T0328315

Viral Gastroenteritis

Molecular Epidemiology and Pathogenesis

Viral Gastroenteritis

Molecular Epidemiology and Pathogenesis

Edited by

Lennart Svensson

Division of Molecular Virology
Departments of Clinical and Experimental Medicine
Medical Faculty, Linköping University
Linköping, Sweden

Ulrich Desselberger

Department of Medicine, University of Cambridge
Addenbrooke's Hospital, Cambridge, United Kingdom

Harry B. Greenberg

Departments of Microbiology and Immunology
Department of Medicine Stanford University, Stanford;
VA Palo Alto Health Care System
Palo Alto, CA, United States

Mary K. Estes

Departments of Molecular Virology and Microbiology
Baylor College of Medicine
Houston, TX, United States

AMSTERDAM • BOSTON • HEIDELBERG • LONDON
NEW YORK • OXFORD • PARIS • SAN DIEGO
SAN FRANCISCO • SINGAPORE • SYDNEY • TOKYO
Academic Press is an imprint of Elsevier

Academic Press is an imprint of Elsevier
125 London Wall, London EC2Y 5AS, United Kingdom
525 B Street, Suite 1800, San Diego, CA 92101-4495, United States
50 Hampshire Street, 5th Floor, Cambridge, MA 02139, United States
The Boulevard, Langford Lane, Kidlington, Oxford OX5 1GB, UK

Library of Congress Cataloging-in-Publication Data
A catalog record for this book is available from the Library of Congress

British Library Cataloguing-in-Publication Data
A catalogue record for this book is available from the British Library

ISBN: 978-0-12-802241-2

For information on all Academic Press publications
visit our website at https://www.elsevier.com/

Publisher: Sara Tenney
Acquisition Editors: Jill Leonard, Linda Versteeg-Buschman
Editorial Project Manager: Fenton Coulthurst
Production Project Manager: Chris Wortley
Designer: Mark Rogers

Typeset by Thomson Digital

Contents

1.3 Immunodeficiencies: Significance for Gastrointestinal Disease

H. Marcotte, L. Hammarström

1.4 Therapy of Viral Gastroenteritis

G. Kang

2.1 Structure and Function of the Rotavirus Particle

S.C. Harrison, P.R. Dormitzer

2.2 Rotavirus Attachment, Internalization, and Vesicular Traffic

C.F. Arias, D. Silva-Ayala, P. Isa, M.A. Díaz-Salinas, S. López

2.3 Rotavirus Replication and Reverse Genetics

A. Navarro, L. Williamson, M. Angel, J.T. Patton

2.4 Pleiotropic Properties of Rotavirus Nonstructural Protein 4 (NSP4) and Their Effects on Viral Replication and Pathogenesis

N.P. Sastri, S.E. Crawford, M.K. Estes

2.5 Rotavirus Replication: The Role of Lipid Droplets

W. Cheung, E. Gaunt, A. Lever, U. Desselberger

2.6 Rotavirus Disease Mechanisms

M. Hagbom, L. Svensson

2.7 Gnotobiotic Neonatal Pig Model of Rotavirus Infection and Disease

A.N. Vlasova, S. Kandasamy, L.J. Saif

2.8 Innate Immune Responses to Rotavirus Infection

A. Sen, H.B. Greenberg

2.9 Human Acquired Immunity to Rotavirus Disease and Correlates of Protection

D. Herrera, M. Parra, M.A. Franco, J. Angel

4.3 Molecular Epidemiology of Astroviruses
P. Khamrin, N. Maneekarn, H. Ushijima

5.1 Enteric Viral Metagenomics
N.J. Ajami, J.F. Petrosino

5.2 Interactions Between Enteric Viruses and the Gut Microbiota
R.R. Garg, S.M. Karst

Contributors

N.J. Ajami, The Alkek Center for Metagenomics and Microbiome Research, Baylor College of Medicine; Department of Molecular Virology and Microbiology, Baylor College of Medicine, Houston, TX, United States

J. Angel, Instituto de Genética Humana, Facultad de Medicina, Pontificia Universidad Javeriana, Bogotá, Colombia

M. Angel, Laboratory of Infectious Diseases, National Institute of Allergy and Infectious Diseases, National Institutes of Health, Bethesda, MD, United States

C.F. Arias, Departamento de Genética del Desarrollo y Fisiología Molecular, Instituto de Biotecnología, Universidad Nacional Autónoma de México, Cuernavaca, Morelos, México

R.L. Atmar, Departments of Molecular Virology and Microbiology, and Medicine, Baylor College of Medicine, Houston, TX, United States

K. Bányai, Institute for Veterinary Medical Research, Centre for Agricultural Research, Hungarian Academy of Sciences, Budapest, Hungary

P. Brandtzaeg, Laboratory for Immunohistochemistry and Immunopathology (LIIPAT), Centre for Immune Regulation (CIR), University of Oslo and Department of Pathology, Oslo University Hospital Rikshospitalet, Oslo, Norway

W. Cheung, Department of Medicine, University of Cambridge, Addenbrooke's Hospital, Cambridge, United Kingdom

J.-M. Choi, Verna and Marrs McLean Department of Biochemistry and Molecular Biology, Baylor College of Medicine, Houston, TX, United States

S.E. Crawford, Department of Molecular Virology and Microbiology, Baylor College of Medicine, Houston, TX, United States

J.B. Cunha, Department of Microbiology and Immunology, University of Michigan Medical School, Ann Arbor, MI, United States

U. Desselberger, Department of Medicine, University of Cambridge, Addenbrooke's Hospital, Cambridge, United Kingdom

M.A. Díaz-Salinas, Departamento de Género Genetica del Desarrollo y Fisiología Molecular, Instituto de Biotecnología, Universidad Nacional Autónoma de México, Cuernavaca, Morelos, México

P.R. Dormitzer, Novartis Influenza Vaccines, Cambridge, MA, United States

M.D. Elftman, Department of Biomedical and Diagnostic Sciences, University of Detroit Mercy, School of Dentistry, Detroit, MI, United States

M.K. Estes, Departments of Molecular Virology and Microbiology, and Medicine, Baylor College of Medicine, Houston, TX, United States

M.A. Franco, Instituto de Genética Humana, Facultad de Medicina, Pontificia Universidad Javeriana, Bogotá, Colombia

R.R. Garg, College of Medicine, Department of Molecular Genetics and Microbiology, Emerging Pathogens Institute, University of Florida, Gainesville, FL, United States

E. Gaunt, Department of Medicine, University of Cambridge, Addenbrooke's Hospital, Cambridge, United Kingdom

I. Goodfellow, Division of Virology, Department of Pathology, University of Cambridge, Addenbrooke's Hospital, Cambridge, United Kingdom

K.Y. Green, Caliciviruses Section, Laboratory of Infectious Diseases, National Institute of Allergy and Infectious Diseases, National Institutes of Health, Bethesda, MD, United States

H.B. Greenberg, Departments of Microbiology and Immunology, Department of Medicine, Stanford University, School of Medicine, Stanford; VA Palo Alto Health Care System, Palo Alto, CA, United States

M. Hagbom, Division of Molecular Virology, Department of Clinical and Experimental Medicine, Medical Faculty, Linköping University, Linköping, Sweden

L. Hammarström, Division of Clinical Immunology and Transfusion Medicine, Department of Laboratory Medicine, Karolinska Institutet at Karolinska University Hospital Huddinge, Stockholm, Sweden

S.C. Harrison, Laboratory of Molecular Medicine, Boston Children's Hospital, Harvard Medical School, and Howard Hughes Medical Institute, Boston, MA, United States

D. Herrera, Instituto de Genética Humana, Facultad de Medicina, Pontificia Universidad Javeriana, Bogotá, Colombia

P. Isa, Departamento de Génetica del Desarrollo y Fisiología Molecular, Instituto de Biotecnología, Universidad Nacional Autónoma de México, Cuernavaca, Morelos, Mexico

S. Kandasamy, Food Animal Health Research Program, The Ohio Agricultural Research and Development Center, Department of Veterinary Preventive Medicine, The Ohio State University, Wooster, OH, United States

G. Kang, The Wellcome Trust Research Laboratory, Division of Gastrointestinal Sciences, Christian Medical College, Vellore, Tamil Nadu, India

S.M. Karst, College of Medicine, Department of Molecular Genetics and Microbiology, Emerging Pathogens Institute, University of Florida, Gainesville, FL, United States

P. Khamrin, Department of Microbiology, Faculty of Medicine, Chiang Mai University, Chiang Mai, Thailand

A.O. Kolawole, Department of Microbiology and Immunology, University of Michigan Medical School, Ann Arbor, MI, United States

G. Larson, Department of Clinical Chemistry and Transfusion Medicine, Sahlgrenska Academy, University of Gothenburg, Gothenburg, Sweden

J. Le Pendu, Inserm, CNRS, Nantes University, IRS UN, Nantes, France

A. Lever, Department of Medicine, University of Cambridge, Addenbrooke's Hospital, Cambridge, United Kingdom

S. López, Departamento de Génetica del Desarrollo y Fisiología Molecular, Instituto de Biotecnología, Universidad Nacional Autónoma de México, Cuernavaca, Morelos, México

T. López, Departamento de Génetica del Desarrollo y Fisiología Molecular, Instituto de Biotecnología, Universidad Nacional Autónoma de México, Cuernavaca, Morelos, México

N. Maneekarn, Department of Microbiology, Faculty of Medicine, Chiang Mai University, Chiang Mai, Thailand

H. Marcotte, Division of Clinical Immunology and Transfusion Medicine, Department of Laboratory Medicine, Karolinska Institutet at Karolinska University Hospital Huddinge, Stockholm, Sweden

S. Marvin, Department of Infectious Diseases, St. Jude Children's Research Hospital, Memphis, TN, United States

V.A. Meliopoulos, Department of Infectious Diseases, St. Jude Children's Research Hospital, Memphis, TN, United States

Z. Muhaxhiri, Verna and Marrs McLean Department of Biochemistry and Molecular Biology, Baylor College of Medicine, Houston, TX, United States

A. Murillo, Departamento de Genética del Desarrollo y Fisiología Molecular, Instituto de Biotecnología, Universidad Nacional Autónoma de México, Cuernavaca, Morelos, México

W. Nasir, Department of Clinical Chemistry and Transfusion Medicine, Sahlgrenska Academy, University of Gothenburg, Gothenburg, Sweden

A. Navarro, Laboratory of Infectious Diseases, National Institute of Allergy and Infectious Diseases, National Institutes of Health, Bethesda, MD, United States

M. Parra, Instituto de Genética Humana, Facultad de Medicina, Pontificia Universidad Javeriana, Bogotá, Colombia

J.T. Patton, Laboratory of Infectious Diseases, National Institute of Allergy and Infectious Diseases, National Institutes of Health, Bethesda, MD, United States; Virginia-Maryland College of Veterinary Medicine, University of Maryland, MD, United States

J.F. Petrosino, The Alkek Center for Metagenomics and Microbiome Research, Baylor College of Medicine, Department of Molecular Virology and Microbiology, Baylor College of Medicine, Houston, TX, United States

V.E. Pitzer, Department of Epidemiology of Microbial Diseases, Yale School of Public Health, New Haven, CT, United States

B.V. Venkataram Prasad, Verna and Marrs McLean Department of Biochemistry and Molecular Biology; Departments of Molecular Virology and Microbiology, and Medicine, Baylor College of Medicine, Houston, TX, United States

S. Ramani, Department of Molecular Virology and Microbiology, Baylor College of Medicine, Houston, TX, United States

G.E. Rydell, Department of Infectious Diseases, Sahlgrenska Academy, University of Gothenburg, Gothenburg, Sweden

L.J. Saif, Food Animal Health Research Program, The Ohio Agricultural Research and Development Center, Department of Veterinary Preventive Medicine, The Ohio State University, Wooster, OH, United States

N.P. Sastri, Department of Molecular Virology and Microbiology, Baylor College of Medicine, Houston, TX, United States

S. Schultz-Cherry, Department of Infectious Diseases, St. Jude Children's Research Hospital, Memphis, TN, United States

A. Sen, Departments of Microbiology and Immunology, Department of Medicine, Stanford University, School of Medicine, Stanford; VA Palo Alto Health Care System, Palo Alto, CA, United States

S. Shanker, Verna and Marrs McLean Department of Biochemistry and Molecular Biology, Baylor College of Medicine, Houston, TX, United States

D. Silva-Ayala, Departamento de Genética del Desarrollo y Fisiología Molecular, Instituto de Biotecnología, Universidad Nacional Autónoma de México, Cuernavaca, Morelos, México

L. Svensson, Division of Molecular Virology, Department of Clinical and Experimental Medicine, Medical Faculty, Linköping University, Linköping, Sweden

S. Taube, Institute of Virology and Cell Biology, University of Lübeck, Lübeck, Germany

H. Ushijima, Division of Microbiology, Department of Pathology and Microbiology, Nihon University School of Medicine, Tokyo, Japan

T. Vesikari, Vaccine Research Center, University of Tampere Medical School, Tampere, Finland

A.N. Vlasova, Food Animal Health Research Program, The Ohio Agricultural Research and Development Center, Department of Veterinary Preventive Medicine, The Ohio State University, Wooster, OH, United States

L. Williamson, Laboratory of Infectious Diseases, National Institute of Allergy and Infectious Diseases, National Institutes of Health, Bethesda, MD, United States

C.E. Wobus, Department of Microbiology and Immunology, University of Michigan Medical School, Ann Arbor, MI, United States

N.C. Zachos, Hopkins Conte Digestive Disease Basic and Translational Research Core Center, Department of Medicine, Division of Gastroenterology and Hepatology, Johns Hopkins University School of Medicine, Baltimore, MD, United States

Introduction

Acute gastroenteritis (AGE) in infants and young children (of <5 years of age) is a very frequent disease worldwide. A decade ago, 1.4 billion cases of AGE were recorded in this age group annually, of which 475 million occurred in children of <1 years of age and 945 million in those of 1–4 years of age (Parashar et al., 2003). For 2010, worldwide 1.7 billion cases of AGE were estimated to have occurred in children of <5 years of age (Fischer Walker et al., 2013). In 2003, most episodes of acute diarrhea (1.3 billion) were cared for at home, 124 million required a visit to a general practitioner, pediatrician or primary care center, and for 9 million children hospitalization was necessary. On average association with rotavirus infection was 1.8% for children in home care and 19–21% for children requiring medical advice and intervention (Parashar et al., 2003). In terms of mortality, in 2008 AGE in children ranged in third place as a cause after neonatal disease and pneumonia worldwide (Table 1; Liu et al., 2012). In absolute numbers, 11.7 million children/annum died from all causes, of those 1.3 million from diarrhea and dehydration, with rotavirus (RV) disease as leading cause of death in 453,000 children (34.8%; Table 2; Tate et al., 2012). Childhood mortality from RV disease was highest in India, Nigeria, Pakistan, DR of Congo, Ethiopia, and Afghanistan, where almost two thirds of all cases occurred (Table 2, Tate et al., 2012). However, for 2011 the number of death from RV-associated diarrhea worldwide was estimated to be lower at about 193,000 (95% CI 133,000–284,000) (Fischer Walker et al., 2013). Since 2006 two live attenuated RV vaccines have been licensed and become part of universal childhood vaccination schedules in over 80 countries. RV vaccination has reduced hospitalization and clinical visits for RV disease significantly in countries of temperate climate (eg, Payne et al., 2013a; Gastañaduy et al., 2013a; Akikusa et al., 2013), less so in countries of the developing world (sub-Saharan Africa, SE Asia) where RV vaccination is needed most (Armah et al., 2010; Madhi et al., 2010; Zaman et al., 2010). In several countries, a very substantial reduction of RV-associated disease and mortality has been recorded (Gastañaduy et al., 2013b; Zhang et al., 2015), contributing to the overall decrease of death from RV-associated AGE (Fischer Walker et al., 2013).

Other viral causes of AGE in childhood are infections with human caliciviruses (noroviruses, sapoviruses), astroviruses, and more rarely, enteric adenoviruses, enteroviruses and picobirnaviruses (Gray and Desselberger, 2009; Estes and Greenberg, 2013). Human noroviruses (NoVs) have become the most frequent viral enteric pathogen in childhood several years after introduction of

TABLE 1 Causes of Childhood Mortality Worldwide in 2010

Childhood mortality 2010	
Global causes	**%**
Neonatal causes	40
Pneumonia	18
Diarrhea	11
Malaria	7
AIDS	2
Meningitis	2
Measles	1
Other	19

From Liu et al., 2012. *Lancet* 379, 2151–2161.

TABLE 2 Mortality in Children <5 Years of Age From Diarrhea Worldwide

		No./annum	
All causes		11,700,000	
Diarrhea (11% of all)		1,300,000	
Rotavirus-associated diarrhea (95% CI 423,000–494,000)		453,000	38.4%[a]
Countries with greatest mortality	**No. of deaths**	**% of all**	
India	98,600	22	
Nigeria	41,100	9	
Pakistan	39,100	9	
DR of Congo	32,600	7	
Ethiopia	28,200	6	
Afghanistan	25,400	6	59%

[a]*Percentage of all deaths from severe diarrhea.*
From Tate et al., 2012. *Lancet Infect. Dis.* 12, 136–141.

universal RV vaccination (Payne et al., 2013b; Koo et al., 2013). Work on developing a vaccine against NoV disease is ongoing (Atmar and Estes, 2012; Richardson et al., 2013).

The facts outlined above emphasize the enormous significance of viruses as cause of diarrheal disease in humans, particularly young children, worldwide. During the last decade substantial progress has been made in characterizing the molecular biology (structure–function relationships, replication), pathogenesis of and immune responses to the viruses causing AGE. In the course of this, a lot has been learned about various functions of gastrointestinal cells and the gastrointestinal immune system, and the opportunities for further discoveries in these research areas are immense. There is a huge potential for applying the knowledge obtained from basic and translational studies of the enteric viral pathogens to counteract the impact of these viruses on acute illness and public health.

The book has been subdivided into five sections. The first (overview) section presents recent advances in gastrointestinal physiology and pathophysiology (Chapter 1.1), mechanisms and functions of gut immunity (Chapter 1.2), primary immunodeficiencies and their significance for gastrointestinal disease (Chapter 1.3), and outlines of present day therapy of AGE in children (Chapter1.4).

The second section is devoted to rotaviruses (RVs). Based on exact knowledge of the RV particle structure at the atomic level and assigned functions (Chapter 2.1), several chapters are committed to viral replication issues: cellular receptors and coreceptors involved in RV entry (Chapter 2.2), structure-function studies of the RV enterotoxin NSP4 (Chapter 2.4), the structural and functional interaction of cellular lipid droplets with RV viroplasms (Chapter 2.5), and an overview of the RV replication mechanisms and of attempts to create a tractable, nucleic acid only-based reverse genetics (RG) system (Chapter 2.3). Newer data on RV pathophysiology (Chapter 2.6), on innate immune responses to RV infection (Chapter 2.8) and on acquired immunity to RV disease including a discussion on correlates of protection (Chapter 2.9) are substantially based on the availability of animal models of RV infection and disease (Chapter 2.7). The advent of whole genome sequencing combined with bio-informatics has advanced RV classification and has revitalized studies of the molecular epidemiology and evolution of human and animal RV strains (Chapter 2.10). Rotavirus vaccine development is lively: in addition to the two licensed RV vaccines mentioned above, new RV vaccines are under investigation or have recently been licensed in various parts of the world (Chapter 2.11).

The successes of RV vaccination programs have directed increasing attention to human caliciviruses and, more specifically, the human NoVs that are major causes of children's AGE besides being the lead agents of food-borne, nonbacterial AGE outbreaks in population groups of all ages worldwide. Norovirus particle structures have been studied in great detail (Chapter 3.1). In contrast to RVs, for certain NoVs fully tractable RG systems have been successfully established (Chapter 3.2), permitting rational genotype–phenotype correlation studies. However, the continued failure to identify a tractable cell culture replication system or animal model system for human NoVs has limited the extent to which the NoV RG system can be optimally used. Human histo-blood group antigens were recognized as cellular receptors for NoVs (Chapter 3.3) and recently also for RVs (Chapter 2.2) (Hu et al., 2012; Tan and Jiang, 2014). Mouse NoVs grow in murine macrophage cultures, have an established RG system, and can be tested in various animal models (Chapter 3.4). The increasing significance of human NoVs has been analysed in numerous molecular epidemiological studies

(Chapter 3.5) and is stimulating work on the development of human NoV vaccines (Chapter 3.6).

Astroviruses (AstV) are also important pathogens that cause AGE in humans (Chapter 4.3). Recent progress in the knowledge of AstV particle structures (Chapter 4.1) is paralleled by the ability to propagate AstVs in vitro and the availability of a RG system (Chapter 4.2).

The recognition of the significance of the gut flora (gut microbiota) for health and disease (various gut functions, immunology, and pathophysiology) has given rise to a rapidly emerging research field (Virgin, 2014; Norman et al., 2014; Lim et al., 2015). The metabolism of gut microbes may affect brain functions; however, a puzzling array of initial data awaits further exploration (Mayer et al., 2015; Smith, 2015). A chapter on the metagenomics of viruses infecting the gut (Chapter 5.1) is complemented by studies on the interaction of enteric viruses with gut microbiota (Chapter 5.2) which fostered a recent report on growing human NoVs in B cells in vitro (Jones et al., 2014).

The editors considered that a book highlighting basic, translational and clinical research of viruses causing AGE in infants and young children is timely and of general interest. The decision to publish it in an *online* format ensures speedy access, allows the contributors actual updating, is geared at individual needs, and maintains the option of acquiring a hard copy of the complete collection of articles. The book is addressed to basic and clinical virologists, molecular biologists, vaccinologists, pediatricians, infectious disease physicians, public health researchers, providers and decision makers, and, last not least, to students of biomedical sciences at various stages of their career.

Last, but not least, the editors gratefully acknowledge the efficient support by the publisher Elsevier Academic Press, initially by Jill Leonard, Acquisition Editor at the Cambridge MA office, Linda Versteeg, Senior Acquisition Editor at the Amsterdam office and during the later stages by Fenton Coulthurst, Editorial Project Manager, and Chris Wortley, Production Project Manager, and their staff at the Kidlington/Oxfordshire office. Their untiring efforts and care for details have made the production phase of the book very easy for the editors and indeed enjoyable.

Lennart Svensson, Ulrich Desselberger, Harry B. Greenberg, Mary K. Estes
Linkoeping, Cambridge, Stanford, Houston
May 2016.

REFERENCES

Armah, G.E., Sow, S.O., Breiman, R.F., Dallas, M.J., Tapia, M.D., Feikin, D.R., Binka, F.N., Steele, A.D., Laserson, K.F., Ansah, N.A., Levine, M.M., Lewis, K., Coia, M.L., Attah-Poku, M., Ojwando, J., Rivers, S.B., Victor, J.C., Nyambane, G., Hodgson, A., Schödel, F., Ciarlet, M., Neuzil, K.M., 2010. Efficacy of pentavalent rotavirus vaccine against severe rotavirus gastroenteritis in infants in developing countries in sub-Saharan Africa: a randomised, double-blind, placebo-controlled trial. Lancet 376 (9741), 606–614.

Akikusa, J.D., Hopper, S.M., Kelly, J.J., Kirkwood, C.D., Buttery, J.P., 2013. Changes in the epidemiology of gastroenteritis in a paediatric short stay unit following the introduction of rotavirus immunisation. J. Paediatr. Child Health 49 (2), 120–124.

Atmar, R.L., Estes, M.K., 2012. Norovirus vaccine development: next steps. Expert. Rev. Vaccines 11 (9), 1023–1025.

Estes, M.K., Greenberg, H.B., 2013. Rotaviruses. In: Knipe, D.M., Howley, P.M. et al., (Eds.), Fields Virology, sixth ed. Wolters Kluwer Health/Lippincott Williams &Wilkins, Philadelphia, PA, pp. 1347–1401.

Fischer Walker, C.L., Rudan, I., Liu, L., Nair, H., Theodoratou, E., Bhutta, Z.A., O'Brien, K.L., Campbell, H., Black, R.E., 2013. Global burden of childhood pneumonia and diarrhoea. Lancet 381 (9875), 1405–1416.

Gastañaduy, P.A., Curns, A.T., Parashar, U.D., Lopman, B.A., 2013a. Gastroenteritis hospitalizations in older children and adults in the United States before and after implementation of infant rotavirus vaccination. JAMA 310 (8), 851–853.

Gastañaduy, P.A., Sánchez-Uribe, E., Esparza-Aguilar, M., Desai, R., Parashar, U.D., Patel, M., Richardson, V., 2013b. Effect of rotavirus vaccine on diarrhea mortality in different socioeconomic regions of Mexico. Pediatrics 131 (4), e1115–e1120.

Gray, J., Desselberger, U., 2009. Viruses other than rotaviruses associated with acute diarrhoeal disease. In: Zuckerman, A.J., Banatvala, J.E., Schoub, B.D., Griffiths, P.D., Mortimer, P. (Eds.), Principles and Practice of Clinical Virology. Wiley-Blackwell, Chichester, pp. 355–372.

Hu, L., Crawford, S.E., Czako, R., Cortes-Penfield, N.W., Smith, D.F., Le Pendu, J., Estes, M.K., Prasad, B.V., 2012. Cell attachment protein VP8* of a human rotavirus specifically interacts with A-type histo-blood group antigen. Nature 485 (7397), 256–259.

Jones, M.K., Watanabe, M., Zhu, S., Graves, C.L., Keyes, L.R., Grau, K.R., Gonzalez-Hernandez, M.B., Iovine, N.M., Wobus, C.E., Vinjé, J., Tibbetts, S.A., Wallet, S.M., Karst, S.M., 2014. Enteric bacteria promote human and mouse norovirus infection of B cells. Science 346 (6210), 755–759.

Koo, H.L., Neill, F.H., Estes, M.K., Munoz, F.M., Cameron, A., Dupont, H.L., Atmar, R.L., 2013. Noroviruses: the most common pediatric viral enteric pathogen at a large university hospital after introduction of rotavirus vaccination. J. Pediatric. Infect. Dis. Soc. 2 (1), 57–60.

Lim, E.S., Zhou, Y., Zhao, G., Bauer, I.K., Droit, L., Ndao, I.M., Warner, B.B., Tarr, P.I., Wang, D., Holtz, L.R., 2015. Early life dynamics of the human gut virome and bacterial microbiome in infants. Nat. Med. 21 (10), 1228–1234.

Liu, L., Johnson, H.L., Cousens, S., Perin, J., Scott, S., Lawn, J.E., Rudan, I., Campbell, H., Cibulskis, R., Li, M., Mathers, C., Black, R.E., Child Health Epidemiology Reference Group of WHO and UNICEF, 2012. Global, regional, and national causes of child mortality: an updated systematic analysis for 2010 with time trends since 2000. Lancet 379 (9832), 2151–2156, [Erratum in: *Lancet.* 2012 Oct 13;**380**(9850):1308.]

Madhi, S.A., Cunliffe, N.A., Steele, D., Witte, D., Kirsten, M., Louw, C., Ngwira, B., Victor, J.C., Gillard, P.H., Cheuvart, B.B., Han, H.H., Neuzil, K.M., 2010. Effect of human rotavirus vaccine on severe diarrhea in African infants. N. Engl. J. Med. 362 (4), 289–298.

Mayer, E.A., Tillisch, K., Gupta, A., 2015. Gut/brain axis and the microbiota. J. Clin. Invest. 125 (3), 926–938.

Norman, J.M., Handley, S.A., Virgin, H.W., 2014. Kingdom-agnostic metagenomics and the importance of complete characterization of enteric microbial communities. Gastroenterology 146 (6), 1459–1469.

Parashar, U.D., Hummelman, E.G., Bresee, J.S., Miller, M.A., Glass, R.I., 2003. Global illness and deaths caused by rotavirus disease in children. Emerg. Infect. Dis. 9 (5), 565–572.

Payne, D.C., Boom, J.A., Staat, M.A., Edwards, K.M., Szilagyi, P.G., Klein, E.J., Selvarangan, R., Azimi, P.H., Harrison, C., Moffatt, M., Johnston, S.H., Sahni, L.C., Baker, C.J., Rench, M.A., Donauer, S., McNeal, M., Chappell, J., Weinberg, G.A., Tasslimi, A., Tate, J.E., Wikswo, M., Curns, A.T., Sulemana, I., Mijatovic-Rustempasic, S., Esona, M.D., Bowen, M.D., Gentsch, J.R., Parashar, U.D., 2013a. Effectiveness of pentavalent and monovalent rotavirus vaccines in concurrent use among US children <5 years of age, 2009–2011. Clin. Infect. Dis. 57 (1), 13–20.

Payne, D.C., Vinjé, J., Szilagyi, P.G., Edwards, K.M., Staat, M.A., Weinberg, G.A., Hall, C.B., Chappell, J., Bernstein, D.I., Curns, A.T., Wikswo, M., Shirley, S.H., Hall, A.J., Lopman, B., Parashar, U.D., 2013b. Norovirus and medically attended gastroenteritis in U.S. children. N. Engl. J. Med. 368 (12), 1121–1130.

Richardson, C., Bargatze, R.F., Goodwin, R., Mendelman, P.M., 2013. Norovirus virus-like particle vaccines for the prevention of acute gastroenteritis. Expert Rev Vaccines. 12 (2), 155–167.

Smith, P.A., 2015. The tantalizing links between gut microbes and the brain (News Feature). Nature 526, 312–314.

Tan, M., Jiang, X., 2014. Histo-blood group antigens: a common niche for norovirus and rotavirus. Expert. Rev. Mol. Med. 16, e5.

Tate, J.E., Burton, A.H., Boschi-Pinto, C., Steele, A.D., Duque, J., Parashar, U.D., WHO-coordinated Global Rotavirus Surveillance Network, 2012. 2008 estimate of worldwide rotavirus-associated mortality in children younger than 5 years before the introduction of universal rotavirus vaccination programmes: a systematic review and meta-analysis. Lancet Infect. Dis. 12 (2), 136–141.

Virgin, H.W., 2014. The virome in mammalian physiology and disease. Cell 157 (1), 142–150.

Zaman, K., Dang, D.A., Victor, J.C., Shin, S., Yunus, M., Dallas, M.J., Podder, G., Vu, D.T., Le, T.P., Luby, S.P., Le, H.T., Coia, M.L., Lewis, K., Rivers, S.B., Sack, D.A., Schödel, F., Steele, A.D., Neuzil, K.M., Ciarlet, M., 2010. Efficacy of pentavalent rotavirus vaccine against severe rotavirus gastroenteritis in infants in developing countries in Asia: a randomised, double-blind, placebo-controlled trial. Lancet 376 (9741), 615–623.

Zhang, J., Duan, Z., Payne, D.C., Yen, C., Pan, X., Chang, Z., Liu, N., Ye, J., Ren, X., Tate, J.E., Jiang, B., Parashar, U.D., 2015. Rotavirus-specific and overall diarrhea mortality in Chinese children younger than 5 years: 2003 to 2012. Pediatr. Infect. Dis. J. 34 (10), e233–e237.

Chapter 1.1

Gastrointestinal Physiology and Pathophysiology

N.C. Zachos

Hopkins Conte Digestive Disease Basic and Translational Research Core Center, Department of Medicine, Division of Gastroenterology and Hepatology, Johns Hopkins University School of Medicine, Baltimore, MD, United States

1 INTRODUCTION

The intestinal epithelium constantly manages the complex regulation of paracellular and transcellular transport of water, electrolytes, and small solutes to promote nutrient absorption and secretion of various compounds (electrolytes, methyl sulfides, benzopyrrole derivatives a.o.) while preventing excess fluid loss. The balance of absorptive and secretory processes is tightly regulated to manage the nearly 9 L/day of fluid in the intestinal lumen, of which nearly 98% are absorbed by the intestinal epithelium. These functions occur through the coordinated interplay of polarized columnar epithelial cells that are aligned in a continuous monolayer contoured by villi and crypts, which in addition to epithelial microvilli, enhance the plain surface area by greater than 600-fold (Montrose et al., 1999). The villous epithelium is predominantly comprised of mature enterocytes along with mucus-secreting Goblet cells, hormone producing entero-endocrine cells, and tuft cells while the crypt epithelium is mostly composed of immature enterocytes with Paneth and stem cells located at the crypt base (Barker et al., 2008; Cheng and Leblond, 1974). Pioneering work by the laboratory of Hans Clevers has identified Lgr5 as the molecular marker for the constantly dividing intestinal stem cells that differentiate into all intestinal epithelial cell types (Sato et al., 2011a). Intestinal stem cells confer segment-specific functions such that the proximal small intestine absorbs carbohydrates and fatty acids while the ileum absorbs bile acids and vitamin B12 (Middendorp et al., 2014). Nutrient and salt uptake by the intestinal epithelium generates an osmotic gradient that drives the active (via apical ion/solute cotransporters) and/or passive (ie, paracellular) absorption of most of the water exposed per day to the intestine. In the absence of nutrients, small intestinal and proximal colonic fluid absorption occurs through the electroneutral coupled absorption of Na^+

and Cl^- while electrogenic absorption of Na^+ predominates in the distal colon (Montrose et al., 1999).

Across all age groups, diarrhea is one of the top five causes of death worldwide; while in children less than 5 years old, diarrhea is the second leading cause of death. Diarrhea etiology includes bacteria, viruses, parasites, toxins, or drugs. Acute viral gastroenteritis remains a major public health concern and a common cause of morbidity and mortality worldwide. Rotaviruses, caliciviruses (particularly norovirus), astroviruses, and enteric adenoviruses are the four predominant causes of viral gastroenteritis, inducing symptoms ranging from self-limiting watery diarrhea that resolves within 7 days to severe dehydration (with complications including cerebral edema, hypovolemic shock, renal failure) and death. Two recently developed rotavirus vaccines (RotaTeq® RV5 and Rotarix® RV1) have reduced rotavirus disease in developed countries but are much less effective in low income settings (Babji and Kang, 2012; Tate et al., 2013). The recent Global Enteric Multicenter Study (GEMS) confirmed that rotavirus is the leading cause of infant diarrhea among the more than 20,000 children studied in seven sites across Asia and Africa (Kotloff et al., 2013). Although Oral Rehydration Solution (ORS) remains the only safe, effective, and low cost therapeutic option to prevent life-threatening dehydration due to acute severe diarrheal diseases (Binder et al., 2014), the GEMS study reported that each episode of severe diarrhea in children increased the risk of delayed physical and intellectual development as well as increased mortality by 8.5-fold (Kotloff et al., 2013). Therefore, there remains a great need to develop additional therapies to treat severe diarrheal diseases.

Diarrhea often involves the dysregulation of absorptive processes and/or loss of epithelial barrier integrity, resulting in nutrient malabsorption and/or activated secretion of fluid into the intestinal lumen. In addition to the contribution of the enteric nervous system to diarrhea, roles for intestinal microbiota and epithelial immunity have been described (Moens and Veldhoen, 2012). While numerous studies of bacterial pathogens have improved our understanding of the pathogenesis of gastroenteritis, other studies suggest that viral diarrhea occurs by distinctly different mechanisms (Lorrot and Vasseur, 2007). This chapter will describe the current understanding of normal human intestinal physiology and compare the molecular mechanisms that are dysregulated by enteric bacterial and viral pathogens in the various model systems studied to date.

2 NORMAL INTESTINAL PHYSIOLOGY

Under normal digestive conditions, the intestine is exposed to ~9 L of fluids daily. The ability of the intestine to efficiently absorb dietary water, electrolytes and nutrients is defined by specific functional processes that occur along the horizontal (ie, proximal to distal intestinal segments) and vertical (ie, crypt/villus) axes. While the entire intestine is capable of absorbing water and salts, uptake of carbohydrates, peptides, amino acids, minerals, vitamins, long/short chain

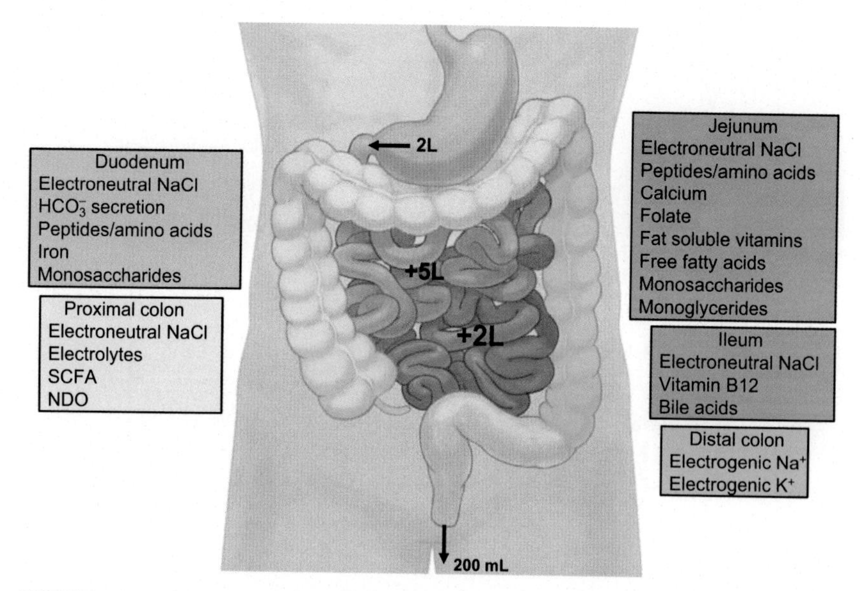

FIGURE 1.1.1 **Segment-specific functions in the human small intestine and colon.** Colored boxes correspond to similar colored intestinal segments and describe the specific absorptive/secretory function of the respective intestinal segment. "+" indicates amount of fluid, including gastric, pancreatic, and intestinal secretions, secreted into intestine daily. *SCFA*, short chain fatty acids; *NDO*, nondigestable oligosaccharides.

fatty acids, and bile acids occurs in distinct intestinal segments that uniquely express the transport proteins to facilitate absorption (Fig. 1.1.1). These proteins are asymmetrically expressed in polarized epithelial cells to perform vectorial transport across the intestinal epithelium resulting in net nutrient and water absorption (Fig. 1.1.2). The intestinal architecture enhances absorption through the increased surface area generated by the villus/crypt axis as well as by the densely packed microvilli of mature enterocytes.

The intestinal epithelium serves as a barrier separating luminal dietary and microbial content from the bloodstream. A single monolayer of polarized columnar epithelial cells is connected by tight junctions that segregate the epithelial apical and basolateral membranes. Tight junctions are formed and maintained by dynamic multiprotein complexes that regulate the paracellular movement of water, ions, and small molecules. Occludin and the claudin family of proteins regulate paracellular movement of ions and small molecules through selectivity based on electrical conductance, size, and charge (Anderson and Van Itallie, 2006). Of these, electrical conductance is the most variable (>fivefold difference) between the "leaky" and "tight" epithelium of the duodenum and distal colon, respectively. The expression of different claudin isoforms varies along the horizontal and vertical axes conferring intestinal segment-specific regulation of paracellular transport (Holmes et al., 2006). Movement through

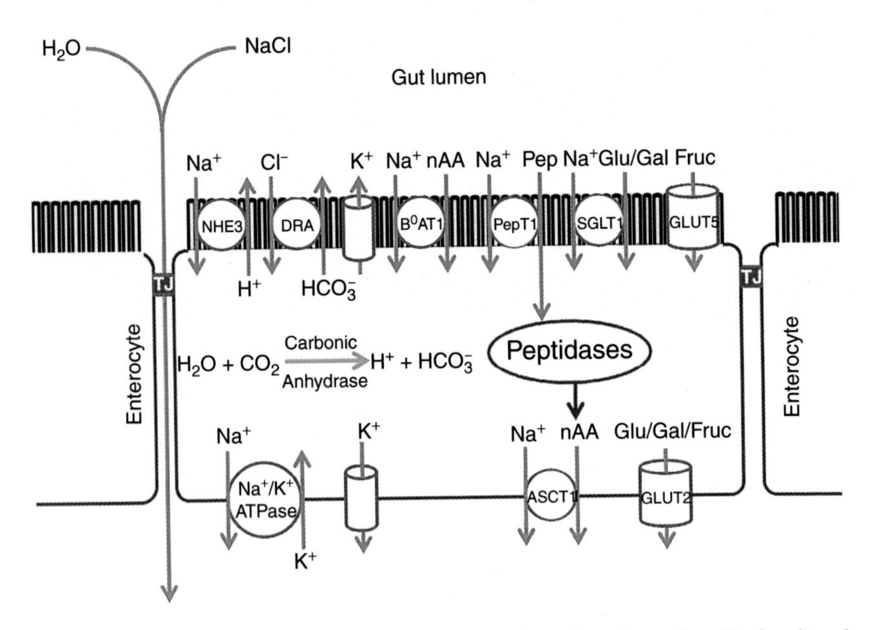

FIGURE 1.1.2 **Normal small intestinal nutrient and electrolyte absorption.** The basolateral Na^+/K^+-ATPase drives the Na^+ gradient across the apical and basolateral membranes. Electroneutral NaCl absorption takes place through the coupled regulation of NHE3 and DRA. Peptides and amino acids are transported by the apical PEPT1 and Na^+-dependent amino acid transporters (eg, B^0AT1), respectively. Intracellular peptidases complete breakdown of absorbed peptides into amino acids that are exported from the epithelium through basolateral amino acid transporters (eg, ASCT1). The dietary sugars, glucose, galactose and fructose are absorbed by apical transporters SGLT-1 and GLUT5. SGLT-1 absorbs glucose in a Na^+-dependent manner while GLUT5 absorbs fructose through facilitative transport. GLUT2 normally resides on the basolateral membrane in between meals to complete transcellular movement of dietary sugars to the portal circulation. However, after a carbohydrate-rich meal, GLUT2 translocates to the apical membrane to handle excess luminal concentrations of glucose that are beyond the capacity of SGLT-1 to absorb. Apical and basolateral potassium channels maintain cellular electrical neutrality during active transport. Carbonic anhydrase maintains physiological pH and fluid balance. *nAA*, neutral amino acids; *Glu*, glucose; *Gal*, galactose; *Fruc*, fructose; *pep*, peptide; *TJ*, tight junction. *Black* denotes Brush border. *Red* denotes basolateral membrane.

tight junctions is a passive process generated by chemical and osmotic gradients that result from active transcellular transport during the postprandial state. Tight junctions regulate the rate of paracellular fluid flow through the opening and closing of the claudin pores (Anderson and Van Itallie, 2006).

The intestinal epithelium is comprised of the major epithelial cell types: enterocytes, entero-endocrine, goblet, tuft and Paneth cells. The leucine-rich G-protein coupled receptor 5, Lgr5, has been established as the molecular marker for the actively dividing intestinal stem cells that give rise to all epithelial cell types (Sato et al., 2011a). In the small intestine, Lgr5+ stem cells reside at the base of the crypt, interspersed among Paneth cells, and undergo asymmetric division producing immature enterocytes which comprise the crypt transit amplifying zone. These enterocytes terminally differentiate into the other epithelial cell types, including mature enterocytes and secretory cell lineages, as they migrate towards the villus tips (Barker et al., 2008). The life cycle of the small intestinal epithelium is about 5–7 days; while Paneth and stem cells turn over approximately every 30 days. This rapid renewal is necessary to handle the harsh luminal environment, and thus the intestinal epithelium is considered the most regenerative region in the body.

In addition to promoting nutrient digestion and absorption, the intestinal epithelial barrier protects the body from luminal pathogens and their secreted products. The intestinal epithelium is exposed to the largest population of microbial species (ranging from $<10^4$/mL in the small intestine to $>10^{15}$/mL in the colon) and serves as the first line of defense against harmful microbial pathogens. Goblet cells constantly secrete highly glycosylated proteins, called mucins (MUC), to produce a mucus layer that serves as a physical barrier, preventing direct interaction of commensal microbes and enteric pathogens with the host epithelium. The mucus layer is a complex organization of transmembrane and gel-forming (ie, secreted) MUC molecules that develop a tightly adhered inner layer and loose outer layer, respectively (Johansson et al., 2013). The small intestine has a single, loose unattached mucus layer comprised exclusively of the gel forming MUC2 while the colon has a two-layered system in which MUC2 serves as the foundation for both layers. The mucus of the small intestine serves to provide a nearly sterile environment through hourly turnover of the mucus layer and antimicrobial peptide secretion by Paneth cells (Bevins et al., 1999). The constant turnover occurs through regulated anion secretion during digestion and between meals. The inner mucus layer of the colon is constantly regenerated and maintained by surface goblet cells. The outer layer of the colonic mucus originates from the inner layer which becomes loose and unattached allowing commensal bacteria to reside there (Johansson et al., 2013). While the intestinal mucus is the first line of defense against enteric pathogens, enteric viruses evade this protective barrier by mechanisms requiring further study. A single report suggested that small intestinal and colonic mucins (colonic mucus had stronger effect) can prevent binding and entry of rotavirus and thus viral replication (Chen et al., 1993). Further studies are required to understand the mechanisms responsible for this effect.

2.1 Fluid Balance in the Intestine

The average dietary intake of fluid, which includes water, electrolytes, and nutrients, is ~2 L; however, due to the secretion of fluids necessary for normal digestion which include saliva, bile, as well as gastric, pancreatic, and intestinal secretions, the total volume of fluid in contact with the intestinal epithelium is ~9 L/day. Together, the small intestine (~7 L) and the colon (~2 L) have the capacity to absorb more than 98% of this fluid load, leaving ~200 mL for excretion in stool (Fig. 1.1.1). Of that absorbed by the small intestine, ~5 L is taken up by the duodenum and jejunum while the ileum is responsible for the remaining ~2 L. Although the small intestinal handles a majority of the fluid absorption, the colon has the capacity to absorb up to 6 L daily (Montrose et al., 1999).

Since the fluid secreted into the intestinal lumen is comprised of gastric, pancreatic and intestinal juices, increased fluid secretion facilitates the spreading of enzymes for normal digestive processes. In addition, increased fluid secretion protects the intestinal mucosa from potential damage as well as allowing stool to pass through the gut. While many studies have elucidated the molecular mechanisms involved in increased water secretion during enteric infections (Hodges and Gill, 2010; Morris and Estes, 2001; Viswanathan et al., 2009), a basic understanding of how water is secreted under basal or normal digestive conditions is still lacking. Moreover, a definitive path (ie, paracellular vs transcellular) for water absorption/secretion in the human intestine has yet to be determined.

The aquaporin (AQP) family of proteins is comprised of 13 isoforms that are categorized into two groups: the AQPs that are only permeable to water, and the aquaglyceroporins that are permeable to water, glycerol, urea, and other small solutes (Agre and Kozono, 2003). AQPs are responsible for osmotic pressure-driven movement of water from the lumen by the intestinal epithelium; however, whether any AQPs participate in active water transport against osmotic gradients remains to be determined. AQP isoforms 1, 3, 7, 10, and 11 are expressed in epithelial cells of both the human small intestine and colon while AQPs 4 and 8 are colonic (Laforenza, 2012). In addition, AQP1 is expressed in the capillary endothelium of the small intestine and may serve as the pathway for water movement from the intestinal mucosa into the circulation. However, detailed characterization of the segment-specific expression and localization of each AQP isoform in the human intestine is incomplete. Understanding the roles of each AQP in water transport under normal digestive conditions and in diarrheal disease in humans is complicated due to conflicting data from animal models in which $AQP^{-/-}$ mice do not exhibit abnormal intestinal phenotypes (Agre and Kozono, 2003). The hypothesis that aquaporins are necessary for normal water transport in humans is supported by correlative data from inflammatory bowel disease, celiac disease, and trichohepatoenteric syndrome patients that have decreased expression of one or more AQP isoforms suggesting that lack of AQPs results in decreased luminal water absorption (Guttman and Finlay, 2008; Hartley et al., 2010). Rodent studies of bacterial pathogenesis observed that decreased AQP activity is due to post translational modifications and to protein trafficking induced by bacterial virulence factors (Guttman et al., 2007). Infection of mice by the attaching and effacing bacterial pathogen, *Citrobacter rodentium*, resulted in mislocalization of AQPs 2 and 3 from the apical membrane to intracellular vesicles. This effect was due to the activity of the bacterial virulence factors EspF and EspG that are injected into host cells by the bacterial type III secretion system. Normal apical membrane expression of AQPs 2 and 3 was restored after *C. rodentium* was cleared from mice (Guttman et al., 2007). One recent report has demonstrated that RV reduces the expression of AQPs 1, 4, and 8 in a mouse model of RV diarrhea (Cao et al., 2014). The roles of AQPs in normal intestinal water movement and in enteric viral pathogenesis remain to be determined.

2.2 Intestinal Absorption

2.2.1 Sugar Transport

Intestinal absorption of the main dietary sugars (ie, glucose, galactose, and fructose) relies on the coordinated function of transporters at the apical and basolateral membranes of enterocytes (Shirazi-Beechey et al., 2011). In between meals, low luminal levels of glucose are absorbed at the apical membrane of enterocytes by the Na^+-dependent glucose transporter, SGLT1, while fructose is absorbed by GLUT5, a facilitative transporter (Mueckler, 1994). SGLT1 has a stoichiometry of 2 Na^+:1 glucose during each transport cycle. Na^+ uptake by SGLT1 is removed from the cell by the basolateral Na^+/K^+ ATPase, and the resulting electrochemical gradient increases paracellular transport of Cl^- from the lumen (Fig. 1.1.2). In addition, activation of SGLT1 leads to increased expression and activity of NHE3 at the brush border (BB) (Hu et al., 2006; Lin et al., 2011). Increased NHE3 activity results in intracellular alkalinization which stimulates BB Cl^-/HCO_3^- exchange activity and apical Cl^- absorption.

The predominantly basolaterally expressed glucose/fructose transporter, GLUT2, provides an exit pathway for glucose and fructose into the bloodstream (Roder et al., 2014). GLUT2 is able to process high sugar concentrations efficiently due to its high maximum rate (V_{max}) of transporting high concentrations of glucose and its high affinity for binding glucose (Michaelis–Menten constant; K_m) compared to SGLT1, which has lower transport rate and affinity for glucose (Raja et al., 2012). SGLT1, GLUT5, and GLUT2 are expressed in the duodenum and jejunum, and at lower levels in the ileum. Upon digestion of carbohydrate-rich meals that produce luminal glucose concentrations above SGLT1 saturating concentrations, unsaturated GLUT2 rapidly (within minutes) translocates to the apical membrane of enterocytes, and is also rapidly retracted back to the basolateral membrane when luminal glucose levels decrease (Gouyon et al., 2003). Different signaling mechanisms have been shown to trigger insertion of GLUT2 into the apical membrane, including high luminal sugar concentrations, stress, corticoids, and enteroendocrine hormones (Au et al., 2002; Chaudhry et al., 2012; Mace et al., 2007; Shepherd et al., 2004). Under normal physiological conditions, insulin triggers GLUT2 internalization, which then limits any increase in serum glucose levels (Tobin et al., 2008).

2.2.2 Amino Acid and Peptide Transport

The small intestine is the major region responsible for amino acid and peptide absorption (Fig. 1.1.1). Dietary proteins are digested to smaller peptide fragments and free amino acids by gastric, pancreatic, and intestinal enzymes.

The BBs of differentiated intestinal epithelial cells express additional digestive enzymes that further hydrolyze oligopeptides to smaller di- and tri-peptides, including aminopeptidase N (APN), carboxypeptidase, and dipeptidylpeptidase IV (DPPIV). All hydrolyzed di- and tri-peptides are absorbed by the BB electrogenic protein/peptide symporter, PEPT1; however, PEPT1 does not transport free amino acids or peptides larger than three amino acids (Daniel, 2004). PEPT1 activity is functionally linked to NHE3 activity through recycling of protons (Kennedy et al., 2002). NHE3 extrudes intracellular protons for PEPT1 symport that in turn lowers intracellular pH and activates NHE3 facilitating Na^+ absorption. Acute second messenger elevation leads to decreased PEPT1 activity similar to that described for NHE3 (Daniel, 2004). Insulin, leptin, growth hormone, and thyroid hormone have all been demonstrated to increase abundance and activity of PEPT1 (Daniel, 2004). Absorbed di- and tri-peptides are further digested by intracellular proteases to free amino acids which are transported to the portal circulation by basolaterally expressed amino acid transporters. Amino acid transporters are expressed at both the apical and basolateral membranes of small intestinal epithelial cells and are characterized by their dependence on a Na^+ gradient for transport activity as well as by the class of amino acids they transport (ie, neutral, anionic, cationic, or imino acids) (Broer, 2008).

2.2.3 *Electroneutral NaCl Absorption*

Electroneutral NaCl absorption occurs throughout the mammalian intestine (except the distal colon) and accounts for most intestinal Na^+ absorption in the period between meals (Fig. 1.1.1). This process contributes more to Na^+ absorption in the ileum and proximal colon post prandially than in the proximal small intestine, where amino acid, peptide, and carbohydrate symporters are linked to Na^+ absorption (Maher et al., 1996, 1997; Zachos et al., 2005). Electroneutral NaCl absorption is both up- and downregulated as part of digestion, appearing to be inhibited initially during and after eating (for spreading digestive enzymes) and then stimulated later in digestion (for fluid absorption) (Donowitz and Tse, 2000). This regulation occurs in response to eating via the complex neural/paracrine/endocrine changes in the intestine. In diarrheal diseases, electroneutral NaCl absorption is inhibited while electrogenic Cl^- secretion is stimulated. These events are responsible for the major loss of water and electrolytes in diarrhea (Donowitz and Tse, 2000). Electroneutral NaCl absorption occurs through the functional linkage of a Na^+ absorptive transporter and a Cl^-/HCO_3^- exchanger. Regulation of the BB Na^+/H^+ exchanger, NHE3 (SLC9a3), accounts for most of the recognized digestive changes in electroneutral NaCl absorption, as well as most of the changes in Na^+ absorption that occur in diarrheal diseases (Zachos et al., 2005). The BB Cl^-/HCO_3^- exchangers, downregulated in adenoma (DRA; SLC26a3) and putative anion transporter (PAT1; SLC26a6), are functionally linked to NHE3 activity by intracellular carbonic anhydrase which provides the H^+ and HCO_3^- for NHE3 and DRA/PAT1 functions, respectively (Melvin et al., 1999; Simpson et al., 2010; Walker et al., 2008). These

Cl^-/HCO_3^- exchangers function in anion secretion as well as Cl^- absorption. DRA is predominantly expressed in the duodenum and colon while PAT1 is expressed in the jejunum and ileum. Elevation of second messengers (ie, cAMP, cGMP, Ca^{2+}) has been shown to acutely regulate NaCl absorption by inhibiting or stimulating the function of NHE3 and/or DRA/PAT1 (Chen et al., 2015; Musch et al., 2009; Xia et al., 2014). In the distal colon, ~50% of luminal Na^+ absorption is electrogenic and occurs through the regulation of the apical epithelial Na^+ channel, ENaC. Examples of regulation include aldosterone (acting via signal transduction) and CFTR, which directly binds and modulates ENaC activity (Wagner et al., 2001).

2.2.4 *Anion Secretion*

Chloride is the major anion responsible for generating the osmotic gradient necessary to drive the active transport of water across the intestinal epithelium and into the lumen (Fig. 1.1.3). HCO_3^- can also increase water transport; however,

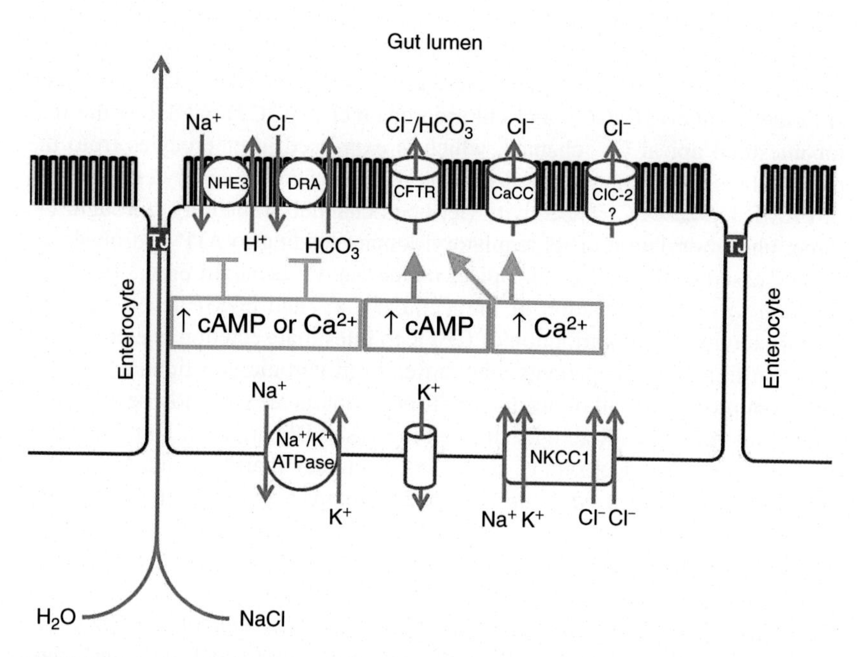

FIGURE 1.1.3 Active fluid secretion is mediated through inhibition of apical NaCl absorption and increased anion secretion. Second messenger elevation (eg, cAMP or Ca^{2+}) inhibits NHE3 and apical Cl^-/HCO_3^- exchange (via DRA/PAT1) while stimulating Cl^- secretion through apical Cl^- channels, CFTR, CaCC, or ClC-2. The basolateral $Na^+/K^+/2Cl^-$ cotransporter provides the Cl^- gradient for secretion. Elevated levels of cAMP and Ca^{2+} can both stimulate CFTR activity by increasing apical membrane expression through increased trafficking as well as open channel probability. The molecular identity of the CaCC involved in intestinal fluid secretion remains to be determined.

to a lesser extent. In intestinal epithelial cells, increased apical Cl^- channel activity and fluid secretion requires the contribution of basolateral transport processes including $Na^+/K^+/2Cl^-$ symporter (NKCC1), Na^+/K^+ ATPase, and K^+ channels (Kunzelmann et al., 2001; Payne et al., 1995; Tabcharani et al., 1991). These three processes collectively contribute to increasing the intracellular concentration of Cl^- stimulating the efflux of Cl^- down its electrochemical gradient through apical chloride channels. NKCC1-mediated transport of Cl^- across the basolateral membrane is driven by the Na^+ gradient (via efflux) established by the Na^+/K^+ ATPase, which exchanges 3 Na^+ (efflux) for 2 K^+ (influx) (Montrose et al., 1999). K^+ ions transported into the cell by both NKCC1 and Na^+/K^+ ATPase are leaked out by basolateral K^+ channels (eg, KCNQ1/KCNE3) to maintain cellular electroneutrality to compensate for the loss of Cl^-. K^+ efflux results in membrane hyperpolarization and is the main driving force for Cl^- secretion (Heitzmann and Warth, 2008). The loss of Cl^- and K^+ establishes the negative charge and the accumulation of luminal Cl^- subsequently drives the paracellular transport of Na^+ through the tight junctions and into the intestinal lumen (Field, 2003). This mucosal to luminal transport of NaCl facilitates the osmotic flow of water and thus net secretion.

In the intestine, the apical Cl^- channels that contribute to fluid secretion include: Cystic Fibrosis Transmembrane Regulator (CFTR), Ca^{2+} activated chloride channels (CaCC), and chloride channel 2 (ClC2). CFTR is the major intestinal apical Cl^- channel, which is expressed in enterocytes from the proximal small intestine to the distal colon. In intestinal enterocytes, elevation of cAMP increases CFTR activity (ie, open channel probability) through: (1) direct phosphorylation of its regulatory domain leading to ATP hydrolysis, or (2) increased trafficking to the apical membrane (Tabcharani et al., 1991). In addition to cAMP, elevation of intracellular Ca^{2+} and cGMP levels increase CFTR activity. The contribution of CFTR to intestinal secretion is emphasized in cystic fibrosis (CF) patients who suffer from chronic constipation (among other symptoms). Small molecule CFTR potentiators are currently being developed to increase CFTR activity through increased BB insertion based on a CF patient's genotype (Dekkers et al., 2013). Secretory diarrheas induced by the enterotoxins of *Vibrio cholerae* (ie, cholera toxin; CTx) and enterotoxigenic *Escherichia coli* (ie, heat stable enterotoxin; STa) involve stimulation of CFTR activity through elevated cAMP and cGMP levels, respectively. Recently, three classes of highly potent (ie, low nM) CFTR inhibitors have been developed and may serve as potential antidiarrheal agents: (1) the thiazolidinone $CFTR_{inh}$-172 binds to cytoplasmic side of CFTR and stabilizes its closed conformational state; (2) the PPQ/BPO compounds (R-BPO-27: IC_{50} ~4 nM) also target the cytoplasmic side of CFTR; and (3) the glycine hydrazide, GlyH-101, inhibits CFTR by binding the channel pore from the extracellular side (Thiagarajah et al., 2014). Both $CFTR_{inh}$-172 and GlyH-101 have demonstrated antisecretory effects in rodent diarrheal models while R-BPO-27 has only been tested in models of polycystic kidney disease. Natural products including the plant extract SB-300 and a Thai herbal remedy have been shown to also effectively inhibit CFTR activity in models of diarrheal disease (Fischer et al., 2004; Tradtrantip et al., 2014).

Luminal purinergic or basolateral adrenergic and cholinergic signaling lead to increases in intracellular Ca^{2+} that evokes a transient Cl^- secretory response. Studies using intestinal epithelial cell lines have observed that the Cl^- secretory response to Ca^{2+} is not as strong or sustained when compared to cAMP stimulated Cl^- secretion. Moreover, the effects of Ca^{2+} and cAMP on Cl^- secretion are synergistic, suggesting that another Ca^{2+}-dependent, cAMP-independent Cl^- channel mediates this secretion; however, the molecular identity of the apical CaCC responsible is not clear (Thiagarajah et al., 2015). The anoctamins (particularly TMEM16A) and bestrophins have been suggested as candidate CaCCs; however, studies have indicated that they do not contribute to intestinal Cl^- secretion. Moreover, CFTR knock-out (KO) mice do not exhibit an intestinal Cl^- secretory response to cAMP or Ca^{2+} agonists (Thiagarajah et al., 2015). An alternate hypothesis suggests that Ca^{2+}-mediated Cl^- secretion occurs due to activation of basolateral K^+ channels which would indirectly stimulate Cl^- secretion due to the change in the electrochemical gradient induced by K^+ efflux.

Similar to the controversy surrounding CaCCs, ClC-2-mediated Cl^- secretion has been measured in intestinal epithelial cell lines and mouse intestine; however, ClC-2 KO mice do not exhibit intestinal obstruction, and the ClC-2/CFTR double KO mice have the same intestinal phenotype as CFTR KO mice (Zdebik et al., 2004). Furthermore, a recent study did not observe altered cAMP and Ca^{2+}-mediated Cl^- secretion in ClC-2 KO mice but rather that ClC-2 was required for electroneutral NaCl and KCl absorption in the colon (Catalan et al., 2012). In addition, ClC-2 localization differs among models tested as well as between gut segments. For example, in the human intestine, ClC-2 localized to a juxtanuclear compartment in the small intestine and basolateral membrane in the colon (Lipecka et al., 2002). However, ClC-2 activity may be coupled with CFTR since lubiprostone stimulation of ClC-2 and Cl^- secretion required WT CFTR (Bijvelds et al., 2009).

2.2.5 *The Enteric Nervous System*

The enteric nervous system (ENS) is a complex network of intrinsic primary afferent neurons, interneurons, and motor neurons that regulate the movement of water and electrolytes between the intestinal lumen and the bloodstream. Intrinsic primary afferent neurons detect physical changes such as mechanical stress/tension in the intestinal mucosa as well as luminal chemical changes that occur during normal digestion (Furness et al., 1999). In response to these changes, intrinsic primary afferent neurons generate reflexes to control motility, secretion and blood flow. Secretomotor (type of motor) neurons innervate the small intestinal and colonic crypts and goblet cells to control ion permeability to balance fluid secretion with absorptive processes (Furness et al., 1999). The release of neurotransmitters acetylcholine and vasoactive intestinal peptide by secretomotor neurons occurs in response to sympathetic innervations to stimulate Cl^- and HCO_3^- secretion (Reddix et al., 1994). Stimulation of epithelial anion secretion by secretomotor neurons is linked with vasodilation, which serves as the source of some of the secreted fluid.

Several studies have demonstrated that activation of the ENS stimulates intestinal fluid secretion during diarrheal diseases. In fact, 50% of fluid loss in secretory diarrheas, including rotavirus, is neuronally mediated (Lundgren, 2002; Lundgren and Jodal, 1997; Lundgren et al., 2000). In response to enteric pathogens, enteroendocrine cells release peptides or hormones (eg, serotonin, secretin, guanylin, etc.) which act on secretomotor neurons to induce fluid loss and increased motility (Furness et al., 1999). Overstimulation of secretion in response to enteric pathogens may lead to dehydration and possibly death.

3 GASTROINTESTINAL PATHOPHYSIOLOGY

Diarrhea is attributed to one or more of the following mechanisms that drive the excessive loss of fluid: osmosis, active secretion, exudation, and abnormal motility (Field, 2003). Osmotic diarrhea may occur when high concentrations of poorly absorbable solutes, such as lactulose, sorbitol, or magnesium, generate an osmotic gradient, driving mucosal to luminal fluid flow. Mutations in nutrient transport or metabolism (eg, lactose deficiency) that prevent digestion of otherwise absorbable nutrients (typically glucose or galactose) can also result in osmotic diarrhea. In the colon, nonabsorbable carbohydrates are metabolized by commensal microbes into short chain fatty acids such as propionate, butyrate, and acetate in concentrations higher than the fluid absorptive capacity of the colon. Secretory diarrhea is characterized by large stool volumes (>1 L) that lack red and white blood cells and an osmotic gap in electrolytes. Na^+, K^+, Cl^-, and HCO_3^- are the major luminal solutes which are either malabsorbed or hypersecreted in response to pathologic secretory stimuli including bacterial toxins, inflammatory mediators, or endocrine tumors (eg, carcinoid syndrome, VIPoma) (Montrose et al., 1999). Exudative diarrhea refers to the disruption of intestinal epithelial barrier integrity that occurs due to either the excess loss of epithelial cells or dysregulation of tight junction function. Under these conditions, increased hydrostatic pressure in blood vessels and lymphatics leak water and electrolytes as well as lose protein and blood cells to the intestinal lumen (Field, 2003). In motility disorders, increased and decreased motor functions have been found to induce diarrhea. Decreased motility can result in bacterial overgrowth that compromises the normal digestive capacity of the small intestine. Increased bacterial colonization results in deconjugation and rapid reabsorption of bile acids which effectively decreases the concentration of bile salts required for fat absorption (Mathias and Clench, 1985). Moreover, bacteria can also induce carbohydrate malabsorption by inactivating BB oligosacchridases necessary for epithelial uptake (Mathias and Clench, 1985).

Comparison of the four major viral causes of gastroenteritis (ie, rotaviruses, noroviruses, astroviruses, and enteric adenoviruses) in humans reveals similar pathologic characteristics such as site of infection, mechanisms of action, and type of diarrhea. These enteric viruses invade the intestinal epithelium of the small intestine, elevate intracellular Ca^{2+} concentrations, decrease nutrient absorption, stimulate Cl^- secretion, and induce epithelial cell loss, which all contribute to the moderate/severe diarrhea phenotype observed in human disease (Karst et al., 2015; Moser and Schultz-Cherry, 2005; Ramig, 2004). Each type of virus infection is associated with various extraintestinal phenotypes (eg, vomiting, nausea, viremia), and some of these are reviewed in Chapters 2.6, 2.9, and 3.4. The ontogeny of viral gastroenteritis in humans has been difficult to characterize due to limited availability of patient tissue samples during the early stages of infection. Thus, much of our understanding of viral pathogenesis has come from animal models, cancer derived intestinal epithelial cell lines, and the infection of the African green monkey kidney cell line, MA104, and other cell lines. The use of multiple model systems has confounded some observations which identified altered cellular and/or molecular events as mechanisms for diarrhea (Lorrot and Vasseur, 2007). For example, some data have indicated that viral infections induce intestinal epithelial cell loss resulting in epithelial lesions and thus a decreased capacity to absorb water, ions, and nutrients due to decreased intestinal surface area (Davidson et al., 1977). However, more recent animal studies of rotavirus, norovirus and astrovirus diarrhea have suggested that diarrhea occurs in the presence of an intact intestinal epithelial barrier and prior to any evidence of epithelial loss (Karst et al., 2015; Moser and Schultz-Cherry, 2005; Ramig, 2004; Lundgren and Svensson, 2001). Furthermore, decreased activity and amount of BB digestive enzymes were also observed in the absence of cell death (Jourdan et al., 1998). Taken together, these data suggest that viral pathogenesis disrupts normal intestinal epithelial physiology that may lead to the subsequent loss of epithelial cells observed in the later stages of viral infection. For further details see Chapter 2.4.

The lack of consistent evidence that epithelial damage is responsible for malabsorption and water loss suggests that viral-induced diarrhea may also result from alterations in intestinal absorption and secretion. Transport physiology studies designed to elucidate the exchangers, symporters, and/or channels responsible for the decreased ion and nutrient absorption and increase Cl^- secretion associated with viral gastroenteritis are incomplete. In rotavirus diarrhea, nutrient malabsorption is correlated with decreased expression of SGLT1 while increased Cl^- secretion has been suggested to involve CaCCs since RV-mediated Cl^- secretion can still occur in the absence of functional CFTR (Halaihel et al., 2000). A recent study reported that rotavirus can inhibit NHE3 activity as early as 30 min post infection in intestinal epithelial cells, suggesting that inhibition of electroneutral Na^+ absorption may contribute to RV diarrhea (Foulke-Abel et al., 2014). Astrovirus, enteric adenovirus, and norovirus diarrhea also present with primary malabsorptive and secretory phenotypes (Karst et al., 2015; Moser and Schultz-Cherry, 2005) and thus, additional transport studies are needed to identify the transporters involved.

Over the last two decades, considerable progress has been made to understand the molecular mechanisms responsible for normal ion and water absorption as well as for each of the previously described contributors to diarrhea. The identification and characterization of bacterial and viral enterotoxins (NSP4 for rotavirus; see Chapter 2.4) have provided valuable insights into the cellular and molecular mechanisms of diarrhea diseases. Mechanisms invoked include pathogen entry/invasion, further characterization of signal transduction pathways, identification of multiprotein regulatory complexes, predicting structure/function relationships, and protein trafficking (Hodges and Gill, 2010; Viswanathan et al., 2009). For example, several intestinal ion transporters and channels, including NHE3 and CFTR, contain PDZ [post synaptic density protein (PSD95), *Drosophila* disc large tumor suppressor (Dlg1), and zonula occludens-1 (ZO-1) protein] protein–protein interacting domains at their carboxy-terminal ends. These PDZ domains directly bind regulatory proteins that increase/decrease transporter/channel activity based on the binding partner. The NHERF (Na^+/H^+ exchanger regulatory factor) family of multi-PDZ domain containing proteins has been shown to regulate the activity of NHE3, CFTR, and DRA (among nearly 70 other targets) by affecting their turnover number and/or trafficking to and from the enterocyte BB (Donowitz et al., 2005). Other PDZ proteins have been shown to regulate tight junction function by affecting tight and adherence junctional complex formation, thereby modulating ion and small solute permeability through the paracellular network (Van Itallie and Anderson, 2014). Studies of bacterial infection, including enterotoxigenic, enteroaggregative, enterohemorrhagic, and enteropathogenic *E. coli*, have demonstrated decreased tight junction function due to mislocalization or extracellular damage of tight junction proteins (Nassour and Dubreuil, 2014; Simonovic et al., 2000; Strauman et al., 2010; Tomson et al., 2004). The effects of enteric viruses on tight junction function are limited as these studies are difficult to assess in vivo. Studies in MDCK and Caco-2 cell lines have reported decreased TER and increase permeability to large cargo (20 kDa) after RV infection (Dickman et al., 2000; Tafazoli et al., 2001). Whether these effects occur in other enteric viral infections in the human intestine remains to be determined.

Since the use of multiple models to study viral pathogens has limited our understanding of human disease, the recent development of the methodologies to culture primary human differentiated epithelial cells from the recently identified Lgr5+ stem cell has reinvigorated the study of gastrointestinal physiology and pathophysiology in general (Jung et al., 2011; Sato et al., 2011b). Primary human intestinal epithelial cultures may be derived from either human crypt based stem cells (termed enteroids/colonoids) or from a single inducible pluripotent stem cell (iPSC) line (termed intestinal organoid) (Jung et al., 2011;

Sato et al., 2011b; McCracken et al., 2011). These cultures are comprised of the complete complement of intestinal epithelial cells including immature enterocytes (that may be differentiated into more mature absorptive enterocytes), enteroendocrine, goblet, Paneth, tuft, and Lgr5+ stem cells. Enteroid/colonoid cultures represent the intestinal segment from which they are derived, and recent modifications generating intestinal organoid cultures have developed organoids with proximal and distal intestinal phenotypes (Watson et al., 2014). Furthermore, the human enteroid model has been shown to perform Na^+ absorptive and Cl^- secretory functions and is being used to understand the pathogenesis of host–pathogen interactions (Foulke-Abel et al., 2014; Kovbasnjuk et al., 2013). Recent work by the laboratory of Mary Estes has demonstrated that intestinal organoids and enteroids may be infected by RV and that each model has the capacity to facilitate viral replication resulting in the production of infectious viral progeny (Finkbeiner et al., 2012; Saxena et al., 2015). The development of enteroids/colonoids and intestinal organoids as models to study the human intestine offer the opportunity to study the pathogenesis of enteric viral infections.

4 CONCLUSIONS

Nearly 50 years of basic and clinical research has advanced our understanding of the complex regulation of fluid absorption that occurs under normal digestive conditions and in diarrheal diseases. However, much of this work has come from studies in animal models and transformed cell lines, and yet the most effective therapy for acute moderate to severe diarrheal diseases remains Oral Rehydration Solution (ORS). While ORS may save patients from life-threatening dehydration, it cannot prevent the long-term effects (eg, stunted growth, delayed intellectual development) that result from episodes of severe diarrhea (especially in children less than 5 years old). Therefore, additional medical therapies are needed to treat severe diarrheal diseases. Currently, our understanding of diarrhea due to enteric pathogens comes from studies of bacteria, and yet the mechanisms responsible for these secretory and enterotoxigenic diarrheas are not completely recapitulated in viral gastroenteritis, based on patient stool analysis and in vitro studies. In comparison to rotavirus diarrhea, similar patterns of pathogenesis (eg, site of infection, primary malabsorption, lack of histologic damage) have emerged relating to diarrhea phenotype, suggesting that similar mechanisms may be common among other types of enteric viruses. However, additional mechanistic studies are required to identify and characterize the membrane transport processes altered in each type of viral infection. The development of normal human intestinal epithelial models that are capable of viral replication now offers an opportunity to advance the field of viral gastroenteritis by providing a more relevant understanding of how enteric viruses affect normal human intestinal physiology and by hopefully uncovering novel therapeutic targets for future drug development.

5 ABBREVIATIONS

APN Aminopeptidase N
AQP Aquaporin
ASCT1 Na^+-dependent neutral amino acid transporter
B^0AT1 Na^+-dependent neutral amino acid transporter
BB Brush border
BPO Benzopyrimido-pyrrolo-oxazine-dione
CaCC Calcium-activated chloride channel
CF Cystic fibrosis
CFTR Cystic fibrosis transmembrane regulator
cGMP Cyclic guanosine monophosphate
ClC2 Chloride channel 2
CTx Cholera toxin
Dlg1 Drosophila disc large tumor suppressor
DPPIV Dipeptidyl peptidase IV
DRA Down regulated in adenoma
ENaC Epithelial Na^+ channel
ENS Enteric nervous system
GLUT2 Glucose transporter 2
GLUT5 Glucose transporter 5
GlyH Glycin hydrazide
iPSC Inducible pluripotent stem cell
KO Knock-out
Lgr5 Leucine-rich G-protein coupled receptor 5
MUC Mucin
nAA Neutral amino acid
NDO Nondigestible oligosaccharides
NHE3 Na^+/H^+ exchanger 3
NHERF Na^+/H^+ exchanger regulatory factor
Na + /K+ ATPase Na^+/K^+ adenosine triphosphatase
NKCC1 $Na^+/K^+/2Cl^-$ symporter
ORS Oral rehydration solution
PAT1 Putative anion transporter 1
PEPT1 Peptide transporter 1
PDZ domain Postsynaptic density protein 95 (PSD-95)/*Drosophila* disc large tumor suppressor (Dlg1)/ZO-1
PPQ Pyrimido-pyrrolo-quinoxalinedione
RV rotavirus
SCFA Short chain fatty acids
SGLT-1 Na^+-dependent glucose cotransporter
STa Heat stable enterotoxin
TER Transepithelial electrical resistance
TJ Tight junction
TMEM16A Anoctamin 1, calcium-activated chloride channel
ZO-1 Zona occludens 1

REFERENCES

Agre, P., Kozono, D., 2003. Aquaporin water channels: molecular mechanisms for human diseases. FEBS Lett. 555 (1), 72–78.

Anderson, J.M., Van Itallie, C.M., 2006. Tight Juction Channels. In: Gonzalez-Mariscal, L. (Ed.), Tight Junctions. Landes Biosciences & Springer Science + Business Media, LLC, New York, pp. 33–42.

Au, A., Gupta, A., Schembri, P., Cheeseman, C.I., 2002. Rapid insertion of GLUT2 into the rat jejunal brush-border membrane promoted by glucagon-like peptide 2. Biochem. J. 367 (Pt 1), 247–254.

Babji, S., Kang, G., 2012. Rotavirus vaccination in developing countries. Curr. Opin. Virol. 2 (4), 443–448.

Barker, N., van de Wetering, M., Clevers, H., 2008. The intestinal stem cell. Genes Dev. 22 (14), 1856–1864.

Bevins, C.L., Martin-Porter, E., Ganz, T., 1999. Defensins and innate host defence of the gastrointestinal tract. Gut 45 (6), 911–915.

Bijvelds, M.J., Bot, A.G., Escher, J.C., De Jonge, H.R., 2009. Activation of intestinal Cl- secretion by lubiprostone requires the cystic fibrosis transmembrane conductance regulator. Gastroenterology 137 (3), 976–985.

Binder, H.J., Brown, I., Ramakrishna, B.S., Young, G.P., 2014. Oral rehydration therapy in the second decade of the twenty-first century. Curr. Gastroenterol. Rep. 16 (3), 376.

Broer, S., 2008. Amino acid transport across mammalian intestinal and renal epithelia. Physiol. Rev. 88 (1), 249–286.

Cao, M., Yang, M., Ou, Z., Li, D., Geng, L., Chen, P., et al., 2014. Involvement of aquaporins in a mouse model of rotavirus diarrhea. Virol. Sin. 29 (4), 211–217.

Catalan, M.A., Flores, C.A., Gonzalez-Begne, M., Zhang, Y., Sepulveda, F.V., Melvin, J.E., 2012. Severe defects in absorptive ion transport in distal colons of mice that lack ClC-2 channels. Gastroenterology 142 (2), 346–354.

Chaudhry, R.M., Scow, J.S., Madhavan, S., Duenes, J.A., Sarr, M.G., 2012. Acute enterocyte adaptation to luminal glucose: a posttranslational mechanism for rapid apical recruitment of the transporter GLUT2. J. Gastrointest. Surg. 16 (2), 312–319.

Chen, C.C., Baylor, M., Bass, D.M., 1993. Murine intestinal mucins inhibit rotavirus infection. Gastroenterology 105 (1), 84–92.

Chen, T., Kocinsky, H.S., Cha, B., Murtazina, R., Yang, J., Tse, C.M., et al., 2015. Cyclic GMP kinase II (cGKII) inhibits NHE3 by altering its trafficking and phosphorylating NHE3 at three required sites: identification of a multifunctional phosphorylation site. J. Biol. Chem. 290 (4), 1952–1965.

Cheng, H., Leblond, C.P., 1974. Origin, differentiation and renewal of the four main epithelial cell types in the mouse small intestine. V. Unitarian Theory of the origin of the four epithelial cell types. Am. J. Anat. 141 (4), 537–561.

Daniel, H., 2004. Molecular and integrative physiology of intestinal peptide transport. Annu. Rev. Physiol. 66, 361–384.

Davidson, G.P., Gall, D.G., Petric, M., Butler, D.G., Hamilton, J.R., 1977. Human rotavirus enteritis induced in conventional piglets. Intestinal structure and transport. J. Clin. Invest. 60 (6), 1402–1409.

Dekkers, J.F., Wiegerinck, C.L., de Jonge, H.R., Bronsveld, I., Janssens, H.M., de Winter-de Groot, K.M., et al., 2013. A functional CFTR assay using primary cystic fibrosis intestinal organoids. Nat. Med. 19 (7), 939–945.

Dickman, K.G., Hempson, S.J., Anderson, J., Lippe, S., Zhao, L., Burakoff, R., et al., 2000. Rotavirus alters paracellular permeability and energy metabolism in Caco-2 cells. Am. J. Physiol. Gastrointest. Liver Physiol. 279 (4), G757–G766.

Donowitz, M., Tse, C.M., 2000. Molecular physiology of mammalian epithelial Na + /H+ exchangers NHE2 and NHE3. Curr. Topics Membr. 50, 437–451.

Donowitz, M., Cha, B., Zachos, N.C., Brett, C.L., Sharma, A., Tse, C.M., et al., 2005. NHERF family and NHE3 regulation. J. Physiol. 567 (Pt 1), 3–11.

Field, M., 2003. Intestinal ion transport and the pathophysiology of diarrhea. J. Clin. Invest. 111 (7), 931–943.

Finkbeiner, S.R., Zeng, X.L., Utama, B., Atmar, R.L., Shroyer, N.F., Estes, M.K., 2012. Stem cell-derived human intestinal organoids as an infection model for rotaviruses. MBio 3 (4), e00159–e00212.

Fischer, H., Machen, T.E., Widdicombe, J.H., Carlson, T.J., King, S.R., Chow, J.W., et al., 2004. A novel extract SB-300 from the stem bark latex of Croton lechleri inhibits CFTR-mediated chloride secretion in human colonic epithelial cells. J. Ethnopharmacol. 93 (2–3), 351–357.

Foulke-Abel, J., In, J., Kovbasnjuk, O., Zachos, N.C., Ettayebi, K., Blutt, S.E., et al., 2014. Human enteroids as an ex-vivo model of host–pathogen interactions in the gastrointestinal tract. Exp. Biol. Med. 239 (9), 1124–1134.

Furness, J.B., Bornstein, J.C., Kunze, W.A.A., Clerc, N., 1999. The enteric nervous system and its intrinsic connections. In: Yamada, T., Alpers, D.H., Laine, L., Owyang, C., Powell, D.W. (Eds.), Textbook of Gastroenterology. Lippincott Williams and Wilkins, Philadelphia, pp. 11–34.

Gouyon, F., Caillaud, L., Carriere, V., Klein, C., Dalet, V., Citadelle, D., et al., 2003. Simple-sugar meals target GLUT2 at enterocyte apical membranes to improve sugar absorption: a study in GLUT2-null mice. J. Physiol. 552 (Pt 3), 823–832.

Guttman, J.A., Finlay, B.B., 2008. Subcellular alterations that lead to diarrhea during bacterial pathogenesis. Trends Microbiol. 16 (11), 535–542.

Guttman, J.A., Samji, F.N., Li, Y., Deng, W., Lin, A., Finlay, B.B., 2007. Aquaporins contribute to diarrhoea caused by attaching and effacing bacterial pathogens. Cell Microbiol. 9 (1), 131–141.

Halaihel, N., Lievin, V., Alvarado, F., Vasseur, M., 2000. Rotavirus infection impairs intestinal brush-border membrane Na(+)-solute cotransport activities in young rabbits. Am. J. Physiol. Gastrointest. Liver Physiol. 279 (3), G587–G596.

Hartley, J.L., Zachos, N.C., Dawood, B., Donowitz, M., Forman, J., Pollitt, R.J., et al., 2010. Mutations in TTC37 cause trichohepatoenteric syndrome (phenotypic diarrhea of infancy). Gastroenterology 138 (7), 2388–2398.

Heitzmann, D., Warth, R., 2008. Physiology and pathophysiology of potassium channels in gastrointestinal epithelia. Physiol. Rev. 88 (3), 1119–1182.

Hodges, K., Gill, R., 2010. Infectious diarrhea: cellular and molecular mechanisms. Gut Microbes 1 (1), 4–21.

Holmes, J.L., Van Itallie, C.M., Rasmussen, J.E., Anderson, J.M., 2006. Claudin profiling in the mouse during postnatal intestinal development and along the gastrointestinal tract reveals complex expression patterns. Gene Expr. Patterns 6 (6), 581–588.

Hu, Z., Wang, Y., Graham, W.V., Su, L., Musch, M.W., Turner, J.R., 2006. MAPKAPK-2 is a critical signaling intermediate in NHE3 activation following Na^+-glucose cotransport. J. Biol. Chem. 281 (34), 24247–24253.

Johansson, M.E., Sjovall, H., Hansson, G.C., 2013. The gastrointestinal mucus system in health and disease. Nat. Rev. Gastroenterol. Hepatol. 10 (6), 352–361.

Jourdan, N., Brunet, J.P., Sapin, C., Blais, A., Cotte-Laffitte, J., Forestier, F., et al., 1998. Rotavirus infection reduces sucrase-isomaltase expression in human intestinal epithelial cells by perturbing protein targeting and organization of microvillar cytoskeleton. J. Virol. 72 (9), 7228–7236.

Jung, P., Sato, T., Merlos-Suarez, A., Barriga, F.M., Iglesias, M., Rossell, D., et al., 2011. Isolation and in vitro expansion of human colonic stem cells. Nat. Med. 17 (10), 1225–1227.

Karst, S.M., Zhu, S., Goodfellow, I.G., 2015. The molecular pathology of noroviruses. J. Pathol. 235 (2), 206–216.

Kennedy, D.J., Leibach, F.H., Ganapathy, V., Thwaites, D.T., 2002. Optimal absorptive transport of the dipeptide glycylsarcosine is dependent on functional Na^+/H^+ exchange activity. Pflugers Arch. 445 (1), 139–146.

Kotloff, K.L., Nataro, J.P., Blackwelder, W.C., Nasrin, D., Farag, T.H., Panchalingam, S., et al., 2013. Burden and aetiology of diarrhoeal disease in infants and young children in developing countries (the Global Enteric Multicenter Study, GEMS): a prospective, case-control study. Lancet 382 (9888), 209–222.

Kovbasnjuk, O., Zachos, N.C., In, J., Foulke-Abel, J., Ettayebi, K., Hyser, J.M., et al., 2013. Human enteroids: preclinical models of non-inflammatory diarrhea. Stem Cell Res. Ther. 4 (Suppl. 1), S3.

Kunzelmann, K., Hubner, M., Schreiber, R., Levy-Holzman, R., Garty, H., Bleich, M., et al., 2001. Cloning and function of the rat colonic epithelial K+ channel KVLQT1. J. Membr. Biol. 179 (2), 155–164.

Laforenza, U., 2012. Water channel proteins in the gastrointestinal tract. Mol. Aspects. Med. 33 (5–6), 642–650.

Lin, R., Murtazina, R., Cha, B., Chakraborty, M., Sarker, R., Chen, T.E., et al., 2011. D-glucose acts via sodium/glucose cotransporter 1 to increase NHE3 in mouse jejunal brush border by a Na^+/H^+ exchange regulatory factor 2-dependent process. Gastroenterology 140 (2), 560–571.

Lipecka, J., Bali, M., Thomas, A., Fanen, P., Edelman, A., Fritsch, J., 2002. Distribution of ClC-2 chloride channel in rat and human epithelial tissues. Am. J. Physiol. Cell Physiol. 282 (4), C805–C816.

Lorrot, M., Vasseur, M., 2007. How do the rotavirus NSP4 and bacterial enterotoxins lead differently to diarrhea? Virol. J. 4, 31.

Lundgren, O., 2002. Enteric nerves and diarrhoea. Pharmacol. Toxicol. 90 (3), 109–120.

Lundgren, O., Jodal, M., 1997. The enteric nervous system and cholera toxin-induced secretion. Comp. Biochem. Physiol. A 118 (2), 319–327.

Lundgren, O., Svensson, L., 2001. Pathogenesis of rotavirus diarrhea. Microbes Infect. 3 (13), 1145–1156.

Lundgren, O., Peregrin, A.T., Persson, K., Kordasti, S., Uhnoo, I., Svensson, L., 2000. Role of the enteric nervous system in the fluid and electrolyte secretion of rotavirus diarrhea. Science 287 (5452), 491–495.

Mace, O.J., Affleck, J., Patel, N., Kellett, G.L., 2007. Sweet taste receptors in rat small intestine stimulate glucose absorption through apical GLUT2. J. Physiol. 582 (Pt 1), 379–392.

Maher, M.M., Gontarek, J.D., Jimenez, R.E., Donowitz, M., Yeo, C.J., 1996. Role of brush border Na^+/H^+ exchange in canine ileal absorption. Dig. Dis. Sci. 41 (4), 651–659.

Maher, M.M., Gontarek, J.D., Bess, R.S., Donowitz, M., Yeo, C.J., 1997. The Na^+/H^+ exchange isoform NHE3 regulates basal canine ileal Na+ absorption in vivo. Gastroenterology 112 (1), 174–183.

Mathias, J.R., Clench, M.H., 1985. Review: pathophysiology of diarrhea caused by bacterial overgrowth of the small intestine. Am. J. Med. Sci. 289 (6), 243–248.

McCracken, K.W., Howell, J.C., Wells, J.M., Spence, J.R., 2011. Generating human intestinal tissue from pluripotent stem cells in vitro. Nat. Protoc. 6 (12), 1920–1928.

Melvin, J.E., Park, K., Richardson, L., Schultheis, P.J., Shull, G.E., 1999. Mouse down-regulated in adenoma (DRA) is an intestinal Cl(-)/HCO(3)(-) exchanger and is up-regulated in colon of mice lacking the NHE3 Na(+)/H(+) exchanger. J. Biol. Chem. 274 (32), 22855–28561.

Middendorp, S., Schneeberger, K., Wiegerinck, C.L., Mokry, M., Akkerman, R.D., van Wijngaarden, S., et al., 2014. Adult stem cells in the small intestine are intrinsically programmed with their location-specific function. Stem Cells 32 (5), 1083–1091.

Moens, E., Veldhoen, M., 2012. Epithelial barrier biology: good fences make good neighbours. Immunology 135 (1), 1–8.

Montrose, M.H., Keely, S.J., Barrett, K.E., 1999. Electrolyte secretion and absorption: small intestine and colon. In: Yamada, T., Alpers, D.H., Laine, L., Owyang, C., Powell, D.W. (Eds.), Textbook of Gastroenterology. Lippincott Williams and Wilkins, Philadelphia, pp. 320–354.

Morris, A.P., Estes, M.K., 2001. Microbes and microbial toxins: paradigms for microbial-mucosal interactions. VIII. Pathological consequences of rotavirus infection and its enterotoxin. Am. J. Physiol. Gastrointest. Liver Physiol. 281 (2), G303–G310.

Moser, L.A., Schultz-Cherry, S., 2005. Pathogenesis of astrovirus infection. Viral Immunol. 18 (1), 4–10.

Mueckler, M., 1994. Facilitative glucose transporters. Eur. J. Biochem. 219 (3), 713–725.

Musch, M.W., Arvans, D.L., Wu, G.D., Chang, E.B., 2009. Functional coupling of the downregulated in adenoma Cl-/base exchanger DRA and the apical Na^+/H^+ exchangers NHE2 and NHE3. Am. J. Physiol. Gastrointest. Liver Physiol. 296 (2), G202–G210.

Nassour, H., Dubreuil, J.D., 2014. Escherichia coli STb enterotoxin dislodges claudin-1 from epithelial tight junctions. PLoS One 9 (11), e113273.

Payne, J.A., Xu, J.C., Haas, M., Lytle, C.Y., Ward, D., Forbush, 3rd., B., 1995. Primary structure, functional expression, and chromosomal localization of the bumetanide-sensitive Na-K-Cl cotransporter in human colon. J. Biol. Chem. 270 (30), 17977–17985.

Raja, M., Puntheeranurak, T., Hinterdorfer, P., Kinne, R., 2012. SLC5 and SLC2 transporters in epithelia-cellular role and molecular mechanisms. Curr. Top. Membr. 70, 29–76.

Ramig, R.F., 2004. Pathogenesis of intestinal and systemic rotavirus infection. J. Virol. 78 (19), 10213–10220.

Reddix, R., Kuhawara, A., Wallace, L., Cooke, H.J., 1994. Vasoactive intestinal polypeptide: a transmitter in submucous neurons mediating secretion in guinea pig distal colon. J. Pharmacol. Exp. Ther. 269 (3), 1124–1129.

Roder, P.V., Geillinger, K.E., Zietek, T.S., Thorens, B., Koepsell, H., Daniel, H., 2014. The role of SGLT1 and GLUT2 in intestinal glucose transport and sensing. PLoS One 9 (2), e89977.

Sato, T., van Es, J.H., Snippert, H.J., Stange, D.E., Vries, R.G., van den Born, M., et al., 2011a. Paneth cells constitute the niche for Lgr5 stem cells in intestinal crypts. Nature 469 (7330), 415–418.

Sato, T., Stange, D.E., Ferrante, M., Vries, R.G., Van Es, J.H., Van den Brink, S., et al., 2011b. Long-term expansion of epithelial organoids from human colon, adenoma, adenocarcinoma, and Barrett's epithelium. Gastroenterology 141 (5), 1762–1772.

Saxena, K., Blutt, S.E., Ettayebi, K., Zeng, X.L., Broughman, J.R., Crawford, S.E., et al., 2015. Human intestinal enteroids: a new model to study human rotavirus infection, host restriction and pathophysiology. J. Virol. 90 (1), 43–56.

Shepherd, E.J., Helliwell, P.A., Mace, O.J., Morgan, E.L., Patel, N., Kellett, G.L., 2004. Stress and glucocorticoid inhibit apical GLUT2-trafficking and intestinal glucose absorption in rat small intestine. J. Physiol. 560 (Pt 1), 281–290.

Shirazi-Beechey, S.P., Moran, A.W., Batchelor, D.J., Daly, K., Al-Rammahi, M., 2011. Glucose sensing and signalling; regulation of intestinal glucose transport. Proc. Nutr. Soc. 70 (2), 185–193.

Simonovic, I., Rosenberg, J., Koutsouris, A., Hecht, G., 2000. Enteropathogenic Escherichia coli dephosphorylates and dissociates occludin from intestinal epithelial tight junctions. Cell Microbiol. 2 (4), 305–315.

Simpson, J.E., Walker, N.M., Supuran, C.T., Soleimani, M., Clarke, L.L., 2010. Putative anion transporter-1 (Pat-1, Slc26a6) contributes to intracellular pH regulation during H + -dipeptide transport in duodenal villous epithelium. Am. J. Physiol. Gastrointest. Liver Physiol. 298 (5), G683–G691.

Strauman, M.C., Harper, J.M., Harrington, S.M., Boll, E.J., Nataro, J.P., 2010. Enteroaggregative Escherichia coli disrupts epithelial cell tight junctions. Infect. Immun. 78 (11), 4958–4964.

Tabcharani, J.A., Chang, X.B., Riordan, J.R., Hanrahan, J.W., 1991. Phosphorylation-regulated Cl^- channel in CHO cells stably expressing the cystic fibrosis gene. Nature 352 (6336), 628–631.

Tafazoli, F., Zeng, C.Q., Estes, M.K., Magnusson, K.E., Svensson, L., 2001. NSP4 enterotoxin of rotavirus induces paracellular leakage in polarized epithelial cells. J. Virol. 75 (3), 1540–1546.

Tate, J.E., Haynes, A., Payne, D.C., Cortese, M.M., Lopman, B.A., Patel, M.M., et al., 2013. Trends in national rotavirus activity before and after introduction of rotavirus vaccine into the national immunization program in the United States, 2000 to 2012. Pediatr. Infect. Dis. J. 32 (7), 741–744.

Thiagarajah, J.R., Ko, E.A., Tradtrantip, L., Donowitz, M., Verkman, A.S., 2014. Discovery and development of antisecretory drugs for treating diarrheal diseases. Clin. Gastroenterol. Hepatol. 12 (2), 204–209.

Thiagarajah, J.R., Donowitz, M., Verkman, A.S., 2015. Secretory diarrhoea: mechanisms and emerging therapies. Nat. Rev. Gastroenterol. Hepatol. 12 (8), 446–457.

Tobin, V., Le Gall, M., Fioramonti, X., Stolarczyk, E., Blazquez, A.G., Klein, C., et al., 2008. Insulin internalizes GLUT2 in the enterocytes of healthy but not insulin-resistant mice. Diabetes 57 (3), 555–562.

Tomson, F.L., Koutsouris, A., Viswanathan, V.K., Turner, J.R., Savkovic, S.D., Hecht, G., 2004. Differing roles of protein kinase C-zeta in disruption of tight junction barrier by enteropathogenic and enterohemorrhagic Escherichia coli. Gastroenterology 127 (3), 859–869.

Tradtrantip, L., Ko, E.A., Verkman, A.S., 2014. Antidiarrheal efficacy and cellular mechanisms of a Thai herbal remedy. PLoS Negl. Trop. Dis. 8 (2), e2674.

Van Itallie, C.M., Anderson, J.M., 2014. Architecture of tight junctions and principles of molecular composition. Semin. Cell Dev. Biol. 36, 157–165.

Viswanathan, V.K., Hodges, K., Hecht, G., 2009. Enteric infection meets intestinal function: how bacterial pathogens cause diarrhoea. Nat. Rev. Microbiol. 7 (2), 110–119.

Wagner, C.A., Ott, M., Klingel, K., Beck, S., Melzig, J., Friedrich, B., et al., 2001. Effects of the serine/threonine kinase SGK1 on the epithelial Na(+) channel (ENaC) and CFTR: implications for cystic fibrosis. Cell Physiol. Biochem. 11 (4), 209–218.

Walker, N.M., Simpson, J.E., Yen, P.F., Gill, R.K., Rigsby, E.V., Brazill, J.M., et al., 2008. Downregulated in adenoma Cl/HCO3 exchanger couples with Na/H exchanger 3 for NaCl absorption in murine small intestine. Gastroenterology 135 (5), 1645–1653 e3.

Watson, C.L., Mahe, M.M., Munera, J., Howell, J.C., Sundaram, N., Poling, H.M., et al., 2014. An in vivo model of human small intestine using pluripotent stem cells. Nat. Med. 20 (11), 1310–1314.

Xia, W., Yu, Q., Riederer, B., Singh, A.K., Engelhardt, R., Yeruva, S., et al., 2014. The distinct roles of anion transporters Slc26a3 (DRA) and Slc26a6 (PAT-1) in fluid and electrolyte absorption in the murine small intestine. Pflugers Arch. 466 (8), 1541–1556.

Zachos, N.C., Tse, M., Donowitz, M., 2005. Molecular physiology of intestinal Na^+/H^+ exchange. Annu. Rev. Physiol. 67, 411–443.

Zdebik, A.A., Cuffe, J.E., Bertog, M., Korbmacher, C., Jentsch, T.J., 2004. Additional disruption of the ClC-2 Cl(-) channel does not exacerbate the cystic fibrosis phenotype of cystic fibrosis transmembrane conductance regulator mouse models. J. Biol. Chem. 279 (21), 22276–22283.

Chapter 1.2

Immunity in the Gut: Mechanisms and Functions

P. Brandtzaeg
Laboratory for Immunohistochemistry and Immunopathology (LIIPAT), Centre for Immune Regulation (CIR), University of Oslo and Department of Pathology, Oslo University Hospital Rikshospitalet, Oslo, Norway

1 INTRODUCTION

Mammalian host defense has successfully handled environmental confrontations over millions of years. To this end, numerous genes involved in innate and adaptive immunity have been subjected to evolutionary modifications, thus being shaped according to microbial pressure and other exogenous impacts. This modulation has been influenced by various ways of living such as hunting, fishing, gathering, agriculture, and animal husbandry.

The major arena for the complex interactions between genes and the exogenous impact is the gut. This is therefore the most crucial organ for communication between the environment and the body (Fig. 1.2.1). In an adult human being, the intestinal epithelium covers a surface area of perhaps 300 m^2 when villi, microvilli, crypts, and folds are taken into account. This barrier has generally only one cell layer and is therefore vulnerable but normally well protected by numerous chemical and physical innate defense mechanisms which cooperate intimately with a local adaptive immune system (Brandtzaeg, 2013a). The dominating component of the latter is an immunoglobulin A (IgA)-generating mucosal B-cell population which basically provides an antiinflammatory first-line defense by giving rise to secretory IgA (SIgA) antibodies performing "immune exclusion". This term is coined for low- and high-affinity antibody functions at the mucosal surface, aiming to control both microbial colonization and penetration of noxious antigens through the epithelial barrier (Brandtzaeg, 2013b).

Acute or protracted diarrhea is a sign of barrier break in the gut and an increasing variety of viral agents are suspected in the etiology (Desselberger, 2000; Glass et al., 2001). Both DNA and RNA viruses are obligate intracellular microorganisms, so those infecting via the gut must in one way or another affect the epithelial lining. However, the inconsistently observed damage

Viral Gastroenteritis

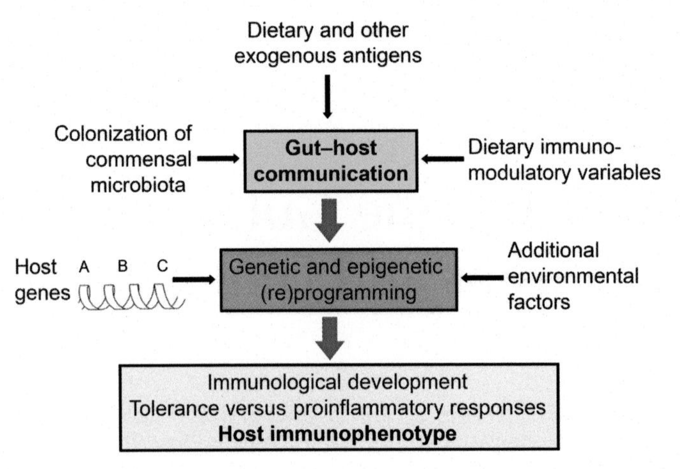

FIGURE 1.2.1 **Influence of exogenous variables on gut–host communication and immunological development.** The gut represents a large contact organ that transmits biological signals from the environment to the host. These signals are essential for the development of adaptive immunity and may determine the balance between tolerance and proinflammatory responses. The interaction of exogenous factors with host genes (A, B, C, etc.) provides opportunities not only for programming but also for reprogramming of the immune system and may be influenced by epigenetic variables. The immunophenotype of the host may therefore be determined at various levels, partly through epigenetic regulation. *(Source: Based on information discussed in Renz et al., 2011.)*

of enterocytes, villous atrophy, and crypt hyperplasia should not necessarily be ascribed to cytopathic effects of the actual viruses but may reflect immunopathology resulting from an immune attack on the infected host cell or from immune-mediated inflammation. Due to ethical considerations, there are only few studies of the pathogenesis of mucosal injury caused by gastroenteritis in humans (Davidson and Barnes, 1979; Philips, 1989). Overt or obscure cellular damage could result from complement- or T-cell-mediated cytotoxicity, antibody-dependent cell-mediated cytotoxicity (ADCC), complement-induced inflammation, or activation of Th1 or Th17 lymphocytes which secrete proinflammatory cytokines. However, similar immune reactions are involved in virus elimination and mucosal healing; the clinical outcome most likely reflects the balance that is continuously evolving in the fine-tuning of the respective defense mechanisms of the pathogen and its host. An inherent biological problem in this field is that host defense against pathogenic microorganisms and immunopathology basically go hand-in-hand (Shacklett, 2010).

The main focus of this chapter is the adaptive defense mechanisms exerted by the intestinal mucosal immune system. Secretory immunity is of special interest because asymptomatic virus infections, or prevention of infections, seems to be related to a relatively high local antibody level—at least for poliovirus (Ogra and Karzon, 1969) and apparently also for rotavirus (Clarke and Desselberger, 2015). Experiments in mice and gnotobiotic pigs support this notion by showing that improved protection correlates with an enhanced specific B-cell response in the gut after oral rotavirus inoculation (Moser and Offit, 2001).

2 NEONATAL ADAPTIVE MUCOSAL IMMUNITY

The enterocytes play a vital role in the defense of the neonate, not only by forming a mechanical barrier but also by transferring breast milk-derived maternal antibodies from the gut lumen, thus providing passive systemic immunity in the newborn period. This enterocytic Ig transmission differs remarkably among species. In the ungulate (horse, cattle, sheep, pig) the whole length of the intestine is involved in a nonselective protein uptake, including all Ig isotypes in a poorly defined pinocytotic process. Since colostrum of these animals is particularly rich in IgG, this antibody class will preferentially reach the circulation of the neonate via its gut epithelium during the two first postnatal days, after which so-called "gut closure" takes place (Mackenzie, 1990). Rodents, on the other hand, express an Fc receptor specific for IgG apically on neonatal enterocytes in the proximal small intestine. This receptor, FcRn, which disappears at weaning, has been particularly well characterized on enterocytes of the neonatal rat; it is a major histocompatibility complex (MHC) class I-related molecule associated with β2-microglobulin (Simister and Mostov, 1989). Complexes of FcRn and IgG are internalized in clathrin-coated pits at the base of the microvilli; binding of the ligand takes place in the acidic luminal environment, and IgG release occurs at physiological pH at the basolateral face of the enterocyte, after which the receptor is recycled.

Contrary to the previously mentioned species, the human fetus acquires maternal IgG via the placenta (Mackenzie, 1990), and perhaps to some extent from swallowed amniotic fluid via FcRn expressed by fetal enterocytes (Israel et al., 1993). Indeed, a bidirectional transport mechanism for IgG has been demonstrated in a human intestinal epithelial cell line (Dickinson et al., 1999), but the functional significance of FcRn on enterocytes in the human newborn remains unknown. Intestinal uptake of SIgA antibodies after breast-feeding appears of little or no importance in the support of systemic immunity (Klemola et al., 1986), except perhaps in the preterm infant (Weaver et al., 1991). Although gut closure in humans normally seems to occur mainly before birth, a patent mucosal barrier function may not be established until after 2 years of age; the different variables involved in this process are poorly defined.

Although breastfeeding initially may reduce the induction of SIgA by antibody shielding of the infant's intestinal immune system, it appears later on in infancy (up to 8 months) to boost secretory immunity (Brandtzaeg, 2013c). One possibility is that SIgA antibodies in mother's milk guide the uptake of cognate luminal antigens via receptors for IgA on M cells of follicle-associated epithelium (FAE) covering the infant's Peyer's patches. Mouse experiments suggest that the antigens may further be targeted to dendritic cells (DCs) which migrate to mesenteric lymph nodes where they induce a homeostatic immune response

(Mathias et al., 2014). One possibility is that SIgA interact with DCs through the specific intercellular adhesion molecule (ICAM)-3 grabbing nonintegrin receptor 1 (SIGNR1)—a mouse homolog of the human C-type lectin receptor DC-SIGN. DC-SIGNR1 interaction has been shown to induce Treg cells and antiinflammatory IL-10 (Monteiro, 2014). Therefore, the remarkable output of SIgA during feeding represents an optimally targeted passive immunization of the breast-fed infant's gut, which might serve as a positive homeostatic feedback loop.

3 POSTNATAL ADAPTIVE MUCOSAL IMMUNITY

Most babies growing up under privileged conditions show remarkably good resistance to infections if their innate nonspecific mucosal defense mechanisms are adequately developed. This can be explained by the fact that immune protection of their mucosae is additionally provided by maternal IgG antibodies, which are distributed in interstitial tissue fluid at a concentration 50–60% of the intravascular level (Offit et al., 2002). In the first postnatal period, only occasional traces of SIgA and secretory IgM (SIgM) normally occur in intestinal juice, whereas some IgG is more often present—either as a result of epithelial FcRn-mediated translocation or, perhaps more likely, mainly passive paracellular leakage from the highly vascularized lamina propria (Persson et al., 1998), which particularly after 34 weeks of gestation contains readily detectable maternal IgG (Brandtzaeg et al., 1991). However, an optimal mucosal barrier function in the neonatal period depends on an appropriate supply of breast milk, as highlighted in relation to mucosal infections, especially in the developing countries (Brandtzaeg, 2013c). Exclusively breast-fed infants are better protected against a variety of infections, and epidemiological data suggest that the risk of dying from diarrhea is reduced 14–24 times in nursed children. In the westernized part of the world, the protective value of breast-feeding is clinically evident in relation to mucosal infections, being most apparent in preterm infants. The role of secretory antibodies for mucosal homeostasis is furthermore supported by the fact that knock-out mice lacking SIgA and SIgM show increased mucosal leakiness and overstimulation of systemic (IgG) immunity to *Escherichia coli* (Johansen et al., 1999) and also to other commensal bacteria (Sait et al., 2007).

After the peak of passive immunity mediated by systemic maternal IgG antibodies and SIgA from breast milk, the survival of the infant will to an increasing extent depend on its own adaptive immune responses (Offit et al., 2002). At mucosal surfaces such responses are largely expressed by local IgA antibody production (Brandtzaeg, 2013b, 2015a). In addition, hypersensitivity due to penetration of exogenous antigens into the mucosa is normally counteracted by adaptive hyporesponsiveness to innocuous agents (Fig. 1.2.2). This phenomenon is traditionally referred to as "oral tolerance" when induced via the gut; it inhibits in particular overreaction to dietary proteins and components of

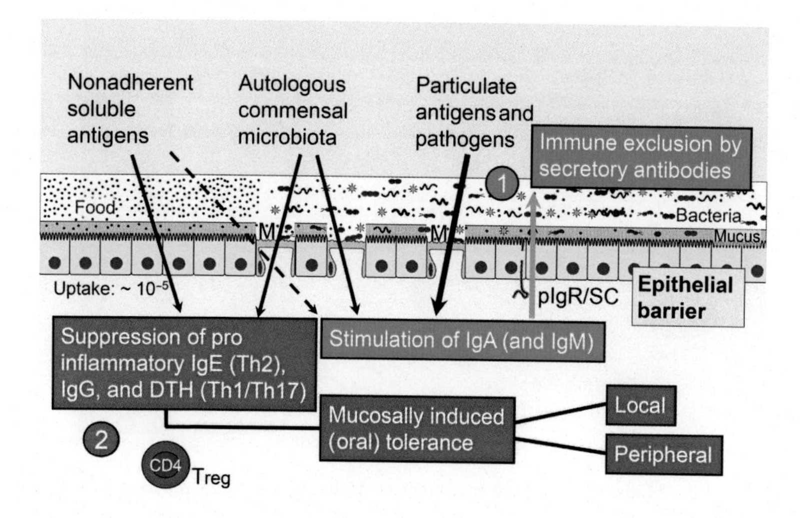

FIGURE 1.2.2 **Two levels of intestinal antiinflammatory immune defense aiming at preserved integrity of the epithelial barrier.** (1) Secretory immunity providing immune exclusion to limit microbial colonization and penetration of harmful agents. This first line of defense is principally mediated by polymeric antibodies of the IgA (and IgM) class in cooperation with various innate defense mechanisms (not shown; for rotavirus see Chapter 2.8). Polymeric antibodies are exported by transcytosis after interaction with, and apical clevage of, the epithelial polymeric Ig receptor *(pIgR)*, also called membrane secretory component *(SC)*. Secretory immunity is preferentially stimulated by pathogens and other particulate antigens taken up through specialized M cells *(M)* located in the dome epithelium covering inductive mucosa-associated lymphoid tissue (Fig. 1.2.4). (2) Innocuous soluble antigens (eg, food proteins; magnitude of uptake after a meal is indicated) and the commensal microbiota are also stimulatory for secretory immunity, but induce additionally suppression of pro-inflammatory Th2-dependent responses (*IgE* antibodies), Th1-dependent delayed-type hypersensitivity *(DTH)*, IgG antibodies, and Th17-dependent neutrophilic reactions (graded arrows indicate presumed importance of stimulatory pathways). This Th-cell balance is regulated by a complex mucosally induced phenomenon called "oral tolerance" in the gut, in which regulatory T *(Treg)* cells are important. Their suppressive effects can be observed both locally and in the periphery.

the commensal microbiota as long as the epithelial barrier remains fairly intact (Brandtzaeg, 2013a).

The cellular basis for the SIgA-mediated defense is the fact that exocrine glands and secretory mucosae contain most of the body's activated B cells—terminally differentiated to Ig-producing plasmablasts and plasma cells (collectively referred to as PCs). Most of these mucosal PCs (70–90%) produce dimers and some larger polymers of IgA (collectively referred to as pIgA) which, along with pentameric IgM, can be transcytosed through serous-type of secretory epithelia to act as SIgA and SIgM in surface defense (Fig. 1.2.3). This function depends on the polymeric Ig receptor (pIgR)—a 110-kDa epithelial glycoprotein also known as membrane SC—and the presence of a small (~15 kDa) joining (J) peptide in the Ig polymers (Brandtzaeg and Prydz, 1984).

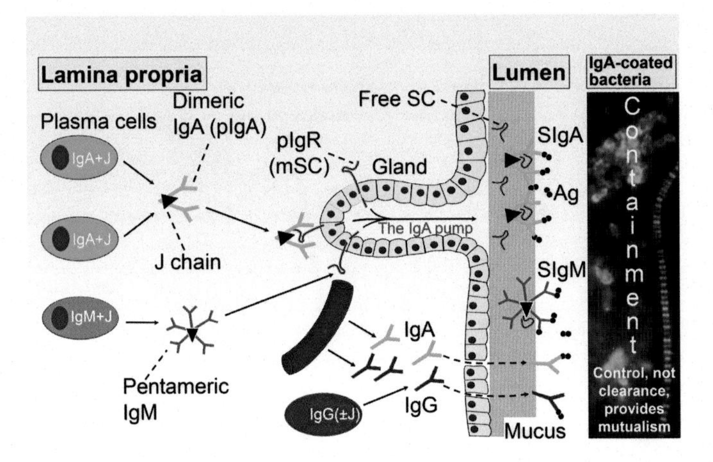

FIGURE 1.2.3 **Model for external transport of J chain-containing polymeric IgA (pIgA, mainly dimers) and pentameric IgM by the polymeric Ig receptor (pIgR), expressed basolaterally on glandular epithelial cells.** This receptor was previously called membrane secretory component *(mSC)*. The polymeric Ig molecules are produced with incorporated J chain (IgA + J and IgM + J) by mucosal plasma cells. The resulting secretory Ig molecules (*SIgA* and *SIgM*) act in a first line of defense by performing immune exclusion of antigens at the mucus layer on the epithelial surface. Although J chain is often (70–90%) produced by mucosal IgG plasma cells (Brandtzaeg, 2015a), it does not combine with this Ig class but is degraded intracellularly as denoted by *(±J)* in the figure. Locally produced and serum-derived IgG is not subjected to active external transport, but can be transmitted paracellularly to the lumen as indicated. Free SC (depicted in mucus) is generated when unoccupied pIgR (top symbol) is cleaved at the apical face of the epithelial cell in the same manner as bound SC in SIgA and SIgM. Most commensal bacteria become coated with SIgA in vivo (right panel) which presumably contains them for mutualism with the host without inhibiting their growth.

IgA^+ PCs are normally undetectable in human intestinal mucosa before 10 days of age but thereafter a rapid increase takes place, although IgM^+ PCs usually remain predominant up to 1 month after birth (Brandtzaeg et al., 1991). Adult salivary IgA levels are reached quite late in childhood, but only a small increase of IgA^+ PCs has been reported in intestinal mucosa after the first year (Brandtzaeg, 2013d). These observations have notably been made in industrialized countries; a faster development of the mucosal IgA system is usually seen in children from developing countries, reflecting the adaptability of local immunity according to the environmental antigenic load (Hoque et al., 2000).

4 MUCOSA-ASSOCIATED LYMPHOID TISSUE

4.1 Induction of Gut Immunity

Intestinal lymphoid cells occur in three distinct tissue compartments: organized gut-associated lymphoid tissue (GALT), the mucosal lamina propria, and the surface epithelium (Fig. 1.2.4). GALT comprises the Peyer's patches, the

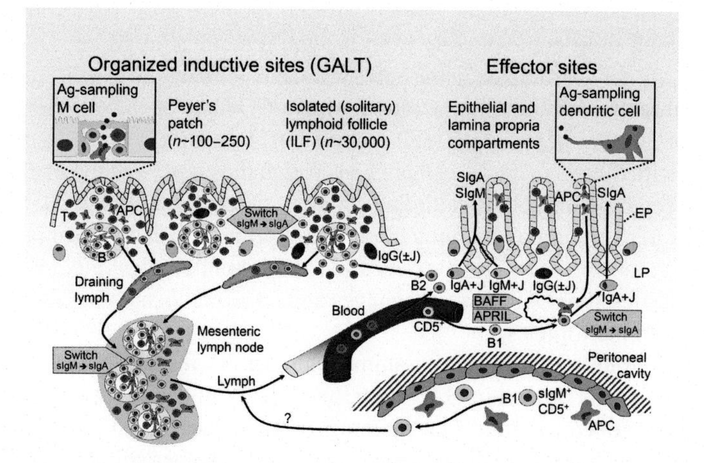

FIGURE 1.2.4 Antigen-sampling and B cell-switching sites for induction of intestinal IgA responses. The classical inductive sites are constituted by gut-associated lymphoid tissue *(GALT)*, here represented by Peyer's patches and isolated (solitary) lymphoid follicles *(ILFs)*. GALT structures are equipped with antigen (Ag)-sampling M cells, T-cell areas *(T)*, B-cell follicles (B), and antigen-presenting cells *(APCs)*. Class switching from surface sIgM to sIgA occurs in GALT and mesenteric lymph nodes; from here primed B and T cells home to the intestinal lamina propria *(LP)* via lymph and blood. T cells mainly end up in the epithelium *(EP)*. Primed B cells may also migrate from ILFs directly into LP where IgG-producing cells occur with downregulated J chain *(±J)*. The sIgA$^+$ cells differentiate to plasma cells that produce dimeric IgA with J chain *(IgA + J)* which becomes secretory IgA *(SIgA)*. Bone marrow-derived sIgM$^+$ B2 cells *(CD5$^-$)* may also give rise to pentameric IgM *(IgM + J)* and secretory IgM (SIgM). B1 cells (CD5$^+$) from the peritoneal cavity reach the LP by an unknown route *(?)*, perhaps via mesenteric lymph nodes. These sIgM$^+$ cells are particularly abundant in mice and may switch to sIgA within the LP under the influence of the cytokines BAFF and APRIL when dendritic cells have sampled luminal antigen within the epithelium. The sIgA$^+$ cells then differentiate to plasma cells that provide SIgA mainly directed against the commensal gut flora. Red dots denote antigen.

appendix and numerous isolated (solitary) lymphoid follicles, especially in the large bowel. All these lymphoid structures are believed to represent inductive sites for gastrointestinal immune responses (Brandtzaeg, 2015a). The lamina propria and the epithelial compartment constitute effector sites but are nevertheless important in terms of cellular expansion and differentiation within the mucosal immune system. GALT and other mucosa-associated lymphoid tissue (MALT) structures (see later) are covered by a characteristic FAE, containing specialized M cells (Fig. 1.2.4) which are effective in the uptake especially of live and dead particulate antigens from the gut lumen. However, many enteropathogenic infectious bacterial (eg, *Salmonella* spp., *Vibrio cholerae*) and viral [eg, poliovirus, reovirus, human immunodeficiency virus (HIV)-1] agents may use the M cells as portals of entry (Hathaway and Kraehenbuhl, 2000).

MALT structures resemble lymph nodes with B-cell follicles, intervening T-cell areas, and a variety of antigen-presenting cells (APCs), but there are no afferent lymphatics supplying antigens (Brandtzaeg, 2015a). Therefore, exogenous stimuli must come directly from the gut lumen via the M cells, although DCs in FAE may participate in the antigen uptake. Among the T cells, the CD4$^+$ helper subset predominates—the ratio between CD4 and CD8 cells being similar to that of other peripheral T-cell populations. In addition, B cells aggregate together with T cells in the M-cell pockets, which thus represent the first contact site between immune cells and luminal antigens (Yamanaka et al., 2001). The B cells may perform important antigen-presenting functions in this compartment, perhaps promoting antibody diversification and immunological memory. Other types of professional APCs, macrophages and DCs, are located below the FAE and between the follicles.

Pioneer studies, performed in animals almost 30 years ago, demonstrated that immune cells primed in GALT are functionally linked to mucosal effector sites by integrated migration or "homing" (Brandtzaeg, 2015a). T cells activated in GALT preferentially differentiate to CD4$^+$ helper cells which—aided by DCs and secretion of cytokines such as transforming growth factor (TGF)-β and interleukin (IL)-10—induce the differentiation of antigen-specific B cells to predominantly IgA expression. Induction of rotavirus-specific B-cell responses, moreover, seems to be promoted by type I interferon derived from plasmacytoid DCs (Deal et al., 2013).

The activated B-cell blasts proliferate and differentiate further on their route through mucosa-draining lymph nodes and the thoracic duct into the blood stream (Fig. 1.2.4). They thereafter home preferentially to the gut mucosa and complete their terminal differentiation to IgA$^+$ PCs locally—most likely under the influence of DCs which may pick up luminal antigens at the secretory effector site. The migration of lymphoid cells into the mucosal lamina propria is facilitated by "homing receptors" that interact with ligands on the microvascular endothelium at the effector site (addressins)—with an additional fine-tuned level of navigation conducted by local chemoattractant cytokines (chemokines). Under normal conditions, therefore, the lamina propria microvasculature exerts a "gatekeeper" function to allow selective extravasation of primed lymphoid cells belonging to the mucosal immune system, but this restriction is not maintained during inflammation (Table 1.2.1). Thus, it has been shown in mice that the selectivity of chemokine-dependent attraction of IgA$^+$ PC to the small intestinal lamina propria becomes redundant during rotavirus infection (Feng et al., 2006). Also, the inductive part of MALT exerts an important impact as shown by the virtual lack of B-cell homing to the small intestinal lamina propria from human nasopharynx-associated lymphoid tissue (NALT) (Johansen et al., 2005), similar to the lack of homing of rotavirus-specific B cells after NALT immunization in mice (Ogier et al., 2005).

The immunological immaturity of the newborn period (Offit et al., 2002) causes very few IgA$^+$ B cells (presumably GALT-derived) to appear in peripheral blood of newborns ($<8/10^6$ lymphocytes), but after 1 month of age this number increases remarkably ($\sim600/10^6$ lymphocytes) in parallel with the

TABLE 1.2.1 Characteristics of the Systemic Versus the Mucosal Immune System

	Systemic immunity	Shared features	Mucosal immunity
Inductive sites			
Antigen uptake and transport	Ordinary surface epithelia		Epithelia with membrane (M) cells
		Dendritic cells (DCs)	
	Blood circulation		
	Peripheral lymph nodes, spleen and bone marrow		Mucosa-associated lymphoid tissue (MALT): Peyer's patches, appendix and isolated (solitary) lymphoid follicles (GALT)
		Tonsils and adenoids Local (regional) lymph nodes	
Influx of circulating lymphoid cells: adhesion molecules and chemokines/chemokine receptors		Postcapillary high endothelial venules (HEVs) PNAd/L-selectin (CD62L) SLC (CCL21), ELC (CCL19)/CCR7	
			GALT: MAdCAM-1/$\alpha4\beta7$
Effector sites			
Homing of memory and effector B and T cells	Peripheral (lymphoid) tissues and sites of chronic inflammation: a variety of adhesion molecules and chemokines/chemokine receptors		Mucosal lamina propria and exocrine glands: MAdCAM-1/$\alpha4\beta7$ (gut), other adhesion molecules (? extraintestinal), TECK (CCL25)/CCR9 (small intestine), MEC (CCL28)/CCR10 (? elsewhere)
		Tonsils and adenoids	
Antibody production	IgG > monomeric IgA > pIgA > pentameric IgM		pIgA > pentameric IgM > >IgG

Abbreviations: *GALT*, gut-associated lymphoid tissue; *PNAd*, peripheral lymph node addressin; *SLC*, secondary lymphoid-tissue chemokine; *ELC*, Epstein Barr virus-induced molecule 1 ligand chemokine; *CCR*, CC chemokine receptor, *MAdCAM-1*, mucosal addressin cell adhesion molecule 1; *TECK*, thymus-expressed chemokine; *MEC*, mucosae-associated epithelial chemokine; *pIgA*, polymeric IgA, mainly dimeric.

progressive environmental stimulation of GALT (Stoll et al., 1993). Later on in childhood, B-cell trafficking between inductive and effector sites is reflected by a significant correlation between the number of IgA^+ cells in the circulation and in the gut lamina propria, as shown for the small intestine of patients with rotavirus infection (Brown et al., 2000).

4.2 Additional Sources of Intestinal B Cells

GALT constitutes the major part of human MALT, but induction of mucosal immunity may also take place in lymphoepithelial structures of Waldeyer's pharyngeal ring, particularly the palatine tonsils and the adenoids (Brandtzaeg, 2015b), and probably also bronchus-associated lymphoid tissue at an early age (Heier et al., 2008). Regionalization exists in the mucosal immune system, especially a dichotomy between the gut and the upper aerodigestive tract with regard to homing properties and terminal differentiation of B cells (Brandtzaeg, 2015a). This disparity may be explained by microenvironmental differences in the antigenic repertoire as well as adhesion molecules and chemokines involved in preferential local leucocyte extravasation (Table 1.2.1). It appears that primed immune cells selectively home to effector sites corresponding to the inductive sites where they initially were triggered by antigens. Such compartmentalization has to be taken into account in the development of local vaccines. However, there is interaction among the various immune compartments, and even the mucosal and systemic cell systems are not completely segregated. Thus, it is possible to obtain a detectable SIgA response also by the parenteral route of immunization, as shown with inactivated poliovirus vaccine when GALT has been previously primed by poliovirus infection (Herremans et al., 1997).

The peritoneal cavity is an additional source of intestinal lamina propria B cells ($CD5^+$ B1 cells) in the mouse. The switching to IgA expression may take place in the intestinal lamina propria and give rise to so-called natural SIgA antibodies (Fig. 1.2.4), which react particularly with commensal bacteria (Macpherson et al., 2000). However, there is no evidence that B1 cells are significantly involved in intestinal IgA production in humans (Brandtzaeg, 2015a), although human secretions contain considerable levels of polyreactive SIgA antibodies recognizing both self and microbial antigens (Bouvet and Fischetti, 1999). Interestingly, however, even in mice the traditional B2 cells, but not the B1 cells, apparently contribute along with $CD4^+$ T cells to the clearance of rotavirus infection by an intestinal IgA response (Kushnir et al., 2001).

5 INTESTINAL IMMUNE-EFFECTOR COMPARTMENTS

5.1 Lamina Propria B Cells and Their Epithelial Cooperation

The intestinal lamina propria of adults contains approximately 10^{10} PCs per meter of intestine, so at least 80% of all antibody-producing cells are located in the gut. Some 80–90% are IgA^+ PCs and a relatively large fraction consists of the IgA2 subclass (17–64%) compared with the proportion (7–25%) seen in peripheral lymphoid tissue, tonsils, and airway mucosae (Brandtzaeg, 2015a,b). A predominance of IgA2$^+$ PCs is regularly seen only in the large bowel but this cannot be explained by isotype switching outside of GALT structures (Lin et al., 2014) as previously proposed (He et al., 2007). The concentration ratios of the two SIgA subclasses in various exocrine secretions are quite similar to the relative distribution of IgA1$^+$ and IgA2$^+$ PCs at the corresponding effector sites, attesting to the fact that pIgA of both isotypes are equally well transported externally (Brandtzaeg, 2013b). The relative increase of the IgA2 subclass in secretions compared with serum may be important for the stability of secretory antibodies because SIgA2, in contrast to SIgA1, is resistant to several IgA1-specific proteases which are produced by a variety of potentially pathogenic bacterial species (Plaut et al., 1974).

More than 90% of the mucosal IgA$^+$ PCs synthesize a small polypeptide called the joining (J) chain (Brandtzaeg, 2015a). The J chain is essential for correct polymerization of pIgA (and also pentameric IgM) and for the subsequent interaction of the Ig polymers with the pIgR which is expressed basolaterally on the secretory epithelial cells (Brandtzaeg, 1985, 2013b; Brandtzaeg and Prydz, 1984). After transcytosis the pIg-receptor complexes are released into the gut lumen by apical pIgR cleavage (Fig. 1.2.3). The extracellular portion of the receptor (~80 kDa) remains as so-called bound SC in SIgA and SIgM, thereby stabilizing the secretory antibodies. Particularly the covalent bonding between SC and one α-chain of pIgA makes SIgA the most stable antibody functioning in external secretions.

5.2 Secretory Immunity

The function of SIg antibodies is principally to perform immune exclusion by complexing with soluble antigens and by binding to the surface of microorganisms, thus preventing their adherence to epithelial cells and penetration of the surface barrier (Fig. 1.2.5, far left panel). Interestingly, during transcytosis of pIgA and pentameric IgM these antibodies may even inactivate infectious agents, for example, rotavirus, influenza virus, and HIV—inside secretory epithelial cells and carry the pathogens and their products back to the lumen, thus avoiding cytolytic damage to the epithelium (Fig. 1.2.5, next to left panel). It has also been proposed that pIgA during transcytosis may neutralize endotoxin from Gram-negative bacteria and thereby inhibit the proinflammatory gene repertoire of the epithelial cell (Fig. 1.2.5, next to right panel). Finally, as previously reviewed (Johansen and Brandtzaeg, 2004), stromal clearance of penetrating antigens has been experimentally documented for pIgA-containing immune complexes (Fig. 1.2.5, far right panel).

Both the agglutinating and virus-neutralizing antibody effect of pIgA is superior compared with monomeric antibodies (Brandtzaeg, 2013b), and SIgA antibodies may block microbial invasion quite efficiently. This has been particularly

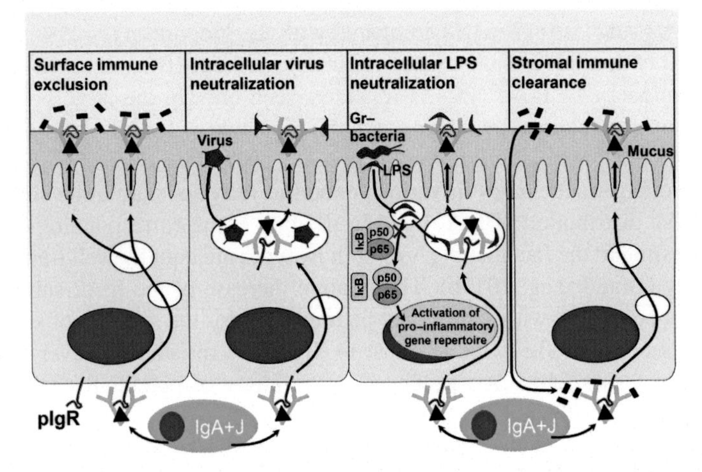

FIGURE 1.2.5 **Schematic representation of four proposed mechanisms for intestinal surface defense performed by dimeric IgA (pIgA) after being produced with J chain (IgA + J) by plasma cells in the lamina propria.** Far left panel: pIgA is transcytosed in vescicles by the polymeric Ig receptor *(pIgR)* across secretory epithelial cells and released into the lumen as secretory IgA *(SIgA)* antibodies which perform immune exclusion by interaction with luminal antigens (▬). Next to left panel: During pIgR-mediated transport pIgA antibodies may interact with viral antigens within apical epithelial endosomes, thereby performing intracellular virus neutralization and removal of viral products. Next to right panel: Intracellular neutralization of endotoxins (lipopolysaccharide, *LPS*) from Gram-negative *(Gr−)* bacteria may inhibit potentially harmful activation of the proinflammatory NFκB pathway in the epithelial cell. Far right panel: Dimeric IgA antibodies interact with paracellularly penetrating antigens reaching the lamina propria and shuttle them back to the lumen by pIgR-mediated export. *(Source: References to primary articles on which the cartoon is partially based, can be found in Johansen and Brandtzaeg, 2004.)*

well documented in relation to HIV (Mazzoli et al., 1997); specific SIgA antibodies isolated from human colostrum were shown to be more efficient in this respect than comparable IgG antibodies (Hocini and Bomsel, 1999). Moreover, serum IgG antibodies mainly depend on a passive paracellular transfer to reach mucosal surfaces (Fig. 1.2.3), and the same is true for the locally produced IgG that is detectable in various secretions (Bouvet and Fischetti, 1999). Also notably, less than 5% of the human intestinal PCs normally produce IgG—a figure that in selective IgA deficiency, however, is increased to ~25% together with ~75% IgM^+ PCs (Brandtzaeg, 2015a). The proportion of IgG^+ PCs is also strikingly increased in inflamed mucosae.

The protective role of SIgA has been questioned by observations in IgA knockout ($IgA^{-/-}$) mice which remain healthy under ordinary laboratory conditions (Mbawuike et al., 1999). When challenged with influenza virus, they show similar pulmonary virus levels and mortality as control wild-type ($IgA^{+/+}$) mice. However, leakage of serum IgG antibodies through an irritated mucosal surface lining (Fig. 1.2.3), according to the classical principle of "pathotopic potentiation of local immunity" (Fazekas de St. Groth, 1951), probably plays a greater protective role in the airways than in the gut (Persson et al., 1998). In addition, the IgA$^{-/-}$ mice usually show a compensatory SIgM response. Interestingly, IgA$^{-/-}$ mice on exclusive parenteral nutrition have reduced IgA antiinfluenza virus titers in the upper respiratory tract and no compensatory SIgM, which most likely explains that they show impaired mucosal immunity (Renegar et al., 2001). The fact that individuals with selective IgA deficiency do not suffer significantly more than others from intestinal virus infections, may largely be ascribed to their consistently enhanced SIgM and IgG1 response in the gut (Brandtzaeg, 2015a). Altogether, a secretory antibody response appears to be essential for adequate mucosal protection. This notion is supported by the finding that systemic IgG antibody production against *E. coli* and other commensals are triggered in pIgR knockout mice lacking both SIgA and SIgM (Johansen et al., 1999; Sait et al., 2007).

The complexity of local immunity is emphasized by the initial finding that clearance of rotavirus after infection, as well as long-term resistance to reinfection, was not impaired in IgA$^{-/-}$ mice (O'Neal et al., 2000). Careful analysis of different variables suggested that in the absence of IgA, the protective effect was primarily mediated by IgG antibodies. Notably, however, both clinical data and animal experiments have suggested that when IgA is present, rotavirus infection is primarily cleared and subsequently prevented by intestinal SIgA antibodies (O'Neal et al., 2000). Rotavirus clearance and protection therefore seems to be critically dependent on IgA (Blutt et al., 2012).

In a similar study, the role of IgA versus IgG in gut immunity to reovirus was investigated (Silvey et al., 2001). Although this virus does not normally cause disease in humans, it constitutes a useful rodent model for enteric virus infections. Both wild-type and IgA$^{-/-}$ mice were competent at clearing a primary infection, but only the wild-type mice were fully protected against entry of reovirus through Peyer's patch M cells upon subsequent infection; partial protection shown by the IgA$^{-/-}$ mice was ascribed to a compensatory SIgM response (Silvey et al., 2001). IgG antibodies to the $\sigma 1$ protein could protect when preincubated with the virus prior to viral challenge, but only the corresponding monoclonal pIgA antibodies protected as SIgA against virus entry when administered via the hybridoma "back-pack" model (Hutchings et al., 2004). Together, these data imply that SIgA antibodies provide the best protection of Peyer's patches against this enteric virus. In addition to local antibodies, mucosal cytotoxic T lymphocytes are likely important to prevent systemic seeding of virus, particularly when the pathogen such as HIV persistently tends to replicate in gut mucosa (Berzofsky et al., 2001).

5.3 Complexity of Mucosal Antigen Clearance

Unlike IgG, antibodies of the IgA class do not cause complement activation by the classical pathway. External transport of pIgA-containing immune complexes has therefore been suggested as an efficient, noninflammatory antigen clearance mechanism (Fig. 1.2.5, far right panel). This possibility is supported by in vivo experiments (Robinson et al., 2001). Pentameric IgM (in contrast to hexameric IgM without J chain) also appears to have poor complement-activating properties and may therefore support the noninflammatory functions of pIgA in competition with corresponding proinflammatory IgG antibodies. Interestingly, monomeric IgA or IgG antibodies, when cross-linked to pIgA via the same antigen, may contribute to this pIgR-mediated stromal immune clearance. Conversely, IgG antibodies against infectious agents and dietary proteins can on their own increase mucosal penetrability of exogenous bystander antigens as suggested by in vivo (Lim and Rowley, 1982) and ex vitro experiments (Brandtzaeg and Tolo, 1977). The mucosal integrity is apparently damaged by lysosomal enzymes released from polymorphonuclear granulocytes which are attracted when complement-activating immune complexes are formed locally.

The proinflammatory potential of IgG is probably less important in the gut of infants who are breast-fed because milk SIgA antibodies will exert a noninflammatory blocking effect (Brandtzaeg, 2013c). Moreover, complement regulatory proteins are expressed by the gastrointestinal epithelium and may counteract immune complex-mediated damage (type III hypersensitivity) of the epithelial lining (Berstad and Brandtzaeg, 1998).

Experimental evidence further suggests that IgA may influence mucosal homeostasis through its binding to the FcαI receptor (CD89) when present on lamina propria leucocytes. In the normal state CD89 is not detectable on human intestinal macrophages (Hamre et al., 2003), but this situation may differ in the inflamed mucosa with extravasation of leucocytes. Both activating and inhibitory signals may then be induced by IgA. Thus, cross-linking of CD89 during infection with IgA-opsonized pathogens causes proinflammatory responses, whereas naturally occurring IgA (not complexed) induces inhibitory signals through CD89, thereby dampening excessive reactions (Monteiro, 2014). Thus, IgA can downregulate the secretion of the proinflammatory cytokine tumour necrosis factor (TNF)-α from activated monocytes and inhibit activation-dependent generation of reactive oxygen intermediates in neutrophils and monocytes (Wolf et al., 1994a,b). On the other hand, perhaps aggregated monomeric IgA and pIgA can trigger monocytes to secrete TNF-α and upregulate the B7 (CD68) costimulatory molecules (Geissmann et al., 2001).

Eosinophil degranulation can also be induced by aggregated IgA (Abu-Ghazaleh et al., 1989). This proinflammatory potential of serum IgA probably reflects the need for reinforcement of mucosal antigen elimination when immune exclusion fails. Moreover, IgA can induce ADCC by interaction with CD89 (Black et al., 1996), and particulary pIgA can activate complement via the mannan-binding lectin pathway (Roos et al., 2001). Therefore, it still remains an open question whether IgA plays a good or bad role in, for instance, HIV infection in humans although a protective effect of mucosally administered pIgA antibodies against HIV challenge has been documented (Zhou and Ruprecht, 2014).

6 DIFFICULTIES IN EVALUATING THE PROTECTIVE EFFECT OF SECRETORY IMMUNITY

As emphasized earlier, the protective effect of SIgA and SIgM antibodies during intestinal infection is blurred by concurrent systemic immunity, both after natural infection (or enteric live vaccines) and when nonproliferating virus-like particles are mucosally applied together with an adjuvant (O'Neal et al., 1998). The effect of serum antibodies, both locally in the gut and systemically such as in chronic HIV infection (Haas et al., 2011), may be to inhibit further spread of the infectious agent by neutralization and immune elimination. In the gut lumen, however, IgG (and probably also monomeric IgA) is less stable than SIgA and may thus be of little protective value, although binding of antibody fragments to the biliary protein Fv (Fv fragment-binding protein) has been suggested to reinforce immune exclusion (Bouvet and Fischetti, 1999).

Methodological problems render evaluation of intestinal antibody induction difficult. Important variables beyond current control are the impact on microbial mucosal challenge caused by differences in mucus layers and mucins produced by goblet cells and enterocytes (Pelaseyed et al., 2014). To maintain mucosal homeostasis, the mucus coat collaborates with SIgA in keeping the microbiota away from the epithelial cells (Rogier et al., 2014a,b). Moreover, mucus contains antimicrobial peptides and can bind DC-SIGN, probably through its content of lactoferrin (Stax et al., 2015). Thus, it has been shown that the capture of HIV-1 via DC-SIGN by dendrites from mucosal DCs, and thereby the transfer of virus to mucosal $CD4^+$ T lymphocytes (so-called *trans*-infection) may be inhibited.

Analysis of intestinal fluid or fecal extracts for secretory antibodies is further jeopardized by leakage of serum IgA antibodies, and even SIgA may be considerably degraded in such samples (Johansen et al., 1999). Thus, it is often possible to measure intestinal IgA antibodies directed against, for instance, rotavirus, whereas a corresponding SIgA activity may be undetectable—depending on whether the employed immunoenzyme assay reveals the α-chain of IgA or bound SC. Another problem is the fact that an unknown fraction of SIgA may remain adsorbed to microbes in feces after the extraction procedure.

In rodent models a significant proportion of intestinal SIgA is derived from bile and therefore does not represent mucosal immunity, which is in important contrast to the human situation where the hepatocytes do not express pIgR/SC (Brandtzaeg, 1985). In addition, it has recently been reported that microbially driven dichotomous fecal IgA levels in mice within the same facility mimic the effect of chromosomal mutations (Moon et al., 2015). Bacteria from mice with low fecal IgA levels degraded bound SC in SIgA as well as IgA itself.

Finally, it is controversial whether serum IgA antibody levels are helpful in the diagnosis of rotavirus infection (Angel et al., 2012). Some studies have suggested that SIgA appearing in plasma to a certain degree may be a correlate for protection of rotavirus infection or attenuated live vaccine (Clarke and

Desselberger, 2015). However, it was shown several decades ago that conditions that increase the permeability of glandular and mucosal tissue may raise the "spill-over" or absorption of SIgA into the circulation (Brandtzaeg, 1971). The presence of SIgA in plasma can therefore not be taken as a reliable proxy of mucosal immunity. SIgA antibody levels in sublingual/submandibular secretion, on the other hand, seem to be a promising noninvasive correlate of immune induction in the gut (Aase et al., 2015).

7 MUCOSAL T CELLS AND THEIR PUTATIVE PROTECTIVE ROLES

7.1 Phenotypic Heterogeneity

The protective effect of live attenuated rotavirus vaccines is apparently mediated not only by IgA antibodies but also by cytotoxic $CD8^+$ T cells (Azegami et al., 2014). The human intestinal lamina propria contains, in addition to IgA^+ PCs and scattered B lymphocytes, an abundance of T cells which resemble other peripheral T lymphocytes in that they virtually always express the T-cell receptor (TCR)-α/β and show a predominance of the CD4 over the CD8 phenotype (Brandtzaeg et al., 1998). This picture is becoming even more complex by the discovery of mucosal natural killer (NK) T cells (Zeissig and Blumberg, 2014) and innate lymphoid cells (Sonnenberg et al., 2012).

Although the $CD4^+$ lamina propria T cells exhibit a memory/effector phenotype, they show little proliferative activity and low expression of CD25 (the high-affinity receptor for IL-2). Also, they are refractory to CD3-triggered activation in vitro, but can be activated through CD2–CD28 costimulation (Abreu-Martin and Targan, 1996). Rotavirus-specific $CD4^+$ T cells with intestinal homing molecules have been shown in peripheral blood of healthy individuals (Parra et al., 2014). Similarly, rotavirus-specific IgA^+ PC are detectable in duodenal mucosa of subjects with no signs of ongoing infection (Di Niro et al., 2010).

Intraepithelial lymphocytes (IELs) are strikingly dominated by the CD8 phenotype (80–90%), and normally amount to 5–10 per 100 epithelial cells (Fig. 1.2.4). B lymphocytes are virtually absent from the intestinal epithelium outside of GALT (Brandtzaeg et al., 1998). In addition, the epithelium contains a small population of non-T non-B lymphocytes of obscure origin and function. Such peculiar subsets consist of T cells expressing the alternative TCR-γ/δ, the αα homodimer of the CD8 molecule (both the TCR-α/β+ and TCR-γ/δ+ subsets contain CD8αα+ subpopulations), as well as CD4 + CD8+ "double positive" and CD4−CD8− "double negative" T cells. These subsets contribute substantially to the IEL pool in rodents, and coexpression of NK-cell markers have been reported also in humans (Cheroutre et al., 2011). It is difficult to decide when IELs are important for mucosal barrier protection or when they are involved in epithelial damage.

Consistent with the expression of CD8 and cytoplasmic granules, human and murine IELs show vivid cytotoxicity in vitro, and this can be elicited through both the CD3 complex, TCR-α/β and TCR-γ/δ· Interestingly, the IELs are potent producers of IFN-γ after in vitro stimulation (Lundqvist et al., 1996), and this capacity may be important for protection of neighbouring epithelial cells against intracellular infectious agents (Chardes et al., 1994).

7.2 Possible Antimicrobial Functions

It is generally believed that IELs play a role in a first line of microbial defense but only few studies have directly demonstrated such a function (Cheroutre et al., 2011). Experiments in mice have suggested that certain infectious intracellular pathogens elicit a dynamic disseminated IEL response which may be quite persistent. Thus, enteric infection with reovirus, rotavirus and *Toxoplasma gondii* resulted in rapid accumulation of cytotoxic TCR-α/β+ IELs that lyzed infected intestinal epithelial cells. The effector cells belonged to the CD8αβ+ subpopulation and recognized intracellular peptides presented on classical MHC class I molecules, thus being similar to the conventional CD8αβ+ cells of systemic lymphoid organs (Helgeland and Brandtzaeg, 2000).

Transfer of germ-free animals to a specific pathogen-free environment converts the composition and functional profile of IELs to that of conventional animals, suggesting that the response of these cells is primarily directed against the commensal microbiota. However, attempts to identify the stimulatory bacteria have been inconclusive. To elucidate the nature of the antigens recognized by IELs, their TCR repertoire has been subjected to detailed investigation. Surprisingly, it has been found in humans, mice and rats that TCR-α/β+ IELs express an oligoclonal antigen receptor repertoire and the same limited number of T-cell clones can be identified over long distances in the gut (Helgeland and Brandtzaeg, 2000). This result is difficult to reconcile with the possibility that IELs recognize diverse microbial antigens. Furthermore, inbred mice and rats express individually distinct oligoclonal repertoires, indicating that TCR-α/β+ IELs recognize different antigens in different animals. Microbial colonization of previously germ-free rats induces random oligoclonal expansion of CD8α/β+ and CD8γ/δ+ IELs (Helgeland et al., 2004). This may exclude specific recognition of indigenous microorganisms, which are believed to be quite comparable but not phenotypically identical among animals of the same species. Moreover, TCR-α/β+ IELs in germ-free mice also express oligoclonal repertoires, further speaking against a role of the intestinal microflora in the clonal T-cell selection process (Regnault et al., 1996).

The possibility remains that different T-cell clones recognize different epitopes of the same microbial species, and certain features of the TCR repertoire suggest that it is shaped by the intestinal flora. Thus, the widely disseminated clones were found to vary markedly in size at different levels of the gut and some clones occurred only locally (Dogan et al., 1996). Also, it has been shown that microbial colonization of rats leads to selective expansion of TCR-α/β+ clones bearing certain variable β gene segments (Helgeland and Brandtzaeg, 2000). Finally, data from chicken suggest that the first T cells to arrive in the intestinal epithelium during ontogeny express a polyclonal repertoire (Dunon et al., 1994), which then might be subjected to oligoclonal selection by naturally occurring microbial antigens in postnatal life. Altogether, it is tempting to speculate that IELs recognize conserved microbial antigens.

The γ/δ IEL subset is associated with certain TCR Vγ and Vδ gene segments in defined anatomical sites. Thus, in the human gut, Vδ1 in expressed by some 70% of the TCR-γ/δ^+ IELs, apparently in the main together with Vγ8 (Brandtzaeg et al., 1998). This may reflect a preferential specificity directed against selected gut antigens, and molecular analyses of Vδ1 sequences have suggested an individually imprinted oligoclonality along extensive gut segments for considerable time periods. Interestingly, Groh et al. (1998) showed that Vδ1 TCR-γ/δ^+ IELs could recognize stress-induced nonclassical MHC class I (or class Ib) molecules on epithelial cells, mainly MICA and MICB. This observation might hint to the existence of an immune surveillance mechanism for detection of damaged, infected or transformed intestinal epithelium, and/or the possibility for stimulated secretion of immunoregulatory cytokines from the TCR-γ/δ^+ IELs.

Interestingly, the number of intestinal TCR-γ/δ^+ IELs is generally increased in AIDS patients, although within a wide range, probably reflecting individual responses to opportunistic infections (Nilssen et al., 1996). Conversely, the almost total depletion of mucosal CD4$^+$ cells is not compensated for by a numerical increase of duodenal CD8$^+$ cells. In AIDS patients with particularly short life expectancy ($<$7 months), γ/δ IELs are decreased to virtually normal levels. Longitudinal studies in a few patients supported this observation, suggesting that γ/δ IELs might be involved in prolonging the life of patients with AIDS.

8 CONCLUSIONS

Secretory immunity is desirable in the defense against intestinal virus infections because it can operate both at the luminal face of the epithelium and intracellularly in infected epithelial cells without causing mucosal damage. In addition, IELs may act in front-line defense by eliminating infected epithelial cells. Several virological studies have shown that natural gut infection and enteric vaccination are more efficient in giving rise to SIgA antibodies than parenteral vaccination. Also, the intraluminal vaccines have been much more efficient when consisting of live than killed viruses. Like natural infections, live vaccines give rise not only to SIgA antibodies, but also to longstanding serum IgG and IgA responses. SIgM and serum IgM antibodies often appear in the acute infection phase but decline after covalescence. In infants and subjects with selective IgA deficiency, however, SIgM may play a major protective role against infectious agents (Brandtzaeg, 2015a).

Despite advanced mechanistic understanding of the protective effects of mucosal immune responses against gut viruses, it is generally difficult to determine unequivocally the importance of SIgA versus serum antibodies. The degree of immune protection may range from complete inhibition of reinfection (poliovirus and murine reovirus) to reduction of symptoms (rotavirus). It is furthermore not clear to what extent T-cell-mediated mechanisms or ADCC are involved in this effect. This complexity is an inherent difficulty in design of vaccines to protect against mucosal infections (Azegami et al., 2014) and makes it difficult to understand how the current live oral rotavirus vaccines function in various parts of the world (Glass et al., 2014).

ACKNOWLEDGMENTS

Supported by the Research Council of Norway through its Centers of Excellence funding scheme (Project No. 179573/40) and by Oslo University Hospital Rikshospitalet. Hege Eliassen is thanked for her excellent assistance with the manuscript.

REFERENCES

Aase, A., Sommerfelt, H., Petersen, L.B., Bolstad, M., Cox, R.J., Langeland, N., Guttormsen, A.B., Steinsland, H., Skrede, S., Brandtzaeg, P., 2015. Salivary IgA from the sublingual compartment as a novel non-invasive proxy for intestinal immune induction. *Mucosal Immunol. Oct 28, 2015. doi: 10.1038/mi.2015.107. [Epub ahead of print].*

Abreu-Martin, M.T., Targan, S.R., 1996. Regulation of immune responses of the intestinal mucosa. Crit. Rev. Immunol. 16, 277–309.

Abu-Ghazaleh, R.I., Fujisawa, T., Mestecky, J., Kyle, R.A., Gleich, G.J., 1989. IgA–induced eosinophil degranulation. J. Immunol. 142, 2393–2400.

Angel, J., Franco, M.A., Greenberg, H.B., 2012. Rotavirus immune responses and correlates of protection. Curr. Opin. Virol. 2, 419–425.

Azegami, T., Yuki, Y., Kiyono, H., 2014. Challenges in mucosal vaccines for the control of infectious diseases. Int. Immunol. 26, 517–528.

Berzofsky, J.A., Ahlers, J.D., Belyakov, I.M., 2001. Strategies for designing and optimizing new generation vaccines. Nature Rev. Immunol. 1, 209–219.

Berstad, A.E., Brandtzaeg, P., 1998. Expression of cell-membrane complement regulatory glycoproteins along the normal and diseased human gastrointestinal tract. Gut 42, 522–529.

Black, K.P., Cummins, Jr., J.E., Jackson, S., 1996. Serum and secretory IgA from HIV-infected individuals mediate antibody-dependent cellular cytotoxicity. Clin. Immunol. Immunopathol. 81, 182–190.

Blutt, S.E., Miller, A.D., Salmon, S.L., Metzger, D.W., Conner, M.E., 2012. IgA is important for clearance and critical for protection from rotavirus infection. Mucosal. Immunol. 5, 712–719.

Bouvet, J.-P., Fischetti, V.A., 1999. Diversity of antibody-mediated immunity at the mucosal barrier. Infect. Immun. 67, 2687–2691.

Brandtzaeg, P., 1971. Human secretory immunoglobulins. V. Occurrence of secretory piece in human serum. J. Immunol. 106, 318–323.

Brandtzaeg, P., 1985. Role of J chain and secretory component in receptor-mediated glandular and hepatic transport of immunoglobulins in man. Scand. J. Immunol. 22, 111–146.

Brandtzaeg, P., 2013a. Gate-keeper function of the intestinal epithelium. Benef. Microbes 4, 67–82.

Brandtzaeg, P., 2013b. Secretory IgA: Designed for anti-microbial defense. Front. Immunol. 4, 222.

Brandtzaeg, P., 2013c. Immune aspects of breast milk: an overview. In: Zibadi, S., Watson, R.R., Preedy, V.R., (Eds.), Handbook of Dietary and Nutritional Aspects of Human Breast Milk, Wageningen Academic Publishers, Wageningen, The Netherlands. Chapter 3, pp. 57–82.

Brandtzaeg, P., 2013d. Secretory immunity with special reference to the oral cavity. J. Oral. Microbiol. 5, 20401.

Brandtzaeg, P., 2015a. The mucosal B-cell system. In: Mestecky, J., Strober, W., Russell, M.W., Kelsall, B.L., Cheroutre, H., Lambrecht, B.N (Eds.), Mucosal Immunology, fourth ed. Academic Press/Elsevier, Amsterdam, pp. 623–681, Chapter 31.

Brandtzaeg, P., 2015b. Immunobiology of tonsils and adenoids. In: Mestecky, J., Strober, W., Russell, M.W., Kelsall, B.L., Cheroutre, H., Lambrecht, B.N. (Eds.), Mucosal Immunology, fourth ed. Academic Press/Elsevier, Amsterdam, pp. 1985–2016, Chapter 103.

Brandtzaeg, P., Prydz, H., 1984. Direct evidence for an integrated function of J chain and secretory component in epithelial transport of immunoglobulins. Nature 311, 71–73.

Brandtzaeg, P., Tolo, K., 1977. Mucosal penetrability enhanced by serum-derived antibodies. Nature 266, 262–263.

Brandtzaeg, P., Nilssen, D.E., Rognum, T.O., Thrane, P.S., 1991. Ontogeny of the mucosal immune system and IgA deficiency. Gastroenterol. Clin. North Am. 20, 397–439.

Brandtzaeg, P., Farstad, I.N., Helgeland, L., 1998. Phenotypes of T cells in the gut. Chem. Immunol. 71, 1–26.

Brown, K.A., Kriss, J.A., Moser, C.A., Wenner, W.J., Offit, P.A., 2000. Circulating rotavirus-specific antibody-secreting cells (ASCs) predict the presence of rotavirus-specific ASCs in the human small intestinal lamina propria. J. Infect. Dis. 182, 1039–1043.

Chardes, T., Buzoni-Gatel, D., Lepage, A., Bernard, F., Bout, D., 1994. Toxoplasma gondii oral infection induces specific cytotoxic CD8 α/β^+ Thy-1^+ gut intraepithelial lymphocytes, lytic for parasite-infected enterocytes. J. Immunol. 153, 4596–4603.

Cheroutre, H., Lambolez, F., Mucida, D., 2011. The light and dark sides of intestinal intraepithelial lymphocytes. Nat. Rev. Immunol. 11, 445–456.

Clarke, E., Desselberger, U., 2015. Correlates of protection against human rotavirus disease and the factors influencing protection in low-income settings. Mucosal Immunol. 8, 1–17.

Davidson, G.P., Barnes, G.L., 1979. Structural and functional abnormalities of the small intestine in infants and young children with rotavirus enteritis. Acta Paediatr. Scand. 68, 181–186.

Deal, E.M., Lahl, K., Narváez, C.F., Butcher, E.C., Greenberg, H.B., 2013. Plasmacytoid dendritic cells promote rotavirus-induced human and murine B cell responses. J. Clin. Invest. 123, 2464–2474.

Desselberger, U., 2000. Gastroenteritis viruses: research update and perspectives. Gastroenteritis viruses, Novartis Foundation Symposium 238, London, UK, May 16–18, 2000. Mol. Med. Today 2000; 6, 383–384.

Dickinson, B.L., Badizadegan, K., Wu, Z., Ahouse, J.C., Zhu, X., Simister, N.E., Blumberg, R.S., Lencer, W.I., 1999. Bidirectional FcRn-dependent IgG transport in a polarized human intestinal epithelial cell line. J. Clin. Invest. 104, 903–911.

Di Niro, R., Mesin, L., Raki, M., Zheng, N.Y., Lund-Johansen, F., Lundin, K.E., Charpilienne, A., Poncet, D., Wilson, P.C., Sollid, L.M., 2010. Rapid generation of rotavirus-specific human monoclonal antibodies from small-intestinal mucosa. J. Immunol. 185, 5377–5383.

Dogan, A., Dunn-Walters, D.K., MacDonald, T.T., Spencer, J., 1996. Demonstration of local clonality of mucosal T cells in human colon using DNA obtained by microdissection of immunohistochemically stained tissue sections. Eur. J. Immunol. 26, 1240–1245.

Dunon, D., Schwager, J., Dangy, J.P., Cooper, M.D., Imhof, B.A., 1994. T cell migration during development: homing is not related to TCR V β 1 repertoire selection. EMBO J. 13, 808–815.

Fazekas de St. Groth, S., 1951. Studies in experimental immunology of influenza. IX. The mode of action of pathotopic adjuvants. Aust. J. Exp. Biol. Med. Sci. 29, 339–352.

Feng, N., Jaimes, M.C., Lazarus, N.H., Monak, D., Zhang, C., Butcher, E.C., Greenberg, H.B., 2006. Redundant role of chemokines CCL25/TECK and CCL28/MEC in IgA$^+$ plasmablast recruitment to the intestinal lamina propria after rotavirus infection. J. Immunol. 176, 5749–5759.

Geissmann, F., Launay, P., Pasquier, B., Lepelletier, Y., Leborgne, M., Lehuen, A., Brousse, N., Monteiro, R.C., 2001. A subset of human dendritic cells expresses IgA Fc receptor (CD89), which mediates internalization and activation upon cross-linking by IgA complexes. J. Immunol. 166, 346–352.

Glass, R.I., Bresee, J., Jiang, B., Gentsch, J., Ando, T., Fankhauser, R., Noel, J., Parashar, U., Rosen, B., Monroe, S.S., 2001. Gastroenteritis viruses: an overview. Novartis Found. Symp. 238, 5–19, discussion 19–25.

Glass, R.I., Parashar, U., Patel, M., Gentsch, J., Jiang, B., 2014. Rotavirus vaccines: successes and challenges. J. Infect. 68 (Suppl 1), S9–S18.

Groh, V., Steinle, A., Bauer, S., Spies, T., 1998. Recognition of stress-induced MHC molecules by intestinal epithelial gammadelta T cells. Science 279, 1737–1740.

Haas, A., Zimmermann, K., Graw, F., Slack, E., Rusert, P., Ledergerber, B., Bossart, W., Weber, R., Thurnheer, M.C., Battegay, M., Hirschel, B., Vernazza, P., Patuto, N., Macpherson, A.J., Günthard, H.F., Oxenius, A, 2011. Swiss HIV Cohort Study. Systemic antibody responses to gut commensal bacteria during chronic HIV-1 infection. Gut 60, 1506–1519.

Hamre, R., Farstad, I.N., Brandtzaeg, P., Morton, H.C., 2003. Expression and modulation of the human immunoglobulin A Fc receptor (CD89) and the FcR gamma chain on myeloid cells in blood and tissue. Scand. J. Immunol. 57, 506–516.

Hathaway, L.J., Kraehenbuhl, J.P., 2000. The role of M cells in mucosal immunity. Cell. Mol. Life Sci. 57, 323–332.

He, B., Xu, W., Santini, P.A., Polydorides, A.D., Chiu, A., Estrella, J., Shan, M., Chadburn, A., Villanacci, V., Plebani, A., Knowles, D.M., Rescigno, M., Cerutti, A., 2007. Intestinal bacteria trigger T cell-independent immunoglobulin A(2) class switching by inducing epithelial-cell secretion of the cytokine APRIL. Immunity 26, 812–826.

Heier, I., Malmström, K., Pelkonen, A.S., Malmberg, L.P., Kajosaari, M., Turpeinen, M., Lindahl, H., Brandtzaeg, P., Jahnsen, F.L., Mäkelä, M.J., 2008. Bronchial response pattern of antigen presenting cells and regulatory T cells in children less than 2 years of age. Thorax 63, 703–709.

Helgeland, L., Brandtzaeg, P., 2000. Development and function of intestinal B and T cells. Microbiol. Ecol. Health Dis. 12 (Suppl. 2), 110–127.

Helgeland, L., Dissen, E., Dai, K.Z., Midtvedt, T., Brandtzaeg, P., Vaage, J.T., 2004. Microbial colonization induces oligoclonal expansions of intraepithelial CD8 T cells in the gut. Eur. J. Immunol. 34, 3389–3400.

Herremans, M.M., van Loon, A.M., Reimerink, J.H., Rümke, H.C., van der Avoort, H.G., Kimman, T.G., Koopmans, M.P., 1997. Poliovirus-specific immunoglobulin A in persons vaccinated with inactivated poliovirus vaccine in The Netherlands. Clin. Diagn. Lab. Immunol. 4, 499–503.

Hocini, H., Bomsel, M., 1999. Infectious human immunodeficiency virus can rapidly penetrate a tight human epithelial barrier by transcytosis in a process impaired by mucosal immunoglobulins. J. Infect. Dis. 179, S448–S453.

Hoque, S.S., Ghosh, S., Poxton, I.R., 2000. Differences in intestinal humoral immunity between healthy volunteers from UK and Bangladesh. Eur. J. Gastroenterol. Hepatol. 12, 1185–1193.

Hutchings, A.B., Helander, A., Silvey, K.J., Chandran, K., Lucas, W.T., Nibert, M.L., Neutra, M.R., 2004. Secretory immunoglobulin A antibodies against the sigma1 outer capsid protein of reovirus type 1 Lang prevent infection of mouse Peyer's patches. J. Virol. 78, 947–957.

Israel, E.J., Simister, N., Freiberg, E., Caplan, A., Walker, W.A., 1993. Immunoglobulin G binding sites on the human foetal intestine: a possible mechanism for the passive transfer of immunity from mother to infant. Immunology 79, 77–81.

Johansen, F.E., Baekkevold, E.S., Carlsen, H.S., Farstad, I.N., Soler, D., Brandtzaeg, P., 2005. Regional induction of adhesion molecules and chemokine receptors explains disparate homing of human B cells to systemic and mucosal effector sites: dispersion from tonsils. Blood 106, 593–600.

Johansen, F.E., Brandtzaeg, P., 2004. Transcriptional regulation of the mucosal IgA system. Trends Immunol. 25, 150–157.

Johansen, F.-E., Pekna, M., Norderhaug, I.N., Haneberg, B., Hietala, M.A., Krajci, P., Betsholtz, C., Brandtzaeg, P., 1999. Absence of epithelial immunoglobulin A transport, with increased mucosal leakiness, polymeric immunoglobulin receptor/secretory component-deficient mice. J. Exp. Med. 190, 915–922.

Klemola, T., Savilahti, E., Leinikki, P., 1986. Mumps IgA antibodies are not absorbed from human milk. Acta Pediatr. Scand. 75, 230–232.

Kushnir, N., Bos, N.A., Zuercher, A.W., Coffin, S.E., Moser, C.A., Offit, P.A., Cebra, J.J., 2001. B2 but not B1 cells can contribute to $CD4^+$ T-cell-mediated clearance of rotavirus in SCID mice. J. Virol. 75, 5482–5490.

Lim, P.L., Rowley, D., 1982. The effect of antibody on the intestinal absorption of macromolecules and on intestinal permeability in adult mice. Int. Arch. Allergy Appl. Immunol. 68, 41–46.

Lin, M., Du, L., Brandtzaeg, P., Pan-Hammarström, Q., 2014. IgA subclass switch recombination in human mucosal and systemic immune compartments. Mucosal Immunol. 7, 511–520.

Lundqvist, C., Melgar, S., Yeung, M.M., Hammarström, S., Hammarström, M.L., 1996. Intraepithelial lymphocytes in human gut have lytic potential and a cytokine profile that suggest T helper 1 and cytotoxic functions. J. Immunol. 157, 1926–1934.

Mackenzie, N.M., 1990. Transport of maternally derived immunoglobulin across the intestinal epithelium. In: MacDonald, T.T. (Ed.), Ontogeny of the Immune System of the Gut. CRC Press, Boca Raton, Florida, pp. 69–81.

Macpherson, A.J., Gatto, D., Sainsbury, E., Harriman, G.R., Hengartner, H., Zinkernagel, R.M., 2000. A primitive T cell-independent mechanism of intestinal mucosal IgA responses to commensal bacteria. Science 288, 2222–2226.

Mathias, A., Pais, B., Favre, L., Benyacoub, J., Corthésy, B., 2014. Role of secretory IgA in the mucosal sensing of commensal bacteria. Gut Microbes 5, 688–695.

Mazzoli, S., Trabattoni, D., Lo Caputo, S., Piconi, S., Ble, C., Meacci, F., Ruzzante, S., Salvi, A., Semplici, F., Longhi, R., Fusi, M.L., Tofani, N., Biasin, M., Villa, M.L., Mazzotta, F., Clerici, M., 1997. HIV-specific mucosal and cellular immunity in HIV-seronegative partners of HIV-seropositive individuals. Nat. Med. 3, 1250–1257.

Mbawuike, I.N., Pacheco, S., Acuna, C.L., Switzer, K.C., Zhang, Y., Harriman, G.R., 1999. Mucosal immunity to influenza without IgA: an IgA knockout mouse model. J. Immunol. 162, 2530–2537.

Monteiro, R.C., 2014. Immunoglobulin A as an anti-inflammatory agent. Clin. Exp. Immunol. 178 (Suppl. 1), 108–110.

Moon, C., Baldridge, M.T., Wallace, M.A., Burnham, C.A., Virgin, H.W., Stappenbeck, T.S., 2015. Vertically transmitted faecal IgA levels determine extra-chromosomal phenotypic variation. Nature 521 (7550), 90–93.

Moser, C.A., Offit, P.A., 2001. Distribution of rotavirus-specific memory B cells in gut-associated lymphoid tissue after primary immunization. J. Gen. Virol. 82, 2271–2274.

Nilssen, D.E., Müller, F., Øktedalen, O., Frøland, S.S., Fausa, O., Halstensen, T.S., Brandtzaeg, P., 1996. Intraepithelial γ/δ T cells in duodenal mucosa are related to the immune state and survival time in AIDS. J. Virol. 70, 3545–3550.

Offit, P.A., Quarles, J., Gerber, M.A., Hackett, C.J., Marcuse, E.K., Kollman, T.R., Gellin, B.G., Landry, S., 2002. Addressing parents' concerns: do multiple vaccines overwhelm or weaken the infant's immune system? Pediatrics 109, 124–129.

Ogier, A., Franco, M.A., Charpilienne, A., Cohen, J., Pothier, P., Kohli, E., 2005. Distribution and phenotype of murine rotavirus-specific B cells induced by intranasal immunization with 2/6 virus-like particles. Eur. J. Immunol. 35, 2122–2130.

Ogra, P.L., Karzon, D.T., 1969. Distribution of poliovirus antibody in serum, nasopharynx and alimentary tract following segmental immunization of lower alimentary tract with poliovaccine. J. Immunol. 102, 1423–1430.

O'Neal, C.M., Clements, J.D., Estes, M.K., Conner, M.E., 1998. Rotavirus 2/6 virus-like particles administered intranasally with cholera toxin, *Escherichia coli* heat-labile toxin (LT), and LT-R192G induce protection from rotavirus challenge. J. Virol. 72, 3390–3393.

O'Neal, C.M., Harriman, G.R., Conner, M.E., 2000. Protection of the villus epithelial cells of the small intestine from rotavirus infection does not require immunoglobulin A. J. Virol. 74, 4102–4109.

Parra, M., Herrera, D., Calvo-Calle, J.M., Stern, L.J., Parra-López, C.A., Butcher, E., Franco, M., Angel, J., 2014. Circulating human rotavirus specific CD4 T cells identified with a class II tetramer express the intestinal homing receptors α4β7 and CCR9. Virology, 452–453, 191–201.

Pelaseyed, T., Bergström, J.H., Gustafsson, J.K., Ermund, A., Birchenough, G.M., Schütte, A., van der Post, S., Svensson, F., Rodríguez-Piñeiro, A.M., Nyström, E.E., Wising, C., Johansson, M.E., Hansson, G.C., 2014. The mucus and mucins of the goblet cells and enterocytes provide the first defense line of the gastrointestinal tract and interact with the immune system. Immunol. Rev. 260, 8–20.

Persson, C.G., Erjefält, J.S., Greiff, L., Erjefält, I., Korsgren, M., Linden, M., Sundler, F., Andersson, M., Svensson, C., 1998. Contribution of plasma-derived molecules to mucosal immune defense, disease and repair in the airways. Scand. J. Immunol. 47, 302–313.

Philips, A.D., 1989. Mechanisms of mucosal injury: human studies. In: Farthing, M.J.G. (Ed.), Viruses and the Gut. Smith Kline & French Laboratories Ltd, London, pp. 30–40.

Plaut, A.G., Wistar, Jr., R., Capra, J.D., 1974. Differential susceptibility of human IgA immunoglobulins to streptococcal IgA protease. J. Clin. Invest. 54, 1295–1300.

Regnault, A., Levraud, J.P., Lim, A., Six, A., Moreau, C., Cumano, A., Kourilsky, P., 1996. The expansion and selection of T cell receptor αβ intestinal intraepithelial T cell clones. Eur. J. Immunol. 26, 914–921.

Renegar, K.B., Johnson, C.D., Dewitt, R.C., King, B.K., Li, J., Fukatsu, K., Kudsk, K.A., 2001. Impairment of mucosal immunity by total parenteral nutrition: requirement for IgA in murine nasotracheal anti-influenza immunity. J. Immunol. 166, 819–825.

Renz, H., von Mutius, E., Brandtzaeg, P., Cookson, W.O., Autenrieth, I.B., Haller, D., 2011. Gene-environment interactions in chronic inflammatory disease. Nat. Immunol. 12, 273–277.

Robinson, J.K., Blanchard, T.G., Levine, A.D., Emancipator, S.N., Lamm, M.E., 2001. A mucosal IgA-mediated excretory immune system in vivo. J. Immunol. 166, 3688–3692.

Rogier, E.W., Frantz, A.L., Bruno, M.E., Kaetzel, C.S., 2014a. Secretory IgA is concentrated in the outer layer of colonic mucus along with gut bacteria. Pathogens 3, 390–403.

Rogier, E.W., Frantz, A.L., Bruno, M.E., Wedlund, L., Cohen, D.A., Stromberg, A.J., Kaetzel, C.S., 2014b. Secretory antibodies in breast milk promote long-term intestinal homeostasis by regulating the gut microbiota and host gene expression. Proc. Natl. Acad. Sci. USA 111, 3074–3079.

Roos, A., Bouwman, L.H., van Gijlswijk-Janssen, D.J., Faber-Krol, M.C., Stahl, G.L., Daha, M.R., 2001. Human IgA activates the complement system via the mannan-binding lectin pathway. J. Immunol. 167, 2861–2868.

Sait, L.C., Galic, M., Price, J.D., Simpfendorfer, K.R., Diavatopoulos, D.A., Uren, T.K., Janssen, P.H., Wijburg, O.L., Strugnell, R.A., 2007. Secretory antibodies reduce systemic antibody responses against the gastrointestinal commensal flora. Int. Immunol. 19, 257–265.

Shacklett, B.L., 2010. Immune responses to HIV and SIV in mucosal tissues: location, location, location'. Curr. Opin. HIV AIDS 5, 128–134.

Silvey, K.J., Hutchings, A.B., Vajdy, M., Petzke, M.M., Neutra, M.R., 2001. Role of immunoglobulin A in protection against reovirus entry into murine Peyer's patches. J. Virol. 75, 10870–10879.

Simister, N.E., Mostov, K.E., 1989. An Fc receptor structurally related to MHC class I antigens. Nature 337, 184–187.

Sonnenberg, G.F., Monticelli, L.A., Alenghat, T., Fung, T.C., Hutnick, N.A., Kunisawa, J., Shibata, N., Grunberg, S., Sinha, R., Zahm, A.M., Tardif, M.R., Sathaliyawala, T., Kubota, M., Farber, D.L., Collman, R.G., Shaked, A., Fouser, L.A., Weiner, D.B., Tessier, P.A., Friedman, J.R., Kiyono, H., Bushman, F.D., Chang, K.M., Artis, D., 2012. Innate lymphoid cells promote anatomical containment of lymphoid-resident commensal bacteria. Science 336, 1321–1325.

Stax, M.J., Mouser, E.E., van Montfort, T., Sanders, R.W., de Vries, H.J., Dekker, H.L., Herrera, C., Speijer, D., Pollakis, G., Paxton, W.A., 2015. Colorectal mucus binds DC-SIGN and inhibits HIV-1 trans-infection of CD4+ T-lymphocytes. PLoS One 10 (3), e0122020.

Stoll, B.J., Lee, F.K., Hale, E., Schwartz, D., Holmes, R., Ashby, R., Czerkinsky, C., Nahmias, A.J., 1993. Immunoglobulin secretion by the normal and the infected newborn infant. J. Pediatr. 122, 780–786.

Weaver, L.T., Wadd, N., Taylor, C.E., Greenwell, J., Toms, G.L., 1991. The ontogeny of serum IgA in the newborn. Pediatr. Allergy Immunol. 2, 72–75.

Wolf, H.M., Fischer, M.B., Pühringer, H., Samstag, A., Vogel, E., Eibl, M.M., 1994a. Human serum IgA downregulates the release of inflammatory cytokines (tumor necrosis factor-α, interleukin-6) in human monocytes. Blood 83, 1278–1288.

Wolf, H.M., Vogel, E., Fischer, M.B., Rengs, H., Schwarz, H.P., Eibl, M.M., 1994b. Inhibition of receptor-dependent and receptor-independent generation of the respiratory burst in human neutrophils and monocytes by human serum IgA. Pediatr. Res. 36, 235–243.

Yamanaka, T., Straumfors, A., Morton, H., Fausa, O., Brandtzaeg, P., Farstad, I., 2001. M cell pockets of human Peyer's patches are specialized extensions of germinal centers. Eur. J. Immunol. 31, 107–117.

Zeissig, S., Blumberg, R.S., 2014. Commensal microbial regulation of natural killer T cells at the frontiers of the mucosal immune system. FEBS Lett. 588, 4188–4194.

Zhou, M., Ruprecht, R.M., 2014. Are anti-HIV IgAs good guys or bad guys? Retrovirology 11, 109.

Chapter 1.3

Immunodeficiencies: Significance for Gastrointestinal Disease

H. Marcotte, L. Hammarström

Division of Clinical Immunology and Transfusion Medicine, Department of Laboratory Medicine, Karolinska Institutet at Karolinska University Hospital Huddinge, Stockholm, Sweden

1 PRIMARY IMMUNODEFICIENCY

Primary immunodeficiency diseases (PIDs) represent a heterogeneous group of genetically determined disorders that influence the development and function of different components of adaptive and innate immunity (Al-Herz et al., 2014). Since the first official scientific publication of a PID patient 60 years ago (Bruton, 1952), more than 250 additional diseases have been described and characterized (Al-Herz et al., 2014). Antibody deficiencies occur when a patient has too few B cells or B cells that do not function properly. Antibody deficiency accounts for nearly half of all PID cases and includes IgA deficiency, X-linked agammaglobulinemia (XLA), the hyper IgM syndrome (HIGM) and common variable immunodeficiency (CVID) (Table 1.3.1). Combined B-cell and T-cell deficiencies comprise around 20% of all PIDs and can arise when the body produces too few B cells and T cells or when the B cells and T cells that are produced do not function correctly. Severe Combined Immunodeficiency (SCID) is the clinically most serious type of combined immunodeficiency. Phagocyte defects, including Chronic Granulomatous Disease (CGD), and complement deficiencies represent around 18 and 5% of all PIDs, respectively (Grumach and Kirschfink, 2014; Song et al., 2011).

PIDs are characterized by various clinical symptoms including a high degree of susceptibility to various types of infections, autoimmunity, inflammation, allergies, and cancer. The main challenge of current immunology is to establish the diagnosis of PID before the onset of clinical symptoms and start therapy as soon as possible in order to prevent the complications and consequences of untreated or nonappropriately treated disease. Patients with antibody deficiency are generally treated either with immunoglobulin replacement (IVIG)

Viral Gastroenteritis

TABLE 1.3.1 Primary Immunodeficiencies and Gastrointestinal Infections

Immunodeficiency	Phenotype	Prevalence	Altered gene	Age of manifestation	Gastrointestinal infections
IgAD	IgA: <0.07g/L	1/600 (1/155–1/18,550)	Multiple, unknown Possibly HLA, B cell (IGHA1, IGHA2, CLEC16A) or T cell (TACI) defect, or viral response (IFH1) defect	All ages	*G. lamblia* *C. jejuni* *Salmonella* Rotavirus
CVID	• IgG deficit: <3 g/L • IgA deficit: <0.05 g/L • IgM concentration normal or low: <0.3 g/L	1/20,000–1/50,000	Multiple, unknown Possibly B cell (CD19, CD20, CLEC16A) or T cell (ICOS, TACI, BAFFR) defect	All ages	*G. lamblia* *C. jejuni* *Salmonella* *C. parvum* *C. difficile* Cytomegalovirus *Helicobacter pylori*
X-linked agammaglobulinemia	• Low to absent B cells • Low to absent immunoglobulin	1/70,000–1/100,000 (male)	Btk (Bruton's tyrosine kinase) in B cells	Childhood	*G. lamblia* *Campylobacter fetus* *Salmonella* spp. *C. difficile* Enterovirus Rotavirus

X-linked hyper IgM syndrome	• IgG, IgA, IgE concentration reduced • IgM normal or elevated	1/1,000,000 (male)	T cells (CD40L)	Childhood	*G. lamblia* *C. parvum* *C. difficile* *Yersinia* *Salmonella* *E. histolytica* Rotavirus
SCID	Defect in T and B cells	1/50,000–1/100,000	>12 genes (including common gamma chain, adenosine deaminase, JAK3, RAG1, RAG2, IL7R)	Infancy Neonatal	*C. jejuni* *G. lamblia* Rotavirus Poliovirus Norovirus

Adapted from Alkhairy and Hammarström (2015).

while bone marrow or stem cell transplantation is the most common treatment for patients with combined deficiencies. Bacterial infections in PID patients are usually controlled with antibiotics.

2 GASTROINTESTINAL DISORDERS IN PID PATIENTS

2.1 Selective IgA Deficiency

IgAD is defined as serum IgA levels equal to or below 0.07 g/L with normal serum IgM and IgG levels in individuals of 4 years of age or older. The threshold of 4 years of age is used to prevent a premature diagnosis of IgAD, which may be transient in children due to a delayed maturation of the immune system. IgAD is the most common PID with an estimated frequency of 1/600 in the Caucasian population but it may vary from 1/155 to 1/18,550, depending on the ethnic background (Wang and Hammarström, 2012). The phenotypic feature of IgAD is a failure of B lymphocyte differentiation into plasma cells producing IgA. The exact etiology is unknown and may result from either a T or B cell defect. However, the major histocompatibility complex (MHC), in particular the ancestral human leukocyte antigen (HLA)-A1, B8, DR3, DQ2 (8.1) haplotype, has previously been reported to be associated with disease development (Alkhairy and Hammarström, 2015; Wang and Hammarström, 2012). In addition, non-MHC genes, such as interferon-induced helicase 1 (*IFH1*) and C-type lectin domain family 16, member A (*CLEC16A*) have also been shown to be associated with IgAD (Ferreira et al., 2010). Mutations in the *TNFRSF13B* gene have been seen in some families with transmembrane activator and cyclophilin ligand interactor (TACI)-related IgAD but no cause–effect relationship has been found (Pan-Hammarström et al., 2007; Salzer et al., 2005). Genetic defects in the *IGHA1* and *IGHA2* genes have been associated with selective IgA1 and IgA2 deficiencies.

IgA normally provides immunologic protection at the mucosal membranes, with a daily production of IgA exceeding that of all other antibody classes combined, presumably reflecting its important role. While the majority of IgAD individuals discovered by screening of blood donors are asymptomatic at the time of diagnosis, some individuals develop recurrent infections, allergic diseases and autoimmune manifestations. IgAD is commonly considered a mild disorder, with approximately 65–70% of the affected individuals being reported as asymptomatic. However, in a recent case study using gender- and age-matched controls, IgAD individuals were found to have a significantly increased proneness to respiratory infections and increased prevalence of allergic diseases and autoimmunity, with a total of 84.4% being affected by any of these diseases, compared to 47.6% of matched controls (Jorgensen et al., 2013). IgA deficiency is also associated with an increased frequency of celiac diseases, Crohn's disease and ulcerative colitis (Jorgensen et al., 2013). Gastrointestinal infections and steatorrhea also occur with increased frequency, with *Giardia lamblia*, *Campylobacter jejuni*, and *Salmonella* spp. being the most implicated (Ammann and Hong, 1971; Spickett et al., 1991).

2.1.1 Selective IgA Deficiency and Gastrointestinal Viral Infections

IgA deficiency has also been associated with an increase in gastrointestinal viral infections. In humans and animals, intestinal rotavirus-specific IgA is a correlate of protection against rotavirus (see Chapter 2.9). This suggests that intestinal IgA may be essential in the immune response to the virus. Protection from rotavirus infection requires B cells; mice lacking B cells (μMt mice on a C57Bl/6 background) are not protected from rotavirus reinfection (Franco and Greenberg, 1995; McNeal et al., 1995). Support for the importance of IgA for rotavirus immunity also comes from studies in J-chain deficient (J chain$^{-/-}$) mice, which cannot transcytose IgA or IgM into the intestinal lumen. J chain$^{-/-}$ mice have difficulty clearing a primary rotavirus infection and are not protected from reinfection (Schwartz-Cornil et al., 2002). Mice lacking IgA (IgA$^{-/-}$) also exhibit a substantial and significant delay in clearance of the initial infection compared to wild type mice (Blutt et al., 2012) and excrete rotavirus in the stool up to 3 weeks after the initial exposure as compared to 10 days in wild type mice. Importantly, IgA$^{-/-}$ mice fail to develop protective immunity against multiple repeat exposures to the virus. All IgA$^{-/-}$ mice excreted virus in the stool upon reexposure to rotavirus while wild type mice were completely protected against reinfection, indicating a critical role for IgA in the establishment of immunity against this pathogen.

There are very few well controlled studies that address the question of whether or not IgA deficiency predisposes individuals to an increased susceptibility to and recurrence of gastrointestinal viral infections. IgAD patients show significantly elevated serum levels of rotavirus-specific IgG antibodies, suggesting a compensatory role of IgG in systemic rotavirus infection and a possible protraction of rotaviral persistence/shedding after infection although the disease is ultimately resolved (Istrate et al., 2008; Günaydın et al., 2014). To date, there has not been any report of an increased frequency or severity of infection by other enteropathogenic viruses (ie, capable of producing disease in the intestinal tract) such as norovirus or astrovirus in IgAD individuals. Elucidation of the role of the gastrointestinal IgA response to norovirus and astrovirus has, however, been restricted by the absence of small animal models in which pathogenesis and immunity is similar to that observed in humans. There is limited evidence that protection from both norovirus and astrovirus infections correlates with mucosal virus-specific IgA in humans (Blutt and Conner, 2013).

Nonenteropathogenic viruses do not infect a significant number of intestinal cells and do not cause gastrointestinal diseases. They invade the body by either breeching or crossing the epithelium of the gastrointestinal tract. However, poliovirus induces a secretory IgA response that appears to neutralize the virus and is associated with protection and decreased virus shedding in the stool (Buisman et al., 2008). An intestinal IgA response is also induced with the live replicating oral polio vaccine (OPV), and it is considered that OPV prevents infection through IgA-mediated viral neutralization in the intestine (Blutt and Conner, 2013). It has been reported that poliovirus can be detected in the stool of IgAD patients for a longer period than in normal individuals after oral vaccination with live attenuated virus, suggesting that IgAD patients have an impaired capacity to eliminate the virus (Savilahti et al., 1988).

Although studies in humans correlate increases in viral-specific IgA levels at the mucosal surface with either the cessation of virus excretion or protection against infection and disease (Chapter 2.9), IgA deficient patients do not appear to present overt clinical manifestations. In most individuals with IgA deficiency, the deficiency is not complete, and thus it is possible that even relatively low levels of IgA produced in patients with deficiency are sufficient to prevent infection. Although IgAD patients have a concomitant lack of IgA in external secretions (Norhagen et al., 1989; Savilahti, 1973), a low proportion of individuals with low levels of serum IgA may actually have sufficient secretory IgA at their mucosal surfaces and thus remain asymptomatic (Ammann and Hong, 1971). We have previously identified a few IgAD and CVID patients with normal level of secretory IgA (Hammarström, unpublished data). Furthermore, other antibody isotypes, in particular IgM transported to the mucosal surface, may compensate for the loss of IgA in selected cases (Savilahti, 1973). This is supported by the fact that patients with CVID are more symptomatic and susceptible to gastrointestinal infections than IgAD patients (see later).

2.2 Common Variable Immunodeficiency

Common variable immunodeficiency (CVID) is the second most common primary immunodeficiency syndrome with an estimated prevalence of 1/20,000 to 1/50,000 in the general population. The diagnosis is based on a marked reduction in serum levels of both IgG (usually <3 g/L) and IgA (<0.07 g/L); IgM is reduced in about half the patients (<0.3 g/L) (Agarwal and Mayer, 2013). Patients have also a poor or absent antibody production to vaccines, such as tetanus or diphtheria toxoids. In a few cases, IgAD can evolve gradually into CVID over a period of months to decades. A linkage of IgAD and CVID was previously observed among family members and patients with these two disorders, suggesting that they may share a common genetic background (Schäffer et al., 2006; Vorechovský et al., 1995).

The pathogenesis of CVID has not been delineated clearly; however, mutations in several genes associated with B-cell development, including autosomal recessive mutations in the genes encoding BAFF-R, TACI, CD20, CD19, CD81, CD21, inducible T cell costimulatory (ICOS) and lipopolysaccharide (LPS)-responsive beige-like anchor (LRBA), have been found in a small number of patients. A recent metaanalysis also suggests involvement of selected HLA alleles and the *CLEC16A* gene (Li et al., 2015). CVID is the most common symptomatic primary antibody deficient syndrome and characterized by upper and lower respiratory tracts infections caused by *Streptococcus pneumoniae, Moraxella catarrhalis,* and *Haemophilus influenzae*. The most common age of the onset of symptoms in patients with CVID generally appears during adolescence or early adulthood (Agarwal and Mayer, 2013). In addition to infection, CVID patients have a wide range of clinical manifestations, including autoimmune disease (mostly immune thrombocytopenic purpura and autoimmune hemolytic anemia), granulomatous/lymphoid infiltrating disease, enteropathy, and an increased incidence of malignancies.

Gastrointestinal symptoms are common in CVID patients and up to 50% of the patients have chronic diarrhea with malabsorption (Hermans et al., 1976; Cunningham-Rundles and Bodian, 1999), with *G. lamblia* being the most common cause, while *C. jejuni*, *Salmonella* spp., *Cryptosporidium parvum*, and *Clostridium difficile* have also be implicated (Agarwal and Mayer, 2009; Daniels et al., 2007; Hermaszewski and Webster, 1993; Sicherer and Winkelstein, 1998). Inflammatory bowel like-diseases resembling Crohn's disease and ulcerative colitis, atrophic gastritis, achlorhydria, gastric carcinoma, and gastrointestinal lymphoma have also been reported (Daniels et al., 2007; Hermaszewski and Webster, 1993). In addition, *H. pylori* has been suggested to be a significant infectious agent causing gastritis in CVID (Daniels et al., 2007; Quinti et al., 2007; Agarwal and Mayer, 2009). In CVID patients with gastrointestinal symptoms, histological lesions are observed in approximately 80% of biopsies sampled from the stomach, small bowel or colon (Malamut et al., 2010). A striking characteristic of CVID patients is the absence of intestinal plasma cells, pointing to a local defect in secretory antibodies (Malamut et al., 2010; van de Ven et al., 2014).

2.2.1 CVID and Viral Infections

Cases of CVID patients infected with echovirus, coxsackievirus, and poliovirus (particularly the vaccine strain), which in some cases may lead to death, have also been reported (Table 1.3.2) (Halliday et al., 2003; de Silva et al., 2012). The age of onset of nonpolio enteroviral infection was very variable, and was not confined to childhood while most polio cases occurred in childhood following immunization with live attenuated oral vaccines (OPV). Cytomegalovirus (CMV) infections are an infrequent complication of this disorder, and very few cases resulting in gastrointestinal diseases have been reported, occasionally occurring after long term high dose steroids (Chapel and Cunningham-Rundles, 2009; Cunningham-Rundles and Bodian, 1999; Daniels et al., 2007; Freeman et al., 1977).

A small proportion of patients (5–15%) suffering from CVID develop a characteristic and severe enteropathy, the cause of which is unknown but might be associated with viral infections (Woodward et al., 2015). A recent study on gastrointestinal viruses in children with antibody deficiencies (48 CVID and 6 XLA

TABLE 1.3.2 Gastrointestinal Viral Infections in Patients with PIDs

Immunodeficiency	Infection	Strain	Patients	Outcome in immunodeficient patients	References
IgAD	Poliovirus	Vaccine strains (OPV)	Control study: • 8 IgAD individuals, 6–23 years • 9 controls, 23–43 years	• Higher virus load in the stool • Longer shedding of virus (>5 weeks)	Savilahti et al. (1988)
	Rotavirus	wt	Control study: • 62 IgAD individuals without gastrointestinal diseases, 23–76 years • 62 controls, 8–69 years	• IgA deficient individuals develop higher serum IgG titers	Istrate et al. (2008)
		wt	Control study: • 783 IgAD individuals • 1009 controls	• Individuals with combined IgA and TLR3 deficiency show increased specific IgG titers as compared to individuals with impaired TLR3 only	Günaydın et al. (2014)
CVID	Various gastrointestinal viruses	wt	Observational study: • 54 antibody deficient patients (48 CVID + CVID-like), 4–18 years • 66 healthy donors, 4–18 years	• Increased prevalence of gastrointestinal viruses (particularly norovirus, parechovirus and adenovirus) • More symptoms (abdominal ache, diarrhea, thin stool)	van de Ven et al. (2014)

	Norovirus	wt	• 8 patients with CVID enteropathy, 25–66 years • 10 patients with CVID but no entheropathy	• Persistent (>22 months) fecal excretion of norovirus in all patients with CVID enteropathy • Clearance of virus in three patients associated with resolution of symptoms	Woodward et al. (2015)
	Cytomegalovirus	wt	20 CVID patients, 9 children (<10 years) and 21 adults	• 2 patients with cytomegalovirus infection	Daniels et al. (2007)
	Enteroviral infections	wt	20 CVID patients from various studies	• 14 with nonpolio enteroviral infection (echoviruses, Coxsackie viruses) and 6 with poliovirus infection • Enterovirus detected in various sites including the stool • Various neurological symptoms • Death occurred in one third of polio cases and half of the nonpolio cases	Halliday et al. (2003)
	Poliovirus	Vaccine strain (OPV)	51 PID patients screened for poliovirus	• A case of CVID patient (8 years old) with prolonged virus excretion (>6 months)	de Silva et al. (2012)
XLA	Rotavirus	wt	201 males with XLA from a registry	• 4 cases with rotavirus induced chronic/recurrent diarrhea • Rotavirus isolated from 8% of the patient with diarrhea	Winkelstein et al. (2006)

(Continued)

TABLE 1.3.2 Gastrointestinal Viral Infections in Patients with PIDs (*cont.*)

Immunodeficiency	Infection	Strain	Patients	Outcome in immunodeficient patients	References
	Enteroviral infections	wt	47 cases from various studies	• 42 with nonpolio infection (echoviruses, poliovirus) and 5 with polio infection • Enterovirus detected in various sites including the stool • Various neurological symptoms • Death occurred in one third of polio cases and half of nonpolio cases	Halliday et al. (2003)
	Poliovirus	Vaccine strain (OPV)	4 cases from various studies	• A 5 year old XLA patient with prolonged virus excretion • A 15 month old XLA patient with prolonged virus excretion (4 months) and acute flaccid paralysis • 2 XLA patients with flaccid paralysis	de Silva et al. (2012), Shahmahmoodi et al. (2008), Winkelstein et al. (2006)
XHIGM	Rotavirus	wt	79 males with XHIGM from a registry	• 2 patients with rotavirus infection • Isolated from 8% of the patients with diarrhea	Winkelstein et al. (2003)
	Enteroviral infections	wt	5 cases from various studies	• 5 with nonpolio infection (echoviruses, Coxsackie viruses)	Halliday et al. (2003)

SCID	Chronic rotavirus infection	wt	6 cases from different studies	• Prolonged diarrhea and persistent fecal excretion of rotavirus (>8 weeks) • Two patients died and had systemic spread of rotavirus at death • In one case, rotavirus infection reversed by HSCT	Saulsbury et al. (1980), Oishi et al. (1991), Gilger et al. (1992), Frange et al. (2012), Patel et al. (2012)
		Vaccine strains (Rotateq or Rotarix)	>10 cases from different studies, 3–9 months old	• Prolonged diarrhea and persistent rotavirus vaccine excretion (up to 6 months) • In one child, the prolonged excretion was resolved following successful cord-blood transplantation	Werther et al. (2009), Bakare et al. (2010), Patel et al. (2010), Uygungil et al. (2010), Donato et al. (2012)
	Norovirus	wt	Prospective study: • 62 children with PID	• Prolonged virus excretion in one SCID (6 months old) patient • Norovirus shedding associated with gastrointestinal symptoms	Frange et al. (2012)
		wt	2 cases from two studies	• Prolonged virus excretion	Chrystie et al. (1982), Xerry et al. (2010)
	Poliovirus	Vaccine strain (OPV)	Screening of 51 PID patients for poliovirus	• Three cases of SCID patients (4–5 months of age) with poliovirus infection • The three patients died before there was a possibility to measure the duration of excretion	de Silva et al. (2012)

IgAD, IgA deficiency; *CVID*, common variable immunodeficiency; *XLA*, X-linked agammaglobulinemia; *XHIGM*, X-linked hyper IgM syndrome; *SCID*, Severe Combined Immunodeficiency.

patients) showed that antibody deficient patients show an increased presence of gastrointestinal viruses (particularly norovirus, adenovirus, parechovirus) and that these patients frequently have gastrointestinal symptoms (abdominal ache, diarrhea and thin stool) (van de Ven et al., 2014) (Table 1.3.2). CVID and XLA patients that tested positive for gastrointestinal viruses also showed diminished levels of serum IgA and secretory IgA in fecal samples. Furthermore, a significant association between mucosal inflammation and the presence of enteric viruses was observed in the patients, but not in healthy controls. These findings suggest that hypogammaglobulinemia, particularly a decreased IgA production, might be associated with prolonged intestinal virus replication and that this may result in an increased risk for chronic mucosal inflammation such as CVID related enteropathy (van de Ven et al., 2014). Similarly, persistent fecal excretion of norovirus was found in CVID patients with duodenal villous atrophy and malabsorption (Table 1.3.2) (Woodward et al., 2015). Clearance of the virus spontaneously or following ribavirin therapy was temporally associated with complete resolution of symptoms, restoration of duodenal architecture, and resolution of intestinal mucosal inflammation. However, ribavirin as an effective antiviral therapy for norovirus infection requires further evaluation.

Severe enteropathy appears to be unique to a subset of patients with CVID and is not reported in other immunodeficiencies (Chapel et al., 2008). Recent evidence described earlier suggests an association between norovirus infection and enteropathy, and patients with CVID enteropathy should be tested for the presence of this virus. Furthermore, since CVID patients with a lower IgA level appear to be more prone to viral infections, secretory IgA concentrations should also be determined in order to better understand the susceptibility to virus infection in these patients. Finally, the development of the microbiota is closely related to intestinal IgA excretion, and it would be relevant to evaluate the microbiome of the antibody deficient patients and healthy controls (van de Ven et al., 2014). The microbiome of antibody deficient patients with low intestinal IgA levels might significantly differ from that of patients with normal intestinal IgA secretion and influence the susceptibility to viral and bacterial infections and development of enteropathy (Lindner et al., 2015; Xiong and Hu, 2015).

2.3 X-Linked Agammaglobulinemia

XLA accounts for 85% of known cases of agammaglobulinemia and is caused by a deficiency of Bruton's tyrosine kinase (BTK), causing a defect in early B-cell development. The disorder has a prevalence of roughly 1/100,000 newborns and has been reported in various ethnic groups worldwide. Being an X-linked disease, only males are affected and females are asymptomatic carriers. The majority of patients develop recurrent or persistent bacterial infections including otitis media, conjunctivitis, sinusitis, pneumonia, diarrhea, and skin infections within the first 2 years of life (Winkelstein et al., 2006; Papadopoulou-Alataki et al., 2012). Over half of the patients developed symptoms referable to their immunodeficiency before 1 year of age, and more than 90% by 5 years of age (Winkelstein et al., 2006). The most frequent clinical manifestations are upper and/or lower respiratory tract infections due to encapsulated bacteria (*S. pneumonia, H. influenza, H. parainfluenza*), *Staphylococcus* or *Pseudomonas* (Winkelstein et al., 2006). Chronic or recurrent diarrhea is observed in more than 20% of the patients with *G. lamblia* as the main causative agent, followed by rotavirus, enterovirus, *Campylobacter fetus, Salmonella* spp., *C. difficile, H. pylori*, and *Shigella* spp. (Winkelstein et al., 2006).

2.3.1 XLA and Viral Infections

Rotavirus is one of the most commonly isolated pathogen (after *G. lamblia*) infecting around 2% of the patient with XLA and 9% of those with diarrhea (Table 1.3.2) (Winkelstein et al., 2006). Severe viral infections are rare in XLA patients, but these patients have a unique susceptibility to acute or chronic meningoencephalitis caused by enteroviruses such as echovirus, Coxsackie virus and poliovirus (Halliday et al., 2003; Winkelstein et al., 2006). The virus enters the host via the oral cavity and respiratory tract, then invades and replicates in the upper respiratory tract and small intestine, with a predilection for lymphoid tissues in these regions (Peyer's patches, mesenteric lymph nodes, tonsils, and cervical lymph nodes). Virus then enters the bloodstream, resulting in a minor viremia and dissemination to a variety of target organs, including the central nervous system (CNS). Enteroviruses have been isolated from a variety of sites, usually from the CSF, and occasionally outside the central nervous system such as the stool and respiratory tract; these sites possibly being the primary site of infection. The infections are chronic in nature and may take the form of meningoencephalitis, dermatomyositis, hepatitis, and/or arthritis. (Winkelstein et al., 2006). Death is frequent and was shown to occur in half of nonpolio cases and a third of polio cases (Halliday et al., 2003; Winkelstein et al., 2006). XLA patients can also show prolonged virus excretion of the polio vaccine strains through vaccination or contact, and in some cases symptoms of polio-like disease can lead to death (Wang et al., 2004; Shahmahmoodi et al., 2008; de Silva et al., 2012; Winkelstein et al., 2006).

Although patients with other PID are also susceptible to enteroviral infections, XLA patients are particularly prone which may partly be due to the fact that they have the most severe antibody deficiency. The death of many of these patients might be prevented with earlier diagnosis of PID (Section 3.2) and immediate institution of high level immunoglobulin replacement therapy (Plebani et al., 2002; Winkelstein et al., 2006).

2.4 Hyper IgM Syndrome

The most common form of the hyper IgM (HIGM) syndrome is inherited as an X-linked trait (XHIGM) which occurs at a frequency of around 1/1,000,000 males. Mutations in the *CD40 ligand* (*CD40L*) gene are known to cause

XHIGM, in which T-cell and B-cell interaction is disrupted, thus preventing T-cell-dependent isotype switching from IgM to IgG or IgA and a subsequent selective increase in IgM level and a reduced level of the other immunoglobulin classes. Intermittent or chronic neutropenia is found in half of the patients and is often associated with oral ulcers, proctitis (inflammation and ulceration of the rectum), skin infections and sepsis (Winkelstein et al., 2003). HIGM syndromes with a pure humoral defect linked to intrinsic B-lymphocyte anomalies are attributed to mutations in genes coding for enzymes involved in isotype switching including activation-induced cytidine deaminase (AID) and uracil-DNA glycosylase (UNG).

Most patients with HIGM develop clinical symptoms, particularly respiratory infection with *Pneumocystis carinii*, during their first or second year of life and nearly all before the age of four (Winkelstein et al., 2003). HIGM patients could have distinct clinical infectious complications based on the type of genetic background and exact type of the syndrome. Protracted or recurrent diarrhea caused by *C. parvum*, *G. lamblia*, *C. difficile*, *Entamoeba histolytica*, *Salmonella* spp., *Yersinia* spp. and rotavirus, is reported in 35–50% of the XHIGM patients although an infectious agent cannot be identified in half of the cases (Winkelstein et al., 2003; Levy et al., 1997).

2.4.1 XHIGM and Viral Infections

Only a few publications are available on gastrointestinal viral infections in patients with the XHIGM syndrome although cases of nonpolio enterovirus (echovirus, Coxsackie virus) infection have been reported (Halliday et al., 2003; Cunningham et al., 1999; Winkelstein et al., 2003). In one study, rotavirus was isolated in 2.5% of the patient and 8% of those with diarrhea (Table 1.3.2) (Winkelstein et al., 2003). No other viruses were isolated and no etiologic agent was detected in 50% of the patients with diarrhea which might explain the apparent low impact of virus in patients with XHIGM (Winkelstein et al., 2003). Another reason might be that in the intestinal mucosa of XHIGM patients, secretory IgM or a low level of secretory IgA antibody originating from T-cell-independent class switching may be sufficient for protection against gastrointestinal viruses. Intestinal bacteria can trigger T-cell-independent IgA2 class switching (Cerutti et al., 2011). LPS can activate B cells through Toll-like receptors (TLRs), and polysaccharides can activate B cells through their B-cell receptors. Furthermore dendritic cells produce IgA class switching recombination-inducing factors, such as APRIL and IL-10, after receiving instructing signals from bacteria and intestinal epithelial cells via TLR ligands and thymic stroma lymphopoietin. In mice, B-1 cells express unmutated IgA antibodies that have not been subjected to somatic hypermutation and recognize multiple specificities with low affinity. In T-cell deficient mice, these antibodies provide limited protection against some pathogens, including rotavirus (Franco and Greenberg, 1997). Although humans seem to lack B-1 cells, they have additional B-cell subsets that might be involved in T-cell-independent IgA responses. The intestinal lamina propria of XHIGM patients includes IgA-producing B cells and plasmablasts that contain AID, a hallmark of ongoing class switching recombination (Cerutti et al., 2011).

2.5 SCID

Severe Combined Immunodeficiency (SCID) represents a group of genetic defects causing severe deficiencies in the number and/or function of T, B, and natural killer cells. The overall incidence is 1 in 50,000–100,000 births and involves at least 12 different molecular causes including the genes of the common gamma chain, adenosine deaminase and JAK3. SCID is fatal if not treated, usually by hematopoietic stem cell transplantation (HSCT). Patients with SCID are highly susceptible to recurrent infections with a variety of pathogens including *P. jiroveci, C. albicans,* cytomegalovirus, respiratory syncytial virus (RSV), herpes simplex virus (HSV), adenoviruses, influenza viruses, parainfluenza viruses, and mycobacteria. They are particularly susceptible to viral infections and viral reactivations after allogeneic HSCT. Gastrointestinal infections are generally caused by *G. lamblia*, *C. jejuni*, norovirus, and rotavirus (Aguilar et al., 2014; Frange et al., 2012).

2.5.1 SCID and Viral Infections

A few cases of SCID patients infected with norovirus have been reported in the literature (Xerry et al., 2010; Frange et al., 2012; Chrystie et al., 1982). Some of the infections occur in hospital settings and the viral shedding may last up to 1 year.

The contribution of the adaptive immune responses in clearance of rotavirus from the stool is well defined and depends on lymphocytes. Mice lacking T and B lymphocytes (SCID and Rag $2^{-/-}$) are unable to clear a primary rotavirus infection from the intestine and chronically shed rotavirus (Franco et al., 1997; Kushnir et al., 2001). Cases of wild type rotavirus infection have indeed been reported in SCID children with prolonged diarrhea and persistent fecal excretion of rotavirus (Table 1.3.2). A 7-month-old male infant with SCID who had persistent rotavirus gastroenteritis and viremia despite oral and intravenous immunoglobulins administration was cured following HSCT (Patel et al., 2012). IVIG alone was not efficient in clearing the infection maybe because both immunoglobulins and $CD8^+$ T cells are important in clearing the rotavirus infection. Clearance of chronic rotavirus infection in SCID mice can be achieved by adoptive transfer of immune $CD8^+$ T lymphocytes in the absence of antibodies (Dharakul et al., 1990).

Children affected by SCID can also become ill from live viruses present in some vaccines including poliovirus and rotavirus (Table 1.3.2). There are several case reports of severe gastroenteritis with prolonged vaccine virus shedding and prolonged diarrhea in infants administered live, oral rotavirus vaccine that were later identified to have SCID (Table 1.3.2). The interval from vaccination to the onset of symptoms ranged from 1 to 33 days (median 12 days)

(Bakare et al., 2010; Patel et al., 2010). Most cases were associated with viral shedding lasting from 1 to at least 7.5 months (median of at least 5.5 months). In one child, prolonged excretion was resolved following successful cord-blood transplantation (Werther et al., 2009).

2.6 Other Immunodeficiency Diseases and Genetic Factors

No increase of gastrointestinal viral infections has been reported in patients with other prevalent immunodeficiency such as phagocyte (CGD) and complement defects. CGD patients show a normal immunity to most viruses, and the role of complement is probably not crucial in the defense against viruses at mucosal surface and in the mucosa. However, there are more than 250 PIDs and some of them have been associated with viral infections. For example, some combined immunodeficiency disorders [deficiency in IL-2-inducible T-cell kinase (ITK), magnesium transporter 1 (MAGT1), signaling lymphocyte activation molecule (SLAM)–associated protein (SH2D1A), macrophage stimulating 1 (MST1), phosphoinositide 3-kinase delta (PI3K-δ), lipopolysaccharide responsive beige-like anchor protein (LRBA), CD27, etc.], diseases of immune dysregulation [deficiency in caspase recruitment domain-containing protein 11 (CARD11), protein kinase C (PRKCδ)], and defects in innate immunity [signal transducer and activator of transcription 1-alpha/beta (STAT1) deficiency] have been associated with Epstein-Barr-Virus and/or cytomegalovirus infection (Al-Herz et al., 2014). Furthermore, mutation in genes encoding IL-10, IL-10 receptors, STAT1, LRBA, forkhead box P3 (FOXP3), autoimmune regulator (AIRE), an E3 ubiquitin ligase (ITCH), and the T-cell receptor alpha constant region gene (TRAC), can lead to chronic diarrhea, enteropathy and/or inflammatory bowel disease (Al-Herz et al., 2014). Those PIDs might also be associated with an increase in the incidence of gastrointestinal viral infections, but little information is available to date.

Innate immunity, more particularly the interferon response, is important in the defense against viruses infecting the gastrointestinal tract (see Chapter 2.8), and defects in genes involved in the interferon pathway might affect susceptibility to infection. A combination of immunodeficiency affecting both the innate and acquired immunity might increase the susceptibility to viral infections. Individuals with combined IgA and TLR3 deficiency show an increased specific IgG antibody titer against rotavirus as compared to individuals with impaired TLR3 only, suggesting an increased susceptibility to severe, prolonged and/or recurrent rotavirus infection in individuals with both of these defects (Günaydın et al., 2014). Other genetic factors might also affect susceptibility to gastrointestinal infections in PID patients. Genetic human variations in human histo-blood group antigen expression on mucosal epithelia strongly affect the risk of norovirus infection, and indications of a similar relationship with rotavirus have recently been suggested in different populations (Nordgren et al., 2014; Payne et al., 2015). The histo-blood group antigens that serve as receptors for norovirus can also act as receptors for rotavirus (see Chapter 3.3) (Tan and Jiang, 2014).

3 CONSIDERATION FOR USE OF LIVE ORAL VACCINE IN IMMUNODEFICIENT PATIENTS

3.1 Oral Rotavirus Vaccine

Two live oral vaccines for rotavirus are currently licensed: a pentavalent bovine-human reassortant vaccine (RV5; marketed as RotaTeq, Merck), and a human monovalent vaccine (Rotarix, GlaxoSmithKline GSK). RotaTeq is administered in a 3-dose series, with doses administered at ages of 2, 4, and 6 months while Rotarix is administered in a 2-dose series, with doses administered at ages of 2 and 4 months (for further details see Chapter 2.11). Immunization with rotavirus vaccine early in life is especially important since rotavirus infection, nearly universal in children by 5 years of age, is typically most severe at first infection in children soon after 3 months of age (CDC, 2015). Both vaccine strains replicate in the human small intestine and can induce shedding of the vaccine strains. After the first doses of vaccine during clinical trials, Rotarix and Rotateq strains were detected in 35–80% and 9% of healthy recipient, respectively (Donato et al., 2012). Live viral vaccines are typically contraindicated in patients with known severe immunodeficiencies due to the risk of prolonged infection. As reported in Section 2.5, several cases of rotavirus vaccine-induced infection have been reported in children with SCID. The rotavirus-vaccine series is recommended to be started at 2 months of age, which is before SCID is typically diagnosed in infants for whom there is no family history of immunodeficiency. Most SCID infants are not diagnosed until after 4 months of age and could potentially have received two doses of the vaccine before a diagnosis is made. It is thus important to consider this diagnosis when treating children with prolonged diarrhea in countries where rotavirus vaccination is implemented.

In the absence of newborn screening, it is a challenge for clinicians to diagnose SCID early enough to withhold the first dose rotavirus vaccine at 2 months of age. In this context, a recent recommendation by the US Health Resources and Services Administration (HRSA) Advisory Committee on Heritable Disorders in Newborns and Children (ACHDNC) to add SCID to the core panel for neonatal universal screening may enable more frequent withholding of rotavirus vaccine from infants with SCID. Pilot programs of universal neonatal screening for SCID are already established in most states of the USA (Kwan et al., 2014). The test is based on detection of T-cell receptor excision circles (TREC assay) in Guthrie card samples from newborns.

Early diagnosis is also important to identify infants with SCID that could be vulnerable to acquiring the vaccine strain from close contacts with children who received rotavirus vaccine and subsequently shed the vaccine strain. Seven cases of horizontal transmission of rotavirus vaccine strains causing symptomatic infection have been reported by the US passive Vaccine Adverse Event Reporting System (VAERS), including transmission to adult relatives and one case of transmission to an immunocompromised father (Donato et al., 2012). Reassortment between vaccine strains of Rotateq vaccine can occur in a small proportion of children and, although a rare event, those reassortants vaccine strains can be transmitted to immunocompetent or immunocompromised individuals (Donato et al., 2012; Payne et al., 2010; Werther et al., 2009). Reassortment between the human rotavirus-derived vaccine (Rotarix) or RotaTeq with rotavirus wild-type strains may also be possible (Hemming and Vesikari, 2014). These reassortants could possess characteristics of increased virulence as compared to the original attenuated strain, including enhanced cell binding and entry into human intestinal cells, and might thus infect immunodeficient individual more readily than the vaccine strain.

There is no question about the substantial benefit from the rotavirus vaccine in reducing morbidity as well as health care costs worldwide (Donato et al., 2012; Patel et al., 2011). Risks of the vaccine to immunocompromised hosts do not negate its widespread use, but rather reinforce the need for neonatal screening measures for PIDs (Gaspar et al., 2014). A method measuring both T-cell receptor excision circles (TRECs) and kappa-deleting recombination excision circles (KRECs) as markers for T or B lymphopenia which can detect both SCID and XLA is currently being implemented in some European countries (Borte et al., 2011). Those excision circles-based methods are inefficient to detect most forms of antibody deficiency (except XLA), and novel methods based on genome- and transcriptome-wide analysis are being developed.

3.2 Oral Poliovirus (OPV)

The OPV is a mixture of three live viruses (types 1, 2, and 3) that is administered at birth (OPV-0 dose) followed by a series of three doses (around 6, 10, and 14 weeks) in endemic regions. Vaccine-derived polioviruses (VDPVs) are rare strains of poliovirus that have genetically mutated from the strains contained in the OPV. Individuals with PID are at a high risk for developing vaccine-associated paralytic poliomyelitis. In addition, prolonged intestinal poliovirus replication in PID patients may lead to development of VDPVs referred to as immunodeficiency-associated vaccine-derived polioviruses (iVDPVs). As reported in Section 2.5, individuals with PID exposed to OPV through vaccination or contact are at increased risk of prolonged (≥6 months) or chronic (>5 years) excretion of VDPVs (de Silva et al., 2012). Prolonged intestinal replication increases the risk for the virus to mutate and regain increased transmissibility to the general population and the neurovirulence of wild type poliovirus. Although it is recommended that persons diagnosed with primary B-cell deficiencies should not receive OPV, the PID is frequently not diagnosed until later in life.

More than 70 individuals with PID, shedding the Sabin VDPVs have been reported to the WHO since 1961 (Diop et al., 2014; Guo et al., 2015). In the majority of these patients, the PIDs were detected only after onset of acute flaccid paralysis. After implementation of intensified surveillance for iVDPV excretion among persons with PID in middle and lower income countries, detection of iVDPV infections increased between 2008 and 2013 (Li et al., 2014). Poliovirus type 2 (Sabin or iVDPV) was isolated most frequently (41%), followed by type 3 (32%), and type 1 (27%). Of the PIDs, only those with antibody deficiency (CVID, XLA) and SCID, which are all associated with lack of mucosal antibodies (SIgA), were found to be excreting poliovirus confirming the importance of secretory IgA in the elimination of the virus.

VDPV outbreaks continue to emerge in settings of conflict, insecurity and poor infrastructure. Over a dozen of such outbreaks have been recorded throughout the world in the last 10 years including Afghanistan, Nigeria, Pakistan, and Somalia (WHO). This poses a challenge to global polio eradication. As the majority of VDPV isolates are type 2 (Guo et al., 2015), the WHO has planned to replace the trivalent OPV with bivalent OPV (types 1 and 3) by 2016, preceded by at least one dose of inactivated poliovirus vaccine. Although the risk for further transmission of iVDPV is relatively low, a potential risk for circulation of iVDPV strains always remains. Circulation of iVDPV has been reported in an Amish community with low immunization coverage in 2005 but it did not result in any case of paralytic disease (Li et al., 2014). Introduction of neonatal screening programs for some immunodeficiencies such as SCID could also help prevent inadvertent exposure of such patients to OPV. This screening program should be easy to implement in nonendemic countries. However, in an endemic zone, this would require delaying 0 dose in order to allow diagnosis of PIDs before administration of the first dose and potentially increasing the risk of polio infection.

The most commonly recommended form of treatment for PID disorders is replacement therapy with IVIG, but due to high cost this treatment is not routinely available in many countries. Furthermore, although appropriate IVIG therapy may protect patients with certain B-cell deficiencies against paralytic disease, current evidence suggests that therapy is unlikely to prevent or clear persistent poliovirus infection or to stop excretion (Halliday et al., 2003; Li et al., 2014). IVIG appears to be somewhat efficient for prophylaxis against gastrointestinal infections and diarrhea in patients with immunoglobulin immunodeficiency (Baris et al., 2011). In humans, neonatal Fc receptor (FcRn) is detected in both fetal and adult intestines and can mediate bidirectional transcytosis of IgG across the intestinal epithelium both in vitro and in vivo (Kuo et al., 2010). Administered immunoglobulin can theoretically be transported within epithelial cells and surface of the mucosa through (FcRn) where they can neutralize viruses. The lower efficacy of IVIG against poliovirus might be because an insufficient amount of immunoglobulins is transported to the mucosal surface. Oral immunoglobulin therapy could also be considered as it could help neutralizing the progeny virus, preventing infection of neighboring cells (Losonsky et al., 1985; Sarker et al., 1998).

4 CONCLUSIONS AND FUTURE PERSPECTIVES

Very few control studies or case reports on gastrointestinal viral infections are available in the literature, possibly because viruses are not routinely screened for or the detection methods are not optimal. In summary, recent data suggest that norovirus is an important pathogen in patients with PID, particularly CVID, and may have a role in exacerbation of immune-mediated enteropathy. It would thus be appropriate to systematically screen PID patients for norovirus in cases of chronic gastrointestinal symptoms or before HSCT. XLA patients have a unique susceptibility to acute or chronic meningoencephalitis caused by enteroviruses, and early diagnosis and treatment may greatly improve the prognosis of the disorder. Furthermore, oral poliovirus and rotavirus vaccines can cause protracted infection in patients with PIDs, reinforcing the need for neonatal screening for SCID. Recurrent respiratory tract infections appear to be a greater problem than intestinal infections in PID patients which might be explained by a difference in local immunoglobulin levels in the respiratory and intestinal tract in these patients. The intestinal microbiota in absence of T cells could induce local B-cell switching to IgA2 that recognize multiple specificities with a low affinity which could give limited protection to gastrointestinal pathogens. In order to better understand the susceptibility of PID patients to gastrointestinal infection, further information should be obtained on local immunoglobulin levels, IgA subclass levels, the intestinal microbiome and blood group antigens. Finally, full genome or exome sequencing could be performed to better detect the underlying genetic defects, potential combined immunodeficiency (including innate immunity deficiency), and other genetic factors increasing the susceptibility to gastrointestinal viral infections.

REFERENCES

Agarwal, S., Mayer, L., 2009. Pathogenesis and treatment of gastrointestinal disease in antibody deficiency syndromes. J. Allergy Clin. Immunol. 124, 658–664.

Agarwal, S., Mayer, L., 2013. Diagnosis and treatment of gastrointestinal disorders in patients with primary immunodeficiency. Clin. Gastroenterol. Hepatol. 11, 1050–1063.

Aguilar, C., Malphettes, M., Donadieu, J., et al., 2014. Prevention of infections during primary immunodeficiency. Clin. Infect. Dis. 59, 1462–1470.

Al-Herz, W., Bousfiha, A., Casanova, J.L., et al., 2014. Primary immunodeficiency diseases: an update on the classification from the international union of immunological societies expert committee for primary immunodeficiency. Front. Immunol. 22, 162.

Alkhairy, O., Hammarström, L., 2015. IgA deficiency and other immunodeficiencies causing mucosal immunity dysfunction. In: Mestecky, J., Strober, W., Russell, M.W., Cheroutre, H., Lambrecht, B.N., Kelsall, B.L. (Eds.), Mucosal Immunology, fourth ed. Academic Press, London, pp. 1441–1459, Chapter 73.

Ammann, A.J., Hong, R., 1971. Selective IgA deficiency: presentation of 30 cases and a review of the literature. Medicine (Baltimore) 50, 223–236.

Bakare, N., Menschik, D., Tiernan, R., Hua, W., Martin, D., 2010. Severe combined immunodeficiency (SCID) and rotavirus vaccination: reports to the Vaccine Adverse Events Reporting System (VAERS). Vaccine 28, 6609–6612.

Baris, S., Ercan, H., Cagan, H.H., et al., 2011. Efficacy of intravenous immunoglobulin treatment in children with common variable immunodeficiency. J. Investig. Allergol. Clin. Immunol. 21, 514–521.

Borte, S., Wang, N., Oskarsdóttir, S., von Döbeln, U., Hammarström, L., 2011. Newborn screening for primary immunodeficiencies: beyond SCID and XLA. Ann. NY Acad. Sci. 1246, 118–130.

Blutt, S.E., Conner, M.E., 2013. The gastrointestinal frontier: IgA and viruses. Front. Immunol. 4, 402.

Blutt, S.E., Miller, A.D., Salmon, S.L., Metzger, D.W., Conner, M.E., 2012. IgA is important for clearance and critical for protection from rotavirus infection. Mucosal Immunol. 5, 712–719.

Bruton, O.C., 1952. Agammaglobulinemia. Pediatrics 9, 722–728.

Buisman, A.M., Abbink, F., Schepp, R.M., Sonsma, J.A., Herremans, T., Kimman, T.G., 2008. Preexisting poliovirus-specific IgA in the circulation correlates with protection against virus excretion in the elderly. J. Infect. Dis. 197, 698–706.

Centers for Disease Control and Prevention (CDC), 2015. Rotavirus. In: Epidemiology and prevention of vaccine-preventable diseases, 13th ed., Hamborsky J, Kroger A, Wolfe S, Eds. Washington D.C. Public Health Foundation, Chapter 19, pp. 311–24.

Cerutti, A., Cols, M., Gentile, M., et al., 2011. Regulation of mucosal IgA responses: lessons from primary immunodeficiencies. Ann. NY Acad. Sci. 1238, 132–144.

Chapel, H., Cunningham-Rundles, C., 2009. Update in understanding common variable immunodeficiency disorders (CVIDs) and the management of patients with these conditions. Br. J. Haematol. 145, 709–727.

Chapel, H., Lucas, M., Lee, M., et al., 2008. Common variable immunodeficiency disorders: division into distinct clinical phenotypes. Blood 112, 277–286.

Chrystie, I.L., Booth, I.W., Kidd, A.H., Marshall, W.C., Banatvala, J.E., 1982. Multiple faecal virus excretion in immunodeficiency. Lancet 1, 282.

Cunningham, C.K., Bonville, C.A., Ochs, H.D., et al., 1999. Enteroviral meningoencephalitis as a complication of X-linked hyper IgM syndrome. J. Pediatr. 134, 584–588.

Cunningham-Rundles, C., Bodian, C., 1999. Common variable immunodeficiency: clinical and immunological features of 248 patients. Clin. Immunol. 92, 34–48.

Daniels, J.A., Lederman, H.M., Maitra, A., Montgomery, E.A., 2007. Gastrointestinal tract pathology in patients with common variable immunodeficiency (CVID): a clinicopathologic study and review. Am. J. Surg. Pathol. 31, 1800–1812.

Dharakul, T., Rott, L., Greenberg, H.B., 1990. Recovery from chronic rotavirus infection in mice with severe combined immunodeficiency: virus clearance mediated by adoptive transfer of immune CD8+ T lymphocytes. J. Virol. 64, 4375–4382.

De Silva, R., Gunasena, S., Ratnayake, D., et al., 2012. Prevalence of prolonged and chronic poliovirus excretion among persons with primary immune deficiency disorders in Sri Lanka. Vaccine 30, 7561–7565.

Diop, O.M., Burns, C.C., Wassilak, S.G., Kew, O.M., 2014. Centers for Disease Control and Prevention (CDC). Update on vaccine-derived polioviruses—worldwide, July 2012-December 2013. MMWR Morb. Mortal. Wkly. Rep. 63, 242–248.

Donato, C.M., Ch'ng, L.S., Boniface, K.F., et al., 2012. Identification of strains of RotaTeq rotavirus vaccine in infants with gastroenteritis following routine vaccination. J. Infect. Dis. 206, 377–383.

Ferreira, R.C., Pan-Hammarström, Q., Graham, R.R., et al., 2010. Association of IFIH1 and other autoimmunity risk alleles with selective IgA deficiency. Nat. Genet. 42, 777–780.

Franco, M.A., Greenberg, H.B., 1995. Role of B cells and cytotoxic T lymphocytes in clearance of and immunity to rotavirus infection in mice. J Virol 69, 7800–7806.

Franco, M.A., Greenberg, H.B., 1997. Immunity to rotavirus in T cell deficient mice. Virology 238, 169–179.

Franco, M.A., Tin, C., Rott, L.S., VanCott, J.L., McGhee, J.R., Greenberg, H.B., 1997. Evidence for CD8+ T-cell immunity to murine rotavirus in the absence of perforin, fas, and gamma interferon. J. Virol. 71, 479–486.

Frange, P., Touzot, F., Debré, M., et al., 2012. Prevalence and clinical impact of norovirus fecal shedding in children with inherited immune deficiencies. J. Infect. Dis. 206, 1269–1274.

Freeman, H.J., Shnitka, T.K., Piercey, J.R., Weinstein, W.M., 1977. Cytomegalovirus infection of the gastrointestinal tract in a patient with late onset immunodeficiency syndrome. Gastroenterology 73, 1397–1403.

Gaspar, H.B., Hammarström, L., Mahlaoui, N., Borte, M., Borte, S., 2014. The case for mandatory newborn screening for severe combined immunodeficiency (SCID). J. Clin. Immunol. 34, 393–397.

Gilger, M.A., Matson, D.O., Conner, M.E., Rosenblatt, H.M., Finegold, M.J., Estes, M.K., 1992. Extraintestinal rotavirus infections in children with immunodeficiency. J. Pediatr. 120, 912–917.

Grumach, A.S., Kirschfink, M., 2014. Are complement deficiencies really rare? Overview on prevalence, clinical importance and modern diagnostic approach. Mol. Immunol. 61, 110–117.

Günaydın, G., Nordgren, J., Svensson, L., Hammarström, L., 2014. Mutations in toll-like receptor 3 are associated with elevated levels of rotavirus-specific IgG antibodies in IgA-deficient but not IgA-sufficient individuals. Clin. Vaccine Immunol. 21, 298–301.

Guo, J., Bolivar-Wagers, S., Srinivas, N., Holubar, M., Maldonado, Y., 2015. Immunodeficiency-related vaccine-derived poliovirus (iVDPV) cases: A systematic review and implications for polio eradication. Vaccine 33, 1235–1242.

Halliday, E., Winkelstein, J., Webster, A.D., 2003. Enteroviral infections in primary immunodeficiency (PID): a survey of morbidity and mortality. J. Infect. 46, 1–8.

Hemming, M., Vesikari, T., 2014. Detection of rotateq vaccine-derived, double-reassortant rotavirus in a 7-year-old child with acute gastroenteritis. Pediatr. Infect. Dis. J. 33, 655–656.

Hermans, P.E., Diaz-Buxo, J.A., Stobo, J.D., 1976. Idiopathic late-onset immunoglobulin deficiency. Clinical observations in 50 patients. Am J Med 61, 221–237.

Hermaszewski, R.A., Webster, A.D., 1993. Primary hypogammaglobulinaemia: a survey of clinical manifestations and complications. Q. J. Med. 86, 31–42.

Istrate, C., Hinkula, J., Hammarström, L., Svensson, L., 2008. Individuals with selective IgA deficiency resolve rotavirus disease and develop higher antibody titers (IgG, IgG1) than IgA competent individuals. J. Med. Virol. 80, 531–535.

Jorgensen, G.H., Gardulf, A., Sigurdsson, M.I., et al., 2013. Clinical symptoms in adults with selective IgA deficiency: a case-control study. J. Clin. Immunol. 33, 742–747.

Kuo, T.T., Baker, K., Yoshida, M., et al., 2010. Neonatal Fc receptor: from immunity to therapeutics. J. Clin. Immunol. 30, 777–789.

Kushnir, N., Bos, N.A., Zuercher, A.W., et al., 2001. B2 but not B1 cells can contribute to CD4+ T-cell-mediated clearance of rotavirus in SCID mice. J. Virol. 75, 5482–5490.

Kwan, A., Abraham, R.S., Currier, R., et al., 2014. Newborn screening for severe combined immunodeficiency in 11 screening programs in the United States. JAMA 312, 729–738.

Levy, J., Espanol-Boren, T., Thomas, C., et al., 1997. Clinical spectrum of X-linked hyper-IgM syndrome. J. Pediatr. 131, 47–54.

Li, J., Jorgensen, S.F., Melkorka Maggadottir, S., et al., 2015. Association of CLEC16A with human common variable immunodeficiency disorder and role in murine B cells. Nat. Commun. 6, 6804.

Li, L., Ivanova, O., Driss, N., et al., 2014. Poliovirus excretion among persons with primary immune deficiency disorders: summary of a seven-country study series. J. Infect. Dis. 210 (Suppl 1), S368–S372.

Lindner, C., Thomsen, I., Wahl, B., et al., 2015. Diversification of memory B cells drives the continuous adaptation of secretory antibodies to gut microbiota. Nat. Immunol. 16, 880–888.

Losonsky, G.A., Johnson, J.P., Winkelstein, J.A., Yolken, R.H., 1985. Oral administration of human serum immunoglobulin in immunodeficient patients with viral gastroenteritis. A pharmacokinetic and functional analysis. J. Clin. Invest. 76, 2362–2367.

Malamut, G., Verkarre, V., Suarez, F., et al., 2010. The enteropathy associated with common variable immunodeficiency: the delineated frontiers with celiac disease. Am. J. Gastroenterol. 105, 2262–2275.

McNeal, M.M., Barone, K.S., Rae, M.N., Ward, R.L., 1995. Effector functions of antibody and CD8+ cells in resolution of rotavirus infection and protection against reinfection in mice. Virology 214, 387–397.

Nordgren, J., Sharma, S., Bucardo, F., et al., 2014. Both Lewis and secretor status mediate susceptibility to rotavirus infections in a rotavirus genotype-dependent manner. Clin. Infect. Dis. 59, 1567–1573.

Norhagen, G., Engström, P.E., Hammarström, L., Söder, P.O., Smith, C.I., 1989. Immunoglobulin levels in saliva in individuals with selective IgA deficiency: compensatory IgM secretion and its correlation with HLA and susceptibility to infections. J. Clin. Immunol. 9, 279–286.

Oishi, I., Kimura, T., Murakami, T., et al., 1991. Serial observations of chronic rotavirus infection in an immunodeficient child. Microbiol. Immunol. 35, 953–961.

Pan-Hammarström, Q., Salzer, U., Du, L., et al., 2007. Reexamining the role of TACI coding variants in common variable immunodeficiency and selective IgA deficiency. Nat. Genet. 39, 429–430.

Papadopoulou-Alataki, E., Hassan, A., Davies, E.G., 2012. Prevention of infection in children and adolescents with primary immunodeficiency disorders. Asian Pac. J. Allergy Immunol. 30, 249–258.

Patel, N.C., Hertel, P.M., Estes, M.K., et al., 2010. Vaccine-acquired rotavirus in infants with severe combined immunodeficiency. N. Engl. J. Med. 362, 314–319.

Patel, M.M., Steele, D., Gentsch, J.R., Wecker, J., Glass, R.I., Parashar, U.D., 2011. Real-world impact of rotavirus vaccination. Pediatr. Infect. Dis J. 30 (Suppl. 1), S1–S5.

Patel, N.C., Hertel, P.M., Hanson, I.C., et al., 2012. Chronic rotavirus infection in an infant with severe combined immunodeficiency: successful treatment by hematopoietic stem cell transplantation. Clin. Immunol. 142, 399–401.

Payne, D.C., Edwards, K.M., Bowen, M.D., et al., 2010. Sibling transmission of vaccine-derived rotavirus (RotaTeq) associated with rotavirus gastroenteritis. Pediatrics 125, e438–e441.

Payne, D.C., Parashar, U.D., Lopman, B.A., 2015. Developments in understanding acquired immunity and innate susceptibility to norovirus and rotavirus gastroenteritis in children. Curr. Opin. Pediatr. 27, 105–109.

Plebani, A., Soresina, A., Rondelli, R., et al., 2002. Italian Pediatric Group for XLA-AIEOP. Clinical, immunological, and molecular analysis in a large cohort of patients with X-linked agammaglobulinemia: an Italian multicenter study. Clin. Immunol. 104, 221–230.

Quinti, I., Soresina, A., Spadaro, G., et al., 2007. Italian Primary Immunodeficiency Network. Long-term follow-up and outcome of a large cohort of patients with common variable immunodeficiency. J. Clin. Immunol. 27, 308–316.

Salzer, U., Chapel, H.M., Webster, A.D., et al., 2005. Mutations in TNFRSF13B encoding TACI are associated with common variable immunodeficiency in humans. Nat. Genet. 37, 820–828.

Sarker, S.A., Casswall, T.H., Mahalanabis, D., et al., 1998. Successful treatment of rotavirus diarrhea in children with immunoglobulin from immunized bovine colostrum. Pediatr. Infect. Dis. J. 17, 1149–1154.

Saulsbury, F.T., Winkelstein, J.A., Yolken, R.H., 1980. Chronic rotavirus infection in immunodeficiency. J. Pediatr. 97, 61–65.

Savilahti, E., 1973. IgA deficiency in children. Immunoglobulin-containing cells in the intestinal mucosa, immunoglobulins in secretions and serum IgA levels. Clin. Exp. Immunol. 13, 395–406.

Savilahti, E., Klemola, T., Carlsson, B., Mellander, L., Stenvik, M., Hovi, T., 1988. Inadequacy of mucosal IgM antibodies in selective IgA deficiency: excretion of attenuated polio viruses is prolonged. J. Clin. Immunol. 8, 89–94.

Schäffer, A.A., Pfannstiel, J., Webster, A.D., Plebani, A., Hammarström, L., Grimbacher, B., 2006. Analysis of families with common variable immunodeficiency (CVID) and IgA deficiency suggests linkage of CVID to chromosome 16q. Hum. Genet. 118, 725–729.

Schwartz-Cornil, I., Benureau, Y., Greenberg, H., Hendrickson, B.A., Cohen, J., 2002. Heterologous protection induced by the inner capsid proteins of rotavirus requires transcytosis of mucosal immunoglobulins. J. Virol. 76, 8110–8117.

Shahmahmoodi, S., Parvaneh, N., Burns, C., et al., 2008. Isolation of a type 3 vaccine-derived poliovirus (VDPV) from an Iranian child with X-linked agammaglobulinemia. Virus Res. 137, 168–172.

Sicherer, S.H., Winkelstein, J.A., 1998. Primary immunodeficiency diseases in adults. JAMA 279, 58–61.

Song, E., Jaishankar, G.B., Saleh, H., Jithpratuck, W., Sahni, R., Krishnaswamy, G., 2011. Chronic granulomatous disease: a review of the infectious and inflammatory complications. Clin. Mol. Allergy 31, 10.

Spickett, G.P., Misbah, S.A., Chapel, H.M., 1991. Primary antibody deficiency in adults. Lancet 337, 281–284.

Tan, M., Jiang, X., 2014. Histo-blood group antigens: a common niche for norovirus and rotavirus. Expert Rev. Mol. Med. 16, e5.

Uygungil, B., Bleesing, J.J., Risma, K.A., McNeal, M.M., Rothenberg, M.E., 2010. Persistent rotavirus vaccine shedding in a new case of severe combined immunodeficiency: a reason to screen. J. Allergy Clin. Immunol. 125, 270–271.

van de Ven, A.A., Janssen, W.J., Schulz, L.S., et al., 2014. Increased prevalence of gastrointestinal viruses and diminished secretory immunoglobulin a levels in antibody deficiencies. J. Clin. Immunol. 34, 962–970.

Vorechovský, I., Zetterquist, H., Paganelli, R., et al., 1995. Family and linkage study of selective IgA deficiency and common variable immunodeficiency. Clin. Immunol. Immunopathol. 77, 185–192.

Wang, N., Hammarström, L., 2012. IgA deficiency: what is new? Curr. Opin. Allergy Clin. Immunol. 12, 602–608.

Wang, L.J., Yang, Y.H., Lin, Y.T., Chiang, B.L., 2004. Immunological and clinical features of pediatric patients with primary hypogammaglobulinemia in Taiwan. Asian Pac. J. Allergy Immunol. 22, 25–31.

Werther, R.L., Crawford, N.W., Boniface, K., Kirkwood, C.D., Smart, J.M., 2009. Rotavirus vaccine induced diarrhea in a child with severe combined immune deficiency. J. Allergy Clin. Immunol. 124, 600.

Winkelstein, J.A., Marino, M.C., Lederman, H.M., et al., 2006. X-linked agammaglobulinemia: report on a United States registry of 201 patients. Medicine 85, 193–202.

Winkelstein, J.A., Marino, M.C., Ochs, H., et al., 2003. The X-linked hyper-IgM syndrome: clinical and immunologic features of 79 patients. Medicine 82, 373–384.

Woodward, J.M., Gkrania-Klotsas, E., Cordero-Ng, A.Y., et al., 2015. The role of chronic norovirus infection in the enteropathy associated with common variable immunodeficiency. Am. J. Gastroenterol. 110, 320–327.

Xerry, J., Gallimore, C.I., Cubitt, D., Gray, J.J., 2010. Tracking environmental norovirus contamination in a pediatric primary immunodeficiency unit. J. Clin. Microbiol. 48, 2552–2556.

Xiong, N., Hu, S., 2015. Regulation of intestinal IgA responses. Cell Mol. Life Sci. 72, 2645–2655.

Chapter 1.4

Therapy of Viral Gastroenteritis

G. Kang
The Wellcome Trust Research Laboratory, Division of Gastrointestinal Sciences, Christian Medical College, Vellore, Tamil Nadu, India

1 INTRODUCTION

It is unlikely that anyone who lives beyond infancy anywhere in the world has not suffered from or will not experience one or more episodes of viral gastroenteritis (Das et al., 2014). Among children, rotaviruses were the leading cause of severe acute gastroenteritis in all parts of the world until the introduction of rotavirus vaccines. The introduction of rotavirus vaccines has resulted in dramatic reductions in disease particularly in industrialized countries, but other viruses, particularly noroviruses still cause significant gastroenteritis in children and adults everywhere (Payne et al., 2013). In lower income countries, the lower efficacy of oral rotavirus vaccines, ranging from 20–65%, leaves between a third to half of all vaccinated children unprotected from severe rotavirus disease (Babji and Kang, 2012).

The morbidity and mortality caused by viral gastroenteritis represent a significant economic and public health burden. Although the total number of deaths is still unacceptably high and disproportionately affects the poorest in low-income countries, there has been a substantial reduction in the past three decades (Das et al., 2014). This reduction can be attributed to several factors, but one important reason is the sustained effort to manage diarrhoeal disease appropriately.

2 BACKGROUND

The mainstay of management of viral gastroenteritis of any severity is rehydration. Treatment of dehydration was first attempted in the 1830s during cholera epidemics. Intravenous fluids were introduced for treatment of dehydration over a century later and they were used for cholera by the 1950s. Given the difficulty of finding a peripheral vein in a dehydrated patient, and safety problems of intravenous solutions and their administration in low resource settings, attempts were made to develop cheap and effective oral solutions. By the late 1960s, oral rehydration solutions (ORS) had been developed and shown to be effective in cholera, and they were deployed on a large scale in 1971–72 at the time when Bangladesh became independent (Mahalanabis et al., 1973). The World Health Organization (WHO) developed guidelines for oral rehydration therapy and established the standards for production of packets of ORS and documents that have subsequently been reviewed and updated, most notably with the recommendation of hypoosmolar ORS (Atia and Buchman, 2009).

2.1 Physiologic Basis of Rehydration

In health, there is a continuous exchange of water through the intestinal wall with secretion of up to 9 L of water from oral intake, salivary, gastric, pancreatic, biliary, and upper intestinal secretion and reabsorption of almost as much every 24 h by the distal ileum and colon, resulting in a stool output of about 250 mL/day (Acra and Ghishan, 1996). The secretion and reabsorption allow soluble metabolites from digested food to be transferred into the bloodstream. When diarrhoea occurs, there is an imbalance between secretion and absorption with much more fluid being secreted, resulting in a net loss of body water of up to several litres a day. In addition to fluid loss, sodium is also lost because sodium ions are held almost entirely extracellularly in blood and body fluids. This differs from the largely intracellular holding of potassium ions (Field, 2003).

Precise control (135–150 mmol/L) of sodium ions in the extracellular fluid is essential for normal metabolism, but in dehydration water is conserved by anuria, and sodium regulation does not work effectively. Continued diarrhoea can cause very rapid depletion of water and sodium. If more than 10% of the body's fluid is lost, it can be fatal.

When sodium ions are not being absorbed by the intestine, then water is not absorbed either, since water passively follows the osmotic gradient generated by transcellular transport of electrolytes and nutrients. Although there are several mechanisms for sodium absorption, the one most important for oral rehydration with a glucose and electrolyte containing solution, is a cotransport mechanism, where glucose transport, unaffected by the diarrhoea, continues across the luminal membrane, facilitated by the protein sodium glucose cotransporter 1, and cotransports sodium in a 1:1 ratio. Glucose-stimulated sodium absorption is cyclic AMP (cAMP) independent while cAMP is likely responsible for stimulation of chloride secretion and inhibition of sodium chloride absorption. The increased concentration of sodium across the intestinal barrier now draws water across, resulting in net retention and increase in fluid (Curran, 1960). Once in the intestinal cell, the glucose is transported through the basolateral membrane via the glucose transporter 2. The $Na^{+}K^{+}$ATPase provides the energy to drive the process (Fig. 1.4.1).

Since there are several mechanism for transport of sodium, including sodium hydrogen exchangers, many additional cotransporters of Na^{+} (eg, of amino acids, products of hydrolysis of cereals or digestion-resistant starch) were targeted

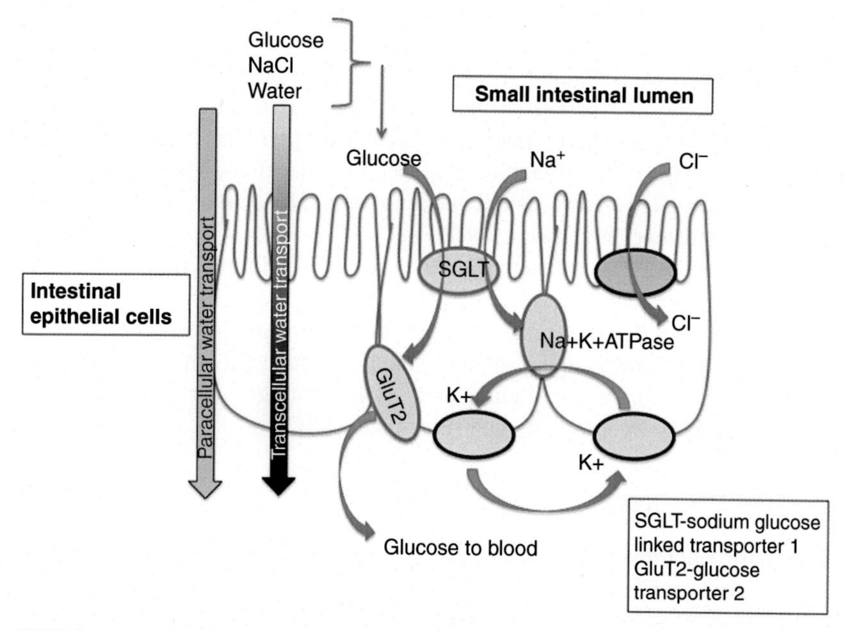

FIGURE 1.4.1 **Physiologic basis of oral rehydration.** (The complexity of all transport mechanisms is not shown. For further details see chapter 1.1.)

(Binder et al., 2014), and some interventions demonstrated promising results, but larger trials have not confirmed their efficacy in gastroenteritis of any aetiology.

2.2 Basis of Pharmacotherapy

Although reliance on pharmacologic agents is not recommended because it might shift the therapeutic focus away from the appropriate fluid, electrolyte and nutritional therapy which is needed for viral gastroenteritis of different aetiologies, there is evidence that some treatment modalities may be appropriate for certain patient groups. In the more industrialized world, recent guidelines for management of acute gastroenteritis include evidence-based recommendations for use of pharmacotherapy with a broad range of activity (Guarino et al., 2014).

Antibiotics have no role to play in viral gastroenteritis, and no specific antiviral agents directed against the causes of gastroenteritis have, as yet, been developed. Antibiotics may be used in children with concomitant bacterial illness or if they develop sepsis. The use of antimotility agents to increase transit time, and thus enable greater absorption of fluid from the gut is not recommended in children less than 8 years of age but is recommended in adults (Farthing et al., 2008). In patients less than 3 years of age, use of antimotility agents such as loperamide can, rarely, result in ileus or death (Li et al., 2007).

Since vomiting is a presenting and common feature of viral gastroenteritis and can result in failure to rehydrate orally, the use of antiemetic agents is recommended in some settings, particularly North America. The agents include 5-HT3 receptor antagonists, such as ondansetron, which can be used safely in children as well as dopamine receptor antagonists, including phenothiazines (prochlorperazine and promethazine), benzamides (metoclopramide and trimethobenzamide), and butyrophenones (such as droperidol). Their use in children is discouraged due to lack of evidence of a beneficial effect and strong association with side effects including extrapyramidal reactions and neuroleptic malignant syndrome (Freedman, 2007).

Antisecretory agents which prevent the excess loss of fluid into the intestinal lumen have multiple modes of action. They include racecadotril, an enkephalinase inhibitor that decreases intestinal hypersecretion and promotes absorption (see later). Other antisecretory agents such as dioctahedral smectite block the activity of enterotoxins by binding them or promoting water and electrolyte absorption across the intestinal wall (Freedman, 2007).

3 CLINICAL ASSESSMENT

Acute gastroenteritis usually refers to an illness with a duration of less than 7 days which is characterized by diarrhoea and/or vomiting. Acute diarrhoea is defined as $\geq$3 loose or watery stools/day. The volume of fluid lost through stools can vary from 5 mL/kg body weight/day to $\geq$200 mL/kg body weight/day, and can result in severe dehydration (Centers for Disease Control and Prevention, 2003).

Diarrhoea can also be seen as an early presenting symptom in illnesses not related to the gastrointestinal tract, such as pneumonia, urinary tract infection, meningitis, and sepsis. Vomiting can result from metabolic disorders, toxin ingestion or trauma. Therefore, a good case history is required to rule out concomitant or other illnesses.

3.1 History

The clinical history should assess the onset, duration, frequency, and quantity of vomiting and diarrhoea. The presence of fever and of bile, blood, or mucus in the stool or vomitus should be noted. A history of recent oral intake, of urine output and mental status is helpful in evaluating dehydration. The past medical history of any underlying medical problems, other recent infections, medications, and human immunodeficiency virus (HIV) infection status is needed.

3.2 Physical Examination

Body weight, temperature, heart rate, respiratory rate, and blood pressure must be measured. A recent premorbid weight is needed to estimate fluid loss, but in its absence expected weight can be calculated from any available prior growth curve data. The general condition of the patient should be assessed, with special attention to activity. The appearance of the eyes, lips, mouth, and tongue are assessed for degree of dehydration. Skin turgor is examined by pinching a small skin fold on the lateral abdominal wall at the level of the umbilicus using the thumb and index finger and measuring the time it takes to return to normal. Deep respirations may indicate metabolic acidosis, and faint or absent bowel sounds can indicate hypokalemia. Capillary refill can help in assessment of dehydration.

3.3 Assessment of Dehydration

Many systems of evaluation can be used to measure the degree of severity of dehydration and several guidelines classify dehydration to indicate appropriate treatment. These published classifications include those of the World Health Organization (WHO, 2005), the American Academy of Pediatrics (Yu et al., 2011) and the European Society of Paediatric Gastroenterology, Hepatology and Nutrition (Guarino et al., 2014). The main variables assessed are shown in Table 1.4.1.

Among infants and children, a decrease in blood pressure is a late sign of dehydration that can correspond to fluid deficits of >10% and signals shock.

TABLE 1.4.1 Clinical Assessment of Dehydration

Symptom	Mild (3–5%)	Moderate (6–9%)	Severe (≥10%)
Urine output	Normal to decreased	Decreased	Minimal
Thirst	Slightly increased	Moderately increased	Drinks poorly, unable to drink
Eyes	Normal	Slightly sunken	Sunken
Tears	Present	Decreased	Absent
Fontanelle	Normal	Sunken	Sunken
Mucous membranes	Moist	Dry	Very dry
Extremities	Warm	Cool	Cold, mottled, cyanotic
Skin pinch	Normal	Return in <2 s	Recoil in >2 s
Mental status	Normal	Normal to irritable	Apathetic, lethargic, unconscious
Capillary refill	Normal	Prolonged	Poor
Breathing	Normal	Normal, fast	Deep
Heart rate	Normal	Increased	Increased
Pulse	Normal	Normal to weak	Weak, thready
Blood pressure	Normal	Normal	Normal to reduced

Increases in heart rate and reduced peripheral perfusion can be more sensitive indicators of moderate dehydration, although both can vary with the degree of fever. Decreased urine output is sensitive but not specific and can be difficult to assess in children with diarrhoea. A finding of increased urine specific gravity can indicate dehydration.

4 THERAPY

Most acute gastroenteritis is mild and as with many minor illnesses, most patients are likely to be treated at home. Stocking a commercially available ORS at home enables parents to start therapy early. There are also recipes for homemade ORS, using salt and sugar, but there is a possibility of error. The goal of home or facility treatment is the need to replace lost fluid and maintain adequate nutrient intake. In breastfed children, more frequent feeds should be encouraged and for nonbreastfed children more age-appropriate fluids should be given (Guarino et al., 2014).

A decision on referral may be needed for very young infants or if a child is dehydrated, and this may require the assessment of a child by a healthcare worker in person or by telephone. When treatment at home is initiated, parents/caregivers should monitor the child carefully and seek help if there is any sign of distress or change in mental status. Infants with acute diarrhoea are more likely to become dehydrated than older children because they have a higher body surface-to-volume ratio, a higher metabolic rate, relatively smaller fluid reserves, and are dependent on others for fluid intake (CDC, 2003).

4.1 Supportive Measures

The mainstay of therapy for acute gastroenteritis is rehydration. Continuation or rapid introduction after rehydration of an age-appropriate diet is a necessary measure to prevent the nutritional consequences of an acute diarrhoeal episode. Supplementation with zinc during acute diarrhoea is an important adjunct to rehydration particularly in developing countries (see later). Finally, probiotics may play a role in restoration of a normal microbial flora after an acute episode of gastroenteritis (see later).

4.1.1 Fluid Replacement and Maintenance

The first approach in diarrhoea with no or mild dehydration should be oral rehydration therapy (Fig. 1.4.2). Treatment with ORS is simple and permits management of uncomplicated cases of diarrhoea at home, regardless of aetiology. Early fluid replacement leads to fewer clinic and emergency care visits and potentially prevents hospitalizations and deaths (Duggan et al., 1999).

Oral rehydration therapy consists of two phases of treatment: (1) a rehydration phase, in which water and electrolytes are given as ORS to replace existing losses, and (2) a maintenance phase, which includes replacement of ongoing fluid and electrolyte losses and adequate dietary intake (CDC, 2003).

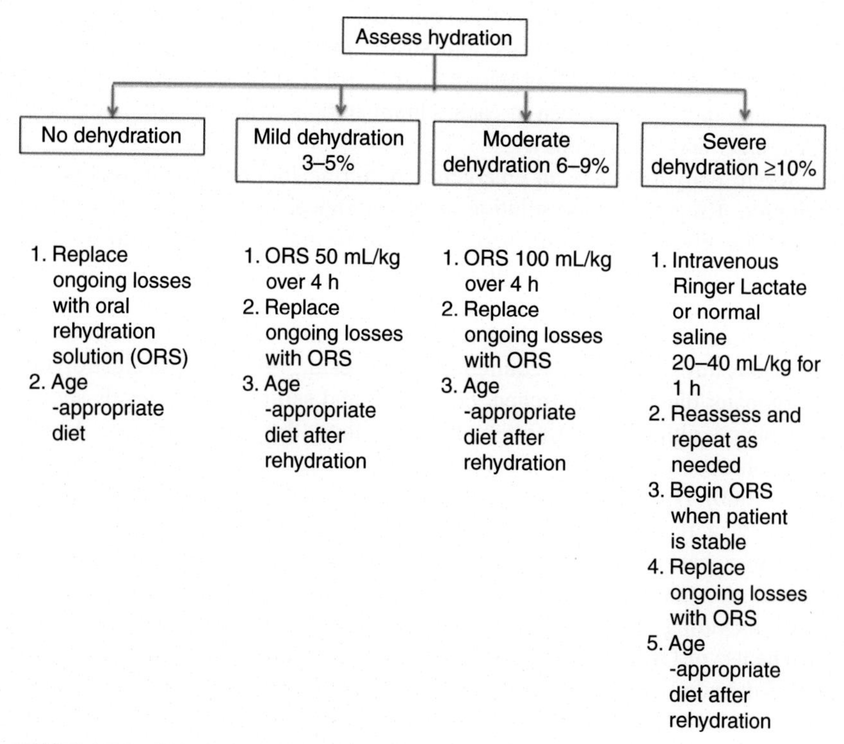

FIGURE 1.4.2 **Rehydration algorithm for acute gastroenteritis.**

In the rehydration phase, the fluid deficit is replaced within 3–4 h, and in the maintenance phase, calories (food) and fluids are given. Rapid realimentation should follow rapid rehydration, with a goal of quickly returning the patient to an age-appropriate unrestricted diet, including solids. Breastfeeding should be continued throughout rehydration. The diet should be increased as soon as tolerated to compensate for lost caloric intake during the acute illness. Lactose restriction is usually not necessary.

For children and adults with no or minimal dehydration, 1 mL of fluid is needed to replace 1 g of stool. When losses are not easily measured, 10 mL/kg body weight of fluid can be given for each watery stool or 2 mL/kg body weight for each emesis. Children with mild to moderate dehydration should have their estimated fluid deficit replaced by administering 50–100 mL of ORS/kg body weight during 2–4 h to replace the estimated fluid deficit, with additional ORS to replace ongoing losses. In case of vomiting, a nasogastric tube can be placed to continue to deliver ORS (Nager and Wang, 2002). Some children may not respond to treatment with ORS, and reassessment may be necessary.

After correction, a guideline for maintenance is a daily fluid requirement of 100 mL/kg for the first 10 kg body weight, 50 mL/kg for the next 10 kg, and 20 mL/kg for each subsequent 1 kg over 20 kg (Canadian Paediatric Society, 2006).

There are few contraindications to oral rehydration; they include patients in haemodynamic shock, with ileus, intussusception, or other bowel obstruction. In patients with vomiting, oral rehydration is still possible in many patients by decreasing the amount given at each administration or by using a nasogastric tube (Nager and Wang, 2002).

Severe dehydration is an emergency requiring immediate intravenous rehydration. Ringer's lactate solution or physiological saline at 20 mL/kg body weight are given until pulse, perfusion, and mental status return to normal. With malnourished infants, smaller amounts (10 mL/kg body weight) are recommended because of the reduced ability of these infants to increase cardiac output. Sometimes, two IV lines or even alternative access sites such as intraosseous infusion may be required (Driggers et al., 1991). Serum electrolytes, bicarbonate, blood urea nitrogen, creatinine, and serum glucose levels should be obtained, although starting rehydration without these results is safe. Hypotonic solutions should not be used for acute parenteral rehydration (Jackson and Bolte, 2000). Hydration status should be reassessed frequently to determine the adequacy of replacement therapy. Boluses of IV fluid may have to be given until pulse, perfusion, and mental status return to normal. A lack of response to fluid administration should raise the suspicion of alternative or concurrent diagnoses, including septic shock and metabolic, cardiac, or neurologic disorders.

The use of ORS resulted in a steady decrease in the number of children dying of dehydration, but uptake of ORS was hampered by low usage (Lenters et al., 2013), in part because there is no effect on stool output (Wadhwa et al., 2011). Studies have shown that the efficacy of ORS solution for treatment of children with acute noncholera diarrhoea is improved by a reduced osmolarity ORS, with reduction in the need for unscheduled supplemental IV therapy, reduction in stool output and reduction in vomiting (Santosham et al., 1996). The WHO recommended the use of reduced osmolarity ORS (Table 1.4.2).

TABLE 1.4.2 Composition of Oral Rehydration Solutions Recommended by the World Health Organization

	Reduced osmolarity ORS (mmol/L)	Standard ORS (mmol/L)	Recommended standards
Sodium	75	90	60–70
Chloride	65	80	60–70
Glucose	75	111	75–90
Potassium	20	20	20
Citrate	10	10	10
Total osmolarity	245	311	210–260

Modified from UNICEF (2004).

In 2004, a joint statement was released by the WHO and the United Nations Children's Emergency Fund (UNICEF) which recommended two changes to treatment protocols for young children with diarrhea (WHO/UNICEF, 2004). The two organizations recommended a switch to a hypoosmolar ORS (UNICEF, 2004) and the introduction of zinc supplementation (see later) for 10–14 days.

Other formulations of ORS with the inclusion of resistant starch or amino acids, designed to reduce stool volume or promote more rapid repair of the gut, have been evaluated in small clinical trials but have not, as yet, built the evidence base to lead to widespread use.

4.1.2 Diet

An age-appropriate diet is critical to the management of acute gastroenteritis. Breastfed and bottle-fed infants should be fed in amounts sufficient to satisfy energy and nutrient requirements. Formula feeds should not be diluted. Except for a subset of malnourished dehydrated infants, lactose-free or soy fibre containing formulas are not necessary (Brown et al., 1994). Without clinical symptoms, low pH and reducing substances in the stool do not prove lactose intolerance. Older children should continue to receive their usual diet during episodes of diarrhoea. Foods high in simple sugars should be avoided to prevent osmotic diarrhoea, but special diets are not necessary. In developing countries, age-appropriate unrestricted diets, including complex carbohydrates, meats, yogurt, fruits, and vegetables are recommended.

In children with severe dehydration, rapid realimentation should follow intravenous rehydration with attempts to reintroduce oral intake as soon as the dehydration is corrected. The promotion of the use of the oral route for rehydration and feeding assists the return to normal activities and prevents nutritional consequences of illness.

4.1.3 Probiotics

Probiotics are live microorganisms that are believed to promote health by improving the intestinal microflora, promoting immunity or repair of damage. Reviews have evaluated their use in preventing or reducing the severity or duration of diarrhoeal illnesses among children, including diarrhoea caused by rotavirus. These products include lactobacilli, bifidobacteria or the nonpathogenic yeast *Saccharomyces boulardii* (Reid et al., 2003). However, many trials were small, and results vary by probiotic strain. A systematic review and two meta-analyses concluded that *Lactobacillus* species, particularly *Lactobacillus rhamnosus* GG, are both safe and effective as treatment for children with infectious diarrhoea (Szajewska and Mrukowicz, 2001, Van Niel et al., 2002, Allen et al., 2004).

Prebiotics are complex carbohydrates used to preferentially stimulate the growth of health-promoting intestinal flora, but data on their efficacy in acute gastroenteritis are limited.

4.1.4 Zinc

Zinc deficiency is prevalent in many developing countries where dietary intake is inadequate. Diarrhoea results in loss of zinc in stool and reduced tissue zinc (Black and Sazawal, 2001). Although severe zinc deficiency is associated with diarrhoea, milder deficiencies of zinc also appear to play a role in childhood diarrhoea. A Cochrane review which analysed data from 24 trials with 9128 children, mainly from Asia, stated that overall in children of all ages with acute diarrhoea, there is currently not enough evidence to say whether zinc supplementation during acute diarrhoea reduces death or hospitalization. In children older than 6 months with acute diarrhoea, zinc supplementation shortened the duration of diarrhoea by around 10 h and probably reduced the number of children whose diarrhoea persisted until day 7. Effects were stronger in children with moderate malnutrition and persistent diarrhoea (Lazzerini and Ronfani, 2013).

Nonetheless, because there are supporting data from large, well-conducted trials in Asia (Bhatnagar et al., 2004, Faruque et al., 1999, Baqui et al., 2002), the WHO has recommended zinc supplementation (20 mg/day for 2 weeks for children >6 months of age and 10 mg/day for children <6 months of age) for treating children with acute and persistent diarrhoea and as a prophylactic supplement for decreasing the incidence of diarrhoeal disease and pneumonia (WHO/UNICEF, 2004). The role of zinc supplements in developed countries is less certain and needs further evaluation.

4.2 Pharmacotherapy

4.2.1 Antimotility Agents

The most widely used antimotility agent, loperamide, is a peripheral opiate receptor agonist with antisecretory and antimotility properties (Schiller, 1995). It slows orocecal transit in normal subjects who take doses of > 4 mg or repetitive doses (Kirby et al., 1989). Although efficacious, the use of loperamide is contraindicated in children of <2 years of age due to reports of drowsiness and ileus. Additionally, deaths due to paralytic ileus were attributed to loperamide administration (Bhutta and Tahir, 1990). However, in adults, 4–6 mg a day is recommended by the World Gastroenterology Organization (Farthing et al., 2008). Lomotil, a combination of diphenoxylate and atropine, is widely used in the United States at doses of 5–20 mg/day in up to 4 divided doses. It is not recommended for children below the age of 2 years and should be used with caution in older children because of the risk of overdosage. In acute gastroenteritis its use should be discontinued, if there is no response within 48 h.

4.2.2 Antiemetics

Ondansetron is a 5-HT3 receptor antagonist available since 1991 and is given by either the oral or IV route. In contrast to most antiemetic drugs, ondansetron does not affect dopamine, histamine, adrenergic, or cholinergic receptors.

Hence, extrapyramidal side effects are extremely uncommon, are usually dose related and occur with repeated dosing (Simpson and Hicks, 1996). Both IV and oral ondansetron have been evaluated in clinical trials in children and shown to reduce vomiting and, in some studies, the need for IV rehydration and admission (Stork et al., 2006, Ramsook et al., 2002). The use of ondansteron in children and adults with vomiting is recommended in several developed countries and is increasingly used in practice in developing countries (Farthing et al., 2008).

Promethazine is a phenothiazine derivative that has pronounced antihistaminic, anticholinergic and sedative effects. It is contraindicated for use in children of <2 years of age and should be used rarely in those >2 years of age. Respiratory depression is common particularly when promethazine is combined with other drugs that also cause respiratory depression (Stork et al., 2006, Ramsook et al., 2002).

4.2.3 Antisecretory Agents

Racecadotril (acetorphan) is an enkephalinase inhibitor that decreases intestinal secretion and promotes absorption, but does not slow intestinal transit. A limited number of randomized controlled trials have shown that oral racecadotril decreased stool output and the duration of diarrhoea, while others have shown no difference (Salazar-Lindo et al., 2000, Mehta et al., 2012). The most recent European recommendations state that there is moderate evidence of benefit based on a systematic review of individual patient data (Lehert et al., 2011, Guarino et al., 2014). In the absence of conclusive evidence, use of racecadotril should be avoided in young children.

Dioctahedral smectite (Diosmectite) is a natural hydrated aluminomagnesium silicate that absorbs endo- and exotoxins, bacteria, and viruses. It increases water and electrolyte absorption and is believed to aid intestinal recovery. A recent metaanalysis showed that diosmectite was effective in all types of acute childhood diarrhoea except dysentery. However, the trials were mainly open-label and had major methodologic limitations, and therefore the evidence is considered to be of low quality (Das et al., 2015). Its routine use is not recommended.

Small studies in Egypt and Bolivia have evaluated nitazoxanide in rotavirus gastroenteritis and showed reduction in the duration of diarrhoea, but there is need for confirmatory studies (Teran et al., 2009, Rossignol et al., 2006).

Octreotide, an analog of somatostatin, has been used subcutaneously and intravenously to control diarrhoea in adults and children, but is recommended mainly in patients with other underlying conditions, such as HIV infections or malignancy (Peeters et al., 2010).

4.2.4 Nonspecific Agents

Bismuth subsalicylate and its degradation products are believed to bind enterotoxin. In trials in Peru and Chile (Figueroa-Quintanilla et al., 1993, Soriano-Brucher et al., 1991), benefits included a reduction in stool frequency and weight and shorter disease duration, but a trial in Bangladesh in patients with acute diarrhoea reported nonsignificant decrease in severity and duration (Chowdhury et al., 2001). There are concerns regarding the toxicity from salicylate absorption, and routine use in children is not recommended (Lewis et al., 2006).

Kaolin-pectin, fibre, and activated charcoal should not be used in the treatment of diarrhoea and dehydration in infants and children. There is no conclusive evidence that they reduce stool losses, duration of diarrhoea or stool frequency. Although nontoxic, disadvantages of their use may include adsorption of nutrients and antibiotics in the intestine and masking the severity of fluid loss. As with other unproven therapies, use should be restricted.

4.2.5 *Immune Active Agents*

In the past few decades there has been interest in passive immune mechanisms to treat or prevent disease due to viral agents of gastroenteritis, particularly rotavirus. Several studies have shown that antirotavirus immunoglobulin, as pooled gamma globulin, bovine colostrum, or human milk, may decrease frequency and duration of diarrhea (eg, Guarino et al., 1994), but none are recommended for routine use.

A recent report of a randomized, double blind placebo controlled trial of a biological molecule consisting of the variable domain of llama heavy chain antibodies directed against rotavirus and expressed in yeast, for the treatment of rotavirus diarrhoea in Bangladesh, showed a decrease in stool output and frequency (Sarker et al., 2013). There appeared to be no effect on virus excretion or duration of illness. Although biological plausibility for the antirotavirus activity is supported by in vitro and animal model experiments, further evaluations of safety and efficacy are needed.

4.3 Special Clinical Scenarios

Special clinical scenarios may include children and adults with underlying disease such as HIV infection, cancers, or malnutrition. Malnutrition, malignancy, and immunodeficient states predispose to more severe episodes of illness or unremitting diarrhoea that can persist until the underlying condition is corrected. In all cases, the principles that underly the use of oral rehydration therapy in managing diarrhoeal disease apply although the specifics of the evaluation, and fluid, electrolyte, and nutritional management differ. Additional management strategies require a lower threshold for admission to the hospital for close observation.

5 CONCLUSIONS

The evidence-based management of viral gastroenteritis indicates that assessment of dehydration is critical to establish severity and monitor treatment. Oral rehydration with hypoosmolar solutions is the mainstay of treatment, and continued or early refeeding feeding is essential in children. Interventions such as antiemetics, probiotics, and antisecretory agents are adjuncts to rehydration, which are recommended in some parts of the world, but their wider use requires further evaluation of effectiveness.

Unfortunately despite the recognition of the value of oral rehydration, ORS is still used in only a subset of cases in most settings around the world. Promotion of oral rehydration and the development of better formulations or adjunct therapies that decrease the volume of stool and shorten the duration of diarrhoea still require research and development of new approaches to improve outcomes in viral gastroenteritis.

REFERENCES

Acra, S.A., Ghishan, G.K., 1996. Electrolyte fluxes in the gut and oral rehydration solutions. Pediatr. Clin. North Am. 43, 433–449.

Allen, S.J., Okoko, B., Martine, E., Gregorio, G., Dans, L.F., 2004. Probiotics for treating infectious diarrhoea. Cochrane Database Syst. Rev. 2, CD003048.

Atia, A.N., Buchman, A.L., 2009. Oral rehydration solutions in non-cholera diarrhea: a review. Am. J. Gastroenterol. 104, 2596–2604.

Babji, S., Kang, G., 2012. Rotavirus vaccination in developing countries. Curr. Opin. Virol. 2 (4), 443–448.

Baqui, A.H., Black, R.E., El Arifeen, S., Yunus, M., Zaman, K., Begum, N., Roess, A.A., Santosham, M., 2002. Effect of zinc supplementation started during diarrhoea on morbidity and mortality in Bangladeshi children: community randomised trial. Brit. Med. J. 325 (7372), 1059.

Bhatnagar, S., Bahl, R., Sharma, P.K., Kumar, G.T., Saxena, S.K., Bhan, M.K., 2004. Zinc with oral rehydration therapy reduces stool output and duration of diarrhea in hospitalized children: a randomized controlled trial. J. Pediatr. Gastroenterol. Nutr. 38, 34–40.

Bhutta, T.I., Tahir, K.I., 1990. Loperamide poisoning in children. Lancet 335 (8685), 363.

Binder, H.J., Brown, I., Ramakrishna, B.S., Young, G.P., 2014. Oral rehydration therapy in the second decade of the twenty-first century. Curr. Gastroenterol. Rep. 16, 376.

Black, R.E., Sazawal, S., 2001. Zinc and childhood infectious disease morbidity and mortality. Brit. J. Nutr. 85 (Suppl. 2), S125–S129.

Brown, K.H., Peerson, J., Fontaine, O., 1994. Use of nonhuman milks in the dietary management of young children with acute diarrhoea: a meta-analysis of clinical trials. Pediatrics 93, 17–27.

Canadian Paediatric Society, 2006. Nutrition and Gastroenterology Committee. Oral rehydration therapy and early re-feeding in the management of childhood gastroenteritis. Paediatr. Child Health 11, 527–531.

Centers for Disease Control and Prevention, 2003. Managing acute gastroenteritis among children: oral rehydration, maintenance, and nutritional therapy. MMWR 52, 1–16.

Chowdhury, H.R., Yunus, M., Zaman, K., Rahman, A., Faruque, S.M., Lescano, A.G., Sack, R.B., 2001. The efficacy of bismuth subsalicylate in the treatment of acute diarrhoea and the prevention of persistent diarrhoea. Acta Paediatr. 90, 605–610.

Curran, P.F., 1960. Na, Cl, and water transport by rat ileum in vitro. J. Gen. Physiol. 43, 1137–1148.

Das, J.K., Salam, R.A., Bhutta, Z.A., 2014. Global burden of childhood diarrhea and interventions. Curr. Opin. Infect. Dis. 27, 451–458.

Das, R.R., Sankar, J., Naik, S.S., 2015. Efficacy and safety of diosmectite in acute childhood diarrhoea: a meta-analysis. Arch. Dis. Child. 100 (7), 704–712.

Driggers, D.A., Johnson, R., Steiner, J.F., Jewell, G.S., Swedberg, J.A., Goller, V., 1991. Emergency resuscitation in children. The role of intraosseous infusion. Postgrad. Med. 89, 129–132.

Duggan, C., Lasche, J., McCarty, M., Mitchell, K., Dershewitz, R., Lerman, S.J., Higham, M., Radzevich, A., Kleinman, R.E., 1999. Oral rehydration solution for acute diarrhoea prevents subsequent unscheduled follow-up visits. Pediatrics 104, e29.

Farthing, M., Lindberg, G., Dite, P., Khalif, I., Salazar-Lindo, E., Ramakrishna, B.S., Goh, K., Thomson, A., Khan, A.G., 2008. World Gastroenterology Organisation practice guideline: Acute diarrhea. http://www.worldgastroenterology.org/assets/downloads/en/pdf/guidelines/01_acute_diarrhea.pdf

Faruque, A.S., Mahalanabis, D., Haque, S.S., Fuchs, G.J., Habte, D., 1999. Double-blind, randomized, controlled trial of zinc or vitamin A supplementation in young children with acute diarrhoea. Acta Paediatr. 88, 154–160.

Field, M., 2003. Intestinal ion transport and the pathophysiology of diarrhoea. J. Clin. Invest. 111, 931–943.

Figueroa-Quintanilla, D., Salazar-Lindo, E., Sack, R.B., León-Barúa, R., Sarabia-Arce, S., Campos-Sánchez, M., Eyzaguirre-Maccan, E., 1993. A controlled trial of bismuth subsalicylate in infants with acute watery diarrheal disease. N. Engl. J. Med. 328, 1653–1658.

Freedman, S.B., 2007. Acute infectious pediatric gastroenteritis: beyond oral rehydration therapy. Expert Opin. Pharmacother. 8, 1651–1665.

Guarino, A., Ashkanzi, S., Gendrel, D., Lo Vecchio, A., Shamir, R., Szajewska, H., 2014. European Society for Pediatric Gastroenterology, Hepatology and Nutrition/European Soceity for Pediatric Infectious Diseases evidence based guidelines for the management of acute gastroenteritis in children in Europe: Update 2014. J. Pediatr. Gastroenterol. Nutr. 59, 132–152.

Guarino, A., Canani, R.B., Russo, S., Albano, F., Canani, M.B., Ruggeri, F.M., Donelli, G., Rubino, A., 1994. Oral immunoglobulins for treatment of acute rotaviral gastroenteritis. Pediatrics 93 (1), 12–16.

Jackson, J., Bolte, R.G., 2000. Risks of intravenous administration of hypotonic fluids for pediatric patients in ED and prehospital settings: let's remove the handle from the pump. Am. J. Emerg. Med. 18, 269–270.

Kirby, M.G., Dukes, G.E., Heizer, W.D., Bryson, J.C., Powell, J.R., 1989. Effect of metoclopramide, bethanechol, and loperamide on gastric residence time, gastric emptying, and mouth-to-cecum transit time. Pharmacotherapy 9, 226–231.

Lazzerini, M., Ronfani, L., 2013. Oral zinc for treating diarrhoea in children. Cochrane Database Syst. Rev. 1, CD005436.

Lehert, P., Chéron, G., Calatayud, G.A., Cézard, J.P., Castrellón, P.G., Garcia, J.M., Santos, M., Savitha, M.R., 2011 Sep. Racecadotril for childhood gastroenteritis: an individual patient data meta-analysis. Dig. Liver Dis. 43 (9), 707–713.

Lewis, T.V., Badillo, R., Schaeffer, S., Hagemann, T.M., McGoodwin, L., 2006. Salicylate toxicity associated with administration of Percy medicine in an infant. Pharmacotherapy 26, 403–409.

Lenters, L.M., Das, J.K., Bhutta, Z.A., 2013. Systematic review of strategies to increase use of oral rehydration solution at the household level. BMC Public Health 13 (Suppl. 3), S28.

Li, S.T.T., Grossman, D.C., Cummings, P., 2007. Loperamide therapy for acute diarrhea in children: Systematic review and meta-analysis. PLoS Med. 4 (3), e98.

Mahalanabis, D., Choudhuri, A.B., Bagchi, N., Bhattacharya, A.K., Simpson, T.W., 1973. Oral fluid therapy of cholera among Bangladesh refugees. Johns Hopkins Med. J. 132, 197–205.

Mehta, S., Khandelwal, P.D., Jain, V.K., Sihag, M., 2012. A comparative study of racecadotril and single dose octreotide as an anti-secretory agent in acute infective diarrhoea. J. Assoc. Physicians India 60, 12–15.

Nager, A.L., Wang, V.J., 2002. Comparison of nasogastric and intravenous methods of rehydration in pediatric patients with acute dehydration. Pediatrics 109, 566–572.

Payne, D.C., Vinjé, J., Szilagyi, P.G., Edwards, K.M., Staat, M.A., Weinberg, G.A., Hall, C.B., Chappell, J., Bernstein, D.I., Curns, A.T., Wikswo, M., Shirley, S.H., Hall, A.J., Lopman, B., Parashar, U.D., 2013. Norovirus and medically attended gastroenteritis in U.S. children. N. Engl. J. Med. 368, 1121–1130.

Peeters, M., Van den Brande, J., Francque, S., 2010. Diarrhea and the rationale to use Sandostatin. Acta Gastroenterol. Belg. 73, 25–36.

Ramsook, C., Sahagun-Carreon, I., Kozinetz, C.A., Moro-Sutherland, D., 2002. A randomized clinical trial comparing oral ondansetron with placebo in children with vomiting from acute gastroenteritis. Ann. Emerg. Med. 39, 397–403.

Reid, G., Jass, J., Sebulsky, M.T., McCormick, J.K., 2003. Potential uses of probiotics in clinical practice. Clin. Microbiol. Rev. 16, 658–672.

Rossignol, J.F., Abu-Zekry, M., Hussein, A., Santoro, M.G., 2006. Effect of nitazoxanide for treatment of severe rotavirus diarrhoea: randomised double-blind placebo-controlled trial. Lancet 368 (9530), 124–129.

Sarker, S.A., Jäkel, M., Sultana, S., Alam, N.H., Bardhan, P.K., Chisti, M.J., Salam, M.A., Theis, W., Hammarström, L., Frenken, L.G., 2013. Anti-rotavirus protein reduces stool output in infants with diarrhea: a randomized placebo-controlled trial. Gastroenterology 145, 740–748.

Salazar-Lindo, E., Santisteban-Ponce, J., Chea-Woo, E., Gutierrez, M., 2000. Racecadotril in the treatment of acute watery diarrhea in children. N. Engl. J. Med. 343, 463–467.

Santosham, M., Fayad, I., Abu Zikri, M., Hussein, A., Amponsah, A., Duggan, C., Hashem, M., el Sady, N., Abu Zikri, M., Fontaine, O., 1996. A double-blind clinical trial comparing World Health Organization oral rehydration solution with a reduced osmolarity solution containing equal amounts of sodium and glucose. J. Pediatr. 128, 45–51.

Schiller, R., 1995. Review article: anti-diarrhoeal pharmacology and therapeutics. Aliment. Pharmacol. Ther. 9, 87–106.

Simpson, K.H., Hicks, F.M., 1996. Clinical pharmacokinetics of ondansetron. A review. J. Pharm. Pharmacol. 48, 774–781.

Soriano-Brucher, H., Avendaño, P., O'Ryan, M., Braun, S.D., Manhart, M.D., Balm, T.K., Soriano, H.A., 1991. Bismuth subsalicylate in the treatment of acute diarrhea in children: a clinical study. Pediatrics 87, 18–27.

Stork, C.M., Brown, K.M., Reilly, T.H., Secreti, L., Brown, L.H., 2006. Emergency department treatment of viral gastritis using intravenous ondansetron or dexamethasone in children. Acad. Emerg. Med. 13, 1027–1033.

Szajewska, H., Mrukowicz, J.Z., 2001. Probiotics in the treatment and prevention of acute infectious diarrhea in infants and children: a systematic review of published randomized, double-blind, placebo-controlled trials. J. Pediatr. Gastroenterol. Nutr. 33 (Suppl. 2), S17–S25.

Teran, C.G., Teran-Escalera, C.N., Villarroel, P., 2009. Nitazoxanide vs. probiotics for the treatment of acute rotavirus diarrhoea in children: a randomized, single-blind, controlled trial in Bolivian children. Int. J. Infect. Dis. 13, 518–523.

UNICEF, 2004, United Nations International Children's Emergency Fund. Technical Bulletin No. 9. New formulation of Oral Rehydration Salts (ORS) with reduced osmolarity. http://www.unicef.org/supply/files/Oral_Rehydration_Salts(ORS)_.pdf

Van Niel, C.W., Feudtner, C., Garrison, M.M., Christakis, D.A., 2002. Lactobacillus therapy for acute infectious diarrhea in children: a meta-analysis. Pediatrics 109, 678–684.

Wadhwa, N., Natchu, U.C., Sommerfelt, H., Strand, T.A., Kapoor, V., Saini, S., Kainth, U.S., Bhatnagar, S., 2011. ORS containing zinc does not reduce duration or stool volume of acute diarrhea in hospitalized children. J. Pediatr. Gastroenterol. Nutr. 53, 161–167.

WHO/UNICEF, 2004. WHO/UNICEF Joint Statement: Clinical Management of Acute Diarrhoea. The United Nations Children's Fund/World Health Organization.

World Health Organization, 2005. Handbook IMCI integrated management of childhood illness. http://whqlibdoc.who.int/publications/2005/9241546441.pdf

Yu, C., Lougee, D., Murno, J.R., 2011. Module 6 Diarrhea and dehydration. American Academy of Pediatrics. https://www.aap.org/en-us/advocacy-and-policy/aap-health-initiatives/children-and-disasters/Documents/MANUAL-06-internacional-2011.pdf

Chapter 2.1

Structure and Function of the Rotavirus Particle

S.C. Harrison*, P.R. Dormitzer**

*Laboratory of Molecular Medicine, Boston Children's Hospital, Harvard Medical School, and Howard Hughes Medical Institute, Boston, MA, United States; **Novartis Influenza Vaccines, Cambridge, MA, United States

1 ROTAVIRUS PARTICLE

The three layers of an infectious rotavirus particle (virion; triple-layered particle, TLP) have differentiated functions. The outer layer (VP4 and VP7) is the agent of attachment and penetration; the inner layer (VP2), the encapsidating shell for genomic RNA; the middle layer (VP6), a reinforcement for the inner layer and an adaptor, to match it to the outer layer (Chen et al., 2009; Li et al., 2008; Settembre et al., 2011).

The so-called double-layered particle (DLP)—the VP6-stabilized inner shell, which holds the 11 genomic, double-stranded (ds) RNA segments and their associated polymerase (VP1) and capping-enzyme (VP3) proteins—is transcriptionally active and has all the information and activities needed to propagate an infection, once introduced into a susceptible cell through the action of the outer-layer proteins (Bass et al., 1992; Lawton et al., 1997a).

The particle structure is correspondingly hierarchical (Fig. 2.1.1). The inner layer templates the overall icosahedral symmetry. The fundamental assembly unit is probably a fivefold-symmetric decamer of VP2: five copies clustered around the symmetry axis and five inserted more peripherally into the gaps between them (Lawton et al., 1997b; McClain et al., 2010). Twelve such decamers then form the complete shell. The 260 VP6 trimers of the middle layer cover the inner layer, packed into a $T = 13$ *laevo* subtriangulated icosahedral lattice. The only location at which the 3 VP6 subunits of a trimer have identical contacts with underlying VP2 subunits is at the icosahedral threefold axis, and those 20 trimers are indeed the most tightly associated; quasi-equivalent lateral interactions then establish the coherence of the middle layer and the structural robustness of the DLP (McClain et al., 2010).

Viral Gastroenteritis

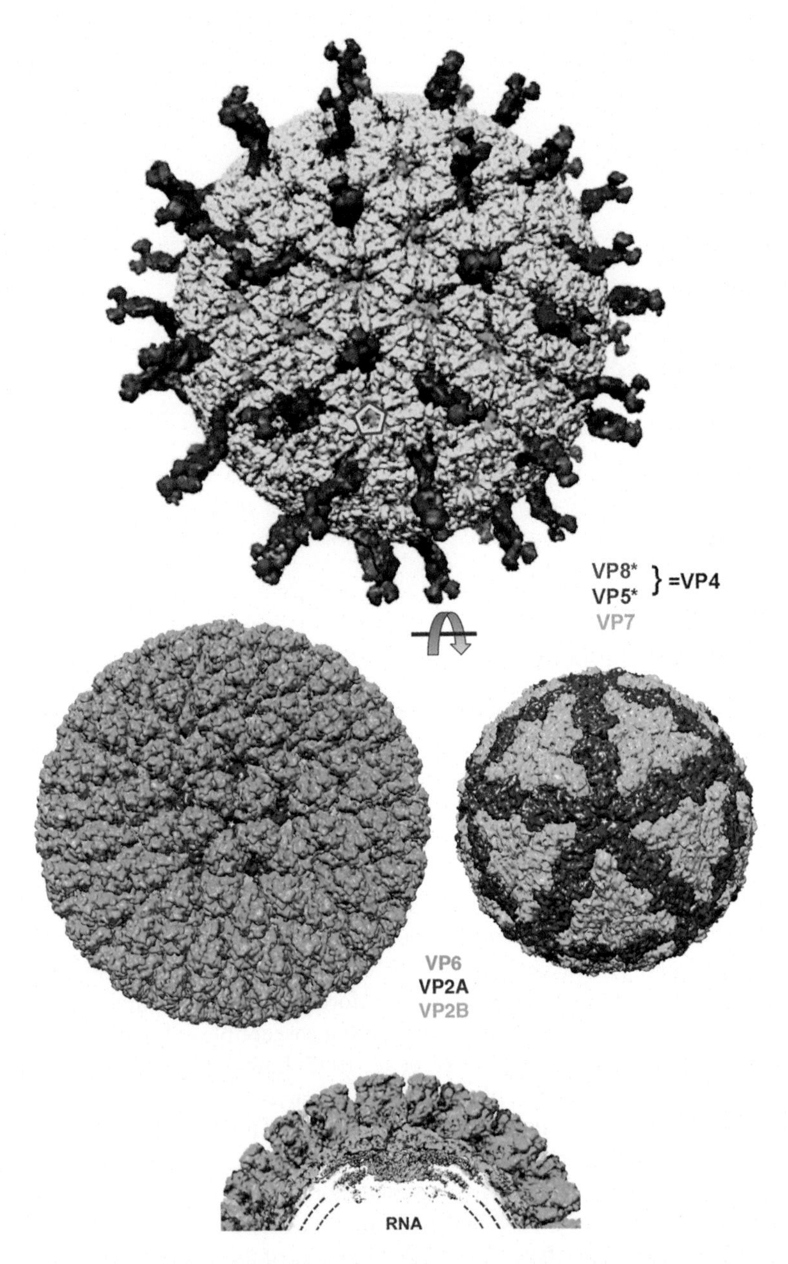

FIGURE 2.1.1 **Rotavirus particles.** Top: the infectious, triple-layer particle (TLP). The $T = 13$ icosahedral surface lattice of VP7 *(yellow)* is evident; likewise, the 60 prominent VP4 spikes (here colored magenta for the VP8* fragment and red for VP5*). Note that the VP7 layer holds the spikes in place. The two layers of the DLP—VP6 *(green)* and VP2 *(blue)*—show through the gaps in the VP7 layer at the local sixfold positions not occupied by VP4 and also through the smaller gaps at the fivefold positions. A pentagon shows the position of one of the icosahedral fivefold axes. Middle left: the DLP. The direction of view is directly along a fivefold symmetry axis. Middle right: the VP2 layer that forms the RNA encapsidating shell. The two shades of blue distinguish the two VP2 conformers. Bottom: DLP viewed in a direction tangential to the VP6 shell, cut away to show VP6, VP2, the polymerase VP1 (for details see Fig. 2.1.7) and the normalized densities of genomic RNA *(purple)*.

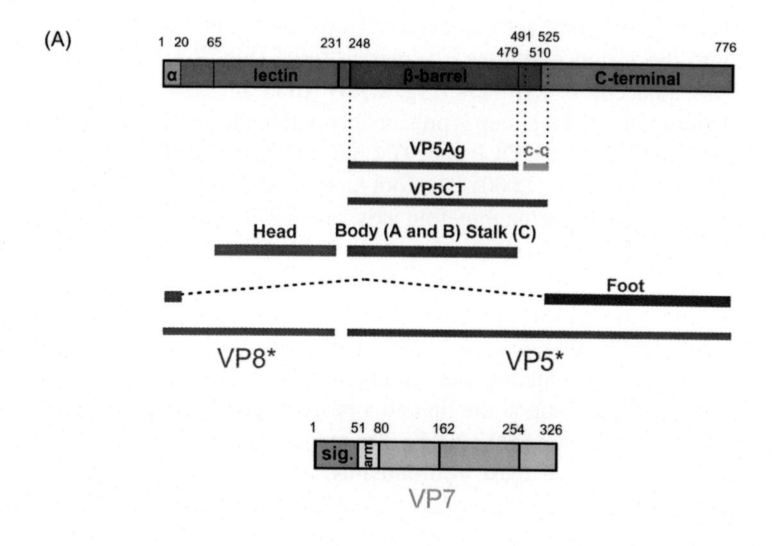

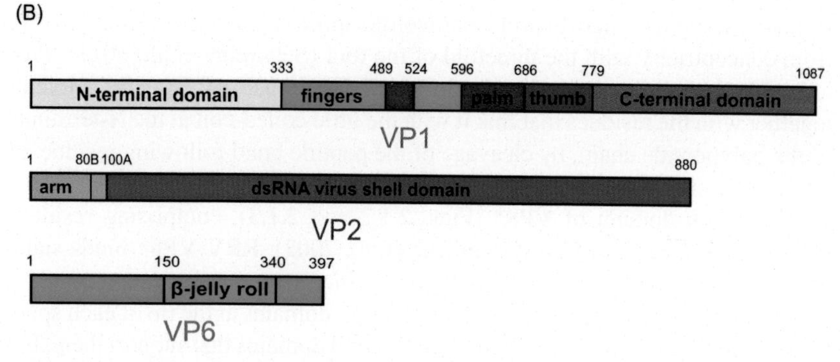

FIGURE 2.1.2 **Domain organization of rotavirus proteins.** (A) Outer-layer proteins. (B) DLP proteins, including the polymerase, VP1.

The middle layer in turn templates the outer layer. A VP7 trimer caps each VP6 trimer, and the $T = 13$ shell thus generated locks down 60 VP4 trimeric "spikes" (Li et al., 2008; Settembre et al., 2011). A T = 13 icosahedral lattice has 2 positions of local sixfold symmetry in each icosahedral asymmetric unit; the VP4 spikes emanate from one of those. The lack of strict threefold symmetry at that site–both in the VP6 array and in the VP7 array matched to it–probably contributes to defining the asymmetric orientation of the spikes.

In the account of specific protein components below, residue numbers refer to rhesus rotavirus (RRV, G3P5B[3]).

2 OUTER LAYER

VP4 has a complex domain organization (Fig. 2.1.2A), and processing to VP8* and VP5* by tryptic cleavage in the gut or in cell culture is necessary for infectivity (Estes et al., 1981; Settembre et al., 2011). VP4 alone is a monomer—at least when made by recombinant expression (Dormitzer et al., 2001). Its incorporation into the virion generates a trimeric "foot" (Fig. 2.1.3), clamped down by the six adjacent VP7 trimers (Fig. 2.1.4) (Li et al., 2008; Settembre et al., 2011). Coordination of this step is presumably part of the process by which NSP4 facilitates budding of the DLP and VP4 into the endoplasmic reticulum (Au et al., 1993; O'Brien et al., 2000). The foot includes a short (29-residue) N-terminal segment, clustered at the threefold axis, and a 277-residue C-terminal domain (Figs. 2.1.2A and 2.1.4) (Settembre et al., 2011). The asymmetric organization of the projecting spike does not, at least for some strains, require proteolytic processing, but the cleavage to VP8* and VP5* promotes spike stability (Crawford et al., 2001; Rodriguez et al., 2014). Two of the VP4 subunits cluster to form a locally twofold symmetric, but acentrically located, projection, which comprises both a lectin domain at the tip and a bean-shaped β-barrel that supports it (Fig. 2.1.3) (Settembre et al., 2011). The tryptic cleavage excises about 18 residues at the junction of these two domains, allowing the newly generated VP5* N-termini to cross-connect the twofold related β-barrels (Fig. 2.1.3) (Yoder and Dormitzer, 2006). The β-barrel of the third VP5* subunit is the diagonal "cantilever" that supports the twofold spike and displaces its axis away from concentricity with the threefold of the foot (Settembre et al., 2011). Proteolytic processing removes the lectin-like VP8* domain of that third subunit, together with the residues that link it with the little coiled-coil at the N-terminus of the polypeptide chain, by cleavage of the peptide bond following residue 29 (Rodriguez et al., 2014).

The lectin domain of VP8* (Figs. 2.1.2 and 2.1.3), comprising residues 65–224, has a galectin fold (Dormitzer et al., 2002). RRV VP8* binds sialic acid, its attachment receptor, in a groove distinct from the site that accepts N-acetyl-lactosamine on galectin 3. The two lectin domains at the tip of each spike have relatively modest contacts with the β-barrel domains that support them, but they are also retained by the tether linking them to the N-terminal coiled-coil in the foot (Settembre et al., 2011). The VP8* receptor-binding site accounts for attachment specificity and at least some degree of tropism (Haselhorst et al., 2009). For example, the human strain, HAL1166 (G8P11[14]), binds the A-type histo-blood group antigen (Hu et al., 2012).

The β-barrels bear a cluster of three hydrophobic loops at the end distal to the DLP (Fig. 2.1.3) (Settembre et al., 2011). The VP8* lectin domains cover these loops on the projecting subunits; the proximal ends of the projecting β-barrels cover the loops on the diagonal "cantilever." The loops resemble the fusion loops of class II and class III viral fusion proteins, and mutation analysis confirms that their hydrophobicity is essential for viral penetration (Kim et al., 2010). The structure of a C-terminally truncated fragment of VP5* (called VP5CT, Figs. 2.1.2, 2.1.4 and 2.1.5) shows a distinct, threefold symmetric conformational state, in which amino-acid residues in the segment connecting the β-barrel to the foot have become a three-chain coiled-coil, with the β-barrels directed toward the C-terminus of the coiled-coil and hence—if this trimer were part of a virion—toward, rather than away from, the DLP (Dormitzer et al., 2004).

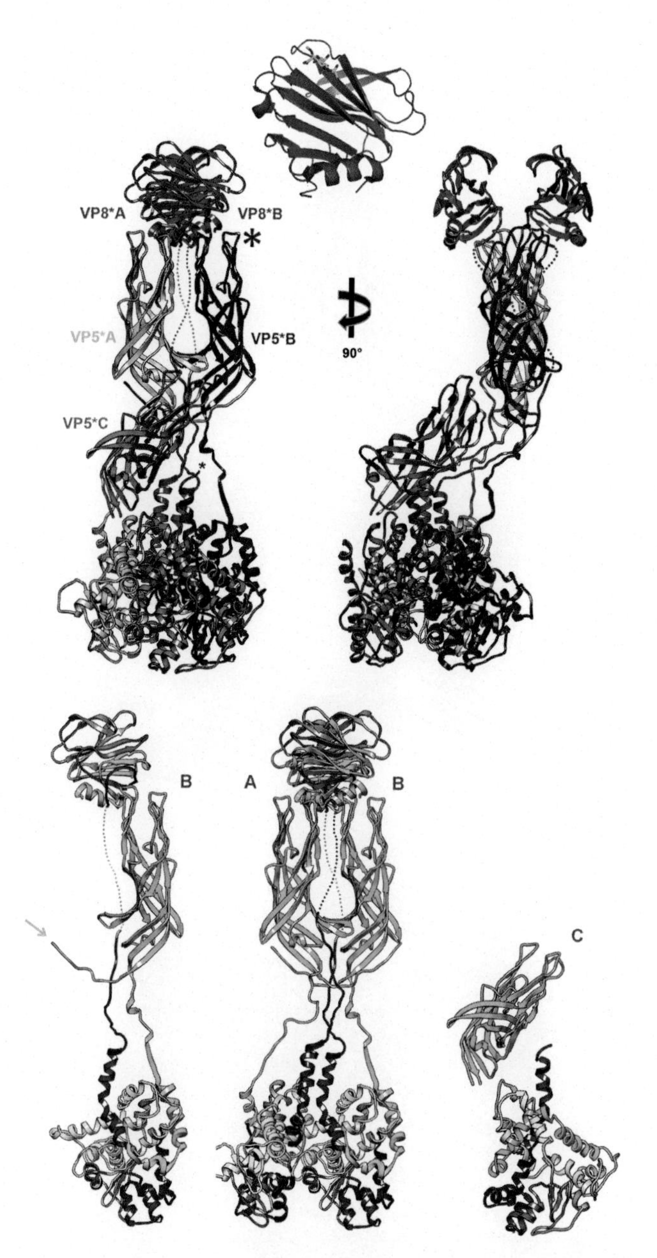

FIGURE 2.1.3 Structure of VP4 and its cleavage fragments. Top: Two of the VP4 subunits (A and B), cleaved to VP8* (shades of *magenta*) and VP5* (light orange and red for A and B, respectively), project as a roughly twofold symmetric spike; the third (C) supports the spike and displaces it from the trimeric "foot" at the base. A separate, magnified ribbon diagram of the lectin domain (from RRV) shows the site for sialic acid *(cyan)*. Note that the N-termini of all three VP8* subunits cluster into a short coiled-coil at the center of the foot and that cleavage just C-terminal to the coiled-coil has released the lectin domain of subunit C. The dotted lines connect the N-terminal helices of VP8* with the lectin domain. *Red asterisk* shows the position of the hydrophobic loops at the tip of the VP5* β-barrel domain. Bottom left: the B subunit, colored from the N-terminus *(blue)* to the C-terminus *(red)*, to illustrate the course of the polypeptide chain. The blue arrow shows the cleavage point at which VP5* has separated from VP8*. Bottom center: the A and B subunits, in the same color gradient, illustrating the twofold clustering of the VP8* lectin domains and the VP5* β-barrel domains. Bottom right: the C subunit, in the same color gradient.

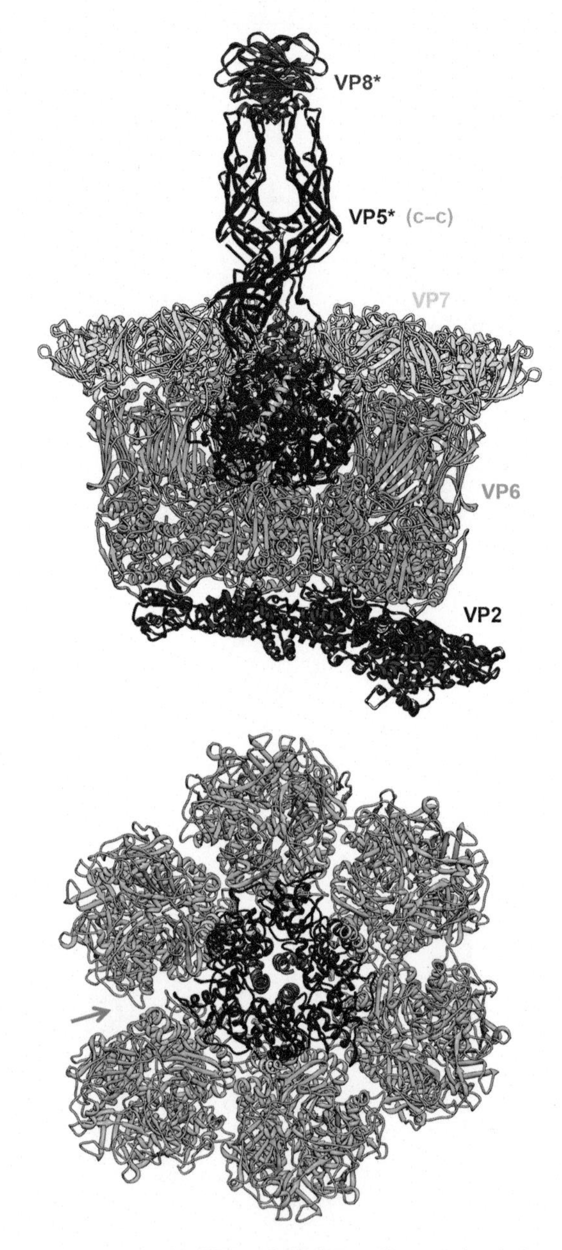

FIGURE 2.1.4 VP4 contacts with VP7 and VP6. Top: Cross-sectional view of part of a TLP, illustrating how the foot of VP4 *(red)* fits within the VP7 shell *(yellow)* and contacts the six surrounding VP6 trimers *(green)*. The designation "c–c" in cyan indicates the segment of VP5* that becomes the three-chain coiled-coil in VP5CT (Figs. 2.1.2A and 2.1.5). Bottom: view along the threefold axis of the VP4 foot, looking toward the center of the particle and illustrating the short, axial coiled-coil of the VP8* N-terminal segments *(magenta)*, the peripheral location of the "c–c" segments *(cyan)*, and the deviation from strict sixfold symmetry that helps establish the asymmetric clustering of the VP4 projections (*arrow*, which shows a wider gap than at other interfaces within the ring of six VP6 trimers).

This inversion resembles the fusion-promoting conformational change in viral fusion proteins, just as the loops resemble the membrane-interacting elements. Fig. 2.1.5 illustrates a proposed mechanism for coupling of this conformational change to membrane perforation during entry.

VP7 is a Ca^{2+}-stabilized trimer, with an N-terminal "arm" appended to a two-lobe globular domain (Fig. 2.1.6) (Aoki et al., 2009; Chen et al., 2009). The globular domain interfaces have two Ca^{2+} binding sites each; removal of Ca^{2+} dissociates the oligomer (Cohen et al., 1979; Dormitzer et al., 2000). The N-terminal arms, which bear a relatively conserved glycan at position 69, clamp the VP7 trimer onto the underlying VP6 trimer (Fig. 2.1.6) (Chen et al., 2009). The hydrated interface between the globular domains and the outer surfaces of VP6 has almost no direct protein-protein contact, and the intimate wrapping of the N-terminal arm accounts for the high affinity of the association. Like other glycoproteins, VP7 is synthesized on membrane-bound ribosomes, which secrete the product into the endoplasmic reticulum (ER) (Stirzaker and Both, 1989). A still puzzling mechanism underlies the final assembly of the triple-layered particle (TLP) in the ER, mediated at least in part by NSP4, the membrane-bound viral protein that recruits the presumed DLP:VP4 complex to the ER membrane, and by VP7, which disrupts a transient envelope that surrounds the DLP:VP4 complex after it buds into the ER lumen (Lopez et al., 2005). DLPs can be recoated with recombinant VP4 and VP7 to form fully infectious particles, provided that VP4 and DLPs are mixed in the recoating reaction before VP7 is added (Trask and Dormitzer, 2006). The required order of addition suggests that VP4 assembles onto DLPs before VP7 during authentic viral assembly.

VP7 holds the VP4 trimer in place, and loss of Ca^{2+} causes both to dissociate from the DLP (Cohen et al., 1979; Settembre et al., 2011). An engineered disulfide bond across the VP7 subunit interface prevents dissociation at low Ca^{2+} concentrations and blocks penetration (Aoki et al., 2011). The TLP structure shows that an intact VP7 layer would interfere with the conformational inversion of VP5* proposed to induce membrane disruption (Settembre et al., 2011). When the rotavirus outer layer is released by calcium chelation in the presence of a lipid bilayer, the VP5* fragment binds the membrane (Trask et al., 2010). Ca^{2+} depletion is thus a likely trigger for membrane penetration by rotavirus. For RRV and BSC-1 cells, this step occurs in a very early compartment, before the engulfed virus reaches a Rab5-labeled endosomc (Abdclhakim et al., 2014). Engulfment in these cells is primarily clathrin independent; it appears to be driven largely by virus–membrane interactions.

3 MIDDLE AND INNER LAYERS: THE DLP

VP6 has a jelly-roll domain, projecting from a largely helical base, composed of residues N- and C-terminal to the jelly-roll (Figs. 2.1.2B and 2.1.6) (Mathieu et al., 2001). It appears to have an exclusively structural function in the rotavirus

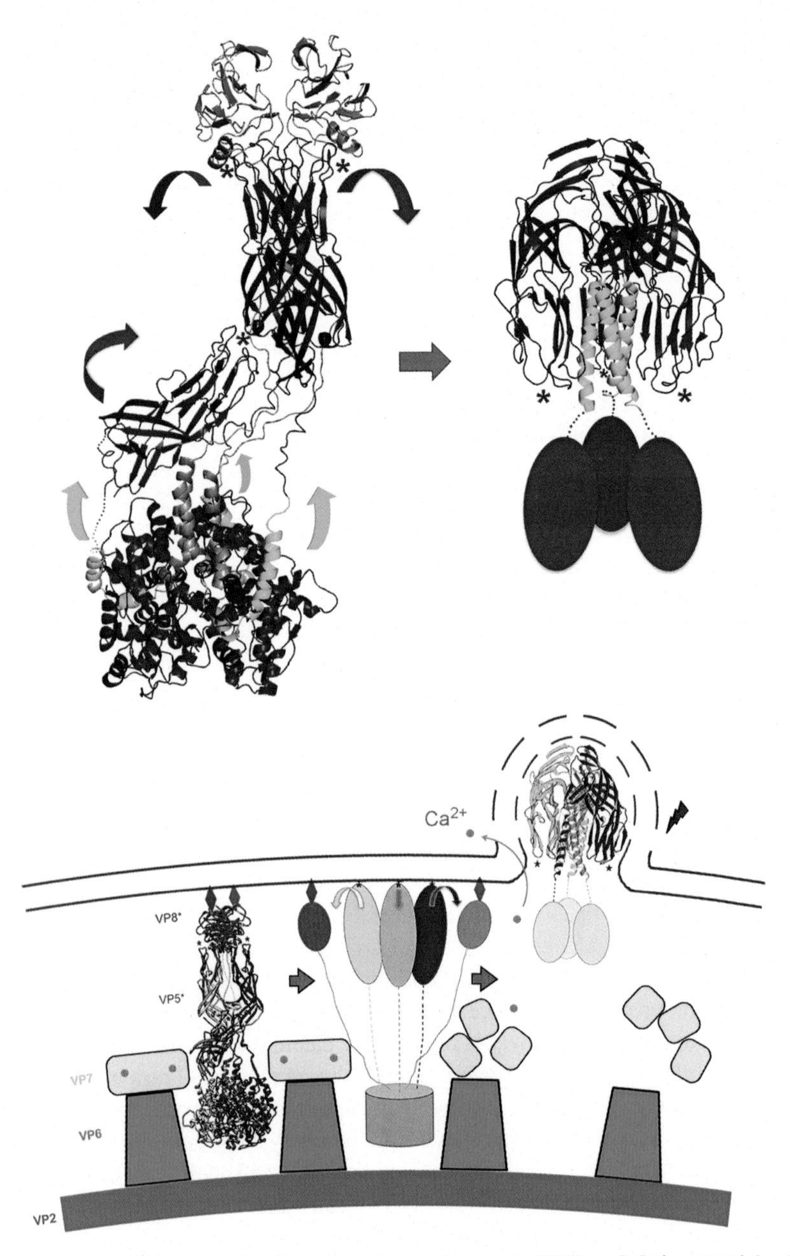

FIGURE 2.1.5 **Entry associated conformational changes in VP5* and their potential role in membrane perforation.** Top: Transition of VP5* from its conformation as part of the mature spike on the TLP to its final conformation after entry as inferred from the structure of VP5CT. VP8* is in magenta; VP5*, in *red. Red asterisks* indicate hydrophobic loops, which are covered either by VP8* (for loops on the two β-barrels that project outward) or by other VP5* subunits (for the β-barrel that creates the diagonal support for the other two). The segment in VP5* that transitions

particle. Its reovirus homolog, μ1, has an even more elaborate helical domain at its base, which has a direct function in membrane perforation and viral penetration (Chandran et al., 2002; Liemann et al., 2002).

VP2 is effectively one very large domain, except for 80–100 residues at the N-terminus (Figs. 2.1.1, 2.1.2B, and 2.1.4) (McClain et al., 2010). The two icosahedrally distinct (but chemically identical) VP2 conformers (VP2A and VP2B) within the inner layer are related by a relatively modest distortion. VP2 has obvious homologs in most other families of dsRNA viruses, but comparison does not give a clear description of how one might have evolved from another. The N-terminal arms cannot be seen in icosahedrally averaged cryoEM reconstructions and X-ray density maps. They project toward the fivefold axis (Fig. 2.1.7), where they may have a role in recruiting VP1 (and presumably VP3, the only protein in the virion for which we have no structure) (Estrozi et al., 2013; McDonald and Patton, 2011). One VP1 polymerase molecule associates with each VP2 decamer (Patton et al., 1997). Its position in the TLP and DLP near their fivefold axes can be deduced from features of cryoEM and X-ray maps of the particles.

VP1, like the polymerases of other dsRNA viruses and of nonsegmented negative-strand RNA viruses, is a "caged" version of the standard fingers-palm-thumb module, common to a very large class of RNA and DNA polymerases (Lu et al., 2008). N- and C-terminal extensions of that module surround it so that four distinct channels (for template entry, template or double-strand product exit, transcript exit, and NTP access, respectively) connect the outside of the protein with the catalytic site at its center (Fig. 2.1.7). A site on the surface of the protein binds the mRNA cap, probably to retain the 5′ end of the plus-sense strand. Figure 2.1.7 shows the position and orientation of VP1 in the particle, the positions of various channels and of the cap-binding site, and a scheme for the threading of RNA through the polymerase during transcription of dsRNA genomic segments. The structures determined to date cannot specify whether each of the possible five positions has equal (20%) occupancy or the VP1 position at one vertex correlates with the VP1 positions at neighboring vertices. In cytoplasmic polyhedrosis virus,

---

◀ to a three-chain coiled-coil in the conformational rearrangement is in cyan. We have no information about the conformation of the C-terminal "foot" region in the postentry state, so that part is represented by a featureless oval. The foot may unfold during the transition because part of the coiled-coil appears to contribute to the folded structure (note *cyan*-colored segment buried within the globular foot domain). Bottom: model for membrane disruption, based on the known structures and on the data in Abdelhakim et al. (2014), showing that penetration occurs from a small vesicle closely surrounding a single virion. Step 1: VP8* binds glycolipid receptor (*blue* diamonds). Step 2: transient dissociation (breathing) of VP8* from the tip of VP5* allows hydrophobic loops of VP5* to engage the membrane, preventing redocking of VP8*. Step 3: Ca^{2+} leak from vesicle into cytosol (perhaps induced by membrane perturbation when VP5* loops insert) leads to reduced Ca^{2+} concentration, dissociation of VP7 *(yellow)*, and release of VP5*. Step 4: Complete foldback of VP5*, driven by formation of stable coiled-coiled and other strong trimer interactions seen in structure and linked through hydrophobic loops to the vesicle bilayer, disrupts the membrane and allows release of DLP (*green* VP6 and *blue* VP2).

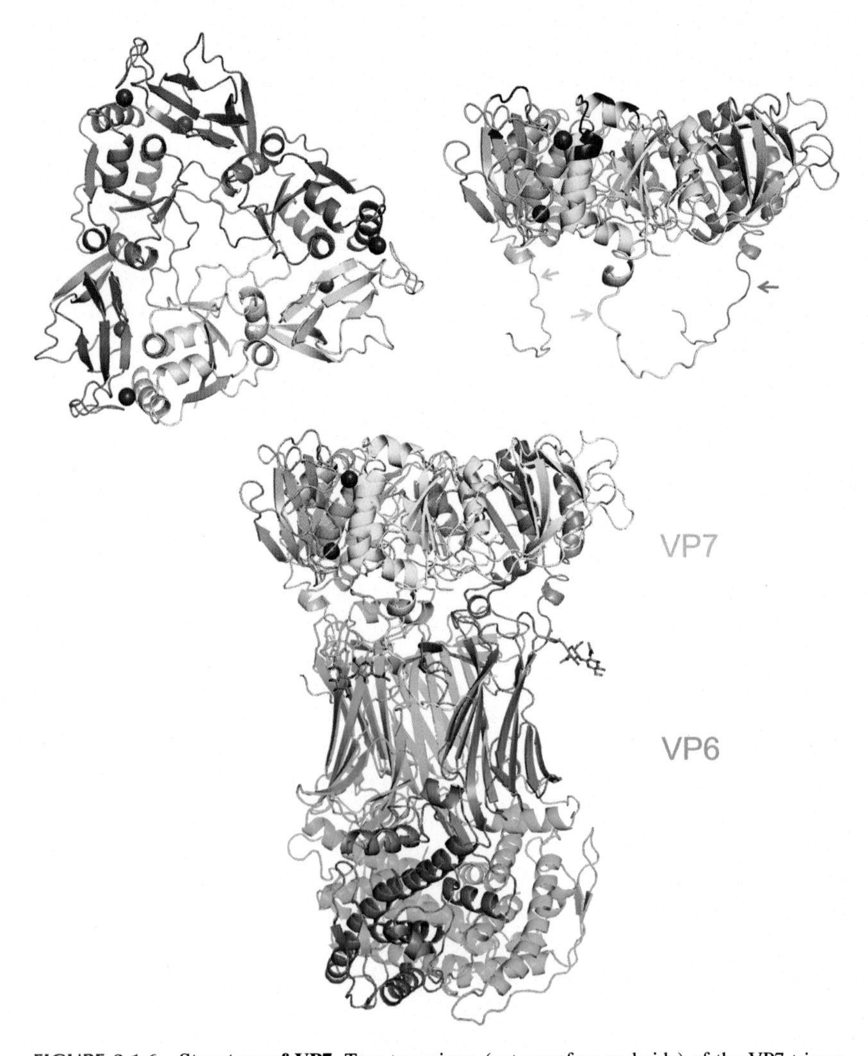

FIGURE 2.1.6 **Structure of VP7.** Top: two views (outer surface and side) of the VP7 trimer. The three subunits are in shades of yellow and orange. The Ca^{2+} ions at the subunit interfaces are in dark gray. The side view shows the N-terminal arms (and the Ca^{2+} ions at only one of the three interfaces). The arrows indicate the conserved glycosylation site. Bottom: VP7 trimer bound with a VP6 trimer. The N-terminal arms of VP7 grip the VP6 trimer. The two initial *N*-acetyl glucosamines of the glycans are shown as sticks.

a single-shelled reovirus with 10 genomic RNA segments, the ten transcription complexes occupy defined positions with respect to each other, so that their full structures emerge from a cryoEM reconstruction that does not impose icosahedral symmetry (Liu and Cheng, 2015; Zhang et al., 2015). The genomic RNA winds among them, traversing the two unoccupied vertices as well. Yet-to-be determined events during capsid assembly must define this global organization.

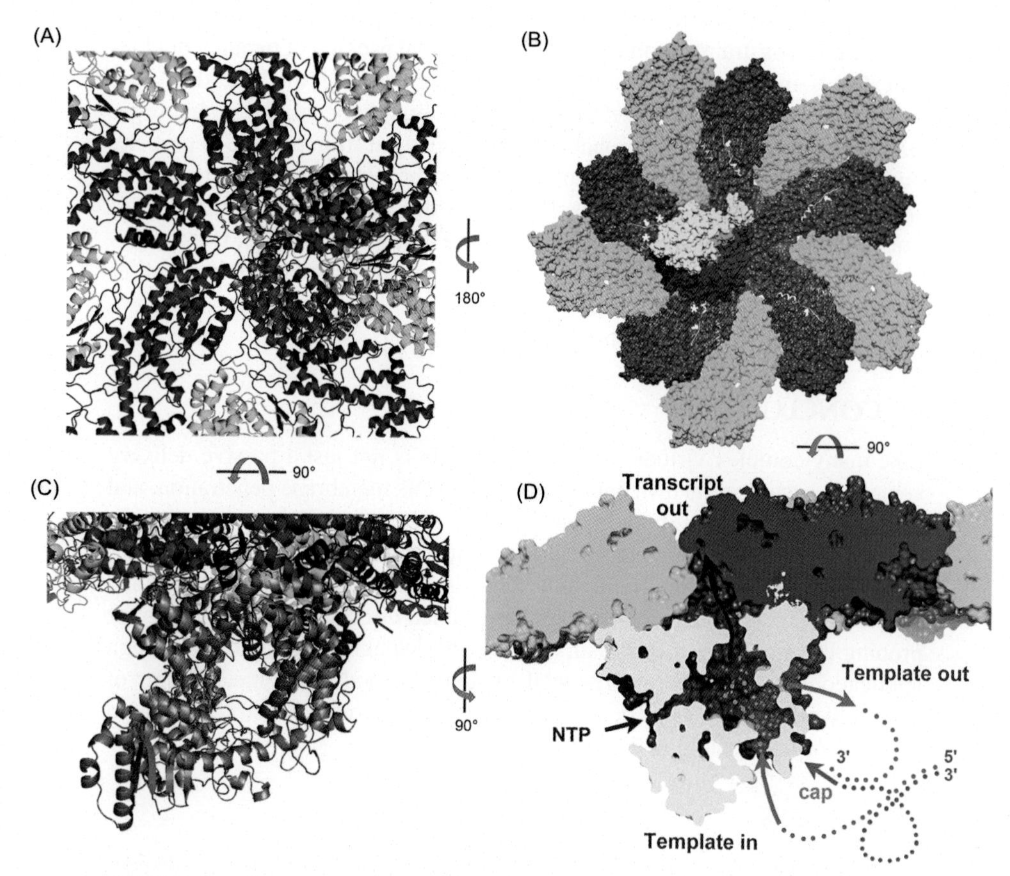

FIGURE 2.1.7 VP1: structure and placement within the DLP. (A) VP1 *(orange)* and its position with respect to VP2A *(blue)* and VP2B *(cyan)*, viewed from outside the VP2 shell along a fivefold axis. (B) Ten VP2 subunits surrounding a fivefold axis, colored as in (A), and the associated VP1, viewed from inside the DLP. The domains of VP1 are colored yellow (N-terminal domain), green (fingers subdomain), red (palm subdomain), blue (thumb subdomain), and magenta (C-terminal domain). The positions at which N-terminal arms of three different VP2 subunits clearly contact the VP1 shown are marked by asterisks; the remaining N-terminal arms may also contact this VP1 molecule. (C) As in (A), but viewed in a direction tangential to the VP2 shell; the opening in the foreground is the exit channel for template (and dsRNA). Arrow: contact between VP1 and a VP2A N-terminal arm. (D) As in (B), but viewed in a direction tangential to the VP2 shell, cut away to show the internal, catalytic-site cavity. Channels leading into this cavity are labeled. The arrows and dotted lines show (not to scale) how the minus-sense strand of the genome *(blue)* threads through the polymerase, leaving a "transcription bubble" in the plus-sense strand *(orange)*; the latter is anchored to the polymerase by its 5′-cap. The transcript exits toward the narrow channel at the fivefold axis of the VP2 shell. It must in some way deviate from this path initially, in order to receive the 5′ cap from VP3, which is probably nearby, but for which there is no clear density signature.

The genomic RNA fits tightly within the volume defined by the inner surface of VP2. This constraint enforces a locally hexagonal packing of parallel, double-helical rods and produces layers of density concentric with the VP2 shell. The ordering of the RNA thus resembles a nematic liquid crystal; the global organization of the 11 segments probably varies somewhat from particle to particle. During transcription, each segment must pass through the polymerase many times, to yield multiple transcripts. Binding of the 5′ cap on the positive-sense strand, which can remain there throughout the transcription cycle, is a mechanism for capturing the complementary 3′ end of the template (minus-sense) strand close to the opening of the template channel and for avoiding entanglement with other RNA segments (Fig. 2.1.7) (Lu et al., 2008).

4 CONCLUSIONS

Like many complex virions, a rotavirus particle is not just a passive delivery vehicle. It organizes its own engulfment, effects its membrane penetration, and synthesizes, caps and exports its mRNA into the cytosol. A rotavirus particle also organizes its own assembly, in an unusual series of steps requiring budding into the endoplasmic reticulum. Knowing the structures of the TLP and DLP and of their protein components has enabled direct experimental tools for probing the molecular mechanism of entry (Abdelhakim et al., 2014), through application of contemporary live-cell imaging methods. The mechanisms of assembly are at least as intricate; one can now contemplate a structure-based strategy for understanding them.

REFERENCES

Abdelhakim, A.H., Salgado, E.N., Fu, X., Pasham, M., Nicastro, D., Kirchhausen, T., Harrison, S.C., 2014. Structural correlates of rotavirus cell entry. PLoS Pathog. 10, e1004355.

Aoki, S.T., Settembre, E.C., Trask, S.D., Greenberg, H.B., Harrison, S.C., Dormitzer, P.R., 2009. Structure of rotavirus outer-layer protein VP7 bound with a neutralizing Fab. Science 324, 1444–1447.

Aoki, S.T., Trask, S.D., Coulson, B.S., Greenberg, H.B., Dormitzer, P.R., Harrison, S.C., 2011. Cross-linking of rotavirus outer capsid protein VP7 by antibodies or disulfides inhibits viral entry. J. Virol. 85, 10509–10517.

Au, K.S., Mattion, N.M., Estes, M.K., 1993. A subviral particle binding domain on the rotavirus nonstructural glycoprotein NS28. Virology 194, 665–673.

Bass, D.M., Baylor, M.R., Chen, C., Mackow, E.M., Bremont, M., Greenberg, H.B., 1992. Liposome-mediated transfection of intact viral particles reveals that plasma membrane penetration determines permissivity of tissue culture cells to rotavirus. J. Clin. Investig. 90, 2313–2320.

Chandran, K., Farsetta, D.L., Nibert, M.L., 2002. Strategy for nonenveloped virus entry: a hydrophobic conformer of the reovirus membrane penetration protein μ1 mediates membrane disruption. J. Virol. 76, 9920–9933.

Chen, J.Z., Settembre, E.C., Aoki, S.T., Zhang, X., Bellamy, A.R., Dormitzer, P.R., Harrison, S.C., Grigorieff, N., 2009. Molecular interactions in rotavirus assembly and uncoating seen by high-resolution cryo-EM. Proc. Natl. Acad. Sci. USA 106, 10644–10648.

Cohen, J., Laporte, J., Charpilienne, A., Scherrer, R., 1979. Activation of rotavirus RNA polymerase by calcium chelation. Arch. Virol. 60, 177–186.

Crawford, S.E., Mukherjee, S.K., Estes, M.K., Lawton, J.A., Shaw, A.L., Ramig, R.F., Prasad, B.V., 2001. Trypsin cleavage stabilizes the rotavirus VP4 spike. J. Virol. 75, 6052–6061.

Dormitzer, P.R., Greenberg, H.B., Harrison, S.C., 2000. Purified recombinant rotavirus VP7 forms soluble, calcium-dependent trimers. Virology 277, 420–428.

Dormitzer, P.R., Greenberg, H.B., Harrison, S.C., 2001. Proteolysis of monomeric recombinant rotavirus VP4 yields an oligomeric VP5* core. J. Virol. 75, 7339–7350.

Dormitzer, P.R., Nason, E.B., Prasad, B.V., Harrison, S.C., 2004. Structural rearrangements in the membrane penetration protein of a non-enveloped virus. Nature 430, 1053–1058.

Dormitzer, P.R., Sun, Z.-Y.J., Wagner, G., Harrison, S.C., 2002. The rhesus rotavirus VP4 sialic acid binding domain has a galectin fold with a novel carbohydrate binding site. EMBO J. 21, 885–897.

Estes, M.K., Graham, D.Y., Mason, B.B., 1981. Proteolytic enhancement of rotavirus infectivity: molecular mechanisms. J. Virol. 39, 879–888.

Estrozi, L.F., Settembre, E.C., Goret, G., McClain, B., Zhang, X., Chen, J.Z., Grigorieff, N., Harrison, S.C., 2013. Location of the dsRNA-dependent polymerase, VP1, in rotavirus particles. J. Mol. Biol. 425, 124–132.

Haselhorst, T., Fleming, F.E., Dyason, J.C., Hartnell, R.D., Yu, X., Holloway, G., Santegoets, K., Kiefel, M.J., Blanchard, H., Coulson, B.S., et al., 2009. Sialic acid dependence in rotavirus host cell invasion. Nat. Chem. Biol. 5, 91–93.

Hu, L., Crawford, S.E., Czako, R., Cortes-Penfield, N.W., Smith, D.F., Le Pendu, J., Estes, M.K., Prasad, B.V., 2012. Cell attachment protein VP8* of a human rotavirus specifically interacts with A-type histo-blood group antigen. Nature 485, 256–259.

Kim, I.S., Trask, S.D., Babyonyshev, M., Dormitzer, P.R., Harrison, S.C., 2010. Effect of mutations in VP5 hydrophobic loops on rotavirus cell entry. J. Virol. 84, 6200–6207.

Lawton, J.A., Estes, M.K., Prasad, B.V., 1997a. Three-dimensional visualization of mRNA release from actively transcribing rotavirus particles [letter]. Nat. Struct. Biol. 4, 118–121.

Lawton, J.A., Zeng, C.Q., Mukherjee, S.K., Cohen, J., Estes, M.K., Prasad, B.V., 1997b. Three-dimensional structural analysis of recombinant rotavirus-like particles with intact and amino-terminal-deleted VP2: implications for the architecture of the VP2 capsid layer. J. Virol. 71, 7353–7360.

Li, Z., Baker, M.L., Jiang, W., Estes, M.K., Prasad, B.V., 2008. Rotavirus architecture at subnanometer resolution. J. Virol.

Liemann, S., Chandran, K., Baker, T.S., Nibert, M.L., Harrison, S.C., 2002. Structure of the reovirus membrane-penetration protein, μ1, in a complex with its protector protein, σ3. Cell 108, 283–295.

Liu, H., Cheng, L., 2015. Cryo-EM shows the polymerase structures and a nonspooled genome within a dsRNA virus. Science 349 (6254), 1347–1350.

Lopez, T., Camacho, M., Zayas, M., Najera, R., Sanchez, R., Arias, C.F., Lopez, S., 2005. Silencing the morphogenesis of rotavirus. J. Virol. 79, 184–192.

Lu, X., McDonald, S.M., Tortorici, M.A., Tao, Y.J., Vasquez-Del Carpio, R., Nibert, M.L., Patton, J.T., Harrison, S.C., 2008. Mechanism for coordinated RNA packaging and genome replication by rotavirus polymerase VP1. Structure 16, 1678–1688.

Mathieu, M., Petitpas, I., Navaza, J., Lepault, J., Kohli, E., Pothier, P., Prasad, B.V., Cohen, J., Rey, F.A., 2001. Atomic structure of the major capsid protein of rotavirus: implications for the architecture of the virion. EMBO J. 20, 1485–1497.

McClain, B., Settembre, E., Temple, B.R., Bellamy, A.R., Harrison, S.C., 2010. X-ray crystal structure of the rotavirus inner capsid particle at 3.8 Å resolution. J. Mol. Biol. 397, 587–599.

McDonald, S.M., Patton, J.T., 2011. Rotavirus VP2 core shell regions critical for viral polymerase activation. J. Virol. 85, 3095–3105.

O'Brien, J.A., Taylor, J.A., Bellamy, A.R., 2000. Probing the structure of rotavirus NSP4: a short sequence at the extreme C terminus mediates binding to the inner capsid particle. J. Virol. 74, 5388–5394.

Patton, J.T., Jones, M.T., Kalbach, A.N., He, Y.W., Xiaobo, J., 1997. Rotavirus RNA polymerase requires the core shell protein to synthesize the double-stranded RNA genome. J. Virol. 71, 9618–9626.

Rodriguez, J.M., Chichon, F.J., Martin-Forero, E., Gonzalez-Camacho, F., Carrascosa, J.L., Caston, J.R., Luque, D., 2014. New insights into rotavirus entry machinery: stabilization of rotavirus spike conformation is independent of trypsin cleavage. PLoS Pathogens 10, e1004157.

Settembre, E.C., Chen, J.Z., Dormitzer, P.R., Grigorieff, N., Harrison, S.C., 2011. Atomic model of an infectious rotavirus particle. EMBO J. 30, 408–416.

Stirzaker, S.C., Both, G.W., 1989. The signal peptide of the rotavirus glycoprotein VP7 is essential for its retention in the ER as an integral membrane protein. Cell 56, 741–747.

Trask, S.D., Dormitzer, P.R., 2006. Assembly of highly infectious rotavirus particles recoated with recombinant outer capsid proteins. J. Virol. 80, 11293–11304.

Trask, S.D., Kim, I.S., Harrison, S.C., Dormitzer, P.R., 2010. A rotavirus spike protein conformational intermediate binds lipid bilayers. J. Virol. 84, 1764–1770.

Yoder, J.D., Dormitzer, P.R., 2006. Alternative intermolecular contacts underlie the rotavirus VP5* two- to three-fold rearrangement. EMBO J. 25, 1559–1568.

Zhang, X., Ding, K., Yu, X., Chang, W., Sun, J., Zhou, Z.H., 2015. In situ structures of the segmented genome and RNA polymerase complex inside a dsRNA virus. Nature 527 (7579), 531–534.

Chapter 2.2

Rotavirus Attachment, Internalization, and Vesicular Traffic

C.F. Arias, D. Silva-Ayala, P. Isa, M.A. Díaz-Salinas, S. López
Departamento de Genética del Desarrollo y Fisiología Molecular, Instituto de Biotecnología, Universidad Nacional Autónoma de México, Cuernavaca, Morelos, México

1 INITIAL INTERACTIONS OF THE VIRUS WITH THE HOST CELL

Rotavirus particles are formed by a triple-layered capsid that surrounds the viral genome composed of eleven double-stranded RNA segments. The outermost layer of the viral particle is made of 780 copies of the glycoprotein VP7 arranged in trimers coordinated by calcium, which form a smooth surface layer from which spikes composed of trimers of VP4 protrude. To be infectious the virus depends on the specific cleavage by trypsin of VP4 into VP8 and VP5. Both cleavage domains of VP4, as well as VP7, have essential roles in the initial steps of the replication cycle of rotaviruses.

1.1 Virus Attachment

Rotavirus cell entry is a multistep process involving cellular glycans for cell binding and several cell surface molecules during postattachment steps (Lopez and Arias, 2006) (Fig. 2.2.1A). The VP8 domain of VP4 mediates the initial interaction of the virus with the cell surface, whereas the VP5 domain of VP4 and the surface glycoprotein VP7 interact with downstream postattachment molecules (Lopez and Arias, 2006). Different rotavirus strains were initially classified as neuraminidase (NA)-sensitive or NA-resistant, depending on their ability to infect cells previously treated with NA. Some animal rotavirus strains require sialic acid (SA) for attachment, whereas other animal and most human rotavirus strains are NA-resistant (Lopez and Arias, 2006). Recently, it was found that some rotavirus strains, which were originally classified as NA-resistant, bind to internal SAs, which are not cleaved by NA (Haselhorst et al., 2007), while others bind to human blood group antigens (HBGAs) (Table 2.2.1).

Viral Gastroenteritis

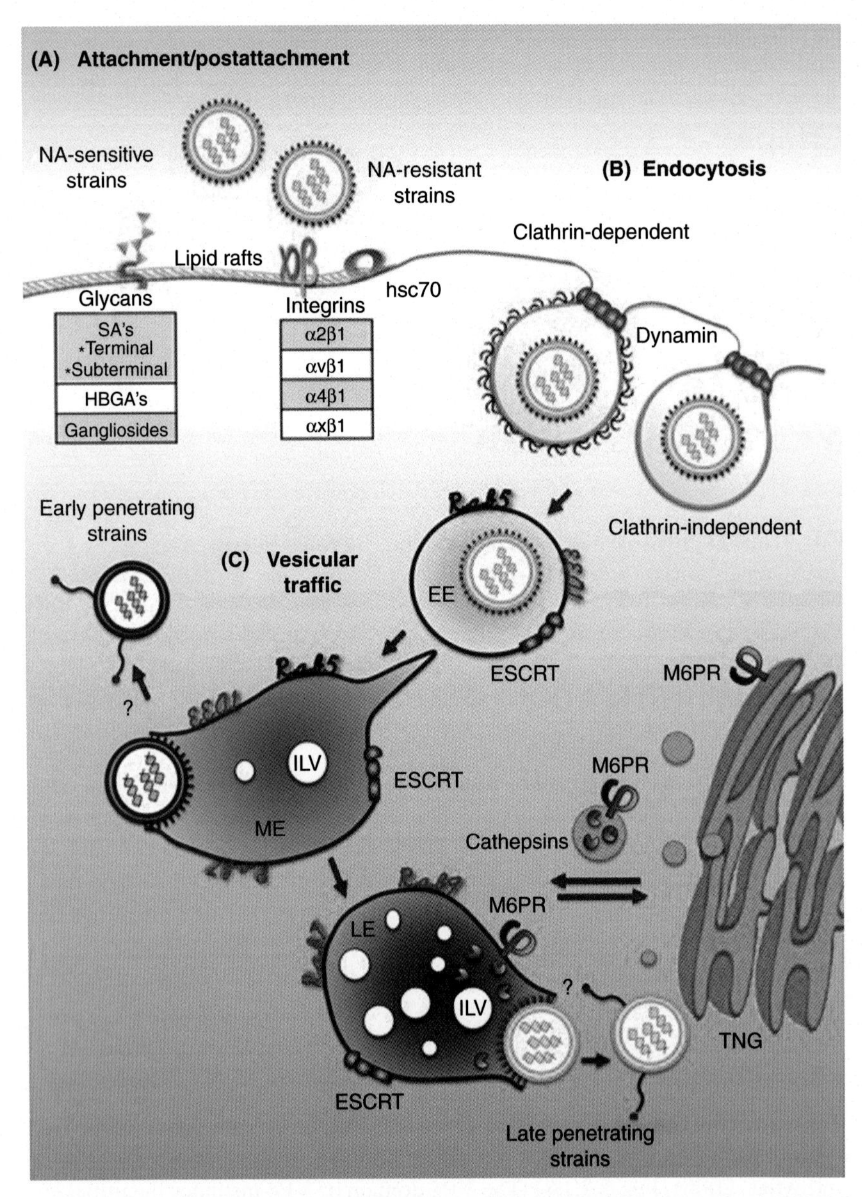

FIGURE 2.2.1 Illustration of the internalization and vesicular traffic of rotavirus in MA104 cells. (A) NA-sensitive and NA-resistant rotavirus strains initially interact with the cell surface through different glycans. After attachment, rotaviruses interact with different integrins and with hsc70, organized in lipid rafts on the cell membrane; (B) viral particles are then internalized either by a clathrin-dependent or –independent endocytosis which is directed by the spike protein VP4 (Diaz-Salinas et al., 2013); and (C) once inside the cell, regardless of their route of entry, all rotavirus strains converge in early endosomes *(EE)*, characterized by the presence of Rab 5 and EEA1, and then proceed to maturing endosomes *(ME)* where intraluminal vesicles *(ILVs)* begin to form with the participation of the ESCRT proteins. Some rotavirus strains, such as RRV and SA11, behave as early penetrating viruses since they escape the endosomal network at this point. Other viral strains continue to LEs, characterized for the presence of Rab 7 and Rab 9; to exit the endosomal network these strains also depend on the presence of CD-M6PR and probably on the activity of cathepsins, which are transported from the trans Golgi network *(TNG)*, behaving as late penetrating strains.

TABLE 2.2.1 Glycans Bound by Different Rotavirus Strains

Virus strain	Origin	Genotype	HA/NA[1]	Ligand	Method	References
NCDV	Bovine	G6P[1]	+/+	NeuGc-GM3	TLC[5]	Delorme et al. (2001)
				NeuGc-GM2	TLC	Delorme et al. (2001)
				NeuGc-GD1a	TLC	Delorme et al. (2001)
				NeuAc-GD1a	TLC	Delorme et al. (2001)
UK	Bovine	G6P[5]	−/−	NeuGc-GM1	TLC	Delorme et al. (2001)
				NeuAc-GM1	TLC	Delorme et al. (2001)
				NeuAca3-neo-LTC[3]	TLC	Delorme et al. (2001)
				NeuGc-GM2	TLC	Delorme et al. (2001)
B223	Bovine	G10P[11]	NT[2]/−	LacNAc	Glycan array	Ramani et al. (2013)
				Neu HMG	HMG SGM[8]	Yu et al. (2011)
DS1	Human	G2P[4]	−/−	Le^b, H type 1	EIA[6]	Huang et al. (2012)
				A-type HBGA	STD NMR[7]	Bohm et al. (2015)
BM5265	Human	P[4]	NT/NT	Le^b, H type 1	EIA	Huang et al. (2012)
BM11596	Human	P[6]	NT/NT	H type 1 antigen	EIA	Huang et al. (2012)
BM151	Human	P[8]	NT/NT	Le^b, H type 1	EIA	Huang et al. (2012)
BM13851	Human	P[8]	NT/NT	Le^b, H type 1	EIA	Huang et al. (2012)

(*Continued*)

TABLE 2.2.1 Glycans Bound by Different Rotavirus Strains (*cont.*)

Virus strain	Origin	Genotype	HA/NA[1]	Ligand	Method	References
BM14113	Human	P[8]	NT/NT	Le^b, H type 1	EIA	Huang et al. (2012)
	Human	G3P[8]	NT/NT	Le^b, H type 1	EIA	Huang et al. (2012)
RV-3	Human	G3P[6]	NT/NT	aceramido-GM1	STD NMR	Bohm et al. (2015)
				A-type HBGA	STD NMR	Bohm et al. (2015)
				Neu and SA HMG[4]	HMG SGM	Yu et al. (2011)
ST3	Human	G4P[6]	NT7/–	H type 1 antigen	EIA	Huang et al. (2012)
Wa	Human	G1P[8]	–/–	aceramido-GM1	STD NMR	Haselhorst et al. (2009)
				Le^b, H type 1	EIA	Huang et al. (2012)
T152	Human	G12P[9]	NT/NT	A type HBGA	EIA	Liu et al. (2012)
K8	Human	G1P[9]	NT7/–	A-type HBGA	STD NMR	Bohm et al. (2015)
			NT/NT	LacNAc Neu HMG	Glycan array	Ramani et al. (2013)
					HMG SGM	Yu et al. (2011)
N1509	Human	G10P[11]	NT/NT	LacNAc	Glycan Array	Ramani et al. (2013)
HAL1166	Human	G8P[14]	NT/NT	A-type HGBA	X-ray[9], Glycan array	Hu et al. (2012)
				A-type HBGA	STD NMR	Bohm et al. (2015)
VAG8.1	Human	G8P[14]	NT/NT	A type HBGA	EIA	Liu et al. (2012)

KTM368	Human	G11P[25]	NT/NT	A type HBGA	EIA	Haselhorst et al. (2009)
CRW-8	Porcine	G3P[7]	NT/+	aceramido-GD1a	STD NMR	Haselhorst et al. (2009)
RRV	Simian	G3P[3]	+/+	Neu5AcGM3	X-ray	Delorme et al. (2001)
SA11	Simian	G3P[2]	+/+	NeuGc-GM3	TLC	Delorme et al. (2001)
				NeuGc-GM2	TLC	Delorme et al. (2001)
				NeuGc-GD1a	TLC	Delorme et al. (2001)
				NeuAc-GD1a	TLC	Delorme et al. (2001)

[1]*HA/NA, hemagglutination activity and neuraminidase sensitivity*
[2]*NT, not tested*
[3]*LTC, lactotetraosylceramide*
[4]*HMG, human milk glycans*
[5]*TLC, thin-layer chromatography binding assay*
[6]*EIA, enzyme immuno assay*
[7]*STD NMR, saturation transfer difference nuclear magnetic resonance spectroscopy*
[8]*HMG SGM, human milk glycans shotgun glycan microarray*
[9]*X-ray crystallography.*

The SA binding domain of the virus is located on the tip of VP8 (Dormitzer et al., 2002; Isa et al., 1997). The crystal structure of several VP8 proteins has been resolved (Dormitzer et al., 2002; Blanchard et al., 2007; Monnier et al., 2006; Yu et al., 2011); this protein has a galectin-like fold with two β-sheets separated by a shallow cleft. The crystal structures of VP8 of NA-sensitive rotavirus strains complexed with SA showed that SA binds near the cleft region. Interestingly, when the VP8 proteins of NA-resistant strains were analyzed, this cleft was found to be wider (Haselhorst et al., 2009). Even though the structure of a human VP8 with a wider cleft in complex with a glycan has not been reported, NMR and modeling studies propose that a wider cleft would allow binding of gangliosides with internal SA (Venkataram Prasad et al., 2014).

Gangliosides and glycosphingolipids having one or more SA residues, are a large and heterogeneous family of lipids present on the extracellular leaflet of mammalian plasma membranes, and have been associated with rotavirus cell entry for some time. Using a thin-layer chromatography binding assay, NA-sensitive rotavirus strains were found to bind to gangliosides with terminal SA, while the NA-resistant strains tested interacted with gangliosides containing subterminal SA (Delorme et al., 2001). The role of gangliosides during the infection of MA104 cells by NA-sensitive and -resistant strains was also analyzed after silencing by RNA interference the expression of two key enzymes involved in ganglioside synthesis: UDP-glucose:ceramide glucosyltransferase and lactosyl ceramide-α-2,3-sialyl transferase 5 (Martinez et al., 2013). Knocking down the expression of these enzymes decreased ganglioside levels, resulting in a diminished infectivity of all rotavirus strains tested. Interestingly, the binding of virus to cells with low ganglioside levels was not affected, suggesting that the rotavirus-ganglioside interaction is not necessary for cell surface binding, but during a later step of the cell entry process.

Recently, it was demonstrated that many human rotavirus strains bind to human histo-blood group antigens (HBGAs). Some of them bind H-type glycans, while others have A-type glycan specificity (Bohm et al., 2015; Hu et al., 2012; Huang et al., 2012; Liu et al., 2012) (Table 2.2.1). In addition, the VP8 of a P[11] neonatal strain, was shown to specifically bind a precursor of the H type II HBGA, which forms the core structure of type II glycans (Ramani et al., 2013). The crystal structure of a [P14] VP8 showed that it has a narrow cleft, like that present in NA-sensitive strains, and that the A-type HBGA binds in the same location in the cleft as SA in the animal VP8 structure (Venkataram Prasad et al., 2014). Of interest, glycan modification is thought to vary during neonatal development, which could explain the age-restricted infectivity of neonatal rotavirus strains. It is also of interest that the HBGA phenotype and the secretory status of children seem to correlate with the VP4 genotype of the infecting rotavirus strain (Nordgren et al., 2014; Van Trang et al., 2014).

1.2 Postattachment Interactions

After the initial attachment to glycans on the cell surface, rotaviruses interact with other surface molecules to gain access into the cell. Among these molecules there are several integrins (α2β1, α4β1, αXβ2, αVβ3) and the heat shock cognate protein 70 (hsc70). Whether all these molecules are used by all rotavirus strains, and whether these interactions are sequential or alternative, is not known. In the particular case of a rhesus rotavirus strain (RRV) it was shown that some of these interactions occur sequentially (Lopez and Arias, 2006). Interestingly, not all rotavirus strains interact with integrins, while all the strains tested require hsc70 for efficient cell infection (Gutierrez et al., 2010).

The interaction of rotavirus with integrin α2β1 is mediated by a DGE motif located towards the amino-terminal end of the VP5 domain of VP4, while the domain I of the integrin subunit α2 is involved in this interaction (Graham et al., 2003; Zarate et al., 2004). The amino acid residues on the I domain that interact with the virus spike protein were identified by expressing α2β1 integrin mutants in CHO cells (Fleming et al., 2011). On the other hand, integrin αVβ3 interacts with rotavirus through a linear sequence in VP7 (Graham et al., 2003; Zarate et al., 2004).

A cellular molecule used by all rotavirus strains is hsc70 (Gutierrez et al., 2010; Guerrero et al., 2002). The interaction between the viral particle and hsc70 is mediated by VP5, through a domain located between amino acids 642 and 659 (Zarate et al., 2003). Since a synthetic peptide corresponding to this region blocks virus infectivity but not cell binding, it has been suggested that the interaction between the virus and hsc70 is at a postattachment step. Apparently the ATPase domain of hsc70 is involved in promoting conformational changes in the viral particle that facilitate virus entry (Perez-Vargas et al., 2006).

Additionally, some of the molecules that participate in attachment and postattachment interactions (gangliosides, integrins α2β1, αVβ3, and hsc70) group together in detergent-resistant membrane domains (lipid rafts), where infectious viral particles are also present during cell infection (Isa et al., 2004). The integrity of these domains is fundamental for viral infection, since their destabilization severely decreases the infectivity of all rotavirus strains tested (Gutierrez et al., 2010; Guerrero et al., 2000) (Fig. 2.2.1A).

Rotaviruses are classical gastrointestinal viruses that infect the mature enterocytes located at the tip of the small intestinal villi. Given the basolateral localization of integrins in these cells, an important question is how rotaviruses reach these cell receptors to enter the cell. One explanation was offered when it was shown that a recombinant VP8 protein of RRV is able to decrease the transepithelial electrical resistance of polarized Madin–Darby canine kidney (MDCK) cells (Nava et al., 2004). Furthermore, this VP8 protein was shown to open the tight junctions, releasing basolateral proteins (integrins αVβ3, β1, and the Na^+-K^+-ATPase) to the apical side of the cells (Nava et al., 2004). In addition, it has recently been found that the tight-junction proteins JAM-A, occludin, and ZO-1 are important for the entry of some rotavirus strains (Torres-Flores et al., 2015).

It remains to be established whether all the described molecules work in concert, or represent alternative routes of entry. However, it is noteworthy that the assays used to block the interaction of rotaviruses with each of these proposed receptors and coreceptors using different approaches, such as proteases, antibodies, peptides, sugar analogues, or siRNAs, only decrease viral infectivity by less than 10-fold, suggesting that either a more relevant entry factor for rotavirus has yet to be found, the virus can use more than one route of entry, or the cellular factors that allow the entry of rotavirus are redundant.

2 VIRUS INTERNALIZATION

Most viruses hijack cellular endocytic pathways to reach the cell's interior. The mechanisms more frequently used by these pathogens to enter the cell are either clathrin-dependent, or caveolae-mediated endocytosis, as well as macropinocytosis (Yamauchi and Helenius, 2013). Also, most endocytic pathways described to date depend on dynamin, a GTPase implicated in several membrane scission events and required during the endocytic process (Praefcke and McMahon, 2004).

It was originally proposed that rotaviruses enter cells via direct penetration at the plasma membrane. However, in recent years the role of the endocytic process in rotavirus cell entry has been carefully reevaluated using pharmacological inhibitors of endocytosis, overexpression of dominant-negative mutant proteins, and RNA interference to knockdown the expression of proteins implicated in different endocytic routes (Table 2.2.2). Using these tools it has been clearly established that rotaviruses enter cells by endocytosis, however, different strains use different endocytic pathways (Gutierrez et al., 2010) (Fig. 2.2.1B). Human rotavirus strains Wa, DS-1, WI69, and animal strains UK, YM, SA11-4S, and nar3 enter through clathrin-mediated endocytosis (Gutierrez et al., 2010; Diaz-Salinas et al., 2013), while RRV uses an atypical endocytic pathway which is clathrin- and caveolin-independent, but depends on dynamin 2, and on the presence of cholesterol (Sanchez-San Martin et al., 2004; Silva-Ayala et al., 2013). The requirement for cholesterol and dynamin is also shared by those rotaviruses that are internalized into MA104 cells by clathrin-dependent endocytosis (Gutierrez et al., 2010) (Fig. 2.2.1B, and Table 2.2.2). In contrast with these observations, it was found that the entry of RRV was not affected by dynasore, a chemical inhibitor of dynamin, in polarized MDCK, BSC-1, and MA104 cells (Abdelhakim et al., 2014; Wolf et al., 2011). Differences in the cells used, and in the methods employed to determine the role of dynamin on rotavirus infectivity might account for these discrepancies. That rotaviruses enter via endocytosis is also supported by the observation that actinin 4 and the small GTPases RhoA and Cdc42, as well as the activator of the latter, CDGAP, which are involved in different types of endocytic processes, have been implicated in

TABLE 2.2.2 Rotavirus Internalization and Vesicular Traffic

	Internalization			Vesicular traffic		
Strain	**Type of endocytosis**	**Experimental approach used[1]**	**References**	**EC[2]**	**Experimental approach used[1]**	**References**
RRV	Clathrin-, caveolin-, and macropinocytosis-independent	Fillipin, nystatin, amiloride, sucrose, chlorpromazine; DN Cav-1 & -3, Eps15; RNAi CHC	Gutierrez et al. (2010), Diaz-Salinas et al. (2013), Sanchez-San Martin et al. (2004)	EE, ME	IFM; DN Rab5, 7, TSG101,VPS4A; LBPA mAb; RNAi EEA1, Rab5, 7, 9, HRS, TSG101, VPS24, 25, and 4A	Silva-Ayala et al. (2013), Wolf et al. (2011), Wolf et al. (2012)
Nar3	Clathrin-dependent	Sucrose, RNAi CHC	Diaz-Salinas et al. (2013)	EE, ME, LE	IFM; DN Rab5, 7, TSG101, VPS4A; RNAi Rab5, 7, 9, HRS, TSG101, VPS24, and 4A	Diaz-Salinas et al. (2014)
SA11	Clathrin-dependent	Sucrose, RNAi CHC.	Diaz-Salinas et al. (2013)	EE	RNAi Rab5, 7, and 9	Diaz-Salinas et al. (2014)
UK	Clathrin-dependent	Amiloride, sucrose; DN Cav-1; RNAi CHC	Gutierrez et al. (2010), Diaz-Salinas et al. (2013)	EE, ME, LE	IFM; DN Rab5, 7, TSG101, VPS4A; LBPA mAb; RNAi EEA1, Rab5, 7, 9, HRS, TSG101, VPS24, and 4A	Diaz-Salinas et al. (2014), Wolf et al. (2012)
DS-1	Clathrin-dependent	Sucrose, RNAi CHC	Diaz-Salinas et al. (2013)	EE, ME, LE	RNAi Rab5, 7 and 9, HRS, TSG101, VPS25, and 4A	Silva-Ayala et al. (2013), Diaz-Salinas et al. (2014)

(*Continued*)

TABLE 2.2.2 Rotavirus Internalization and Vesicular Traffic (*cont.*)

	Internalization			Vesicular traffic		
Strain	Type of endocytosis	Experimental approach used[1]	References	EC[2]	Experimental approach used[1]	References
Wa	Clathrin-dependent	Amiloride, sucrose; DN Cav-1; RNAi CHC	Gutierrez et al. (2010), Diaz-Salinas et al. (2013)	EE, ME, LE	DN TSG101 and VPS4A; LBPA mAb; RNAi Rab5, 7, 9, HRS, TSG101, VPS25, and 4A	Silva-Ayala et al. (2013), Diaz-Salinas et al. (2014)
WI61	Clathrin-dependent	Sucrose, RNAi CHC.	Diaz-Salinas et al. (2013)	EE, ME, LE	RNAi Rab5, 7, and 9	Diaz-Salinas et al. (2014)
TFR-41	Clathrin-dependent	Amiloride, sucrose; DN Cav-1; RNAi CHC	Gutierrez et al. (2010)	NT[3]		
YM	Clathrin-dependent	Sucrose, RNAi CHC	Diaz-Salinas et al. (2013)	EE, LE	RNAi Rab5, 7, and 9	Diaz-Salinas et al. (2014)

[1]*Drugs used in pharmacological approaches:* DN, *over expression of dominant negative mutants of indicated proteins;* Cav, *Caveolin;* CHC, *Clathrin heavy chain;* IFM, *Immunofluorescence microscopy;* RNAi, *RNA interference assays of the indicated proteins;* LBPA MAb, *lysobisophosphatidic acid monoclonal antibody.*
[2]EC, *Endocytic compartment required for the infectivity of the indicated rotavirus strain.* EE, *early endosomes;* ME, *maturing endosomes;* LE, *late endosomes.*
[3]NT, *not tested.*

the entry of all rotavirus tested so far (Silva-Ayala et al., 2013; Diaz-Salinas et al., 2014).

It is interesting to note that interactions of the virus with the receptors characterized so far do not seem to determine the endocytic pathway used, since both NA-resistant and -sensitive strains, as well as rotaviruses that interact with HBGAs, can enter cells using a clathrin-dependent mechanism (Diaz-Salinas et al., 2013). Using reassortant viruses it was recently found that the outer layer protein VP4 determines the endocytic pathway used (Diaz-Salinas et al., 2014). A single amino acid substitution (K187R) in the VP8 protein of RRV was also found to change the pathway of entry of this virus from clathrin-independent to clathrin-dependent (Diaz-Salinas et al., 2014). The RRV variant bearing this mutation, nar3, was found to attach initially to the cell surface through an interaction between VP5 and integrin α2β1, skipping the first interaction with SA (Lopez and Arias, 2004). These findings suggest that the endocytic pathway used by rotaviruses might be dictated by a particular interaction of VP4 with a specific cellular coreceptor not yet characterized.

2.1 The Endosomal Network and the ESCRT Machinery

After cell internalization rotavirus travels along the intracellular vesicle apparatus moving from the cell periphery to the perinuclear space to finally reach the cytoplasm; during this tour the viral particle uncoats. To accomplish this early stage in replication, rotavirus hijacks the endosomal network that is typically used by the cell for the uptake and transport of molecules such as hormones, growth, and membrane factors, and lipids (Marsh and Helenius, 2006). The endosomal network is composed of early endosomes (EEs), maturing endosomes (MEs), late endosomes (LEs), recycling endosomes (REs), and lysosomes. These compartments are characterized by their luminal pH, protein and lipid composition, localization, and structure, and are often defined by the presence of different Rab GTPases on their surface. Rab proteins are a large family of small GTPases that control membrane identity and vesicle budding, uncoating, motility, and fusion through the recruitment of effector proteins (Mercer et al., 2010).

An interesting feature of the endosomal pathway is the formation of characteristic intraluminal vesicles (ILVs); these ILVs are the product of the endosomal sorting complex required for transport (ESCRT) machinery. The ESCRT machinery consists of four complexes, ESCRT-0, -I, -II and -III, plus several accessory components. Besides their participation in the endocytosis of specific cargos, ESCRT complexes play critical roles in receptor downregulation, retroviral budding, and other normal and pathological cellular processes (Stenmark, 2009).

Independently of the nature of the cell surface receptor and the endocytic pathway used for cell internalization (Gutierrez et al., 2010; Diaz-Salinas et al., 2013; Sanchez-San Martin et al., 2004; Silva-Ayala et al., 2013;

Diaz-Salinas et al., 2014) all rotavirus strains tested converge in EEs during entry, since their infectivity depends on the presence of Rab5 and EEA1 (Fig. 2.2.1C) (Silva-Ayala et al., 2013; Diaz-Salinas et al., 2014; Wolf et al., 2012). In addition, rotaviruses depend on functional ESCRT machinery in MA104 and Caco-2 cells, since knocking down the expression of components of each of the four ESCRT complexes by RNA interference reduces virus infectivity (Silva-Ayala et al., 2013) (Fig. 2.2.1C). ILVs play an important role during virus entry since silencing the expression of VPS4A, the ESCRT-associated ATPase involved in membrane fission, or blocking by specific antibodies of the phospholipid lysobisphosphatidic acid, crucial for ILV formation, inhibit viral infectivity (Silva-Ayala et al., 2013) (Table 2.2.2). It is puzzling to understand why the entry of rotaviruses depends on the ESCRT machinery, and what is the role of ILVs in this process. The possibilities to be explored are either that the ILVs activate or attenuate a signaling cascade triggered by the internalized virions, as has been previously observed for other ligand-activated receptor systems, or that ILVs serve as sites where a given coreceptor clusters, which might be important to induce a conformational change in the viral particle necessary to disrupt the endosomal membrane (discussed in (Silva-Ayala et al., 2013)). In contrast with these results, it has been recently described that live-cell imaging of RRV DLPs recoated with fluorophore-tagged VP4 and VP7 enter the cytoplasmic space in a Rab5-independent manner, without reaching EEs (Abdelhakim et al., 2014) (see Chapter 2.1), and the kinetics of RRV entry reported in this system were also different to those previously described (Gutierrez et al., 2010; Silva-Ayala et al., 2013; Wolf et al., 2012) (Table 2.2.2). These contradictory results might be due to the different cell line (BSC-1) used in these studies.

For rotaviruses RRV and SA11 the intracellular traffic comes to an end at MEs, since none of the cellular components required downstream of MEs are required for their productive infection. Based on this observation, these strains have been considered as early-penetrating viruses (Table 2.2.2, Fig. 2.2.1B). In contrast, the infectivity of all rotavirus strains tested other than RRV, including nar3, depends on the expression of Rab7 (Silva-Ayala et al., 2013; Diaz-Salinas et al., 2014), suggesting that these viruses continue their travel through the endosomal network to reach LEs (Diaz-Salinas et al., 2014). Rab7 localizes to both maturing and late endosomes and has been shown to be critical for trafficking through the endocytic pathway (Huotari and Helenius, 2011). Rab7 colocalizes by confocal microscopy with viral particles during virus entry, suggesting the presence of viral particles in LEs (Silva-Ayala et al., 2013; Diaz-Salinas et al., 2014). LEs might provide the optimal environment for the Rab7-dependent strains to enter the cytosol. In this regard, Rab7-dependent rotaviruses behave as late-penetrating viruses. Interestingly, the differential exit from the endosomal network followed by rotaviruses RRV and UK, is also determined by the spike protein VP4 (Diaz-Salinas et al., 2013).

3 M6PR AND CATHEPSINS IN ROTAVIRUS ENTRY

The small GTPase Rab9 is a key component of LEs and orchestrates the transport of mannose-6-phosphate receptors (M6PRs) from LEs to the trans-Golgi network. Rotavirus strains UK, Wa, WI61, nar3, DS-1, and YM, all of which reach LEs, depend on a functional Rab9 to infect the cell, and all of them, with the exception of nar3, also require the activity of the cation-dependent (CD) M6PR. Lysosomal acid hydrolases are delivered from the trans-Golgi network to endosomes by M6PRs, and the recycling of these receptors to the Golgi depends on Rab9 (Braulke and Bonifacino, 2009) (Fig. 2.2.1C). Among the transported hydrolases are cysteine cathepsins, which are important components of lysosomes (Turk et al., 2012). It has been observed that these enzymes are required for the processing of viruses, such as reovirus, Ebola virus, SARS virus, and Nipah virus, during their penetration to the cytoplasmic space (Chandran et al., 2005; Diederich et al., 2012; Ebert et al., 2002; Schornberg et al., 2006; Simmons et al., 2005). Of interest, it was recently shown that the infectivity of rotavirus UK, but not that of RRV or nar3, decreased in MA104 cells treated either with pharmacological inhibitors of cathepsins B and L, or when the expression of cathepsins B, L, or S were silenced by RNA interference (Diaz-Salinas et al., 2014), suggesting that rotavirus UK requires the activity of these hydrolases for cell entry. It remains to be determined whether the proteolytic activity is required directly on the viral particle and/or to process a cellular factor, and whether other rotavirus strains also depend on the activity of cathepsins.

4 PENETRATION AND UNCOATING

During their vesicular transit, viruses are exposed to changes in luminal pH, calcium concentration, membrane components, and lysosomal enzymes, which could induce conformational changes in the viral particle to promote the exit of the viral nucleocapsid into the cytoplasm (Tsai, 2007). Little is known about the mechanism by which rotaviruses exit the endosomal compartment and the changes that viral particles undergo to achieve this task. On the basis of cryo-electron microscopy and crystallography data of the VP5 domain of VP4 (Yoder and Dormitzer, 2006) it has been proposed that to exit the endosomal compartment the spike protein VP4 undergoes an initial conformational change, triggered by an unknown factor, that allows the intraendosomal calcium to diffuse to the cytosol. This decrease in the endosomal calcium concentration then promotes the disassembly of the VP7 trimers, with the subsequent release of the smooth surface layer. In turn, calcium decrease, and possibly the lower pH of this compartment, causes a drastic rearrangement of VP5 to a fold-back conformation that exposes a hydrophobic domain on VP5 and allows its interaction with the endosomal membrane to disrupt it, a process that is followed by the escape of the double-layered virus particles into the cytosol to begin transcription the virus genome (Settembre et al., 2011) (Fig. 2.2.1C). The colocalization of Rab5 and the "fold-back" conformation of RRV VP5 by confocal microscopy strongly suggest that the environment in the endosomes triggers the conformational change that facilitates the penetration of RRV strain into the cytoplasm (Wolf et al., 2011)

Taken together, these observations suggest that the exit of rotavirus particles from the endosomal compartment depends on a combination of factors, such as the interplay between pH and calcium concentration, as well as on cysteine proteases, and possibly other cellular factors. Thus, it is likely that the particular vesicular compartment from which different rotavirus strains penetrate into the cytoplasm provide the specific luminal conditions needed to induce the conformational changes of the viral particle. The outer layer proteins VP5 and VP7 seem to act as environment sensors that detect the conditions present at the endosomal compartments that trigger uncoating and engage the viral particle in membrane penetration. The net result of the processes involving attachment, uncoating, and penetration of the endosomal membrane is the release of the transcriptionally competent double-layered particle into the cytosol. For subsequent steps of rotavirus replication see Chapter 2.3.

5 PERSPECTIVES

Even though our knowledge on rotavirus entry and vesicular traffic has advanced a great deal in the last years, several aspects of these processes remain to be elucidated both in cell culture models as well as in cells that more closely resemble the mature enterocytes, where the virus replicates in a natural infection. The recent development of human intestinal organoids (Spence et al., 2011), and the possibility to use the CRISPR/Cas9 technology (Ran et al., 2013) in these cells, together with assays where structural studies of the surface proteins of the virus and their interaction with cellular receptors/coreceptors can be followed by live cell imaging systems (Sun et al., 2013) coupled with novel super-resolution microscopy techniques (Viswanathan et al., 2015), and the current possibility of carrying out single cell studies, will be important tools to advance this field.

ACKNOWLEDGMENTS

The work in our laboratory was supported by grants 153639 and 221019 from the National Council for Science and Technology (CONACyT) (Mexico), and IG-200114 from DGAPA-UNAM. We acknowledge the excellent technical assistance of Rafaela Espinosa that supported several of these studies. We also would like to apologize to our colleagues whose work was not cited due to space limitations.

REFERENCES

Abdelhakim, A.H., Salgado, E.N., Fu, X., et al., 2014. Structural correlates of rotavirus cell entry. PLoS Pathog. 10, e1004355.

Blanchard, H., Yu, X., Coulson, BS, von Itzstein, M, 2007. Insight into host cell carbohydrate-recognition by human and porcine rotavirus from crystal structures of the virion spike associated carbohydrate-binding domain (VP8*). J. Mol. Biol. 367, 1215–1226.

Bohm, R., Fleming, F.E., Maggioni, A., et al., 2015. Revisiting the role of histo-blood group antigens in rotavirus host-cell invasion. Nat. Commun. 6, 5907.

Braulke, T., Bonifacino, J.S., 2009. Sorting of lysosomal proteins. Biochim. Biophys. Acta 1793, 605–614.

Chandran, K., Sullivan, N.J., Felbor, U., Whelan, S.P., Cunningham, J.M., 2005. Endosomal proteolysis of the Ebola virus glycoprotein is necessary for infection. Science 308, 1643–1645.

Delorme, C., Brussow, H., Sidoti, J., et al., 2001. Glycosphingolipid binding specificities of rotavirus: identification of a sialic acid-binding epitope. J. Virol. 75, 2276–2287.

Diaz-Salinas, M.A., Romero, P., Espinosa, R., Hoshino, Y., Lopez, S., Arias, C.F., 2013. The spike protein VP4 defines the endocytic pathway used by rotavirus to enter MA104 cells. J. Virol. 87, 1658–1663.

Diaz-Salinas, M.A., Silva-Ayala, D., Lopez, S., Arias, C.F., 2014. Rotaviruses reach late endosomes and require the cation-dependent mannose-6-phosphate receptor and the activity of cathepsin proteases to enter the cell. J. Virol. 88, 4389–4402.

Diederich, S., Sauerhering, L., Weis, M., et al., 2012. Activation of the Nipah virus fusion protein in MDCK cells is mediated by cathepsin B within the endosome-recycling compartment. J. Virol. 86, 3736–3745.

Dormitzer, P.R., Sun, Z.Y., Wagner, G., Harrison, S.C., 2002. The rhesus rotavirus VP4 sialic acid binding domain has a galectin fold with a novel carbohydrate binding site. Embo J. 21, 885–897.

Ebert, D.H., Deussing, J., Peters, C., Dermody, T.S., 2002. Cathepsin L and cathepsin B mediate reovirus disassembly in murine fibroblast cells. J. Biol. Chem. 277, 24609–24617.

Fleming, F.E., Graham, K.L., Takada, Y., Coulson, B.S., 2011. Determinants of the specificity of rotavirus interactions with the alpha2beta1 integrin. J. Biol. Chem. 286, 6165–6174.

Graham, K.L., Halasz, P., Tan, Y., et al., 2003. Integrin-using rotaviruses bind alpha2beta1 integrin alpha2 I domain via VP4 DGE sequence and recognize alphaXbeta2 and alphaVbeta3 by using VP7 during cell entry. J. Virol. 77, 9969–9978.

Guerrero, C.A., Zarate, S., Corkidi, G., Lopez, S., Arias, C.F., 2000. Biochemical characterization of rotavirus receptors in MA104 cells. J. Virol. 74, 9362–9371.

Guerrero, C.A., Bouyssounade, D., Zarate, S., et al., 2002. Heat shock cognate protein 70 is involved in rotavirus cell entry. J. Virol. 76, 4096–4102.

Gutierrez, M., Isa, P., Sanchez-San Martin, C., et al., 2010. Different rotavirus strains enter MA104 cells through different endocytic pathways: the role of clathrin-mediated endocytosis. J. Virol. 84, 9161–9169.

Haselhorst, T., Blanchard, H., Frank, M., et al., 2007. STD NMR spectroscopy and molecular modeling investigation of the binding of *N*-acetylneuraminic acid derivatives to rhesus rotavirus VP8* core. Glycobiology 17, 68–81.

Haselhorst, T., Fleming, F.E., Dyason, J.C., et al., 2009. Sialic acid dependence in rotavirus host cell invasion. Nat. Chem. Biol. 5, 91–93.

Hu, L., Crawford, S.E., Czako, R., et al., 2012. Cell attachment protein VP8* of a human rotavirus specifically interacts with A-type histo-blood group antigen. Nature 485, 256–259.

Huang, P., Xia, M., Tan, M., et al., 2012. Spike protein VP8* of human rotavirus recognizes histo-blood group antigens in a type-specific manner. J. Virol. 86, 4833–4843.

Huotari, J., Helenius, A., 2011. Endosome maturation. Embo J. 30, 3481–3500.

Isa, P., Lopez, S., Segovia, L., Arias, C.F., 1997. Functional and structural analysis of the sialic acid-binding domain of rotaviruses. J. Virol. 71, 6749–6756.

Isa, P., Realpe, M., Romero, P., Lopez, S., Arias, C.F., 2004. Rotavirus RRV associates with lipid membrane microdomains during cell entry. Virology 322, 370–381.

Liu, Y., Huang, P., Tan, M., et al., 2012. Rotavirus VP8*: phylogeny, host range, and interaction with histo-blood group antigens. J. Virol. 86, 9899–9910.

Lopez, S., Arias, C.F., 2004. Multistep entry of rotavirus into cells: a Versaillesque dance. Trends Microbiol. 12 (6), 271–278.

Lopez, S., Arias, C.F., 2006. Early steps in rotavirus cell entry. Curr. Top. Microbiol. Immunol. 309, 39–66.

Marsh, M., Helenius, A., 2006. Virus entry: open sesame. Cell 124, 729–740.

Martinez, M.A., Lopez, S., Arias, C.F., Isa, P., 2013. Gangliosides have a functional role during rotavirus cell entry. J. Virol. 87, 1115–1122.

Mercer, J., Schelhaas, M., Helenius, A., 2010. Virus entry by endocytosis. Annu. Rev. Biochem. 79, 803–833.

Monnier, N., Higo-Moriguchi, K., Sun, Z.Y., Prasad, B.V., Taniguchi, K., Dormitzer, P.R., 2006. High-resolution molecular and antigen structure of the VP8* core of a sialic acid-independent human rotavirus strain. J. Virol. 80, 1513–1523.

Nava, P., Lopez, S., Arias, C.F., Islas, S., Gonzalez-Mariscal, L., 2004. The rotavirus surface protein VP8 modulates the gate and fence function of tight junctions in epithelial cells. J. Cell Sci. 117, 5509–5519.

Nordgren, J., Sharma, S., Bucardo, F., et al., 2014. Both lewis and secretor status mediate susceptibility to rotavirus infections in a rotavirus genotype-dependent manner. Clin. Infect. Dis. 59, 1567–1573.

Perez-Vargas, J., Romero, P., Lopez, S., Arias, C.F., 2006. The peptide-binding and ATPase domains of recombinant hsc70 are required to interact with rotavirus and reduce its infectivity. J. Virol. 80, 3322–3331.

Praefcke, G.J., McMahon, H.T., 2004. The dynamin superfamily: universal membrane tubulation and fission molecules? Nat. Rev. Mol. Cell Biol. 5, 133–147.

Ramani, S., Cortes-Penfield, N.W., Hu, L., et al., 2013. The VP8* domain of neonatal rotavirus strain G10P[11] binds to type II precursor glycans. J. Virol. 87, 7255–7264.

Ran, F.A., Hsu, P.D., Lin, C.Y., et al., 2013. Double nicking by RNA-guided CRISPR Cas9 for enhanced genome editing specificity. Cell 154, 1380–1389.

Sanchez-San Martin, C., Lopez, T., Arias, C.F., Lopez, S., 2004. Characterization of rotavirus cell entry. J. Virol. 78, 2310–2318.

Schornberg, K., Matsuyama, S., Kabsch, K., Delos, S., Bouton, A., White, J., 2006. Role of endosomal cathepsins in entry mediated by the Ebola virus glycoprotein. J. Virol. 80, 4174–4178.

Settembre, E.C., Chen, J.Z., Dormitzer, P.R., Grigorieff, N., Harrison, S.C., 2011. Atomic model of an infectious rotavirus particle. Embo J. 30, 408–416.

Silva-Ayala, D., Lopez, T., Gutierrez, M., Perrimon, N., Lopez, S., Arias, C.F., 2013. Genome-wide RNAi screen reveals a role for the ESCRT complex in rotavirus cell entry. Proc. Natl. Acad. Sci. USA 110, 10270–10275.

Simmons, G., Gosalia, D.N., Rennekamp, A.J., Reeves, J.D., Diamond, S.L., Bates, P., 2005. Inhibitors of cathepsin L prevent severe acute respiratory syndrome coronavirus entry. Proc. Natl. Acad. Sci. USA 102, 11876–11881.

Spence, J.R., Mayhew, C.N., Rankin, S.A., et al., 2011. Directed differentiation of human pluripotent stem cells into intestinal tissue in vitro. Nature 470, 105–109.

Stenmark, H., 2009. Rab GTPases as coordinators of vesicle traffic. Nat. Rev. Mol. Cell Biol. 10, 513–525.

Sun, E., He, J., Zhuang, X., 2013. Live cell imaging of viral entry. Curr. Opin. Virol. 3, 34–43.

Torres-Flores, J.M., Silva-Ayala, D., Espinoza, M.A., Lopez, S., Arias, C.F., 2015. The tight junction protein JAM-A functions as coreceptor for rotavirus entry into MA104 cells. Virology 475, 172–178.

Tsai, B., 2007. Penetration of nonenveloped viruses into the cytoplasm. Annu. Rev. Cell Dev. Biol. 23, 23–43.

Turk, V., Stoka, V., Vasiljeva, O., et al., 2012. Cysteine cathepsins: from structure, function and regulation to new frontiers. Biochim. Biophys. Acta 1824, 68–88.

Van Trang, N., Vu, H.T., Le, N.T., Huang, P., Jiang, X., Anh, D.D., 2014. Association between norovirus and rotavirus infection and histo-blood group antigen types in Vietnamese children. J. Clin. Microbiol. 52, 1366–1374.

Venkataram Prasad, B.V., Shanker, S., Hu, L., et al., 2014. Structural basis of glycan interaction in gastroenteric viral pathogens. Curr. Opin. Virol. 7, 119–127.

Viswanathan, S., Williams, M.E., Bloss, E.B., et al., 2015. High-performance probes for light and electron microscopy. Nat. Methods 12, 568–576.

Wolf, M., Vo, P.T., Greenberg, H.B., 2011. Rhesus rotavirus entry into a polarized epithelium is endocytosis dependent and involves sequential VP4 conformational changes. J. Virol. 85, 2492–2503.

Wolf, M., Deal, E.M., Greenberg, H.B., 2012. Rhesus rotavirus trafficking during entry into MA104 cells is restricted to the early endosome compartment. J. Virol. 86, 4009–4013.

Yamauchi, Y., Helenius, A., 2013. Virus entry at a glance. J. Cell Sci. 126, 1289–1295.

Yoder, J.D., Dormitzer, P.R., 2006. Alternative intermolecular contacts underlie the rotavirus VP5* two- to three-fold rearrangement. Embo J 25, 1559–1568.

Yu, X., Coulson, B.S., Fleming, F.E., Dyason, J.C., von Itzstein, M., Blanchard, H., 2011. Novel structural insights into rotavirus recognition of ganglioside glycan receptors. J. Mol. Biol. 413, 929–939.

Zarate, S., Cuadras, M.A., Espinosa, R., et al., 2003. Interaction of rotaviruses with Hsc70 during cell entry is mediated by VP5. J. Virol. 77, 7254–7260.

Zarate, S., Romero, P., Espinosa, R., Arias, C.F., Lopez, S., 2004. VP7 mediates the interaction of rotaviruses with integrin alphavbeta3 through a novel integrin-binding site. J. Virol. 78, 10839–10847.

Chapter 2.3

Rotavirus Replication and Reverse Genetics

A. Navarro*, L. Williamson*, M. Angel*, J.T. Patton*,**
*Laboratory of Infectious Diseases, National Institute of Allergy and Infectious Diseases, National Institutes of Health, Bethesda, MD, United States; **Virginia-Maryland College of Veterinary Medicine, University of Maryland, MD, United States

1 INTRODUCTION

The establishment of efficient reverse genetics (RG) systems opens a window to studies on the biology and pathogenesis of viruses that cannot be accomplished through other means. Efficient complete plasmid-based RG systems have been developed for four members of the *Reoviridae*: mammalian reovirus (MRV), bluetongue virus (BTV), African horse sickness virus (AHSV), and epizootic hemorrhagic disease virus (EHDV). These RG systems all rely on the principle that viral (+)RNAs are "infectious" because viral (+)RNAs function both as templates for protein and genome synthesis in the *Reoviridae* life cycle. Indeed, the MRV, BTV, AHSV, and EHDV RG systems all use the introduction of recombinant viral (+)RNAs produced from T7 transcription vectors into uninfected cells as a mechanism to launch replication of recombinant viruses. Despite extensive efforts, attempts to develop similar plasmid-only RV RG systems have yet to be successful. However, several single-gene replacement systems have been developed that enable reverse engineering of selected RV genome segments. In this chapter, we review the RV replication cycle and consider issues that may be hampering the development of fully recombinant RG systems for these viruses.

2 ROTAVIRUS REPLICATION

2.1 Virus Structure

The RV virion is a nonenveloped icosahedral triple-layered particle (TLP) with a diameter of approximately 100 nm (Li et al., 2009). Enclosed within the virion is a genome consisting of 11 dsRNA segments ranging in size from 0.7 to 3.3 kbp (Estes et al., 1989). The (+) strand of each segment contains a

Viral Gastroenteritis

5′-cap, but lacks a 3′-poly(A) tail. In contrast, the 5′-γ-phosphate of the (−) strand is missing, reflecting a hydrolysis event occurring for an unknown reason (Imai et al., 1983). The smooth outer layer of the virion is formed by the glycoprotein VP7; emanating from it are spikes of the viral attachment protein VP4 (Settembre et al., 2011). Exposure of the virus to trypsin-like proteases cleaves VP4 into the subunits VP5* and VP8*, resulting in conformational changes that enhance virus infectivity (Estes et al., 1981; Arias et al., 1996). The intermediate layer of the virion is composed of elongate twisted columns formed by VP6 trimers (Mathieu et al., 2001), and the inner layer is composed of 12 thin decamers of the core shell protein VP2 (McClain et al., 2010). Each decamer forms one of 12 corners (vertices) of the core and serves as an attachment site for the viral RNA-dependent RNA polymerase (RdRP), VP1, and the RNA-capping enzyme, VP3 (Estrozi et al., 2013). Within the core, the genome is likely organized such that each dsRNA segment is associated with one specific VP1/VP3/VP2 decamer complex (Prasad et al., 1996; Periz et al., 2013).

2.2 Viral Transcription

RV entry is an endosome-dependent process that leads to the loss of the VP4-VP7 outer capsid layer and the release of a double-layered particle (DLP) into the cytosol (Abdelhakim et al., 2014; Arias et al., 2015) (see Chapter 2.2). The DLP acts as transcription machine, with VP1 and VP3 in the core directing synthesis of 11 species of RV (+)RNAs (Cohen, 1977). The 11 species are made at nonequimolar levels in the infected cell (Stacy-Phipps and Patton, 1987). Newly made transcripts extrude through channels at the 5-fold axes of the DLP, passing through both the VP2 and VP6 protein layers (Lawton et al., 1997). The 5′-ends of most (+)RNAs contain cap-1 structures (Imai et al., 1983); these modifications result from the guanyltransferase, N-7-methyltransferase and 2′-O-methyltransferase activities associated with the VP3 RNA capping enzyme (Ogden et al., 2014). Some RV (+)RNAs recovered from infected cells lack 5′-caps, ending instead with 5′-terminal phosphates (Uzri and Greenberg, 2013). Whether capped and uncapped viral transcripts have functional differences is not clear, although presumably the lack of caps would affect recruitment of translational initiation factors and ribosomal subunits to the 5′-end of (+)RNAs. The lack of 5′-caps increases the likelihood that RV (+)RNAs will be recognized by retinoic acid-inducible gene 1 (RIG-I), a intracellular pattern recognition receptor (PRR) that can promote IFN expression and an antiviral response (Reikine et al., 2014). The effect of RIG-I can be countered in RV-infected cells by the action of the viral protein, NSP1, a suspected viral E3 ubiquitin ligase that inhibits IFN expression by targeting key host factors for proteasomal degradation (Morelli et al., 2015).

RV (+)RNAs are generally monocistronic, encoding either a single structural (VP1–VP4, VP6–VP7) or nonstructural protein (NSP1–NSP4) (Estes et al., 2007). The exception is the transcript made from RNA segment 11, which contains an open reading frame (ORF) for NSP5 and a second ORF for the smaller NSP6 (Mattion et al., 1991). Comparison of the 11 segments of RV (+) RNAs shows that the only conserved sequences among them are limited to a few residues at the 5′-(5′-GGC(A/U)n-3′) and 3′-(5′-UGUGACC-3′) termini. Notably, the 5′ and 3′ untranslated regions (UTRs) of RV (+)RNAs contain long stretches of complementary sequences, often extending into the ORF. Based on secondary structure predictions, the complementary sequences are believed to mediate long-range interactions between the ends of viral (+)RNAs, producing 5′–3′ panhandle structures (Chen and Patton, 1998; Tortorici et al., 2003). These predictions also indicate that one or more stable stem-loop structures project from the panhandles (Li et al., 2010). The panhandle and stem-loop structures and conserved 5′ and 3′ terminal sequences likely represent *cis*-acting signals that are critical for replication and packaging of viral RNAs (Biswas et al., 2014; Suzuki, 2014). These highly ordered structural elements of RV (+) RNAs are potentially recognized by several intracellular PRRs, including the host dsRNA sensors melanoma differentiation-associated protein 5 (MDA5), protein kinase R (PKR), and 2′–5′ oligoadenylate synthetase (OAS) (Wu and Chen, 2014). Within infected cells, RV NSP1, VP3, and other proteins likely subvert the antiviral functions of these PRRs (Morelli et al., 2015).

Transcriptionally active DLPs extrude transcripts in a 5′ to 3′ direction (Lawton et al., 1997, 2000), an orientation that may allow interaction of translation initiation factors and ribosomal subunits with the 5′-terminus before synthesis of the (+)RNA is complete. As a result, RV DLPs may promote (+)RNA-polysome formation in a manner that cannot be replicated by full-length folded RNAs containing 5′–3′ panhandles. The extent to which concurrent transcription–translation impacts the efficiency of viral protein synthesis in infected cells is not known, but could have bearing on the poor translatability noted for viral (+)RNAs transfected into cells (Richards, 2012; Richards et al., 2013). Concurrent loading of translation factors and ribosomes onto the 5′-ends of viral (+) RNAs during transcription may prevent the formation of 5′–3′ panhandle structures that are recognized by PRRs and trigger antiviral responses.

2.3 Viroplasm Formation and Function

RV (+)RNAs not only direct protein synthesis in infected cells, but also serve as templates for the synthesis of dsRNA genome segments through a process that is coordinated with the packaging and assembly of progeny cores (Trask et al., 2012a,b). Replication and assembly take place in electron-dense cytoplasmic inclusion bodies, termed viroplasms (Altenburg et al., 1980; Eichwald et al., 2012). Viroplasms are dynamic structures initially appearing in infected cells as small punctate structures that then fuse and grow in size over the course of infection (Eichwald et al., 2004a). Although viroplasms are not membrane enclosed, they are often located adjacent to vesicles formed by the endoplasmic reticulum (ER) (see Chapter 2.5).

Two RV nonstructural proteins, NSP2 and NSP5, are essential for the formation of viroplasms (Fabbretti et al., 1999). NSP2 exists in the infected cell as large doughnut-shaped octamers with deep diagonal grooves that mediate strong sequence-independent binding of single-stranded (ss)RNA (Jayaram et al., 2002; Taraporewala et al., 1999). This activity is an essential component of the helix-destabilizing activity detected for the octamer (Taraporewala and Patton, 2001). The octamer is also an enzyme, displaying phosphatase activities able to hydrolyze the γ-phosphate from NTP and RNA substrates (*vis-à-vis*, NTPase and RTPase activities) (Vasquez-Del Carpio et al., 2006). In contrast, NSP5 forms dimers, which may self-associate into larger decameric structures (Martin et al., 2011). NSP5 also binds RNA (Vende et al., 2002), although the activity is weak compared to that of NSP2. NSP5 undergoes two types of posttranslational modification: O-linked glycosylation and phosphorylation; the latter is believed to result from casein kinase activities (Eichwald et al., 2004a,b; Gonzalez and Burrone, 1991). NSP2 stimulates the hyperphosphorylation of NSP5 and such forms of NSP5 preferentially accumulate in viroplasms. Binding sites for NSP5 and ssRNA are located close together on the NSP2 octamer (Jiang et al., 2006), potentially allowing competition between the two to regulate the function of the octamer and the formation of viroplasms. NSP2 and NSP5 have affinity for core structural proteins, and these interactions are probably key to recruiting VP1, VP2, VP3, and VP6 to viroplasms (Arnoldi et al., 2007; Berois et al., 2003; Kattoura et al., 1994; Viskovska et al., 2014). Moreover, interactions with NSP2 and NSP5 may be important for regulating the intrinsic self-assembly tendencies of the structural proteins, thereby suppressing the formation of empty capsids. Of interest, all the nonstructural and structural proteins that accumulate in viroplasms, with the exception of VP6, have affinity for ssRNA (Patton, 1995; Patton et al., 2006).

Transcriptionally active DLPs are proposed to serve as nucleation sites for the formation of viroplasms (Silvestri et al., 2004). In this scenario, transcripts produced by DLPs soon after infection are incorporated into nearby polysomes that direct viral protein synthesis. Due to the affinity of the viroplasm proteins for RV (+)RNAs and for each other, many will interact with and concentrate around transcriptionally active DLPs, ultimately embedding the particle within small protein-rich inclusions. With the continued synthesis of transcripts by DLPs and viral proteins by nearby polysomes, the inclusions grow into larger electron-dense structures that capture large amounts of (+)RNAs through the activity of their RNA-binding proteins. An important aspect of this scenario is the assumption that (+)RNAs used as templates for dsRNA synthesis may originate from transcriptionally active DLPs contained within the viroplasm. Consistent with this hypothesis are data from BrUTP-labeling studies indicating that transcriptionally active DLPs are present within viroplasms and that newly made transcripts accumulate in viroplasms (Silvestri et al., 2004). On a related note, studies of other members of the *Reoviridae* (ie, MRV, rice dwarf virus) indicate that transcriptionally active particles generated from their incoming virions are recruited into viroplasmic structures (Miller et al., 2010; Wei et al., 2006).

Transfection of RV-infected cells with gene 5- and gene 7-specific small interfering (si)RNAs knock down expression of NSP1 and NSP3, respectively, without reducing levels of genome replication and virion assembly (Silvestri et al., 2004; Montero et al., 2006). Not only do such data demonstrate that NSP1 and NSP3 have nonessential roles in virus replication, but they also reveal the presence of two pools of (+)RNAs in infected cells. One pool, which directs protein synthesis, is sensitive to siRNA-induced degradation, while the other pool, which directs dsRNA synthesis, is not subject to degradation. Moreover, these data indicate that the pool of (+)RNAs that directs protein synthesis is not the source of (+)RNAs used for dsRNA synthesis. These data also suggest that (+)RNAs contained in viroplasms—the site of genome replication—are not targeted by siRNA-induced degradation pathways, consistent with the concept that viroplasms operate as "safe houses" walled off from host antiviral effectors. Given that (+)RNAs present in the cytosol [ie, polysomal (+)RNAs] are sensitive to siRNA-induced degradation, it can be argued that the more likely source of (+)RNAs used for dsRNA synthesis are transcriptionally active DLPs contained within viroplasms. If this argument holds, then viral (+)RNAs transfected into infected cells may be poorly integrated into viroplasms (Silvestri et al., 2004).

2.4 Genome Replication and Particle Assembly

Steps in RV genome replication and particle assembly are poorly understood, but some insight has been gleaned from the study of (1) viral replication intermediates (Boudreaux et al., 2015; Gallegos and Patton, 1989), (2) the structure and function of the core proteins VP1, VP2, and VP3, and the viroplasm-building blocks NSP2 and NSP5 (Boudreaux et al., 2013; Liu et al., 1988; McDonald and Patton, 2011), and (3) the location and activities of *cis*-acting replication signals in template RNAs (Chen et al., 2001; Patton et al., 1996). Fig. 2.3.1 summarizes a model for replication and assembly that we believe is most consistent with available data. The model proposes that the 11 different segments of RV (+)RNAs accumulate in the viroplasm, folded in such a way that their 5′–3′ panhandle structures are present. The viral RdRP, VP1, is recruited to the 3′-end of (+)RNA panhandles due to the affinity of its template entry tunnel for the conserved 3′-terminal sequence UGUGACC (Lu et al., 2008). Due to the affinity of VP3 for the 5′-end of the RNA and possibly for VP1, VP1–VP3–(+)RNA complexes are formed (Patton and Chen, 1999). The model predicts that these complexes undergo assortment via RNA–RNA interactions mediated by conserved sequences and/or structural elements associated with 5′–3′ panhandles (Li et al., 2010). VP2 decameric plates assemble around the assorted complexes due to the protein's affinity for VP1 and VP3, and ssRNA. Contacts made between the assembled VP2 core shell and VP1 trigger conformational

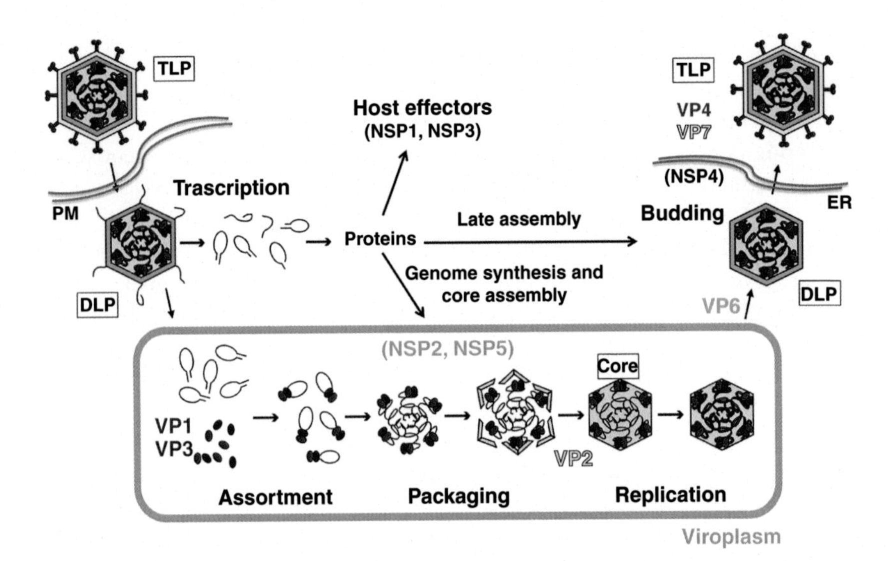

FIGURE 2.3.1 **Model for the rotavirus (RV) replication cycle.** The VP4–VP7 outer capsid layer is lost from the RV triple-layered particle *(TLP)* during virus entry, yielding a double-layered particle *(DLP)*. Within the cytosol, the DLP directs the synthesis of 11 types of capped (+)RNAs. These are translated, producing proteins that influence host cell processes, direct viroplasm formation, and support genome replication and virus assembly. In the viroplasm, viral (+)RNAs interact with the viral polymerase, VP1, and RNA-capping enzyme, VP3, and then undergo assortment. The core shell, consisting of VP2, encloses VP1-VP3-(+)RNA complexes, that, after forming core-like structures, support dsRNA synthesis. Newly formed cores interact with VP6 to form DLPs; these migrate to the endoplasmic reticulum *(ER)* due to the affinity of the DLP VP6 protein for ER-transmembrane protein, NSP4. During budding into the ER, assembly of the VP4–VP7 outer capsid layer around the DLP results in the formation of the TLP. PM, plasma membrane; ER, endoplasmic reticulum.

changes in the priming loop of the polymerase, resulting in the initiation of (−) strand RNA synthesis (Lu et al., 2008; Gridley and Patton, 2014). This polymerase activity yields the formation of progeny cores that contain 11 dsRNA genome segments. Interaction of cores with VP6 concentrated around the periphery of viroplasms leads to DLP formation (Chen and Ramig, 1993). Newly formed DLPs can amplify the replication cycle by serving as sources of secondary transcripts or migrate to the ER due the affinity of the VP6 capsid component for the ER-transmembrane protein NSP4. The VP4–VP7 outer capsid of the TLP is acquired during or after budding of DLPs into the ER (Trask and Dormitzer, 2006). TLPs characteristically accumulate within ER vesicles during the course of infection and are then released by cell lysis. To some extent, TLPs may be released early in infection by exocytosis from the apical surface of polarized cells (Chwetzoff and Trugnan, 2006).

This model indicates that genome replication and core assembly are coordinated via the VP2-dependent polymerase activity of VP1, a mechanism assuring that dsRNA synthesis does not take place unless progeny cores are available to capture the RNA product. The model also predicts that (+)RNA assortment is precise and occurs prior to the synthesis of dsRNA by newly formed cores. Indeed, the 11 dsRNA genome segments are made in equimolar levels in infected cells (Patton, 1990), and based on CsCl gradient ultracentrifugation studies, incompletely packaged virus particles do not accumulate. There can be little doubt that the nonstructural proteins NSP2 and NSP5 play significant roles in genome replication and core assembly, as these proteins are major components of RV replication intermediates (Boudreaux et al., 2015; Gallegos and Patton, 1989; Campagna et al., 2005; Vascotto et al., 2004). In particular, NSP2 may prepare viral (+)RNAs for assortment and packaging into cores and facilitate (–) strand initiation (Jayaram et al., 2002).

3 REVERSE GENETICS SYSTEMS

3.1 Fully Recombinant Systems

Mechanisms of genome replication are similar for the *Reoviridae*, with viral (+)RNAs serving as templates for protein synthesis and dsRNA synthesis. This implies that for all members of the family, development of fully recombinant RG systems should be straightforward, simply requiring the introduction of a complete cohort of synthetic viral (+)RNAs into permissive cells. Indeed, this has been the pathway for development of RG systems for MRV, BTV, ASHV, and EHDV (Table 2.3.1). In the case of MRV, viral (+)RNAs are made in vivo by transfection of T7 transcription vectors into cells producing T7 RNA polymerase. For BTV, ASHV, and EHDV, viral (+)RNAs are made in vitro from T7 transcription vectors and then cotransfected into cells. Both approaches have yielded tractable RG systems that allow the study of virus biology and pathogenesis through mutagenesis (Boehme et al., 2013; Sandekian and Lemay, 2015), the directed formation of reassortant viruses (Celma et al., 2014; Kobayashi et al., 2010), the production of recombinant viruses that express nonviral proteins (Demidenko et al., 2013; Kobayashi et al., 2007; Shaw et al., 2013), and the exploration of next generation vaccines (Celma et al., 2013; Matsuo et al., 2011). Despite extensive efforts, a fully recombinant RV RG system has yet to be developed, even though efforts have closely patterned after those used for MRV, BTV, AHSV, and EHDV.

3.1.1 MRV

The plasmid-based MRV RG system has advanced through a number of improvements since its initial description in 2007. The original system required transfection of 10 plasmids—each a T7 transcription vector with a select MRV cDNA—into cells infected with a recombinant vaccinia virus encoding T7 RNA polymerase (rDIs-T7pol) (Kobayashi et al., 2007). The vectors were constructed by placing cDNAs of MRV genome segments immediately downstream of a T7 promoter and upstream from a *cis*-acting hepatitis delta virus (HDV) ribozyme (Rz). These T7(MRV-cDNA)Rz cassettes allowed synthesis of viral (+)RNA

TABLE 2.3.1 Fully Recombinant Reverse Genetics Systems for the Reoviridae

Virus	Source of T7 RNA polymerase	Approach for generating recombinant virus	Key references
Mammalian reovirus	Vaccinia virus rDIs-T7pol	Transfection of T7 transcription vectors (10-plasmid)	Kobayashi et al. (2007)
	BHK cell line expressing T7 RNA polymerase	Transfection of T7 transcription vectors (4-plasmid)	Kobayashi et al. (2010), Boehme et al. (2011)
	Plasmid encoding T7 RNA polymerase	Transfection of T7 transcription vectors (10+1-plasmid)	Komoto et al. (2014)
Bluetongue virus	Cell-free T7 transcription system	Single transfection of synthetic viral (+)RNAs	Boyce et al. (2008)
	Cell-free T7 transcription system	Double transfection of synthetic viral (+)RNAs	Matsuo and Roy (2009)
	Cell-free T7 transcription system	Sequential transfection of protein expression vectors and synthetic viral (+)RNAs	Matsuo and Roy (2013)
	Cell-free T7 transcription system	Transfection of reconstituted cores containing *in vitro* replicated viral (+)RNAs	Lourenco and Roy (2011)
African horse sickness virus	Cell-free T7 transcription system	Double transfection of synthetic viral (+)RNAs	Matsuo et al. (2010), Kaname et al. (2013)
Epizootic hemorrhagic disease virus	Cell-free T7 transcription system	Sequential transfection of protein expression vectors and synthetic viral (+)RNAs	Yang et al. (2015)

transcripts with authentic 5′ and 3′-termini. The efficiency of the system was subsequently improved by constructing multicistronic plasmids that contained two to four T7(MRV-cDNA)Rz cassettes, reducing the number of plasmids that needed to be transfected, from 10 to 4 (Kobayashi et al., 2010). A particularly important advance was the discovery that transfection of the MRV RG plasmids into BHK cell lines expressing T7 RNA polymerase (BHK-T7) also yielded recombinant virus, thus negating the need for continued use of rDIs-T7pol (Kobayashi et al., 2010). Interestingly, this discovery also indicated that viral transcripts lacking 5′ caps supported the formation of recombinant MRV. More recent studies show that recombinant MRV can be made in any of several cell lines by cotransfecting the MRV RG plasmid set with a plasmid encoding T7 RNA polymerase; this is an important finding for generating candidate vaccine viruses in qualified cell substrates (Komoto et al., 2014).

3.1.2 Orbiviruses: BTV, AHDV, and EHDV

Fully recombinant RG systems have been developed for three orbiviruses: BTV, AHDV, and EHDV. Key to their development was the initial discovery that viral (+)RNAs synthesized by BTV core particles in vitro are infectious when transfected into permissive cells (Boyce and Roy, 2007). Boyce et al. (2008) advanced this finding by creating a set of 10 plasmids, each containing a cDNA of one of the 10 BTV genome segments. The cDNAs were positioned with an upstream T7 promoter and a unique downstream restriction enzyme site, allowing linearized plasmids to be used as templates for the synthesis of BTV (+)RNAs in vitro. Transfection of capped T7 transcripts made from the plasmids resulted in the recovery of infectious virus, creating the first fully recombinant BTV RG system. Subsequent studies revealed that the efficiency of the system was significantly improved if cells were transfected twice with BTV transcripts, spaced 18-h apart (Matsuo and Roy, 2009). These studies went on to show that the principal function of the early transfected transcripts was to promote synthesis of BTV proteins required to support RNA synthesis, the assembly of subcore structures, and the formation of viroplasms. Based on this information, the components of the early transfection were adjusted to include only those viral (+)RNAs necessary for expression of the required BTV proteins (Matsuo and Roy, 2013). Further studies showed that the early transfected transcripts could be replaced with expression vectors encoding the required BTV proteins (Matsuo and Roy, 2013). Transcription of these vectors is driven by host RNA polymerase II and generates capped (+)RNAs that lack the terminal sequences found on BTV RNAs, stressing their importance as templates for protein synthesis and not for viral genome replication. Consistently, the results showed that viral (+)RNAs introduced into cells with the second transfection served as the source of templates for genome replication. It is interesting that there have been no reports indicating that recombinant BTV can be generated using T7 expression vectors as a source of viral (+)RNAs that serve as templates for genome replication.

Lourenco and Roy (2011) developed an alternative system for producing recombinant BTV, relying on the in vitro reconstitution of infectious BTV core particles containing synthetic viral (+)RNAs. Key to the success of the system was the sequential incubation of BTV (+)RNAs with wheat-germ expressed subcore and core proteins, in a process emulating virus-assembly steps in the infected cell. Clearly, the core-reconstitution RG system is a more technically challenging approach for generating recombinant virus than are infectious RNA-based RG systems. However, in situations where RNA transfections do not yield recombinant viruses, perhaps due to the poor translation or toxicity of viral (+)RNAs (Richards et al., 2013; Trask et al., 2012a; Wentzel, 2014), core-reconstitution systems may provide a useful alternative route for creating RG systems.

The approaches taken to generate RG systems for AHSV and EHDV mirror those used for BTV. In the case of AHSV, initial experiments showed that double transfection of viral (+)RNAs into permissive cells was more efficient than a single transfection in generating recombinant virus (Matsuo et al., 2010). The AHSV system was subsequently modified, relying on the early transfection of expression vectors instead of (+)RNAs to drive the synthesis of viral proteins (Kaname et al., 2013). As with the BTV system, only six vectors were required, those encoding viral proteins associated with genome replication, subcore assembly, and the formation of viroplasms. The EHDV RG system is like the modified AHSV system: the early transfection comprised of pCI-expression vectors for six viral proteins and the later transfection comprised of 10 full-length viral (+)RNAs synthesized in vitro by T7 transcription (Yang et al., 2015).

3.1.3 RV—Attempts at Developing Fully Recombinant RG Systems

Many laboratories have attempted to establish fully recombinant RV RG systems, extrapolating from methods used to develop the MRV and orbivirus systems. The best documented efforts are described in the publication by Richards et al. (2013) and in the PhD theses of Richards (2012), Mlera (2013), and Wentzel (2014). These scientists attempted to develop RV infectious RNA systems using viral (+)RNAs produced in vitro by the polymerase activity of purified DLPs or by T7 transcription of viral cDNAs cloned into plasmids. Their work included transfecting viral transcripts of a variety of virus strains (eg, SA11, Wa, DS-1, RF) into a range of cell substrates (eg, MA104, COS-7, BSR, Caco-2, Vero, 293T), under optimized transfection conditions (Richards, 2012; Wentzel, 2014; Mlera, 2013). In their attempts to establish an infectious RNA synthesis, they also evaluated the double-transfection approach that, for orbiviruses, significantly enhanced the production of recombinant virus. In their trials, the early transfection step was varied to include the complete cohort of viral (+)RNAs or only those viral (+)RNAs encoding proteins essential for genome replication, DLP assembly, and viroplasm formation (Richards, 2012; Richards et al., 2013). In other trials, the early transfected RNAs were replaced with viral protein expression vectors, driven either by cellular RNA polymerase II (CMV promoter) or by T7 RNA polymerase expressed by recombinant fowlpox virus (FPV-T7) (Richards, 2012; Wentzel, 2014). None of the efforts resulted in the recovery of recombinant RV. A common observation has been that transfection of viral (+)RNAs results in extensive cellular cytopathic effects (CPE) (Richards, 2012; Richards et al., 2013; Wentzel, 2014; Mlera, 2013). The basis for the CPE is not clear, but may result from the recognition of the RNAs by PRRs, inducing innate immune pathways that lead to apoptosis and necrosis.

The failure of transfected viral (+)RNAs to be efficiently translated may limit the expression of viral antagonists that are needed to suppress innate immune responses (Richards, 2012; Richards et al., 2013; Wentzel, 2014).

Although the efforts are poorly documented, several laboratories have attempted to develop a plasmid-based RV RG system. In our laboratory, we cloned full-length cDNAs of the 11 genome segments of SA11-4F RV into the same T7 transcription vectors used in the MRV RG system. These 11 RV plasmids were transfected into COS-7 cells previously infected with rDIs-T7pol. No recombinant RVs were recovered, even though in parallel transfection experiments using the 10-plasmid MRV RG system, we were able to recover recombinant MRV. We also reduced the number of RV plasmids transfected into cells by creating multicistronic bacmids; this allowed us to transfect cells with 4 plasmids instead of 11. This modified approach also failed to generate recombinant RV. In analyzing cells transfected with SA11-4F transcription vectors and infected with rDIs-T7pol, we detected little or no expression of RV proteins by either immunofluorescence or western blot assay. Although this raises questions as to whether the expression of RV proteins was adequate to launch infection, we noted that protein expression in transfected cells was similarly difficult to detect with the MRV RG system. In our experiments we also observed that cells transfected with RV plasmids and infected with rDIs-T7pol displayed considerable cytopathic effect (CPE) within 24 h, even though after transfection, the cells were maintained in media containing fetal bovine serum. The basis for the CPE is not fully understood, but could reflect the impact of rDIs-T7pol, transfection reagent, or viral transcripts or proteins produced by the expression vector. Further damage to the cells was noted when the cells were maintained in serum-free medium containing trypsin, conditions anticipated to promote spread of recombinant viruses.

In our attempts to develop a plasmid-based RV RG system, we found that viral protein expression was enhanced by adding extra G residues to the 5′-end of RV cDNAs inserted into T7 transcription vectors. Indeed, the level of protein expression by the modified vectors progressively increased as the number of G residues increased from one to three. Due to the importance of precise terminal sequences on RV RNAs for genome replication, it is doubtful that the modified vectors can be used in the development of functional plasmid-based RG systems. However, such modified vectors may prove useful for expressing viral proteins at levels adequate to launch the replication of viral (+)RNAs with authentic termini. Computer modeling indicated that the secondary structures of viral (+)RNAs with and without extra G residues are the same, containing 5′–3′ panhandles and associated stem loop structures. The fact that viral (+)RNAs with extra G residues were efficiently translated indicates that the highly ordered structural elements in the RNAs did not trigger innate immune responses that suppressed protein expression. From these results, we suggest that the failure of viral (+)RNAs made by unmodified T7 transcription vectors to drive efficient protein expression is not due to an innate immune response, but rather reflects an issue with the recruitment of the (+)RNAs into polysomes. In considering why authentic viral (+)RNAs are not efficiently translated, it is worth noting that RVs encoding defective NSP1 proteins induce very high levels of IFN expression in infected cells, yet these mutant viruses replicate quite well (Barro and Patton, 2005). It was also observed that transfection of BTV (+)RNAs into uninfected cells gives rise to strong innate immune responses (Wentzel, 2014), yet such methods are used to generate recombinant BTV (Boyce et al., 2008). Thus, although RV infection and RNA transfections may activate antiviral pathways, this does not provide sufficient information to explain the poor translation of RV (+)RNAs or the failure create a RV RG system.

3.2 Rotavirus Single-Gene Replacement Systems

The failure to establish a fully tractable recombinant RV RG system has led to efforts to generate single-gene RG systems that allow replacement of a single genome segment in a helper virus with an RNA of recombinant origin. Four such single-gene systems have been described, using four different helper viruses and a variety of selection methods to recover recombinant viruses from a background of parental viruses (Table 2.3.2).

3.2.1 Gene 4 (VP4)

The first single-gene replacement system described for RV was used to introduce a recombinant SA11 gene 4 (VP4) RNA into human KU RV (Komoto et al., 2006). The recombinant virus [recKU(SA11g4)] was produced by infecting COS-7 cells with rDIs-T7pol, which were then transfected with a plasmid containing a T7(SA11g4-cDNA)Rz cassette and infected with KU helper virus. To recover recombinant virus, the COS-7 cell lysates were passaged serially on MA104 cells in the presence of neutralizing antibody against KU VP4, followed by triple plaque purification. The basis for producing the recKU(SA11g4) reassortant came from earlier studies showing that KU(SA11g4) reassortant viruses grew better than the KU parent in cell culture. In a later study, the gene 4 RG system was used to replace one of the antibody neutralization epitopes in SA11 VP4 with the corresponding epitope in DS1 VP4 (Komoto et al., 2008). Interestingly, there have been no reports of gene 9 (VP7) single-gene replacement systems, despite the availability of numerous VP7 neutralizing antibody that could be used in selecting recombinant viruses.

3.2.2 Gene 8 (NSP2)

The most extensively used RV single-gene replacement system allows substitution of the mutant gene 8 (NSP2) RNA in the SA11 temperature-sensitive (*ts*) mutant *tsE* with recombinant RNA (Trask et al., 2010). A critical element of this system is that *tsE* NSP2 functions at low temperature (30°C), but not at elevated temperature (39°C). To generate SA11 viruses using this system, COS-7 cells were infected with rDIs-T7pol and then transfected with a plasmid containing

TABLE 2.3.2 Single-Gene Reverse Genetics Systems for the Reoviridae

Virus	Target segment (protein)	Helper virus (species)	Source of RNA polymerase	Source of RNA transcripts	Selection and recovery of recombinant virus	Key references
Rotavirus	g4 (VP4)	KU (human)	Vaccinia virus rDIs-T7pol	Transfection of g4 transcription vector	Passage with neutralizing αVP4 antibody	Komoto et al. (2006), Komoto et al. (2008)
	g7 (NSP3)	RF (bovine)	Vaccinia virus rDIs-T7pol	Transfection of g7 transcription vector	Serial passage at high MOI	Troupin et al. (2010)
	g8 (NSP2)	SAll*tsE* (simian)	Vaccinia virus rDIs-T7pol	Transfection of g8 transcription vector	Passage in g8 shRNA-expressing cell line at non-permissive temperature	Trask et al. (2010), Navarro et al. (2013)
	g10 (NSP4)	CHLY (bovine)	Plasmid encoding T7 RNA polymerase	Transfection of g10 transcription vector	Passage in g10 shRNA-expressing cell line	Yang et al. (2012)
Mammalian reovirus	S1 (σl)	T3D	Endogenous RNA pol II	Integrated S1 expression cassette	Binding of virion His-tagged s1 to cell-surface αHis-tag antibody	van den Wollenberg et al. (2008)
	S2 (σ2)	ST3 (ST2)	Cell-free T7 transcription system	Transfection of synthetic transcripts	Plaque assay	Roner and Joklik (2001)

a T7(g8-cDNA)Rz cassette that directs the synthesis of gene 8 transcripts that encode non-*ts* NSP2. Afterward, the cells were infected with the *tsE* helper virus and maintained at 30°C. To select for the recovery of viruses containing recombinant gene 8 RNA, COS-7 cell lysates were passaged serially ~3-times at elevated temperature (39°C) in MA104 cells expressing an siRNA (MA104/g8D) that targets the *tsE* gene 8 RNA of the helper virus. Recombinant viruses were then isolated by triple plaque purification. The gene 8 RG system has been used to produce more than 20 recombinant viruses, including those that contain chimeric gene 8 RNAs (Trask et al., 2010), gene 8 RNAs with viral sequence duplications, and gene 8 RNAs with insertions of nonrotaviral sequences (Navarro et al., 2013) (Table 2.3.3). Notably, genetically stable recombinant RVs were produced that contained gene 8 RNAs with sequences for the FLAG tag, Hepatitis C Virus E2 epitope (HCV2), and Cricket Paralysis Virus internal ribosome entry site (CrPV IRES) inserted into the 3′ UTR (Navarro et al., 2013). The ability to engineer foreign sequences into the RV genome opens for the door for the potential development of these viruses as expression vectors, potentially leading to the generation of next generation RV vaccines that can induce protection against other enteric pathogens. Given that mutant strains of RVs have been described with *ts* lesions that map to nearly all eleven genome segments, it seems likely that *ts*-based RG systems could be developed for other RV genes (Criglar et al., 2011).

3.2.3 Gene 7 (NSP3)

RVs with rearranged genome segments, resulting from head-to-tail sequence duplications, can be generated in cell culture by serial passage of virus at high multiplication of infection (MOI) (Hundley et al., 1985; Patton et al., 2001; Arnold et al., 2012). Viruses with rearranged genome segments have also been detected in immunocompromised children (Hundley et al., 1987; Gault et al., 2001). Analysis of viral progeny produced in cells coinfected with wild type viruses and viruses with rearranged segments have indicated that there can be a selective advantage favoring packaging of the mutant RNAs (Troupin et al., 2011). Troupin et al. (2010) used this concept as the basis for developing a single-gene replacement system that allowed the introduction of recombinant gene 7 (NSP3) RNA into bovine RF RV. These investigators engineered two plasmids containing T7(g7-cDNA)Rz cassettes, one directing transcription of the rearranged gene 7 RNA of the human M1 strain and, the second, directing transcription of the rearranged gene 7 RNA segment of the human M3 strain (Gault et al., 2001). To generate recombinant viruses, the plasmids were transfected into COS-7 cells, which were later infected with rDIs-T7pol and RF viruses. Recombinant viruses contained in COS-7 infected cell lysates were amplified by serial passage 18-times at high MOI in MA104 cells. RF viruses containing M1 and M3 rearranged gene 7 (NSP3) RNAs were purified from 18th passage lysates by triple plaque purification. In addition to viruses containing rearranged gene 7 segments, viruses have been isolated that have rearranged

TABLE 2.3.3 Recombinant SA11 Rotaviruses Made From the Temperature-Sensitive Mutant *tsE* (Navarro et al., 2013)

Recombinant virus	Features of the gene 8 RNA
rSA11/g8SA11	Represents an siRNA-resistant SA11 RNA
rSA11/g8DC1	Represents a DCI/SA11 chimera
rSA11/g8DS-1	Represents a DS-1/SA11 chimera
rSA11/g8OSU	Represents an OSU/SA11 chimera
rSA11/g8UK	Represents a UK/SA11 chimera
rSA11/g8-3′PacI	Contains a Pac I site at the NSP2 ORF/3′-UTR junction
rSA11/g8-3′25D	3′-UTR contains a 25-nt sequence duplication
rSA11/g8-3′50D	3′-UTR contains a 50-nt sequence duplication
rSA11/g8-3′100D	3′-UTR contains a 100-nt sequence duplication
rSA11/g8-3′200D	3′-UTR contains a 200-nt sequence duplication
rSA11/g8-3′Flag	3′-UTR contains a nonexpressing Flag sequence
rSA11/g8-3′HCV2	3′-UTR contains a nonexpressing HCV2-epitope sequence
rSA11/g8-3′Flag-25D	3′-UTR contains a nonexpressing Flag sequence and 25-nt g8 sequence duplication
rSA11/g8-3′Flag-50D	3′-UTR contains a nonexpressing Flag sequence and 50-nt g8 sequence duplication
rSA11/g8-3′HCV2-25D	3′-UTR contains a nonexpressing HCV2-epitope sequence and 25-nt g8 sequence duplication
rSA11/g8-3′HCV2-50D	3′-UTR contains a nonexpressing HCV2-epitope sequence and 50-nt g8 sequence duplication
rSA11/g8-3′CrPV	3′-UTR contains a CrPV IRES
rSA11/g8-3′CrPV-Flag	3′-UTR contains a FLAG expression sequence downstream of a CrPV IRES
rSA11/g8-3′PP7	3′-UTR contains a recognition element for PP7 protein
rSA11/g8-3′(2x)PP7	3′-UTR contains two tandem recognition elements for PP7 protein
rSA11/g8-3′MS2	3′-UTR contains a recognition element for MS2 protein
rSA11/g8-3′MS2-PP7	3′-UTR contains recognition elements for MS2 and PP7 proteins
rSA11/g8-3′TAR	3′-UTR contains HIV TAR element

gene 5 (NSP1), 6 (VP6), 8 (NSP2), 10 (NSP4), and 11 (NSP5/NSP6) segments (Patton et al., 2001; Arnold et al., 2012; Desselberger, 1996). Based on experience with the gene 7 reverse genetics system, it may be possible to use the packaging preference of rearranged RNAs to generate RG systems for these other segments as well. A limitation of this packaging preference is that it cannot be

used to produce viruses with wild type genome segments; instead, it can only be used to generate viruses with rearranged segments.

3.2.4 Gene 10 (NSP4)

A single-gene replacement system has been described that allows modification of gene 10 (NSP4) of bovine CHLY RV (Yang et al., 2012) (see http://www.chinaagrisci.com/CN/abstract/abstract17260.shtml) In this system, MA104 cells were infected with CHLY helper virus and then transfected with two plasmids, one expressing T7 RNA polymerase and the other expressing gene 10 transcripts via a T7 promoter. Selection for CHLY virus containing a recombinant gene 10 RNA was accomplished by serial passage of transfected-infected cell lysates on MA104 cells expressing an siRNA that targeted the CHLY gene 10 RNA. Recombinant viruses were recovered by triple plaque purification, and the gene 10 sequences of the viruses were confirmed by sequencing. The gene 10 RG system is technically important because it indicates that the formation of recombinant viruses can be driven by T7 RNA polymerase expressed by a plasmid instead of by a vaccinia virus (rDIs-T7pol). Also, the gene 10 system suggests that viral transcripts need not be capped by vaccinia virus capping enzymes to serve as templates for the production of recombinant viruses. However, because the gene 10 transcripts may have been capped in *trans* by the VP3 capping enzymes of the CHLY helper virus, the results do not exclude the possibility that capped gene 10 transcripts are essential for recovery of recombinant virus.

3.3 Summary and Future Directions

The absence of a fully tractable recombinant RG system for RVs is a significant impediment to the pursuit of studies required to gain a clearer understanding of RV biology. The reasons for the lack of progress in creating such a system are not known, and indeed may be multifactorial, but a principal issue seems to be the inability of synthetic RV (+)RNAs to undergo/elicit efficient translation when introduced into uninfected cells. This holds true for in vitro synthesized (+)RNAs transfected into cells and for (+)RNAs made within cells by T7 transcription vectors. In contrast, when BTV (+)RNAs are transfected into cells, they are translated into readily detectable levels of viral proteins. In fact, viroplasm-like structures can be detected in many cells transfected with BTV (+)RNAs, quite unlike what is seen when cells are transfected with RV (+)RNAs. The 5′-terminal sequences of RV, BTV, and MRV (+)RNAs are different (GGC, GUU, and GCU, respectively); it is possible that these differences influence the ability of translation initiation factors in uninfected cells to engage the (+)RNAs and facilitate their introduction into polysomes. Only when the ends of RV (+)RNAs are modified, by the addition of a few extra 5′-terminal G-residues, does expression of RV proteins in transfected cells become easily detected. Unfortunately, the extra G residues create template (+)RNAs that are not effectively replicated in RG systems.

In the RV replication cycle, the outer capsid protein layer of virions is lost, producing DLPs that express low levels of viral (+)RNA at early times of infection. Translation of the (+)RNAs produces proteins that promote the formation of viroplasms, alter the host translational apparatus, and inhibit the development of an antiviral state. The synthesis of viral proteins during early times of infection when intracellular levels of viral (+)RNAs are low probably allows the virus to gain control of intracellular processes before viral RNAs accumulate to levels that are readily sensed by host PRRs and induce IFN and ISG expression. Given the large amount of RV (+)RNA introduced into cells as part of most RG experiments, it is probably unavoidable that the RNAs will be sensed by PRRs, triggering innate immune pathways that if not rapidly counteracted by viral antagonists will induce an antiviral state. The failure of transfected viral (+)RNAs to undergo efficient translation creates a situation where there is not only a lack of viral proteins needed to support genome replication and virus assembly, but also a lack of viral proteins necessary to control and redirect intracellular processes to establish an environment favorable for virus replication.

It is perplexing that transfection of synthetic (+)RNAs does not launch productive infection, while transfection of DLPs does. This contrast suggests that the DLP itself, or the process by which the particle produces transcripts, represents a variable that contributes importantly to initiation of infection. Since DLPs extrude nascent (+)RNAs in a 5′ to 3′ direction during transcription, the translational machinery may have an opportunity to recruit the 5′-end of the growing transcript into polysomes before the RNA is fully elongated. Such cotranscription-translation could prevent viral (+)RNAs from folding into structures that are poorly translated and/or highly immunogenic. It is also possible that DLPs localize to sites in the cytosol that favor translation of its (+)RNA products, or indeed, physically engage host factors that effect the structure of (+)RNAs or influence the loading of translation initiation factors onto the 5′-end of (+)RNAs. Clearly, studies are needed that focus on DLP transcription in the early stages of infection, and how this process connects to effective translation of viral (+)RNAs. Not only would these experiments provide important information of the RV biology, but could be key to the development of RV plasmid only-based RG systems. Given the importance of RG systems for probing mechanisms of virus virulence and pathogenesis, and for rational vaccine design, the continued pursuit of a fully tractable recombinant RV RG system is warranted.

ACKNOWLEDGMENTS

The unrecognized efforts of Anne Kalbach, Zenobia Taraporewala, Shane Trask, Aitor Navarro, and Lauren Williamson to develop RV RG systems are greatly appreciated. This work was supported in part by the Intramural Research Program of the National Institute of Allergy and Infectious Diseases, National Institutes of Health.

NOTE ADDED IN PROOF

Recently, a reverse genetics system in which viral (+)RNAs are made in vivo by transfection of T7 transcription vectors into cells producing T7 RNA polymerase has also been established for bluetongue virus.

[Pretorius, J.M., Huismans, H., Theron, J., 2015. Establishment of an entirely plasmid-based reverse genetics system for Bluetongue virus. Virology 486, 71–77.]

REFERENCES

Abdelhakim, A.H., Salgado, E.N., Fu, X., et al., 2014. Structural correlates of rotavirus cell entry. PLoS Pathog. 10 (9), e1004355.

Altenburg, B.C., Graham, D.Y., Estes, M.K., 1980. Ultrastructural study of rotavirus replication in cultured cells. J. Gen. Virol. 46 (1), 75–85.

Arias, C.F., Romero, P., Alvarez, V., Lopez, S., 1996. Trypsin activation pathway of rotavirus infectivity. J. Virol. 70 (9), 5832–5839.

Arias, C.F., Silva-Ayala, D., Lopez, S., 2015. Rotavirus entry: a deep journey into the cell with several exits. J. Virol. 89 (2), 890–893.

Arnold, M.M., Brownback, C.S., Taraporewala, Z.F., Patton, J.T., 2012. Rotavirus variant replicates efficiently although encoding an aberrant NSP3 that fails to induce nuclear localization of poly(A)-binding protein. J. Gen. Virol. 93 (Pt 7), 1483–1494.

Arnoldi, F., Campagna, M., Eichwald, C., Desselberger, U., Burrone, O.R., 2007. Interaction of rotavirus polymerase VP1 with nonstructural protein NSP5 is stronger than that with NSP2. J. Virol. 81 (5), 2128–2137.

Barro, M., Patton, J.T., 2005. Rotavirus nonstructural protein 1 subverts innate immune. response by inducing degradation of IFN regulatory factor 3. Proc. Natl. Acad. Sci. USA. 102 (11), 4114–4119.

Berois, M., Sapin, C., Erk, I., Poncet, D., Cohen, J., 2003. Rotavirus nonstructural protein NSP5 interacts with major core protein VP2. J. Virol. 77 (3), 1757–1763.

Biswas, S., Li, W., Manktelow, E., et al., 2014. Physicochemical analysis of rotavirus segment 11 supports a "modified panhandle" structure and not the predicted alternative tRNA-like structure (TRLS). Arch. Virol. 159 (2), 235–248.

Boehme, K.W., Ikizler, M., Kobayashi, T., Dermody, T.S., 2011. Reverse genetics for mammalian reovirus. Methods 55 (2), 109–113.

Boehme, K.W., Hammer, K., Tollefson, W.C., Konopka-Anstadt, J.L., Kobayashi, T., Dermody, T.S., 2013. Nonstructural protein sigma1s mediates reovirus-induced cell cycle arrest and apoptosis. J. Virol. 87 (23), 12967–12979.

Boudreaux, C.E., Vile, D.C., Gilmore, B.L., Tanner, J.R., Kelly, D.F., McDonald, S.M., 2013. Rotavirus core shell subdomains involved in polymerase encapsidation into virus-like particles. J. Gen. Virol. 94 (Pt 8), 1818–1826.

Boudreaux, C.E., Kelly, D.F., McDonald, S.M., 2015. Electron microscopic analysis of rotavirus assembly-replication intermediates. Virology 477, 32–41.

Boyce, M., Roy, P., 2007. Recovery of infectious bluetongue virus from RNA. J. Virol. 81 (5), 2179–2186.

Boyce, M., Celma, C.C., Roy, P., 2008. Development of reverse genetics systems for bluetongue virus: recovery of infectious virus from synthetic RNA transcripts. J. Virol. 82 (17), 8339–8348.

Campagna, M., Eichwald, C., Vascotto, F., Burrone, O.R., 2005. RNA interference of rotavirus. segment 11 mRNA reveals the essential role of NSP5 in the virus replicative cycle. J. Gen. Virol. 86 (Pt 5), 1481–1487.

Celma, C.C., Boyce, M., van Rijn, P.A., et al., 2013. Rapid generation of replication-deficient monovalent and multivalent vaccines for bluetongue virus: protection against virulent virus challenge in cattle and sheep. J. Virol. 87 (17), 9856–9864.

Celma, C.C., Bhattacharya, B., Eschbaumer, M., Wernike, K., Beer, M., Roy, P., 2014. Pathogenicity study in sheep using reverse-genetics-based reassortant bluetongue viruses. Vet. Microbiol. 174 (1–2), 139–147.

Chen, D., Patton, J.T., 1998. Rotavirus RNA replication requires a single-stranded 3' end for efficient minus-strand synthesis. J. Virol. 72 (9), 7387–7396.

Chen, D., Ramig, R.F., 1993. Rescue of infectivity by sequential in vitro transcapsidation of rotavirus core particles with inner capsid and outer capsid proteins. Virology 194 (2), 743–751.

Chen, D., Barros, M., Spencer, E., Patton, J.T., 2001. Features of the 3'-consensus sequence of rotavirus mRNAs critical to minus strand synthesis. Virology 282 (2), 221–229.

Chwetzoff, S., Trugnan, G., 2006. Rotavirus assembly: an alternative model that utilizes an atypical trafficking pathway. Curr. Top. Microbiol. Immunol. 309, 245–261.

Cohen, J., 1977. Ribonucleic acid polymerase activity associated with purified calf rotavirus. J. Gen. Virol. 36 (3), 395–402.

Criglar, J., Greenberg, H.B., Estes, M.K., Ramig, R.F., 2011. Reconciliation of rotavirus temperature-sensitive mutant collections and assignment of reassortment groups D, J, and K to genome segments. J. Virol. 85 (10), 5048–5060.

Demidenko, A.A., Blattman, J.N., Blattman, N.N., Greenberg, P.D., Nibert, M.L., 2013. Engineering recombinant reoviruses with tandem repeats and a tetravirus 2A-like element for exogenous polypeptide expression. Proc. Natl. Acad. Sci. USA 110 (20), E1867–E1876.

Desselberger, U., 1996. Genome rearrangements of rotaviruses. Adv. Virus Res. 46, 69–95.

Eichwald, C., Rodriguez, J.F., Burrone, O.R., 2004a. Characterization of rotavirus NSP2/NSP5 interactions and the dynamics of viroplasm formation. J. Gen. Virol. 85 (Pt 3), 625–634.

Eichwald, C., Jacob, G., Muszynski, B., Allende, J.E., Burrone, O.R., 2004b. Uncoupling substrate. and activation functions of rotavirus NSP5: phosphorylation of Ser-67 by casein kinase 1 is essential for hyperphosphorylation. Proc. Natl. Acad. Sci. USA 101 (46), 16304–16309.

Eichwald, C., Arnoldi, F., Laimbacher, A.S., et al., 2012. Rotavirus viroplasm fusion and perinuclear localization are dynamic processes requiring stabilized microtubules. PLoS One 7 (10), e47947.

Estes, M.K., Graham, D.Y., Mason, B.B., 1981. Proteolytic enhancement of rotavirus infectivity: molecular mechanisms. J. Virol. 39 (3), 879–888.

Estes, M.K., Cohen, J., 1989. Rotavirus gene structure function. Microbiol. Rev. 53 (4), 410–449.

Estes, M.K., Kapikian, A.Z., 2007. Rotaviruses. In: Knipe, D., Griffin, D., Lamb, R., Martin, M., Roizman, B., Strauss, S. (Eds.), Fields Virology. fifth ed. Lippincott Williams and Wilkins, Philadelphia, pp. 1917–1974.

Estrozi, L.F., Settembre, E.C., Goret, G., et al., 2013. Location of the dsRNA-dependent polymerase, VP1, in rotavirus particles. J. Mol. Biol. 425 (1), 124–132.

Fabbretti, E., Afrikanova, I., Vascotto, F., Burrone, O.R., 1999. Two non-structural rotavirus proteins, NSP2 and NSP5, form viroplasm-like structures in vivo. J. Gen. Virol. 80 (Pt 2), 333–339.

Gallegos, C.O., Patton, J.T., 1989. Characterization of rotavirus replication intermediates: a model for the assembly of single-shelled particles. Virology 172 (2), 616–627.

Gault, E., Schnepf, N., Poncet, D., Servant, A., Teran, S., Garbarg-Chenon, A., 2001. A human rotavirus with rearranged genes 7 and 11 encodes a modified NSP3 protein and suggests an additional mechanism for gene rearrangement. J. Virol. 75 (16), 7305–7314.

Gonzalez, S.A., Burrone, O.R., 1991. Rotavirus NS26 is modified by addition of single O-linked residues of N-acetylglucosamine. Virology 182 (1), 8–16.

Gridley, C.L., Patton, J.T., 2014. Regulation of rotavirus polymerase activity by inner capsid proteins. Curr. Opin. Virol. 9, 31–38.

Hundley, F., Biryahwaho, B., Gow, M., Desselberger, U., 1985. Genome rearrangements of bovine rotavirus after serial passage at high multiplicity of infection. Virology 143 (1), 88–103.

Hundley, F., McIntyre, M., Clark, B., Beards, G., Wood, D., Chrystie, I., Desselberger, U., 1987. Heterogeneity of genome rearrangements in rotaviruses isolated from a chronically infected immunodeficient child. J. Virol. 61 (11), 3365–3372.

Imai, M., Akatani, K., Ikegami, N., Furuichi, Y., 1983. Capped and conserved terminal structures in human rotavirus genome double-stranded RNA segments. J. Virol. 47 (1), 125–136.

Jayaram, H., Taraporewala, Z., Patton, J.T., Prasad, B.V., 2002. Rotavirus protein involved in genome replication and packaging exhibits a HIT-like fold. Nature 417 (6886), 311–315.

Jiang, X., Jayaram, H., Kumar, M., Ludtke, S.J., Estes, M.K., Prasad, B.V., 2006. Cryoelectron microscopy structures of rotavirus NSP2-NSP5 and NSP2-RNA complexes: implications for genome replication. J. Virol. 80 (21), 10829–10835.

Kaname, Y., Celma, C.C., Kanai, Y., Roy, P., 2013. Recovery of African horse sickness virus from synthetic RNA. J. Gen. Virol. 94 (Pt 10), 2259–2265.

Kattoura, M.D., Chen, X., Patton, J.T., 1994. The rotavirus RNA-binding protein NS35 (NSP2) forms 10S multimers and interacts with the viral RNA polymerase. Virology 202 (2), 803–813.

Kobayashi, T., Antar, A.A., Boehme, K.W., et al., 2007. A plasmid-based reverse genetics system for animal double-stranded RNA viruses. Cell Host Microbe 1 (2), 147–157.

Kobayashi, T., Ooms, L.S., Ikizler, M., Chappell, J.D., Dermody, T.S., 2010. An improved reverse genetics system for mammalian orthoreoviruses. Virology 398 (2), 194–200.

Komoto, S., Sasaki, J., Taniguchi, K., 2006. Reverse genetics system for introduction of site-specific mutations into the double-stranded RNA genome of infectious rotavirus. Proc. Natl. Acad. Sci. USA 103 (12), 4646–4651.

Komoto, S., Kugita, M., Sasaki, J., Taniguchi, K., 2008. Generation of recombinant rotavirus with an antigenic mosaic of cross-reactive neutralization epitopes on VP4. J. Virol. 82 (13), 6753–6757.

Komoto, S., Kawagishi, T., Kobayashi, T., et al., 2014. A plasmid-based reverse genetics system for mammalian orthoreoviruses driven by a plasmid-encoded T7 RNA polymerase. J. Virol. Methods 196, 36–39.

Lawton, J.A., Estes, M.K., Prasad, B.V., 1997. Three-dimensional visualization of mRNA release from actively transcribing rotavirus particles. Nat. Struct. Biol. 4 (2), 118–121.

Lawton, J.A., Estes, M.K., Prasad, B.V., 2000. Mechanism of genome transcription in segmented dsRNA viruses. Adv. Virus Res. 55, 185–229.

Li, Z., Baker, M.L., Jiang, W., Estes, M.K., Prasad, B.V., 2009. Rotavirus architecture at subnanometer resolution. J. Virol. 83 (4), 1754–1766.

Li, W., Manktelow, E., von Kirchbach, J.C., Gog, J.R., Desselberger, U., Lever, A.M., 2010. Genomic analysis of codon, sequence and structural conservation with selective biochemical-structure mapping reveals highly conserved and dynamic structures in rotavirus RNAs with potential cis-acting functions. Nucleic Acids Res. 38 (21), 7718–7735.

Liu, M., Offit, P.A., Estes, M.K., 1988. Identification of the simian rotavirus SA11 genome segment 3 product. Virology 163 (1), 26–32.

Lourenco, S., Roy, P., 2011. In vitro reconstitution of bluetongue virus infectious cores. Proc. Natl. Acad. Sci. USA 108 (33), 13746–13751.

Lu, X., McDonald, S.M., Tortorici, M.A., et al., 2008. Mechanism for coordinated RNA packaging and genome replication by rotavirus polymerase VP1. Structure 16 (11), 1678–1688.

Martin, D., Ouldali, M., Menetrey, J., Poncet, D., 2011. Structural organisation of the rotavirus nonstructural protein NSP5. J. Mol. Biol. 413 (1), 209–221.

Mathieu, M., Petitpas, I., Navaza, J., et al., 2001. Atomic structure of the major capsid protein of rotavirus: implications for the architecture of the virion. EMBO J. 20 (7), 1485–1497.

Matsuo, E., Roy, P., 2009. Bluetongue virus VP6 acts early in the replication cycle and can form the basis of chimeric virus formation. J. Virol. 83 (17), 8842–8848.

Matsuo, E., Roy, P., 2013. Minimum requirements for bluetongue virus primary replication in vivo. J. Virol. 87 (2), 882–889.

Matsuo, E., Celma, C.C., Roy, P., 2010. A reverse genetics system of African horse sickness virus reveals existence of primary replication. FEBS Lett. 584 (15), 3386–3391.

Matsuo, E., Celma, C.C., Boyce, M., et al., 2011. Generation of replication-defective virus-based vaccines that confer full protection in sheep against virulent bluetongue virus challenge. J. Virol. 85 (19), 10213–10221.

Mattion, N.M., Mitchell, D.B., Both, G.W., Estes, M.K., 1991. Expression of rotavirus proteins encoded by alternative open reading frames of genome segment 11. Virology 181 (1), 295–304.

McClain, B., Settembre, E., Temple, B.R., Bellamy, A.R., Harrison, S.C., 2010. X-ray crystal structure of the rotavirus inner capsid particle at 3 8 A resolution. J. Mol. Biol. 397 (2), 587–599.

McDonald, S.M., Patton, J.T., 2011. Rotavirus VP2 core shell regions critical for viral polymerase activation. J. Virol. 85 (7), 3095–3105.

Miller, C.L., Arnold, M.M., Broering, T.J., Hastings, C.E., Nibert, M.L., 2010. Localization of mammalian orthoreovirus proteins to cytoplasmic factory-like structures via nonoverlapping regions of microNS. J. Virol. 84 (2), 867–882.

Mlera, L., 2013. Preparatory investigations for developing a transcript-based rotavirus reverse genetics system. PhD Thesis, North-West University, Potchefstroom, South Africa.

Montero, H., Arias, C.F., Lopez, S., 2006. Rotavirus nonstructural protein NSP3 is not required for viral protein synthesis. J. Virol. 80 (18), 9031–9038.

Morelli, M., Ogden, K.M., Patton, J.T., 2015. Silencing the alarms: Innate immune antagonism by rotavirus NSP1 and VP3. Virology 479–480, 75–84.

Navarro, A., Trask, S.D., Patton, J.T., 2013. Generation of genetically stable recombinant rotaviruses containing novel genome rearrangements and heterologous sequences by reverse genetics. J. Virol. 87 (11), 6211–6220.

Ogden, K.M., Snyder, M.J., Dennis, A.F., Patton, J.T., 2014. Predicted structure and domain organization of rotavirus capping enzyme and innate immune antagonist VP3. J. Virol. 88 (16), 9072–9085.

Patton, J.T., 1990. Evidence for equimolar synthesis of double-strand RNA and minus-strand RNA in rotavirus-infected cells. Virus Res. 17 (3), 199–208.

Patton, J.T., 1995. Structure and function of the rotavirus RNA-binding proteins. J. Gen. Virol. 76 (Pt 11), 2633–2644.

Patton, J.T., Chen, D., 1999. RNA-binding and capping activities of proteins in rotavirus open cores. J. Virol. 73 (2), 1382–1391.

Patton, J.T., Wentz, M., Xiaobo, J., Ramig, R.F., 1996. cis-Acting signals that promote genome replication in rotavirus mRNA. J. Virol. 70 (6), 3961–3971.

Patton, J.T., Taraporewala, Z., Chen, D., Chizhikov, V., Jones, M., Elhelu, A., Collins, M., Kearney, K., Wagner, M., Hoshino, Y., Gouvea, V., 2001. Effect of intragenic rearrangement and changes in the 3' consensus sequence on NSP1 expression and rotavirus replication. J. Virol. 75 (5), 2076–2086.

Patton, J.T., Silvestri, L.S., Tortorici, M.A., Vasquez-Del Carpio, R., Taraporewala, Z.F., 2006. Rotavirus genome replication and morphogenesis: role of the viroplasm. Curr. Top Microbiol. Immunol. 309, 169–187.

Periz, J., Celma, C., Jing, B., Pinkney, J.N., Roy, P., Kapanidis, A.N., 2013. Rotavirus mRNAs are. released by transcript-specific channels in the double-layered viral capsid. Proc. Natl. Acad. Sci. USA 110 (29), 12042–12047.

Prasad, B.V., Rothnagel, R., Zeng, C.Q., et al., 1996. Visualization of ordered genomic RNA and localization of transcriptional complexes in rotavirus. Nature 382 (6590), 471–473.

Reikine, S., Nguyen, J.B., Modis, Y., 2014. Pattern recognition and signaling mechanisms of RIG-I and MDA5. Front. Immunol. 5, 342.

Richards, J.E., 2012. Engineering a helper virus-free reverse genetics system for rotavirus. PhD Thesis, University of Cambridge.

Richards, J.E., Desselberger, U., Lever, A.M., 2013. Experimental pathways towards developing a rotavirus reverse genetics system: synthetic full length rotavirus ssRNAs are neither infectious nor translated in permissive cells. PLoS One 8 (9), e74328.

Roner, M.R., Joklik, W.K., 2001. Reovirus reverse genetics: Incorporation of the CAT gene into the reovirus genome. Proc. Natl. Acad. Sci. USA 98 (14), 8036–8041.

Sandekian, V., Lemay, G., 2015. A single amino acid substitution in the mRNA capping enzyme lambda2 of a mammalian orthoreovirus mutant increases interferon sensitivity. Virology 483, 229–235.

Settembre, E.C., Chen, J.Z., Dormitzer, P.R., Grigorieff, N., Harrison, S.C., 2011. Atomic model of an infectious rotavirus particle. EMBO J. 30 (2), 408–416.

Shaw, A.E., Ratinier, M., Nunes, S.F., et al., 2013. Reassortment between two serologically unrelated bluetongue virus strains is flexible and can involve any genome segment. J. Virol. 87 (1), 543–557.

Silvestri, L.S., Taraporewala, Z.F., Patton, J.T., 2004. Rotavirus replication: plus-sense templates for double-stranded RNA synthesis are made in viroplasms. J. Virol. 78 (14), 7763–7774.

Stacy-Phipps, S., Patton, J.T., 1987. Synthesis of plus- and minus-strand RNA in rotavirus-infected cells. J. Virol. 61 (11), 3479–3484.

Suzuki, Y., 2014. A possible packaging signal in the rotavirus genome. Genes Genet. Syst. 89 (2), 81–86.

Taraporewala, Z.F., Patton, J.T., 2001. Identification and characterization of the helix-destabilizing activity of rotavirus nonstructural protein NSP2. J. Virol. 75 (10), 4519–4527.

Taraporewala, Z., Chen, D., Patton, J.T., 1999. Multimers formed by the rotavirus nonstructural protein NSP2 bind to RNA and have nucleoside triphosphatase activity. J. Virol. 73 (12), 9934–9943.

Tortorici, M.A., Broering, T.J., Nibert, M.L., Patton, J.T., 2003. Template recognition and formation of initiation complexes by the replicase of a segmented double-stranded RNA virus. J. Biol. Chem. 278 (35), 32673–32682.

Trask, S.D., Dormitzer, P.R., 2006. Assembly of highly infectious rotavirus particles. recoated with recombinant outer capsid proteins. J. Virol. 80 (22), 11293–11304.

Trask, S.D., Taraporewala, Z.F., Boehme, K.W., Dermody, T.S., Patton, J.T., 2010. Dual selection mechanisms drive efficient single-gene reverse genetics for rotavirus. Proc. Natl. Acad. Sci. USA 107 (43), 18652–18657.

Trask, S.D., McDonald, S.M., Patton, J.T., 2012a. Structural insights into the coupling of virion assembly and rotavirus replication. Nat. Rev. Microbiol. 10 (3), 165–177.

Trask, S.D., Ogden, K.M., Patton, J.T., 2012b. Interactions among capsid proteins orchestrate rotavirus particle functions. Curr. Opin. Virol. 2 (4), 373–379.

Troupin, C., Dehee, A., Schnuriger, A., Vende, P., Poncet, D., Garbarg-Chenon, A., 2010. Rearranged genomic RNA segments offer a new approach to the reverse genetics of rotaviruses. J. Virol. 84 (13), 6711–6719.

Troupin, C., Schnuriger, A., Duponchel, S., Deback, C., Schnepf, N., Dehee, A., Garbarg-Chenon, A., 2011. Rotavirus rearranged genomic RNA segments are preferentially packaged into viruses despite not conferring selective growth advantage to viruses. PLoS One 6 (5), e20080.

Uzri, D., Greenberg, H.B., 2013. Characterization of rotavirus RNAs that activate innate immune signaling through the RIG-I-like receptors. PLoS One 8 (7), e69825.

van den Wollenberg, D.J., van den Hengel, S.K., Dautzenberg, I.J., Cramer, S.J., Kranenburg, O., Hoeben, R.C., 2008. A strategy for genetic modification of the spike-encoding segment of human reovirus T3D for reovirus targeting. Gene Ther. 15 (24), 1567–1578.

Vascotto, F., Campagna, M., Visintin, M., Cattaneo, A., Burrone, O.R., 2004. Effects of intrabodies specific for rotavirus NSP5 during the virus replicative cycle. J. Gen. Virol. 85 (Pt 11), 3285–3290.

Vasquez-Del Carpio, R., Gonzalez-Nilo, F.D., Riadi, G., Taraporewala, Z.F., Patton, J.T., 2006. Histidine triad-like motif of the rotavirus NSP2 octamer mediates both RTPase and NTPase activities. J. Mol. Biol. 362 (3), 539–554.

Vende, P., Taraporewala, Z.F., Patton, J.T., 2002. RNA-binding activity of the rotavirus phosphoprotein NSP5 includes affinity for double-stranded RNA. J. Virol. 76 (10), 5291–5299.

Viskovska, M., Anish, R., Hu, L., et al., 2014. Probing the sites of interactions of rotaviral proteins involved in replication. J. Virol. 88 (21), 12866–12881.

Wei, T., Shimizu, T., Hagiwara, K., et al., 2006. Pns12 protein of Rice dwarf virus is essential for formation of viroplasms and nucleation of viral-assembly complexes. J. Gen. Virol. 87 (Pt 2), 429–438.

Wentzel, J.F., 2014. Investigating the importance of co-expressed rotavirus proteins in the development of a selection-free rotavirus reverse genetics system. PhD Thesis, North-West University, Potchefstroom, South Africa.

Wu, J., Chen, Z.J., 2014. Innate immune sensing and signaling of cytosolic nucleic acids. Annu. Rev. Immunol. 32, 461–488.

Yang, S.H., He, H.B., Yang, H.J., Chen, F.Y., Gao, Y.D., Zhong, J.F., 2012. Rescue and identification of the recombinant bovine rotavirus with mutational NSP4 gene. China Agric. Sci. 45 (19), 4102–4108.

Yang, T., Zhang, J., Xu, Q., et al., 2015. Development of a reverse genetics system for epizootic hemorrhagic disease virus and evaluation of novel strains containing duplicative gene rearrangements. J. Gen. Virol. 96 (9), 2714–2720.

Chapter 2.4

Pleiotropic Properties of Rotavirus Nonstructural Protein 4 (NSP4) and Their Effects on Viral Replication and Pathogenesis

N.P. Sastri*, S.E. Crawford*, M.K. Estes*,**

**Department of Molecular Virology and Microbiology, Baylor College of Medicine, Houston, TX, United States; **Department of Medicine, Baylor College of Medicine, Houston, TX, United States*

1 INTRODUCTION

Rotaviruses (RV) remain major causes of life-threatening diarrheal disease in children under 5 years of age, resulting in nearly half a million deaths annually. Although RV vaccines are available and successful in many developed countries, vaccine efficacy and delivery are suboptimal in developing countries where they are needed most. RV-induced disease is life-threatening compared to disease caused by most other enteric microbes, emphasizing the need to more fully understand the pathophysiology of RV gastroenteritis. RV infection, replication and pathogenesis are carried out by the viral encoded six structural and six nonstructural proteins and the viral genome. One intriguing protein encoded in the viral genome is nonstructural protein 4 (NSP4). RV NSP4 has been the focus of much attention in our laboratory since it was first described as a nonstructural glycoprotein with an uncleaved signal sequence and a glycosylation pattern that indicated it does not traffic to the Golgi apparatus (Arias et al., 1982; Both et al., 1983; Ericson et al., 1983). At that time, few viral glycoproteins with uncleaved signal sequences were known. More importantly, the existence of a nonstructural glycoprotein, which contained only high mannose glycans that are trimmed and produced by a nonenveloped virus was certainly intriguing.

Broadly speaking, NSP4 is involved in at least three different functions: pathogenesis, replication, and morphogenesis. This chapter summarizes our current understanding of how NSP4 regulates these processes and new information about how it affects host cell physiologic pathways to enhance viral replication and cause disease. NSP4's functions in multiple critical pathways are consistent with it being a master regulator of RV replication. NSP4's central role in virus replication is further highlighted by the consequences of knockdown of NSP4, which affects viroplasm formation that is central to viral replication, modulates viral transcription, inhibits genome packaging into particles and reduces infectious particle assembly (Silvestri et al., 2005). The role of NSP4 in these events critical to replication may explain the inability to obtain mutants of the NSP4 gene using methods that have successfully selected mutants of the other 10 RV genes (Vende et al., 2013; Criglar et al., 2014). This chapter focuses on NSP4s from Group A rotaviruses (RVA) and updates information reviewed in recent years (Ruiz et al., 2000; Estes et al., 2001; Morris and Estes, 2001; Ball et al., 2005; Greenberg and Estes, 2009; Hagbom et al., 2012; Hu et al., 2012).

2 CLASSIFICATION AND PHYLOGENY OF NSP4

The epidemiology and evolutionary patterns of RVs are being documented by studies of viruses from humans and a variety of animal species. RVs are broadly classified based on the serologic reactivity and genetic variability of the middle capsid structural protein VP6 into eight groups/species (RVA-RVH). RVA viruses are the most extensively studied and they are further classified based on antigenic and sequence differences of the two outer capsid proteins VP7 and VP4 into G (Glycosylated) and P (Protease-sensitive) types, respectively. Due to the unavailability of serological reagents for these neutralization antigens, and the increasing ease and availability of sequencing, including whole genome sequencing, each genomic segment of RVA is now classified based on sequencing data. A Rotavirus Classification Working Group (RCWG) established a system to identify and differentiate each genome segment according to particular cut-off points of nucleotide sequence identities. Each RV genome is designated by listing the segments in the order of VP7-VP4-VP6-VP1-VP2-VP3-NSP1-NSP2-NSP3-NSP4-NSP5/6 with their genotypes being represented by Gx-P[x]-Ix-Rx-Cx-Mx-Ax-Nx-Tx-Ex-Hx (Matthijnssens et al., 2008a,b, 2011) as discussed more thoroughly in Chapter 2.10.

Here, we review the current NSP4 genotype (called E for enterotoxin) classification and phylogenetic analysis. The initial classification of NSP4 into genotypes was based on amino acid clusters, and six groups (A–F) were identified (Horie et al., 1997; Ciarlet et al., 2000; Mori et al., 2002). The current method of classification based on nucleotide sequence has characterized 19 E genotypes, which supersede those earlier groupings of NSP4 protein sequences (Table 2.4.1).

The results of analyses at both the nucleotide and amino acid levels are in complete concordance with each other but use of nucleotide sequences makes it clearer to assign different genotypes without overlaps as accepted by the

TABLE 2.4.1 Summary of Properties of Different Enterotoxin Genotypes

Genotype/ reference strain	#AA	Early group	Diarrhea in mice (strain)	References
E1/Human-tc/ USA/Wa/1974/ G1P1A[8]	175	B	Yes (Wa, ST3, 116E)	Sastri et al. (2011)
E2/Human-tc/ USA/DS-1/1976/ G2P1B[8]	175	A	Yes (SA11, BRV, CHLY, Hg18, I321, NCDV, S2, 1040, N136, N138, 2KD-851, 99-D/214)	Ball et al. (1996), Zhang et al. (1998), Seo et al. (2008), Chen et al. (2011), Sastri et al. (2011)
E3/Human-tc/ JPN/AU-1/1982/ G3P3[9]	175	C	Yes (RRV)	Sastri et al. (2011)
E4/Pigeon-tc/ JPN/PO-13/1983/ G18P[17]	169	E	Yes (PO-13, TY-1)	Mori et al. (2002)
E5/Rabbit-wt/USA/ Alabama/2000/ G3P[14]	175	A		
E6/Human-wt/ BGD/N26/2000/ G12P[6]	175			
E7/Mouse-wt/ JPN/EW/1999/ G16P[16]	175	D	Yes (EW, EHP) No (EC)	Horie et al. (1997); Tsugawa et al. (2014), Sastri et al. (2011)
E8/Cow-lab/GBR/ PP-1/1976/G3P[7]	175	B		
E9/Pig-wt/THA/ CMPP034/2000/ G2P[27]	175			
E10/Chicken-wt/ JPN/Ch-1/2001/ G19P[17]	168	F	Yes (Ch-1)	Mori et al. (2002)
E11/Turkey-tc/ IRL/Ty-3/1979/ G7P[17]	169	E	Yes (Ty-3)	Mori et al. (2002)
E12/Guanaco-wt/ARG/ Chubut/1999/ G8P[14]	175			

(*Continued*)

TABLE 2.4.1 Summary of Properties of Different Enterotoxin Genotypes (*cont.*)

Genotype/ reference strain	#AA	Early group	Diarrhea in mice (strain)	References
E13/Human-tc/ KEN/B10/1987/ G3P[2]	175			
E14/Horse-tc/ GBR/L338/1991/ G13P[18]	175			
E15/Camel/ KUW/s21/2010/ G10P[15]	181			
E16/Vicuna-wt/ ARG/C75/2010/ G8P[14]	175			
E17/Sugar glider-tc/JPN/ SG385/2012/ G27P[36]	175			
E18/Rat-wt/GER/ KS-11-573/2011/ G3P[3]	175			
E19/Fox/ ITA/288356/2010/ G18P[17]	169			

RCWG. The cut-off value for designating a new E genotype is defined as 85% nucleotide identity. Analysis and understanding the genotype differences and distribution of all the RV genes provide a framework to analyze interspecies evolutionary relationships, gene reassortment events, functional gene linkage with other genome segments in reassortant progeny and the emergence of new RVs by interspecies transmission (Matthijnssens et al., 2008a) (see Chapter 2.10).

A phylogenetic tree (Fig. 2.4.1A) using nucleotide sequences from each representative genotype constructed with MEGA software (version 6.0) (Tamura et al., 2013) shows there are three distinct NSP4 clusters from genes from human viruses (E1, E2, E3) that are in agreement with the previously described gene constellations by RNA-RNA hybridization (called Wa- DS-1- and KUN-/AU-1-like genogroups). These three E genotypes are commonly detected among viruses from humans that contain common G and P types, whereas E6 contains a human NSP4 gene from unusual G12 strains (Rahman et al., 2007). E types from avian species (E4, E10, and E11) branch out and form a distinct constellation. Full genome classification of multiple genes has revealed genetic relatedness of human RV genes with those from animals supporting the occurrence of interspecies transmission and penetration of animal virus genes into human RVs. A common origin between human Wa-like and porcine RV strains is now recognized (Matthijnssens et al., 2008a). A close genetic relatedness between E1 sequences and human strains and an evolutionarily positively selected site in NSP4 have also been described (Malik et al., 2014). Interestingly, the most recently identified E genotype (E19) was isolated from a fox in Italy (Fox/Italy/288356/2010) and is closely related to avian strains (Boniotti, Unpublished), which further supports the transmission and circulation of new RV species in wild animals. To our knowledge, this is the first report that suggests the circulation of avian RVs in other animal species. This raises the question of whether a fox ate a bird. Full genotypic characterization of the fox isolate from Italy will allow conclusions about its evolution.

Other phylogenetic analyses of whole RV genome sequences have sought to determine if there are genetic linkages between specific RV gene segments, which might support dependent functions. The first reported genetic linkage was between the NSP4 gene and the VP6 gene that seems to correlate with the host

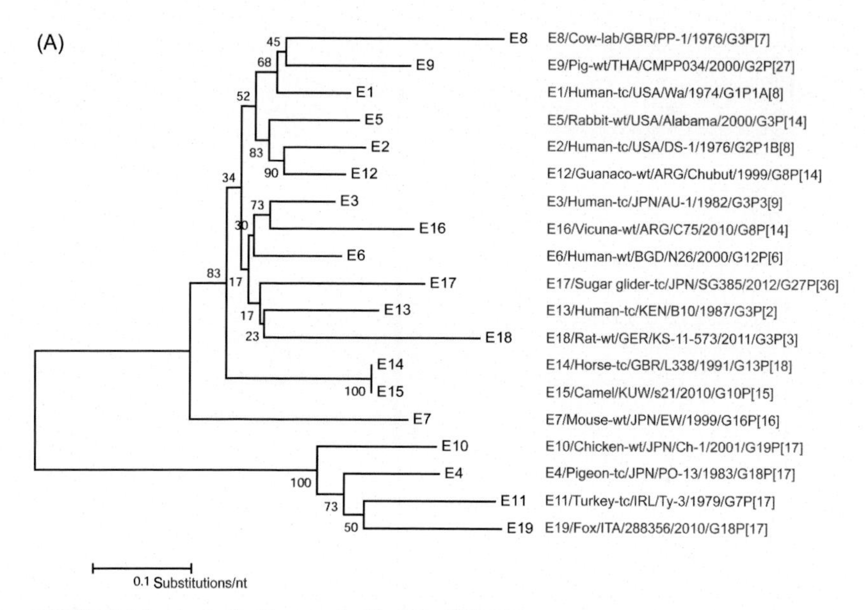

FIGURE 2.4.1 Analysis of currently identified NSP4 E genotypes. (A) *Phylogenetic analysis of NSP4 nucleotide sequences representing each genotype.* A phylogenetic tree was constructed by the neighbor-joining method using MEGA software (Version 6.0) and boot strap values (2000 replicates) are shown at the nodes. Representative strains from each genotype are labeled to the right. GenBank accession numbers for the NSP4 nucleotide sequences used in this analysis are as follows: E1 (K02032), E2 (AF174305), E3 (D89873), E4 (AB009627), E5 (AB005472), E6 (DQ146691), E7 (AB003805), E8 (AF427521), E9 (KM820684), E10 (AB065287), E11 (AB065286), E12 (FJ347109), E13 (HM627562), E14 (JF712564), E15 (JX968472), E16 (JX070055), E17 (AB971769), E18 (KJ879457), E19 (unpublished and kindly provided by Beatrice Boniotti).

(Continued)

(B)

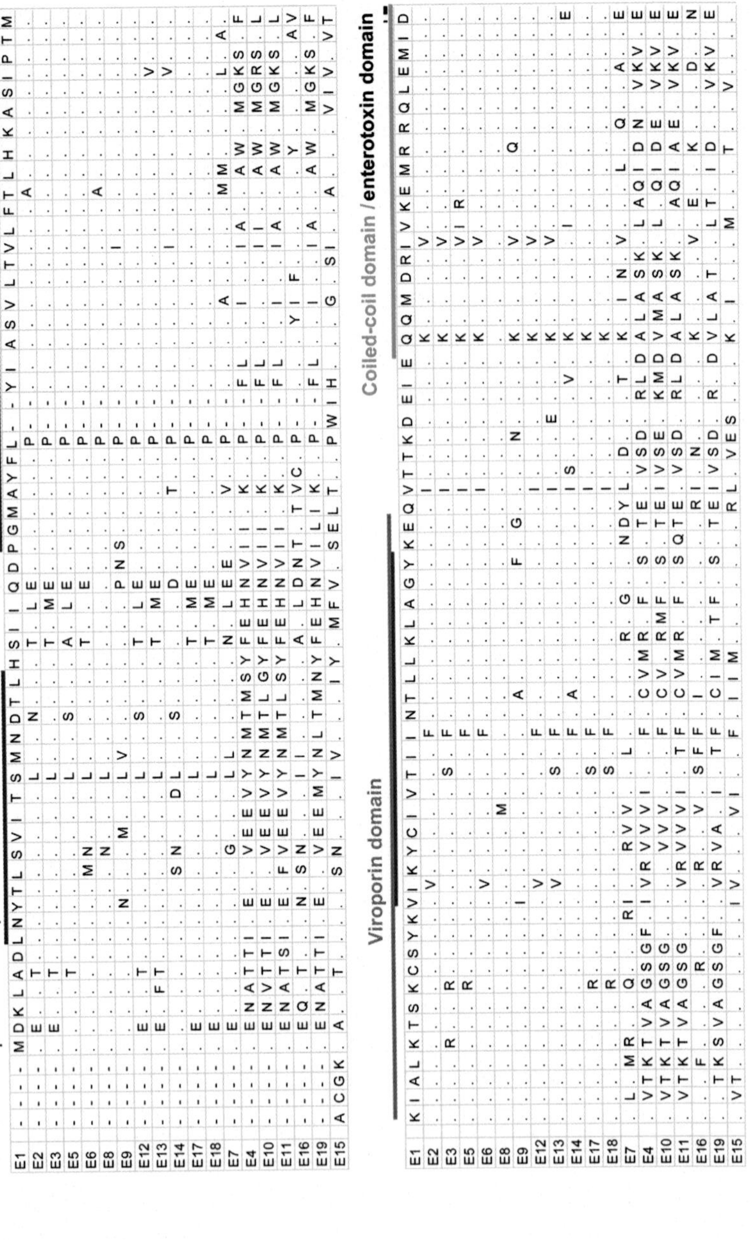

1 Y Y

Strain	Sequence
E1	- - - - M D K L A D L N Y T L S V I T S M N D T L H S I I Q D P G M A Y F L - - Y I A S V L T V L F T L H K A S I P T M
E2	- - - - . E . . T . . . . L . . N . . . T . L E . . . . P - - . . . . A . . . .
E3	- - - - . E . . T . . . . L . . . . . . T . M E . . . . P - - . . . .
E5	- - - - . . . . T . . . . L . . S . . . A . L E . . . . P - - . . . .
E6	- - - - . . . . M N . . . L . . . . . . T . . E . . . . P - - . . . .
E8	- - - - . . . . N . . . L . . . . P - - . . . . A . . . .
E9	- - - - . . . . N . . . M . . L V . . . . P N S . . . . . P - - . . . . I . . . .
E12	- - - - . E . . T . . . . L . . S . . . T . L E . . . . P - - . . . . V . . .
E13	- - - - . E . F T . . . . L . . . . . . T . M E . . . . P - - . . . . V . . .
E14	- - - - . . . . S N . . D L . . S . . . . . . D . . . . T . . P - - . . . . I . . . . V . . .
E17	- - - - . E . . . . L . . . . . . T . M E . . . . P - - . . . .
E18	- - - - . E . . . . L . . . . . . T . M E . . . . P - - . . . .
E7	- - - - . E . . . . G . . . L L . . . . . N . L E E . . . V . . P - - . . . . A . . . . . M M . . . . L . A .
E4	- - - - . E N A T T I . E . . V E E V Y N M T M S Y F E H N V I I . K . . P - - F L . . I . . I A . . A W . M G K S . F
E10	- - - - . E N V T T I . E . . V E E V Y N M T L G Y F E H N V I I . K . . P - - F L . . . . . I I . . A W . M G R S . L
E11	- - - - . E N A T S I . E . F V E E V Y N M T L S Y F E H N V I I . K . . P - - F L . . I . . I A . . A W . M G K S . L
E16	- - - - . E Q . T . . . N . S N . . . I I . . . . . A . L D N T . T V C . P - - . . . Y I F . . . . . . Y . . . . . A V
E19	- - - - . E N A T T I . E . . V E E M Y N L T M N Y F E H N V I L I K . . P - - F L . . I . . I A . . A W . M G K S . F
E15	A C G K . A . . T . . . . . S N . . . I V . . . I Y . M F V . S E L T . . P W I H . . . G . S I . . A . . . V I V . V T

Viroporin domain

Coiled-coil domain / enterotoxin domain

Strain	Sequence
E1	K I A L K T S K C S Y K V I K Y C I V T I I N T L L K L A G Y K E Q V T T K D E I E Q Q M D R I V K E M R R Q L E M I D
E2	. . . . V . . . . F . . . . I . . . . K . . . . V . . . .
E3	. . . . R . . R . . . . S . F . . . . I . . . . K . . . . V . . . .
E5	. . . . R . . . . F . . . . I . . . . K . . . . V I R . . . .
E6	. . . . V . . . . F . . . . I . . . . K . . . . V . . . .
E8	. . . . M . . . .
E9	. . . . I . . . . A . . . . . . F . G . . . . N . . . . K . . . . V . . . . Q . . . .
E12	. . . . V . . . . F . . . . I . . . . K . . . . V . . . .
E13	. . . . V . . . . . S . F . . . . I . . . E . . . K . . . . V . . . .
E14	. . . . F . A . . . . I S . . . . V . K . . . . . I . . . . E
E17	. . . . R . . . . S . F . . . . I . . . . K . . . .
E18	. . . . R . . . . S . F . . . . I . . . . K . . . .
E7	. L . M R . . Q . . . R I . . R V V . . L . . . . . R . G . . N D Y L . D . . . T . K . I N . V . . . L . Q . . A . . E
E4	. V T K T V A G S G F . I V R V V V I . . F . C V M R . F . S . T E . V S D . R L D A L A S K . L A Q I D N . V K V . E
E10	. V T K T V A G S G . . . . . V V V . . . F . C V . R M F . S . T E I V S E . K M D V M A S K . L . Q I D E . V K V . E
E11	. V T K T V A G S G . . . V R V V V I . T F . C V M R . F . S Q T E . V S D . R L D A L A S K . . A Q I A E . V K V . E
E16	. . . F . . . . R . . . . . R . . V . S F F . I . . . . R I . N . . . . . K . . . . V . E . . K . . . D . . N
E19	. . T K S V A G S G F . . V R V A . I . T F . C I M . T F . S . T E I V S D . R . D V L A T . . L T . I D . . V K V . E
E15	. V T . . . . I V . . . V I . . F . I I M . . . . R L . V E S . . . K . I . . . M . . . T . . V . . . .

Coiled-coil domain / **enterotoxin domain**

175

E1	KLTTREIEQVELLKRIHDNLITRPVDVIDMSKEFNQKNIKTLDEWESGKNPYEPSEVTASM
E2	................Y.K.MV.ST.E...T..I....VR..E...N.......K....A.
E3	..................M..IK...K....Q....RQF...N..AE.E.....K.....L
E5	.............R..Y.R.TV.KT.E......I.....R......N.............L
E6	R.................M..IK.IEK..........QF......TD.E.....K.....L
E8	..................K.V..
E9	......V.........Y.K.MVKNA.A............R......N.R............
E12	................Y.K.VV..I.GV..T..I....VR..E...N.......K....T.
E13	................Y.MM.V..TEKV...L.T...YF..SG..KN.E.....K.....L
E14	......V.........Y.K..IQRT.D.............SIE...E.R......K.A...
E17	..................MMVVN.PEK....Q.I...FY..MAD.AE.E.....K.....L
E18	..................MMVVN.PEK....Q.I...FY..MAD.AE.E.....K.....L
E7	................Y.MMVV.QNRE......T...AF...HD.GNDR.YDDNTD.I.PL
E4	Q..K..L...K..AD.YEM.KFK-K.E....F.T.K.EYEKWVK-----D..Q.TRAVSLD
E10	Q..K..L...K..AD.YEL.KYK-SEG-NI.S.T.R.AYE.WSK-----D..Q.TRAVSLN
E11	Q..K..L...K..AD.YEM.KMK-KNDV...F.T.K.EYEKWMK-----D..Q.TRAVSLD
E16	......M...........M.TIK.IGK...TQ....RHF...N..TE.E....SG......
E19	Q..K..L...K..AD.YEL.KFK-K.G....F.T.K.AYEQWIQ-----D..K.TRAVSLD
E15	.............RN.YNK.MAKSI.DMEIMERTKH..MEI.EMKKDRESIN.MRK.....

FIGURE 2.4.1 (*cont.*) (B) *Amino acid sequence alignment of representative NSP4 genotypes.* NSP4 amino acid sequences corresponding to the representative E genotypes shown in Fig. 2.4.1A were aligned by ClustalW. Mammalian strains (E1-E3, E5, E6, E8, E9, E12-E15, E17, E18) are shown on the top followed by mouse (E7), avian [pigeon (E4), chicken (E10), turkey (E11)], vicuna (E16), fox (E19), and camel (E15). Domains are indicated by lines over the alignment: hydrophobic domains H1, H2, and H3: blue line; viroporin domain: orange line; coiled-coil domain: solid green line; enterotoxin peptide: black dashed line showing the well-characterized aa 114–135 followed by a black dotted line that represents an extended domain with enterotoxin activity whose function might depend on regions other than the core domain. Conserved glycosylation sites at aa 8 and 18 are represented by "Y"; avian and the new fox genotypes have a conserved predicted glycosylation site at aa 17 (not shown). The interspecies variable domain (ISVD) is boxed with a blue dashed line. The conserved calcium-binding residues E120 and Q123 are indicated by asterisks. Residues that are identical to the top sequence (E1 genotype) are indicated by a dot. GenBank accession numbers used for this amino acid alignment are as follows: E1 (AAA47309), E2 (AAG09190), E3 (BAA24413), E4 (BAA24144), E5 (BAA36286), E6 (ABA34245), E7 (BAA33948), E8 (AAL31532), E9 (AIY56440), E10 (BAB83747), E11 (BAB83746), E12 (ACN86097), E13 (ADP68540), E14 (AEH96577), E15 (AGC24703), E16 (AFO84069), E17 (BAP91310), E18 (AIL24115), E19 (unpublished and kindly provided by Beatrice Boniotti).

(*Continued*)

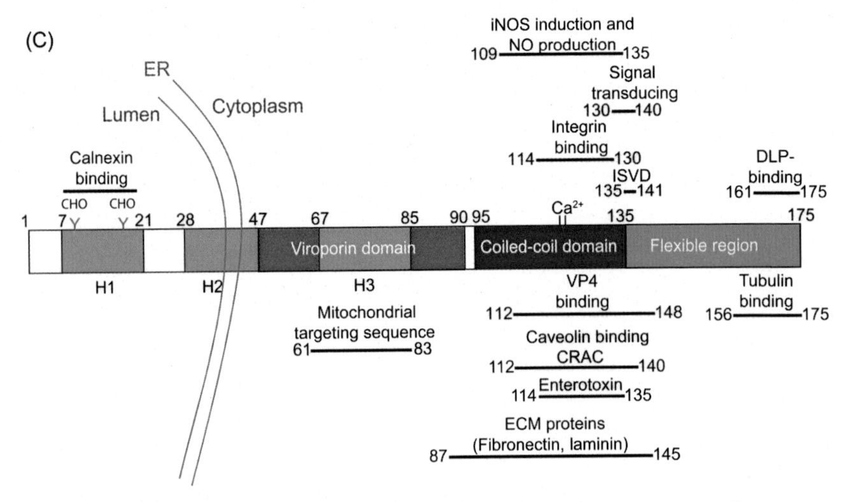

FIGURE 2.4.1 (cont.) (C) *Schematic diagram of NSP4 domain organization and interacting partner proteins.* A linear diagram of SA11 NSP4 with amino acid numbers marking the N- and C-terminal amino acids of the following domains: hydrophobic domains H1, H2, and H3 are indicated by blue boxes; the viroporin domain, shown in orange, overlaps the H3 domain; the coiled-coil domain is shown in red; and the flexible region is shown in green. The H2 domain traverses the ER membrane (shown by blue lines) followed by an extended C-terminus present in the cytoplasm. Conserved glycosylation sites (CHO), indicated with a "Y," reside in the ER lumen. The conserved calcium binding residues (E120 and Q123) are shown within the coiled-coil domain. NSP4 functional domains and cellular and viral protein interaction domains are shown above and below the linear schematic. *ISVD*,: interspecies variable domain; *DLP*, double-layered particle; *CRAC*, cholesterol recognition amino acid consensus; *ECM*, extracellular matrix.

species of origin (Iturriza-Gomara et al., 2003). This linkage may have biological relevance because it is known that NSP4 functions as an intracellular receptor that binds VP6 as part of virus morphogenesis (Au et al., 1989; Bergmann et al., 1989; Meyer et al., 1989). Early studies of receptor interactions suggested there might be specific interactions between NSP4s and VP6s of different subgroups (Au et al., 1989). Other studies have confirmed this linkage in human RVs, and in some cases, also reported linkages between VP4, VP7, VP6, NSP4, and NSP5 (Araujo et al., 2007; Tavares Tde et al., 2008; Benati et al., 2010; Khamrin et al., 2010; Chaimongkol et al., 2012). By contrast, VP6 and NSP4 reportedly segregate independently for porcine RV strains (Ghosh et al., 2007).

3 NSP4 DOMAIN ORGANIZATION

A multiple amino acid sequence alignment of the current 19 E genotypes (Fig. 2.4.1B) provides information about similarities that may be related to protein function. Most mammalian NSP4 proteins contain 175 amino acids while NSP4s from avian, camel and fox species have smaller and larger proteins, respectively (Table 2.4.1 and Fig. 2.4.1B). This figure aligns NSP4 proteins from mammalian viruses on the top followed by those from mouse (E7), avian [pigeon

(E4), chicken (E10), turkey (E11)], vicuna (E16), fox (E19) and camel (E15) at the bottom. Highly conserved residues in the mammalian viruses are evident that diverge in the NSP4 proteins from rodent, avian and camel viruses. Fig. 2.4.1C shows a schematic of the known functional domains in NSP4 in more detail. These figures illustrate that NSP4, which is synthesized as an ER transmembrane glycoprotein, consists of three hydrophobic domains (H1–H3) with two N-linked high mannose glycosylation sites oriented to the luminal side of the ER in the H1 domain (Chan et al., 1988; Bergmann et al., 1989). The H2 transmembrane domain traverses the ER bilayer and serves as an uncleaved signal sequence and aa 85–123 are reported to be involved in ER retention although this is likely to be an indirect involvement because these residues are located in the cytoplasmic domain of NSP4 (Mirazimi et al., 2003). The H3 domain consists of an amphipathic α-helix, which is part of the viroporin domain (Hyser et al., 2010). The remainder of NSP4 is an extended cytoplasmic domain containing a highly conserved coiled-coil region and a flexible C-terminus (Estes and Kapikian, 2007). The coiled-coil domain contains a calcium ion binding site coordinated by amino acids E120 and Q123, which are 100% conserved among all RVA viruses. This is consistent with the evidence showing that this region plays a critical role in conserved functions of NSP4 across different species. The number of multiple overlapping binding domains with distinct partners is striking (Fig. 2.4.1C). An interspecies variable domain (ISVD, aa 135–141) is clearly evident in the amino acid alignment. This alignment highlights that there are multiple other highly conserved domains without current functions, which suggest areas for future research.

The fact that NSP4 has been found in dimeric, tetrameric, pentameric, and higher ordered multimeric structures may explain how NSP4 is able to interact with multiple partners within the same region of the protein (Fig. 2.4.1C) (Maass and Atkinson, 1990; Bowman et al., 2000; Chacko et al., 2011; Sastri et al., 2011, 2014). The coiled-coil domain interacts with multiple proteins including integrin I domains, caveolin, extracellular matrix proteins, a cholesterol recognition/interaction amino acid consensus sequence (CRAC) and VP4 (Boshuizen et al., 2004; Parr et al., 2006; Hyser et al., 2008; Seo et al., 2008; Schroeder et al., 2012). In addition, the flexible C-terminal domain interacts with both tubulin and newly formed, immature double-layered particles (DLPs) (Au et al., 1989; O'Brien et al., 2000; Xu et al., 2000). Whether specific structural forms are associated with these different interactions remains to be determined.

4 NSP4 STRUCTURE

The coiled-coil domain (aa 95–137) is the only domain whose structure has been determined by crystallographic studies. Challenges for obtaining crystals of the full-length protein are due to the presence of the N-terminal hydrophobic domains and a C-terminal flexible region, which leads to full-length NSP4 being highly insoluble in aqueous solution. Thus, crystallization studies have focused on the soluble domains. An initial study, which used a synthetic peptide composed of aa 95–137 of the simian strain SA11 (Bowman et al., 2000), and later studies using bacterially expressed protein (aa 95–146) from SA11 and a human I321 strain reported that the coiled-coil domain forms a tetramer. In spite of crystallizing longer proteins, only aa 95–137 have been resolved in the structures, and the amino acids after 137 are disordered (Bowman et al., 2000; Deepa et al., 2007). The first structure from the SA11 NSP4 peptide showed a tetramer with a calcium ion bound and coordinated by aa E120 and Q123. In contrast, the structure of the NSP4 coiled-coil domain from a human RV ST3 strain formed a pentameric coiled-coil that lacked a calcium ion at its core (Chacko et al., 2011). In these studies, the tetramer was crystallized at neutral pH while the pentamer was crystallized at low pH. It was unclear whether the two oligomeric forms of the coiled-coil domain from the two different strains represented a calcium-dependent, strain-dependent, or pH-dependent phenomenon. This was tested by determining the crystal structure of the NSP4 coiled-coil domain from a single strain SA11 (Sastri et al., 2014); these studies showed that at neutral pH this domain exists as a tetramer, which binds a calcium ion at its core (Fig. 2.4.2,

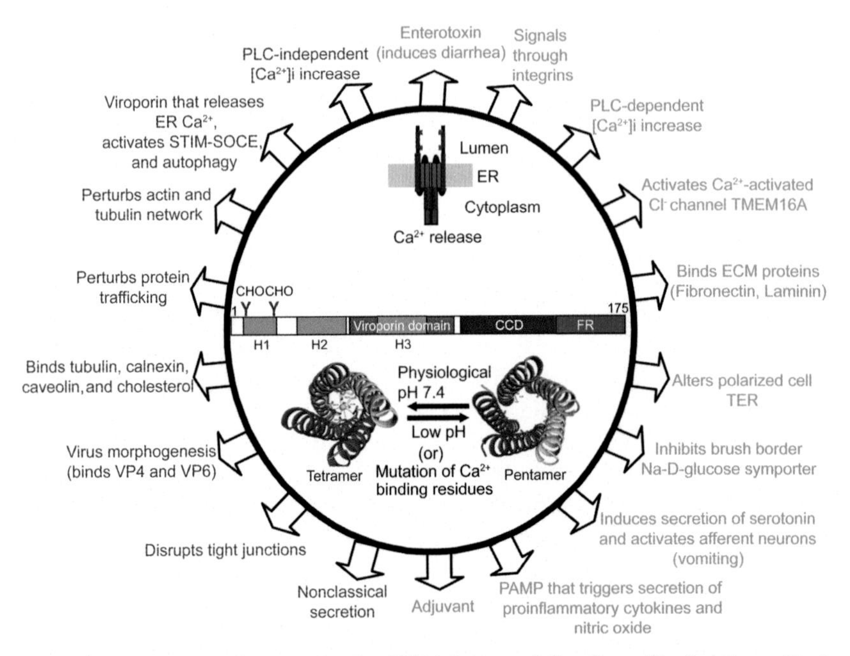

FIGURE 2.4.2 **Wheel diagram showing NSP4 forms and functions.** The functions of both intracellular NSP4 (iNSP4, shown in *blue*) and extracellular NSP4 (eNSP4, shown in *green*) are indicated with arrows. Inside the wheel at the top: a model of the topology of the NSP4 viroporin, in which the pentalysine domain is predicted to mediate viroporin domain insertion into the ER membrane, NSP4 oligomerization and the amphipathic α-helix formation of a pore that allows the release of calcium into the cytoplasm. In the center: NSP4 domain organization (refer to Fig. 2.4.1C for details). At the bottom: two structures of the NSP4 coiled-coil domain. The calcium-bound tetramer that forms at pH 7.4 (left) and a pentamer that forms as a result of mutation of the calcium-coordinating amino acids (E120 and Q123) or at low pH 5.6, which abolishes calcium binding.

structure on the left). However, mutation of the calcium- coordinating amino acids resulted in formation of a pentamer that lacks a calcium ion at its core (Fig. 2.4.2, structure on the right). Further solution studies showed that the NSP4 coiled-coil domain can directly bind calcium at physiological pH and exists as a tetramer. In contrast, at low pH, the coiled-coil domain does not bind calcium and forms a pentamer. This structural plasticity of NSP4 regulated by pH and calcium may be the basis for its pleiotropic functions during RV replication and pathogenesis.

5 THE PATHOPHYSIOLOGY OF ROTAVIRUS-INDUCED DISEASE AND CHARACTERIZATION OF THE ROLE OF NSP4 IN DIARRHEA INDUCTION

The spectrum of responses to RV infection varies and can be asymptomatic, mild or severe, which results in a life-threatening dehydrating illness. The incubation time is short (<48 h), and illness often begins with a sudden onset of nausea and vomiting, followed by diarrhea that leads to a high frequency of dehydration and a mean duration of illness lasting 5–6 days. Several studies have shown infection leads to extra-intestinal spread of virus with antigenemia and viremia, and limited systemic replication in a variety of sites occurs frequently (Blutt et al., 2007). It remains unclear whether this systemic spread and replication cause any specific pathology in the normal host, but fever has been associated with antigenemia (Sugata et al., 2008), and antigenemia is reported to be associated with more severe clinical manifestations of acute gastroenteritis (Hemming et al., 2014).

The mechanisms causing these symptoms remain incompletely understood although our understanding of the pathophysiological processes of RV-induced disease has improved in recent years. The outcome of infection is more complex than initially appreciated and is affected by a complex interplay of host and viral factors. Chapter 2.6 covers this topic in detail; so later we focus on information related to NSP4 in pathogenesis.

RVs infect intestinal enterocytes and the early events in infection are mediated by virus-epithelial cell interactions (see Chapter 2.2). The cause of RV-induced diarrhea is multifactorial, and mechanisms include (1) increased intestinal secretion stimulated by infection, (2) villus ischemia and activation of the enteric nervous system that may be evoked by release of a vasoactive agent from infected epithelial cells in the absence of significant pathologic lesions or enterocyte damage, (3) malabsorption that occurs secondary to the down regulation of host proteins or destruction of differentiated absorptive enterocytes, and (4) alterations in transepithelial fluid balance caused by loss of polarized epithelial cell tight junctions and cell integrity. Each of these mechanisms is associated with functions of NSP4 either expressed intracellularly (iNSP4) or secreted extracellularly (eNSP4). Fig. 2.4.2 shows the functions attributed to iNSP4 (blue) or eNSP4 (green).

A possible role for NSP4 in pathogenesis (diarrhea induction) was initially suggested in studies of a virulent porcine RV (SB-1A strain) and an avirulent human RV (DS-1) in gnotobiotic pigs (Hoshino et al., 1995). Pigs administered a single-gene reassortant virus that derived its 10th gene (that encodes NSP4) from the avirulent human DS-1 virus failed to develop diarrhea, and these pigs were also protected from subsequent challenge with a different virulent porcine RV. These results were unexpected and a mechanistic explanation for why the gene encoding NSP4 was a virulence gene associated with diarrhea was unclear.

Discovery that administration of NSP4 alone or synthetic peptides from NSP4 cause age-dependent diarrhea in mice without causing histologic changes in the intestine led to the recognition that NSP4 functions as an enterotoxin (Ball et al., 1996). Subsequent work showed that NSP4 is a novel secretory agonist eliciting both paracrine and autocrine signaling. Both trigger (by different mechanisms) calcium-dependent signaling pathways that alter epithelial cell permeability and augment endogenous secretory pathways (Estes and Morris, 1999; Morris et al., 1999; Estes et al., 2001). This led to a new explanation for RV-induced diarrhea and why diarrhea can be seen *before* or *in the absence* of histologic changes in several species (Mebus et al., 1975; Collins et al., 1989; Burns et al., 1995; Ward et al., 1996; Ciarlet and Estes, 1999). Such observations had been made previously in several settings, and the production of endogenous, neuroactive, hormonal substances of pathophysiological importance had been proposed (Osborne et al., 1988).

The ability of NSP4 from a number of mammalian and avian virus strains to induce diarrhea in neonatal mice has been demonstrated by multiple groups (Table 2.4.1). The enterotoxin domain was initially characterized using both purified protein and a synthetic peptide that contained aa 114–135 (Ball et al., 1996). Further studies revealed that this region has two distinct domains: amino acids 114–130 are essential for binding to α1 and α2 integrin I domains while amino acids 131–140, which are not associated with the initial binding to the I domain, elicit signaling (Fig. 2.4.3). NSP4 mutants that fail to bind or signal through integrin α2 are attenuated in diarrhea induction in neonatal mice (Seo et al., 2008). It is unclear whether interaction of NSP4 with other proteins is involved in diarrhea induction. In addition, the diarrheagenic peptide contains only a part of the active domain of the protein based on comparisons of the effective dose of the peptide and the full-length protein to cause diarrhea, but the exact size of the active toxin domain remains unknown.

Based on the hypothesis that diarrhea induction should be predicated on a secreted form of NSP4, studies were performed to detect such forms, and several secreted forms of NSP4 have been described. In a first study, a 7 kDa molecular weight secreted form of NSP4, spanning amino acids 112–175, was isolated from the media of rotavirus-infected MA104 and intestinal HT-29 cells. This product was secreted in a nonclassical, Golgi apparatus-independent mechanism that utilizes the microtubule and actin microfilament network (Zhang et al., 2000). Diarrhea was induced in neonatal mice using an expressed

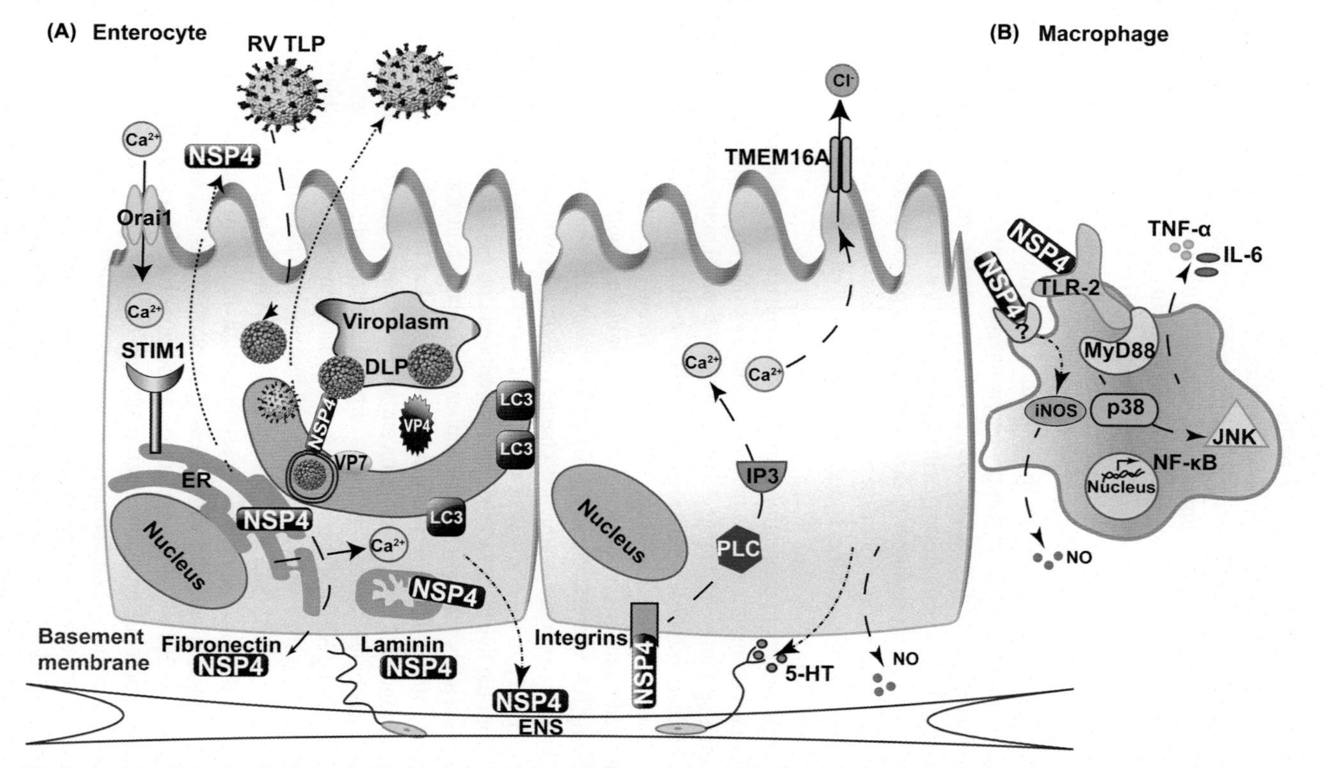

FIGURE 2.4.3 **NSP4 is a master regulator that modulates cellular signaling pathways during rotavirus infection.** (A) Following RV infection of an intestinal enterocyte, iNSP4 is cotranslationally inserted into the ER membrane. Functioning as a viroporin, iNSP4 mediates the release of calcium from ER into the cytoplasm. iNSP4 viroporin-mediated loss of ER calcium activates and translocates STIM1, which interacts and activates Orai1 for SOCE. Elevated cytoplasmic calcium initiates the autophagy process, which is necessary for RV replication and production of infectious particles. iNSP4 also induces apoptosis by activating intrinsic apoptotic pathway and releases cytochrome c, which activates procaspase-9 that triggers caspase-3 cleavage. Secreted eNSP4 binds integrins and triggers a PLC-mediated pathway that elevates cytoplasmic calcium. The increased cytoplasmic calcium activates calcium-activated chloride channels such as TMEM16A, which results in cellular Cl^- secretion. (B) Macrophage depicting paracrine signaling pathway elicted by eNSP4, which triggers the release of the proinflammatory cytokines TNF-α and IL-6 and activates p38 and JNK MAP kinases and NF-κB. NSP4 also induces NO release.

and purified form of this protein. Another secreted, fully glycosylated form of NSP4 was isolated from Caco-2 cells as discrete detergent-sensitive oligomers in a complex with phospholipids. Secretion of this form of NSP4 was dramatically inhibited by brefeldin A and monensin, suggesting that a Golgi-dependent pathway is involved in release of the protein, which was supported by its partial resistance to deglycosylation by endoglycosidase H. This form of NSP4 was found to bind cell surface glycosaminoglycans (Didsbury et al., 2011). A third full-length, glycosylated, endoglycosidase H-sensitive form of NSP4 was detected in the media of two cell types (MDCK and HT-29 cells) (Gibbons et al., 2011). Others detected impairment of lactase activity in the brush border of Caco-2 cells caused by secreted NSP4 in the media from virus-infected cells although the exact form of the protein was not characterized (Beau et al., 2007). It remains unclear whether the different forms of NSP4 detected in these studies relate to the use of different virus strains, cells or methods of isolation, and this issue remains to be resolved, perhaps by designing new experiments that take into account the new knowledge about the plasticity of the structure of NSP4 that can be influenced by pH and ionic conditions of buffers used in the isolation and purification of NSP4.

One continuing impediment preventing complete characterization of the activity of NSP4 is the lack of any mutants in gene 10 that encodes NSP4 and the lack of a reverse genetics system (see Chapter 2.3) to study site-directed mutants in the context of a viral infection. These limitations have forced studies of the functions of NSP4 to focus on examining the effects of purified protein or expressed proteins with engineered mutations by using a variety of cell biology and biochemical studies including RNA interference (RNAi) knockdowns and pharmacologic treatments. In another approach, sequence changes associated with modified function can be identified in the NSP4 gene of isogenic virulent/avirulent pairs of viruses. NSP4s from the avirulent viruses showed reduced enterotoxin activity; in one example, amino acids 136 and 138 were implicated in reduced enterotoxin activity (Zhang et al., 1998). These results also demonstrate that the enterotoxin domain extends beyond aa 114–135. More recent studies indicate that the diarrheagenic domain may extend to the C-terminus of NSP4 (Sastri et al., 2011). It is interesting to note that the NSP4 of the porcine/human reassortant viruses studied by Hoshino had sequence changes at aa 135 and 138 that might explain their phenotypes (Hoshino et al., 1995). Some studies have tried to examine sequences of RVs that are associated with symptomatic or asymptomatic infections and failed to find distinct sequences associated with illness (Lee et al., 2000; Rajasekaran et al., 2008; Sastri et al., 2011). This is not surprising because both host and virus properties affect infection outcome, and clear results are unlikely to be obtained unless isogenic virus strains are studied and compared.

Another novel approach to define virulence-associated genome mutations of murine RVs evaluated a cell culture-adapted murine EB RV strain that was serially passaged in mouse pups or in cell cultures, alternately and repeatedly, and then the genomes of viruses with different virulence phenotypes were fully sequenced. Mouse-passaged virus that regained virulence had aa substitutions in VP4 and at aa 37 in NSP4 (Tsugawa et al., 2014). This is an unexpected but interesting result because this aa is present in a highly conserved domain (aa 35–41) of the NSP4s of mammalian RVs that currently does not have a known function (Fig. 2.4.1B). A human vaccine strain of lamb RV also has a change at aa 36 in this highly conserved region of NSP4 (Mohan et al., 2003). These results raise the question of whether this domain may be correlated with virulence and suggest that it should be examined in additional functional studies.

A correlation of RV-induced diarrhea in neonatal mice and NSP4 has been confirmed using RNAi technology, specifically using a short hairpin RNA (shRNA) reactive with the NSP4 gene engineered into a lentivirus. No suckling mice developed diarrhea induced by bovine RV when the NSP4 was silenced by the shRNA while 100% of suckling mice without shRNA treatment or with treatment with a control shRNA developed diarrhea (Chen et al., 2011). These results provide an independent approach to confirm that NSP4 is associated with diarrhea induction in neonatal mice. The ability to prevent or reduce RV-induced diarrhea by treating mice with NSP4-specific antibody is additional independent evidence of the role of NSP4 in diarrhea induction (Ball et al., 1996; Estes et al., 2001; Hou et al., 2008).

6 MECHANISM OF NSP4 IN DIARRHEA INDUCTION

The mechanism of diarrhea induction is best characterized through studies of the paracrine action of eNSP4 on uninfected intestinal epithelial cells. NSP4 binding to intestinal epithelial cell lines has been shown to initiate a signaling pathway that involves activation of phospholipase C (PLC), elevation in inositol 1,4,5-trisphophate ($InsP_3$) and release of ER Ca^{2+} stores that increases cytoplasmic Ca^{2+} resulting in Ca^{2+}-dependent Cl^- secretion (Tian et al., 1995; Ball et al., 1996; Dong et al., 1997). Crypt cells isolated from mice also respond in a similar manner (Morris et al., 1999). Studies in mice lacking the cystic fibrosis transmembrane regulator (CFTR) channel showed that age-dependent diarrheal disease results from an event downstream from increased cytoplasmic Ca^{2+} (Morris et al., 1999). Since the CFTR channel is a c-AMP-regulated Cl^- channel, the finding that diarrhea is observed in CFTR knockout mice following RV infection or NSP4 treatment indicated that a Cl^- channel different than CFTR mediates this effect (Angel et al., 1998; Morris et al., 1999). Cl^- secretion is age-dependent in CFTR mice, indicating that age-dependent disease may result from an age-dependent induction, activation or regulation of the Cl^- channel, which was proposed to be the calcium-activated chloride channel (CaCC). This pathway would be one that is activated when eNSP4 is released from virus-infected cells and the affected cells are neighboring, noninfected secretory crypt or epithelial cells (Fig. 2.4.3). These studies also illustrate that NSP4 is a novel secretory agonist because CFTR knockout mice do not respond to classical secretory agonists (Morris et al., 1999). Confirmatory data for this proposed model came from showing NSP4 induces diarrhea by activation of the proposed CaCC TMEM16A and inhibition of Na^+ absorption (Ousingsawat et al., 2011) and that CaCC inhibitors prevent RV secretory diarrhea in neonatal mice (Ko et al., 2014). NSP4 also does not have a direct, specific effect on chloride transport in brush border membranes, confirming the hypothesis that its actions are through triggering signal transduction pathways (Lorrot and Vasseur, 2006).

Further studies using Ussing chamber experiments showed that the enteric nervous system (ENS) also plays a role in the fluid secretion evoked by RV in neonatal mice (Lundgren et al., 2000). Drugs that inhibit ENS functions attenuate the intestinal secretory response to RV in vitro and in vivo, suggesting that the ENS participates in RV-induced electrolyte and fluid secretion as shown previously for bacterial enterotoxins (Lundgren and Jodal, 1997; Farthing et al., 2004; Kordasti et al., 2004). Activation of secretory reflexes in the ENS system represents another proposed mechanism for RV fluid secretion. This hypothesis implies that nerve afferent fibers, located just underneath the intestinal epithelium, are activated by amines/peptides secreted from the enteroendocrine cells and/or by chemokines, prostaglandins, or nitric oxide (NO) released from enterocytes exposed to microorganisms (Fig. 2.4.3) (Lundgren and Svensson, 2001). In line with this, NSP4 and RV can stimulate the release of NO from HT-29 cells, and increased levels of NO_2/NO_3 are detected in the urine of RV-infected mice and young children (Rodriguez-Diaz et al., 2006). Finally, NSP4 can induce nitric oxide synthase (iNOS) and stimulate NO release from macrophages (Borghan et al., 2007). It is in this interesting context that Rollo et al. reported that RV evokes an age-dependent release of chemokines from epithelial cells, occurring only in mice less than 2 weeks old (Rollo et al., 1999). Thus, age-dependent diarrhea may be multifactorial, and mechanisms may include an infection-mediated intestinal chemokine response as well as through a NSP4-mediated activation of an age- and Ca^{2+}-dependent plasma membrane Cl^- channel distinct from CFTR.

Subsequent studies indicated that the neurotransmitters serotonin and vasoactive intestinal peptide (VIP) are involved in RV diarrhea (Kordasti et al., 2004). RV can infect cultured human entero-chromaffin (EC) cells and stimulate serotonin secretion, which activates the vagal afferent nerves connected to the brain stem structures associated with nausea and vomiting; NSP4 alone also was shown to evoke release of serotonin from EC cells in a calcium-dependent manner, suggesting new mechanisms to explain the nausea and vomiting associated with RV infections (Fig. 2.4.3) (Hagbom et al., 2011, 2012) (see Chapter 2.6). It is evident that NSP4 activates and triggers multiple signaling pathways that affect pathogenesis.

Another mechanism of RV-induced pathogenesis is loss of cell integrity and polarized epithelial cell tight junctions. A calcium-dependent depolymerization of microvillar actin occurs in RV-infected polarized human intestinal Caco-2 cells that results in alterations of protein trafficking (Jourdan et al., 1998; Brunet et al., 2000) and disruption of tight junctions with loss of cell monolayer integrity (Obert et al., 2000). Silencing NSP4 provided confirmation that disruption of actin in RV-infected MA104 cells is directly related to changes in the intracellular calcium concentration mediated by the expression of NSP4 (Zambrano et al., 2012). One mechanism of iNSP4-mediated, calcium-dependent actin reorganization appears to be through decreased phosphorylation of the actin remodeling protein cofilin (Berkova et al., 2007). In addition to iNSP4 disruption of actin, addition of eNSP4 to the apical media of polarized MDCK-1 cells causes filamentous actin (F-actin) redistribution accompanied by a reduction of transepithelial resistance (Tafazoli et al., 2001). These alterations of the actin cytoskeleton may contribute to RV pathogenesis.

7 iNSP4-MEDIATED APOPTOSIS

Many viruses induce apoptosis for release and dissemination of viral progeny. Apoptosis may also be a contributing factor to pathogenesis through destruction of differentiated absorptive enterocytes. In murine EDIM RV-infected mouse pups, apoptosis, shown by detection of caspase-3 in infected mature small intestinal enterocytes, was hypothesized to be responsible for the observed villous atrophy (Boshuizen et al., 2003). Further in vitro studies in human colon carcinoma HT-29 and fully differentiated Caco-2 cells indicated that RV infection induces apoptosis as demonstrated by peripheral condensation of chromatin and fragmentation of the nuclei (Superti et al., 1998; Chaibi et al., 2005). Apoptosis in RV-infected MA104 cells has been attributed to the activation of Bax, a proapoptotic member of the Bcl-2 family (Martin-Latil et al., 2007).

More recently, NSP4 and NSP1 have been proposed to regulate apoptosis during infection; NSP4 plays a proapoptotic role and NSP1 an antiapoptotic role (Bhowmick et al., 2012, 2013). NSP4 has been implicated in activation of the intrinsic apoptosis pathway by depolarizing the mitochondria; this leads to the release of cytochrome c into the cytosol, causing the autocatalytic activation of procaspase-9 that triggers caspase-3 cleavage (Bhowmick et al., 2012). However, apoptosis is inhibited by NSP1 signaling through PI3K/AKT to upregulate XIAP, which inhibits caspase-3 and -9. This NSP1-activated signaling pathway is downregulated later in infection, resulting in apoptosis of the infected cell. In RV-infected and NSP4-expressing cells, nonglycosylated NSP4 (20 kDa) was detected in both the outer and inner mitochondria membranes. The mitochondrial targeting signal sequence was identified as the amphipathic α-helix within the viroporin domain, amino acids 61–83 (Bhowmick et al., 2012, 2013).

To further confirm NSP4 localization in both the inner and outer mitochondrial membranes, NSP4 was shown to coimmunoprecipitiate with the outer mitochondrial membrane voltage-dependent anion channel (VDAC) and the inner mitochondrial membrane adenine nucleotide translocator (ANT) (Bhowmick et al., 2012). VDAC facilitates the exchange of ions and molecules between the mitochondria and cytosol and ANT exports ATP from the mitochondrial matrix. NSP4-mediated mitochondria depolarization is proposed to occur through

NSP4 interaction with VDAC that may impair the ion supply to the mitochondria and/or interaction with ANT to inhibit ATP exportation. Future studies are needed to address these proposed mechanisms of nonglycosylated NSP4-mediated activation of the intrinsic apoptotic pathway.

8 iNSP4 VIROPORIN-MEDIATED ELEVATION OF CYTOPLASMIC CALCIUM

Perturbing the host Ca^{2+} signaling pathway is not only important in pathogenesis but also for viral replication and morphogenesis. RV infection induces dramatic changes in cellular calcium homeostasis causing a two- to fourfold elevation in cytoplasmic calcium ([Ca^{2+}]cyto). Cellular calcium homeostasis is tightly regulated as calcium ions are ubiquitous intracellular signaling molecules responsible for controlling a plethora of cellular processes. To determine the viral protein(s) responsible for the elevation in [Ca^{2+}]cyto, individual RV proteins were expressed in Sf9 insect cells or a variety of mammalian cell lines and [Ca^{2+}]cyto was measured. NSP4 was identified as the sole RV protein responsible for the increase in [Ca^{2+}]cyto by a PLC-independent mechanism (Tian et al., 1994, 1995; Berkova et al., 2007; Diaz et al., 2008, 2012). NSP4 expression in cells recapitulates all of the changes in calcium homeostasis observed in RV-infected cells, while silencing of NSP4 expression in infected cells inhibits these changes (Tian et al., 1994; Berkova et al., 2007; Diaz et al., 2008, 2012; Zambrano et al., 2008). A breakthrough in identifying the mechanism of iNSP4-mediated global disruption in cellular calcium homeostasis was attained when NSP4 was characterized as a viroporin in the ER (Fig. 2.4.2) (Hyser et al., 2010). Viroporins are small, hydrophobic proteins that contain a cluster of basic residues (Lys or Arg) and an amphipathic-α-helix that oligomerizes to create a transmembrane aqueous pore. The NSP4 viroporin domain is comprised of amino acids 47–90, which are critical for elevation of [Ca^{2+}]cyto, since mutation of either the cluster of basic residues or amphipathic α-helix abolishes the observed elevation in [Ca^{2+}]cyto (Hyser et al., 2010). Whether NSP4 viroporin functions as a channel or a pore to elevate cytoplasmic calcium requires further study. Although NSP4 mediates the release of calcium from the ER store into the cytoplasm, this alone is not sufficient to explain the dramatic increase in [Ca^{2+}]cyto or the progressive increase of calcium permeability of the plasma membrane.

The mechanism of NSP4-mediated [Ca^{2+}]cyto was further illuminated when NSP4-mediated depletion of calcium from the ER store was shown to involve store-operated calcium entry (SOCE) through activation of the ER calcium sensor stromal interaction molecule 1 (STIM1) (Hyser et al., 2013). STIM1 is an ER single transmembrane glyco/phosphoprotein that senses ER calcium levels through a low-affinity EF-hand calcium binding site located within its N-terminal domain in the ER lumen. Loss of calcium from the EF-hand induces a conformational change in STIM1 and formation of STIM1 oligomers. The

STIM1 oligomers are retained in the ER but rapidly move to ER–PM junctions to activate a variety of calcium-release-activated calcium (CRAC) channels, including Orai1 and TRPC channels, for SOCE. In RV-infected and NSP4-expressing cells, STIM1 colocalizes with and activates Orai1 at the PM for SOCE. In contrast, a NSP4 viroporin mutant failed to induce STIM1 activation and did not activate the PM calcium entry pathway. These studies provided a mechanism for and confirmed early work indicating iNSP4 release of intracellular Ca^{2+} stimulated Ca^{2+} influx from the extracellular medium through the capacitative calcium entry pathway (Tian et al., 1995).

NSP4 viroporin-mediated STIM1 activation of SOCE (Fig. 2.4.3) is supported by pharmacological studies showing that ion channel inhibitors block calcium entry into RV-infected or NSP4-expressing cells. Due to the constitutive activation of STIM1 in RV-infected or NSP4-expressing cells, it is likely that a variety of calcium entry channels are activated, and Orai1, voltage-gated calcium channels and the sodium/calcium exchanger NCX have been implicated. The Orai1 inhibitor 2-APB partially blocks NSP4-mediated calcium entry (Diaz et al., 2012). Similarly, methoxyverapamil (D600), a blocker of L-type voltage-gated channels, partially inhibits the entry of calcium in virus-infected MA104 and HT-29 cells (Perez et al., 1999). Under normal conditions, NCX pumps Ca^{2+} out of cells using the Na^{+} gradient as the driving force, but if intracellular Na^{+} is elevated, as seen in RV-infected cells, NCX will pump Na^{+} out and bring Ca^{2+} into the cytoplasm, thus functioning in reverse mode. The inhibitory effect of KB-R7943, a commonly used blocker of NCX functioning in reverse mode, raises the possibility of Ca^{2+} entry through NCX in its reverse mode in RV-infected or NSP4-expressing cells (Diaz et al., 2012). Thus, while RV may ultimately activate multiple calcium channels in the PM, calcium influx is predicated on NSP4 viroporin-mediated activation of STIM1 in the ER.

Another consequence of NSP4 viroporin-mediated increase in $[Ca^{2+}]$cyto is the activation of the initial stages of autophagy. Autophagy is a cellular degradation process involving an intracellular membrane trafficking pathway that recycles cellular components or eliminates intracellular microbes in lysosomes. RV infection initiates autophagy, characterized by an increase in lipidated LC3 II, a marker of autophagy (Crawford et al., 2012; Crawford and Estes, 2013; Arnoldi et al., 2014). Autophagy initiation is required for the production of infectious virus; pharmacological inhibition of autophagy, RNAi knockdown of autophagy genes and infection of cells genetically deficient in autophagy proteins results in reduced virus yields (Crawford et al., 2012; Crawford and Estes, 2013; Arnoldi et al., 2014). The mechanism of RV-initiated autophagy is through NSP4 viroporin-mediated increases in $[Ca^{2+}]$cyto that activates a calcium signaling pathway involving CAMKK2 and AMPK (Crawford et al., 2012; Crawford and Estes, 2013). NSP4-initiated autophagy is inhibited by RNAi knockdown of NSP4, expression of a viroporin mutant that does not increase $[Ca^{2+}]$cyto, and chelation of $[Ca^{2+}]$cyto following NSP4 expression (Crawford et al., 2012; Crawford and Estes, 2013). Furthermore, a specific inhibitor of

CAMKK2, STO-609, not only inhibits AMPK phosphorylation and autophagy induction, but also significantly reduces the yield of infectious virus (Crawford et al., 2012; Crawford and Estes, 2013). Induction of autophagy also leads to the colocalization of NSP4 and LC3 in puncta that traffic to viroplasms that is inhibited by STO-609 (Crawford et al., 2012; Crawford and Estes, 2013).

Although RV infection initiates autophagy, autophagy maturation is inhibited as demonstrated by a lack of autophagosomes and autophagolysosomes (Crawford et al., 2012; Crawford and Estes, 2013; Arnoldi et al., 2014). Instead of forming autophagosomes, LC3 colocalizes with NSP4 surrounding viroplasms (Berkova et al., 2006; Crawford et al., 2012; Crawford and Estes, 2013). Furthermore, colocalization analysis of NSP4 and endogenous LC3 following RV infection indicates that NSP4 colocalizes with LC3 in puncta and traffics to viroplasms. Induction of autophagy and trafficking of ER-localized NSP4 and VP7 to viroplasms is hindered by STO-609 (Crawford et al., 2012; Crawford and Estes, 2013). These results suggest that RV exploits the membrane trafficking function of autophagy to transport the ER-localized viral glycoproteins NSP4 and VP7 to viroplasms to facilitate infectious particle assembly, thus disrupting autophagosome and autophagolysosome formation. These results were confirmed by Arnoldi et al., (Arnoldi et al., 2014) who found that RV replication: (1) induces accumulation of LC3 II; (2) does not lead to accumulation of autophagosomes in spite of significant accumulation of lipidated LC3; and (3) takes advantage of LC3 II for improved production of infectious progeny virus. However, they did not detect LC3 colocalization with NSP4 surrounding viroplasms. We predict that the use of different LC3 antibodies or staining conditions are responsible for the lack of detecting LC3 colocalization with NSP4 surrounding viroplasms by Arnoldi et al. (2014). Nonetheless, RV infection induces autophagy, which is required for infectious particle assembly. Questions remain as to whether immature particles bud through ER membranes or ER-derived membranes.

9 ROLE OF iNSP4 IN VIRAL MORPHOGENESIS

The ability of NSP4 to interact with multiple viral and cellular proteins may rely on the differential localization of this protein in infected cells during the RV replication process. In the ER, NSP4 interacts with the ER transmembrane chaperone calnexin, which accelerates NSP4 carbohydrate trimming by glucosidase I and II as well as affects the stability of infectious particles. TLPs produced in calnexin-silenced cells are less infectious, less dense in CsCl gradients (1.35 g/cm^3) compared to control (1.36 g/cm^3) and do not assemble properly suggesting that NSP4–calnexin interactions and ER carbohydrate quality control affect TLP assembly (Mirazimi et al., 1998; Maruri-Avidal et al., 2008).

NSP4 is essential to RV morphogenesis by serving as an intracellular receptor to DLPs (Au et al., 1989; O'Brien et al., 2000). RV genome replication and nascent particle assembly occurs in electron dense viroplasms located in the cytoplasm of the infected cell. NSP4-containing membranes are detected adjacent to viroplasms. The C-terminal cytoplasmic domain of NSP4, amino acids 161–175, binds the inner coat protein (VP6) of DLPs in viroplasms. This interaction triggers the budding of the DLP into the NSP4-containing membranes where the particles become transiently enveloped. The transient envelope is removed by an unknown mechanism and the outer capsid proteins, VP7 and VP4, are assembled onto the particle.

NSP4 has been detected in the ER-Golgi intermediate compartment (ERGIC) and in the plasma membrane (Berkova et al., 2006; Storey et al., 2007; Crawford et al., 2012; Ball et al., 2013; Crawford and Estes, 2013). The form and functional significance of detecting NSP4 in the ERGIC is unknown. However at the plasma membrane, a glycosylated, endoglycosidase H-sensitive form of NSP4 is detected at the apical surface or apical and basolateral surfaces of infected polarized MDCK or HT-29.F8 cells, respectively (Gibbons et al., 2011). It has been proposed that NSP4 is transported to the plasma membrane by interacting with both caveolin and cholesterol; caveolin forms a complex with chaperone proteins that deliver cholesterol from the ER to the plasma membrane, bypassing the Golgi apparatus (Uittenbogaard and Smart, 2000; Storey et al., 2007; Ball et al., 2013). NSP4 colocalizes with caveolin in the ER and the plasma membrane and a domain of NSP4, the hydrophobic face of the NSP4 amphipathic helix, interacts with both the N- and C-terminal domains of caveolin-1. NSP4 contains a cholesterol recognition/interaction amino acid consensus sequence (CRAC) and binds cholesterol at a 1:1 ratio (Fig. 2.4.1C). Transport of NSP4 to the plasma membrane is altered by treatment of cells with the cholesterol disrupting drugs filipin and nystatin. Since reactivity with different NSP4 antibodies revealed that the cytoplasmic tail of NSP4 is exposed on the exofacial surface of the plasma membrane, it was hypothesized that caveolin/cholesterol transport may be the mechanism by which glycosylated, endoglycosidase H-sensitive NSP4 is secreted into the media of infected cells (Gibbons et al., 2011; Schroeder et al., 2012). Further studies are needed to verify this hypothesis.

NSP4 also was found to colocalize with the enterocyte basement membrane during RV infection in suckling mice suggesting this may be a secreted form of NSP4 (Boshuizen et al., 2004). Although the form of NSP4 detected remains uncharacterized, protein interaction and mutation analysis studies mapped domains on NSP4 that interact with the extracellular matrix proteins laminin-3 and fibronectin (Boshuizen et al., 2004). The functional significance of such interactions remains to be fully understood. A nonglycosylated 20K form of NSP4 can also be detected in RV-infected cells (Ericson et al., 1982) and this may be the form recently reported to be associated with mitochondria (see earlier).

10 NSP4-ACTIVATED IMMUNE RESPONSES

Roles for NSP4 in innate immune responses have been described. NSP4 has adjuvant properties when coadministered with model antigens in mice (Kavanagh et al., 2010). This response may be through NSP4 functioning through pathogen-associated molecular pattern (PAMP) activity (Ge et al., 2013). Treatment of macrophages with NSP4 purified from the media of RV-infected Caco-2 cells induces secretion of the pro-inflammatory cytokines TNF-α and IL-6 as well as activation of p38 and JNK mitogen-activated protein kinases (MAPKs) and nuclear factor NF-κB (Ge et al., 2013) (Fig. 2.4.3). Secretion of cytokines was mediated through Toll-like receptor 2 (TLR-2) and may provide a mechanism for the production of proinflammatory cytokines associated with the clinical symptoms of infection in humans and animals.

In addition, a role for NSP4 in injury to the extrahepatic biliary epithelium that may be associated with biliary atresia in neonatal mice has been found (Feng et al., 2011; Zheng et al., 2014). Biliary atresia is an obliterative cholangiopathy with progressive hepatobiliary disease. Although the pathogenic mechanisms of biliary atresia remain largely unknown, interferon (IFN)-γ-driven and CD8+ T-cell-dependent inflammatory injury to extrahepatic biliary epithelium is likely to be involved in the development of biliary atresia in a neonatal mouse model of this disease. Injection of NSP4 peptides, amino acids 144–152 or 157–170, into neonatal mice increased IFN-γ release by CD8+ T cells, elevated the population of hepatic memory CD8+ T cells, and augmented cytotoxicity of CD8+ T cells to rhesus RV-infected or naïve EHBE cells (Zheng et al., 2014). Furthermore, immunization of mouse dams with GST-NSP4, or the NSP4 peptides 144–152 or 157–170 decreased the incidence of rhesus RV-induced biliary atresia in their offspring. These results support a role for autoimmune responses to NSP4 in the pathogenesis of experimental biliary atresia.

11 CONCLUDING REMARKS AND SUMMARY

Due to their limited coding capacities, most viruses encode multifunctional proteins that accomplish a variety of tasks during infection. NSP4 is a pleiotropic protein that performs many functions as described in this review. Our knowledge of the molecular mechanisms by which iNSP4 and eNSP4 mediate RV-induced pathogenesis, replication, and morphogenesis has increased over the last years. However, additional studies are needed to elucidate the different structural forms of NSP4 in order to augment biochemical and functional studies addressing the many functions that NSP4 performs during RV infection. Many questions remain for the structure of NSP4 and how it facilitates multiple functions of NSP4 including the functions of eNSP4 and iNSP4 in infection and pathogenesis (Table 2.4.2).

Knowledge of RV-induced disease pathogenesis is based primarily on studies using animal rotaviruses, which grow well in cultured cells and in animal models, or in vitro studies in immortalized or cancer cell lines. New models to understand human rotavirus infection are needed. New models that show promise are human intestinal organoids or enteroids produced from approved stem cell lines or from human intestinal biopsies or tissue, respectively; these

TABLE 2.4.2 Questions for Future Research on NSP4

Structure–function	• What is structure of full-length NSP4? • What are the functions of the conserved domains identified in the amino acid alignment of the E genotypes? • How do the different NSP4 interacting domains bind to so many different partners? Structure-dependent? Spatiotemporal-dependent? pH-dependent? Cooperative-binding?
eNSP4	• How does eNSP4 traffic to the plasma membrane for secretion? • Is NSP4-integrin interaction necessary and sufficient for diarrhea induction? • Are there unknown functions for eNSP4?
iNSP4	• How does iNSP4 traffic to different subcellular compartments? • Is the NSP4 viroporin an ion channel? • Does binding of other viral or cellular proteins to domains of iNSP4 modulate any of the iNSP4 functions? • Are other signaling pathways activated by the NSP4-mediated increase in cytoplasm calcium? • How does NSP4 interaction with DLPs trigger the budding process? • Does NSP4 play a role in removal of the transient envelope? • Does NSP4 mediate the assembly of the outer capsid proteins onto DLPs? • Are there unknown functions for iNSP4?

models are the first physiologically active ex vivo model of the human intestine for studying RV-induced pathophysiology (Finkbeiner et al., 2012; Kovbasnjuk et al., 2013; Foulke-Abel et al., 2014; Saxena et al., 2015; Zachos et al., 2015). Use of these nontransformed human miniguts that include all epithelial cell types normally present in the human small intestine may identify the form of eNSP4 produced in the infected human intestinal cells and clarify whether the form of eNSP4 produced and its functions are strain-dependent. These systems will help us better understand the roles of eNSP4 and iNSP4 following infection with human RV strains and may identify new functions for these proteins.

ACKNOWLEDGMENTS

We gratefully acknowledge partial support of our rotavirus research from NIH grants R01 AI080656 and P30 DK56338 (M.K.E.) and R37 AI36040 (B.V.V. Prasad). We also thank Dr Beatrice Boniotti, Istituto Zooprofilattico Sperimentale della Lombardia e dell' Emilia Romagna "Bruno Ubertini," Brescia, Italy for sharing the sequence of the NSP4 from a fox prior to its publication.

REFERENCES

Angel, J., Tang, B., Feng, N., Greenberg, H.B., Bass, D., 1998. Studies of the role for NSP4 in the pathogenesis of homologous murine rotavirus diarrhea. J. Infect. Dis. 177 (2), 455–458.

Araujo, I.T., Heinemann, M.B., Mascarenhas, J.D., Assis, R.M., Fialho, A.M., Leite, J.P., 2007. Molecular analysis of the NSP4 and VP6 genes of rotavirus strains recovered from hospitalized children in Rio de Janeiro, Brazil. J. Med. Microbi. 56 (Pt 6), 854–859.

Arias, C.F., Lopez, S., Espejo, R.T., 1982. Gene protein products of SA11 simian rotavirus genome. J. Virol. 41 (1), 42–50.

Arnoldi, F., De Lorenzo, G., Mano, M., Schraner, E.M., Wild, P., Eichwald, C., et al., 2014. Rotavirus increases levels of lipidated LC3 supporting accumulation of infectious progeny virus without inducing autophagosome formation. PloS One 9 (4), e95197.

Au, K.S., Chan, W.K., Burns, J.W., Estes, M.K., 1989. Receptor activity of rotavirus nonstructural glycoprotein NS28. J. Virol. 63 (11), 4553–4562.

Ball, J.M., Tian, P., Zeng, C.Q., Morris, A.P., Estes, M.K., 1996. Age-dependent diarrhea induced by a rotaviral nonstructural glycoprotein. Science 272 (5258), 101–104.

Ball, J.M., Mitchell, D.M., Gibbons, T.F., Parr, R.D., 2005. Rotavirus NSP4: a multifunctional viral enterotoxin. Viral Immunol. 18 (1), 27–40.

Ball, J.M., Schroeder, M.E., Williams, C.V., Schroeder, F., Parr, R.D., 2013. Mutational analysis of the rotavirus NSP4 enterotoxic domain that binds to caveolin-1. Virol. J. 10, 336.

Beau, I., Cotte-Laffitte, J., Geniteau-Legendre, M., Estes, M.K., Servin, A.L., 2007. An NSP4-dependant mechanism by which rotavirus impairs lactase enzymatic activity in brush border of human enterocyte-like Caco-2 cells. Cell. Microbiol. 9 (9), 2254–2266.

Benati, F.J., Maranhao, A.G., Lima, R.S., da Silva, R.C., Santos, N., 2010. Multiple-gene characterization of rotavirus strains: evidence of genetic linkage among the VP7-, VP4-, VP6-, and NSP4-encoding genes. J. Med. Virol. 82 (10), 1797–1802.

Bergmann, C.C., Maass, D., Poruchynsky, M.S., Atkinson, P.H., Bellamy, A.R., 1989. Topology of the non-structural rotavirus receptor glycoprotein NS28 in the rough endoplasmic reticulum. EMBO J. 8 (6), 1695–1703.

Berkova, Z., Crawford, S.E., Trugnan, G., Yoshimori, T., Morris, A.P., Estes, M.K., 2006. Rotavirus NSP4 induces a novel vesicular compartment regulated by calcium and associated with viroplasms. J. Virol. 80 (12), 6061–6071.

Berkova, Z., Crawford, S.E., Blutt, S.E., Morris, A.P., Estes, M.K., 2007. Expression of rotavirus NSP4 alters the actin network organization through the actin remodeling protein cofilin. J. Virol. 81 (7), 3545–3553.

Bhowmick, R., Halder, U.C., Chattopadhyay, S., Chanda, S., Nandi, S., Bagchi, P., et al., 2012. Rotaviral enterotoxin nonstructural protein 4 targets mitochondria for activation of apoptosis during infection. J. Biol. Chem. 287 (42), 35004–35020.

Bhowmick, R., Halder, U.C., Chattopadhyay, S., Nayak, M.K., Chawla-Sarkar, M., 2013. Rotavirus-encoded nonstructural protein 1 modulates cellular apoptotic machinery by targeting tumor suppressor protein p53. J. Virol. 87 (12), 6840–6850.

Blutt, S.E., Matson, D.O., Crawford, S.E., Staat, M.A., Azimi, P., Bennett, B.L., et al., 2007. Rotavirus antigenemia in children is associated with viremia. PLoS Med. 4 (4), e121.

Borghan, M.A., Mori, Y., El-Mahmoudy, A.B., Ito, N., Sugiyama, M., Takewaki, T., et al., 2007. Induction of nitric oxide synthase by rotavirus enterotoxin NSP4: implication for rotavirus pathogenicity. J. Gen. Virol. 88 (Pt 7), 2064–2072.

Boshuizen, J.A., Reimerink, J.H., Korteland-van Male, A.M., van Ham, V.J., Koopmans, M.P., Buller, H.A., et al., 2003. Changes in small intestinal homeostasis, morphology, and gene expression during rotavirus infection of infant mice. J. Virol. 77 (24), 13005–13016.

Boshuizen, J.A., Rossen, J.W., Sitaram, C.K., Kimenai, F.F., Simons-Oosterhuis, Y., Laffeber, C., et al., 2004. Rotavirus enterotoxin NSP4 binds to the extracellular matrix proteins laminin-beta3 and fibronectin. J. Virol. 78 (18), 10045–10053.

Both, G.W., Siegman, L.J., Bellamy, A.R., Atkinson, P.H., 1983. Coding assignment and nucleotide sequence of simian rotavirus SA11 gene segment 10: location of glycosylation sites suggests that the signal peptide is not cleaved. J. Virol. 48 (2), 335–339.

Bowman, G.D., Nodelman, I.M., Levy, O., Lin, S.L., Tian, P., Zamb, T.J., et al., 2000. Crystal structure of the oligomerization domain of NSP4 from rotavirus reveals a core metal-binding site. J. Mol. Biol. 304 (5), 861–871.

Brunet, J.P., Cotte-Laffitte, J., Linxe, C., Quero, A.M., Geniteau-Legendre, M., Servin, A., 2000. Rotavirus infection induces an increase in intracellular calcium concentration in human intestinal epithelial cells: role in microvillar actin alteration. J. Virol. 74 (5), 2323–2332.

Burns, J.W., Krishnaney, A.A., Vo, P.T., Rouse, R.V., Anderson, L.J., Greenberg, H.B., 1995. Analyses of homologous rotavirus infection in the mouse model. Virology 207 (1), 143–153.

Chacko, A.R., Arifullah, M., Sastri, N.P., Jeyakanthan, J., Ueno, G., Sekar, K., et al., 2011. Novel pentameric structure of the diarrhea-inducing region of the rotavirus enterotoxigenic protein NSP4. J. Virol. 85 (23), 12721–12732.

Chaibi, C., Cotte-Laffitte, J., Sandre, C., Esclatine, A., Servin, A.L., Quero, A.M., et al., 2005. Rotavirus induces apoptosis in fully differentiated human intestinal Caco-2 cells. Virology 332 (2), 480–490.

Chaimongkol, N., Khamrin, P., Malasao, R., Thongprachum, A., Ushijima, H., Maneekarn, N., 2012. Genotypic linkages of gene segments of rotaviruses circulating in pediatric patients with acute gastroenteritis in Thailand. Infect. Genet. Evol. 12 (7), 1381–1391.

Chan, W.K., Au, K.S., Estes, M.K., 1988. Topography of the simian rotavirus nonstructural glycoprotein (NS28) in the endoplasmic reticulum membrane. Virology 164 (2), 435–442.

Chen, F., Wang, H., He, H., Song, L., Wu, J., Gao, Y., et al., 2011. Short hairpin RNA-mediated silencing of bovine rotavirus NSP4 gene prevents diarrhoea in suckling mice. J. Gen. Virol. 92 (Pt 4), 945–951.

Ciarlet, M., Estes, M.K., 1999. Human and most animal rotavirus strains do not require the presence of sialic acid on the cell surface for efficient infectivity. J. Gen. Virol. 80 (Pt 4), 943–948.

Ciarlet, M., Liprandi, F., Conner, M.E., Estes, M.K., 2000. Species specificity and interspecies relatedness of NSP4 genetic groups by comparative NSP4 sequence analyses of animal rotaviruses. Archiv. Virol. 145 (2), 371–383.

Collins, J.E., Benfield, D.A., Duimstra, J.R., 1989. Comparative virulence of two porcine group-A rotavirus isolates in gnotobiotic pigs. Ame. J. Vet. Res. 50 (6), 827–835.

Crawford, S.E., Estes, M.K., 2013. Viroporin-mediated calcium-activated autophagy. Autophagy 9 (5), 797–798.

Crawford, S.E., Hyser, J.M., Utama, B., Estes, M.K., 2012. Autophagy hijacked through viroporin-activated calcium/calmodulin-dependent kinase kinase-beta signaling is required for rotavirus replication. Proc. Natl. Acad. Sci. USA 109 (50), E3405–E3413.

Criglar, J.M., Hu, L., Crawford, S.E., Hyser, J.M., Broughman, J.R., Prasad, B.V., et al., 2014. A novel form of rotavirus NSP2 and phosphorylation-dependent NSP2-NSP5 interactions are associated with viroplasm assembly. J. Virol. 88 (2), 786–798.

Deepa, R., Durga Rao, C., Suguna, K., 2007. Structure of the extended diarrhea-inducing domain of rotavirus enterotoxigenic protein NSP4. Arch. Virol. 152 (5), 847–859.

Diaz, Y., Chemello, M.E., Pena, F., Aristimuno, O.C., Zambrano, J.L., Rojas, H., et al., 2008. Expression of nonstructural rotavirus protein NSP4 mimics Ca^{2+} homeostasis changes induced by rotavirus infection in cultured cells. J. Virol. 82 (22), 11331–11343.

Diaz, Y., Pena, F., Aristimuno, O.C., Matteo, L., De Agrela, M., Chemello, M.E., et al., 2012. Dissecting the Ca(2)(+) entry pathways induced by rotavirus infection and NSP4-EGFP expression in Cos-7 cells. Virus Res. 167 (2), 285–296.

Didsbury, A., Wang, C., Verdon, D., Sewell, M.A., McIntosh, J.D., Taylor, J.A., 2011. Rotavirus NSP4 is secreted from infected cells as an oligomeric lipoprotein and binds to glycosaminoglycans on the surface of non-infected cells. Virol. J. 8, 551.

Dong, Y., Zeng, C.Q., Ball, J.M., Estes, M.K., Morris, A.P., 1997. The rotavirus enterotoxin NSP4 mobilizes intracellular calcium in human intestinal cells by stimulating phospholipase C- mediated inositol 1,4,5-trisphosphate production. Proc. Natl. Acad. Sci. USA 94 (8), 3960–3965.

Ericson, B.L., Graham, D.Y., Mason, B.B., Estes, M.K., 1982. Identification, synthesis, and modifications of simian rotavirus SA11 polypeptides in infected cells. J. Virol. 42 (3), 825–839.

Ericson, B.L., Graham, D.Y., Mason, B.B., Hanssen, H.H., Estes, M.K., 1983. Two types of glycoprotein precursors are produced by the simian rotavirus SA11. Virology 127 (2), 320–332.

Estes, M.K., Kapikian, A.Z., 2007. Rotaviruses. In: Knipe, D.M.H.P. (Ed.), Fields Virology. fifth ed. Lippincott Williams & Wilkins, Philadelphia.

Estes, M.K., Morris, A.P., 1999. A viral enterotoxin. A new mechanism of virus-induced pathogenesis. Adv. Exp. Med. Biol. 473, 73–82.

Estes, M.K., Kang, G., Zeng, C.Q., Crawford, S.E., Ciarlet, M., 2001. Pathogenesis of rotavirus gastroenteritis. Novartis Foundation Symposium, 238, 82–96; discussion-100.

Farthing, M.J., Casburn-Jones, A., Banks, M.R., 2004. Enterotoxins, enteric nerves, and intestinal secretion. Curr. Gastroenterol. Rep. 6 (3), 177–180.

Feng, J., Yang, J., Zheng, S., Qiu, Y., Chai, C., 2011. Silencing of the rotavirus NSP4 protein decreases the incidence of biliary atresia in murine model. PloS One 6 (8), e23655.

Finkbeiner, S.R., Zeng, X.L., Utama, B., Atmar, R.L., Shroyer, N.F., Estes, M.K., 2012. Stem cell-derived human intestinal organoids as an infection model for rotaviruses. mBio 3 (4), e00159–e00212.

Foulke-Abel, J., In, J., Kovbasnjuk, O., Zachos, N.C., Ettayebi, K., Blutt, S.E., et al., 2014. Human enteroids as an ex-vivo model of host-pathogen interactions in the gastrointestinal tract. Exp. Biol. Med. 239 (9), 1124–1134.

Ge, Y., Mansell, A., Ussher, J.E., Brooks, A.E., Manning, K., Wang, C.J., et al., 2013. Rotavirus NSP4 triggers secretion of proinflammatory cytokines from macrophages via toll-Like receptor 2. J. Virol. 87 (20), 11160–11167.

Ghosh, S., Varghese, V., Samajdar, S., Bhattacharya, S.K., Kobayashi, N., Naik, T.N., 2007. Evidence for independent segregation of the VP6- and NSP4- encoding genes in porcine group A rotavirus G6P[13] strains. Arch. Virol. 152 (2), 423–429.

Gibbons, T.F., Storey, S.M., Williams, C.V., McIntosh, A., Mitchel, D.M., Parr, R.D., et al., 2011. Rotavirus NSP4: Cell type-dependent transport kinetics to the exofacial plasma membrane and release from intact infected cells. Virol. J. 8, 278.

Greenberg, H.B., Estes, M.K., 2009. Rotaviruses: from pathogenesis to vaccination. Gastroenterology 136 (6), 1939–1951.

Hagbom, M., Istrate, C., Engblom, D., Karlsson, T., Rodriguez-Diaz, J., Buesa, J., et al., 2011. Rotavirus stimulates release of serotonin (5-HT) from human enterochromaffin cells and activates brain structures involved in nausea and vomiting. PLoS Pathog. 7 (7), e1002115.

Hagbom, M., Sharma, S., Lundgren, O., Svensson, L., 2012. Towards a human rotavirus disease model. Curr. Opin. Virol. 2 (4), 408–418.

Hemming, M., Huhti, L., Rasanen, S., Salminen, M., Vesikari, T., 2014. Rotavirus antigenemia in children is associated with more severe clinical manifestations of acute gastroenteritis. Pediatr. Infect. Dis. J. 33 (4), 366–371.

Horie, Y., Masamune, O., Nakagomi, O., 1997. Three major alleles of rotavirus NSP4 proteins identified by sequence analysis. J. Gen. Virol. 78 (Pt 9), 2341–2346.

Hoshino, Y., Saif, L.J., Kang, S.Y., Sereno, M.M., Chen, W.K., Kapikian, A.Z., 1995. Identification of group A rotavirus genes associated with virulence of a porcine rotavirus and host range restriction of a human rotavirus in the gnotobiotic piglet model. Virology 209 (1), 274–280.

Hou, Z., Huang, Y., Huan, Y., Pang, W., Meng, M., Wang, P., et al., 2008. Anti-NSP4 antibody can block rotavirus-induced diarrhea in mice. J. Pediatr. Gastroenterol. Nutr. 46 (4), 376–385.

Hu, L., Crawford, S.E., Hyser, J.M., Estes, M.K., Prasad, B.V., 2012. Rotavirus non-structural proteins: structure and function. Curr. Opin. Virol. 2 (4), 380–388.

Hyser, J.M., Zeng, C.Q., Beharry, Z., Palzkill, T., Estes, M.K., 2008. Epitope mapping and use of epitope-specific antisera to characterize the VP5* binding site in rotavirus SA11 NSP4. Virology 373 (1), 211–228.

Hyser, J.M., Collinson-Pautz, M.R., Utama, B., Estes, M.K., 2010. Rotavirus disrupts calcium homeostasis by NSP4 viroporin activity. mBio 1 (5).

Hyser, J.M., Utama, B., Crawford, S.E., Broughman, J.R., Estes, M.K., 2013. Activation of the endoplasmic reticulum calcium sensor STIM1 and store-operated calcium entry by rotavirus requires NSP4 viroporin activity. J. Virol. 87 (24), 13579–13588.

Iturriza-Gomara, M., Anderton, E., Kang, G., Gallimore, C., Phillips, W., Desselberger, U., et al., 2003. Evidence for genetic linkage between the gene segments encoding NSP4 and VP6 proteins in common and reassortant human rotavirus strains. J. Clin. Microbiol. 41 (8), 3566–3573.

Jourdan, N., Brunet, J.P., Sapin, C., Blais, A., Cotte-Laffitte, J., Forestier, F., et al., 1998. Rotavirus infection reduces sucrase-isomaltase expression in human intestinal epithelial cells by perturbing protein targeting and organization of microvillar cytoskeleton. J. Virol. 72 (9), 7228–7236.

Kavanagh, O.V., Ajami, N.J., Cheng, E., Ciarlet, M., Guerrero, R.A., Zeng, C.Q., et al., 2010. Rotavirus enterotoxin NSP4 has mucosal adjuvant properties. Vaccine 28 (18), 3106–3111.

Khamrin, P., Maneekarn, N., Malasao, R., Nguyen, T.A., Ishida, S., Okitsu, S., et al., 2010. Genotypic linkages of VP4, VP6, VP7, NSP4, NSP5 genes of rotaviruses circulating among children with acute gastroenteritis in Thailand. Infect. Genet. Evol. MEEGID 10 (4), 467–472.

Ko, E.A., Jin, B.J., Namkung, W., Ma, T., Thiagarajah, J.R., Verkman, A.S., 2014. Chloride channel inhibition by a red wine extract and a synthetic small molecule prevents rotaviral secretory diarrhoea in neonatal mice. Gut 63 (7), 1120–1129.

Kordasti, S., Sjovall, H., Lundgren, O., Svensson, L., 2004. Serotonin and vasoactive intestinal peptide antagonists attenuate rotavirus diarrhoea. Gut 53 (7), 952–957.

Kovbasnjuk, O., Zachos, N.C., In, J., Foulke-Abel, J., Ettayebi, K., Hyser, J.M., et al., 2013. Human enteroids: preclinical models of non-inflammatory diarrhea. Stem Cell Res. Ther. 4 (Suppl. 1), S3.

Lee, C.N., Wang, Y.L., Kao, C.L., Zao, C.L., Lee, C.Y., Chen, H.N., 2000. NSP4 gene analysis of rotaviruses recovered from infected children with and without diarrhea. J. Clin. Microbiol. 38 (12), 4471–4477.

Lorrot, M., Vasseur, M., 2006. Rotavirus NSP4 114-135 peptide has no direct, specific effect on chloride transport in rabbit brush-border membrane. Virol. J. 3, 94.

Lundgren, O., Jodal, M., 1997. The enteric nervous system and cholera toxin-induced secretion. Comp. Biochem. Physiol. Part A 118 (2), 319–327.

Lundgren, O., Svensson, L., 2001. Pathogenesis of rotavirus diarrhea. Microbes Infect. 3 (13), 1145–1156.

Lundgren, O., Peregrin, A.T., Persson, K., Kordasti, S., Uhnoo, I., Svensson, L., 2000. Role of the enteric nervous system in the fluid and electrolyte secretion of rotavirus diarrhea. Science 287 (5452), 491–495.

Maass, D.R., Atkinson, P.H., 1990. Rotavirus proteins VP7, NS28, and VP4 form oligomeric structures. J. Virol. 64 (6), 2632–2641.

Malik, Y.S., Kumar, N., Sharma, K., Ghosh, S., Banyai, K., Balasubramanian, G., et al., 2014. Molecular analysis of non structural rotavirus group A enterotoxin gene of bovine origin from India. Infect. Genet. Evol. MEEGID 25, 20–27.

Martin-Latil, S., Mousson, L., Autret, A., Colbere-Garapin, F., Blondel, B., 2007. Bax is activated during rotavirus-induced apoptosis through the mitochondrial pathway. J. Virol. 81 (9), 4457–4464.

Maruri-Avidal, L., Lopez, S., Arias, C.F., 2008. Endoplasmic reticulum chaperones are involved in the morphogenesis of rotavirus infectious particles. J. Virol. 82 (11), 5368–5380.

Matthijnssens, J., Ciarlet, M., Heiman, E., Arijs, I., Delbeke, T., McDonald, S.M., et al., 2008a. Full genome-based classification of rotaviruses reveals a common origin between human Wa-Like and porcine rotavirus strains and human DS-1-like and bovine rotavirus strains. J. Virol. 82 (7), 3204–3219.

Matthijnssens, J., Ciarlet, M., Rahman, M., Attoui, H., Banyai, K., Estes, M.K., et al., 2008b. Recommendations for the classification of group A rotaviruses using all 11 genomic RNA segments. Arch. Virol. 153 (8), 1621–1629.

Matthijnssens, J., Ciarlet, M., McDonald, S.M., Attoui, H., Banyai, K., Brister, J.R., et al., 2011. Uniformity of rotavirus strain nomenclature proposed by the Rotavirus Classification Working Group (RCWG). Arch. Virol. 156 (8), 1397–1413.

Mebus, C.A., Newman, L.E., Stair, Jr., E.L., 1975. Scanning electron, light, and immunofluorescent microscopy of intestine of gnotobiotic calf infected with calf diarrheal coronavirus. Am. J. Vet. Res. 36 (12), 1719–1725.

Meyer, J.C., Bergmann, C.C., Bellamy, A.R., 1989. Interaction of rotavirus cores with the nonstructural glycoprotein NS28. Virology 171 (1), 98–107.

Mirazimi, A., Nilsson, M., Svensson, L., 1998. The molecular chaperone calnexin interacts with the NSP4 enterotoxin of rotavirus in vivo and in vitro. J. Virol. 72 (11), 8705–8709.

Mirazimi, A., Magnusson, K.E., Svensson, L., 2003. A cytoplasmic region of the NSP4 enterotoxin of rotavirus is involved in retention in the endoplasmic reticulum. J. Gen. Virol. 84 (Pt 4), 875–883.

Mohan, K.V., Kulkarni, S., Glass, R.I., Zhisheng, B., Atreya, C.D., 2003. A human vaccine strain of lamb rotavirus (Chinese) NSP4 gene: complete nucleotide sequence and phylogenetic analyses. Virus Genes 26 (2), 185–192.

Mori, Y., Borgan, M.A., Ito, N., Sugiyama, M., Minamoto, N., 2002. Sequential analysis of nonstructural protein NSP4s derived from Group A avian rotaviruses. Virus Res. 89 (1), 145–151.

Morris, A.P., Estes, M.K., 2001. Microbes and microbial toxins: paradigms for microbial-mucosal interactions. VIII. Pathological consequences of rotavirus infection and its enterotoxin. Am. J. Physiol. Gastrointest. Liver Physiol. 281 (2), G303–G310.

Morris, A.P., Scott, J.K., Ball, J.M., Zeng, C.Q., O'Neal, W.K., Estes, M.K., 1999. NSP4 elicits age-dependent diarrhea and Ca(2 +)mediated I(-) influx into intestinal crypts of CF mice. Am. J. Physiol. 277 (2 Pt 1), G431–g444.

Obert, G., Peiffer, I., Servin, A.L., 2000. Rotavirus-induced structural and functional alterations in tight junctions of polarized intestinal Caco-2 cell monolayers. J. Virol. 74 (10), 4645–4651.

O'Brien, J.A., Taylor, J.A., Bellamy, A.R., 2000. Probing the structure of rotavirus NSP4: a short sequence at the extreme C terminus mediates binding to the inner capsid particle. J. Virol. 74 (11), 5388–5394.

Osborne, M.P., Haddon, S.J., Spencer, A.J., Collins, J., Starkey, W.G., Wallis, T.S., et al., 1988. An electron microscopic investigation of time-related changes in the intestine of neonatal mice infected with murine rotavirus. J. Pediatr. Gastroenterol. Nutr. 7 (2), 236–248.

Ousingsawat, J., Mirza, M., Tian, Y., Roussa, E., Schreiber, R., Cook, D.I., et al., 2011. Rotavirus toxin NSP4 induces diarrhea by activation of TMEM16A and inhibition of Na+ absorption. Pflugers Archiv: Eur. J. Physiol. 461 (5), 579–589.

Parr, R.D., Storey, S.M., Mitchell, D.M., McIntosh, A.L., Zhou, M., Mir, K.D., et al., 2006. The rotavirus enterotoxin NSP4 directly interacts with the caveolar structural protein caveolin-1. J. Virol. 80 (6), 2842–2854.

Perez, J.F., Ruiz, M.C., Chemello, M.E., Michelangeli, F., 1999. Characterization of a membrane calcium pathway induced by rotavirus infection in cultured cells. J. Virol. 73 (3), 2481–2490.

Rahman, M., Matthijnssens, J., Yang, X., Delbeke, T., Arijs, I., Taniguchi, K., et al., 2007. Evolutionary history and global spread of the emerging g12 human rotaviruses. J. Virol. 81 (5), 2382–2390.

Rajasekaran, D., Sastri, N.P., Marathahalli, J.R., Indi, S.S., Pamidimukkala, K., Suguna, K., et al., 2008. The flexible C terminus of the rotavirus non-structural protein NSP4 is an important determinant of its biological properties. J. Gen. Virol. 89 (Pt 6), 1485–1496.

Rodriguez-Diaz, J., Banasaz, M., Istrate, C., Buesa, J., Lundgren, O., Espinoza, F., et al., 2006. Role of nitric oxide during rotavirus infection. J. Med. Virol. 78 (7), 979–985.

Rollo, E.E., Kumar, K.P., Reich, N.C., Cohen, J., Angel, J., Greenberg, H.B., et al., 1999. The epithelial cell response to rotavirus infection. J. Immunol. 163 (8), 4442–4452.

Ruiz, M.C., Cohen, J., Michelangeli, F., 2000. Role of Ca^{2+} in the replication and pathogenesis of rotavirus and other viral infections. Cell Calcium 28 (3), 137–149.

Sastri, N.P., Pamidimukkala, K., Marathahalli, J.R., Kaza, S., Rao, C.D., 2011. Conformational differences unfold a wide range of enterotoxigenic abilities exhibited by rNSP4 peptides from different rotavirus strains. Open Virol. J. 5, 124–135.

Sastri, N.P., Viskovska, M., Hyser, J.M., Tanner, M.R., Horton, L.B., Sankaran, B., et al., 2014. Structural plasticity of the coiled-coil domain of rotavirus NSP4. J. Virol. 88 (23), 13602–13612.

Saxena, K., Blutt, S.E., Ettayebi, K., Zeng, X.L., Broughman, J.R., Crawford, S.E., et al., 2015. Human intestinal enteroids: a new model to study human rotavirus infection, host restriction, and pathophysiology. J. Virol. 90 (1), 43–56.

Schroeder, M.E., Hostetler, H.A., Schroeder, F., Ball, J.M., 2012. Elucidation of the rotavirus NSP4-caveolin-1 and -cholesterol interactions using synthetic peptides. J. Amino Acids 2012, 575180.

Seo, N.S., Zeng, C.Q., Hyser, J.M., Utama, B., Crawford, S.E., Kim, K.J., et al., 2008. Integrins alpha1beta1 and alpha2beta1 are receptors for the rotavirus enterotoxin. Proc. Natl. Acad. Sci. USA 105 (26), 8811–8818.

Silvestri, L.S., Tortorici, M.A., Vasquez-Del Carpio, R., Patton, J.T., 2005. Rotavirus glycoprotein NSP4 is a modulator of viral transcription in the infected cell. J. Virol. 79 (24), 15165–15174.

Storey, S.M., Gibbons, T.F., Williams, C.V., Parr, R.D., Schroeder, F., Ball, J.M., 2007. Full-length, glycosylated NSP4 is localized to plasma membrane caveolae by a novel raft isolation technique. J. Virol. 81 (11), 5472–5483.

Sugata, K., Taniguchi, K., Yui, A., Miyake, F., Suga, S., Asano, Y., et al., 2008. Analysis of rotavirus antigenemia and extraintestinal manifestations in children with rotavirus gastroenteritis. Pediatrics 122 (2), 392–397.

Superti, F., Amici, C., Tinari, A., Donelli, G., Santoro, M.G., 1998. Inhibition of rotavirus replication by prostaglandin A: evidence for a block of virus maturation. J. Infect. Dis. 178 (2), 564–568.

Tafazoli, F., Zeng, C.Q., Estes, M.K., Magnusson, K.E., Svensson, L., 2001. NSP4 enterotoxin of rotavirus induces paracellular leakage in polarized epithelial cells. J. Virol. 75 (3), 1540–1546.

Tamura, K., Stecher, G., Peterson, D., Filipski, A., Kumar, S., 2013. MEGA6: Molecular Evolutionary Genetics Analysis version 6.0. Mol. Biol. Evol. 30 (12), 2725–2729.

Tavares Tde, M., de Brito, W.M., Fiaccadori, F.S., Parente, J.A., da Costa, P.S., Giugliano, L.G., et al., 2008. Molecular characterization of VP6-encoding gene of group A human rotavirus samples from central west region of Brazil. J. Med. Virol. 80 (11), 2034–2039.

Tian, P., Hu, Y., Schilling, W.P., Lindsay, D.A., Eiden, J., Estes, M.K., 1994. The nonstructural glycoprotein of rotavirus affects intracellular calcium levels. J. Virol. 68 (1), 251–257.

Tian, P., Estes, M.K., Hu, Y., Ball, J.M., Zeng, C.Q., Schilling, W.P., 1995. The rotavirus nonstructural glycoprotein NSP4 mobilizes Ca2+ from the endoplasmic reticulum. J. Virol. 69 (9), 5763–5772.

Tsugawa, T., Tatsumi, M., Tsutsumi, H., 2014. Virulence-associated genome mutations of murine rotavirus identified by alternating serial passages in mice and cell cultures. J. Virol. 88 (10), 5543–5558.

Uittenbogaard, A., Smart, E.J., 2000. Palmitoylation of caveolin-1 is required for cholesterol binding, chaperone complex formation, and rapid transport of cholesterol to caveolae. J. Biol. Chem. 275 (33), 25595–25599.

Vende, P., Gratia, M., Duarte, M.D., Charpilienne, A., Saguy, M., Poncet, D., 2013. Identification of mutations in the genome of rotavirus SA11 temperature-sensitive mutants D, H, I and J by whole genome sequences analysis and assignment of tsI to gene 7 encoding NSP3. Virus Res. 176 (1–2), 144–154.

Ward, L.A., Rosen, B.I., Yuan, L., Saif, L.J., 1996. Pathogenesis of an attenuated and a virulent strain of group A human rotavirus in neonatal gnotobiotic pigs. J. Gen. Virol. 77 (Pt 7), 1431–1441.

Xu, A., Bellamy, A.R., Taylor, J.A., 2000. Immobilization of the early secretory pathway by a virus glycoprotein that binds to microtubules. EMBO J. 19 (23), 6465–6474.

Zachos, N.C., Kovbasnjuk, O., Foulke-Abel, J., 2015. In: J, Blutt SE, deJonge HR, et al., 2015. Human enteroids/colonoids and intestinal organoids functionally recapitulate normal intestinal physiology and pathophysiology. J. Biol. Chem.

Zambrano, J.L., Diaz, Y., Pena, F., Vizzi, E., Ruiz, M.C., Michelangeli, F., et al., 2008. Silencing of rotavirus NSP4 or VP7 expression reduces alterations in Ca2+ homeostasis induced by infection of cultured cells. J. Virol. 82 (12), 5815–5824.

Zambrano, J.L., Sorondo, O., Alcala, A., Vizzi, E., Diaz, Y., Ruiz, M.C., et al., 2012. Rotavirus infection of cells in culture induces activation of RhoA and changes in the actin and tubulin cytoskeleton. PloS One 7 (10), e47612.

Zhang, M., Zeng, C.Q., Dong, Y., Ball, J.M., Saif, L.J., Morris, A.P., et al., 1998. Mutations in rotavirus nonstructural glycoprotein NSP4 are associated with altered virus virulence. J. Virol. 72 (5), 3666–3672.

Zhang, M., Zeng, C.Q., Morris, A.P., Estes, M.K., 2000. A functional NSP4 enterotoxin peptide secreted from rotavirus-infected cells. J. Virol. 74 (24), 11663–11670.

Zheng, S., Zhang, H., Zhang, X., Peng, F., Chen, X., Yang, J., et al., 2014. CD8+ T lymphocyte response against extrahepatic biliary epithelium is activated by epitopes within NSP4 in experimental biliary atresia. Am. J. Physiol. Gastrointest. Liver Physiol. 307 (2), G233–G240.

Chapter 2.5

Rotavirus Replication: the Role of Lipid Droplets

W. Cheung, E. Gaunt, A. Lever, U. Desselberger
Department of Medicine, University of Cambridge, Addenbrooke's Hospital, Cambridge, United Kingdom

1 INTRODUCTION

Rotaviruses (RVs), a genus of the *Reoviridae* family, are a major cause of acute gastroenteritis (AGE) in infants and young children and in the young of a large variety of other mammalian and avian species. The genome of *Rotavirus* consists of 11 segments of double-stranded (ds) RNA, which encode 6 structural (VP) and 5–6 nonstructural (NSP) proteins. Upon entry into gut epithelial cells, the infectious, triple-layered RV particles (TLPs) are processed to become double-layered particles (DLPs), which actively transcribe single-stranded (ss) RNAs of (+) polarity from all 11 genomic segments to be released into the cytoplasm. These RNAs act either as mRNAs and are translated into viral proteins, or they become templates for progeny genomes and are replicated into dsRNAs to be packaged into newly formed virion particles. Early virion morphogenesis and viral RNA replication occur in cytoplasmic inclusion bodies termed "viroplasms". DLPs released from viroplasms "mature" to TLPs in the endoplasmic reticulum (ER) before they are released from cells, either by lysis or a budding process (for details see Estes and Greenberg, 2013). The RV nonstructural proteins NSP2 and NSP5 are essential components of viroplasms. The transfection of NSP2- and NSP5-expressing plasmids into uninfected cells leads to the formation of viroplasm-like structures (VLS) (Fabbretti et al., 1999). Disruption of NSP2 or NSP5 synthesis in RV-infected cells by specific siRNAs, intrabodies or the use of NSP2- or NSP5-specific temperature sensitive mutants at the nonpermissive temperature abolishes viroplasm formation and the production of infectious viral progeny (Silvestri et al., 2004; Vascotto et al., 2004; Campagna et al., 2005).

Throughout its replication cycle, RV depends on and interacts with various components of the host cell. Here the interaction of the RV viroplasms with the cellular organelles lipid droplets is described.

2 VIROPLASMS

RV-encoded NSP2 and NSP5 are essential components of viroplasms, which also contain several other viral proteins including VP1, VP2, VP3, VP6, and NSP4. NSP2 forms octamers (Taraporewala et al., 2002); grooves in the NSP2 octamer are binding sites for which NSP5 dimers and ssRNAs compete, possibly regulating the balance between RV RNA replication and translation (Jiang et al., 2006). In addition NSP2 interacts with tubulin (Martin et al., 2010; Criglar et al., 2014) and induces microtubule depolymerization and stabilization by acetylation (Eichwald et al., 2012). Cellular components of the autophagy pathway are also involved in viroplasm formation and RV replication (Berkova et al., 2006; Berkova et al., 2007; Contin et al., 2011; López et al., 2011; Crawford et al., 2012; Arnoldi et al., 2014).

3 LIPID DROPLETS

Lipid droplets (LDs) are the principal cellular storage sites of triacylglycerols and sterol esters (Herker and Ott, 2012) and are surrounded by a phospholipid monolayer into which numerous proteins including perilipin A and adipose differentiation-related protein (ADRP) are inserted (Martin and Parton, 2006; Walther and Farese, 2012; Brasaemle and Wolins, 2012). LDs are dynamic organelles with functions in the cell other than energy storage including protein trafficking and interaction with the other cellular components such as mitochondria, lysosomes, and the plasma membrane (Heaton and Randall, 2011; Saka and Valdivia, 2012). Examination of cellular components associated with viroplasms revealed proteins characteristic of cellular LDs.

4 INTERACTION OF VIROPLASMS WITH LIPID DROPLETS

Viroplasms interact with and possibly trigger the formation of LDs, inducing and recruiting them during the replication cycle (Cheung et al., 2010; Saxena et al., 2015). The initial discovery of the interaction of RV viroplasms with LDs was based on the observation, in RV-infected MA104 and Caco-2 cells, of colocalization of viroplasm-associated viral proteins (eg, NSP2 and NSP5) with cellular proteins typically found inserted into the LD-surface phospholipid monolayer (perilipin A and ADRP) (Cheung et al., 2010; Cheung, 2010; Fig. 2.5.1). LDs were also found to colocalize with VLS (Fabbretti et al., 1999) in uninfected cells (Cheung et al., 2010), indicating that complexes of coexpressed NSP2 and NSP5 are sufficient to interact with LDs. In addition, MA104 cells constitutively expressing the fusion protein NSP5-EGFP and infected with RV showed—as expected—concentration of NSP5-EGFP in viroplasms with close spatial proximity to LD-associated proteins as evidenced by fluorescence resonance energy transfer (FRET) (Cheung et al., 2010).

LD recruitment (measured by perilipin A localization) was found to depend on viroplasm formation in rotavirus-infected cells. MA104 cells pretreated with

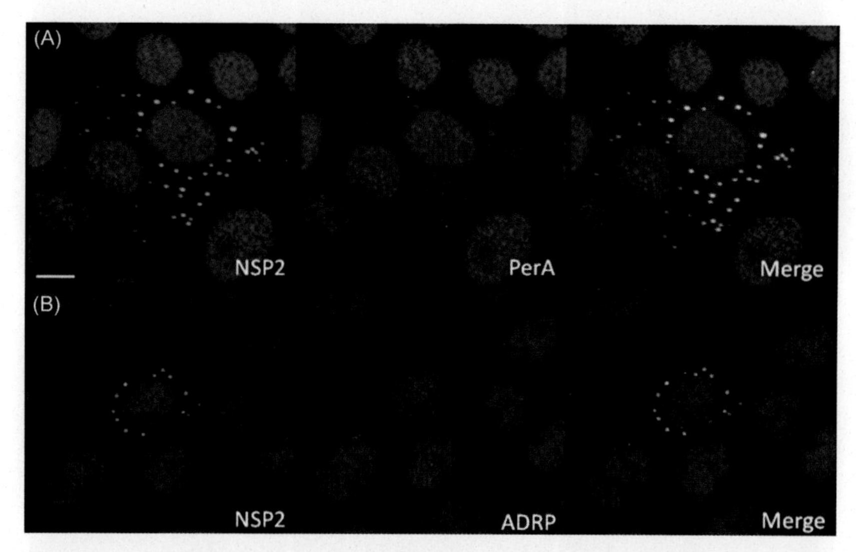

FIGURE 2.5.1 **NSP2 colocalizes with lipid droplet-associated proteins perilipin A and ADRP in viroplasms of rotavirus-infected Caco-2 cells.** Confocal images of rotavirus-infected Caco-2 cells at 8 h p.i. Viroplasms were detected with anti-NSP2 antibodies followed by visualization with Alexa Fluor 488 *(green)*-labeled species-specific secondary antibody, while LD-associated proteins were detected with anti-perilipin A and anti-ADRP antibodies followed by reaction with Alexa Fluor 633 *(red)*-labeled species-specific secondary antibody. Scale bar: 10 μm. *(Source: From Cheung W, PhD Thesis, University of Cambridge, Cambridge, 2010.)*

siRNAs targeting NSP5 mRNAs [of either the porcine OSU (siOSU) or the simian SA11 (siSA11) strains] (Campagna et al., 2005) were infected with the OSU rotavirus strain for 6 h, and the localization of NSP5 and perilipin A was investigated using immunofluorescence and confocal microscopy. As shown in Fig. 2.5.2, siOSU largely blocked both, viroplasm formation and perilipin A recruitment, in comparison to cells treated with an irrelevant siRNA (siSA11) (Cheung, 2010). The yield of infectious progeny from the siOSU treated cells was 100 times lower than that of the siSA11-treated or the untreated cells (data not shown) (Cheung, 2010).

RV-infected cells contain significantly higher concentrations of lipids than do uninfected cells (Kim and Chang, 2011; Gaunt et al., 2013b). The link between viroplasms and LDs was further strengthened by equilibrium ultracentrifugation through iodixanol gradients of RV-infected cell extracts (detergent-free): viral dsRNA sedimented in the same low-density fractions (1.11–1.15 g/mL) as NSP5 and perilipin A, markers of viroplasms and LDs, respectively. By contrast, purified RV DLPs spiked into uninfected cell extracts passed through the gradient without localization in the low density fractions (Cheung et al., 2010). Lipid components preferentially found in LDs (triacylglycerol, ceramide, sphingomyelin, phosphatidylinositol, phosphatidic acid) were also concentrated in the low-density fractions, further supporting the physical interaction between viroplasms and LDs (Gaunt et al., 2013b).

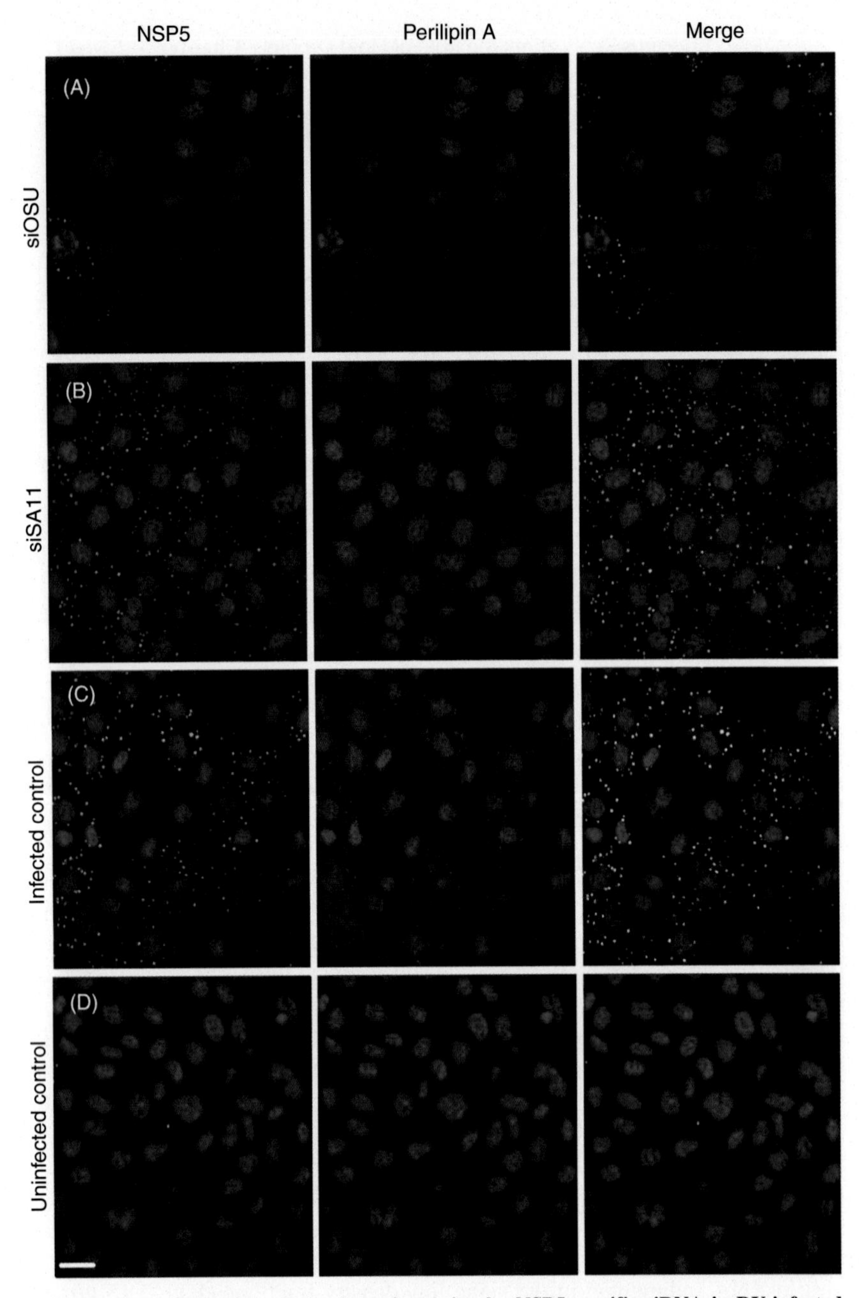

FIGURE 2.5.2 **Inhibition of viroplasm formation by NSP5-specific siRNA in RV-infected cells affects recruitment of perilipin A and by implication of LDs.** MA104 cells pretreated with siRNAs directed against NSP5 of RV strains OSU (panel A) or SA11 (panel B) were infected with the OSU RV strain at a m.o.i. of 5 and fixed at 6 h p.i., followed by staining with NSP5- *(green)* and perilipin A- *(red)* specific antibodies (as indicated in Legend of Figure 2.5.1). siRNA directed against the OSU strain inhibits viroplasm formation in most cells (panel A), while siRNA directly against the SA11 strain has not effect (panel B). Untreated, infected (panel C) and uninfected (panel D) cells served as controls. Scale bar: 20 μm. *(Source: From Cheung W, PhD Thesis, University of Cambridge, Cambridge, 2010.)*

5 LIPID DROPLET HOMOEOSTASIS AND ROTAVIRUS REPLICATION

The association between RV "factories" and the cellular energy reservoirs prompted exploration of whether disturbance of the homoeostasis of cellular lipid metabolism might also affect RV replication. Treatment of cells with a combination of isoproterenol, a beta-adrenergic agonist, and isobutylmethylxanthine (IBMX), a phosphodiesterase inhibitor, raises the level of intracellular cAMP, leading to phosphorylation of hormone-dependent lipase, which catalyses fragmentation of LDs (lipolysis) (Gross et al., 2006; Marcinkiewicz et al., 2006). LD formation can also be inhibited by antagonists of various enzymes involved in fatty acid synthesis. Triacsin C [*N*-(((2*E*,4*E*,7*E*)-undeca-2,4,7-trienylidene) amino) nitrous amide] is a specific inhibitor of long chain acyl coenzyme A synthetases (Igal et al., 1997; Namatame et al., 1999) and prevents LD formation in Huh7 cells (Zou et al., 2010); C75 (tetrahydro-4-methylene-2R-octyl-5-oxo-3S-furancarboxylic acid) is an inhibitor of the fatty acid synthase (FAS) complex and significantly reduces LD accumulation (Schmid et al., 2005); TOFA [5-(tetradecyloxy)-2-furoic acid] inhibits the enzyme acetyl-CoA carboxylase 1 (ACC1), acting early in the fatty acid biosynthesis pathway (Parker et al., 1977; Halvorson and McCune, 1984; Fukuda and Ontko, 1984).

Pretreatment of cells with nontoxic concentrations of all of these: a combination of [isoproterenol + IBMX], or triacsin C, or TOFA alone reduced the number and size of viroplasms generated by a subsequent RV infection (data not shown). The amounts of newly synthesized RV dsRNA declined by 3.8- to 5.9-fold, and the infectivity of viral progeny by 20- to 50-fold (Table 2.5.1) (Cheung et al., 2010; Gaunt et al., 2013a). Interestingly, treatment appeared to protect cells

TABLE 2.5.1 Comparison of Inhibitory Effects of Different Compounds Affecting Lipid Droplet Homoeostasis on Rotavirus Replication

Treatment of cells		Viral dsRNA		Infectivity of progeny	
		Relative values[a]	Diff[b]	log TCID50/mL[c]	Diff[b]
Isoproterenol	–	1.00		8.2	
+ IBMX[d]	+	0.25	4.0-fold	6.5	50-fold
Triacsin C[d]	–	1.00		7.5	
	+	0.26	3.8-fold	6.2	20-fold
TOFA[e]	–	1.00		8.4	
	+	0.17	5.9-fold	6.7	50-fold

[a]*Calculated from densitometric values of RNA gels (Cheung, 2010)*
[b]*Underlining indicates statistical difference.*
[c]*S.E. values not shown*
[d]*From: Cheung et al., J. Virol. 2010; 84: 6782–6798.*
[e]*From: Gaunt et al., J. Gen. Virol. 2013a: 94: 1310–1317.*

from RV-induced cytopathicity since a higher percentage of infected, drug-treated cells was viable at 16 h p.i. compared to those infected but not treated (Cheung et al., 2010). TOFA still had an inhibitory effect when added to RV-infected cells as late as at 4 h after infection. Specific siRNA knockdown of ACC1 (verified by Western blot) produced a qualitatively similar effect to chemical ACC1 inhibition (Gaunt et al., 2013a). Triacsin C and analogs were also found to inhibit RV replication by Kim et al. (2012). Inhibition of RV replication by C75 was only marginal, most likely due to the low chemotherapeutic index (Gaunt et al., 2013a). More recently it was shown that inhibitors of lipolysis and of transport of fatty acids to mitochondria decreased RV replication by 94–97% (Crawford et al., 2013), but the mechanism of action remains to be fully understood.

Interestingly, the decrease of RV RNA production in TOFA-treated cells was disproportionately smaller than the decrease in infectivity (Gaunt et al., 2013a). This effect was analysed further by CsCl gradient ultracentrifugation of RVs (Arnoldi et al., 2007) synthesized after infection of untreated or TOFA-treated cells. TOFA treatment led to a 2-fold reduction in double-layered particles (DLPs) but a 20-fold reduction in triple-layered particles (TLPs, the infectious virions), compared to untreated cells (Table 2.5.2). These data were confirmed by electrophoresis of DLPs and TLPs on nondenaturing agarose gels and subsequent densitometry; the infectivity of RV progeny from the TOFA-treated cells was also decreased by 10-fold (data not shown). The results suggest that TOFA treatment, apart from interfering with DLP synthesis in viroplasms, may also affect the lipid composition of the endoplasmic reticulum (ER) where RV particle maturation from DLPs to TLPs occurs.

Farnesoid X receptor (FXR) and its natural ligands bile acids [such as chenodeoxycholic acid (CDCA)] play major roles in cholesterol and lipid homeostasis (Makishima et al., 1999; Parks et al., 1999; Trauner and Boyer, 2003; Watanabe et al., 2004). Treatment of MA104 cells with CDCA, deoxycholic acid (DCA), and other FXR agonists led to a reduction of cellular triglyceride content (while RV infection, as mentioned, increased it); after RV infection, this treatment was correlated with a significant reduction of viral replication in a dose-dependent

TABLE 2.5.2 Effect of TOFA on the Production of Rotavirus DLPs and TLPs ug/mL Protein Ratio TOFA/Untr

ug/mL protein				Ratio TOFA/Untr	
TOFA		Untr			
DLP	TLP	DLP	TLP	DLP	TLP
7.6	1.2	21.9	18.2	0.48	0.05
±2.6[a]	±1.1	±13.1	±10.4	± 0.19	±0.03

[a]*Arithmetic mean ± standard error (N = 2)*
TOFA, [5-(Tetradecyloxy)-2-furoic acid]; *DLP*, double-layered particles; *TLP*, triple-layered RV particles

manner, with downregulation of cellular lipids as a possible contributing factor. In a mouse model of RV infection, oral administration of CDCA significantly reduced fecal RV shedding (Kim and Chang, 2011), suggesting that FXR agonists could become a treatment option against RV disease. However, more extensive experiments with animal models of RV infection are required to pursue this idea.

Infection of HT29 cells with RV in the presence of nontoxic concentrations of stilbenoids, cannabinoid receptor antagonists, led to a 10- to 20-fold decrease of the infectivity of viral progeny; one of the suggested mechanisms of action was interference with lipid homoeostasis, since stilbenoids are very lipophilic (Ball et al., 2015). However, more work is needed to explore this interesting observation in more detail.

6 LIPID DROPLETS, LIPID HOMOEOSTASIS AND REPLICATION OF VIRUSES AND OTHER MICROBES

Rotaviruses have joined the growing list of viruses and microbes that interact with LDs during their replication and depend on cellular lipid homoeostasis, including hepatitis C virus (Miyanari et al., 2007; Boulant et al., 2007; Shavinskaya et al., 2007; Salloum et al., 2013; Paul et al., 2014; Liefhebber et al., 2014; Shahidi et al., 2014; Filipe et al., 2015), dengue virus (Samsa et al., 2009; Jain et al., 2014; Soto-Acosta et al., 2014), GB virus B (Hope et al., 2002), bunyavirus (Wu et al., 2014) and the intracellular parasites *Chlamydia* (Kumar et al., 2006), *Mycobacterium tuberculosis* (Daniel et al., 2011) and *Mycobacterium leprae* (Mattos et al., 2011). On a wider scale, FA biosynthesis has been recently recognized as being essential for the replication of a larger number of viruses such as enteroviruses, West Nile virus, human cytomegalovirus, Kaposi sarcoma-associated herpes virus and Epstein Barr virus (Chukkapalli et al., 2012).

7 FUTURE WORK ON LIPID DROPLETS AND ROTAVIRUS REPLICATION

Future work should be aimed at identifying which genes/proteins involved in LD homoeostasis (Guo et al., 2008) are important for the interaction of LDs with viroplasms and what is the basis for the observed inhibition of TLP compared to DLP production in TOFA-treated cells. Most importantly, the in vivo effects of (isoproterenol + IBMX) and TOFA should be explored in animal models of RV infection.

8 OTHER CELLULAR PROTEINS INVOLVED IN ROTAVIRUS REPLICATION

Throughout its replication cycle RV depends on the interaction with many different cellular proteins. The RV surface proteins VP4 (VP8*) and VP7 interact with various cellular receptor molecules and undergo structural alterations

(López and Arias, 2004; Hu et al., 2012; Díaz-Salinas et al., 2014; Abdelhakim et al., 2014) as described in detail in Chapters 2.1 and 2.2.

For viral translation the N terminus of the RV-encoded NSP3 interacts with the 3′ terminus of viral (+) ssRNA and the C terminus of NSP3 with the translation factor eIF4G which in turn interacts with the 5′ terminus of the mRNA, leading to RNA circularization (Groft and Burley, 2002). More importantly, NSP3 can displace the cellular poly A binding protein (PABP) from cellular mRNAs and thus prevent their translation (Piron et al., 1998; Montero et al., 2006; Harb et al., 2008; Rubio et al., 2013). Interestingly, after transfection into susceptible cells full length RV (+) ssRNAs are translated very poorly (or not at all), despite them being functional in in vitro translation systems. This suggests that viral mRNAs, after being extruded from DLPs into the cytoplasm, encounter ribosomes immediately and are translated without delay (Richards et al., 2013).

The cytoskeleton protein actin interacts with RV VP4 to be remodeled into actin bodies (Gardet et al., 2006; Gardet et al. 2007). Since actin rearrangements are Ca^{2+} dependent, they are also regulated by RV NSP4 (Berkova et al., 2007). NSP4 is a RV-encoded transmembrane glycoprotein which acts as a viroporin (Hyser et al., 2010, 2012) and also with the autophagy pathway (Crawford et al., 2012; Crawford and Estes, 2013). The cellular ubiquitin-proteasome system (UPS) plays a significant role in the replication of many viruses (Isaacson and Ploegh, 2009). For RVs it has been shown that UPS inhibition reduces recruitment of viral proteins to viroplasms and viral RNA replication (Contin et al., 2011; López et al., 2011; Crawford et al., 2012). These and other activities of NSP4 are described in detail in Chapter 2.4.

Rotavirus infection activates different cascades of the cellular innate immune response (IIR) (Sen et al., 2011; Sen et al., 2012; Angel et al., 2012). The viral NSP1 is an IIR antagonist (Barro and Patton, 2005; Barro and Patton, 2007; Arnold et al., 2013). The details of the virus–host relationship regarding the IIR are presented in Chapter 2.8.

9 CONCLUSIONS

Rotavirus replication in cells depends on an intense interplay of viral and cellular factors throughout the viral growth cycle. Here, details of the interaction of viroplasms with lipid droplets have been presented, including the demonstration that interference with lipid droplet homoeostasis reduces the amounts of synthesized rotavirus significantly. These observations may provide intriguing new therapeutic approaches to combat this ubiquitous pathogen.

REFERENCES

Abdelhakim, A.H., Salgado, E.N., Fu, X., Pasham, M., Nicastro, D., Kirchhausen, T., Harrison, S.C., 2014. Structural correlates of rotavirus cell entry. PLoS Pathog. 10 (9), e1004355.

Angel, J., Franco, M.A., Greenberg, H.B., 2012. Rotavirus immune responses and correlates of protection. Curr. Opin. Virol. 2 (4), 419–425.

Arnold, M.M., Sen, A., Greenberg, H.B., Patton, J.T., 2013. The battle between rotavirus and its host for control of the interferon signaling pathway. PLoS Pathog. 9 (1), e1003064.

Arnoldi, F., Campagna, M., Eichwald, C., Desselberger, U., Burrone, O.R., 2007. Interaction of rotavirus polymerase VP1 with nonstructural protein NSP5 is stronger than that with NSP2. J. Virol. 81 (5), 2128–2137.

Arnoldi, F., De Lorenzo, G., Mano, M., Schraner, E.M., Wild, P., Eichwald, C., Burrone, O.R., 2014. Rotavirus increases levels of lipidated LC3 supporting accumulation of infectious progeny virus without inducing autophagosome formation. PLoS One 9 (4), e95197.

Ball, J.M., Medina-Bolivar, F., Defrates, K., Hambleton, E., Hurlburt, M.E., Fang, L., Yang, T., Nopo-Olazabal, L., Atwill, R.L., Ghai, P., Parr, R.D., 2015. Investigation of stilbenoids as potential therapeutic agents for rotavirus gastroenteritis. Adv. Virol. 2015, 293524.

Barro, M., Patton, J.T., 2005. Rotavirus nonstructural protein 1 subverts innate immune response by inducing degradation of IFN regulatory factor 3. Proc. Natl. Acad. Sci. USA 102 (11), 4114–4119.

Barro, M., Patton, J.T., 2007. Rotavirus NSP1 inhibits expression of type I interferon by antagonizing the function of interferon regulatory factors IRF3, IRF5, and IRF7. J. Virol. 81 (9), 4473–4481.

Berkova, Z., Crawford, S.E., Trugnan, G., Yoshimori, T., Morris, A.P., Estes, M.K., 2006. Rotavirus NSP4 induces a novel vesicular compartment regulated by calcium and associated with viroplasms. J. Virol. 80 (12), 6061–6071.

Berkova, Z., Crawford, S.E., Blutt, S.E., Morris, A.P., Estes, M.K., 2007. Expression of rotavirus NSP4 alters the actin network organization through the actin remodeling protein cofilin. J. Virol. 81 (7), 3545–3553.

Boulant, S., Targett-Adams, P., McLauchlan, J., 2007. Disrupting the association of hepatitis C virus core protein with lipid droplets correlates with a loss in production of infectious virus. J. Gen. Virol. 88, 2204–2213.

Brasaemle, D.L., Wolins, N.E., 2012. Packaging of fat: an evolving model of lipid droplet assembly and expansion. J. Biol. Chem. 287 (4), 2273–2279.

Campagna, M., Eichwald, C., Vascotto, F., Burrone, O.R., 2005. RNA interference of rotavirus segment 11 mRNA reveals the essential role of NSP5 in the virus replicative cycle. J. Gen. Virol. 86 (Pt 5), 1481–1487.

Cheung, W., 2010. Rotavirus inclusion bodies (viroplasms) are structurally and functionally associated with lipid droplet components. PhD Thesis, University of Cambridge, Cambridge.

Cheung, W., Gill, M., Esposito, A., Kaminski, C.F., Courousse, N., Chwetzoff, S., Trugnan, G., Keshavan, N., Lever, A., Desselberger, U., 2010. Rotaviruses associate with cellular lipid droplet components to replicate in viroplasms, and compounds disrupting or blocking lipid droplets inhibit viroplasm formation and viral replication. J. Virol. 84 (13), 6782–6798.

Chukkapalli, V., Heaton, N.S., Randall, G., 2012. Lipids at the interface of virus-host interactions. Curr. Opin. Microbiol. 15, 512–518.

Contin, R., Arnoldi, F., Mano, M., Burrone, O.R., 2011. Rotavirus replication requires a functional proteasome for effective assembly of viroplasms. J. Virol. 85 (6), 2781–2792.

Crawford, S.E., Estes, M.K., 2013. Viroporin-mediated calcium-activated autophagy. Autophagy 9 (5), 797–798.

Crawford, S.E., Hyser, J.M., Utama, B., Estes, M.K., 2012. Autophagy hijacked through viroporin-activated calcium/calmodulin-dependent kinase kinase-β signaling is required for rotavirus replication. Proc. Natl. Acad. Sci. USA 109 (50), E3405–E3413.

Crawford, S.E., Utama, B., Hyser, J.M., Broughman, J.R., Estes, M.K., 2013. Rotavirus exploits lipid metabolism and energy production for replication. American Society for Virology Annual Meeting, Pennsylvania State University, University Park, PA., p. 74 (W2–6).

Criglar, J.M., Hu, L., Crawford, S.E., Hyser, J.M., Broughman, J.R., Prasad, B.V., Estes, M.K., 2014. A novel form of rotavirus NSP2 and phosphorylation-dependent NSP2–NSP5 interactions are associated with viroplasm assembly. J. Virol. 88 (2), 786–798.

Daniel, J., Maamar, H., Deb, C., Sirakova, T.D., Kolattukudy, P.E., 2011. Mycobacterium tuberculosis uses host triacylglycerol to accumulate lipid droplets and acquires a dormancy-like phenotype in lipid loaded macrophages. PLoS Pathog. 7, e1002093.

Díaz-Salinas, M.A., Silva-Ayala, D., López, S., Arias, C.F., 2014. Rotaviruses reach late endosomes and require the cation-dependent mannose-6-phosphate receptor and the activity of cathepsin proteases to enter the cell. J. Virol. 88 (8), 4389–4402.

Eichwald, C., Arnoldi, F., Laimbacher, A.S., Schraner, E.M., Fraefel, C., Wild, P., Burrone, O.R., Ackermann, M., 2012. Rotavirus viroplasm fusion and perinuclear localization are dynamic processes requiring stabilized microtubules. PLoS One 7 (10), e47947.

Estes, M.K., Greenberg, H.B., 2013. Rotaviruses. In: Knipe, D.M., Howley, P.M. et al., (Eds.), Fields Virology, sixth ed. Wolters Kluwer Health/Lippincott Williams & Wilkins, Philadelphia, PA, pp.1347–1401.

Fabbretti, E., Afrikanova, I., Vascotto, F., Burrone, O.R., 1999. Two non-structural rotavirus proteins, NSP2 and NSP5, form viroplasm-like structures in vivo. J. Gen. Virol. 80 (Pt 2), 333–339.

Filipe, A., McLauchlan, J., 2015. Hepatitis C virus and lipid droplets: finding a niche. Trends Mol. Med. 21 (1), 34–42.

Fukuda, N., Ontko, J.A., 1984. Interactions between fatty acid synthesis, oxidation, and esterification in the production of triglyceride-rich lipoproteins by the liver. J. Lipid Res. 25 (8), 831–842.

Gardet, A., Breton, M., Fontanges, P., Trugnan, G., Chwetzoff, S., 2006. Rotavirus spike protein VP4 binds to and remodels actin bundles of the epithelial brush border into actin bodies. J. Virol. 80 (8), 3947–3956.

Gardet, A., Breton, M., Trugnan, G., Chwetzoff, S., 2007. Role for actin in the polarized release of rotavirus. J. Virol. 81 (9), 4892–4894.

Gaunt, E.R., Cheung, W., Richards, J.E., Lever, A., Desselberger, U., 2013a. Inhibition of rotavirus replication by downregulation of fatty acid synthesis. J. Gen. Virol. 94 (Pt 6), 1310–1317, Erratum in: J. Gen. Virol. 2013; 94 (Pt 9), 2140.

Gaunt, E.R., Zhang, Q., Cheung, W., Wakelam, M.J., Lever, A.M., Desselberger, U., 2013b. Lipidome analysis of rotavirus-infected cells confirms the close interaction of lipid droplets with viroplasms. J. Gen. Virol. 94 (Pt 7), 1576–1586.

Groft, C.M., Burley, S.K., 2002. Recognition of eIF4G by rotavirus NSP3 reveals a basis for mRNA circularization. Mol. Cell 9 (6), 1273–1283.

Gross, D.N., Miyoshi, H., Hosaka, T., Zhang, H.H., Pino, E.C., Souza, S., Obin, M., Greenberg, A.S., Pilch, P.F., 2006. Dynamics of lipid droplet-associated proteins during hormonally stimulated lipolysis in engineered adipocytes: stabilization and lipid droplet binding of adipocyte differentiation-related protein/adipophilin. Mol. Endocrinol. 20 (2), 459–466.

Guo, Y., Walther, T.C., Rao, M., Stuurman, N., Goshima, G., Terayama, K., Wong, J.S., Vale, R.D., Walter, P., Farese, R.V., 2008. Functional genomic screen reveals genes involved in lipid-droplet formation and utilization. Nature 453 (7195), 657–661.

Halvorson, D.L., McCune, S.A., 1984. Inhibition of fatty acid synthesis in isolated adipocytes by 5-(tetradecyloxy)-2-furoic acid. Lipids 19 (11), 851–856.

Harb, M., Becker, M.M., Vitour, D., Baron, C.H., Vende, P., Brown, S.C., Bolte, S., Arold, S.T., Poncet, D., 2008. Nuclear localization of cytoplasmic poly(A)-binding protein upon rotavirus infection involves the interaction of NSP3 with eIF4G and RoXaN. J. Virol. 82 (22), 11283–11293.

Heaton, N.S., Randall, G., 2011. Multifaceted roles for lipids in viral infection. Trends Microbiol. 19 (7), 368–375.

Herker, E., Ott, M., 2012. Emerging role of lipid droplets in host/pathogen interactions. J. Biol. Chem. 287 (4), 2280–2287.

Hope, R.G., Murphy, D.J., McLauchlan, J., 2002. The domains required to direct core proteins of hepatitis C virus and GB virus-B to lipid droplets share common features with plant oleosin proteins. J. Biol. Chem. 277 (6), 4261–4270.

Hu, L., Crawford, S.E., Czako, R., Cortes-Penfield, N.W., Smith, D.F., Le Pendu, J., Estes, M.K., Prasad, B.V., 2012. Cell attachment protein VP8* of a human rotavirus specifically interacts with A-type histo-blood group antigen. Nature 485 (7397), 256–259.

Hyser, J.M., Collinson-Pautz, M.R., Utama, B., Estes, M.K., 2010. Rotavirus disrupts calcium homeostasis by NSP4 viroporin activity. MBio 1 (5.), e00265–e00310.

Hyser, J.M., Utama, B., Crawford, S.E., Estes, M.K., 2012. Genetic divergence of rotavirus nonstructural protein 4 results in distinct serogroup-specific viroporin activity and intracellular punctate structure morphologies. J. Virol. 86 (9), 4921–4934.

Igal, R.A., Wang, P., Coleman, R.A., 1997. Triacsin C blocks de novo synthesis of glycerolipids and cholesterol esters but not recycling of fatty acid into phospholipid: evidence for functionally separate pools of acyl-CoA. Biochem. J. 324 (Pt 2), 529–534.

Isaacson, M.K., Ploegh, H.L., 2009. Ubiquitination, ubiquitin-like modifiers, and deubiquitination in viral infection. Cell Host Microbe. 5 (6), 559–570.

Jain, B., Chaturvedi, U.C., Jain, A., 2014. Role of intracellular events in the pathogenesis of dengue; an overview. Microb. Pathog. 69–70, 45–52.

Jiang, X., Jayaram, H., Kumar, M., Ludtke, S.J., Estes, M.K., Prasad, B.V., 2006. Cryoelectron microscopy structures of rotavirus NSP2-NSP5 and NSP2-RNA complexes: implications for genome replication. J. Virol. 80 (21), 10829–10835.

Kim, Y., Chang, K.O., 2011. Inhibitory effects of bile acids and synthetic farnesoid X receptor agonists on rotavirus replication. J. Virol. 85 (23), 12570–12577.

Kim, Y., George, D., Prior, A.M., Prasain, K., Hao, S., Le, D.D., Hua, D.H., Chang, K.O., 2012. Novel triacsin C analogs as potential antivirals against rotavirus infections. Eur. J. Med. Chem. 50, 311–318.

Kumar, Y., Cocchiaro, J., Valdivia, R.H., 2006. The obligate intracellular pathogen Chlamydia trachomatis targets host lipid droplets. Curr. Biol. 16 (16), 1646–1651.

Liefhebber, J.M., Hague, C.V., Zhang, Q., Wakelam, M.J., McLauchlan, J., 2014. Modulation of triglyceride and cholesterol ester synthesis impairs assembly of infectious hepatitis C virus. J. Biol. Chem. 289 (31), 21276–21288.

López, S., Arias, C.F., 2004. Multistep entry of rotavirus into cells: a Versaillesque dance. Trends Microbiol. 12 (6), 271–278.

López, T., Silva-Ayala, D., López, S., Arias, C.F., 2011. Replication of the rotavirus genome requires an active ubiquitin-proteasome system. J. Virol. 85 (22), 11964–11971.

Makishima, M., Okamoto, A.Y., Repa, J.J., Tu, H., Learned, R.M., Luk, A., Hull, M.V., Lustig, K.D., Mangelsdorf, D.J., Shan, B., 1999. Identification of a nuclear receptor for bile acids. Science 284 (5418), 1362–1365.

Marcinkiewicz, A., Gauthier, D., Garcia, A., Brasaemle, D.L., 2006. The phosphorylation of serine 492 of perilipin A directs lipid droplet fragmentation and dispersion. J. Biol. Chem. 281 (17), 11901–11909.

Martin, D., Duarte, M., Lepault, J., Poncet, D., 2010. Sequestration of free tubulin molecules by the viral protein NSP2 induces microtubule depolymerization during rotavirus infection. J. Virol. 84 (5), 2522–2532.

Martin, S., Parton, R.G., 2006. Lipid droplets: a unified view of a dynamic organelle. Nat. Rev. Mol. Cell Biol. 7 (5), 373–378.

Mattos, K.A., Lara, F.A., Oliveira, V.G., Rodrigues, L.S., D'Avila, H., Melo, R.C., Manso, P.P., Sarno, E.N., Bozza, P.T., Pessolani, M.C., 2011. Modulation of lipid droplets by Mycobacterium leprae in Schwann cells: a putative mechanism for host lipid acquisition and bacterial survival in phagosomes. Cell Microbiol. 13, 259–273.

Miyanari, Y., Atsuzawa, K., Usuda, N., Watashi, K., Hishiki, T., Zayas, M., Bartenschlager, R., Wakita, T., Hijikata, M., Shimotohno, K., 2007. The lipid droplet is an important organelle for hepatitis C virus production. Nat. Cell Biol. 9 (9), 1089–1097.

Montero, H., Arias, C.F., Lopez, S., 2006. Rotavirus Nonstructural Protein NSP3 is not required for viral protein synthesis. J. Virol. 80 (18), 9031–9038.

Namatame, I., Tomoda, H., Arai, H., Inoue, K., Omura, S., 1999. Complete inhibition of mouse macrophage-derived foam cell formation by triacsin C. J. Biochem. 125 (2), 319–327.

Parker, R.A., Kariya, T., Grisar, J.M., Petrow, V., 1977. 5-(Tetradecyloxy)-2-furancarboxylic acid and related hypolipidemic fatty acid-like alkyloxyarylcarboxylic acids. J. Med. Chem. 20 (6), 781–791.

Parks, D.J., Blanchard, S.G., Bledsoe, R.K., Chandra, G., Consler, T.G., Kliewer, S.A., Stimmel, J.B., Willson, T.M., Zavacki, A.M., Moore, D.D., Lehmann, J.M., 1999. Bile acids: natural ligands for an orphan nuclear receptor. Science 284 (5418), 1365–1368.

Paul, D., Madan, V., Bartenschlager, R., 2014. Hepatitis C virus replication and assembly: living on the fat of the land. Cell Host Microbe 16 (5), 569–579.

Piron, M., Vende, P., Cohen, J., Poncet, D., 1998. Rotavirus RNA-binding protein NSP3 interacts with eIF4GI and evicts the poly(A) binding protein from eIF4F. EMBO J. 17 (19), 5811–5821.

Richards, J.E., Desselberger, U., Lever, A.M., 2013. Experimental pathways towards developing a rotavirus reverse genetics system: synthetic full length rotavirus ssRNAs are neither infectious nor translated in permissive cells. PLoS One 8 (9), e74328.

Rubio, R.M., Mora, S.I., Romero, P., Arias, C.F., López, S., 2013. Rotavirus prevents the expression of host responses by blocking the nucleocytoplasmic transport of polyadenylated mRNAs. J. Virol. 87 (11), 6336–6345.

Saka, H.A., Valdivia, R., 2012. Emerging roles for lipid droplets in immunity and host-pathogen interactions. Annu. Rev. Cell Dev. Biol. 28, 411–437.

Salloum, S., Wang, H., Ferguson, C., Parton, R.G., Tai, A.W., 2013. Rab18 binds to hepatitis C virus NS5A and promotes interaction between sites of viral replication and lipid droplets. PLoS Pathog. 9 (8), e1003513.

Samsa, M.M., Mondotte, J.A., Iglesias, N.G., Assunção-Miranda, I., Barbosa-Lima, G., Da Poian, A.T., Bozza, P.T., Gamarnik, A.V., 2009. Dengue virus capsid protein usurps lipid droplets for viral particle formation. PLoS Pathog. 5 (10), e1000632.

Saxena, K., Blutt, S.E., Ettayebi, K., Zeng, X.L., Broughman, J.R., Crawford, S.E., Karandikar, U.C., Sastri, N.P., Conner, M.E., Opekun, A., Graham, D.Y., Qureshi, W., Sherman, V., Foulke-Abel, J., In, J., Kovbasnjuk, O., Zachos, N.C., Donowitz, M., Estes, M.K., 2015. Human intestinal enteroids: a new model to study human rotavirus infection, host restriction and pathophysiology. J. Virol. 90 (1), 43–56.

Schmid, B., Rippmann, J.F., Tadayyon, M., Hamilton, B.S., 2005. Inhibition of fatty acid synthase prevents preadipocyte differentiation. Biochem. Biophys. Res. Commun. 328 (4), 1073–1082.

Sen, A., Pruijssers, A.J., Dermody, T.S., García-Sastre, A., Greenberg, H.B., 2011. The early interferon response to rotavirus is regulated by PKR and depends on MAVS/IPS-1, RIG-I, MDA-5, and IRF3. J. Virol. 85 (8), 3717–3732.

Sen, A., Rothenberg, M.E., Mukherjee, G., Feng, N., Kalisky, T., Nair, N., Johnstone, I.M., Clarke, M.F., Greenberg, H.B., 2012. Innate immune response to homologous rotavirus infection in the small intestinal villous epithelium at single-cell resolution. Proc. Natl. Acad. Sci. USA 109 (50), 20667–20672.

Shahidi, M., Tay, E.S., Read, S.A., Ramezani-Moghadam, M., Chayama, K., George, J., Douglas, M.W., 2014. Endocannabinoid CB1 antagonists inhibit hepatitis C virus production, providing a novel class of antiviral host-targeting agents. J. Gen. Virol. 95 (Pt 11), 2468–2479.

Shavinskaya, A., Boulant, S., Penin, F., McLauchlan, J., Bartenschlager, R., 2007. The lipid droplet binding domain of hepatitis C virus core protein is a major determinant for efficient virus assembly. J. Biol. Chem. 282, 37158–37169.

Silvestri, L.S., Taraporewala, Z.F., Patton, J.T., 2004. Rotavirus replication: plus-sense templates for double-stranded RNA synthesis are made in viroplasms. J. Virol. 78 (14), 7763–7774.

Soto-Acosta, R., Bautista-Carbajal, P., Syed, G.H., Siddiqui, A., Del Angel, R.M., 2014. Nordihydroguaiaretic acid (NDGA) inhibits replication and viral morphogenesis of dengue virus. Antiviral Res. 109, 132–140.

Taraporewala, Z.F., Schuck, P., Ramig, R.F., Silvestri, L., Patton, J.T., 2002. Analysis of a temperature-sensitive mutant rotavirus indicates that NSP2 octamers are the functional form of the protein. J. Virol. 76 (14), 7082–7093.

Trauner, M., Boyer, J.L., 2003. Bile salt transporters: molecular characterization, function, and regulation. Physiol. Rev. 83 (2), 633–671.

Vascotto, F., Campagna, M., Visintin, M., Cattaneo, A., Burrone, O.R., 2004. Effects of intrabodies specific for rotavirus NSP5 during the virus replicative cycle. J. Gen. Virol. 85 (Pt 11), 3285–3290.

Walther, T.C., Farese, Jr., R.V., 2012. Lipid droplets and cellular lipid metabolism. Annu. Rev. Biochem. 81, 687–714.

Watanabe, M., Houten, S.M., Wang, L., Moschetta, A., Mangelsdorf, D.J., Heyman, R.A., Moore, D.D., Auwerx, J., 2004. Bile acids lower triglyceride levels via a pathway involving FXR, SHP, and SREBP-1c. J. Clin. Invest. 113 (10), 1408–1418.

Wu, X., Qi, X., Liang, M., Li, C., Cardona, C.J., Li, D., Xing, Z., 2014. Roles of viroplasm-like structures formed by nonstructural protein NSs in infection with severe fever with thrombocytopenia syndrome virus. FASEB J. 28 (6), 2504–2516.

Zou, J., Ganji, S., Pass, I., Ardecky, R., Peddibhotla, M., Loribelle, M., Heynen-Genel, S., Sauer, M., Pass, I., Vasile, S., Suyama, E., Malany, S., Mangravita-Novo, A., Vicchiarelli, M., McAnally, D., Cheltsov, A., Derek, S., Shi, S., Su, Y., Zeng, F.Y., Pinkerton, A.B., Smith, L.H., Kim, S., Ngyuen, H., Zeng, F.Y., Diwan, J., Heisel, A.J., Coleman, R., McDonough, P.M., Chung, T.D.Y., 2010. Potent inhibitors of lipid droplet formation. Probe Reports from the NIH Molecular Libraries Program. Bethesda, MD: National Center for Biotechnology Information, US; 2010–2011.

Chapter 2.6

Rotavirus Disease Mechanisms

M. Hagbom, L. Svensson
Division of Molecular Virology, Department of Clinical and Experimental Medicine, Medical Faculty, Linköping University, Linköping, Sweden

1 INTRODUCTION

Rotavirus (RV) infections are associated with approximately 450,000 deaths each year worldwide, mainly occurring in developing countries among children under 5 years of age (Tate et al., 2012). The infection can be asymptomatic or symptomatic, and the outcome is affected by both viral and host factors. One host factor is age, with infected neonates rarely responding with symptomatic disease. This protection is thought to result primarily from transplacental transfer of maternal antibodies (Ray et al., 2007). Reductions in these antibodies coincide with the age of maximum susceptibility of infants to severe RV-induced diarrhoea. Another host factor consists of histo-blood group antigens, with the Lewis and secretor antigens contributing to RV susceptibility (Nordgren et al., 2014; Lopman et al., 2015; Imbert-Marcille et al., 2014). Libonati et al. (2014) found that the genomic sequences of RVs infecting symptomatic and asymptomatic neonates are virtually identical, further supporting the notion that host factors contribute to the modulation of the clinical response.

While the clinical importance of human RV disease is well recognized and potent vaccines have been developed, our current understanding of RV disease mechanisms is to a major part derived from in vitro studies and animal models and to a minor part from pathology and treatment interventions in humans (Lundgren and Svensson, 2001, 2003; Michelangeli and Ruiz, 2003; Ramig, 2004). While in vitro and animal studies have generated a significant amount of important information regarding virus–cell interactions, certain issues of human RV illness still remain to be understood. For example, emesis, a hallmark of RV disease, which contributes to dehydration, cannot be studied in rodents as they lack the emetic-gastric reflex (Horn, 2008). However, a recent study (Hagbom et al., 2011) has shown that RV stimulates release of serotonin [5-hydroxytryptamine (5-HT)] from human enterochromaffin (EC) cells and activates structures in the central nervous system (CNS) involved in nausea and vomiting, all suggesting that the enteric nervous system (ENS) and vagal

Viral Gastroenteritis

afferent stimuli are involved in RV-induced vomiting and diarrhoea. Furthermore, the mechanisms behind clinical symptoms such as fever, malaise, social withdrawal, fatigue, and anorexia, collectively called "sickness responses" have either scantily or not at all been addressed in respect to RV illness. Four mechanisms have been implicated in RV diarrhoeal disease; increased secretion, decreased absorption (ions and water), altered motility and permeability (Fig. 2.6.1). This review will discuss our current understanding of the mechanisms behind the clinical symptoms associated with RV disease.

2 CLINICAL SYMPTOMS

Human RV infection results in a spectrum of responses that vary and can either be asymptomatic, mild or severe, sometimes resulting in lethal dehydrating illness. The incubation period is less than 48 h with a sudden onset of vomiting, a high frequency of dehydration and a mean duration of diarrhoea lasting 5–6 days (Uhnoo et al., 1986).

2.1 Sickness Response

The clinical symptoms are not limited to life-threatening diarrhoea, but also include systemic responses collectively called "sickness responses." The acute phase response, or "sickness" refers to an initial response of the innate immune system to a broad range of potentially infectious agents. It comprises an inflammatory reaction mediated by pro-inflammatory factors such as interleukin-1 (IL-1), IL-6, and tumour necrosis factor alpha (TNF-α). In contrast to many invasive bacterial gastrointestinal infections, both C-reactive protein (CRP) and calprotectin, two clinical biomarkers of inflammation, are not elevated during human RV infection (Greenberg and Estes, 2009; Chen et al., 2012; Weh et al., 2013). Uhnoo et al. (1986) found that bloody stools, prolonged diarrhoea, and leukocytosis was significantly associated with pathogenic bacteria. On the other hand viral infections are more associated with nausea and vomiting compared to bacterial infections (Uhnoo et al., 1986; Weh et al., 2013). Collectively this suggests that RV infection in humans, albeit with significant pathology, results in a modest clinical inflammatory response compared to infections with pathogenic bacteria.

Fever, a response of the body's thermostat, located in hypothalamus, is a part of the acute-phase response by the immune system (Brodal, 2010) and is usually accompanied by sickness behavior, such as inactivity, sleepiness, depression, and reduced intake of food and water (Hart, 1988; Hennessy et al., 2014). IL-1β, TNF-α, and IL-6 have been shown to be the cytokines responsible for the induction of fever, with IL-1β seemingly depending on IL-6 to induce fever (Brodal, 2010; Eskilsson et al., 2014). RV infection is commonly associated with less fever than bacteria (Kutukculer and Caglayan, 1997; Elliott, 2007). Furthermore, children with RV infection have been shown to have elevated levels of

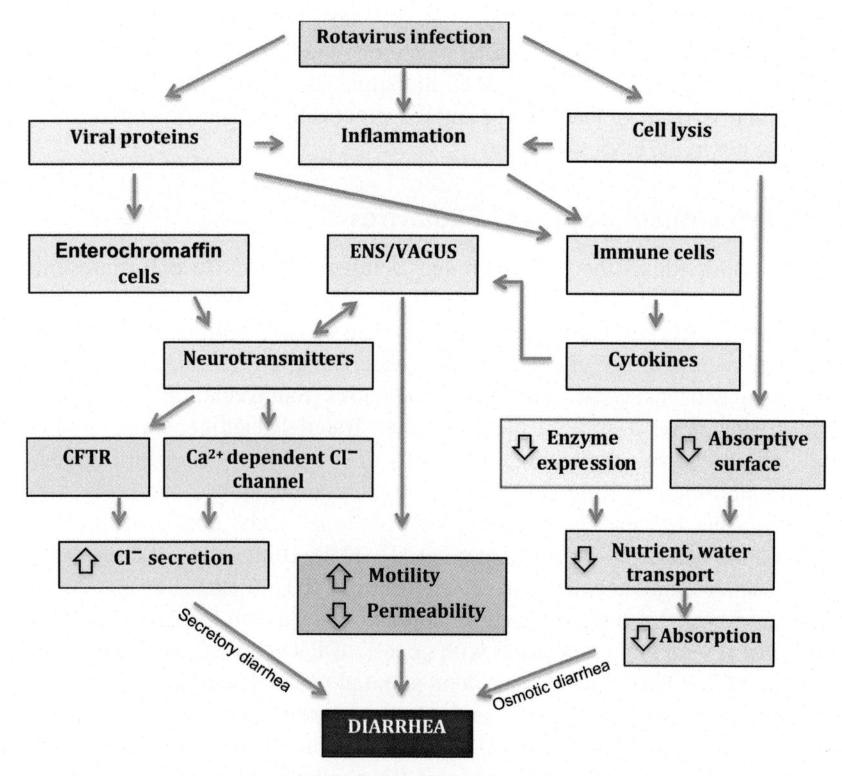

FIGURE 2.6.1 **Schematic mechanisms of rotavirus diarrhoea.** At least four mechanisms have been proposed to be associated with RV diarrhoea; secretory diarrhoea, altered motility, and permeability and osmotic diarrhoea. It is reasonable to assume that rotavirus diarrhoea includes more than one of these mechanisms. High concentration of poorly absorbable compounds may create an osmotic force causing a loss of fluid across the intestinal epithelium (osmotic diarrhoea). Other types of diarrhoea are characterized by an overstimulation of the intestinal tract's secretory capacity (secretory diarrhoea) by nerves not coupled to an inhibition of fluid absorptive mechanisms. Furthermore, if the barrier function of the epithelium is compromised by loss of epithelial cells or disruption of tight junctions, hydrostatic pressure in blood vessels and lymphatic's will cause water and electrolytes to accumulate in lumen. Motility of the small intestine, as in all parts of the digestive tract, is controlled predominantly by excitatory and inhibitory signals from the ENS with inputs from the CNS. Loperamide reduces intestinal motility by acting on opioid receptors of the myenteric plexa. The association of enteric nerves with RV-induced disease mechanism is not only limited to electrolyte secretion, intestinal motility and vomiting but may also include intestinal permeability. In several observations it have been found that intestinal permeability is controlled by the vagus nerve (Costantini et al., 2010; Hu et al., 2013). *(Source: The illustration is modified from Michelangeli and Ruiz, 2003.* CFTR*; cystic fibrosis transmembrane regulator.)*

cytokines in serum and children with fever had significantly higher levels of IL-6 (Jiang et al., 2003).

Application of noxious stimuli to the gastrointestinal (GI) tract may activate peripheral nerve receptors that are sensitive to chemical, mechanical or inflammatory stimuli and may in turn result in abdominal pain. Abdominal pain is associated with RV infection (Uhnoo et al., 1986) but the underlying mechanisms of how RV induces abdominal pain remain to be determined. In general, abdominal pain is supposed to include alterations in smooth muscle and enteric nerves and is likely related to the altered processing of sensory information from the gut to the CNS.

2.2 Extra Mucosal Spread of Rotavirus

In a few individuals the primary mucosal infection leads to extra-intestinal spread and RV RNA has been detected in cerebrospinal fluid (Medici et al., 2011a; Iturriza-Gomara et al., 2002; Liu et al., 2009), possibly associated with meningitis (Wong et al., 1984), encephalopathy (Keidan et al., 1992; Nakagomi and Nakagomi, 2005), and encephalitis (Salmi et al., 1978; Ushijima et al., 1986). Several studies have also demonstrated that antigenemia, viremia, and limited systemic replication in a variety of sites is likely to occur frequently, although there is little evidence that this systemic spread and replication is responsible for any specific pathologic findings in the normal host (Nakagomi and Nakagomi, 2005; Ramani et al., 2010; Blutt et al., 2003; Fischer et al., 2005; Blutt and Conner, 2007; Ramig, 2007; Fenaux et al., 2006). Studies in neonatal mice indicate a lymphatic mechanism for extra-intestinal spread of RV and an association with gene 7 of RRV RV strain (Mossel and Ramig, 2002, 2003). Studies have demonstrated that in severely immune compromised infants, RV can replicate and cause abnormalities in the liver and other organs (Gilger et al., 1992). In suckling mice without an intact interferon signalling system some strains of RV replicate in the biliary tract and pancreas and cause biliary atresia and pancreatic disease (Feng et al., 2008). A liver function test in children with RV gastroenteritis found that 20% had elevated ALT or AST levels (Teitelbaum and Daghistani, 2007). However, the pathophysiological importance of these changes remains controversial. RV is also a special threat to individuals who are immunosuppressed following bone marrow transplantation. In a bone marrow transplant unit, 8 of 78 patients with gastroenteritis shed RV as the sole pathogen, and 5 of these individuals died (Yolken et al., 1982). Moreover, RV infections acquired nosocomially have been associated with severe diarrhoea in adult renal transplant recipients (Peigue-Lafeuille et al., 1991). In conclusion, the importance of extra-intestinal spread remains to be determined.

2.2.1 Intussusception

Intussusception (IS) has been associated with RV infections/vaccines. IS is a process in which a segment of the intestine invaginates into the adjoining intestinal lumen, thereby resulting in bowel obstruction and infarction, which may require clinical or surgical intervention. The first licensed RV vaccine, RRV-TV (Rotashield®) was withdrawn in USA following reports of IS among the vaccinated children (Centers for Disease Control and Prevention, 1999;

Murphy et al., 2001). Recent postmarketing surveillance after wide application of Rotarix® and Rotateq® vaccines in Latin America and Australia has reported lower risks of IS in comparison to those observed with the RRV-TV vaccine (Patel et al., 2011; Buttery et al., 2011). Two earlier phase III clinical trials evaluating safety and efficacy of Rotarix® and Rotateq® vaccines did not observe a significant correlation with IS (Ruiz-Palacios et al., 2006; Vesikari et al., 2006). Besides in vaccinated children, IS has also been reported in a few young children after natural RV infections (Konno et al., 1978; Dallar et al., 2009; Mulcahy et al., 1982). Robinson et al. (2004) using ultrasound observed increased thickness of distal ileum and lymphadenopathy in RV infected children compared to controls, which may be the cause of IS. However, data from epidemiological studies did not demonstrate a significant increase of gut IS after natural RV infection (Velazquez et al., 2004; Bahl et al., 2009; Bines et al., 2006).

2.3 Rotavirus Infection Delays Gastric Emptying

The prominence of nausea and vomiting during RV illness suggests abnormal gastric function. A marked delay of gastric emptying has been observed not only after ingestion of norovirus (Meeroff et al., 1980) but also after RV infection. Bardhan et al. (1992) found that RV infection is accompanied by abnormal gastric motor function as manifested by delayed emptying of liquid. Gastric emptying of liquids is believed to be primary a function of the pressure gradient between the stomach and the duodenum. The mechanisms of delay in gastric emptying is proposed to include neural pathways and gastrointestinal hormones such as secretin, gastrin, glucagon (Cooke, 1975), and cholecystokinin (Debas et al., 1975). The neural pathways influencing gastric emptying may include noncholinergic, nonadrenergic, and dopaminergic vagal neurons (Minami and McCallum, 1984). The response is mediated by 5-HT_3 receptors and also sodium glucose cotransporter (SGLT-1) expressed by EC cells (Raybould, 2002) further supporting the observation that RV infection stimulates the ENS. It should also be mentioned that gastric emptying and food intake are related and reduction in food intake has been observed in children with acute RV illness (Molla et al., 1983) as part of the sickness response.

3 NITRIC OXIDE IN RV ILLNESS

Nitric oxide (NO) is synthesized from L-arginine by NO synthases (NOS) and is secreted by cells involved in host defence, homeostatic, and development functions. NOS exists in three isoforms; endothelial NOS (eNOS), neuronal NOS (nNOS), and inducible NOS (iNOS). While eNOS activity is increased in response to an increased blood flow (the augmented NO causing a relaxation of vascular smooth muscles), the iNOS is activated during infection and inflammation. However, immunohistochemical investigations of the intestinal wall have demonstrated nNOS in the myenteric plexus, indicating that NO is involved in the control of intestinal motility and/or as a transmitter in interneurons of local reflexes of the ENS (Furness, 2006). Accumulating evidence suggests that NO plays a role in the modulation of aqueous secretion and the barrier function of intestinal cells. Several recent studies have reported elevated levels of NO in patients with gastroenteritis (Kawashima et al., 2004; Kukuruzovic et al., 2002; Rodriguez-Diaz et al., 2006; Sowmyanarayanan et al., 2009). A study by Rodriguez-Diaz et al. (2006) demonstrated NSP4 to induce rapid release of NO from intestinal epithelial cells (HT-29). Moreover, time kinetics studies showed release of NO in RV-infected mice peaking between days 6 and 9 after infection, thereby suggesting participation of iNOS. The authors further observed elevated levels of NO in RV infected diarrhoeal children, thereby confirming the results of both in vitro and animal experiments. A similar study from India (Sowmyanarayanan et al., 2009) reported elevated levels of NO metabolites among diarrhoeal children infected with RV and norovirus. Borghan et al. (2007) found that ex vivo NSP4 treatment of ileum excised from CD-1 suckling mice resulted in up regulation of ileal iNOS mRNA expression within 4 h. Furthermore, NSP4 was able to induce iNOS expression and NO production in murine peritoneal macrophages (see Chapter 2.4). Kawashima et al. (2004) observed elevated levels of NO metabolites in both serum and cerebrospinal fluid of RV infected gastroenteritis patients having febrile convulsions in comparison with patients with purulent meningitis, encephalitis, and febrile convulsion and of a control group. Additionally, they also observed a relative correlation between IL-6 levels and NO metabolites in some cases. The functional consequences of an increased epithelial NO production is not fully established. Several effects seem possible. Locally produced NO increases the permeability of the intestinal epithelia to hydrophilic solutes. Furthermore, the produced NO may directly influence epithelial transport mechanisms. The control of epithelial sodium and hence water transport by nerve-mediated NO release was recently reviewed (Althaus, 2012).

4 ROLE OF PROSTAGLANDINS AND ACETYLSALICYLIC ACID IN ROTAVIRUS DIARRHOEA

Prostaglandins (PGs) are lipid compounds, enzymatically derived from fatty acids located in the cell lipid bilayer, and can elicit a wide range of physiological responses in the body (Scher and Pillinger, 2009). Cyclooxygenases (COXs) are essential enzymes in the biosynthesis of PGs, converting the arachidonic acid to PGH_2; subsequently specific isomerases transform PGH_2 to biologically active PGs (Scher and Pillinger, 2009). It has been shown that PGEs are produced under the influence of microorganisms and have immunomodulatory, antiinflammatory as well as pro-inflammatory actions (Scher and Pillinger, 2009). Moreover it has been shown that PGE_2 can stimulate water secretion (Sandhu et al., 1981), an effect that can be blocked by drugs attenuating nervous activity, such as hexamethonium (nicotinic receptor blocker) or lidocaine (local anesthetic) (Brunsson et al., 1987) thus indicating that nerves are involved. Studies carried out among children with RV gastroenteritis found elevated levels of PGE_2 and PGF_2 in both plasma and stool (Yamashiro et al., 1989) and treatment with the COX-inhibitor acetylsalicylic acid (aspirin) (Vane, 1971) reduces the duration of diarrhoea (Yamashiro et al., 1989; Gracey et al., 1984). PGs can be converted to cyclopentenone PGs (cyPGs), which have antiviral properties through NF-$_{k}\beta$ activation and inhibit viral replication of both DNA and RNA viruses (Santoro, 1997). Interestingly, in vitro studies have shown that RV replication is inhibited by the cyPGA1 (Superti et al., 1998; Suzuki and Oshitani, 1999).

In conclusion, these observations suggest that RV infection stimulates release of PGs from epithelial cells, which may act directly on the epithelium and/or indirectly via an activation of nerves. The positive therapeutic effect of aspirin on RV diarrhoea is interesting, but needs to be confirmed in larger studies. However, the side effects of aspirin should be considered before giving the drug to small children (Litalien and Jacqz-Aigrain, 2001).

5 MECHANISMS OF DIARRHOEA

The intestinal epithelium consists of absorptive cells, EC cells, goblet cells, paneth cells, intraephithelial lymphocytes, and undifferentiated cells. Cell division takes place in the undifferentiated cells of the crypts and as these cells migrate upwards toward the villi, they differentiate into different cell types, for example, absorptive enterocytes. Chloride secreting cells, which are undifferentiated cells, are located in the crypts. The villi of the small intestine are the site where most of the absorption of nutrients such as minerals, sugars, and amino acids occurs. After food has been digested in the stomach by strong hydrochloric acid and enzymes, the pyloric sphincter opens, and food gets pushed into small intestine by peristalsis.

The fluid loss in diarrhoea may be caused by several mechanisms (Field, 2003), see Fig. 2.6.1. High concentration of poorly absorbable water-soluble compounds may create an osmotic force causing a loss of fluid across the intestinal epithelium (osmotic diarrhoea). Other types of diarrhoea are characterized by an overstimulation of the intestinal tract's secretory capacity (secretory diarrhoea) not coupled to an inhibition of fluid absorptive mechanisms. Furthermore, if the barrier function of the epithelium is compromised by loss of epithelial cells or disruption of tight junctions, hydrostatic pressure in blood vessels, and lymphatic's will cause water and electrolytes to accumulate in lumen (exudative diarrhoea) (Field, 2003). Finally, motility and permeability disturbances may cause an intestinal fluid loss. In most types of diarrhoea more than one of these pathophysiological mechanisms are involved. Fluid loss and electrolyte disturbances can cause the death of the diarrhoeic patient due to collapse of the circulatory system when too much fluid has been lost. Electrolyte changes such as potassium loss and acid-base disturbances may also contribute to the patient's death (Sachdeva, 1996).

6 PATHOLOGY

Most knowledge of pathological changes that occur during RV infections comes from animal studies. The severity, localization, and histological findings of intestinal infection vary among animal species, between strains in a single species and between studies. However, almost all pathological changes due to RV infection are limited to the small intestine. Most severe intestinal damage is observed in piglets, where intestinal villi can be completely eroded (Shepherd et al., 1979a). Macroscopic changes in the pig intestine also include thinning of the intestinal wall, and microscopic changes consist of the conversion from columnar to a cuboidal epithelium. In calves, infection also leads to a change in the villus epithelium from columnar to cuboidal, and villi become stunted and shortened. Histological changes in calves are not limited to a symptomatic response; in fact asymptomatically infected calves show also villus blunting (Reynolds et al., 1985). Compared with infections in piglets and calves, RV infection in lambs results in less severe histopathological changes and mild clinical disease (Greenberg et al., 1994).

RV infections have been studied most extensively in mice, either in the infant mouse model (Offit et al., 1984) or an adult model (Ward et al., 1990). The pathology of murine RV is generally similar to that of lambs, pigs, and calves, but differs in certain aspects. During infection, histological changes are characterized by swollen and vacuolated enterocytes (Osborne et al., 1988). Vacuolization of enterocytes is most prominent on the villus tip, but can occur in enterocytes throughout the villus (Kordasti et al., 2006). While vacuolization can also be observed in other species, it is most extensive in mice (Offit et al., 1984; Osborne et al., 1988) and absent in calves, lambs, and piglets. These differences have been attributed to the specific nature of the host response and not to the virus, since vacuolization is also a characteristic feature of heterologous infections in mice (Feng et al., 2008). Thus far, the origin and nature of the vacuoles remain ambiguous. Mice usually do not develop any symptoms when infection occurs beyond 2 weeks of age, and infection of adult mice occurs without disease (Offit et al., 1984; Ward et al., 1990). Unlike in other species, villus blunting is limited in mice. A most important observation is that no clear correlation exists between the degree of histopathology changes and the severity of diarrhoeal disease (Lundgren and Svensson, 2001). In cows and pigs significant diarrhoea may occurs before the signs of intestinal pathology.

While intestinal histological investigations are difficult to perform in fatal cases, due to the rapid autolysis and bacterial overgrowth post mortem, only a few biopsies from RV-infected children have been performed. Bishop et al. (1973) first observed the presence of RV in humans and carried out by histological investigations of the small intestine of the infected patients. All nine children exhibited mucosal changes varying from mild to severe. The changes included shortening and blunting of villi and increased infiltration of inflammatory cells in the *lamina propria*. The same year, 1973, RV particles were discovered by electron microscopy in stools of children with acute gastroenteritis (Flewett et al., 1973). Moreover, Davidson, and Barnes found structural and functional alterations in the duodenal mucosa of 17 children and observed that the patients with the most severe mucosal damage were most likely to require intravenous therapy (Davidson et al., 1979), suggesting that RV caused structural lesions in the upper small intestine. However, structural changes of the small intestine have not always been observed. Kohler et al. (1990) investigated intestinal biopsies of 40 RV-infected children who were less than 18 months old. Biopsies were taken between 2 and 10 days after onset of acute gastroenteritis. 95% of the children (38/40) had normal histological findings of the small intestinal mucosa, and no correlation was found between clinical findings, morphological results and therapy. Mavromichalis et al. (1977) addressed the question of impaired cellular function by conducting a blood xylose test in 2–16 months-old children with acute RV gastroenteritis. They observed low blood xylose levels (range 0.15–0.78 mmol/L) in children in whom RV was found in the intestinal aspirate as opposed to children with RV present in faeces only who had significantly higher xylose levels (>1.26 mmol/L). Noone et al. (1986) studied 18 RV-positive patients with abnormal intestinal lactose hydrolysis and observed raised urine lactulose/L-rhamnose excretion. Moreover, Davidson et al. (1979) investigated 17 children and found diminished disaccharidase activity in duodenal homogenates of 14 children, but tests for sugar in faeces were rarely positive. Taken together, these findings suggest that RV infection causes patchy epithelial lesions (Davidson et al., 1979) in the small intestine and impaired diasaccharidase function, but the clinical significance of these remains unclear as oral rehydration therapy (ORT) is most successful. Furthermore, it was proposed that management of mild-to-moderately dehydrated children should consist of reintroduction of normal feeding, continued breast feeding and that presumably lactose-free formula is not justified in most cases (Tormo et al., 2008), all suggesting a rather intact absorptive and digestive capacity of the small intestine.

6.1 Pathology of Fatal Cases of Rotavirus Infection

Deaths occur mainly among children with poor access to medical care, and children die presumably due to dehydration and electrolyte imbalance. Clinical and pathological investigations of patients who died of RV infection are rare. Twenty-one fatal cases were reported in Canada with death occurring within 3 days after onset of symptoms in all cases. Sixteen of the subjects were dehydrated and had sodium levels in excess concentration. It was suggested that the rapid fluid depletion contributed significantly to death (Carlson et al., 1978). In Italy, two fatal cases occurred in children that were severely dehydrated and death was related to severe cerebral oedema. Histological examination demonstrated extensive damage of the intestinal epithelium, villous atrophy, and macrophage infiltration (Medici et al., 2011b). Lynch et al. (2003) performed pathologic investigations on various tissue samples of three children who died of RV infection using immunohistochemistry, in situ hybridization and reverse transcription polymerase chain reaction (RT-PCR) assays. Besides at intestinal sites, RV genomes were detected by RT-PCR in spleen, heart, lung, kidney, testis, and bladder of one patient; and spleen, adrenal gland, pancreas, and kidney of a second patient while no RV genome was detected in the CNS (the only tissue examined) of the third patient. In none of the patients was it possible to determine the exact cause of death based on histological findings. In conclusion, several reports suggest that children who have died during RV illness may have had some extra mucosal manifestations of the illness (Gilger et al., 1992; Pager et al., 2000; Lynch et al., 2001; Morrison et al., 2001; Grech et al., 2001) but the significance of these as a contributor to death remains to be clarified.

7 PATHOPHYSIOLOGY OF ROTAVIRUS-INDUCED DIARRHOEA

RV replicate primarily, but not exclusively in nondividing mature enterocytes of the small intestine suggesting that differentiated enterocytes express factors required for efficient infection and replication. A paradox is that, while colon is resistant to infection in vivo, colon-derived cell lines such as Caco2 and HT29 cells are highly susceptible for infection and replication in vitro.

Mechanisms that have been proposed to explain the diarrhoea include malabsorption secondary to impaired transport of electrolytes and/or glucose/amino acids (Davidson et al., 1977; Graham et al., 1984), a toxin-like effect of the nonstructural NSP4 protein (Morris et al., 1999; Ball et al., 1996) (for further details see Chapter 2.4) and the stimulation of the ENS (Lundgren and Svensson, 2003; Lundgren et al., 2000; Kordasti et al., 2004). Changes in microcirculation have been also proposed to participate in the genesis of RV diarrhoea (Osborne et al., 1991). According to this hypothesis, infection of villus cells by RV would trigger the release of "neuroactive/hormonal substances" which would cause ischemia by vasoconstriction and in turn villus shortening and decrease of absorptive capacity. Details of this mechanism need to be further investigated.

7.1 Effect of Rotavirus Infection on Electrolyte and Fluid Transport

Several reports showed that RV infection provokes a net secretion of fluid in vivo, caused by an attenuated uptake of sodium and secretion of chloride. When perfusing the rat intestinal lumen with an isotonic electrolyte solution devoid of glucose, net fluid secretion and decreased net sodium uptake were observed after RV inoculation (Salim et al., 1995). In contrast, net fluid and sodium uptake was observed in control rats. At the peak of net fluid secretion, villus height of infected rats was decreased to one third of controls, whereas crypt depth was unaltered. Mouse intestinal segments perfused with an isotonic electrolyte solution containing mannitol showed a significant secretion of fluid, sodium and chloride ions at 72 h post RV infection, when clinical signs of diarrhoea were most pronounced (Lundgren and Svensson, 2001; Osborne et al., 1988; Starkey et al., 1990). In accordance with this finding, Lundgren et al. showed that RV induced a net fluid secretion and increased electrolyte secretion at 48–60 h after infecting newborn mice (Lundgren and Svensson, 2001).

A mechanism that may in part explain the virus-induced attenuation of electrolyte and fluid absorption is an inhibition of the symporters for sodium and glucose/amino acids. Apical symporters in villus enterocytes transport sodium ions together with glucose or amino acids and numerous studies indicate that the cotransport of glucose and sodium is impaired in intestinal segments exposed to RV. Most of the studies of electrolyte transport in virus-infected intestinal segments have been performed in vitro with the so-called Ussing chamber technique. The intestinal segment is then almost always stripped of its muscle layer to improve oxygenation of the tissue and furthermore lacks influences of hormones and nerves. Thus, McClung et al. (1976) found reduction in glucose-stimulated sodium absorption in intestinal segments isolated from pigs with viral enteritis. A factor of importance for the absorption of glucose and other sugars consists of disaccharidases localized to the brush border region of enterocytes. In several reports of viral gastroenteritis it has been demonstrated that the activities of all mucosal disaccharidases (sucrase, lactase, maltase) are markedly attenuated (Shepherd et al., 1979a,b; Davidson et al., 1977; Kerzner et al., 1977; Collins et al., 1988).

7.2 Oral Rehydration Corrects Rotavirus-Induced Loss of Electrolytes and Water

It has been suggested that severe lesion and atrophy of the mature absorptive enterocytes could result in impaired sodium-glucose cotransporter 1 (SGLT-1) function and thus explain RV-diarrhoea (Halaihel et al., 2000). A strong inhibition of both SGLT-1 and sodium leucine symport activities have been found in brush-border membrane vesicles (Halaihel et al., 2000) and impaired transport of electrolytes and/or glucose/amino acids (Davidson et al., 1977; Graham et al., 1984). The clinical importance of these observations for humans remains to be confirmed, particularly as oral rehydration solutions (ORS) designed to replace and maintain fluid levels of electrolytes and water have saved the lives of millions of children and are a standard treatment to replace electrolytes and water in RV disease.

In one of the earliest studies, Sack et al. (1978) evaluated the response of glucose and electrolyte solution given orally in comparison to the intravenous rehydration therapy to RV infected children and observed no differences. Although the secretory nature of diarrhoea results in substantial loss of water and electrolytes, an intact sodium-coupled solute cotransporter system allows efficient absorption of salt and water (Tormo et al., 2008). Water passively follows through the osmotic gradient generated by the transcellular transport of electrolytes and nutrients (Hallback et al., 1979). Thus, the coupled transport of sodium and glucose at the intestinal brush border is essential for oral rehydration to function. Cotransport across the luminal membrane is facilitated by the SGLT-1. This mechanism remains intact, even in patients with severe diarrhoea (Acra and Ghishan, 1996), including RV-induced diarrhoea (Rautanen et al., 1997), all suggesting that even if RV infection is lytic to enterocytes and impairs the sodium–glucose transport in these cells, a sufficient number of enterocytes in the small intestine maintain their functional capacity to transport sodium and glucose.

8 SECRETORY DIARRHOEA

Secretory diarrhoea is caused by electrogenic exit of Cl^- across the apical plasma membrane of epithelial cells. The additional paracellular movement of Na^+ and H_2O leads to accumulation of fluid within the lumen of the GI tract and subsequently diarrhoea. Clinically secretory diarrhoea is diagnosed when faecal fluid Na^+, K^+, and accompanying anion concentrations are in balance with those in plasma. By contrast, in osmotic diarrhoea an osmotic gap exists as nonabsorbed solutes within the lumen prevent fluid absorption.

The small intestine is able to secrete a large amount of fluid, mainly chloride from plasma to lumen. The chloride secretory mechanism occurs in crypt cells and the main apical chloride pathway in the intestine is the cystic fibrosis transmembrane conductance regulator (CFTR) chloride channel activated mainly by cAMP-dependent phosphorylation resulting from adenylate cyclase activation (Furness, 2006). Mutations of the CFTR gene affecting chloride ion channel function lead to dysregulation of epithelial fluid transport in the lung, intestines, pancreas, and other organs, resulting in cystic fibrosis (Cutting, 2015). Complications include thickened mucus in the lungs with frequent respiratory infections, obstipation, and pancreatic insufficiency giving rise to malnutrition and diabetes. These conditions lead to chronic disability and reduced life expectancy. Homozygous CFTR (−/−) knockout mice have shown an absence of intestinal Cl^- secretion after treatment with a variety of drugs that normally provoke Cl^- secretion. There is a secretory component of RV diarrhoea that is thought to be mediated by activation of the ENS and the effects of the virus-encoded enterotoxin NSP4. In search of the chloride secretion pathway, RV and NSP4 studies have been performed in CFTR knockout mice (Angel et al., 1998). Morris et al. (1999) found that NSP4 or its active peptide may induce diarrhoea in neonatal mice through the activation of an age- and Ca^{2+}-dependent plasma membrane anion distinct from CFTR. Moreover Angel et al. (1998) found that murine RV and NSP4 from virulent and attenuated strains responded with diarrhoea, all together suggesting that diarrhoea, either by RV infection or by NSP4, is not mediated through CFTR. It is noteworthy that patients with cystic fibrosis may experience dehydrating diarrhoea with RV (Angel et al., 1998). Thus it remains uncertain if results from the age-dependent infant CFTR mouse model can be directly translated to human. Using a different approach,

Kordasti et al. (2004) used the vasoactive intestinal peptide (VIP) receptor antagonist (4Cl-D-Phe^{6}, Leu^{17})-VIP, to attenuate the effects of VIP-ergic secretomotor neurones and thus provided information of the role of VIP in RV diarrhoea. Mice treated with the VIP receptor antagonist responded with significantly less diarrhoea (Kordasti et al., 2004), suggesting that VIP and VIP-ergic secretomotor neurons and indirectly CFTR participate in RV electrolyte secretion. VIP is a hormone that stimulates the secretion and inhibits the absorption of sodium, chloride, potassium, and water within the small intestine and increases bowel motility through activation of cellular adenylate cyclase and cAMP production (Furness, 2006). These actions lead to secretory diarrhoea, hypokalaemia, and dehydration.

9 THE NSP4 ENTEROTOXIN

Several observations in mice suggest that the N-linked glycosylated nonstructural NSP4 protein of RV may function as an enterotoxin. The NSP4 and a fragment containing amino acids 114–135 has been shown to induce diarrhoea in infant mice (Ball et al., 1996). The NSP4 protein increases intracellular Ca^{2+} concentration mainly by mobilizing Ca^{2+} from the endoplasmic reticulum (ER) (Tian et al., 1995). It is not clear how intracellular NSP4 releases Ca^{2+} from the ER, but this is presumably by a phospholipase C (PLC)-dependent mechanism (Tian et al., 1995; Berkova et al., 2003) (Fig. 2.6.2). It has also been shown that NSP4 and a cleavage fragment of NSP4 is secreted from infected cells (Zhang et al., 2000; Bugarcic and Taylor, 2006; Didsbury et al., 2011) and that NSP4 can trigger pro-inflammatory cytokines (Ge et al., 2013) and stimulate 5-HT release from human EC cells (Hagbom et al., 2011). Moreover, Seo et al. (2008) reported that integrin $\alpha 1\beta 1$ and $\alpha 2\beta 1$ activate intracellular signalling pathways that regulate cell spreading and may cause diarrhoea (Fig. 2.6.3). Further details are presented in Chapter 2.4.

10 THE ENTERIC NERVOUS SYSTEM AND ENTEROCHROMAFFIN CELLS

In humans the ENS contains 500 million neurons (Furness, 2006). Due to its size, complexity and certain structure similarities, it has been named the second brain. The ENS controls motility, secretion, and blood supply in the gastrointestinal tract. There are two major plexa, the myenteric plexus and submucosal plexus (Fig. 2.6.4). The myenteric plexus (Auerbach's plexus) provides motoric innervation to the circular muscle layer and the longitudinal muscle (Hansen, 2003). The myenteric plexus regulate intestinal motility including peristalsis. The principal role of the submucosal plexus is to coordinate reflexes such as secretion and absorption as well as motor control of smooth muscles. The myenteric plexus is a network of small ganglia between the outer longitudinal and inner circular muscle layer of the intestine (Fig. 2.6.4). Neurons

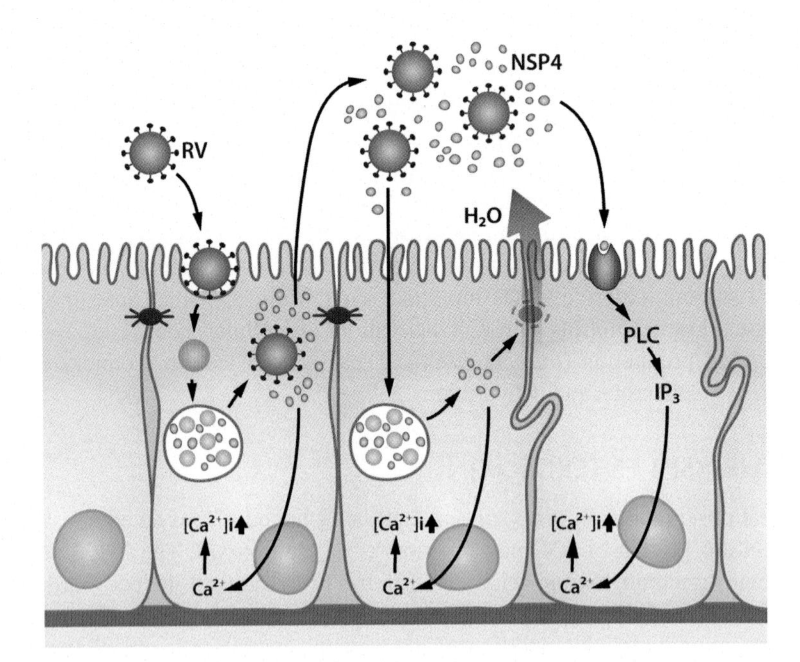

FIGURE 2.6.2 **Proposed models of how NSP4 disrupts tight junctions and induces water loss through paracellular leakage.** RV infect mature enterocytes (left cell) with replication resulting in formation of viroplasms, containing subviral particles and viral proteins. Intracellularly expressed NSP4 *(green dots)* releases intracellular calcium from internal stores, mainly from the ER. NSP4 and a peptide fragment (aa 112–175) of NSP4 are secreted from cells by a Golgi-independent pathway. Another model (middle cell) proposes that intracellularly expressed NSP4 disrupts tight junctions, allowing paracellular flow of water. A third model proposes (right cell) that extracellularly released NSP4 from a previously infected cell binds to a receptor triggering a signalling cascade through PLC and inositol phosphate $(IP)_3$ that results in release of Ca^{2+} and paracellular leakage with water loss. *(Source: Modified from Ramig, 2004 and Greenberg and Estes, 2009 for further details see Chapter 2.4.)*

can be grouped by their functions as (1) intrinsic primary afferent neurons (IPANs), which are sensory neurons, (2) interneurons that connect neurons between the two layers, and (3) motor neurons. The IPANs are located within the submucosal and myenteric plexuses and can activate enteric reflexes that regulate motility, secretion, and blood flow. Transmission from vagal input neurons to enteric neurons is mediated principally by acetylcholine acting on nicotinic cholinergic receptors, but several other transmitters are also involved in these processes (Hansen, 2003).

The role of ENS in RV diarrhoea was first described in 2000 by Lundgren et al (Lundgren et al., 2000). They perfused mouse intestinal segments, with and without treatment with 4 different drugs acting on nerves within the ENS. The observation that all four drugs significantly attenuated the intestinal secretory response to RV infection strongly suggested that the ENS participated in the RV-induced electrolyte and fluid secretion. The involvement of the ENS

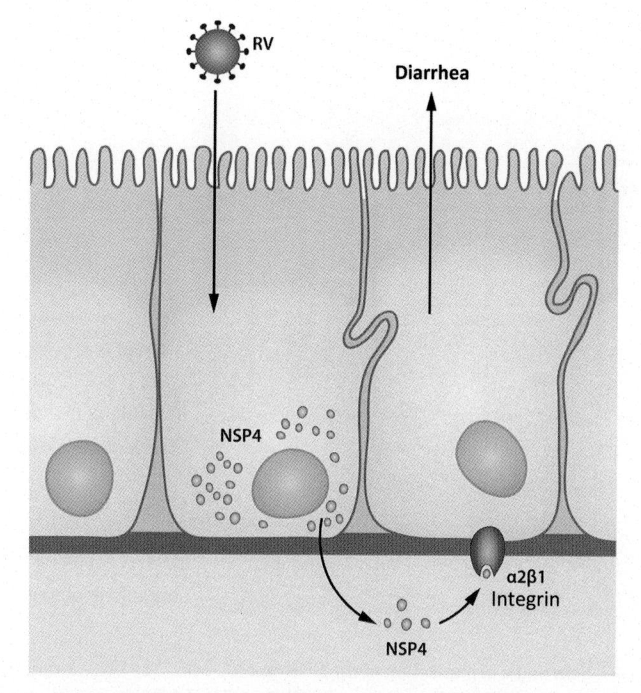

FIGURE 2.6.3 Intracellularly expressed NSP4 recognize α2β1 integrin. One hypothesis is that following infection of mature enterocytes NSP4 is expressed and targeted to the basolateral domain, released and further interacts with integrin α2β1 on the basolateral surface of surrounding intestinal cells. NSP4 mutants that fail to bind or signal through α2β1 are attenuated in diarrhoea induction in neonatal mice (Greenberg and Estes, 2009; Seo et al., 2008). At present it is unclear how the integrin signaling can give raise to diarrhoea. *(Source: Illustration modified from Greenberg and Estes, 2009.)*

may explain how comparatively very few virus-infected cells at the villus tips can cause the intestinal crypt cells to augment their secretion of electrolytes and water. Following the original observation it has been shown that ENS and VIP are associated with RV diarrhoea (Kordasti et al., 2004). Moreover loperamide, a drug acting on myenteric neuron and atropine, acting on muscarinic acetylcholine receptors of muscle cells, both attenuated diarrhoea of infant mice (Istrate et al., 2014).

EC cells are specialized sensory cells in the intestinal epithelium, located predominately in the lower part of villi. These cells are thought to "taste" and "sense" the intestinal lumen and can respond by releasing transmitters to underlying neurons (Hansen, 2003).

EC cells are widely distributed in the GI tract and are found in the mucosa of the gastric antrum, duodenum, jejunum, ileum, appendix, colon, and rectum (Ahlman and Nilsson, 2001). The cytoplasm of the EC cells contains a large number of secretory granules, which are storage sites of secretory products.

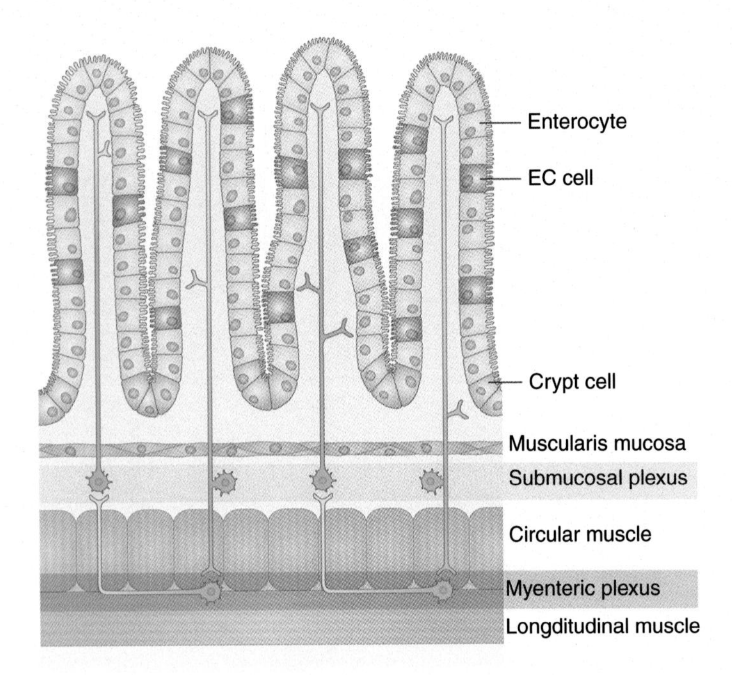

FIGURE 2.6.4 **Schematic organization of the ENS and EC cells.** The ENS in vertebrates is part of the autonomic nervous system that controls the GI tract. The ENS consists of several-hundred million neurons, one thousandth of the number of neurons in the brain, and is essentially equal in size to the 100 million neurons in the spinal cord. The ENS is composed of two major nerve plexus, the myenteric plexus (Auerbach's plexus) provides motor innervation to the circular muscle layer and the longitudinal muscle. Auerbach's plexus originates in the medulla oblongata as a collection of neurons from the ventral part of the brain stem. The vagus nerve then carries the axons to their destination in the GI tract. The myenteric plexus regulate intestinal motility including peristalsis. The submucosal plexus is located in the submucosa, and its principal role is to coordinate reflexes such as secretion and absorption as well as motor control of smooth muscles. The EC cells "taste" and sense" the luminal contents and release mediators such as serotonin (5-hydroxytryptamine, 5-HT) to activate the ENS as well as to stimulate extrinsic vagal afferents to the brain. 5-HT is contained in secretory granules of the EC cells and is released following stimulation by hyperosmolarity, carbohydrates, mechanical distortion, and stimulation by RV of the mucosa (Hagbom et al., 2011; Hansen and Witte, 2008; Spiller et al., 2008; Hagbom et al., 2012).

The main secretory product of EC cells is 5-hydroxytryptamine (5-HT) and the EC cells account for more than 90% of all 5-HT synthesized in the body. Minor amounts of peptide hormones, for example, tachykinins, enkephalins (ENK), motilin (Ahlman and Nilsson, 2001), and PGs (Hansen, 2003) may also be synthesized in EC cells. Serotonin is synthesized from the amino acid tryptophan and transported into granules (Ahlman and Nilsson, 2001). Following stimulation by several agents (eg, hyperosmolarity, carbohydrates, mechanical distortion of the mucosa, cytostatic drugs) including the cholera toxin (Hansen and Witte, 2008; Spiller et al., 2008) and rotavirus (Hagbom et al., 2011), EC cells mobilize intracellular Ca^{2+} and release 5-HT (Cetin et al., 1994). 5-HT is involved in the regulation of gut motility, intestinal secretion, blood flow, and several GI disorders (Coates et al., 2004; Belai et al., 1997; Bearcroft et al., 1998; Dunlop et al., 2005; Gershon, 1999) including rotavirus diarrhoea (Hagbom et al., 2011), chemotherapy-induced nausea and vomiting (Aapro, 2004; Moreno et al., 2005) and *Staphylococcal* enterotoxin-induced vomiting (Hu et al., 2007). RV infection and extracellular NSP4 have been shown to stimulate human EC cells in vitro and ex vivo, and to release 5-HT (Hagbom et al., 2011). Moreover, rotavirus has been shown to be able to infect EC cells in mice small intestine (Hagbom et al., 2011). Nerve stimulation by 5-HT may lead to release of VIP from nerve endings adjacent to crypt cells and presumably increase of cAMP, water and Cl^- secretion. The VIP receptor antagonist (4Cl-D-Phe^6, Leu^{17})-VIP can attenuate RV-induced diarrhoea in mice (Kordasti et al., 2004) and VIP alone can stimulate secretion of water and electrolytes (Krejs et al., 1978). ENK are endogenous morphine like substances (opiates) and function by activating opiate receptors, thus reducing the level of cAMP, and may hence prevent fluid secretion (Turvill and Farthing, 1997). In a clinical study racecadotril has been shown to significantly reduce stool volume and duration of RV diarrhoea and thus reduces the need for oral rehydration (Salazar-Lindo et al., 2000) an observation also supported by a metaanalysis of 1384 patients with acute gastroenteritis (Lehert et al., 2011). The efficiency of racecadotril in the treatment of acute diarrhoea is thought to be due to its inhibition of enkephalinase (Lecomte et al., 1986). Based on these observations an illustration of the proposed cross-talk between EC cells, ENS, VIP, and crypt cells is presented in Fig. 2.6.5.

11 ROTAVIRUS EFFECT ON INTESTINAL MOTILITY

Alterations in intestinal transit/motility have been associated with several forms of gastrointestinal diseases, Motility of the small intestine, as in all parts of the digestive tract, is controlled predominantly by excitatory and inhibitory signals from the ENS with inputs from the CNS. Only a few studies have addressed the question whether RV infection induces intestinal motility and thereby contributes to diarrhoea. In general, infectious diarrhoeal diseases are frequently treated with the pharmacological drug loperamide (Hanauer, 2008), an opioid-receptor agonist that attenuates the activity of the myenteric plexus and thus reduces intestinal motility (Ooms et al., 1984). However, there are contrasting observations whether loperamide actively reduces RV-induced diarrhoea (Yamashiro et al., 1989; Owens et al., 1981). Yamashiro et al. (1989) found a significant effect of loperamide by evaluation of the stool score after 3, 4, and 5 days of treatment. A metaanalysis showed that patients treated with loperamide were less likely to have diarrhoea 24 h after infection, had a shorter duration of diarrhoea and had less stools compared to patients in the placebo group (Li et al., 2007; Dalby-Payne and Elliott, 2011). Istrate et al. (2014) found that

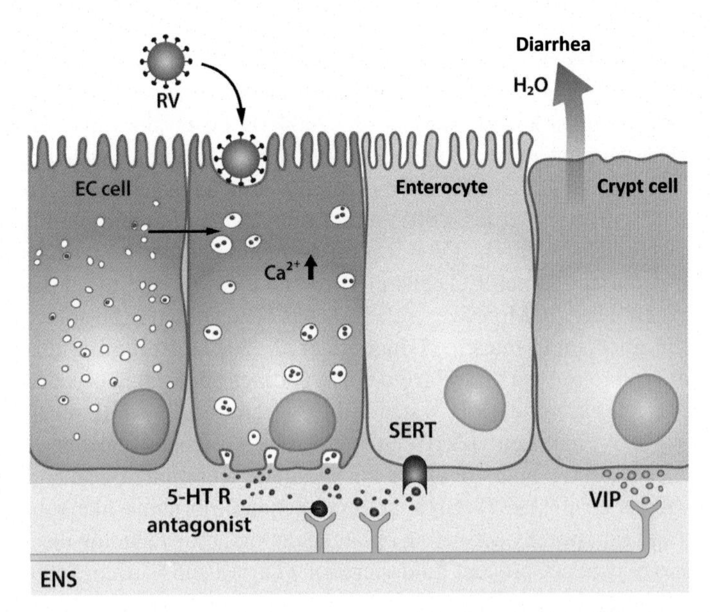

FIGURE 2.6.5 **Schematic hypothesis how rotavirus infection stimulates EC cells and trigger electrolyte and water secretion from crypt cells, via release of 5-HT and vasoactive intestinal peptide *(VIP)*.** RV infection of EC cells *(red cells)* leads to expression of NSP4 which increase intracellular calcium (Figure 2.6.2), and accumulation of 5-HT into secretory granules and release of 5-HT (Hagbom et al., 2011, 2012). The 5-HT reuptake transporter *(SERT)* regulates the 5-HT content and availability of 5-HT along the GI tract, and 5-HT alone can cause secretory diarrhoea (Hagbom et al., 2011), which can be blocked by 5-HT_3 receptor antagonists. Most interesting, the same drug can attenuate RV–induced diarrhoea in mice (Kordasti et al., 2004) and vomiting in children with acute gastroenteritis (Levine, 2011). Stimulation by 5-HT of nerves leads to release of VIP from nerve endings, which presumably increase of cAMP, water and Cl^- secretion from adjacent crypt cells. The VIP receptor antagonist (4Cl-D-Phe^6, Leu^{17})-VIP can attenuate RV-induced diarrhoea in mice (Kordasti et al., 2004), suggesting that VIP and VIPergic secretomotor neurons participate in RV-induced electrolyte secretion.

stimulation of μ-opioid receptors in the myenteric plexa attenuated RV-induced diarrhoea and motility in mice; they found that loperamide almost completely reduced diarrhoea by reducing intestinal motility, further supporting not only the human observations but also the finding that RV stimulates the ENS via an effect on the myenteric plexus (Fig. 2.6.6). Furthermore, they noticed (Istrate et al., 2014) that the muscarinic receptor antagonist atropine significantly reduced RV-diarrhoea. The fact that the antidiarrhoeal effect was associated with attenuating intestinal motility also suggests that ENS is involved in the motility response to RV infection.

Adult mice were investigated (Istrate et al., 2014) to determine whether motility changes were associated with water and electrolyte loss (ie, diarrhoea) from the small intestine or whether motility changes occurred independently of diarrhoea. Previous studies have shown that the small intestine of infant mice

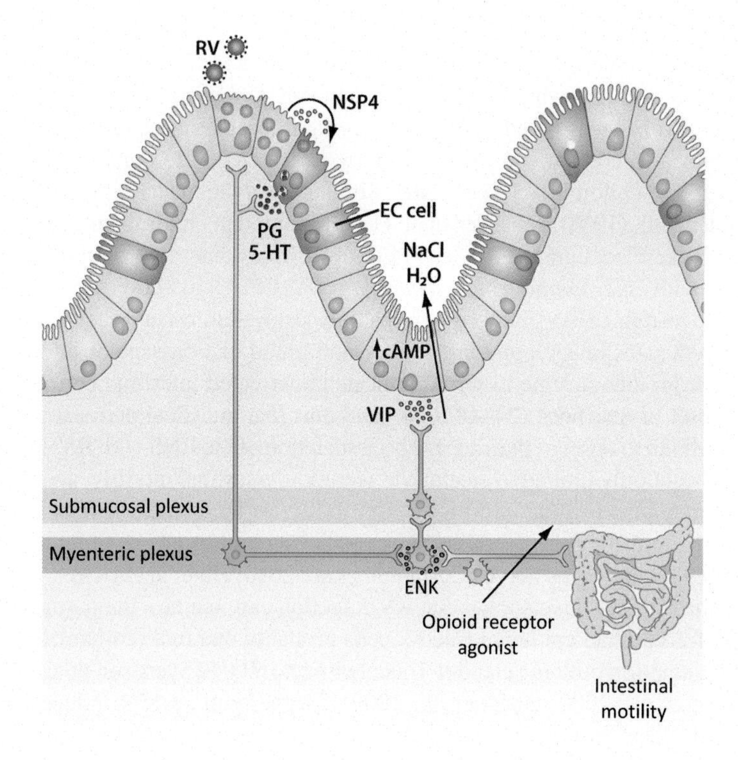

FIGURE 2.6.6 **Schematic model of RV/NSP4-induced intestinal motility and water and electrolyte secretion from crypt cells.** RV/NSP4 interact with EC cells which give raise to secretion of 5-HT (Hagbom et al., 2011) and maybe PG (Yamashiro et al., 1989; Gracey et al., 1984), which stimulates 5-HT_3 receptors on afferent nerves. 5-HT alone can induce diarrhoea (Hagbom et al., 2011). Afferent nerves activate the myenteric plexus, increasing intestinal motility. This motility activity including RV diarrhoea can be attenuated by a opioid receptor antagonist (Istrate et al., 2014). As also illustrated in Fig. 2.6.5, nerve stimulation by 5-HT leads to release of VIP from nerve endings adjacent to crypt cells and presumably increases secretion of cAMP, water, and Cl^- electrolytes.

but not adult mice (Kordasti et al., 2006) responds with diarrhoea following infection with RV. However, in diarrhoea-resistant adult mice no significant increase of intestinal motility was observed at 24 h p.i. or during the later course of infection (Kordasti et al., 2006).

12 PERMEABILITY

Alterations in intestinal permeability and thus possible leak of electrolytes and water as a mechanism of diarrhoea have only briefly been investigated in RV infection. Stintzing et al. (1986) investigated intestinal permeability in young children with RV infection using polyethylene glycol (PEG) of different molecular weight. PEG was given orally to nine children with acute RV diarrhoea, and after 6 h urine was collected and the PEG concentration determined. The same children served as their own control 3–5 weeks later. A significantly lower urinary recovery of PEG was noted in the urine during the acute phase of RV-associated diarrhoea in comparison to high urinary recovery of PEG among the same (then healthy) children 3–5 weeks later. Similar observations have been made by Serrander et al. (1984). Johansen et al. (1989) found that children with acute RV infection excreted significantly less PEG of all sizes than control children and children with *enteropathogenic Escherichia coli* (EPEC) infections. Istrate et al. (2014) investigated RV-induced permeability using fluorescent-dextran probes of different sizes in adult and infant mice and found that passage of fluorescent dextran from the intestine to serum indicated unaffected intestinal permeability at the onset of diarrhoea (24–48 h p.i.) and thus that intestinal permeability did not contribute to onset of diarrhoea. The association of the ENS with RV-induced disease is not only limited to electrolyte secretion, intestinal motility, and vomiting but may also include intestinal permeability. In the context it is of interest to note that intestinal permeability in several observations has been found to be controlled by the vagus nerve (Costantini et al., 2010; Hu et al., 2013).

In vitro studies have shown that RV infection can increase the permeability of polarized human epithelial Caco-2 cells probably due to a reorganization of the tight junction proteins claudin-1, occludin and ZO-1 (Svensson et al., 1991; Dickman et al., 2000; Obert et al., 2000). Furthermore, NSP4 induces paracellular leakage in polarized epithelial cells and prevents lateral targeting of ZO-1 (Tafazoli et al., 2001). In line with this, it has also been demonstrated in Ussing chamber experiments that the electrical tissue conductance is increased in RV-infected intestines (Isolauri et al., 1993). In conclusion, in vitro studies have suggested permeability leak in contrast to in vivo studies, which shows decreased permeability. These contradictory results may reflect the contribution of nerves and hormones in the in vivo model.

13 VOMITING AND SEROTONIN RECEPTOR ANTAGONIST TREATMENT

In human and most animals, vomiting is a protective mechanism that serves to remove noxious agents, such as toxins from the GI tract prior to absorption, but in some species it is also used to eliminate the gut of indigestible material (Andrews and Horn, 2006). Emesis might also be a response generated by cognitive, visual, flavour, or stimuli including chemotherapy, stress, intracranial pressure, and motion. The vagal nerve is an important pathway in the detection of emetic stimuli and generation of vomiting (Andrews and Horn, 2006). Vomiting is a hallmark of RV disease and contributes not only to dehydration but also hampers the effectiveness of ORT (Leung and Robson, 2007). The antiemetic drugs, Ondansetron and Granisetron (5-HT_3 receptor antagonists), once developed to attenuate vomiting and nausea in cancer patients undergoing chemotherapy and radiotherapy (Aapro, 2004, 2005; Moreno et al., 2005) are now widely used as effective antiemetic drugs in children with acute gastroenteritis (Leung and Robson, 2007; Freedman et al., 2006, 2011a,b; Reeves et al., 2002; Ramsook et al., 2002; DeCamp et al., 2008). The effect of the drug is not limited to attenuation of vomiting but can also attenuate RV-induced diarrhoea in mice (Kordasti et al., 2004). The proposed action is blocking of 5-HT_3 receptors on vagal afferent nerves. Noxious agents such as microorganisms or toxins can stimulate intestinal EC cells to release 5-HT which then interact with 5-HT_3 receptors on vagal afferents that project to the vomiting centre of the brain (Endo et al., 2000), a mechanism recently proposed for RV-induced vomiting (Hagbom et al., 2011) (Fig. 2.6.7).

In conclusion, while 5-HT_3 receptor antagonists are frequently used to attenuate vomiting in young children with acute gastroenteritis (Freedman et al., 2011a), and RV has been shown to stimulate the vomiting centre in mice

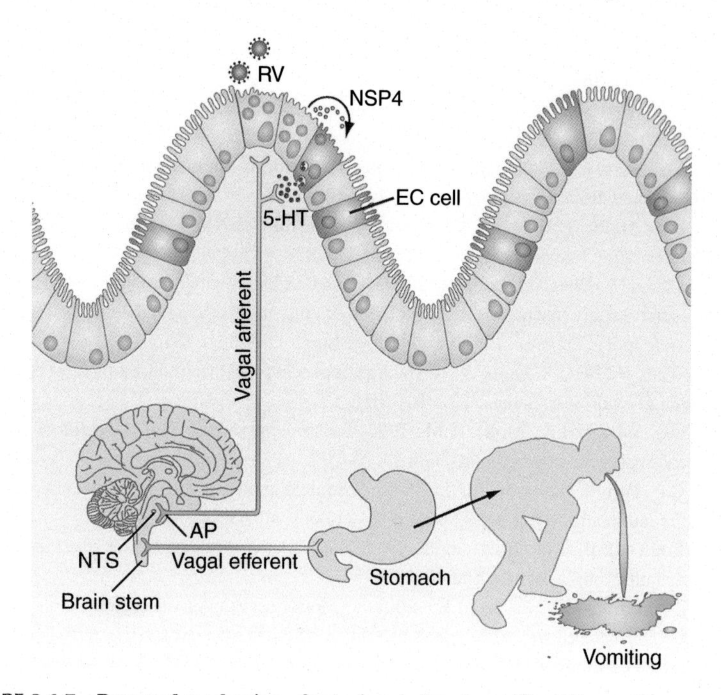

FIGURE 2.6.7 **Proposed mechanism of rotavirus induced vomiting.** The vomiting center lies in the medulla oblongata of the brain stem and comprises the reticular formation and the nucleus of the *tractus solitarii (NTS)* and area postrema *(AP)*. In the GI tract potentially harmful chemical or infectious stimuli activate receptors on EC cells. These cells then respond by release of transmitters *(5-HT)* that can stimulate the vagus nerve to activate the vomiting center. Briefly, RV or NSP4 stimulates vagal afferents to the NTS and AP of the vomiting centre by release of 5-HT from EC cells in the gut (Hagbom et al., 2011). Efferent vagal signalling stimulates a nerve-muscle vomiting reflex in the stomach. 5-HT_3 receptor antagonists are used to attenuate vomiting in children with acute gastroenteritis (Leung and Robson, 2007; Freedman et al., 2006, 2011a,b; Reeves et al., 2002; Ramsook et al., 2002; DeCamp et al., 2008).

(Hagbom et al., 2011), there is yet no clinical proof that this drug can attenuate RV-induced vomiting in humans.

In summary, while the mechanism of diarrhoea is relatively well understood, the mechanisms of vomiting, the low inflammatory response and the "sickness response" are not only poorly understood, but have in certain aspects not even been investigated. To obtain a more complete picture of rotavirus induced illnesses and host response, future research should focus on the latter subjects.

REFERENCES

Aapro, M., 2004. Granisetron: an update on its clinical use in the management of nausea and vomiting. Oncologist 9, 673–686.

Aapro, M.S., 2005. Medical oncology. Rev. Med. Suisse 1, 59–67.

Acra, S.A., Ghishan, G.K., 1996. Electrolyte fluxes in the gut and oral rehydration solutions. Pediatr. Clin. North Am. 43, 433–449.

Ahlman, H., Nilsson, 2001. The gut as the largest endocrine organ in the body. Ann. Oncol. 12 (Suppl. 2), S63–S68.

Althaus, M., 2012. Gasotransmitters: novel regulators of epithelial na(+) transport? Front. Physiol. 3, 83.

Andrews, P.L., Horn, C.C., 2006. Signals for nausea and emesis: Implications for models of upper gastrointestinal diseases. Auton Neurosci. 125, 100–115.

Angel, J., Tang, B., Feng, N., et al., 1998. Studies of the role for NSP4 in the pathogenesis of homologous murine rotavirus diarrhea. J. Infect. Dis. 177, 455–458.

Bahl, R., Saxena, M., Bhandari, N., et al., 2009. Population-based incidence of intussusception and a case-control study to examine the association of intussusception with natural rotavirus infection among indian children. J. Infect. Dis. 200 (Suppl. 1), S277–S281.

Ball, J.M., Tian, P., Zeng, C.Q., et al., 1996. Age-dependent diarrhea induced by a rotaviral nonstructural glycoprotein. Science 272, 101–104.

Bardhan, P.K., Salam, M.A., Molla, A.M., 1992. Gastric emptying of liquid in children suffering from acute rotaviral gastroenteritis. Gut 33, 26–29.

Bearcroft, C.P., Perrett, D., Farthing, M.J., 1998. Postprandial plasma 5-hydroxytryptamine in diarrhoea predominant irritable bowel syndrome: a pilot study. Gut 42, 42–46.

Belai, A., Boulos, P.B., Robson, T., et al., 1997. Neurochemical coding in the small intestine of patients with Crohn's disease. Gut 40, 767–774.

Berkova, Z., Morris, A.P., Estes, M.K., 2003. Cytoplasmic calcium measurement in rotavirus enterotoxin-enhanced green fluorescent protein (NSP4-EGFP) expressing cells loaded with Fura-2. Cell Calcium 34, 55–68.

Bines, J.E., Liem, N.T., Justice, F.A., et al., 2006. Risk factors for intussusception in infants in Vietnam and Australia: adenovirus implicated, but not rotavirus. J. Pediatr. 149, 452–460.

Bishop, R.F., Davidson, B.L., Holmes, I.H., et al., 1973. Virus particles in epithelial cells of duodenal mucosal from children with viral gastroenteritis. Lancet 2, 1281–1283.

Blutt, S.E., Conner, M.E., 2007. Rotavirus: to the gut and beyond! Curr. Opin. Gastroenterol. 23, 39–43.

Blutt, S.E., Kirkwood, C.D., Parreno, V., et al., 2003. Rotavirus antigenaemia and viraemia: a common event? Lancet 362, 1445–1449.

Borghan, M.A., Mori, Y., El-Mahmoudy, A.B., et al., 2007. Induction of nitric oxide synthase by rotavirus enterotoxin NSP4: implication for rotavirus pathogenicity. J. Gen. Virol. 88, 2064–2072.

Brodal, P., 2010. The Central Nervous System; structure and function. Fourth edition. Oxford University Press, New York.

Brunsson, I., Sjoqvist, A., Jodal, M., et al., 1987. Mechanisms underlying the small intestinal fluid secretion caused by arachidonic acid, prostaglandin E1 and prostaglandin E2 in the rat in vivo. Acta Physiol. Scand. 130, 633–642.

Bugarcic, A., Taylor, J.A., 2006. Rotavirus nonstructural glycoprotein NSP4 is secreted from the apical surfaces of polarized epithelial cells. J. Virol. 80, 12343–12349.

Buttery, J.P., Danchin, M.H., Lee, K.J., et al., 2011. Intussusception following rotavirus vaccine administration: post-marketing surveillance in the National Immunization Program in Australia. Vaccine 29, 3061–3066.

Carlson, J.A., Middleton, P.J., Szymanski, M.T., et al., 1978. Fatal rotavirus gastroenteritis: an analysis of 21 cases. Am. J. Dis. Child 132, 477–479.

Centers for Disease Control and Prevention, 1999. Intussusception among recipients of rotavirus vaccine—United States, 1998–1999. JAMA 282, 520–521.

Cetin, Y., Kuhn, M., Kulaksiz, H., et al., 1994. Enterochromaffin cells of the digestive system: cellular source of guanylin, a guanylate cyclase-activating peptide. Proc. Natl. Acad. Sci. USA 91, 2935–2939.

Chen, C.C., Huang, J.L., Chang, C.J., et al., 2012. Fecal calprotectin as a correlative marker in clinical severity of infectious diarrhea and usefulness in evaluating bacterial or viral pathogens in children. J. Pediatr. Gastroenterol. Nutr. 55, 541–547.

Coates, M.D., Mahoney, C.R., Linden, D.R., et al., 2004. Molecular defects in mucosal serotonin content and decreased serotonin reuptake transporter in ulcerative colitis and irritable bowel syndrome. Gastroenterology 126, 1657–1664.

Collins, J., Starkey, W.G., Wallis, T.S., et al., 1988. Intestinal enzyme profiles in normal and rotavirus-infected mice. J. Pediatr. Gastroenterol. Nutr. 7, 264–272.

Cooke, A.R., 1975. Control of gastric emptying and motility. Gastroenterology 68, 804–816.

Costantini, T.W., Bansal, V., Peterson, C.Y., et al., 2010. Efferent vagal nerve stimulation attenuates gut barrier injury after burn: modulation of intestinal occludin expression. J. Trauma. 68, 1349–1354, discussion 1354–1356.

Cutting, G.R., 2015. Cystic fibrosis genetics: from molecular understanding to clinical application. Nat. Rev. Genet. 16, 45–56.

Dalby-Payne, J.R., Elliott, E.J., 2011. Gastroenteritis in children. Clin. Evid., 2011.

Dallar, Y., Bostanci, I., Bozdayi, G., et al., 2009. Rotavirus-associated intussusception followed by spontaneous resolution. Turk. J. Gastroenterol. 20, 209–213.

Davidson, G.P., Gall, D.G., Petric, M., et al., 1977. Human rotavirus enteritis induced in conventional piglets. Intestinal structure and transport. J. Clin. Invest. 60, 1402–1409.

Davidson, G.P., Barnes, G.L., 1979. Structural and functional abnormalities of the small intestine in infants and young children with rotavirus enteritis. Acta Paediatr. Scand. 68, 181–186.

Debas, H.T., Farooq, O., Grossman, M.I., 1975. Inhibition of gastric emptying is a physiological action of cholecystokinin. Gastroenterology 68, 1211–1217.

DeCamp, L.R., Byerley, J.S., Doshi, N., et al., 2008. Use of antiemetic agents in acute gastroenteritis: a systematic review and meta-analysis. Arch. Pediatr. Adolesc. Med. 162, 858–865.

Dickman, K.G., Hempson, S.J., Anderson, J., et al., 2000. Rotavirus alters paracellular permeability and energy metabolism in Caco-2 cells. Am. J. Physiol. Gastrointest. Liver Physiol. 279, G757–G766.

Didsbury, A., Wang, C., Verdon, D., et al., 2011. Rotavirus NSP4 is secreted from infected cells as an oligomeric lipoprotein and binds to glycosaminoglycans on the surface of non-infected cells. Virol. J. 8, 551.

Dunlop, S.P., Coleman, N.S., Blackshaw, E., et al., 2005. Abnormalities of 5-hydroxytryptamine metabolism in irritable bowel syndrome. Clin. Gastroenterol. Hepatol. 3, 349–357.

Elliott, E.J., 2007. Acute gastroenteritis in children. BMJ 334, 35–40.

Endo, T., Minami, M., Hirafuji, M., et al., 2000. Neurochemistry and neuropharmacology of emesis - the role of serotonin. Toxicology 153, 189–201.

Eskilsson, A., Mirrasekhian, E., Dufour, S., et al., 2014. Immune-induced fever is mediated by IL-6 receptors on brain endothelial cells coupled to STAT3-dependent induction of brain endothelial prostaglandin synthesis. J. Neurosci. 34, 15957–15961.

Fenaux, M., Cuadras, M.A., Feng, N., et al., 2006. Extraintestinal spread and replication of a homologous EC rotavirus strain and a heterologous rhesus rotavirus in BALB/c mice. J. Virol. 80, 5219–5232.

Feng, N., Kim, B., Fenaux, M., et al., 2008. Role of interferon in homologous and heterologous rotavirus infection in the intestines and extraintestinal organs of suckling mice. J. Virol. 82, 7578–7590.

Field, M., 2003. Intestinal ion transport and the pathophysiology of diarrhea. J. Clin. Invest. 111, 931–943.

Fischer, T.K., Ashley, D., Kerin, T., et al., 2005. Rotavirus antigenemia in patients with acute gastroenteritis. J. Infect. Dis. 192, 913–919.

Flewett, T.H., Bryden, A.S., Davies, H., Letter:, 1973. Virus particles in gastroenteritis. Lancet 2, 1497.

Freedman, S.B., Adler, M., Seshadri, R., et al., 2006. Oral ondansetron for gastroenteritis in a pediatric emergency department. N. Engl. J. Med. 354, 1698–1705.

Freedman, S.B., Sivabalasundaram, V., Bohn, V., et al., 2011a. The treatment of pediatric gastroenteritis: a comparative analysis of pediatric emergency physicians' practice patterns. Acad. Emerg. Med. 18, 38–45.

Freedman, S.B., Gouin, S., Bhatt, M., et al., 2011b. Prospective assessment of practice pattern variations in the treatment of pediatric gastroenteritis. Pediatrics 127, e287–e295.

Furness, J., 2006. In: JB, F. (Ed.), The enteric nervous system. Blackwell Publishing, Inc, Oxford, pp. 1–267.

Ge, Y., Mansell, A., Ussher, J.E., et al., 2013. Rotavirus NSP4 triggers secretion of pro-inflammatory cytokines from macrophages via Toll-Like Receptor-2. J. Virol. 87 (20), 11160–11167.

Gershon, M.D., 1999. Review article: roles played by 5-hydroxytryptamine in the physiology of the bowel. Aliment. Pharmacol. Ther. 13 (Suppl. 2), 15–30.

Gilger, M.A., Matson, D.O., Conner, M.E., et al., 1992. Extraintestinal rotavirus infections in children with immunodeficiency. J. Pediatr. 120, 912–917.

Gracey, M., Phadke, M.A., Burke, V., et al., 1984. Aspirin in acute gastroenteritis: a clinical and microbiological study. J. Pediatr. Gastroenterol. Nutr. 3, 692–695.

Graham, D.Y., Sackman, J.W., Estes, M.K., 1984. Pathogenesis of rotavirus-induced diarrhea. Preliminary studies in miniature swine piglet. Dig. Dis. Sci. 29, 1028–1035.

Grech, V., Calvagna, V., Falzon, A., et al., 2001. Fatal, rotavirus-associated myocarditis and pneumonitis in a 2-year-old boy. Ann. Trop. Paediatr. 21, 147–148.

Greenberg, H.B., Estes, M.K., 2009. Rotaviruses: from pathogenesis to vaccination. Gastroenterology 136, 1939–1951.

Greenberg, H.B., Clark, H.F., Offit, P.A., 1994. Rotavirus pathology and pathophysiology. Curr. Top. Microbiol. Immunol. 185, 255–283.

Hagbom, M., Istrate, C., Engblom, D., et al., 2011. Rotavirus stimulates release of serotonin (5-HT) from human enterochromaffin cells and activates brain structures involved in nausea and vomiting. PLoS Pathog. 7, e1002115.

Hagbom, M., Sharma, S., Lundgren, O., et al., 2012. Towards a human rotavirus disease model. Curr. Opin. Virol. 2, 408–418.

Halaihel, N., Lievin, V., Alvarado, F., et al., 2000. Rotavirus infection impairs intestinal brush-border membrane Na(+)-solute cotransport activities in young rabbits. Am. J. Physiol. Gastrointest. Liver Physiol. 279, G587–G596.

Hallback, D.A., Jodal, M., Sjoqvist, A., et al., 1979. Villous tissue osmolality and intestinal transport of water and electrolytes. Acta Physiol. Scand. 107, 115–126.

Hanauer, S.B., 2008. The role of loperamide in gastrointestinal disorders. Rev. Gastroenterol. Disord. 8, 15–20.

Hansen, M.B., 2003. The enteric nervous system I: organisation and classification. Pharmacol. Toxicol. 92, 105–113.

Hansen, M.B., Witte, A.B., 2008. The role of serotonin in intestinal luminal sensing and secretion. Acta Physiol. 193, 311–323.

Hart, B.L., 1988. Biological basis of the behavior of sick animals. Neurosci. Biobehav. Rev. 12, 123–137.

Hennessy, M.B., Deak, T., Schiml, P.A., 2014. Sociality and sickness: have cytokines evolved to serve social functions beyond times of pathogen exposure? Brain Behav. Immun. 37, 15–20.

Horn, C.C., 2008. Why is the neurobiology of nausea and vomiting so important? Appetite 50, 430–434.

Hu, D.L., Zhu, G., Mori, F., et al., 2007. Staphylococcal enterotoxin induces emesis through increasing serotonin release in intestine and it is downregulated by cannabinoid receptor 1. Cell Microbiol. 9, 2267–2277.

Hu, S., Du, M.H., Luo, H.M., et al., 2013. Electroacupuncture at Zusanli (ST36) prevents intestinal barrier and remote organ dysfunction following gut ischemia through activating the cholinergic anti-inflammatory-dependent mechanism. Evid. Based Complement. Alternat. Med. 2013, 592127.

Imbert-Marcille, B.M., Barbe, L., Dupe, M., et al., 2014. A FUT2 gene common polymorphism determines resistance to rotavirus A of the P^8 genotype. J. Infect. Dis. 209, 1227–1230.

Isolauri, E., Kaila, M., Arvola, T., et al., 1993. Diet during rotavirus enteritis affects jejunal permeability to macromolecules in suckling rats. Pediatr. Res. 33, 548–553.

Istrate, C., Hagbom, M., Vikstrom, E., et al., 2014. Rotavirus infection increases intestinal motility but not permeability at the onset of diarrhea. J. Virol. 88, 3161–3169.

Iturriza-Gomara, M., Auchterlonie, I.A., Zaw, W., et al., 2002. Rotavirus gastroenteritis and central nervous system (CNS) infection: characterization of the VP7 and VP4 genes of rotavirus strains isolated from paired fecal and cerebrospinal fluid samples from a child with CNS disease. J. Clin. Microbiol. 40, 4797–4799.

Jiang, B., Snipes-Magaldi, L., Dennehy, P., et al., 2003. Cytokines as mediators for or effectors against rotavirus disease in children. Clin. Diagn. Lab. Immunol. 10, 995–1001.

Johansen, K., Stintzing, G., Magnusson, K.E., et al., 1989. Intestinal permeability assessed with polyethylene glycols in children with diarrhea due to rotavirus and common bacterial pathogens in a developing community. J. Pediatr. Gastroenterol. Nutr. 9, 307–313.

Kawashima, H., Inage, Y., Ogihara, M., et al., 2004. Serum and cerebrospinal fluid nitrite/nitrate levels in patients with rotavirus gastroenteritis induced convulsion. Life Sci. 74, 1397–1405.

Keidan, I., Shif, I., Keren, G., et al., 1992. Rotavirus encephalopathy: evidence of central nervous system involvement during rotavirus infection. Pediatr. Infect. Dis. J. 11, 773–775.

Kerzner, B., Kelly, M.H., Gall, D.G., et al., 1977. Transmissible gastroenteritis: sodium transport and the intestinal epithelium during the course of viral enteritis. Gastroenterology 72, 457–461.

Kohler, T., Erben, U., Wiedersberg, H., et al., 1990. Histological findings of the small intestinal mucosa in rotavirus infections in infants and young children. Kinderarztl Prax 58, 323–327.

Konno, T., Suzuki, H., Kutsuzawa, T., et al., 1978. Human rotavirus infection in infants and young children with intussusception. J. Med. Virol. 2, 265–269.

Kordasti, S., Sjovall, H., Lundgren, O., et al., 2004. Serotonin and vasoactive intestinal peptide antagonists attenuate rotavirus diarrhoea. Gut 53, 952–957.

Kordasti, S., Istrate, C., Banasaz, M., et al., 2006. Rotavirus infection is not associated with small intestinal fluid secretion in the adult mouse. J. Virol. 80, 11355–11361.

Krejs, G.J., Barkley, R.M., Read, N.W., et al., 1978. Intestinal secretion induced by vasoactive intestinal polypeptide. A comparison with cholera toxin in the canine jejunum in vivo. J. Clin. Invest. 61, 1337–1345.

Kukuruzovic, R., Robins-Browne, R.M., Anstey, N.M., et al., 2002. Enteric pathogens, intestinal permeability and nitric oxide production in acute gastroenteritis. Pediatr. Infect. Dis. J. 21, 730–739.

Kutukculer, N., Caglayan, S., 1997. Tumor necrosis factor-alpha and interleukin-6 in stools of children with bacterial and viral gastroenteritis. J. Pediatr. Gastroenterol. Nutr. 25, 556–557.

Lecomte, J.M., Costentin, J., Vlaiculescu, A., et al., 1986. Pharmacological properties of acetorphan, a parenterally active "enkephalinase" inhibitor. J. Pharmacol. Exp. Ther. 237, 937–944.

Lehert, P., Cheron, G., Calatayud, G.A., et al., 2011. Racecadotril for childhood gastroenteritis: an individual patient data meta-analysis. Dig. Liver Dis. 43, 707–713.

Leung, A.K., Robson, W.L., 2007. Acute gastroenteritis in children: role of anti-emetic medication for gastroenteritis-related vomiting. Paediatr. Drugs 9, 175–184.

Levine, D.A., 2011. Oral ondansetron decreases vomiting, as well as the need for intravenous fluids and hospital admission, in children with acute gastroenteritis. Evid. Based Med. 17 (4), 112–113.

Li, S.T., Grossman, D.C., Cummings, P., 2007. Loperamide therapy for acute diarrhea in children: systematic review and meta-analysis. PLoS Med. 4, e98.

Libonati, M.H., Dennis, A.F., Ramani, S., et al., 2014. Absence of genetic differences among G10P[11] rotaviruses associated with asymptomatic and symptomatic neonatal infections in Vellore. India. J. Virol. 88, 9060–9071.

Litalien, C., Jacqz-Aigrain, E., 2001. Risks and benefits of nonsteroidal anti-inflammatory drugs in children: a comparison with paracetamol. Paediatr. Drugs 3, 817–858.

Liu, B., Fujita, Y., Arakawa, C., et al., 2009. Detection of rotavirus RNA and antigens in serum and cerebrospinal fluid samples from diarrheic children with seizures. Jpn. J. Infect Dis. 62, 279–283.

Lopman, B.A., Trivedi, T., Vicuna, Y., et al., 2015. Norovirus infection and disease in an ecuadorian birth cohort: association of certain norovirus Genotypes with host FUT2 secretor status. J. Infect. Dis. 211 (11), 1813–1821.

Lundgren, O., Svensson, L., 2001. Pathogenesis of rotavirus diarrhea. Microbes Infect. 3, 1145–1156.

Lundgren, O., Svensson, L., 2003. The enteric nervous system and infectious diarrhea. In: Desselberger, U., Gray, J. (Eds.), Viral Gastroenteritis, vol. 9, Elsevier, Amsterdam, pp. 51–67.

Lundgren, O., Peregrin, A.T., Persson, K., et al., 2000. Role of the enteric nervous system in the fluid and electrolyte secretion of rotavirus diarrhea. Science 287, 491–495.

Lynch, M., Lee, B., Azimi, P., et al., 2001. Rotavirus and central nervous system symptoms: cause or contaminant? Case reports and review. Clin. Infect. Dis. 33, 932–938.

Lynch, M., Shieh, W.J., Tatti, K., et al., 2003. The pathology of rotavirus-associated deaths, using new molecular diagnostics. Clin. Infect. Dis. 37, 1327–1333.

Mavromichalis, J., Evans, N., McNeish, A.S., et al., 1977. Intestinal damage in rotavirus and adenovirus gastroenteritis assessed by d-xylose malabsorption. Arch. Dis. Child 52, 589–591.

McClung, H.J., Butler, D.G., Kerzner, B., et al., 1976. Transmissible gastroenteritis. Mucosal ion transport in acute viral enteritis. Gastroenterology 70, 1091–1095.

Medici, M.C., Abelli, L.A., Guerra, P., et al., 2011a. Case report: detection of rotavirus RNA in the cerebrospinal fluid of a child with rotavirus gastroenteritis and meningism. J. Med. Virol. 83, 1637–1640.

Medici, M.C., Abelli, L.A., Martinelli, M., et al., 2011b. Clinical and molecular observations of two fatal cases of rotavirus-associated enteritis in children in Italy. J. Clin. Microbiol. 49, 2733–2739.

Meeroff, J.C., Schreiber, D.S., Trier, J.S., et al., 1980. Abnormal gastric motor function in viral gastroenteritis. Ann. Intern. Med. 92, 370–373.

Michelangeli, F., Ruiz, M.C., 2003. Physiology and pathophysiology of the gut in relation to viral diarrahea. In: Desselberger, U., Gray, J. (Eds.), Viral Gastroenteritis, vol. 9, Elsevier, Amsterdam, pp. 23–50.

Minami, H., McCallum, R.W., 1984. The physiology and pathophysiology of gastric emptying in humans. Gastroenterology 86, 1592–1610.

Molla, A., Molla, A., Sarker, S., et al., 1983. Food intake during and after recovery from diarrhoea in children. In: Chen, L., Scrinushaw, N. (Eds.), Diarrhoea and Malnutrion. Plenum Press, New York, pp. 113–123.

Moreno, J., Sahade, M., del Giglio, A., 2005. Low-dose granisetron for prophylaxis of acute chemotherapy-induced nausea and vomiting: a pilot study. Support Care Cancer 13, 850–853.

Morris, A.P., Scott, J.K., Ball, J.M., et al., 1999. NSP4 elicits age-dependent diarrhea and Ca(2 +) mediated I(-) influx into intestinal crypts of CF mice. Am. J. Physiol. 277, G431–G444.

Morrison, C., Gilson, T., Nuovo, G.J., 2001. Histologic distribution of fatal rotaviral infection: an immunohistochemical and reverse transcriptase in situ polymerase chain reaction analysis. Hum. Pathol. 32, 216–221.

Mossel, E.C., Ramig, R.F., 2002. Rotavirus genome segment 7 (NSP3) is a determinant of extraintestinal spread in the neonatal mouse. J. Virol. 76, 6502–6509.

Mossel, E.C., Ramig, R.F., 2003. A lymphatic mechanism of rotavirus extraintestinal spread in the neonatal mouse. J. Virol. 77, 12352–12356.

Mulcahy, D.L., Kamath, K.R., de Silva, L.M., et al., 1982. A two-part study of the aetiological role of rotavirus in intussusception. J. Med. Virol. 9, 51–55.

Murphy, T.V., Gargiullo, P.M., Massoudi, M.S., et al., 2001. Intussusception among infants given an oral rotavirus vaccine. N. Engl. J. Med. 344, 564–572.

Nakagomi, T., Nakagomi, O., 2005. Rotavirus antigenemia in children with encephalopathy accompanied by rotavirus gastroenteritis. Arch. Virol. 150, 1927–1931.

Noone, C., Menzies, I.S., Banatvala, J.E., et al., 1986. Intestinal permeability and lactose hydrolysis in human rotaviral gastroenteritis assessed simultaneously by non-invasive differential sugar permeation. Eur. J. Clin. Invest. 16, 217–225.

Nordgren, J., Sharma, S., Bucardo, F., et al., 2014. Both lewis and secretor status mediate susceptibility to rotavirus infections in a rotavirus genotype-dependent manner. Clinical Infect. Dis. 59, 1567–1573.

Obert, G., Peiffer, I., Servin, A.L., 2000. Rotavirus-induced structural and functional alterations in tight junctions of polarized intestinal Caco-2 cell monolayers. J. Virol. 74, 4645–4651.

Offit, P.A., Clark, H.F., Kornstein, M.J., et al., 1984. A murine model for oral infection with a primate rotavirus (simian SA11). J. Virol. 51, 233–236.

Ooms, L.A., Degryse, A.D., Janssen, P.A., 1984. Mechanisms of action of loperamide. Scand. J. Gastroenterol Suppl. 96, 145–155.

Osborne, M.P., Haddon, S.J., Spencer, A.J., et al., 1988. An electron microscopic investigation of time-related changes in the intestine of neonatal mice infected with murine rotavirus. J. Pediatr. Gastroenterol. Nutr. 7, 236–248.

Osborne, M.P., Haddon, S.J., Worton, K.J., et al., 1991. Rotavirus-induced changes in the microcirculation of intestinal villi of neonatal mice in relation to the induction and persistence of diarrhea. J. Pediatr. Gastroenterol. Nutr. 12, 111–120.

Owens, J.R., Broadhead, R., Hendrickse, R.G., et al., 1981. Loperamide in the treatment of acute gastroenteritis in early childhood. Report of a two centre, double-blind, controlled clinical trial. Ann. Trop. Paediatr. 1, 135–141.

Pager, C., Steele, D., Gwamanda, P., et al., 2000. A neonatal death associated with rotavirus infection—detection of rotavirus dsRNA in the cerebrospinal fluid. S. Afr. Med. J. 90, 364–365.

Patel, M.M., Lopez-Collada, V.R., Bulhoes, M.M., et al., 2011. Intussusception risk and health benefits of rotavirus vaccination in Mexico and Brazil. N. Engl. J. Med. 364, 2283–2292.

Peigue-Lafeuille, H., Henquell, C., Chambon, M., et al., 1991. Nosocomial rotavirus infections in adult renal transplant recipients. J. Hosp. Infect. 18, 67–70.

Ramani, S., Paul, A., Saravanabavan, A., et al., 2010. Rotavirus antigenemia in Indian children with rotavirus gastroenteritis and asymptomatic infections. Clin. Infect. Dis. 51, 1284–1289.

Ramig, R.F., 2004. Pathogenesis of intestinal and systemic rotavirus infection. J. Virol. 78, 10213–10220.

Ramig, R.F., 2007. Systemic rotavirus infection. Expert Rev. Anti. Infect. Ther. 5, 591–612.

Ramsook, C., Sahagun-Carreon, I., Kozinetz, C.A., et al., 2002. A randomized clinical trial comparing oral ondansetron with placebo in children with vomiting from acute gastroenteritis. Ann. Emerg. Med. 39, 397–403.

Rautanen, T., Kurki, S., Vesikari, T., 1997. Randomised double blind study of hypotonic oral rehydration solution in diarrhoea. Arch. Dis. Child 76, 272–274.

Ray, P.G., Kelkar, S.D., Walimbe, A.M., et al., 2007. Rotavirus immunoglobulin levels among Indian mothers of two socio-economic groups and occurrence of rotavirus infections among their infants up to six months. J. Med. Virol. 79, 341–349.

Raybould, H.E., 2002. Visceral perception: sensory transduction in visceral afferents and nutrients. Gut 51 (Suppl. 1), i11–i14.

Reeves, J.J., Shannon, M.W., Fleisher, G.R., 2002. Ondansetron decreases vomiting associated with acute gastroenteritis: a randomized, controlled trial. Pediatrics 109, e62.

Reynolds, D.J., Hall, G.A., Debney, T.G., et al., 1985. Pathology of natural rotavirus infection in clinically normal calves. Res. Vet. Sci. 38, 264–269.

Robinson, C.G., Hernanz-Schulman, M., Zhu, Y., et al., 2004. Evaluation of anatomic changes in young children with natural rotavirus infection: is intussusception biologically plausible? J. Infect. Dis. 189, 1382–1387.

Rodriguez-Diaz, J., Banasaz, M., Istrate, C., et al., 2006. Role of nitric oxide during rotavirus infection. J. Med. Virol. 78, 979–985.

Ruiz-Palacios, G.M., Perez-Schael, I., Velazquez, F.R., et al., 2006. Safety and efficacy of an attenuated vaccine against severe rotavirus gastroenteritis. N. Engl. J. Med. 354, 11–22.

Sachdeva, D., 1996. Oral rehydration therapy. J. Indian Med. Assoc. 94, 298–305.

Sack, D.A., Chowdhury, A.M., Eusof, A., et al., 1978. Oral hydration rotavirus diarrhoea: a double blind comparison of sucrose with glucose electrolyte solution. Lancet 2, 280–283.

Salazar-Lindo, E., Santisteban-Ponce, J., Chea-Woo, E., et al., 2000. Racecadotril in the treatment of acute watery diarrhea in children. N. Engl. J. Med. 343, 463–467.

Salim, A.F., Phillips, A.D., Walker-Smith, J.A., et al., 1995. Sequential changes in small intestinal structure and function during rotavirus infection in neonatal rats. Gut 36, 231–238.

Salmi, T.T., Arstila, P., Koivikko, A., 1978. Central nervous system involvement in patients with rotavirus gastroenteritis. Scand. J. Infect. Dis. 10, 29–31.

Sandhu, B.K., Tripp, J.H., Candy, D.C., et al., 1981. Loperamide: studies on its mechanism of action. Gut 22, 658–662.

Santoro, M.G., 1997. Antiviral activity of cyclopentenone prostanoids. Trends Microbiol. 5, 276–281.

Scher, J.U., Pillinger, M.H., 2009. The anti-inflammatory effects of prostaglandins. J. Investig. Med. 57, 703–708.

Seo, N.S., Zeng, C.Q., Hyser, J.M., et al., 2008. Integrins alpha1beta1 and alpha2beta1 are receptors for the rotavirus enterotoxin. Proc. Natl. Acad. Sci. USA 105, 8811–8818.

Serrander, R., Magnusson, K.E., Sundqvist, T., 1984. Acute infections with Giardia lamblia and rotavirus decrease intestinal permeability to low-molecular weight polyethylene glycols (PEG 400). Scand. J. Infect. Dis. 16, 339–344.

Shepherd, R.W., Butler, D.G., Cutz, E., et al., 1979a. The mucosal lesion in viral enteritis. Extent and dynamics of the epithelial response to virus invasion in transmissible gastroenteritis of piglets. Gastroenterology 76, 770–777.

Shepherd, R.W., Gall, D.G., Butler, D.G., et al., 1979b. Determinants of diarrhea in viral enteritis. The role of ion transport and epithelial changes in the ileum in transmissible gastroenteritis in piglets. Gastroenterology 76, 20–24.

Sowmyanarayanan, T.V., Natarajan, S.K., Ramachandran, A., et al., 2009. Nitric oxide production in acute gastroenteritis in Indian children. Trans. R Soc. Trop. Med. Hyg. 103, 849–851.

Spiller, R., Serotonin, 2008. GI clinical disorders. Neuropharmacology 55, 1072–1080.

Starkey, W.G., Collins, J., Candy, D.C., et al., 1990. Transport of water and electrolytes by rotavirus- infected mouse intestine: a time course study. J. Pediatr. Gastroenterol. Nutr. 11, 254–260.

Stintzing, G., Johansen, K., Magnusson, K.E., et al., 1986. Intestinal permeability in small children during and after rotavirus diarrhoea assessed with different-size polyethyleneglycols (PEG 400 and PEG 1000). Acta Paediatr. Scand. 75, 1005–1009.

Superti, F., Amici, C., Tinari, A., et al., 1998. Inhibition of rotavirus replication by prostaglandin A: evidence for a block of virus maturation. J. Infect. Dis. 178, 564–568.

Suzuki, H., Oshitani, H., 1999. Effect of prostaglandin A and tunicamycin on rotavirus assembly. J. Infect. Dis. 179, 522.

Svensson, L., Finlay, B.B., Bass, D., et al., 1991. Symmetric infection of rotavirus on polarized human intestinal epithelial (Caco-2) cells. J. Virol. 65, 4190–4197.

Tafazoli, F., Zeng, C.Q., Estes, M.K., et al., 2001. NSP4 enterotoxin of rotavirus induces paracellular leakage in polarized epithelial cells. J. Virol. 75, 1540–1546.

Tate, J.E., Burton, A.H., Boschi-Pinto, C., et al., 2012. 2008 Estimate of worldwide rotavirus-associated mortality in children younger than 5 years before the introduction of universal rotavirus vaccination programmes: a systematic review and meta-analysis. Lancet Infect. Dis. 12, 136–141.

Teitelbaum, J.E., Daghistani, R., 2007. Rotavirus causes hepatic transaminase elevation. Dig. Dis. Sci. 52, 3396–3398.

Tian, P., Estes, M.K., Hu, Y., et al., 1995. The rotavirus nonstructural glycoprotein NSP4 mobilizes Ca^{2+} from the endoplasmic reticulum. J. Virol. 69, 5763–5772.

Tormo, R., Polanco, I., Salazar-Lindo, E., et al., 2008. Acute infectious diarrhoea in children: new insights in antisecretory treatment with racecadotril. Acta paediatrica 97, 1008–1015.

Turvill, J., Farthing, M., 1997. Enkephalins and enkephalinase inhibitors in intestinal fluid and electrolyte transport. Eur. J. Gastroenterol. Hepatol. 9, 877–880.

Uhnoo, I., Olding-Stenkvist, E., Kreuger, A., 1986. Clinical features of acute gastroenteritis associated with rotavirus, enteric adenoviruses, and bacteria. Arch. Dis. Child 61, 732–738.

Ushijima, H., Bosu, K., Abe, T., et al., 1986. Suspected rotavirus encephalitis. Arch. Dis. Child 61, 692–694.

Vane, J.R., 1971. Inhibition of prostaglandin synthesis as a mechanism of action for aspirin-like drugs. Nat. New Biol. 231, 232–235.

Velazquez, F.R., Luna, G., Cedillo, R., et al., 2004. Natural rotavirus infection is not associated to intussusception in Mexican children. Pediatr. Infect. Dis. J. 23, S173–S178.

Vesikari, T., Matson, D.O., Dennehy, P., et al., 2006. Safety and efficacy of a pentavalent human-bovine (WC3) reassortant rotavirus vaccine. N. Engl. J. Med. 354, 23–33.

Ward, R.L., McNeal, M.M., Sheridan, J.F., 1990. Development of an adult mouse model for studies on protection against rotavirus. J. Virol. 64, 5070–5075.

Watkins, L.R., Goehler, L.E., Relton, J.K., et al., 1995. Blockade of interleukin-1 induced hyperthermia by subdiaphragmatic vagotomy: evidence for vagal mediation of immune-brain communication. Neurosci. Lett. 183, 27–31.

Weh, J., Antoni, C., Weiss, C., et al., 2013. Discriminatory potential of C-reactive protein, cytokines, and fecal markers in infectious gastroenteritis in adults. Diagn. Microbiol. Infect. Dis. 77, 79–84.

Wong, C.J., Price, Z., Bruckner, D.A., 1984. Aseptic meningitis in an infant with rotavirus gastroenteritis. Pediatr. Infect. Dis. 3, 244–246.

Yamashiro, Y., Shimizu, T., Oguchi, S., et al., 1989. Prostaglandins in the plasma and stool of children with rotavirus gastroenteritis. J. Pediatr. Gastroenterol. Nutr. 9, 322–327.

Yolken, R.H., Bishop, C.A., Townsend, T.R., et al., 1982. Infectious gastroenteritis in bone-marrow-transplant recipients. N. Engl. J. Med. 306, 1009–1012.

Zhang, M., Zeng, C.Q., Morris, A.P., et al., 2000. A functional NSP4 enterotoxin peptide secreted from rotavirus-infected cells. J. Virol. 74, 11663–11670.

Chapter 2.7

Gnotobiotic Neonatal Pig Model of Rotavirus Infection and Disease

A.N. Vlasova, S. Kandasamy, L.J. Saif
Food Animal Health Research Program, The Ohio Agricultural Research and Development Center, Department of Veterinary Preventive Medicine, The Ohio State University, Wooster, OH, United States

1 INTRODUCTION

Poor sanitation and water treatment systems, and lack of medical care and efficacious vaccines translate into nearly 1,600,000 deaths among children <5 years old and several million hospitalizations due to severe, dehydrating diarrhea annually (Tagbo et al., 2014). Rotavirus (RV) has been recognized as a major cause of severe diarrhea in infants and young children, associated with ~453,000 deaths worldwide annually (232,000 deaths in sub-Saharan Africa) (Tate et al., 2012). Besides supportive care and fluid/electrolyte replacement therapy, no specific antiviral treatments are available.

Rotaviruses belong to a genus within the *Reoviridae* family. They possess a genome of 11 segments of double-stranded RNA, encoding 6 structural viral proteins (VPs) and 6 nonstructural proteins (NSPs). The glycoprotein VP7 (G type) and the hemagglutinin VP4 (P type) independently induce neutralizing antibodies (Abs) (Estes and Kapikan, 2007).

Two RV vaccines are currently licenced in many countries: Rotarix (GlaxoSmithKline), a monovalent vaccine of sero-/genotype G1P1A[8] and RotaTeq (Merck), a pentavalent vaccine containing genes encoding human RV (HRV) VP7 proteins of serotypes G1, G2, G3, and G4 and genotype P1A[8] on the G6P7[5] bovine RV genetic background (WC3 strain) (Vesikari et al., 2006). Although, these vaccines effectively prevent severe RV gastroenteritis in developed countries (>80%), they show reduced efficacy (~30–50%) in impoverished countries, where RV diarrhea is most severe (Gray, 2011). The reduced efficacy of oral vaccines and increased mortality rates from RV diarrhea in children are often associated with their poor macro- and micronutrient status (specifically vitamin A deficiency, VAD) and/or intestinal dysbiosis. Vaccine efficacy may be further affected by the significant genetic diversity of RV strains. In developed countries, direct and indirect medical costs related to RV disease are estimated to be €400 million in Europe and over $1 billion in the United States (Desselberger and Huppertz, 2011). Therefore, the need for animal experimentation to study HRV pathogenesis, vaccine-induced immunity and possible interventions is paramount.

A number of laboratory and agricultural animal models (reviewed in Desselberger and Huppertz, 2011) have been developed to study the mechanisms of RV pathogenesis and immunity, including piglets, rabbits, rats, lambs, and calves. However, the majority of the mechanistic studies have been conducted with the adult mouse model established in 1990 (Ward et al., 1990). Although of great convenience, mouse models have significant limitations because of their homogenous and inbred genetic background, physiology, anatomy, and immunity, all of which are distinct from those of humans. Additionally, a major limitation of the mouse models is that they do not support the efficient replication of human RVs. Also mice only have a short-term (up to 15 days of life) susceptibility to diarrhea induced by most murine RVs (Franco and Greenberg, 1999, 2000). Consequently, the murine (as well as rabbit) models reproduce only HRV infection but not disease; while studies using lamb and calf models were done only with homologous (ovine and bovine, respectively) RVs.

Further, several investigators have reported that oral inoculation of different nonhuman primates with either cell culture-adapted simian (SA11) or HRVs (including strain Wa) resulted in diarrheal illness during the first week of life (Kalter et al., 1983; Leong and Awang, 1990; Mitchell et al., 1977; Petschow et al., 1992; Wyatt et al., 1976). However, no illness, virus shedding or seroconversions were observed in the older animals to evaluate active immunity postchallenge. Additionally, the transplacental transfer of maternal Abs together with the economic and ethical issues associated with the use of primate models greatly decreases their value and convenience.

The use of swine in biomedical research has been widely accepted because pigs and humans share numerous similarities in anatomy, physiology, metabolism, feeding patterns, and immunity (Meurens et al., 2012) (Table 2.7.1). Apart from the primate and murine immune systems, the porcine immune system is the best characterized, offering a wide range of established research protocols and tools (Summerfield, 2009). Additionally, as outbred animals, pigs are more representative of human population heterogeneity than inbred mouse strains. Piglets are born immunologically immature and devoid of circulating maternal Abs because the sow's placenta (epitheliochorial type) blocks the transfer in utero of immunoglobulins (Hammerberg et al., 1989; Friess et al., 1981). Neonatal pigs resemble infants in their physiology, anatomy and in the development of mucosal immunity (Butler et al., 2007; Yuan et al., 1996; Yuan and Saif, 2002). Gnotobiotic (Gn) pigs are a unique animal model to assess HRV

TABLE 2.7.1 Advantages of Porcine, Murine, and Nonhuman Primate Models for Biomedical Research

Advantages of the porcine model for biomedical research	Murine models	Nonhuman primate models
Availability (most important meat-producing livestock species worldwide)	Yes	No
Size similar to human infant	No	Yes
Possibility of performing analogous surgical procedures and of collecting many samples	No	Yes
Similar anatomy	No	Yes
Omnivorous (similar gastrointestinal physiology)	No	Yes
Closely resemble humans for >80% of immune parameters analyzed (Dawson et al., 2013)	No (<10%)	Yes
Cheaper and ethically more acceptable than nonhuman primates	Yes	N/A
Various breeds (541), outbred and inbred	Yes	Yes/No
Large litter size (10–12 piglets/litter)	Yes	No
Standardized breeding conditions	Yes	No
High pig genome and protein sequence homologies with human counterparts (up to 95%) (Wernersson et al., 2005)	No	Yes
Prolonged susceptibility (up to 8 weeks of life) to some human pathogens, including HRVs	No	No

disease pathogenesis and to evaluate RV vaccines in the presence or absence of commensal gut microbiota. HRV-infected Gn pigs exhibit diarrhea, transient viremia, and intestinal lesions (villous atrophy) mimicking those in children (Azevedo et al., 2005; Ward et al., 1996). The Gn pig is susceptible to HRV diarrhea for at least 8 weeks of age, which is the time necessary to assess disease pathogenesis and protective immunity (Azevedo et al., 2004). The correlates of protection against challenge with HRVs are serum and intestinal RV-specific IgA Abs but not the serum levels of RV-neutralizing Abs (predominantly IgG) (Saif et al., 1996; Azevedo et al., 2004). However, VP7- or VP4-specific Abs protect against challenge with homologous or heterologous RVs (Yuan and Saif, 2002; Saif et al., 1996; Hoshino et al., 1988; Yuan et al., 1996); whereas, VP6-specific IgA Abs do not (Saif et al., 1996; Yuan et al., 2001). Further, being omnivores, pigs represent a translatable model for human research to study the effects of macro- and micronutrient deficiencies and supplementation on HRV pathogenesis and immunity via dietary manipulations (Zijlstra et al., 1997; Kandasamy et al., 2014a; Vlasova et al., 2013a; Chattha et al., 2013a). Finally, the complete pig genome sequence has provided an important resource for transcriptomic research to understand the host-microbiota-pathogen interactions at the molecular level (Groenen et al., 2012; Wernersson et al., 2005). Thus, the Gn pig model represents a unique system and may serve as a preclinical pipeline to study the various aspects of interactions among the immunologically immature neonatal host, dietary components, lactogenic immune factors, commensal/probiotic bacteria and HRV under strictly controlled manipulative conditions.

1.1 Lactobacilli and Bifidobacteria Modulate Innate and Adaptive Immune Responses to Human Rotavirus Infection in Neonatal Gnotobiotic Pigs

Intestinal commensals regulate development of gut immunity (Macpherson and Harris, 2004) and influence the outcome of viral infections (Wilks and Golovkina, 2012) including RV infections (Uchiyama et al., 2014) (Fig. 2.7.1). An understanding of the complex interplay among host immunity, intestinal commensals/probiotics and viral pathogens will provide novel strategies to prevent enteric viral infections including RV. Enhancing the efficacy and effectiveness of RV vaccines in impoverished countries is essential for prevention of the disease.

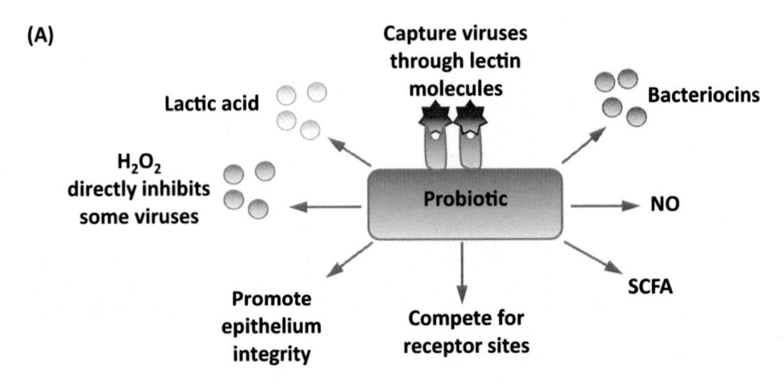

FIGURE 2.7.1 **(A) Probiotic-associated immune-mediators/immune mechanisms, and (B) interactions between probiotics and the immune system modelled in mice and Gn pigs. (A)** In the intestinal lumen, probiotics (1) inhibit certain viruses directly by producing lactic acid, H_2O_2, bacteriocins, and other inhibitory agents; (2) probiotics can also preserve the integrity of the epithelium and compete with pathogens for intestinal epithelial cell (IEC) receptors; (3) Lactobacilli could also capture viruses by lectin-mediated binding to viral glycoproteins and in this way prevent infection; (4) Lactobacilli/Bifidobacteria enhance the local immune system during health and disease and thereby inhibit infection; (5) nitric oxide (NO) produced by Lactobacilli plays a role in microbicidal and tumoricidal activities and in immunopathology; (6) Bifidobacteria-derived short chain fatty acids (SCFA) have immunomodulatory effects: inhibiting dendritic cell development, decreasing IL-12 levels, but increasing LI-23 production by DCs, and inducing Fas-mediated T cell apoptosis.

(Continued)

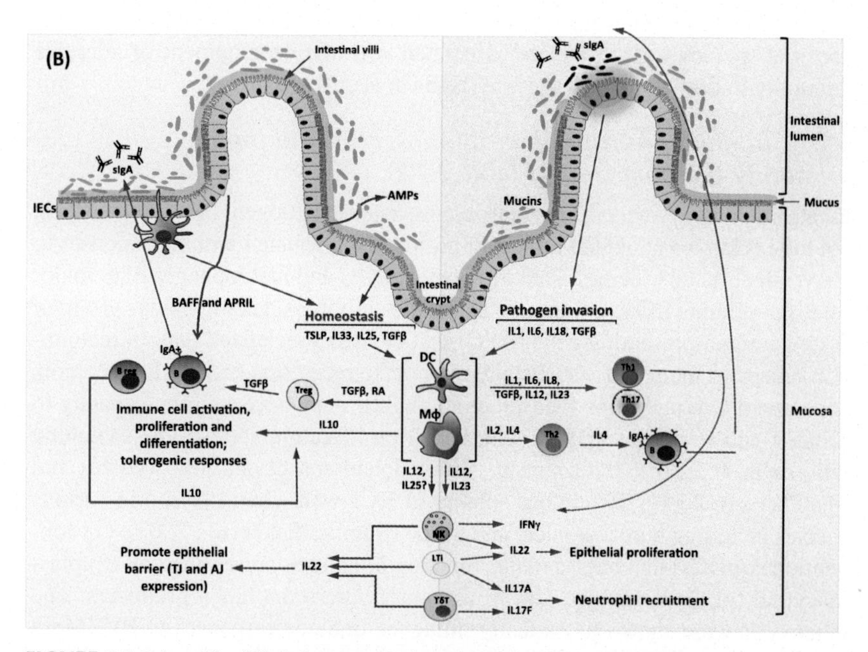

FIGURE 2.7.1 (*cont.*) **(B)** *In homeostasis*, intestinal epithelial cells (IECs) secrete mucins and AMPs in response to the commensal microbiota, regulating microbial replication, and interaction with intestinal mucosa. Additionally, IECs produce BAFF and APRIL factors, stimulating activated B (plasma) cells that produce secretory IgA (sIgA) in the lumen that further limits microbial interaction with the epithelium. Under homeostatic conditions, commensal microbiota stimulate the secretion of cytokines [including thymus stimulating lymphoprotein *(TSLP)*, IL-33, IL-23, IL-25, and *TGFβ*] by IECs that promote development of antigen presenting cells [macrophages *(Mϕ)* and dendritic cells *(DCs)*]. Antigen presenting cells induce regulatory T *(Treg)* cell generation through TGFβ- and retinoic acid (RA)-dependent mechanisms. Through APC and Treg derived TGFβ and IL-10, the antiinflammatory nature of the intestine is maintained by inhibiting or reducing effector responses. Intestinal innate lymphoid cells (ILCs), including natural killer (NK) cells, lymphoid tissue inducer (LTi) cells, and γδ T cells, produce IL-22 that regulates expression of tight and adherent junction *(TJ and AJ)* proteins by IECs, regulating intestinal barrier function. *Upon pathogen invasion, mucosal injury, or dysbiosis*, microbe associated molecular patterns (MAMPs) stimulate the secretion of proinflammatory and pluripotent cytokines by IECs (including, IL-6, IL-1, and IL-18) and APCs (including IL-6, IL-23, and IL-12) that induce effector CD4+ T cells *(Th1, Th2, and Th17)* via IL-23 or IL-12 signalling leading to generation of pathogen specific IgA+ B cells and diverse interactions with intestinal ILCs. Intestinal innate lymphoid cells respond to proinflammatory cytokines to upregulate IL-22, which helps to maintain the epithelial barrier, and IL-17A and IL-17F, which are involved in neutrophil recruitment and inflammatory responses. (*AMP*, antimicrobial peptides; *BAFF*, B-cell activating factor; *APRIL*, A proliferation-inducing ligand).

In our initial experiments, we focused on delineating the immunomodulatory effects of commonly used probiotics, such as Lactobacilli and Bifidobacteria on immune responses to RV. Lactobacillus spp. are the components of normal intestinal microflora of humans and pigs (Ahrne et al., 2005), and Lactobacilli and Bifidobacteria are the two major phyla in naturally delivered and breastfed infants. Also a recent study showed that the presence of certain commensal bacterial species was negatively associated with the development of adaptive immunity to oral poliovirus vaccine (Huda et al., 2014).

1.1.1 Interactions Among Probiotics, HRV and Innate Immunity (Summarized in Table 2.7.2)

Innate immunity provides the initial defense against pathogens (see Chapter 2.8), but little is known about the impact of probiotics on innate immune responses to HRV infection or vaccines. Dendritic cells (DCs) and macrophages play major roles in initiating innate immunity. Dendritic cells are a heterogeneous group of potent antigen-presenting cells (APC) that express specialized pattern recognition receptors including various antigen uptake receptors, such as Fc receptors and certain C-type lectins (Kelsall et al., 2002). They have a strong capacity to prime T-cell responses and regulate B-cell proliferation and isotype switching (Hivroz et al., 2012; Balazs et al., 2002; Johansson et al., 2000; Wykes and MacPherson, 2000). Two major subsets of DCs were characterized in various species including humans, mice, and swine (Summerfield et al., 2003): (1) conventional (cDCs) that have a major function of antigen presenting and (2) plasmacytoid (pDCs) DCs also known as natural interferon alpha producers. The latter cells were shown to critically influence immune responses to RV (Mesa et al., 2007; Deal et al., 2013). Further, DCs and other innate immune cells express Toll-like receptors (TLRs), a family of pattern recognition receptors, which are central to activation of innate immunity (Medzhitov, 2001). Among the various TLRs, TLR2 (expressed on cell surface) in combination with TLR1/TLR6 recognizes gram positive bacterial components, such as peptidoglycans, lipoteichoic acids, and lipoproteins (Song and Lee, 2012). Intracellular TLR3 recognizes double-stranded (ds)RNA of microbes and intracellular TLR9 recognizes unmethlylated CpG motifs that are commonly found in microbial genomes (Song and Lee, 2012).

Recently, we assessed the impact of *Lactobacillus rhamnosus* GG (LGG) and *Bifidobacterium lactis* Bb12 (Bb12) on innate and adaptive (see Section 1.1.2) immunity in RV vaccinated piglets (Vlasova et al., 2013b) (Table 2.7.2). Dual colonization with LGG and Bb12 resulted in less severe diarrhea and reduced virus shedding titers compared to uncolonized piglets and differentially modulated mucosal and systemic innate immunity during HRV infection of Gn pigs (Vlasova et al., 2013b). These probiotics exerted inhibitory effects on DC populations at the systemic level as evident by lower frequencies of activated splenic conventional and plasmacytoid DC (cDC and pDC) in probiotic colonized vaccinated piglets compared to uncolonized vaccinated piglets post-HRV challenge. Alternatively, this finding may also be a result of the reduced HRV replication in the probiotic colonized piglets. Consistent with reduced HRV shedding/diarrhea, LGG + Bb12 colonized piglets had lower IFNα responses compared to uncolonized piglets post-HRV challenge. However, post-challenge, probiotic colonized, vaccinated piglets had higher frequencies of

TABLE 2.7.2 Summary of Probiotic (*Lactobacillus spp.* and *Bifidobacterium lactis* Bb12) Effects on Various Immune Parameters, and Responses to AttHRV Vaccine and Virulent HRV Infection Studied in Gn Pigs.

Probiotic/probiotic combination	Commensal microbiota	AttHRV vaccine	VirHRV infection/challenge[a]	Observed effects of the probiotics	References
LGG + Bb12	No No	Yes No	Yes Yes	*Postchallenge, clinical parameters*: decreased severity of HRV infection and disease; *postchallenge, innate immune parameters*: decreased systemic, but promoted intestinal innate immune responses and immune trafficking, differentially affected TLR responses (decreased pro-inflammatory, increased B-cell promoting); *postchallenge, adaptive immune responses*: promoted adaptive immune (including B, effector and regulatory T cell) responses	Vlasova et al. (2013b), Kandasamy et al. (2014b), Chattha et al. (2013b)
LA + LR	No	No	Yes	*Innate immune parameters:* differentially affected APC frequencies in HRV infected and noninfected piglets, increased TLR expression by blood cDCs; *Adaptive immune responses*: promoted T cell responses and decreased pro-inflammatory cytokine production	Zhang et al. (2008a), Zhang et al. (2008c), Wen et al. (2009), Azevedo et al. (2012), Wen et al. (2011)
LGG	No	No	Yes	*Clinical parameters*: decreased severity of HRV infection and disease; *Innate immune parameters*: decreased intestinal damage and other effects of HRV infection	Liu et al. (2013), Wu et al. (2013)
LA	No	Yes	No	*Adaptive immune responses*: increased adaptive B and T cell immune responses	Zhang et al. (2008b)
LGG	Yes	No	Yes	Moderated HRV effects on intestinal microbiota	Zhang et al. (2014)

[a]*In the experiments that involved VirHRV infection/challenge, the immune/clinical parameters were assessed after the VirHRV infection/challenge.*

activated pDCs/cDCs in ileum and blood compared to uncolonized vaccinated piglets. We observed a synergistic interaction between the attenuated HRV (AttHRV) vaccine and LGG and Bb12 colonization as evident by increased frequencies of ileal TLR9+ mononuclear cells (MNCs) in intestinal tissues of probiotic colonized, RV vaccinated piglets compared to uncolonized AttHRV vaccinated piglets prechallenge. The probiotics alone had little or no long-term stimulatory effect on frequencies of TLR9 + MNCs. Further, the increased TLR9+ MNC frequencies prechallenge coincided with a higher protective effect against virus shedding and diarrhea observed postvirulent HRV challenge. In contrast, the LGG and Bb12 colonized, vaccinated piglets had decreased frequencies of ileal TLR2+ and TLR4+ MNCs compared to uncolonized vaccinated piglets. An earlier study of adult human subjects reported increased TLR2 and TLR4 expression in submucosal immune cells of inflamed intestinal mucosa compared to healthy mucosa (Hausmann et al., 2002). Thus, regulating the expression of specific TLRs by these probiotics in the small intestine might play a role in intestinal immune homeostasis and also prevent excessive inflammatory responses during viral infection.

Investigators have reported that immunomodulatory effects vary with strain (Medina et al., 2007) and composition of the probiotic bacteria (Gackowska et al., 2006). Thus, we also assessed the impact of two other lactic acid producing probiotic bacteria (LAB), *Lactobacillus acidophilus* (LA) and *Lactobacillus reuteri* (LR), on intestinal and systemic innate immune responses (Zhang et al., 2008a,b,c; Wen et al., 2009). Compared to uninfected negative control piglets, HRV infection alone significantly increased monocytes/macrophages, but not the cDC population in ileum. However, LA + LR colonized HRV infected piglets had lower frequencies of monocytes/macrophages compared to HRV only infected piglets in ileum. Additionally, probiotic colonized piglets had lower frequency of activated macrophages post-HRV infection. The antigen presenting cell (APC) populations in spleen were significantly reduced in LA + LR colonized, compared to uncolonized piglets, postvirulent HRV infection. Previous studies by others showed that LR reduced TNFα production in lipopolysaccharide (LPS) treated macrophages (Pena et al., 2004) and suppressed proinflammatory cytokines in macrophages from children with Crohn's disease (Lin et al., 2008). Similarly, colonization of piglets with LA + LR significantly reduced TNFα cytokine secreting cells in ileum and spleen post-HRV challenge (Azevedo et al., 2012). Thus, reduction in total, as well as activated intestinal monocyte/macrophage populations, and decreased inflammatory cytokine production in LAB colonized piglets during HRV infection indicates that these probiotics have a protective effect on inflammatory damage during HRV infection.

Colonization of Gn piglets with LAB alone resulted in significant modulation of innate immunity. LA + LR dual colonization significantly increased both monocytes/macrophages and cDC populations in ileum in comparison to uncolonized Gn piglets (Zhang et al., 2008c). Further, in the absence of HRV infection, probiotic colonization alone increased the frequencies of TLR2 and

TLR9 positive cDC in blood (Wen et al., 2009). These results indicate that LA and LR alone had significant stimulatory effects on the innate immune system.

Intestinal epithelial cells are the target cells for RV, and their anatomic location facilitates interactions with probiotics and intestinal commensals. In a recent study, LGG colonization modulated HRV effects on the levels of tight junction and adherent junction proteins (Liu et al., 2013) and downregulated autophagy in ileal epithelium after HRV infection (Wu et al., 2013). Thus, it appears that probiotics can alleviate the RV induced pathological changes in intestinal epithelial cells and supplementation of probiotics might be a potential strategy to reduce the severe consequences of RV infection on intestinal epithelial cells.

1.1.2 Functional Effects of Probiotics on Adaptive Immunity to RV (Table 2.7.2)

Virus specific B-cell responses play an important role in clearing RV infection and are critical for development of antiviral T-cell responses implicated in controlling primary viral infections (Franco and Greenberg, 1995). Further, a significant correlation was also observed between virus specific serum IgA Ab levels and RV vaccine efficacy in children (Patel et al., 2013). Since RV vaccines lack efficacy in impoverished countries, where diarrhea mortality is highest, we have focused our studies on enhancing the immunogenicity and protective efficacy of RV vaccines using economical approaches, such as probiotics and intestinal commensals.

Dual colonization of LGG and Bb12 probiotics had significant effects on HRV vaccine induced B- and T-cell responses. B-cell responses, including activation of intestinal B cells and RV specific IgA Ab titers were enhanced in vaccinated, probiotic colonized piglets compared to uncolonized, vaccinated piglets postvirulent HRV challenge (Kandasamy et al., 2014b). Further, T-cell responses, specifically ileal T regulatory cells, and systemic IFNγ producing T-cell responses, were increased in probiotic colonized, vaccinated compared to uncolonized vaccinated piglets (Chattha et al., 2013b). Importantly, the probiotic induced immunomodulatory effects on adaptive immune responses coincided with decreased diarrhea severity and reduced fecal virus shedding.

Similar to LGG and Bb12 effects on B-cell responses, LA probiotic significantly enhanced the immunogenicity of AttHRV vaccine responses as indicated by higher numbers of ileal HRV specific IgA and IgG antibody secreting cells (ASCs) and increased intestinal IFNγ producing T cells compared to uncolonized piglets post HRV inoculation (Zhang et al., 2008b). Apart from individual effects of LA on adaptive vaccine-specific immunity, dual-colonization of LA and LR significantly modulated the types of γδ T-cell responses (critical for early responses to infections at epithelial surfaces) during HRV infection of Gn piglets without vaccination (Wen et al., 2011). There were lower numbers of inflammatory type CD2 + CD8 − γδ T cells and higher regulatory type CD2 + CD8+ γδ T cells (Saalmuller et al., 1990) in LA + LR probiotic colonized piglets in comparison to uncolonized piglets postvirulent HRV infection. Additionally, higher systemic IFNγ and IL4 cytokine responses in LA + LR colonized compared to uncolonized HRV infected piglets suggest that LAB modulated both Th1 and Th2 immunity, respectively (Wen et al., 2009). Thus, the probiotics tested had measurable beneficial effects on AttHRV vaccine protective efficacy and immunogenicity, and they moderated the severity of HRV diarrhea when given at least 21 days prior to HRV challenge (Chattha et al., 2013b). However, whether these observed beneficial effects can be reproduced by these probiotics in the presence of complex microbiota has yet to be determined.

1.1.3 Neonatal Pig Models Colonized With Complex Intestinal Microbiota of Human or Swine Origin

The Gn piglet model can also be utilized to address important questions of how intestinal microbiota of human infants modulate or are modulated by host immunity, intestinal health, and RV infections (Pang et al., 2007; Wen et al., 2014). Initial studies using humanized piglets, that is, piglets transplanted with intestinal microbiota from a breast-fed infant (delivered by Caesarean section), revealed that RV infection shifted bacterial abundance from phylum *Fimicutes* to phylum *Proteobacteria* (Zhang et al., 2014), whereas LGG supplementation prevented the HRV infection-induced changes in the microbial community. By contrast, our preliminary data demonstrate that in Gn pigs colonized with defined microbiota of swine origin, HRV infection increased the relative abundance of phyla *Firmicutes* and *Bacteroidetes* and decreased that of *Proteobacteria* in the large and small intestine (Saif, Rajashekara, Vlasova et al., unpublished). Consistent with our findings, alterations in the composition of the *Bacteroides* phylum after RV infection were observed in human subjects (Zhang et al., 2009), suggesting that the composition of intestinal microbiota also influences the outcome of RV infections. Finally, in an ongoing study, we are evaluating the interactions among intestinal microflora, malnutrition, and RV infection to identify efficacious dietary interventions using Gn piglets transplanted with fecal microbiota from a healthy vaginally delivered breast-fed infant (Saif, Vlasova, Rajashekara et al., unpublished). Thus, humanized piglets or piglets colonized with other complex defined microbiota are a valuable model to study the effects of diet, RV infection and vaccines on microbial ecology and host immunity.

1.2 Interactions Between Lactogenic Immune Factors, Probiotics, Neonatal Immune System and Human Rotavirus Vaccine in a Gnotobiotic Pig Model

There are few studies on the impact of selected probiotics on responses to oral vaccines in neonates in the context of colostrum/milk (col/milk) feeding. We have recently examined how LGG + Bb12 colonization with or without col/milk (to mimic breastfed versus formula-fed infants) affects development of B cell responses to an oral AttHRV vaccine in the relevant Gn pig model.

In agreement with previous findings that breast-milk promotes growth of *Bifidobacteria* and *Lactobacilli*, supplementation of col/milk (naturally containing TGFβ and other growth factors) in our study increased fecal probiotic shedding suggesting that milk containing regulatory cytokines (such as TGFβ) and other soluble factors (glycans etc.) can promote establishment and extended colonization by probiotics (LGG + Bb12) (Ahrne et al., 2005; Yoshioka et al., 1983; Rinne et al., 2005). Breast milk is a major source of TGFβ for neonates when intrinsic production is limited (Penttila, 2010; Nguyen et al., 2007) promoting intestinal immune responses, including class-switch to IgA, induction of regulatory T lymphocytes, attenuation of proinflammatory responses, and reduction of immune-mediated and allergic conditions (Kalliomaki et al., 1999). Unexpectedly, an increase in probiotic fecal shedding was not observed in col/milk fed vaccinated pigs, possibly due to the AttHRV vaccine related reduction in serum TGFβ and increase in proinflammatory and Th1 cytokines (IL6 and IL12), resulting in conditions less favorable for commensal colonization (Azevedo et al., 2006).

Lower counts of probiotics detected in cecum/colon of col/milk fed pigs, irrespective of vaccination, suggested a differential impact of col/milk on fecal bacterial shedding versus intestinal distribution or mucosal adherence. Maternal Abs in sow col/milk to bacterial components including peptidoglycan may prevent mucosal adhesion of probiotics resulting in lower mucosa-associated bacterial counts as observed in suckling Gn mice previously (Kramer and Cebra, 1995). Another study demonstrated that in humans, bacteria from *B. animalis* subsp. *lactis* taxon are rarely found in intestinal biopsy samples (mucosa-adherent), whereas they are frequently present in fecal samples (luminal), suggesting that this taxon may not be an abundant component of the mucosa-adherent bacteria in humans or pigs (Turroni et al., 2009).

Combined probiotic colonization and col/milk supplementation in vaccinated pigs enhanced serum IgA HRV Ab titers and intestinal IgA HRV ASC levels, which was not observed in vaccinated pigs that did not receive col/milk, suggesting complex interactions between probiotics and col/milk components. Col/milk containing HRV Abs transiently suppressed serum IgA Ab responses after two vaccine doses irrespective of probiotic colonization, but this effect was ameliorated after the three dose vaccine regimen. Thus, colonization with LGG + Bb12 in breast fed vaccinated infants (with preexisting maternal HRV Abs) may overcome the suppressive effects of maternal Abs at least on IgA Ab responses. Similar to our study, Isolauri et al. (1995) showed enhanced RV IgA Ab responses in LGG fed infants of unknown breastfeeding status after oral immunization with live oral RV vaccine. In contrast, supplementation of *Bifidobacterium breve* strain Yakult (BBG-01) in breast fed children resulted in no significant difference in vibriocidal Ab responders following oral cholera vaccination (Matsuda et al., 2011). The quantity of maternal Abs received in utero (humans) or through colostrum (pigs) and the duration of breast milk feeding are critical factors in determining active Ab responses of neonates to oral vaccines.

Probiotics alone did not enhance IgA Ab responses to oral HRV vaccine in noncol/milk supplemented pigs, suggesting lack of adjuvancy of LGG + Bb12 for primary IgA Ab responses under the conditions tested. Similarly, Perez et al. (2010) have shown a lack of adjuvant effect of probiotics in enhancement of tetanus and pneumococcal IgA and IgG Abs in children, but this was after parenteral immunizations. In contrast, use of *Bifiobacterium* spp. in formula fed infants enhanced poliovirus and RV IgA Abs to the respective vaccines (Holscher et al., 2012; Mullie et al., 2004). Probiotic effects vary with strain, dose, vaccine type and route, level of maternal Abs, duration of breast feeding, age, the existing microflora, and other factors. Further studies are needed to elucidate reasons for inconsistent results.

Probiotic colonization resulted in significantly lower serum HRV IgG Ab titers and IgG HRV ASC in HRV vaccinated Gn pigs that did not receive col/milk. This suggests that LGG + Bb12 may enhance gut barrier integrity in multiple ways, similar to other probiotics (Ohland and Macnaughton, 2010). This would reduce systemic translocation and exposure to AttHRV and reduce IgG HRV ASC and Ab responses. However, this hypothesis was not investigated directly in our study. Thus, our results using the Gn pig model suggest that feeding LGG + Bb12 in breastfed infants may be advantageous, not only by directly enhancing IgA HRV Abs, but also by reducing systemic exposure and preventing adverse clinical effects of HRV gastroenteritis.

1.3 Gut Transcriptome Responses to *Lactobacillus rhamnosus* GG and *Lactobacillus acidophilus* in Neonatal Gnotobiotic Piglets

Molecular mechanisms of probiotic action on neonatal intestinal mucosal immunity are largely undefined. Using a transcriptomic approach, we assessed mucosal tissue responses to LGG or LA monocolonization in comparison to uncolonized Gn piglets (Kumar et al., 2014). Results suggest that transcriptomic responses vary with the strain of probiotic, duration of probiotic colonization and region of the intestinal tract. Immediately after probiotic colonization (day 1), both LA and LGG induced higher transcriptional responses in ileum, whereas at later stages (7 days), LGG, but not LA, induced profound changes in expression of transcripts in duodenum. Both of these probiotics seem to polarize mucosal immunity toward Th1 type as indicated by higher expression of chemokine (C–C motif) ligand 9 [CCL9; macrophage inflammatory protein-1 gamma (MIP-1γ)] in LA and LGG piglets and higher granzyme (serine proteases involved in apoptosis) expression in LA piglets.

Compared to LA, LGG significantly modulated genes associated with the following pathways: inflammatory response, immune cell trafficking, and hematological system development in the duodenum. Pathways associated with immune modulation and carbohydrate metabolism were highly altered by LGG, whereas LA predominantly induced changes in energy and lipid metabolism-related trancriptomic responses. Further LA, but not LGG, induced prominent changes in transcription of vitamin A related genes in duodenum. Thus, LA and LGG differentially modulated major pathways in intestinal tissues. Further, both LGG and LA colonization resulted in higher expression of glucagon-like peptide 2 receptor (GLP2R) which regulates villus height and crypt depth in the small intestine (Jeppesen et al., 2001). LGG colonization also increased expression of claudin-8, a tight junction protein that regulates paracellular permeability (Ulluwishewa et al., 2011). Collectively, our intestinal tissue transcriptomic study revealed that Lactobacilli have prominent impacts on the host immune and metabolic functions. Our future studies will assess the effect of commonly used probiotics on transcriptomic responses of individual cell types in the small intestine. These studies may illuminate the precise mechanisms of probiotic action on mucosal immunity and suggest strategies to tailor preventive and therapeutic probiotic therapy for RV infections.

1.4 Prenatal Vitamin A Deficiency Alters Immune Responses to Virulent Human Rotavirus/Human Rotavirus Vaccines in a Gnotobiotic Pig Model

Vitamin A deficiency (VAD) in children is a significant health concern in impoverished countries. Even marginal (subclinical) VAD may compromise various aspects of innate and adaptive immune responses, resulting in enhanced susceptibility to infectious diseases (Fig. 2.7.2). Prophylactic supplemental vitamin A recommended by WHO has reduced the incidence of diarrhea associated morbidity and mortality in these countries (Humphrey et al., 1996; Imdad et al., 2011). To investigate interactions among VAD, supplemental vitamin A, AttHRV vaccine and virulent HRV infection, we have established a VAD Gn pig model with significantly decreased liver vitamin A levels and positive serum 30-day dose-response test (Ferraz et al., 2004) (Fig. 2.7.3). We showed that VAD increases the severity of HRV-induced diarrhea and impairs innate, mucosal, and systemic adaptive immune responses to HRV infection and AttHRV vaccines. Two separate trials using G1P1A[8] HRV Wa (genetically similar to commercial monovalent HRV vaccine, Rotarix) and pentavalent RotaTeq (Merck) yielded similar results indicating that VAD compromised the protective efficacy of both vaccines (Kandasamy et al., 2014a; Chattha et al., 2013a). HRV challenge resulted in decreased vitamin A levels in serum early postchallenge as was previously shown for measles virus (West, 2000), suggesting that HRV-induced intestinal villous atrophy (Ward et al., 1996) may affect vitamin A metabolism or its absorption from the milk diet.

We showed that porcine retinol binding protein 4 (RBP4) and retinoic acid receptor-alpha (RARα) mRNAs (reflective of respective protein levels) are expressed by splenic MNCs and that their levels are affected by VAD, vitamin A supplementation and HRV challenge. Vlasova et al. (2013a) VAD decreased RARα mRNA levels and lowered relative responses to supplemental vitamin A. This suggests that in VAD pigs, supplemental vitamin A may not be

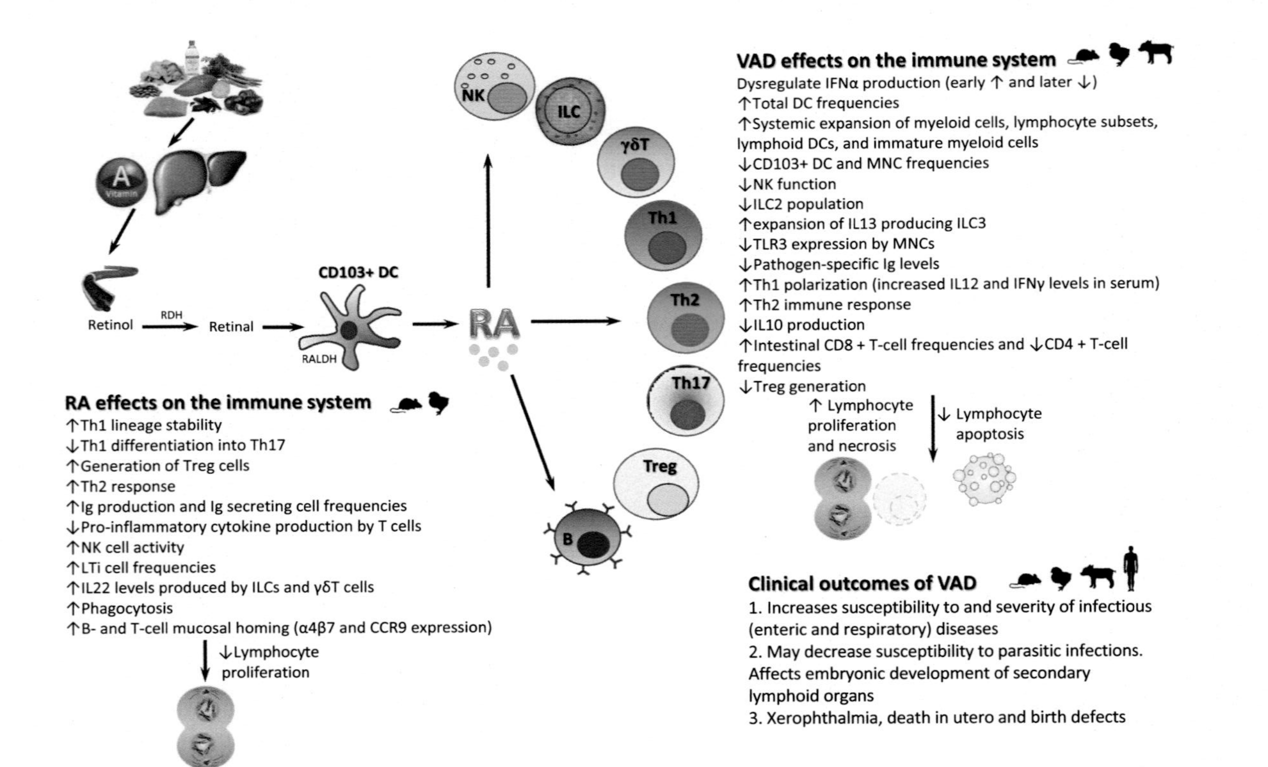

FIGURE 2.7.2 **Vitamin A (Retinoic Acid) and VAD effects on the immune system.** Retinol is taken up from the blood and oxidized to retinal by retinol dehydrogenases *(RDH)* and then to all-trans-retinoic acid *(RA)* by retinal dehydrogenases *(RALDH)*, expressed predominantly by dendritic cells *(DCs)*. The evidence generated in various animal models (rodent and avian species mostly) indicates that RA affects most major immune cell subsets including natural killer *(NK)* cells, innate lymphoid cells *(ILC)*, T cells [*Th1, Th2, Th17*, and regulatory *(Treg)*], B cells; and it regulates Ig and cytokine production by the immune cells. Additionally, RA affects the cell cycle by decreasing lymphocyte (mononuclear cell) proliferation rates and increasing lymphocyte differentiation. Studies in rodent, avian, and porcine models demonstrated that VAD has multiple and varied effects on the immune system, including increased lymphocyte proliferation rates and decreased apoptosis that result in systemic expansion of various immune cells of lymphoid and myeloid origin (immature or incompletely differentiated), dysregulated cytokine responses and decreased IgA, CD103+ DC, Treg and toll-like receptor (TLR) 3 responses.

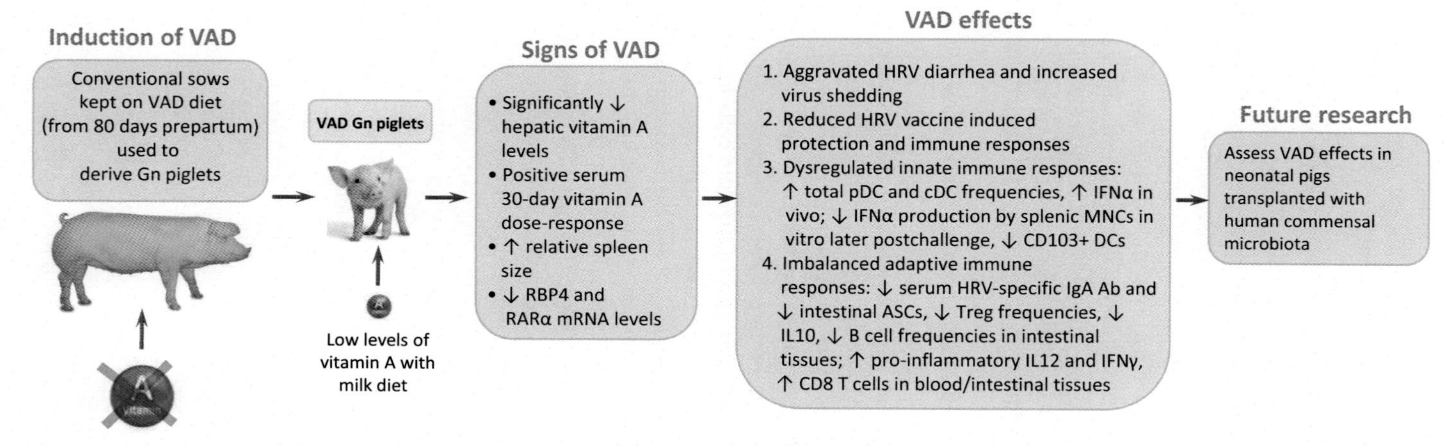

FIGURE 2.7.3 **Prenatal VAD effects on the immune responses in gnotobiotic (Gn) pig model.** Prenatal VAD in Gn piglets resulted in increased severity of human rotavirus (HRV) disease and infection, dysregulated [increased total dendritic cell (DC), but decreased CD103+ DC frequencies, increased early IFNα responses, followed by a significant decrease in MNC capacity to produce IFNα in response to HRV restimulation in vitro, etc.] innate immune responses and decreased adaptive [including IgA and regulatory T (Treg) cell] immune responses. Vitamin A supplementation at the treatment concentrations and times tested did not reverse the VAD effects. Retinoic acid is required for lymphoid tissue inducer (LTi) cell development in the prenatal period; thus, its lack or insufficiency prenatally may result in impaired LTi cell function and the associated permanent or long-term deficiency of the innate and adaptive immune responses.

metabolized to all-*trans* retinoic acid (ATRA) efficiently, or insufficient RARα expression results in aberrant cell signalling and the observed lack of consistent compensatory effects of vitamin A supplementation.

1.4.1 VAD Effects on the Innate Immune Responses

Numerous studies have demonstrated that VAD decreases resistance to and aggravates respiratory and enteric infections by depressing immune function. In our experiments, we observed not just a decrease, but VAD-induced dysregulation of IFNα production in response to AttHRV vaccine or HRV infection. Significantly higher levels of circulating IFNα in VAD piglets early postchallenge were consistent with higher HRV replication titers. However, later postchallenge, the capacity for IFNα production by MNCs from VAD piglets ex vivo was significantly decreased as compared to prechallenge and to MNCs from vitamin A sufficient (VAS) piglets postchallenge. This, together with higher amounts of IFNα in VAD piglets' sera, suggests that imbalanced IFNα release by immune cells may contribute to increased inflammation and does not efficiently control HRV infection. Although RVs are known to be potent IFNα inducers, the role of IFNα in RV clearance is uncertain and varies between homologous and heterologous infections (Feng et al., 2008). Moreover, type I and II interferons were not demonstrated to be major inhibitors of RV replication in mice (Angel et al., 1999).

VAD also resulted in higher total frequencies of pDCs (gut tissues) and cDCs (all tissues) prechallenge in agreement with previous findings by others that VAD causes a systemic expansion of myeloid cells in *sen*sitive to skin *car*cinogenesis (SENCAR) mice and an increase in different lymphocytes/lymphoid DCs in C57BL/6J mice (Duriancik et al., 2010; Kuwata et al., 2000).

Small intestinal CD103+ DCs are imprinted with an ability to metabolize vitamin A (retinol) and generate gut-tropic T cells (expressing CC chemokine receptor-9 and α4β7). Significantly lower CD103+ pDC and cDC frequencies in ileum, duodenum, and spleen of VAD piglets compared to VAS piglets suggest that DC-associated vitamin A metabolism was altered in VAD piglets. The loss of CD103 integrin by colonic DCs during experimentally induced colitis was described in mice (Strauch et al., 2010), suggesting that VAD-induced MNC necrosis could contribute to, or be the result of the significantly decreased CD103+ DC frequencies. Ultimately, disruption of signaling between DCs and T/B-lymphocytes may be due to lack of proper antigen presentation to the latter cells, resulting in the lower HRV specific IgA Ab titers that we observed (Kandasamy et al., 2014a; Chattha et al., 2013a) in VAD piglets (see later). Finally, increased frequencies of apoptotic MNCs in VAS piglets prechallenge confirm that adequate vitamin A levels control apoptosis, a mechanism involved in regulation of autoimmunity and T-cell tolerance (Jin et al., 2010).

1.4.2 VAD Effects on the Adaptive Immune Responses

Our Gn pig VAD model demonstrated that reduced protection against HRV challenge coincided with significantly higher CD8 T-cell frequencies in the blood and intestinal tissues, higher proinflammatory (IL12) and 2–3-fold lower antiinflammatory (IL10) cytokines, in VAD compared to VAS control pigs. Similar to our study, lower frequencies of CD4 T cells and increased frequencies of CD8 T cells in intestinal tissues and spleen have been reported in VAD mouse and rat models (Duriancik et al., 2010; Bjersing et al., 2002). The AttHRV vaccinated VAD pigs had significantly higher serum IL12 (PID2) and IFNγ (PID6) compared to vaccinated VAS groups suggesting higher Th1 responses in VAD conditions (Chattha et al., 2013a). Furthermore, regulatory T-cell responses were compromised in VAD pigs (Chattha et al., 2013a).

Vaccinated VAD pigs had lower serum IgA HRV Ab titers and significantly lower intestinal IgA ASCs postchallenge suggesting lower anamnestic responses. Additionally, VAS piglets had higher frequencies of B cells in duodenum postchallenge suggesting effective homing of activated B cells to effector sites in the gut mucosa.

In conclusion, experimental VAD in neonatal Gn pigs resulted in innate immune system overactivity suggesting that one of the major vitamin A functions is to maintain the immunoregulatory profile in the gut and at systemic immune sites. CD103+ DC frequencies were significantly decreased in VAD piglets, suggesting their critical role in efficient immune responses to HRV. The impaired vaccine-specific intestinal Ab responses and decreased immunoregulatory cytokine responses coincided with reduced protective efficacy of the AttHRV vaccine against virulent HRV challenge in VAD piglets. Finally, a high dose vitamin A supplementation concurrent with the AttHRV vaccine did not compensate for the VAD effects or act as an adjuvant for AttHRV vaccine, possibly indicating a need for maternal, longer, or higher levels of postnatal vitamin A supplementation.

2 CONCLUDING REMARKS AND FUTURE DIRECTIONS

It is widely speculated that micronutrient deficiencies (specifically vitamin A) and intestinal dysbiosis may play a role in reduced efficacy of RV vaccines in developing countries. Our experimental studies with the Gn piglet-HRV disease model provided the first experimental evidence that VAD impairs both innate and adaptive immunity to HRV vaccine. Apart from the vitamin A research, our pioneering studies with probiotics in the Gn piglet model elucidated their direct role in modulating HRV infection, the immunogenicity of HRV vaccines and their interactions with lactogenic immune factors. Further, we advanced probiotic-related transcriptomic research in the neonatal Gn pig model and proved that this model can contribute to understanding of probiotic/commensal functions in humans. Regarding future directions, a key area to be addressed is determining functional interactions among RV, selected probiotics, intestinal commensals, intestinal epithelial cells, and immune cells. Elucidating the complex interactions among them and modulating the individual or combined functionality of those factors by probiotics/commensals and micronutrients are active areas of research in our laboratory. Future directions include: (1) evaluating the impact of selected probiotics on HRV vaccine and diarrhea in Gn piglets transplanted with human infant microbiome; (2) identifying intestinal commensals of human infants that are either positively or negatively associated with immunogenicity and efficacy of HRV vaccines in the Gn piglet model; (3) delineating the molecular mechanisms for how intestinal commensals affect host responses to HRV with the aim of enhancing HRV vaccine efficacy, or treating HRV diarrhea; (4) assessing the role of micronutrients and the impact of malnutrition in modulating HRV infections and the efficacy of HRV vaccines. Another critical research area is finding ways to promote intestinal epithelial cell repair and strengthening of intestinal barrier function following RV infection. Due to HRV vaccine failures in impoverished countries, it is essential to provide other cost effective treatments, such as probiotic or effective micronutrient supplementation in order to reduce the severe consequences of RV diarrhea in children. Finally, the Gn piglet provides a highly relevant and tractable model to assess the complex interactions among the intestinal microbiota, probiotics, enteric pathogens, and vaccines and neonatal immunity in the context of the vitamin deficiencies and malnutrition so common in children in impoverished countries where vaccines often fail.

REFERENCES

Ahrne, S., Lonnermark, E., Wold, A.E., et al., 2005. Lactobacilli in the intestinal microbiota of Swedish infants. Microbes Infect. 7, 1256–1262.

Angel, J., Franco, M.A., Greenberg, H.B., et al., 1999. Lack of a role for type I and type II interferons in the resolution of rotavirus-induced diarrhea and infection in mice. J. Interferon Cytokine Res. 19, 655–659.

Azevedo, M.S., Yuan, L., Iosef, C., et al., 2004. Magnitude of serum and intestinal antibody responses induced by sequential replicating and nonreplicating rotavirus vaccines in gnotobiotic pigs and correlation with protection. Clin. Diagn. Lab. Immunol. 11, 12–20.

Azevedo, M.S., Yuan, L., Jeong, K.I., et al., 2005. Viremia and nasal and rectal shedding of rotavirus in gnotobiotic pigs inoculated with Wa human rotavirus. J. Virol. 79, 5428–5436.

Azevedo, M.S., Yuan, L., Pouly, S., et al., 2006. Cytokine responses in gnotobiotic pigs after infection with virulent or attenuated human rotavirus. J. Virol. 80, 372–382.

Azevedo, M.S., Zhang, W., Wen, K., et al., 2012. Lactobacillus acidophilus and Lactobacillus reuteri modulate cytokine responses in gnotobiotic pigs infected with human rotavirus. Benef. Microbes 3, 33–42.

Balazs, M., Martin, F., Zhou, T., et al., 2002. Blood dendritic cells interact with splenic marginal zone B cells to initiate T-independent immune responses. Immunity 17, 341–352.

Bjersing, J.L., Telemo, E., Dahlgren, U., et al., 2002. Loss of ileal IgA+ plasma cells and of CD4+ lymphocytes in ileal Peyer's patches of vitamin A deficient rats. Clin. Exp. Immunol. 130, 404–408.

Butler, J.E., Lemke, C.D., Weber, P., et al., 2007. Antibody repertoire development in fetal and neonatal piglets: XIX. Undiversified B cells with hydrophobic HCDR3s preferentially proliferate in the porcine reproductive and respiratory syndrome. J. Immunol. 178, 6320–6331.

Chattha, K.S., Kandasamy, S., Vlasova, A.N., et al., 2013a. Vitamin A deficiency impairs adaptive B and T cell responses to a prototype monovalent attenuated human rotavirus vaccine and virulent human rotavirus challenge in a gnotobiotic piglet model. PLoS One 8, e82966.

Chattha, K.S., Vlasova, A.N., Kandasamy, S., et al., 2013b. Divergent immunomodulating effects of probiotics on T cell responses to oral attenuated human rotavirus vaccine and virulent human rotavirus infection in a neonatal gnotobiotic piglet disease model. J. Immunol. 191, 2446–2456.

Dawson, H.D., Loveland, J.E., Pascal, G., et al., 2013. Structural and functional annotation of the porcine immunome. BMC Genomics 14, 332.

Deal, E.M., Lahl, K., Narvaez, C.F., et al., 2013. Plasmacytoid dendritic cells promote rotavirus-induced human and murine B cell responses. J. Clin. Invest. 123, 2464–2474.

Desselberger, U., Huppertz, H.I., 2011. Immune responses to rotavirus infection and vaccination and associated correlates of protection. J. Infect. Dis. 203, 188–195.

Duriancik, D.M., Hoag, K.A., Vitamin A, 2010. deficiency alters splenic dendritic cell subsets and increases CD8(+)Gr-1(+) memory T lymphocytes in C57BL/6J mice. Cell Immunol. 265, 156–163.

Estes, M.K., Kapikan, A.Z., 2007. Rotaviruses. In: Knipe, D.M.H.P. et al., (Ed.), Fields Virology. 5th ed. Wolters Kluwer Health/Lippincott Williams&Wilkins, Philadelphia, PA, pp. 1917–1974.

Feng, N., Kim, B., Fenaux, M., et al., 2008. Role of interferon in homologous and heterologous rotavirus infection in the intestines and extraintestinal organs of suckling mice. J. Virol. 82, 7578–7590.

Ferraz, I.S., Daneluzzi, J.C., Vannucchi, H., et al., 2004. Detection of vitamin A deficiency in Brazilian preschool children using the serum 30-day dose-response test. Eur. J. Clin. Nutr. 58, 1372–1377.

Franco, M.A., Greenberg, H.B., 1995. Role of B cells and cytotoxic T lymphocytes in clearance of and immunity to rotavirus infection in mice. J. Virol. 69, 7800–7806.

Franco, M.A., Greenberg, H.B., 1999. Immunity to rotavirus infection in mice. J. Infect. Dis. 179 (Suppl. 3), S466–S469.

Franco, M.A., Greenberg, H.B., 2000. Immunity to homologous rotavirus infection in adult mice. Trends Microbiol. 8, 50–52.

Friess, A.E., Sinowatz, F., Skolek-Winnisch, R., et al., 1981. The placenta of the pig. II. The ultrastructure of the areolae. Anat. Embryol. 163, 43–53.

Gackowska, L., Michalkiewicz, J., Krotkiewski, M., et al., 2006. Combined effect of different lactic acid bacteria strains on the mode of cytokines pattern expression in human peripheral blood mononuclear cells. J. Physiol. Pharmacol. 57 (Suppl. 9), 13–21.

Gray, J., 2011. Rotavirus vaccines: safety, efficacy and public health impact. J. Intern. Med. 270, 206–214.

Groenen, M.A., Archibald, A.L., Uenishi, H., et al., 2012. Analyses of pig genomes provide insight into porcine demography and evolution. Nature 491, 393–398.

Hammerberg, C., Schurig, G.G., Ochs, D.L., 1989. Immunodeficiency in young pigs. Am. J. Vet. Res. 50, 868–874.

Hausmann, M., Kiessling, S., Mestermann, S., et al., 2002. Toll-like receptors 2 and 4 are up-regulated during intestinal inflammation. Gastroenterology 122, 1987–2000.

Hivroz, C., Chemin, K., Tourret, M., et al., 2012. Crosstalk between T lymphocytes and dendritic cells. Crit. Rev. Immunol. 32, 139–155.

Holscher, H.D., Czerkies, L.A., Cekola, P., et al., 2012. Bifidobacterium lactis Bb12 enhances intestinal antibody response in formula-fed infants: a randomized, double-blind, controlled trial. JPEN J. Parenter. Enteral. Nutr. 36, 106S–117S.

Hoshino, Y., Saif, L.J., Sereno, M.M., et al., 1988. Infection immunity of piglets to either VP3 or VP7 outer capsid protein confers resistance to challenge with a virulent rotavirus bearing the corresponding antigen. J. Virol. 62, 744–748.

Huda, M.N., Lewis, Z., Kalanetra, K.M., et al., 2014. Stool microbiota and vaccine responses of infants. Pediatrics 134, e362–e372.

Humphrey, J.H., Agoestina, T., Wu, L., et al., 1996. Impact of neonatal vitamin A supplementation on infant morbidity and mortality. J. Pediatr. 128, 489–496.

Imdad, A., Yakoob, M.Y., Sudfeld, C., et al., 2011. Impact of vitamin A supplementation on infant and childhood mortality. BMC Public Health 11 (Suppl. 3), S20.

Isolauri, E., Joensuu, J., Suomalainen, H., et al., 1995. Improved immunogenicity of oral D x RRV reassortant rotavirus vaccine by Lactobacillus casei GG. Vaccine 13, 310–312.

Jeppesen, P.B., Hartmann, B., Thulesen, J., et al., 2001. Glucagon-like peptide 2 improves nutrient absorption and nutritional status in short-bowel patients with no colon. Gastroenterology 120, 806–815.

Jin, C.J., Hong, C.Y., Takei, M., et al., 2010. All-trans retinoic acid inhibits the differentiation, maturation, and function of human monocyte-derived dendritic cells. Leuk. Res. 34, 513–520.

Johansson, B., Ingvarsson, S., Bjorck, P., et al., 2000. Human interdigitating dendritic cells induce isotype switching and IL-13-dependent IgM production in CD40-activated naive B cells. J. Immunol. 164, 1847–1854.

Kalliomaki, M., Ouwehand, A., Arvilommi, H., et al., 1999. Transforming growth factor-beta in breast milk: a potential regulator of atopic disease at an early age. J. Allergy Clin. Immunol. 104, 1251–1257.

Kalter, S.S., Heberling, R.L., Rodriguez, A.R., et al., 1983. Infection of baboons ("Papio cynocephalus") with rotavirus (SA11). Dev. Biol. Stand. 53, 257–261.

Kandasamy, S., Chattha, K.S., Vlasova, A.N., et al., 2014a. Prenatal vitamin A deficiency impairs adaptive immune responses to pentavalent rotavirus vaccine (RotaTeq(R)) in a neonatal gnotobiotic pig model. Vaccine 32, 816–824.

Kandasamy, S., Chattha, K.S., Vlasova, A.N., et al., 2014b. Lactobacilli and Bifidobacteria enhance mucosal B cell responses and differentially modulate systemic antibody responses to an oral human rotavirus vaccine in a neonatal gnotobiotic pig disease model. Gut Microbes 5, 639–651.

Kelsall, B.L., Biron, C.A., Sharma, O., et al., 2002. Dendritic cells at the host-pathogen interface. Nat. Immunol. 3, 699–702.

Kramer, D.R., Cebra, J.J., 1995. Role of maternal antibody in the induction of virus specific and bystander IgA responses in Peyer's patches of suckling mice. Int. Immunol. 7, 911–918.

Kumar, A., Vlasova, A.N., Liu, Z., et al., 2014. In vivo gut transcriptome responses to Lactobacillus rhamnosus GG and Lactobacillus acidophilus in neonatal gnotobiotic piglets. Gut Microbes 5, 152–164.

Kuwata, T., Wang, I.M., Tamura, T., et al., 2000. Vitamin A deficiency in mice causes a systemic expansion of myeloid cells. Blood 95, 3349–3356.

Leong, Y.K., Awang, A., 1990. Experimental group A rotaviral infection in cynomolgus monkeys raised on formula diet. Microbiol. Immunol. 34, 153–162.

Lin, Y.P., Thibodeaux, C.H., Pena, J.A., et al., 2008. Probiotic Lactobacillus reuteri suppress proinflammatory cytokines via c-Jun. Inflamm. Bowel Dis. 14, 1068–1083.

Liu, F., Li, G., Wen, K., et al., 2013. Lactobacillus rhamnosus GG on rotavirus-induced injury of ileal epithelium in gnotobiotic pigs. J. Pediatr. Gastroenterol. Nutr. 57, 750–758.

Macpherson, A.J., Harris, N.L., 2004. Interactions between commensal intestinal bacteria and the immune system. Nat. Rev. Immunol. 4, 478–485.

Matsuda, F., Chowdhury, M.I., Saha, A., et al., 2011. Evaluation of a probiotics, Bifidobacterium breve BBG-01, for enhancement of immunogenicity of an oral inactivated cholera vaccine and safety: a randomized, double-blind, placebo-controlled trial in Bangladeshi children under 5 years of age. Vaccine 29, 1855–1858.

Medina, M., Izquierdo, E., Ennahar, S., et al., 2007. Differential immunomodulatory properties of Bifidobacterium logum strains: relevance to probiotic selection and clinical applications. Clin. Exp. Immunol. 150, 531–538.

Medzhitov, R., 2001. Toll-like receptors and innate immunity. Nat. Rev. Immunol. 1, 135–145.

Mesa, M.C., Rodriguez, L.S., Franco, M.A., et al., 2007. Interaction of rotavirus with human peripheral blood mononuclear cells: plasmacytoid dendritic cells play a role in stimulating memory rotavirus specific T cells in vitro. Virology 366, 174–184.

Meurens, F., Summerfield, A., Nauwynck, H., et al., 2012. The pig: a model for human infectious diseases. Trends Microbiol. 20, 50–57.

Mitchell, J.D., Lambeth, L.A., Sosula, L., et al., 1977. Transmission of rotavirus gastroenteritis from children to a monkey. Gut 18, 156–160.

Mullie, C., Yazourh, A., Thibault, H., et al., 2004. Increased poliovirus-specific intestinal antibody response coincides with promotion of Bifidobacterium longum-infantis and Bifidobacterium breve in infants: a randomized, double-blind, placebo-controlled trial. Pediatr. Res. 56, 791–795.

Nguyen, T.V., Yuan, L., Azevedo, M.S., et al., 2007. Transfer of maternal cytokines to suckling piglets: in vivo and in vitro models with implications for immunomodulation of neonatal immunity. Vet. Immunol. Immunopathol. 117, 236–248.

Ohland, C.L., Macnaughton, W.K., 2010. Probiotic bacteria and intestinal epithelial barrier function. Am. J. Physiol. Gastrointest. Liver Physiol. 298, G807–G819.

Pang, X., Hua, X., Yang, Q., et al., 2007. Inter-species transplantation of gut microbiota from human to pigs. ISME J. 1, 156–162.

Patel, M., Glass, R.I., Jiang, B., et al., 2013. A systematic review of anti-rotavirus serum IgA antibody titer as a potential correlate of rotavirus vaccine efficacy. J. Infect. Dis. 208, 284–294.

Pena, J.A., Li, S.Y., Wilson, P.H., et al., 2004. Genotypic and phenotypic studies of murine intestinal lactobacilli: species differences in mice with and without colitis. Appl. Environ. Microbiol. 70, 558–568.

Penttila, I.A., 2010. Milk-derived transforming growth factor-beta and the infant immune response. J. Pediatr. 156, S21–S25.

Perez, N., Iannicelli, J.C., Girard-Bosch, C., et al., 2010. Effect of probiotic supplementation on immunoglobulins, isoagglutinins and antibody response in children of low socio-economic status. Eur. J. Nutr. 49, 173–179.

Petschow, B.W., Litov, R.E., Young, L.J., et al., 1992. Response of colostrum-deprived cynomolgus monkeys to intragastric challenge exposure with simian rotavirus strain SA11. Am. J. Vet. Res. 53, 674–678.

Rinne, M.M., Gueimonde, M., Kalliomaki, M., et al., 2005. Similar bifidogenic effects of prebiotic-supplemented partially hydrolyzed infant formula and breastfeeding on infant gut microbiota. FEMS Immunol. Med. Microbiol. 43, 59–65.

Saalmuller, A., Hirt, W., Reddehase, M.J., 1990. Porcine gamma/delta T lymphocyte subsets differing in their propensity to home to lymphoid tissue. Eur. J. Immunol. 20, 2343–2346.

Saif, L.J., Ward, L.A., Yuan, L., et al., 1996. The gnotobiotic piglet as a model for studies of disease pathogenesis and immunity to human rotaviruses. Arch. Virol. Suppl. 12, 153–161.

Song, D.H., Lee, J.O., 2012. Sensing of microbial molecular patterns by Toll-like receptors. Immunol. Rev. 250, 216–229.

Strauch, U.G., Grunwald, N., Obermeier, F., et al., 2010. Loss of CD103+ intestinal dendritic cells during colonic inflammation. World J. Gastroenterol. 16, 21–29.

Summerfield, A., 2009. Special issue on porcine immunology: an introduction from the guest editor. Dev. Comp. Immunol. 33, 265–266.

Summerfield, A., Guzylack-Piriou, L., Schaub, A., et al., 2003. Porcine peripheral blood dendritic cells and natural interferon-producing cells. Immunology 110, 440–449.

Tagbo, B.N., Mwenda, J.M., Armah, G., et al., 2014. Epidemiology of rotavirus diarrhea among children younger than 5 years in Enugu, South East, Nigeria. Pediatr. Infect. Dis. J. 33 (Suppl. 1), S19–S22.

Tate, J.E., Burton, A.H., Boschi-Pinto, C., et al., 2012. 2008 estimate of worldwide rotavirus-associated mortality in children younger than 5 years before the introduction of universal rotavirus vaccination programmes: a systematic review and meta-analysis. Lancet Infect. Dis. 12, 136–141.

Turroni, F., Foroni, E., Pizzetti, P., et al., 2009. Exploring the diversity of the bifidobacterial population in the human intestinal tract. Appl. Environ. Microbiol. 75, 1534–1545.

Uchiyama, R., Chassaing, B., Zhang, B., et al., 2014. Antibiotic treatment suppresses rotavirus infection and enhances specific humoral immunity. J. Infect. Dis. 210, 171–182.

Ulluwishewa, D., Anderson, R.C., McNabb, W.C., et al., 2011. Regulation of tight junction permeability by intestinal bacteria and dietary components. J. Nutr. 141, 769–776.

Vesikari, T., Matson, D.O., Dennehy, P., et al., 2006. Safety and efficacy of a pentavalent human-bovine (WC3) reassortant rotavirus vaccine. N. Engl. J. Med. 354, 23–33.

Vlasova, A.N., Chattha, K.S., Kandasamy, S., et al., 2013a. Prenatally acquired vitamin A deficiency alters innate immune responses to human rotavirus in a gnotobiotic pig model. J. Immunol. 190, 4742–4753.

Vlasova, A.N., Chattha, K.S., Kandasamy, S., et al., 2013b. Lactobacilli and bifidobacteria promote immune homeostasis by modulating innate immune responses to human rotavirus in neonatal gnotobiotic pigs. PLoS One 8, e76962.

Ward, R.L., McNeal, M.M., Sheridan, J.F., 1990. Development of an adult mouse model for studies on protection against rotavirus. J. Virol. 64, 5070–5075.

Ward, L.A., Rosen, B.I., Yuan, L., et al., 1996. Pathogenesis of an attenuated and a virulent strain of group A human rotavirus in neonatal gnotobiotic pigs. J. Gen. Virol. 77 (Pt 7), 1431–1441.

Wen, K., Azevedo, M.S., Gonzalez, A., et al., 2009. Toll-like receptor and innate cytokine responses induced by lactobacilli colonization and human rotavirus infection in gnotobiotic pigs. Vet. Immunol. Immunopathol. 127, 304–315.

Wen, K., Li, G., Zhang, W., et al., 2011. Development of gammadelta T cell subset responses in gnotobiotic pigs infected with human rotaviruses and colonized with probiotic lactobacilli. Vet. Immunol. Immunopathol. 141, 267–275.

Wen, K., Tin, C., Wang, H., et al., 2014. Probiotic Lactobacillus rhamnosus GG enhanced Th1 cellular immunity but did not affect antibody responses in a human gut microbiota transplanted neonatal gnotobiotic pig model. PLoS One 9, e94504.

Wernersson, R., Schierup, M.H., Jorgensen, F.G., et al., 2005. Pigs in sequence space: a 0.66X coverage pig genome survey based on shotgun sequencing. BMC Genomics 6, 70.

West, C.E., 2000. Vitamin A and measles. Nutr. Rev. 58, S46–S54.

Wilks, J., Golovkina, T., 2012. Influence of microbiota on viral infections. PLoS Pathog. 8, e1002681.

Wu, S., Yuan, L., Zhang, Y., et al., 2013. Probiotic Lactobacillus rhamnosus GG mono-association suppresses human rotavirus-induced autophagy in the gnotobiotic piglet intestine. Gut Pathog. 5, 22.

Wyatt, R.G., Sly, D.L., London, W.T., et al., 1976. Induction of diarrhea in colostrum-deprived newborn rhesus monkeys with the human reovirus-like agent of infantile gastroenteritis. Arch. Virol. 50, 17–27.

Wykes, M., MacPherson, G., 2000. Dendritic cell-B-cell interaction: dendritic cells provide B cells with CD40-independent proliferation signals and CD40-dependent survival signals. Immunology 100, 1–3.

Yoshioka, H., Iseki, K., Fujita, K., 1983. Development and differences of intestinal flora in the neonatal period in breast-fed and bottle-fed infants. Pediatrics 72, 317–321.

Yuan, L., Saif, L.J., 2002. Induction of mucosal immune responses and protection against enteric viruses: rotavirus infection of gnotobiotic pigs as a model. Vet. Immunol. Immunopathol. 87, 147–160.

Yuan, L., Ward, L.A., Rosen, B.I., et al., 1996. Systematic and intestinal antibody-secreting cell responses and correlates of protective immunity to human rotavirus in a gnotobiotic pig model of disease. J. Virol. 70, 3075–3083.

Yuan, L., Iosef, C., Azevedo, M.S., et al., 2001. Protective immunity and antibody-secreting cell responses elicited by combined oral attenuated Wa human rotavirus and intranasal Wa 2/6-VLPs with mutant Escherichia coli heat-labile toxin in gnotobiotic pigs. J. Virol. 75, 9229–9238.

Zhang, W., Azevedo, M.S., Gonzalez, A.M., et al., 2008a. Influence of probiotic Lactobacilli colonization on neonatal B cell responses in a gnotobiotic pig model of human rotavirus infection and disease. Vet. Immunol. Immunopathol. 122, 175–181.

Zhang, W., Azevedo, M.S., Wen, K., et al., 2008b. Probiotic Lactobacillus acidophilus enhances the immunogenicity of an oral rotavirus vaccine in gnotobiotic pigs. Vaccine 26, 3655–3661.

Zhang, W., Wen, K., Azevedo, M.S., et al., 2008c. Lactic acid bacterial colonization and human rotavirus infection influence distribution and frequencies of monocytes/macrophages and dendritic cells in neonatal gnotobiotic pigs. Vet. Immunol. Immunopathol. 121, 222–231.

Zhang, M., Zhang, M., Zhang, C., et al., 2009. Pattern extraction of structural responses of gut microbiota to rotavirus infection via multivariate statistical analysis of clone library data. FEMS Microbiol. Ecol. 70, 21–29.

Zhang, H., Wang, H., Shepherd, M., et al., 2014. Probiotics and virulent human rotavirus modulate the transplanted human gut microbiota in gnotobiotic pigs. Gut Pathog. 6, 39.

Zijlstra, R.T., Donovan, S.M., Odle, J., et al., 1997. Protein-energy malnutrition delays small-intestinal recovery in neonatal pigs infected with rotavirus. J. Nutr. 127, 1118–1127.

Chapter 2.8

Innate Immune Responses to Rotavirus Infection

A. Sen, H.B. Greenberg
Departments of Microbiology and Immunology, Department of Medicine, Stanford University, School of Medicine, Stanford; VA Palo Alto Health Care System, Palo Alto, CA, United States

1 INTRODUCTION

Hosts have evolved highly sophisticated mechanisms to prevent and control viral infections. These include the innate immune response, hardwired into most eukaryotic cells and triggered soon after virus entry as well as the more specialized adaptive immune response, defined by aspects of antigen specificity and memory. Following host cell infection, most viruses trigger one or more pattern recognition receptors (PRRs), which have evolved to recognize virus-specific signatures, or pathogen associated molecular patterns (PAMPs) (eg, 5′-triphosphate RNA), and upon ligand binding, trigger conserved signaling pathways that culminate in the functional activation of critical host transcription factors (TFs) involved in the initiation of various antiviral responses. Several virus stress-induced genes (vSIGs) are transcribed as a result of this early activation, including those encoding the secretory type I and III interferons (IFNs). Expression and secretion of IFNs result in their binding to cognate surface receptors on cells in both autocrine and paracrine patterns, followed by ligand-stimulated activation of the JAK-STAT signaling cascade. Activated STATs, together with other accessory factors, then translocate to the nucleus and initiate a second wave of transcription—resulting in the expression of hundreds of genes that encode antiviral proteins targeting multiple aspects of viral replication, assembly, maturation, and spread. This second phase of the IFN response is also crucial for ensuring that initial IFN expression is amplified through positive feedback, thus ensuring robust establishment of an antiviral state in different host cell types, both infected and uninfected. Co-evolution of hosts and their viruses [also known as the Red Queen Hypothesis (Muraille, 2013)] likely resulted not only in the present complexity of the host innate response, but also in the emergence of viral strategies to block the innate response at multiple steps in order to ensure evolutionary advantage (Medzhitov and Janeway, 1997). Accordingly, most pathogenic viruses, including the rotaviruses (RVs), encode factors that target redundant steps of the IFN induction and amplification signaling pathways. Since many of the basic aspects of RV structure, replication, and pathophysiology are reviewed elsewhere in this book (Chapters 2.1–2.4 and 2.6), they will not be covered here. Certain facets of RV biology may however be specifically pertinent to the host innate immune response to viruses. Rotaviruses encapsidate an 11-segmented dsRNA genome and during replication capped mRNA is generated that serves as a template for nonconservative genome replication. Together, such RV nucleic acids represent extremely potent stimuli for components of the host PRR machinery, including RIG-I, MDA-5, and TLR3. However replication, transcription, and assembly steps in the RV life cycle (according to the current paradigm) occur within subviral particles and specialized sites of assembly within the cell (viroplasms) and likely preclude any significant exposure of the viral dsRNA to innate sensors in the cytoplasm, particularly early during infection (Arnold et al., 2013a). Interferon induction by RVs in several but not all cell types in vitro requires viral replication (Sen et al., 2009; Deal et al., 2010), and the sites of RV assembly (viroplasms) may play a protective role by sequestering not only potential viral PAMPs (Sen et al., 2011; Uzri and Greenberg, 2013), but also critical host innate response factors, such as NF-kB (Holloway et al., 2009). Although in vitro cell culture models continue to be vital in deciphering RV interactions with the innate immune response, crucial differences exist between in vitro cell culture systems and replication in vivo due to unique aspects of RV tropism and the heterogeneity of innate immune responses to RV in different intestinal cell types. Specifically, RVs predominantly infect mature villous enterocytes in the small intestine and replicate poorly, if at all, in intestinal immune cells. However, shortly after infection with RV both intestinal epithelial and immune cells mount innate responses, and recent studies (Frias et al., 2012; Sen et al., 2012; Deal et al., 2013; Feng et al., 2013; Pane et al., 2014) indicate crucial roles for both cell types in restricting the replication of certain strains of RV.

2 HOMOLOGOUS AND HETEROLOGOUS ROTAVIRUSES: CONTRASTING PARADIGMS OF INNATE IMMUNE REGULATION

Rotavirus infection of the immunocompetent host follows an acute course and provided the host survives severe dehydration, is resolved in a matter of days. Early work established that both in vitro and in vivo, although RV is a potent inducer of IFN, the replication of several RV strains is efficient and relatively insensitive to IFN's effects (La Bonnardiere and Laude, 1983; De Boissieu et al., 1993). The lack of a substantial antiviral effect of IFN on RV replication in vivo is reflected by its highly contagious nature; remarkably, a single rotaviral infectious particle in cell culture can constitute the minimal infectious dose in several host species (Graham et al., 1987; Burns et al., 1995). In contrast to these findings for homologous RVs (ie, RV frequently isolated from host species where they cause disease), the replication (and infectivity) of rotaviruses in a heterologous host species (eg, of simian RV or bovine RV in suckling mice) is severely restricted (Arnold et al., 2013a; Sen et al., 2012; Feng et al., 2008; Greenberg and Estes, 2009; Angel et al., 2012). This natural "species barrier," or host range restriction (HRR), has been exploited for the development of several attenuated rotavirus vaccines and vaccine candidates, and previous studies from this laboratory established a prominent role for the RV nonstructural protein NSP1 and the host IFN response in the HRR phenomenon (Feng et al., 2008, 2013; Burns et al., 1995; Greenberg and Estes, 2009; Angel et al., 2012; Bass et al., 1992; Franco et al., 1996a,b; Rose et al., 1998; Franco and Greenberg, 1999). In the suckling mouse model of RV infection, infection with heterologous RVs (such as simian or bovine RV) results in poor viral replication, limited disease when the inoculum is not very large, and poor transmissibility. Genetic ablation of the receptor specific for type I IFN (IFNAR1$^{-/-}$) results in a modest (but not statistically significant) increase in intestinal replication of RRV in IFNAR1 mice (Feng et al., 2008). Combined knockouts of the type I and II IFN receptors (IFNAGR$^{-/-}$) result in significantly more prolonged and extensive intestinal replication of RRV compared to WT mice. Similar gains in replication can be seen in mice lacking STAT1, which likely disables cumulative IFN-$\alpha/\beta/\gamma$- and IFN-λ-dependent responses (Sen et al., 2012; Feng et al., 2008). In addition to higher intestinal replication, RRV replicates substantially better at systemic sites in mice lacking types I and II IFN receptors, or STAT1. Notably, in the IFNAGR$^{-/-}$ or STAT1$^{-/-}$ mice RRV infection leads to the development of a severe systemic disease, which results in morbidity or mortality of nearly 85–90% of animals (Feng et al., 2008). A milder self-limiting illness was observed in IFNAR$^{-/-}$ mice infected with RRV, but was not seen in the IFNGR$^{-/-}$ mice (in contrast to IFNAGR$^{-/-}$ mice as noted earlier). The combined effect of IFNAR1 and IFNGR1, and the absence of systemic illness in IFNGR-deficient mice, is interesting and in line with findings that during RV infection, intestinal immune cells (which are the major IFNGR1 responders) secrete significant type I IFNs (Sen et al., 2012). Studies examining the nature of IFN receptor cross talk found that in MEFs from IFNAR1- and IFNGR1-deficient mice, STAT1-mediated responses to IFN-γ are impaired, while STAT1-responses to IFN-α in IFNGR1$^{-/-}$ cells are not affected (Takaoka et al., 2000). Therefore, cross-talk between the two types of IFNs appears to be unidirectional in that IFN-γ-signaling is dependent on IFN-α/β signaling, but not vice-versa.

In marked contrast to the findings with the simian RRV strain, infection of IFN-receptor/STAT1 knockout suckling mice with the homologous murine EW RV is not substantially more efficient or pathogenic from that seen in WT mice, and IFNAGR- or STAT1-deficient mice show no discernible systemic illness following murine RV infection (Sen et al., 2012; Feng et al., 2008, 2013; Greenberg and Estes, 2009). Thus, susceptibility of homologous and heterologous RV strains to the effects of IFN is markedly different in suckling mice. An important distinction was also observed in these studies between simian RRV and the heterologous bovine and porcine strains (NCDV and OSU), which failed to replicate to detectable levels even in the IFN-receptor or STAT1 deficient mice (Feng et al., 2008, 2013). The basis for these differences may be related to replication restrictions mediated by VP4-mediated viral entry, in addition to IFN-based restrictions, as discussed later. In contrast, another heterologous simian strain (SA11) was similar to RRV, and showed $\sim 10^3$-fold increased intestinal replication at 5 dpi in mice that lacked IFN-α/γ responses (Feng et al., 2008). More recently, in other studies we have reproduced and extended these earlier findings on the substantial differential IFN susceptibility ($> 10^4$) of murine EW and simian RRV intestinal replication, using highly quantitative (and comparable) measurement of viral loads by qRT-PCR (Sen et al., 2012; Feng et al., 2013). Intestinal RRV replication, which is quite efficient compared to several other heterologous viruses in suckling mice, was still severely restricted (more than 10,000-fold) when compared to the homologous murine EW strain. The restricted replication of RRV could be rescued by a factor of $\sim 10^3$ in STAT1$^{-/-}$ mice. In contrast, EW RV replication did not vary substantially between WT and STAT1-deficient mice (less than 10-fold). The RRV VP4 protein, when expressed on a murine RV genetic background, only restricted viral replication by $\sim$1-log (Sen et al., 2012; Feng et al., 2013), whereas substitution of the murine NSP1 encoding gene with its RRV counterpart resulted in RRV-like replication restriction (Feng et al., 2013), indicating that the IFN sensitivity of RRV is not related to VP4-associated entry requirements but instead depends crucially on NSP1. Other comparisons revealed that unlike RRV VP4, the VP4 gene from the heterologous bovine UK RV strain severely restricts replication in the murine intestine, suggesting that the heterologous host may also be refractory to infection by certain RVs due to VP4-mediated inefficient viral entry (Feng et al., 2013). Thus, considerable evidence exists to support the conclusion that in the suckling mouse model, homologous murine RVs are potent inhibitors of IFN-mediated antiviral responses, including all STAT1-dependent responses (types I, II, and III IFNs). In contrast, mouse intestinal cell entry-competent heterologous RVs such as the simian RRV and SA11 strains are highly sensitive to the antireplication effects of IFN, and the roles of different IFNs (particularly of IFN-λ) in restricting their replication will need to be carefully examined in future studies. In a recently published study (Lin et al., 2016) using suckling mice deficient in IFNAR1, IFNLR, or both IFNALR we found that the absence of the IFNAR or IFNLR singly or in combination did not significantly enhance the replication of murine EW RV. Similar studies with heterologous simian RV infection reveal IFNAR1- and IFNLR-dependent viral replication, and enable a more detailed comparison of the role of these IFNRs in regulating heterologous RV replication in vivo (Lin et al., 2016).

3 DETECTION OF ROTAVIRUS INFECTION: HOST INNATE SENSORS INVOLVED IN ROTAVIRUS RECOGNITION

Virus entry and replication result in the appearance of several virus-specific PAMPs in the host intestinal epithelial cell including viral glycoprotein, genomic RNA/DNA, viral transcripts, and nucleic acid replication intermediates. Host cells express a repertoire of structurally related PRRs that enable detection of infection and signal transduction of IFN induction pathways. Eukaryotic PRRs include the *cell-intrinsic RIG-I-like receptors* (RLRs) and NOD-like receptors (NLRs), which are activated purely within infected cells, as well as *cell-extrinsic Toll-like receptors* (TLRs) whose expression on cell surfaces and vesicular membranes extends their functionality to uninfected bystander cells, such as dendritic cells and macrophages. Several studies have examined interactions between RV PAMPs and host PRRs (Sen et al., 2011; Broquet et al., 2011; Qin et al., 2011; Pott et al., 2012; Nandi et al., 2014), leading to a better understanding of this component of host innate immunity to RV.

RV genomic RNA: Remarkably, as early as in 1982 it was noted that "infectious RV does not appear to present its dsRNA in a form suitable for IFN induction" (McKimm-Breschkin and Holmes, 1982). Regardless of their ability to subvert IFN, the replication of diverse RV strains results in an early and significant induction of the IFN pathway, which is accompanied by the increased expression of several transcripts (ISG15, ISG20, IFIT1, IFIT2, etc.), proteins (IFITs, IRF7), and by the phosphorylation of signaling intermediates (IRF3, STAT1, etc.) (Sen et al., 2009, 2011, 2012, 2014; Holloway et al., 2009; Frias et al., 2010, 2012; Broquet et al., 2011; Qin et al., 2011; Nandi et al., 2014; Rollo et al., 1999; Graff et al., 2009; Bagchi et al., 2010; Liu et al., 2010; Pott et al., 2011; Arnold et al., 2013b; Bagchi et al., 2013; Di Fiore et al., 2015; Morelli et al., 2015). This induction occurs as early as 3–5 hpi (Sen et al., 2009, 2011) and differs from IFN pathway stimulation by UV-irradiated "inactivated" virus—which occurs much later (~12 hpi) (McKimm-Breschkin and Holmes, 1982), likely via release and exposure of the viral dsRNA genome following degradation of capsid proteins. As discussed later, by these later times in the infected cell, RV encoded nonstructural proteins are synthesized that can potently inhibit IFN induction (and that are absent in experiments using inactivated RV). Studies using purified RV dsRNA have found it to potently induce the activation of RLRs, TLRs, and NLRs leading to expression of IFNβ and inflammatory cytokines, such as IL1β and IL18 (Kanneganti et al., 2006; Sato et al., 2006). Although TLR3, the primary PRR for dsRNA detection, is poorly expressed on IECs in suckling mice and human infants, there is an age-dependent increase in its expression in these cells in older mice and adults (Pott et al., 2012). Experiments using siRNA-mediated knockdown of TLR3 have failed to uncover a convincing role in early IFN responses to RV in MEFs (Sen et al., 2011; Broquet et al., 2011).

In contrast, a cell type where RV dsRNA is a potent PAMP is the plasmacytoid dendritic cell (pDC), although the cognate PRR(s) involved is not known (Deal et al., 2010, 2013; Douagi et al., 2007; Gonzalez et al., 2010; Lopez-Guerrero et al., 2010). Highly purified human primary pDCs that are exposed to inactivated RV strongly express IFNα as early as 6 h postexposure while pDCs exposed to virus-like particles containing VP2, VP6, VP4, and VP7 but no dsRNA do not express type 1 interferons (Deal et al., 2010). Thus, early dsRNA-mediated activation of IFN may not be particularly relevant in intestinal epithelial cells supporting RV replication, but could be quite important for the IFN induction that occurs in vivo in nonreplicating compartments, such as intestinal immune cells (Sen et al., 2012), and possibly in TLR3 expressing IECs from adult mice infected with RV (Pott et al., 2012).

RV replication by-products: Early activation of PRRs in infected cells is dependent on viral replication. Studies on the bovine UK RV in murine embryonic fibroblasts (MEFs) revealed that despite encoding a full complement of both structural and nonstructural viral proteins, UK infection of MEFs resulted in robust IFNβ secretion (Sen et al., 2009). It is worth noting that in several other cell lines, such as COS7 and HT29, UK RV is very capable of efficiently inhibiting IFNβ responses (Sen et al., 2009), demonstrating that RV regulation of IFN expression is contextual rather than absolute—an important factor to consider when classifying RV IFN phenotypes in single cell types. In MEFs, early IFN induction was abrogated with UV-inactivated UK virus, indicating that a by-product(s) of replication acted as the PAMP(s) for early PRR activation (Sen et al., 2009). Subsequent biochemical and enzymatic characterization of RV RNAs from infected cells and in vitro transcription systems revealed that RV mRNA species with exposed 5′-phosphate groups and those with incompletely 5′-O-methylated "cap" structures were particularly effective at stimulating PRRs (Uzri and Greenberg, 2013) (Fig. 2.8.1). Using genetic knockout MEFs lacking different PRRs, their relative contributions to early IFN activation by RVs have been determined (Sen et al., 2011; Broquet et al., 2011). From such studies, it is likely that RV transcription triggers the mitochondrion-associated RLRs RIG-I (retinoic acid induced gene I) and MDA-5 (melanoma differentiation-associated antigen 5) early after infection. In accordance with the function of RLRs, activation of RIG-I and MDA-5 by RV is completely dependent on the mitochondrial adaptor MAVS. While absence of MAVS results in a complete loss of RV recognition (and subsequent IFN secretion later in infection) by the host cell, this defect is not fully recapitulated by the removal of either RIG-I or MDA5 alone, indicating that both PRRs function redundantly in sensing the presence of RV, perhaps by targeting 5′-phosphate moieties and incompletely 5′-O-capped ends of viral transcripts, respectively. Interestingly, the RLR LGP2 (laboratory of genetics and physiology 2), which lacks the caspase activation and recruitment (CARD) domains required for signaling to MAVS, is important for IFN induction by RV (Broquet et al., 2011), although the mechanism involved is not understood. Apart from viral nucleic acid, RV also encodes a

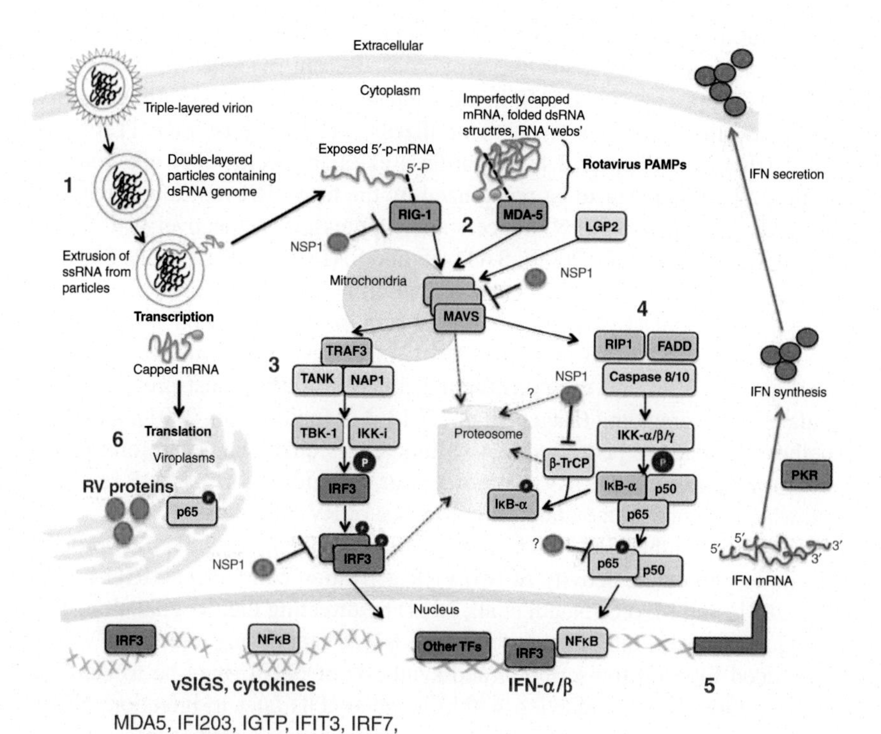

FIGURE 2.8.1 The early host interferon response to rotavirus infection. (1) RV entry into host cells is accompanied by removal of the outermost capsid layer and transcriptional activation of the double-layered particle. Capped viral mRNA is extruded from within particles containing the viral dsRNA genome. (2) Viral transcription (and replication) derivatives, including molecules with exposed 5′-phosphate moieties, imperfectly capped RNA, local hairpin structures, and higher order RNA oligomers, are potential RV PAMPs. Such PAMPs trigger the activation of host PRRs—primarily RIG-I, MDA-5, and LGP2—followed by prion-like oligomerization of the mitochondrial factor MAVS. Pathway activation in this manner leads to assembly of specific signaling complexes that phosphorylate and functionally activate the transcription factors IRF3 (3) and NF-κB (4). Nuclear translocation of IRF3, NF-κB, and other transcription factors results in occupancy of promoter elements on several antiviral genes, including the type I IFNs, and the initiation of a transcriptional program that is characteristic of the initial IFN induction phase (5). Specific examples of antiviral genes induced exclusively within infected cells have been identified from single cell measurements in vivo and are shown. The dsRNA-dependent kinase PKR plays a critical role during RV infection in the expression of type I IFNs from induced transcripts. Rotavirus encodes several inhibitory mechanisms to prevent the establishment of this initial antiviral state. Depending on the specific host cell infected and the viral strain, viral translation of the RV NSP1 protein (6) results in the degradation of the following factors: RIG-I, MAVS, IRF3, and β-TrCP. Interestingly, NSP1 itself can be proteasomally degraded although the significance of this phenomenon is not clear. In addition, RV-mediated sequestration of the NF-κB subunit p65 in viroplasms (by unknown viral factors) (6) occurs during infection. Abbreviations used: RV, rotavirus; dsRNA, double-stranded RNA; PAMP, pathogen associated molecular pattern; PRR, pattern recognition receptor; RIG-I, retinoic acid-inducible gene I; MDA-5, melanoma differentiation associated protein 5; LGP2, laboratory of genetics and physiology 2; MAVS, mitochondrial antiviral signaling protein; IRF3, interferon regulatory factor 3; NF-κB, nuclear factor kappa B; IFN, interferon; PKR, dsRNA-dependent protein kinase; NSP1, nonstructural protein 1; β-TrCP, beta transducin repeat-containing E3 ubiquitin protein ligase.

nonstructural glycoprotein, NSP4 that is secreted into the extracellular milieu and exhibits certain adjuvant-like properties. Recent evidence indicates that NSP4 may represent a RV PAMP in macrophages and trigger pro-inflammatory cytokine expression by a pathway that utilizes the cell-extrinsic PRR TLR2 (Ge et al., 2013) (see Chapter 2.4). Regardless of their specific IFN antagonistic abilities, all RVs appear to be recognized by the host PRR machinery [eg, the following strain pairs of IFN inducers and suppressors, respectively—UK and RRV in MEFs (Sen et al., 2009), SA11-5S and SA11-4F in HT29 cells (Arnold and Patton, 2011)]. PAMP-mediated activation results in recruitment of adaptor TRAF molecules by MAVS—a crucial step in MAVS-dependent activation of both IRF3 and NF-κB (Liu et al., 2013). Interestingly, both the RV outer capsid protein VP4 and the nonstructural interferon antagonist protein NSP1 regulate TRAF signaling (Bagchi et al., 2013; LaMonica et al., 2001), although whether this leads to perturbed MAVS function is unknown. Therefore, based on the available evidence, RVs do not appear to regulate early PAMP recognition, which may be an inevitable consequence of RV infection.

Following synthesis of viral proteins during infection, the viral NSP1 protein appears to reduce the activity of two PRR activation components, RIG-I (Qin et al., 2011) and MAVS (Nandi et al., 2014) by directing their degradation. The consequence of this inhibition (occurring ~8 h or later during infection), which is preceded by viral transcription and synthesis of NSP1, may be to dampen the magnitude of IFN induction within infected cells later in infection. NSP1 proteins from both OSU and SA11 strains have been reported to interact with RIG-I in a transient overexpression system, and this interaction does not require the NSP1 C-terminal (~170-aa) IRF3-binding domain (Qin et al., 2011). Expression of NSP1 also led to inhibition of RIG-I mediated IFNβ expression, and a decrease in RIG-I protein expression by a proteasome-independent pathway (Qin et al., 2011). Recently it was also reported that NSP1s from porcine OSU and human Wa, DS-1, and KU RV strains are able to degrade MAVS in a proteasome-dependent manner (Nandi et al., 2014). NSP1 interacted with MAVS, and the C-terminal 395-aa alone was sufficient to mediate this interaction, although MAVS degradation required the full-length protein (Nandi et al., 2014). Although these observations are interesting, several key questions should be resolved in future studies. Whether NSP1-mediated inhibition of RIG-I occurs during RV infection is not known presently. Somewhat paradoxically, although OSU-NSP1 protein is able to degrade both RIG-I and MAVS, infection with the porcine RV OSU (and the related RV SB1A) has been shown to lead to a robust and sustained activation of IRF3, presumably directed by MAVS (Graff et al., 2009; Sen et al., 2014). Similarly, UK RV infection leads to IRF3 S396 activation during infection of 3T3 cells that persists until later times (12–16 hpi) in a MAVS-dependent manner (Sen et al., 2009, 2011). Finally, definitive evidence is lacking on whether NSP1 interacts with either RIG-I or MAVS directly, and similar to most other potential NSP1 interactions, this interaction has not been analyzed using the relevant purified binary components. NSP1 may instead target a common component of this signaling complex, leading to the secondary observed associations of NSP1 with the other members (through indirect interactions). Such indirect interactions with NSP1 are a possibility since multiple NSP1 host partners that have already been identified are a part of common signaling complexes.

4 INTERMEDIARIES: INNATE FACTORS RELAYING INTERFERON INDUCTION DURING ROTAVIRUS INFECTION

Following pathogen detection, PRRs undergo conformational changes, resulting in the recruitment of adaptors that then recruit and activate protein kinases and downstream transcription factors (TFs) (Arnold et al., 2013a). In the case of the RLRs RIG-I and MDA-5, activation leads to conformational and higher-order rearrangements, a process in which RNA binding is followed by PRR interaction with unanchored K63-ubiquitin chains. Such activated PRRs rapidly induce prion-like oligomerization of the mitochondrial adaptor MAVS, which then recruits TRAFs-2, 5, and 6, and two kinase complexes (IKK-α/β/γ and TBK1-IKKi), resulting in activation of the TFs NF-κB and IRF3 (Fig. 2.8.1). Seminal studies by Hardy and coworkers concluded that, depending on the RV strain studied, the RV NSP1 protein targets IRF3 (Graff et al., 2002) or β-TrCP (an essential cofactor for NF-κB activation) (Graff et al., 2009) in the IFN induction cascade. Several studies have uncovered details of these interactions resulting in fascinating insights into RV regulation of the IFN response (Sen et al., 2009, 2011, 2012; Arnold et al., 2013b; Douagi et al., 2007; Arnold and Patton, 2011; Barro and Patton, 2005, 2007; Graff et al., 2007; Feng et al., 2009). In the context of infection, two main approaches have proved useful in deciphering the component intermediate factors involved: the use of mutant RV strains encoding truncated NSP1 that lack IFN antagonistic function, and the use of wild type RV strains encoding full-length NSP1, which induce IFN in some cell types (but not in others). It is important to note that these 2 approaches are likely to query distinct NSP1 functions during RV infection.

TRAFs: In addition to VP4-encoded TRAF binding motifs, NSP1 has also been identified as an inhibitor of TRAF2, an important component of MAVS and noncanonical NF-kB signaling (Bagchi et al., 2013). The NSP1 protein from different RV strains interacts with and targets TRAF2 for degradation. Interestingly, both IRF3 degrading (simian SA11) and NF-kB inhibiting (porcine OSU, bovine A5-13) RV strains degrade TRAF2. The degradation of TRAF2 by NSP1 was shown to inhibit the noncanonical NF-kB pathway triggered by exogenous IFN (Bagchi et al., 2013). Specifically, NSP1 blocks the nuclear translocation of p52 (which, unlike "canonical" p65, is a repressor lacking a transcription activation domain, and is nonessential for IFN induction) in response to exogenous IFN stimulation. Thus far, NSP1-directed TRAF2 degradation does not appear to be important for IFN induction, but may instead play a role in p52-dependent regulation of cytokine expression triggered by IFN (or other cytokines), as well as in sensitizing cells to potential p52-mediated pro-apoptotic effects of IFNβ, although these possibilities have not been directly examined.

NF-kB: Timely and robust induction of type I IFNs requires formation of a complex that includes IRF3 and NF-kB, both of which are critical for gene induction with an important caveat. Specifically, once viruses activate IRF3/IRF7, whether NF-kB is essential for IFNβ induction remains controversial, and NF-kB has alternately been proposed as essential for maintaining basal autocrine expression of IFNβ and ISGs in the absence of infection (or in uninfected bystander cells) (Wang et al., 2010). Of note, infection with RV strains such as NCDV and OSU that inhibit NF-kB activation results in efficient NSP1-dependent blockage of IFNβ induction and secretion, although IRF3 is activated and nuclear (Holloway et al., 2009; Graff et al., 2009). In addition, IECs infected with murine RV in vivo induce several IRF3-dependent transcripts, but neither NF-kB-dependent or IFNβ transcripts (Sen et al., 2012). Thus RV-mediated activation of IRF3 per se is not sufficient to induce IFNβ, and specifically targeting of NF-kB may lead to effective IFN inhibition as well.

The RV NSP1 protein from certain strains (primarily porcine and human strains) targets β-TrCP, an F-box protein and essential NF-kB activating factor, for degradation (Graff et al., 2009; Di Fiore et al., 2015; Morelli et al., 2015). In the canonical pathway, NF-kB subunits are held in an inactive configuration by inhibitory IκB-α molecules. Upon RLR stimulation, IκB-α is phosphorylated within a phosphodegron motif (DSGxS, where x is generally a hydrophobic residue) by IKKs, and is subsequently recognized and degraded by the F-box protein β-TrCP which is the substrate-directing component of the E3 ubiquitin ligase complex Skp1-Cul1-F box protein-Rbx1 [(SCF), the key E3 ligase enzymatic component of this complex being Rbx1], thereby releasing NF-kB for transcriptional duties. The NSP1 protein from porcine and human RV strains contains an IκB-α-like phosphodegron sequence (called PDL, or phosphodegron-like motif) at the C-terminus—effectively constituting a "β-TrCP trap" (Di Fiore et al., 2015; Morelli et al., 2015). The PDL motif DSGIS, occurs within the carboxyl-8-aa region of NSP1 from human and porcine strains, and is absent in other NSP1 proteins, including the bovine strains. Notably, not all PDL-containing human and porcine NSP1s degrade β-TrCP, although they are all potent inhibitors of NF-kB, indicating that β-TrCP interaction (rather than degradation) is critical for RV inhibition of NF-kB (Morelli et al., 2015). The sequence determinants that predict whether NSP1-mediated β-TrCP degradation is likely to occur are presently not known but are unlikely to reside in the NSP1 RING domain, which is highly conserved. Mutation of two predicted casein kinase II phosphorylation serines in the NSP1 phosphodegron motif (positions 480 and 483 in OSU NSP1) resulted in complete abrogation of NF-kB inhibition, indicating an essential requirement for priming phosphorylation of NSP1 for subsequent β-TrCP interaction (Morelli et al., 2015). These findings are exciting as they reveal the determinants of NSP1's ability to block NF-kB- dependent IFN responses, by mimicking a cellular motif. However, certain puzzling questions about RV-mediated NF-kB regulation vis-à-vis IFN induction remain unanswered. Bovine RV NCDV, whose NSP1 does not contain the PDL motif is still able to degrade β-TrCP in a proteasome-dependent manner, and actively inhibits p-IκB-α degradation and NF-kB function during infection indicating that PDL-motif independent mechanisms may exist (Graff et al., 2009). Indeed, several lines of evidence point to additional RV strategies to usurp NF-kB functions related to the innate response (Holloway et al., 2009; Sen et al., 2012; Graff et al., 2009; Arnold and Patton, 2011; Holloway et al., 2014). The bovine RV A5-16, which encodes a severely truncated ~50-aa long NSP1 lacking both the PDL motif and the RING finger, is able to sequester NF-kB p65 within viroplasms at 6 hpi. Interestingly, inhibition of p65 following TNF-α-mediated activation has also been reported during infection with RRV and Wa RV strains (which have contrasting effects on β-TrCP) as early as 6 hpi, and may represent a alternate β-TrCP-independent mechanism by which RV strains block NF-kB function (Holloway et al., 2009). In infected cells, p65 is held in viroplasms instead of translocating into the nucleus, although the viral proteins underlying this interaction remain unknown (Holloway et al., 2009, 2014). Murine EW RV, whose encoded NSP1 also does not contain the β-TrCP interacting motif, is associated with perturbation of NF-kB function (Sen et al., 2012). Intestinal lysates from mice infected with murine EW (but not with simian RRV strain) exhibit accumulation of IκB-α protein at 16hpi, which is accompanied in virus infected IECs by an increase in IRF3- but not NF-kB-dependent ISG transcripts, and within bystander (virus-negative) IECs by significantly reduced basal transcription of NF-kB target genes such as Peli1 and A20 (Sen et al., 2012). It is thus likely that other RV proteins, and/or secreted cellular factors, lead to NF-kB inhibition during infection.

IRF3 and other IRFs: Activation of PRRs, and subsequent assembly of adaptor-kinase complexes leads to phosphorylation of IRF3 at multiple serine/threonine residues and to conformational changes that mediate IRF3 dimerization. Dimeric IRF3 then translocates to the nucleus, where it participates in the expression of several virus stress-induced genes (vSIGs) and IFN. The first indication that RV blocks early IRF3-dependent IFN responses was from findings that RV NSP1 interacts with IRF3, and mediates its proteasomal degradation (Graff et al., 2002; Barro and Patton, 2005). Using yeast 2-hybrid analysis, recombinant GST-IRF3 pull-downs from infected cell lysates, and immunoprecipitation of endogenous IRF3 from infected cells, it was shown that RV NSP1 interacts with IRF3. Mutagenesis approaches have determined that this interaction requires the NSP1 carboxy-terminal, and mutants containing only the last 326-aa, a region rich in helical content, can interact with IRF3 (Graff et al., 2007). A conserved N-terminal RING finger motif in NSP1 is also important, but not sufficient, for the interaction. The NSP1–IRF3 interaction is likely a direct interaction due to its initial identification in a binary yeast 2-hybrid screen, although it is important to note that IRF3 regulation (ie, degradation) itself could involve other unidentified essential intermediary factors. The IFN-inducing ability of a phospho-mimetic mutant of IRF3 in which five critical S/T residues are mutated to aspartic acid resulting in constitutive activation and dimerization is efficiently inhibited by NSP1, indicating that NSP1 targets the dimeric form of IRF3 (Sen et al., 2009). This notion is strengthened by analysis of different IRF mutants (Arnold et al., 2013b), which demonstrates that degradation of IRFs by NSP1 requires the conserved carboxyl IRF-dimerization domain. We have observed previously that NSP1-mediated degradation of IRF3 is in fact more efficient when IRF3 is activated (Sen et al., 2009), pointing to the possibility that the initial PAMP-mediated stimulation of IRF3 during RV infection may serve to generate a more suitable target for degradation by NSP1. This also offers an interesting explanation for the activation of IRF3 (at least as inferred from its transcriptional activity) at 16 hpi within RV-infected intestinal villous epithelial cells in vivo (Sen et al., 2012). Certain RV strains, notably those isolated from porcine and human hosts encode NSP1 proteins that are unable to degrade IRF3, at least in the cell types examined so far (Arnold and Patton, 2011). Infection with such strains generally results in robust and sustained IRF3 activation during the course of infection, although how such strains cope with IRF3-induced antiviral vSIGs is still unknown (Sen et al., 2009; Graff et al., 2009). As discussed, such NSP1 proteins nevertheless inhibit IFN induction by targeting NF-kB activation in the IFN induction pathway.

IFN regulation by NSP1 is likely to be more complex than a framework where all NSP1 proteins are either IRF3 or NF-kB inhibitors, and such functions can be highly dependent on the cell type studied. Wild type RV strains (as opposed to mutant strains encoding truncated NSP1 proteins) have been identified that induce IRF3-dependent IFNβ secretion in certain cell types and degrade IRF3 efficiently in others (Sen et al., 2009). Bovine RV UK, or its encoded NSP1 expressed singly, can target endogenous IRF3 for degradation in simian COS7 cells (and in human 293/HT29 cells). However, UK RV infection or expression of UK NSP1 does not result in IRF3 degradation in murine 3T3 fibroblasts (or in primary MEFs), and instead triggers IFNβ secretion. Interestingly, UK NSP1 efficiently targets recombinant simian IRF3 exogenously expressed in 3T3 cells for proteasomal degradation. Conversely, recombinant murine IRF3 expressed in COS7 cells is still refractory to UK NSP1-mediated degradation (Sen et al., 2009). Thus, the ability of NSP1 to degrade IRF3 depends on the host cell, and a comparison of different NSP1s in 3T3 cells would likely identify UK NSP1 as "non-IRF3 degrading." Alternate NSP1 degradation-independent mechanisms for inhibiting IRF3, which depend on the nature of PRR stimulation, are also likely to exist. Specifically, in 3T3 cells, UK NSP1 (or UK RV infection) fails to inhibit activation of an IRF3-responsive (PRDIII) luciferase reporter stimulated by overexpression of murine IRF3. However, UK NSP1 efficiently inhibits PRDIII activity in the absence of IRF3 degradation when the pathway is stimulated with liposome-complexed intracellular poly(I:C) instead (Sen et al., 2009). Thus, NSP1 can also inhibit IRF3 function despite lack of degradation under specific contexts of PRR stimulation, which is likely relevant in vivo where stimulation can be cell type- and timing-specific. Adding to this complexity, lack of IRF3 inhibition by an "IRF3-degrading" NSP1, as reported for UK NSP1 in vitro—has also been noted in vivo (Sen et al., 2012). Measurement of IRF3-dependent transcription in isolated murine IECs 16 h after infection with murine RV EW in vivo indicates that IRF3-dependent vSIGs are significantly upregulated. The IRF3 activity is likely pathway-specific, as it is not accompanied by simultaneous increases in NF-kB targets, or of the IFN genes themselves in these single cells. One possibility is that during infection in vivo, RV inhibits NF-kB (and consequently its transcriptional target IFN) earlier than IRF3, leading to the transcriptional changes observed. An alternate possibility is that such transcripts are a result of EW NSP1 failure to inhibit IRF1 and/or IRF7, both of which have transcriptional "footprints" largely overlapping with that of IRF3. Clearly, more studies are needed on NSP1–IRF interactions, particularly in the context of in vivo infection where IRF1/IRF3/IRF7 signaling states are likely influenced by crosstalk from immune cells in the intestine and where IRFs may be regulated temporally, in order to deepen our understanding.

Apart from IRF3, RV NSP1 also targets other IRFs for degradation. Detailed analysis of different NSP1s and IRFs have demonstrated that apart from IRF3, NSP1 can also target IRF5, IRF7, and IRF9 for degradation (Arnold et al., 2013b; Arnold and Patton, 2011), and this regulation involves a conserved C-terminal IRF interaction domain (IAD), which mediates IRF dimerization following its activation. In most cell types, IRF7 is constitutively expressed at low levels, and initial IRF3 activation and IFN secretion increase IRF7 expression and activation, resulting in IRF7-mediated expression of the antiviral program. IRF7 has also been noted to be induced in an IFN-independent manner, and could thus be also considered as a vSIG (Schmid et al., 2010). Interestingly although human RVs are unable to degrade IRF3 unlike most animal RVs, they share the ability to degrade IRF7. Thus, as noted earlier (Arnold et al., 2013b), compared to IRF3, IRF7 inhibition may be more evolutionarily conserved among NSP1s. Single infected enterocytes in vivo contain significantly elevated levels of IRF7 transcripts, but this is not accompanied by an increase in IFN transcripts (Sen et al., 2012)—an expected consequence of increased IRF7 expression. Elevated IRF7 transcription is also not observed in adjacent bystander cells, indicating that IRF7 transcripts are induced directly within infected cells (as a vSIG). It is not known whether murine EW NSP1 targets IRF7 function in vivo. Interestingly, none of the NSP1 proteins examined so far have acquired the ability to degrade IRF1 (Arnold et al., 2013b), which lacks the IAD and is activated by peroxisomal MAVS to drive expression of a subset of early vSIGs. Porcine RVs seem to be rather unique in encoding NSP1 that does not degrade any of the IRFs examined so far. Perhaps this inability of RVs to block certain IRFs is compensated by NSP1's ability to target MAVS and/or RIG-I—it remains to be seen whether RV also targets peroxisomal MAVS during infection.

5 THE AMPLIFIERS: ROTAVIRUS REGULATION OF THE EFFECTS OF INTERFERONS

Following induction, secreted IFNs bind to their cognate receptors on the cell surface and trigger signaling via the JAK-STAT pathway, resulting in an extensive transcriptional program and expression of over 300 distinct antiviral genes Fig. 2.8.2 (Arnold et al., 2013a). Binding of type I IFN to the IFNAR1 and IFNAR2 subunits, results in dimerization of the receptor chains, and phosphorylation of Tyk2, which is associated with IFNAR1, by Janus Kinase 2 (JAK2). The activated Tyk2 subsequently phosphorylates JAK1, which is associated with IFNAR2. The activated JAK in turn phosphorylates STAT2, which is associated with IFNAR2, resulting in STAT1 activation at a critical residue (Y701) and formation of STAT1-STAT2 dimers that translocate to the nucleus and form a TF complex with IRF9 called ISGF3 (IFN stimulated gene factor 3). ISGF3 binds to DNA sequence motifs called IFN sensitive response elements (ISRE), while STAT1 homodimers bind to the IFN-γ-activated site (GAS) sequence motif, to induce subsets of hundreds of antiviral genes. Most viral strategies have evolved to prevent the establishment of this antiviral state in the first instance, as its negation by viruses would require targeting numerous downstream host factors. Although RVs encode a potent inhibitor of IFN induction, in order to replicate successfully they would also seem to need to encode inhibitory factors to specifically negate the effects of secreted IFNs.

Murine RV infection in vivo results in significant induction of IFNβ when bulk intestinal tissues are analyzed. However, using single cell and population-specific measurements of transcription in mice, we observed that this induction does not originate in the intestinal epithelium—either in infected or bystander IECs. Instead, a significant source of IFN is the intestinal hematopoietic cell compartment, where RV replicates poorly (Sen et al., 2012). Examination of selected strains of genetic knockout mice demonstrates that such IFNs are clearly important for restricting heterologous RV. Intestinal replication and extra-intestinal spread of heterologous (host range restricted) RV strains is significantly restricted by STAT1 (which is critical for signaling by types I, II, and III IFNs), and by IFN-αβγ-receptor mediated signaling. Additional potential contributors to restriction of RV replication in mice are the type III IFNs, which are expressed predominantly by IECs. Their effect on homologous murine RV replication is unclear with conflicting data in the literature (Sen et al., 2012; Pott et al., 2011). The effects of type III IFNs on replication of heterologous RV have not yet been examined (Pott et al., 2011). Secreted IFNs from both epithelial and immune cells are likely to trigger antiviral responses and restrict RV replication and spread in the IEC compartment unless actively counteracted by virus. In vivo, homologous RVs are likely to encode potent strategies to negate the actions of such IFNs since their replication is remarkably insensitive to signaling dependent on STAT1, IFNAR1, IFNGR1, or IFNLR (unpublished data). Unpublished studies in MEFs have demonstrated that pretreatment with IFNβ

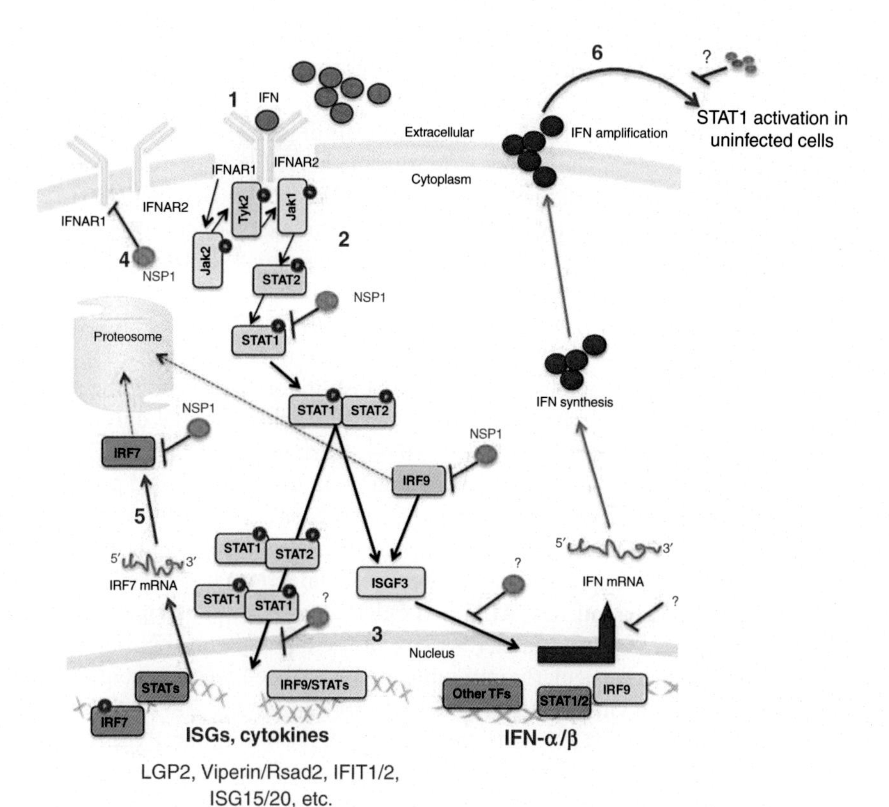

FIGURE 2.8.2 **IFN signaling and amplification during RV infection.** (1, 2) Binding of IFN to its cognate receptor results in receptor dimerization and sequential activation of the JAK-STAT pathway in both an autocrine and paracrine manner. Although the type I IFN signaling pathway is depicted, similar signal transduction occurs following IFN lambda stimulation. (3) Pathway activation results in the transcriptional activation of STAT1and STAT2 and formation of ISGF3—a complex consisting of STAT1/2 and IRF9—which translocates to the nucleus and mediates expression of hundreds of antiviral genes (IFN-stimulated genes, ISGs), including the amplification of IFN-encoding genes. Examples of ISGs induced by RV in vivo in single intestinal epithelial cells are shown. RVs inhibit this stage of the host IFN response in several ways. Relatively later during infection, NSP1 blocks STAT1 activation in response to exogenous IFN (2) and, in unpublished studies, depletes IFN receptors (4) by an unknown mechanism. (5) IRF7, an ISG that acts as a key mediator of antiviral gene expression, can be targeted in infected cells by NSP1 for proteasomal degradation. (3) Early during infection, STAT1 activation by exogenous IFN proceeds unhindered but RV inhibits the nuclear translocation of activated STAT1 and STAT2 by an unknown process. (6) RV also exerts inhibitory effects on this pathway in uninfected bystander cells by blockage of both STAT1 activation and the expression of IFN transcripts by unknown mechanisms. Abbreviations used: JAK, janus kinase; STAT, signal transducer and activator of transcription; ISGF3, interferon stimulated gene factor 3; IRF9, interferon regulatory factor 9; ISG, interferon stimulated gene; IRF7, interferon regulatory factor 7; for additional abbreviations used, see legend to Fig. 2.8.1.

does not suppress the replication of murine RV significantly, while heterologous RV strains with opposing replication phenotypes have also been identified in such experiments

Bystander effects of RV: Transcriptional analysis of isolated infected and bystander single IECs during homologous murine in vivo RV infection has provided new insights on RV regulation of the IFN signaling phase (Sen et al., 2012). At 16 hpi, IECs that are infected with RV contain significant induction of several IRF3-dependent and IRF3-augmented genes, but no increase in transcripts encoding NF-kB-responsive factors or the type I IFNs (IFNα4, IFNα5, and IFNβ). Induced IEC transcripts include the antiviral genes IFIT1, IFIT2, RSAD2, ISG15, ISG20, IGTP, and SHISA5, which are also significantly increased in uninfected bystander IECs, and likely represent transcription due to paracrine and/or autocrine IFN signaling. A second set of genes is induced by RV only in infected cells, and likely represents direct vSIGs—these include IRF7, CXCL10, STAT1, and IRF9. Of these, RV mediated inhibition of the expressed proteins from IRF7, IRF9, and STAT1 transcripts has also been reported to take place in vitro. A final set of transcripts represents significantly decreased expression (compared to baseline expression in the absence of infection) in bystanders and contains the NF-kB targets TNFAIP3 and Peli1, as well as transcripts encoding the type I IFNs. Such "bystander" effects of RV infection on IFN signaling in the epithelium are particularly intriguing, and have been recently confirmed and extended in an in vitro context (Sen et al., 2014).

STAT1 inhibition: In HT29 cells, the porcine SB1A strain robustly activates pS396-IRF3, lacks IRF3 degradative ability, and degrades βTrCP—correlating with its PDL-motif. In these cells, SB1A is able to potently inhibit STAT1 phosphorylation at Y701 in response to saturating doses of exogenous IFNβ (Sen et al., 2014). Interestingly, almost total STAT1-pY701 inhibition also occurs in uninfected bystander cells that do not express the SB1A VP6 antigen. Inhibition of pY701-STAT1 by SB1A in bystander cells occurs later in infection (10–16 hpi), and is accompanied by pY701 regulation in virus-infected cells. Among infected cells, 2 populations can be clearly identified—upto about 50% of infected cells contain significant levels of activated STAT1 in the absence of exogenous stimulation with IFN, while the remainder of SB1A infected cells do not contain any constitutive pY701-STAT1. Following exogenous IFNβ stimulation, none of the infected cells show an IFN-mediated increase in pY701-STAT1 levels. Thus, SB1A shares an interesting similarity with the murine EW strain in being able to inhibit IFN signaling in bystander cells in the absence of detectable IRF3 inhibition. Similar experiments with simian RRV highlight some interesting differences. Unlike SB1A, RRV infection is not accompanied by detectable STAT1 activation in the absence of exogenous IFN stimulation, and at later times during infection only the virus-infected cells are refractory to exogenous IFN-mediated STAT1 activation, while bystander cells respond to exogenous IFN similarly to uninfected control cells. The ability to inhibit STAT1-Y701 activation in response to IFN was mapped to the NSP1 protein using mutant RV SA11-5S encoding a truncated NSP1, and transient expression of recombinant NSP1 from RRV, UK, and EW RV strains resulted in STAT1 inhibition in cells positive for NSP1 expression, indicating that this is an evolutionarily conserved function across a number of RV NSP1s (Sen et al., 2014). Interestingly, the NSP1 RING motif, which is critical for IRF3 degradation, was not essential for STAT1 inhibition indicating that the regulation involves a novel NSP1 domain and possibly underlying mechanism.

In addition to the NSP1-mediated inhibition of IFN-receptor-mediated STAT1 phosphorylation that occurs later in infection, other RV mechanisms also target antiviral functions of STAT1 earlier in infection. The observation that RV regulates STAT1 was first reported by Holloway et al. (2009), who found that in MA104 cells, different RV strains (RRV, Wa, UK, SA11, and several others) could inhibit STAT1 and STAT2 nuclear translocation in response to exogenous IFN stimulation at a step after its phosphorylation as early as 6 hpi. In these studies, Wa RV infection led to STAT1 activation in the absence of any IFN stimulation as measured by immunoblot analysis of bulk cell lysates, resembling the pattern found with SB1A in HT29 cells. It remains to be seen whether similar to SB1A, Wa RV can inhibit STAT1 activation in bystander cells at later times during infection. The cytoplasmic sequestration of STAT1 could not be recapitulated by singly expressing any of the 12 RRV proteins as FLAG-tagged constructs in HEK293T cells (Holloway et al., 2009, 2014), and thus the viral factor(s) involved in this inhibition are presently not known.

RV VP3 inhibits RNAse L: Despite encoding several mechanisms for inhibiting the induction and amplification phases of the IFN response, RVs likely also negate the function of downstream effectors of IFNs (or ISGs). Potential stimuli for such ISGs include the induction of IFN early during infection when NSP1 synthesis has yet to begin or is nascent, and the early secretion of IFNs from intestinal hematopoietic cells that can induce ISGs before STAT1 functions are inhibited. Although not much attention has been focused on the effector aspect of the antiviral response to RV, a notable recent finding is the inhibition of the IFN-inducible RNAse L pathway by the RV VP3 protein (Zhang et al., 2013; Ogden et al., 2015). The group A RV VP3 protein, which is the capping enzyme of the polymerase complex, contains a C-terminal 2H-phosphodiesterase domain that can cleave 2′-5′-oligoadenylates, preventing the activation of RNAse L. Interestingly, we previously observed that efficient enteric replication of RVs in the mouse requires a constellation of murine RV genes including VP3. Whether the replication disadvantage associated with a heterologous VP3 is due to differences in inhibition of RNAse L is not certain, and a subject of ongoing studies in the laboratory using RNAse L deficient mice.

The regulatory crosstalk between rotaviruses and the host innate immune response is complex, and involves strain- and cell-specific differences that necessitate studies at the level of single cells in vivo in order to be fully understood. Interestingly, our recent (unpublished) findings suggest that at a single cell level, the activation of pY701-STAT1 in (a fraction of) SB1A infected cells is accompanied by the simultaneous activation of pY705-STAT3, which in some contexts is known to counteract the antiviral effector functions of STAT1. Current studies in our laboratory also indicate that RVs efficiently deplete expression of surface receptors for multiple IFN subtypes from the host cell, and this inhibition can be observed by expressing the NSP1 protein alone. The depletion of IFN receptors by NSP1 during infection offers a potential mechanism to explain the similar replication phenotype of homologous RV in WT- and IFN-receptor deficient mice. These findings also provide a plausible explanation for the lack of IFN secretion and amplification observed during infection with RVs that apparently activate IRF3 and/or STAT1 (AS and HBG, unpublished data). A working model based on these observations is that, depending on the cell type considered, specific RV strains directly trigger activation of distinct TFs (IRF3, STAT3, and STAT1) in the absence of significant IFN induction. Such RV strains (in the relevant cell type) are able to potently exert inhibition of NF-kB and exogenous IFN-mediated STAT1 signaling both in infected and bystander cells (exogenous IFNs likely originating from both epithelial and immune cells in the intestine), thus allowing IFN-resistant RV replication and spread later in infection. In contrast, or depending on the cell type in which infection occurs in vitro, RVs that avoid activating IRF3 and STAT1, and degrade IRF3 effectively, seem to lack the ability to suppress IFN signaling in adjacent uninfected cells. Never the less, it seems likely that such viruses (bovine UK or simian RRV for example), as is the case with murine RVs, must be able to suppress IFN responsiveness in bystander cells in vivo during infection of their homologous hosts. The mechanisms regulating host range restriction in species other than mice need further characterization and study. Unpublished studies in our laboratory using comparative proteomics indicate that several cullins and the E3 ligase Rbx1 (Really interesting gene box protein) interact differentially with NSP1s from individual rotavirus strains. Inhibition of either cullin-3 or Rbx1 appears to disrupt NSP1-directed β-TrCP degradation. These findings strongly suggest that NSP1 uses cellular E3 ligases to interfere with the NF-κB signaling pathway and identify the mechanism by which this takes place.

ACKNOWLEDGMENTS

AS and HGB were supported by a VA Merit Review Award GRH0022 and NIH/NIAID grants RO1 AI021362, U19 AI057229, UO1 AI115715, and U19 AI090019.

REFERENCES

Angel, J., Franco, M.A., Greenberg, H.B., 2012. Rotavirus immune responses and correlates of protection. Curr. Opin. Virol. 2 (4), 419–425.

Arnold, M.M., Patton, J.T., 2011. Diversity of interferon antagonist activities mediated by NSP1 proteins of different rotavirus strains. J. Virol. 85 (5), 1970–1979.

Arnold, M.M., Sen, A., Greenberg, H.B., Patton, J.T., 2013a. The battle between rotavirus and its host for control of the interferon signaling pathway. PLoS Pathog. 9 (1), e1003064.

Arnold, M.M., Barro, M., Patton, J.T., 2013b. Rotavirus NSP1 mediates degradation of interferon regulatory factors through targeting of the dimerization domain. J. Virol. 87 (17), 9813–9821.

Bagchi, P., Dutta, D., Chattopadhyay, S., et al., 2010. Rotavirus nonstructural protein 1 suppresses virus-induced cellular apoptosis to facilitate viral growth by activating the cell survival pathways during early stages of infection. J. Virol. 84 (13), 6834–6845.

Bagchi, P., Bhowmick, R., Nandi, S., Kant Nayak, M, 2013. Chawla-Sarkar M. Rotavirus NSP1 inhibits interferon induced noncanonical NFkappaB activation by interacting with TNF receptor associated factor 2. Virology 444 (1-2), 41–44.

Barro M., Patton J.T., 2005. Rotavirus nonstructural protein 1 subverts innate immune response by inducing degradation of IFN regulatory factor 3. Proc. Natl. Acad. Sci. USA; 102(11): 4114-4119.

Barro, M., Patton, J.T., 2007. Rotavirus NSP1 inhibits expression of type I interferon by antagonizing the function of interferon regulatory factors IRF3, IRF5, and IRF7. J. Virol. 81 (9), 4473–4481.

Bass, D.M., Baylor, M., Broome, R., Greenberg, H.B., 1992. Molecular basis of age-dependent gastric inactivation of rhesus rotavirus in the mouse. J. Clin. Investig. 89 (6), 1741–1745.

Broquet, A.H., Hirata, Y., McAllister, C.S., Kagnoff, M.F., 2011. RIG-I/MDA5/MAVS are required to signal a protective IFN response in rotavirus-infected intestinal epithelium. J. Immunol. 186 (3), 1618–1626.

Burns, J.W., Krishnaney, A.A., Vo, P.T., Rouse, R.V., Anderson, L.J., Greenberg, H.B., 1995. Analyses of homologous rotavirus infection in the mouse model. Virology 207 (1), 143–153.

De Boissieu, D., Lebon, P., Badoual, J., Bompard, Y., Dupont, C., 1993. Rotavirus induces alpha-interferon release in children with gastroenteritis. J. Pediatric Gastroenterol. Nutr. 16 (1), 29–32.

Deal, E.M., Jaimes, M.C., Crawford, S.E., Estes, M.K., Greenberg, H.B., 2010. Rotavirus structural proteins and dsRNA are required for the human primary plasmacytoid dendritic cell IFNalpha response. PLoS Pathog. 6 (6), e1000931.

Deal, E.M., Lahl, K., Narvaez, C.F., Butcher, E.C., Greenberg, H.B., 2013. Plasmacytoid dendritic cells promote rotavirus-induced human and murine B cell responses. J. Clin. Investig. 123 (6), 2464–2474.

Di Fiore, I.J., Pane, J.A., Holloway, G., Coulson, B.S., 2015. NSP1 of human rotaviruses commonly inhibits NF-kappaB signaling by inducing beta-TrCP degradation. J. Gen. Virol. 96, 1768–1776.

Douagi, I., McInerney, G.M., Hidmark, A.S., et al., 2007. Role of interferon regulatory factor 3 in type I interferon responses in rotavirus-infected dendritic cells and fibroblasts. J. Virol. 81 (6), 2758–2768.

Feng, N., Kim, B., Fenaux, M., et al., 2008. Role of interferon in homologous and heterologous rotavirus infection in the intestines and extraintestinal organs of suckling mice. J. Virol. 82 (15), 7578–7590.

Feng, N., Sen, A., Nguyen, H., et al., 2009. Variation in antagonism of the interferon response to rotavirus NSP1 results in differential infectivity in mouse embryonic fibroblasts. J. Virol. 83 (14), 6987–6994.

Feng, N., Yasukawa, L.L., Sen, A., Greenberg, H.B., 2013. Permissive replication of homologous murine rotavirus in the mouse intestine is primarily regulated by VP4 and NSP1. J. Virol. 87 (15), 8307–8316.

Franco, M.A., Greenberg, H.B., 1999. Immunity to rotavirus infection in mice. J. Infect. Dis. 179 (Suppl. 3), S466–S469.

Franco, M.A., Feng, N., Greenberg, H.B., 1996a. Molecular determinants of immunity and pathogenicity of rotavirus infection in the mouse model. J. Infect. Dis. 174 (Suppl. 1), S47–S50.

Franco, M.A., Feng, N., Greenberg, H.B., 1996b. Rotavirus immunity in the mouse. Archiv. Virol. 12 (Suppl.), 141–152.

Frias, A.H., Vijay-Kumar, M., Gentsch, J.R., et al., 2010. Intestinal epithelia activate anti-viral signaling via intracellular sensing of rotavirus structural components. Mucosal Immunol. 3 (6), 622–632.

Frias, A.H., Jones, R.M., Fifadara, N.H., Vijay-Kumar, M., Gewirtz, A.T., 2012. Rotavirus-induced IFN-beta promotes anti-viral signaling and apoptosis that modulate viral replication in intestinal epithelial cells. Innate Immun. 18 (2), 294–306.

Ge, Y., Mansell, A., Ussher, J.E., et al., 2013. Rotavirus NSP4 Triggers Secretion of Proinflammatory Cytokines from Macrophages via Toll-Like Receptor 2. J. Virol. 87 (20), 11160–11167.

Gonzalez, A.M., Azevedo, M.S., Jung, K., Vlasova, A., Zhang, W., Saif, L.J., 2010. Innate immune responses to human rotavirus in the neonatal gnotobiotic piglet disease model. Immunology 131 (2), 242–256.

Graff, J.W., Mitzel, D.N., Weisend, C.M., Flenniken, M.L., Hardy, M.E., 2002. Interferon regulatory factor 3 is a cellular partner of rotavirus NSP1. J. Virol. 76 (18), 9545–9550.

Graff, J.W., Ewen, J., Ettayebi, K., Hardy, M.E., 2007. Zinc-binding domain of rotavirus NSP1 is required for proteasome-dependent degradation of IRF3 and autoregulatory NSP1 stability. J. Gen. Virol. 88 (Pt 2), 613–620.

Graff, J.W., Ettayebi, K., Hardy, M.E., 2009. Rotavirus NSP1 inhibits NFkappaB activation by inducing proteasome-dependent degradation of beta-TrCP: a novel mechanism of IFN antagonism. PLoS Pathog. 5 (1), e1000280.

Graham, D.Y., Dufour, G.R., Estes, M.K., 1987. Minimal infective dose of rotavirus. Archiv. Virol. 92 (3-4), 261–271.

Greenberg, H.B., Estes, M.K., 2009. Rotaviruses: from pathogenesis to vaccination. Gastroenterology 136 (6), 1939–1951.

Holloway, G., Truong, T.T., Coulson, B.S., 2009. Rotavirus antagonizes cellular antiviral responses by inhibiting the nuclear accumulation of STAT1, STAT2, and NF-kappaB. J. Virol. 83 (10), 4942–4951.

Holloway, G., Dang, V.T., Jans, D.A., Coulson, B.S., 2014. Rotavirus inhibits IFN-induced STAT nuclear translocation by a mechanism that acts after STAT binding to importin-alpha. J. Gen. Virol. 95 (Pt 8), 1723–1733.

Kanneganti, T.D., Body-Malapel, M., Amer, A., et al., 2006. Critical role for Cryopyrin/Nalp3 in activation of caspase-1 in response to viral infection and double-stranded RNA. J. Biol. Chem. 281 (48), 36560–36568.

La Bonnardiere, C., Laude, H., 1983. Interferon induction in rotavirus and coronavirus infections: a review of recent results. A Annals Vet. Res. 14 (4), 507–511.

LaMonica, R., Kocer, S.S., Nazarova, J., et al., 2001. VP4 differentially regulates TRAF2 signaling, disengaging JNK activation while directing NF-kappa B to effect rotavirus-specific cellular responses. J. Biol. Chem. 276 (23), 19889–19896.

Lin, J., Feng, N., Sen, A., Balan, M., Tseng, H., McElrath, C., Smirnov, S., Peng, J., Yasukawa, L.L., Durbin, R.K., Durbin, J.E., Greenberg, H.B., and Kotenko, S.V., 2016. Distinct roles of type I and type III interferons in intestinal immunity to homologous and heterologous rotavirus infections. Plos Pathog. 12 (4), e1005600.

Liu, F., Li, G., Wen, K., et al., 2010. Porcine small intestinal epithelial cell line (IPEC-J2) of rotavirus infection as a new model for the study of innate immune responses to rotaviruses and probiotics. Viral Immunol. 23 (2), 135–149.

Liu, S., Chen, J., Cai, X., et al., 2013. MAVS recruits multiple ubiquitin E3 ligases to activate antiviral signaling cascades. eLife 2, e00785.

Lopez-Guerrero, D.V., Meza-Perez, S., Ramirez-Pliego, O., et al., 2010. Rotavirus infection activates dendritic cells from Peyer's patches in adult mice. J. Virol. 84 (4), 1856–1866.

McKimm-Breschkin, J.L., Holmes, I.H., 1982. Conditions required for induction of interferon by rotaviruses and for their sensitivity to its action. Infect. Immun. 36 (3), 857–863.

Medzhitov, R., Janeway, Jr., C.A., 1997. Innate immunity: the virtues of a nonclonal system of recognition. Cell 91 (3), 295–298.

Morelli, M., Dennis, A.F., Patton, J.T., 2015. Putative E3 ubiquitin ligase of human rotavirus inhibits NF-kappaB activation by using molecular mimicry to target beta-TrCP. mBio 6 (1), e02490–e02514.

Muraille, E., 2013. Redefining the immune system as a social interface for cooperative processes. PLoS Pathog. 9 (3), e1003203.

Nandi, S., Chanda, S., Bagchi, P., Nayak, M.K., Bhowmick, R., Chawla-Sarkar, M., 2014. MAVS protein is attenuated by rotavirus nonstructural protein 1. PloS One 9 (3), e92126.

Ogden, K.M., Hu, L., Jha, B.K., Sankaran, B., Weiss, S.R., Silverman, R.H., Patton, J.T., Prasad, B.V., 2015. Structural basis for 2′-5′-oligoadenylate binding and enzyme activity of a viral RNase L antagonist. J. Virol. 89 (13), 6633–6645.

Pane, J.A., Webster, N.L., Coulson, B.S., 2014. Rotavirus activates lymphocytes from nonobese diabetic mice by triggering toll-like receptor 7 signaling and interferon production in plasmacytoid dendritic cells. PLoS Pathog. 10 (3), e1003998.

Pott J., Mahlakoiv T., Mordstein M., et al., 2011. IFN-lambda determines the intestinal epithelial antiviral host defense. Proc. Natl. Acad. Sci. USA; 108(19), 7944–7949.

Pott, J., Stockinger, S., Torow, N., et al., 2012. Age-dependent TLR3 expression of the intestinal epithelium contributes to rotavirus susceptibility. PLoS Pathog. 8 (5), e1002670.

Qin, L., Ren, L., Zhou, Z., et al., 2011. Rotavirus nonstructural protein 1 antagonizes innate immune response by interacting with retinoic acid inducible gene I. Virol. J. 8, 526.

Rollo, E.E., Kumar, K.P., Reich, N.C., et al., 1999. The epithelial cell response to rotavirus infection. J. Immunol. 163 (8), 4442–4452.

Rose, J., Franco, M., Greenberg, H., 1998. The immunology of rotavirus infection in the mouse. Adv. Virus Res. 51, 203–235.

Sato, A., Iizuka, M., Nakagomi, O., et al., 2006. Rotavirus double-stranded RNA induces apoptosis and diminishes wound repair in rat intestinal epithelial cells. J. Gastroenterol. Hepatol. 21 (3), 521–530.

Schmid, S., Mordstein, M., Kochs, G., Garcia-Sastre, A., Tenoever, B.R., 2010. Transcription factor redundancy ensures induction of the antiviral state. J. Biol. Chem. 285 (53), 42013–42022.

Sen, A., Feng, N., Ettayebi, K., Hardy, M.E., Greenberg, H.B., 2009. IRF3 inhibition by rotavirus NSP1 is host cell and virus strain dependent but independent of NSP1 proteasomal degradation. J. Virol. 83 (20), 10322–10335.

Sen, A., Pruijssers, A.J., Dermody, T.S., Garcia-Sastre, A., Greenberg, H.B., 2011. The early interferon response to rotavirus is regulated by PKR and depends on MAVS/IPS-1, RIG-I, MDA-5, and IRF3. J. Virol. 85 (8), 3717–3732.

Sen A., Rothenberg M.E., Mukherjee G., et al., 2012. Innate immune response to homologous rotavirus infection in the small intestinal villous epithelium at single-cell resolution. Proc. Natl. Acad. Sci. USA 2012; 109 (50): 20667–20672.

Sen, A., Rott, L., Phan, N., Mukherjee, G., Greenberg, H.B., 2014. Rotavirus NSP1 protein inhibits interferon-mediated STAT1 activation. J. Virol. 88 (1), 41–53.

Takaoka, A., Mitani, Y., Suemori, H., et al., 2000. Cross talk between interferon-gamma and -alpha/beta signaling components in caveolar membrane domains. Science 288 (5475), 2357–2360.

Uzri, D., Greenberg, H.B., 2013. Characterization of rotavirus RNAs that activate innate immune signaling through the RIG-I-like receptors. PloS One 8 (7), e69825.

Wang, J., Basagoudanavar, S.H., Wang, X., et al., 2010. NF-kappa B RelA subunit is crucial for early IFN-beta expression and resistance to RNA virus replication. J. Immunol. 185 (3), 1720–1729.

Zhang R., Jha B.K., Ogden K.M., Dong B., Zhao L., Elliott R., Patton J.T., Silverman R.H., Weiss S.R., 2013. Homologous 2′,5′-phosphodiesterases from disparate RNA viruses antagonize antiviral innate immunity. Proc. Natl. Acad. Sci. USA;110(32):13114–13119.

Chapter 2.9

Human Acquired Immunity to Rotavirus Disease and Correlates of Protection

D. Herrera, M. Parra, M.A. Franco, J. Angel
Instituto de Genética Humana, Facultad de Medicina, Pontificia Universidad Javeriana, Bogotá, Colombia

1 INTRODUCTION

Vaccines against rotavirus (RV), until recently the main etiological agent of severe acute gastroenteritis (GE) in children, have been very successful (Yen et al., 2014). However, because of their lower efficacy in developing countries, where they are most needed, they still require improvement, and this will depend, in part, on our understanding of the adaptive immune response against the virus, a main interest of this research group (Angel et al., 2014). Immunity (Angel et al., 2012) and correlates of protection (Angel et al., 2014) to RV have been recently reviewed, and in the present chapter we will focus on our recent investigations regarding the immunity and correlates of protection to RV infection.

Multiple RV infections in the first few years of life result in virtually complete protection against moderate-to-severe gastroenteritis, but sterilizing immunity is not achieved (Angel et al., 2012). Recent data indicate that the number of RV infections (either symptomatic or asymptomatic) necessary to generate this protection are greater in children of low-income settings (Gladstone et al., 2011). These children also get their first infection earlier in life, which may contribute to the delayed protection because at least neutralizing antibodies responses are age dependent (Ward et al., 2006). Of importance for the understanding of RV immunity is the observation that RV induces both a transitory mucosal intestinal IgA response (likely related to protection) and persistent serological RV-specific IgM, IgG, and IgA responses (likely related to the antigenaemia and viraemia that acompany infection in children) (Angel et al., 2012). Moreover, although both serotype-specific and heterotypic immune responses are significant components of protective immunity against RV the latter probably plays a more important role (Angel et al., 2012).

Our studies of the adaptative immune response against RV in children and adults have shown that circulating human RV-specific memory lymphocytes have two unique characteristics: (1) both B (Jaimes et al., 2004; Rojas et al., 2007) and T (Rojas et al., 2003; Parra et al., 2014a) cells which are RV-specific preferentially express intestinal homing receptors. (2) Both B (Rojas et al., 2008; Narváez et al., 2012; Herrera et al., 2014) and T (Mesa et al., 2010; Parra et al., 2014b) RV-specific cells seem to have peculiar functional attributes: while the former are enriched in cells that have innate features, the latter have a poor functional profile. These characteristics are probably related to the fact that RV predominantly replicates in the intestinal microenvironment that is tolerogenic (Lamichhane et al., 2014), making this an excellent model to study lymphocytes from this compartment.

2 ROTAVIRUS-SPECIFIC T CELLS

2.1 Rotavirus-Specific CD4 T Cells

The understanding of RV-specific CD4 T-cell responses is a key issue because most of RV-specific antibody responses in the mouse model seem to be dependent on CD4 T-cells help (Franco and Greenberg, 1997). Although RV predominantly replicates in mature enterocytes of the small intestine, it also has a systemic dissemination and consequently, both intestinally and systemically primed T cells are expected to be elicited (Franco et al., 2006). Our main objectives for the study of human RV-specific T cells have been to characterize their phenotype and function, and to develop in vitro models to understand how they develop in an intestinal context.

Many functional experiments of RV-specific T cells included ELISPOT and intracellular cytokine staining assays (Rojas et al., 2003; Mesa et al., 2010; Jaimes et al., 2002; Narváez et al., 2005). We found that healthy adults have circulating RV-specific CD4 T cells that secrete IFN-γ or IL-2, whereas cells producing IL-4, IL-13, IL-10, or IL-17 were below detection limits (Rojas et al., 2003; Mesa et al., 2010; Jaimes et al., 2002; Narváez et al., 2005). The frequencies of RV-specific CD4 T cells producing IFN-γ in these subjects are comparable to those specific for other mucosal respiratory viruses (Mesa et al., 2010). However, the majority of RV-specific CD4 T cells were IFN-γ single producers, followed by a low percentage of double IFN-γ/IL-2 producers cells (Mesa et al., 2010), suggesting that T cells found in healthy adults are probably terminally differentiated effector cells, unable to provide long-term immunity. Analysis of the RV-specific CD4 T-cell stimulation with different RV antigens demonstrated that cellular responses of healthy adults were similar for cell culture adapted RV and reassortant strains, whether live or UV inactivated, and when tested either individually or pooled in an IFN-γ ELISPOT (Kaufhold et al., 2005).

During acute and convalescence phases of RV infection intestinally primed RV-specific lymphocytes are expected to circulate in blood, on their way back to the intestine (see later). Our studies of RV-specific CD4 T cells circulating in children with acute gastroenteritis showed very low or below detection limit frequencies of circulating cells producing IFN-γ, compared to RV infection in healthy adults (Rojas et al., 2003; Mesa et al., 2010; Jaimes et al., 2002). Additional cytokine producing RV-specific CD4 cells, such as IL-2$^+$, IL-4$^+$, IL-10$^+$, IL-13$^+$, or IL-17$^+$, were below detection levels (Rojas et al., 2003; Mesa et al., 2010; Jaimes et al., 2002). However, other research groups found increased levels of IL-6, IL-10, IFN-γ (Jiang et al., 2003), and TNF-α (Azim et al., 1999) in serum of acutely infected children. In blood of adults with RV gastroenteritis, virus specific CD4 T cells producing IFN-γ and low frequencies of those producing IL-10 and IL-2 were identified during acute and convalescence phases, respectively (Mesa et al., 2010; Jaimes et al., 2002).

In most of the studies from this laboratory, children with RV gastroenteritis were compared with children with gastroenteritis of other etiologies and only recently we typified circulating RV-specific CD4 T cells of a small number of healthy children (Parra et al., 2014b): most cells identified produced IFN-γ or TNF-α.

Multifunctional CD4 and CD8 T cells (proliferating and secreting various cytokines) are associated with protection against different pathogens (Mahnke et al., 2013; Seder et al., 2008). Since our results showed that the frequencies of RV-specific T cells producing IFN-γ are similar to those specific for mucosal respiratory viruses (Mesa et al., 2010), it was considered that the RV-specific T-cell responses may be poor mainly in terms of quality rather than quantity. To evaluate this hypothesis we recently compared proliferation and production of IFN-γ, TNF-α, and IL-2 by RV-specific T cells to cells specific for tetanus toxoid and influenza virus circulating in healthy adults and children (Parra et al., 2014b). Whereas tetanus toxoid- and influenza virus antigen-specific CD4 T cells are enriched in cells that produce two or more cytokines, RV-specific CD4 T cells are enriched in those producing only one cytokine (IFN-γ or TNF-α), (Parra et al., 2014b) (Fig. 2.9.1A). Besides, the frequencies of CD4 T cells producing cytokines and the proliferative responses were significantly higher after stimulation with tetanus toxoid and influenza virus antigens, compared to RV. Moreover, in the context of the linear functional differentiation model for Th1 CD4 T cells proposed by Seder et al. (Mahnke et al., 2013; Seder et al., 2008), RV-specific CD4 T cells are enriched in terminal effector memory populations, whereas those specific of tetanus toxoid and influenza virus antigens are more related with effector and central memory populations (Fig. 2.9.1B). Notably, the predominance of monofunctional RV-specific T cells was also detected in healthy children, signifying that this characteristic appears after infection at an early age, lasts until adult life, and may partially explain why immunity to RV is unable to provide long lasting protection (Parra et al., 2014b).

2.2 Rotavirus-Specific CD8 T Cells

In mice, RV-specific CD8 T cells provide the first mechanism, although not the only one, to clear a primary RV infection (Franco et al., 1997). Assessment of

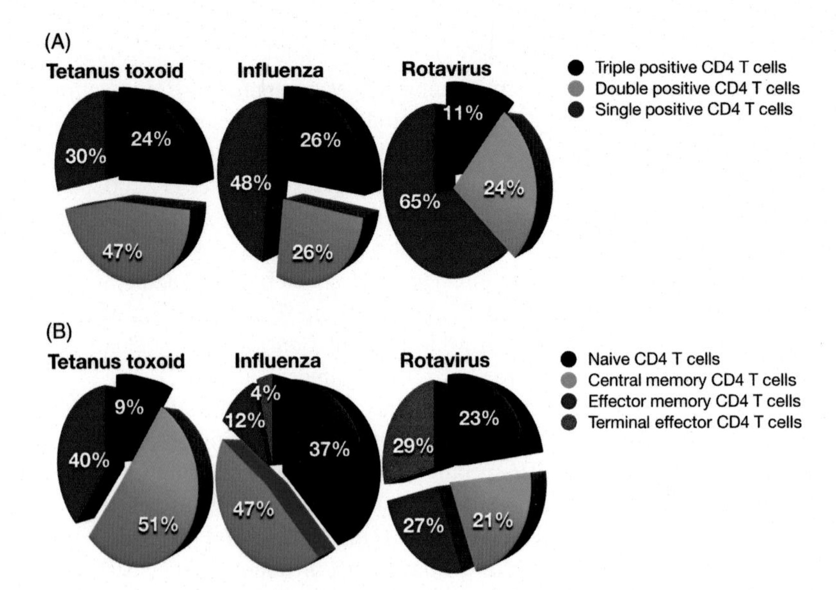

FIGURE 2.9.1 **Comparison of the cytokine and differentiation profiles of antigen specific CD4 T cells.** Fresh PBMC from healthy adults were cultured with tetanus toxoid, influenza virus antigen, and RV antigens in the presence of anti-CD28 and anti-CD49d monoclonal antibodies (MAbs) for 10 h; brefeldin A was added for the last 5 h of incubation. (A) After that, frequencies of live CD4 T cells producing one, two or three cytokines (IL-2, IFN-γ, and/or TNF-α) were evaluated by intracellular cytokine staining and flow cytometry (n = 13). (B) Frequencies of cytokine producing CD4 T cells in response to each antigen with a naïve (CCR7+ CD45RA+), central memory (CCR7+ CD45RA−), effector memory (CCR7− CD45RA−), and terminal effector memory (CCR7− CD45RA+) are represented (n = 3–5). *(Source: Data were taken from publication Parra et al., 2014b.)*

human circulating RV-specific CD8 T cells of healthy adults using ELISPOT and intracellular cytokine staining assays showed that the majority of these cells produced only IFN-γ, whereas cells producing both IFN-γ/IL-2 and those only producing IL-2 were rarely observed (Rojas et al., 2003; Mesa et al., 2010; Jaimes et al., 2002; Narváez et al., 2005). In blood of adults with RV gastroenteritis virus-specific CD8 T cells producing IFN-γ were identified during acute and convalescence phases (Mesa et al., 2010; Jaimes et al., 2002). Compared to RV-infected and healthy adults, very low or below detection limit frequencies of circulating RV-specific CD8 T cells producing IFN-γ were identified in children with RV gastroenteritis (Rojas et al., 2003; Mesa et al., 2010; Jaimes et al., 2002), and they were below detection limit in healthy children (Parra et al., 2014b).

As for CD4 T cells, both the quality and magnitude of RV-specific CD8 T-cell responses in healthy adults were significantly lower than those specific for tetanus toxoid and influenza virus antigens (Parra et al., 2014b). In addition, RV-specific CD8 T cells also seem to be related to terminally differentiated memory cells (Mahnke et al., 2013) because the majority of cells produced only IFN-γ and had low proliferation capacity (Parra et al., 2014b).

2.3 Epitopes Recognized by Rotavirus-Specific T Cells

The fine specificity of the RV-specific T-cell response, either at a protein or epitope level, has been rarely investigated. In animal models, some epitopes have been identified in mice (Jaimes et al., 2005; Banos et al., 1997) and monkeys (Zhao et al., 2008).

Responses of T cells from healthy adults to a peptide pool based on the human VP6 sequence were compared with an IFN-γ ELISPOT (Kaufhold et al., 2005). In many cases the ELISPOT response to a human VP6 peptide pool was predictive of the subject's overall response to individual RV strains. Thus, VP6 seems to be an important target of human RV-specific T cells. A RV-specific human T-cell epitope from VP7 protein (aa 40–52) restricted by HLA-DR4 (DRB1*0401) has been described, which stimulates proliferation of peripheral blood mononuclear cells (PBMC) from healthy adults and may be involved in molecular mimicry that could promote autoimmunity to pancreatic islet antigens (Honeyman et al., 2010).

We have recently identified three RV peptides restricted by HLA-DR1 (DRB1*0101) from viral proteins NSP2 (NSP2-3 SGNVIDFNLLDQRIIWQNWYA), VP3 (VP3-4 YNALIYYRYNYAFDLKRWIYL), and VP6 (VP6-7 DTIRLLFQLMRPPNMTPAVNA) that induce proliferation or production of IL-2, TNF-α, and/or IFN-γ by circulating CD4 T cells from healthy adults (Parra et al., 2014a). Of these, VP6-7 has all the characteristics of an epitope and totally overlaps with one previously found in mice (Banos et al., 1997) and partially with a VP6 epitope found in Rhesus macaques (Zhao et al., 2008), implying that this region is particularly prone to be recognized by CD4 T cells.

2.4 Markers of Intestinal Homing on Rotavirus-Specific T Cells

Since RV replication is highly restricted to enterocytes in vivo, the immune response against it originates in and exhibits its effector function directly at the intestinal mucosa. The homing of lymphocytes stimulated by antigens first encountered in intestinal Peyer's patches back to the same or other Peyer's patches or the intestinal lamina propria is mediated by interactions between the integrin α4β7 and CCR9 expressed on B and T lymphocytes and the cell adhesion molecule MadCAM 1 and CCL25, respectively, expressed on the vascular endothelium of the postcapillary venules in the intestine (Franco et al., 2006; Lee et al., 2014). Intestinal dendritic cells secreting retinoic acid imprint these receptors on T cells, which support trafficking properties based on the site-specific expression of their ligands and define a unique subset of CD4 memory T cells (Sigmundsdottir and Butcher, 2008). Initial studies of the expression of intestinal homing receptors by RV-specific T cells of healthy adults involved the

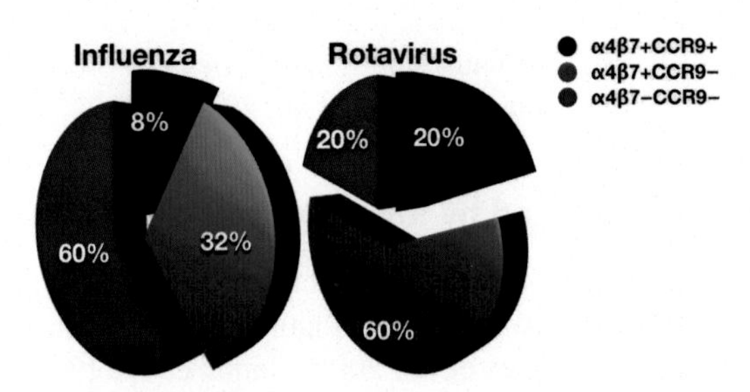

FIGURE 2.9.2 **Intestinal migration markers on circulating influenza- and RV-specific CD4 T cells.** RV- and influenza-specific CD4 T cells from eight healthy adults were identified using tetramers coupled with specific peptides and simultaneously stained with antibodies against α4β7 and CCR9. *(Source: Data were taken from publication Parra et al., 2014a.)*

proliferation and IFN-γ responses of circulating purified CD4 T cells expressing or lacking α4β7: Rott et al. (1997) found that purified $\alpha 4\beta 7^{+}$ $CD45RA^{-}$ cells proliferated more than 2.3-fold over $\alpha 4\beta 7^{-}$ $CD45RA^{-}$ CD4 T cells to RV antigen. Subsequently we found that around 80% of RV-specific CD4 T cells producing IFN-γ expressed α4β7 (Rojas et al., 2003). Only recently, with the identification of DR1 restricted RV peptides and further production of HLA class II tetramers (Parra et al., 2014a), we were able to characterize the expression of both intestinal homing receptors, α4β7 and CCR9, on unpurified circulating RV peptide-specific CD4 T cells. As shown in Fig. 2.9.2, compared to influenza virus-specific $tetramer^{+}$ CD4 T cells, RV-specific $tetramer^{+}$ antigen experienced cells expressed both α4β7 and CCR9. Moreover, such cells were also detected in blood of two RV vaccinated children (Parra et al., 2014a). These results support the notion that small intestine originated cells are, indeed, a unique subset of CD4 T cells (Mahnke et al., 2013).

2.5 Intestinal Human Rotavirus-Specific T Cells

Memory CD4 and CD8 T cells expressing intestinal homing receptor accumulate in the lamina propria, where they can reside for a long time. For this reason, it is possible that a bulk of RV-specific T cells remains in the intestine. However, studies with human intestines are difficult to perform because of the relatively high background CD4 and CD8 T cells present in an intracellular cytokine assay and only RV-specific CD8 T cells have been significantly detected (Narváez et al., 2010). Future studies with human intestinal RV-specific T cells will be important to establish whether they also have a poor functional profile similar to their circulating counterparts.

2.6 In Vitro Models to Study Human Rotavirus-Specific T Cells

We developed in vitro models to study the human systemic and intestinal RV immune response. The evaluation of systemic antigen presenting cells (APC) showed that monocyte derived dendritic cells (moDC) in contact with infectious RV are able to stimulate a strong CD4 Th1 allogeneic response (Narváez et al., 2005). Besides, circulating plasmacytoid dendritic cells are necessary to stimulate RV-specific memory T cells to produce IFN-γ (Mesa et al., 2007). Because moDC generally reflect the function of circulating peripheral blood myeloid dendritic cells, these findings predict that an important systemic Th1 response against RV should be generated under the viremic phase of infection. Since, as previously described, this is not the case, our results suggest that either the systemic antigen/virus that circulates during an acute infection with RV is presented to T cells in a tolerogenic context or that the amount of virus present systemically is unsuited for T-cell immunity induction.

The tolerogenic gut environment (Lamichhane et al., 2014) may substantially influence the T-cell response against RV. For the development of an in vitro model of human intestinal immune response against RV, we initially cultivated polarized Caco-2 cells in transwells and identified the "danger signals" released by cells infected by RV (Rodríguez et al., 2009). After infection, Caco-2 cells released IL-8, PGE_2, small quantities of TGF-β1, and the constitutive and inducible heat shock proteins HSC70 and HSP70, which are known to induce a noninflammatory (non-Th-1) immune response (Rodríguez et al., 2009). Furthermore, HSC70, HSP70, and TGF-β1 were released, in part, associated with membrane vesicles (MV) obtained from filtrated Caco-2 supernatants concentrated by ultracentrifugation (Barreto et al., 2010). These MV were heterogeneous, with characteristics of exosomes and probably also of apoptotic bodies, and had immunomodulatory functions: MV from RV-infected cells induced death and inhibited proliferation of polyclonally stimulated CD4 T cells, and these effects were in part due to TGF-β (Barreto et al., 2010).

We next studied the effect of these intestinal immunomodulators in relation to the interaction of RV with moDC (Rodriguez et al., 2012): moDC treated with supernatants from RV-infected Caco-2 cells promoted a significantly lower Th1 response, in comparison with those treated with purified RV. Moreover, TGF-β, unlike thymic stromal lymphopoietin (TSLP), was an importat mediator of this modulation, suggesting that TGF-β could be an immune evasion mechanism (Rodriguez et al., 2012). In agreement with this hypothesis, in PBMC from healthy adults the inhibition of the TGF-β signaling pathway increased the frequency of RV-specific CD4 T cells that produce IFN-γ (Mesa et al., 2010). However, this inhibition was undetected in children, suggesting that RV-specific CD4 T cells could be modulated by other tolerogenic mechanisms, such as anergy.

We used three anergy inhibitors to assess the hypothesis of the presence of circulating anergic T cells in the response against RV: after stimulation of PBMC from healthy adults with RV in the presence of IL-2—unlike IL-12 or R59949 (a pharmacological diacyl- glycerol kinase alpha inhibitor)—increased frequencies of RV-specific CD4 and CD8 T cells producing cytokines were identified (Parra et al., 2014b). This finding depicts a poor functional T-cell profile that may be partially reversed in vitro by the addition of rIL-2.

The role of regulatory T cells (Treg) in RV infection has been investigated in few reports. In RV-infected mice, the numbers of $FoxP3^+$ Treg cells are increased, but RV clearance or Abs levels are unsignificantly modified in their absence (Miller et al., 2014). In PBMC from healthy adults the depletion of $CD25^+$ T cells (probably containing Treg cells) increases the frequency of RV-specific CD4 T cells that produce IFN-γ, suggesting that, at least systemically, Treg may modulate the function of RV-specific CD4 T cells (Mesa et al., 2010).

3 ROTAVIRUS-SPECIFIC B CELLS

A review of RV-specific B cells (Bc) has been recently published (Franco and Greenberg, 2013). Key to our studies of RV-specific Bc is a flow cytometry assay that detects specific binding of a fluorescent RV antigen to a B cell that expresses RV-specific Ig (Franco et al., 2006). In vitro, RV has been shown to induce activation and differentiation of human Bc, present in PBMC, into antibody secreting cells (ASC) (Narváez et al., 2010). However, this effect was undetected with purified Bc, suggesting the participation of other cells in activating the Bc; most likely dendritic cells that produce IFN-α (Deal et al., 2010).

Unexpectedly, a significant number (approximately 1–2%) of naïve IgD^+ $CD27^-$ RV-specific Bc are detectable in cord blood (Parez et al., 2004). Although clearly specific, these cells secrete antibodies with a low affinity for VP6 (Kallewaard et al., 2008). Infants that have or lack serum RV-specific IgA (considered a hallmark of primary RV infection) have circulating RV-specific Bc that express IgM, IgD, and CD27, a phenotype compatible with memory B cells (mBc) (Rojas et al., 2007). In healthy adults, RV-specific mBc are enriched in all IgM Bc subsets, but in particular in the IgD^{low} IgM^{hi} $CD27^+$ subgroup of Bc (Narváez et al., 2012; Herrera et al., 2014) (Fig. 2.9.3). This phenotype is reminiscent of spleen marginal zone Bc, a subset that has been postulated to develop (by an unknown mechanism) a prediversified Ig repertoire and to participate in "innate" Ig responses to pathogens (Cerutti et al., 2013). Recently, in humans, it has been shown that circulating IgM mBc contains a population of cells that have characteristics of marginal zone Bc (Descatoire et al., 2014). Although the function of these cells remains unknown, in experiments in which human IgM mBc (mostly IgD +) are passively transferred to RV infected immunodeficient mice, they are able to switch to IgG ASC and mediate immunity against RV antigenemia and viremia (Narváez et al., 2012).

As for memory T cells, both RV-specific ASC and mBc predominantly expressed α4β7 and CCR9. In children with an acute RV infection, the great majority (~70%) of circulating virus specific Bc are ASC that coexpress both intestinal homing receptors (Jaimes et al., 2004). In the convalescent phase of viral infection, approximately one-third of RV-specific mBc express both homing receptors and are presumably targeted to the small intestine. Another third of RV-specific mBc only express α4β7 (presumably targeted to other parts of the intestine and other mucosal surfaces) and the final third express

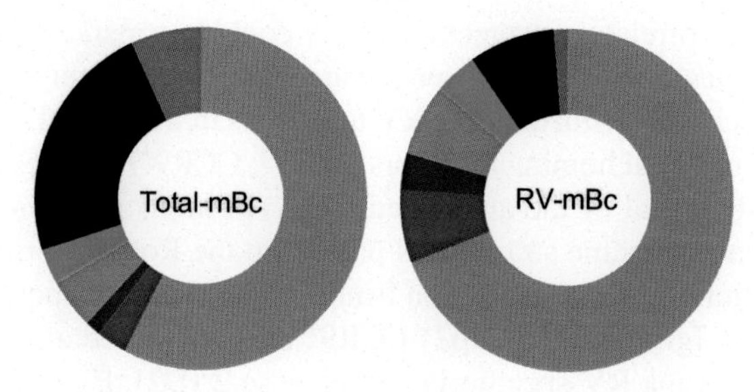

Total mBc %	mBc subset	RV-mBc %
58.1	Naïve	66.1
2.9	$IgM^{hi}IgD^{low}$	6.5*
1.5	$IgM^{low}IgD^{hi}$	3.3*
4.8	IgM^{+} only	6.4*
3.4	IgD^{+} only	3.9
23.2	Sw_{mBc}	7.9*
7.0	$CD27^{-}_{mBc}$	1.3*

FIGURE 2.9.3 **RV-specific memory B cells (mBc) are enriched in the IgM^{hi} IgD^{low}, IgM^{low} IgD^{hi}, and IgM^{+} only mBc subsets.** Summary of the median frequencies of seven subsets of total and RV-specific mBc assessed by multiparametric flow cytometry from 10 healthy adults. Donuts denote the median frequencies of $CD19^{+}$ B cells (%) subsets. * Statistically significant differences ($p < 0.05$, two-tailed Wilcoxon test) between total and RV-specific mBc subsets. *(Source: Data were taken from publication Herrera et al., 2014.)*

neither receptor (presumably targeted to the spleen and other systemic organs) (Jaimes et al., 2004; Gonzalez et al., 2003). Notably, like for RV-specific T cells described earlier, IgD^{-} mBc that express these two markers are enriched compared to the total nonantigen specific lymphocytes [10 times for T cells (Parra et al., 2014a) and approximately 1.5 times for IgD^{-} mBc (Jaimes et al., 2004)].

4 CORRELATES OF PROTECTION

Correlates of protection for RV vaccines have been recently reviewed (Angel et al., 2014), and we focus here only on our most recent studies on this subject (Rojas et al., 2007; Herrera et al., 2013). Serum RV-specific IgA transiently reflect the intestinal RV-specific IgA and are, so far, the best practically measured correlate of protection against RV gastroenteritis in humans (Angel et al., 2012). Nevertheless, they are unsuitable to predict individual protection and therefore can be only used as an epidemiological tool at the population level (Cheuvart et al., 2014). The use of a validated correlate of protection as a surrogate endpoint after vaccination would contribute to the faster development of a new generation of RV vaccines and the assessment of its efficacy in a wider number of settings (Angel et al., 2014).

An adequate correlate of protection for RV could be a marker able to precisely reflect the intestinal immune response induction. Consistent with this is the finding that in children with an acute RV infection circulating IgD^- RV-specific mBc express intestinal homing receptors ($\alpha4\beta7^+$, $CCR9^+$) (Jaimes et al., 2004). In a double blind trial of the attenuated RIX4414 human RV vaccine (which contains the same vaccine strain virus present in the Rotarix formulation), we found correlations between protection from disease and frequencies of circulating RV-specific IgD^- $CD27^+$ $\alpha4\beta7^+$ $CCR9^+$ mBc measured after dose 1 (D1) and levels of plasma RV-specific IgA after dose 2 (D2). However, other factors may be relevant for conferring protection from disease because correlation coefficients for both tests were low and protection against disease was significantly higher in vaccinees that lacked RV-specific IgA (titer $< 1:100$) compared to placebo recipients that also lacked these antibodies (Rojas et al., 2007).

Secretory Ig (SIg) in serum has been proposed as another method for indirectly measuring intestinal Ig (Grauballe et al., 1981). The polymeric immunoglobulin receptor on the basolateral membrane of the mucosal epithelial cells captures polymeric IgA and IgM (Mantis et al., 2011). This complex is endocytosed and transported to the apical membrane, where it is cleaved and part of it (the secretory component [SC]) remains attached to the Ig (IgA or IgM), forming SIg, which may be retro-transported across epithelial cells and ultimatelly enter the circulation (Mantis et al., 2011).

Given that RV-specific SIgs have been detected in serum of children with acute RV infection (Grauballe et al., 1981; Hjelt et al., 1985) and correlated with the amounts detected in duodenal fluid 1 week after the infection (Hjelt et al., 1986), we sought to confirm the presence of plasma RV-specific SIg in children with natural RV infection and to determine if circulating RV-specific SIg could more precisely reflect the intestinal protective immune response induced by the RIX4414 RV vaccine, and be a better correlate of protection than circulating RV-specific IgA after vaccination. Plasma samples collected from children with natural RV infection and from children who had received two doses of the RIX4414 RV vaccine or placebo were assessed by an in-house ELISA designed for meassuring plasma RV-specific SIg (Fig. 2.9.4). As shown in Figure 2.9.4, for this ELISA an antihuman SC monoclonal antibody is used as a capture antibody and thus, the assay does not discriminate between RV-specific SIgA and SIgM (Herrera et al., 2013).

We detected plasma RV-specific SIg in naturally infected children with RV infection and in the great majority of children with evidence of previous RV infection without an ongoing RV gastroenteritis. Children vaccinated with the attenuated RIX4414 human RV vaccine presented higher RV-specific SIg titers than placebo recipients. Furthermore, plasma RV-specific SIg seemed to be a better correlate of protection than plasma RV-specific IgA because, in contrast to RV-specific IgA (Fig. 2.9.5A), protection rates increased as a function of RV-specific SIg titers (Fig. 2.9.5B). Additionally, RV-specific SIg measured after D2 also correlated with protection in vaccinees and placebo recipients analyzed

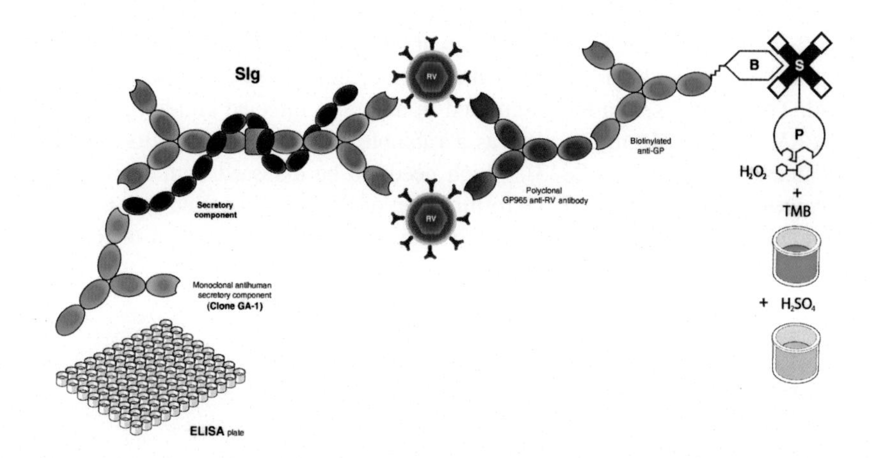

FIGURE 2.9.4 Schematic representation of the ELISA for measuring plasma RV-specific secretory immunoglobulin. ELISA plates are coated with an antihuman SC mAb, and plasma samples are deposited in each well. After incubation, dilutions of a supernatant from RF (bovine RV) virus-infected MA104 cells or the supernatant of mock-infected cells (negative control) are included. Then, the following sequence of reagents is added: guinea pig antirhesus RV hyperimmune serum, biotinylated goat anti-guinea pig serum, peroxidase-labeled streptavidin, and tetramethyl benzidine substrate. The reaction is stopped by the addition of sulfuric acid, and absorbance is read on an ELISA plate reader (at 650 nm wavelength).

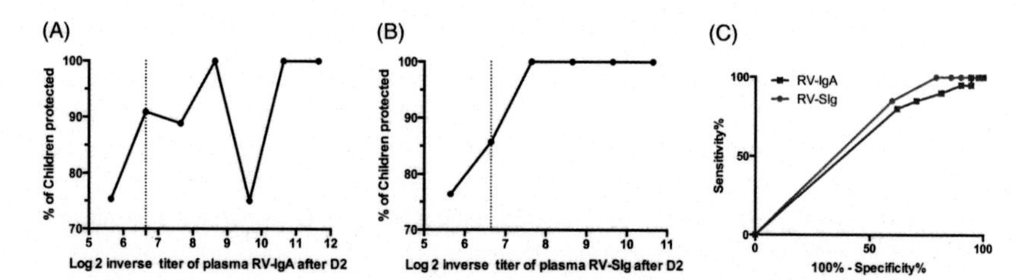

FIGURE 2.9.5 Correlation between RV-specific IgA and SIg titers after D2 and protection against any RV gastroenteritis for vaccinees and placebo recipients analyzed jointly. (A) and (B) The reported values correspond to the $\log_2$ of the inverse titer measured by ELISA. Vertical dashed lines represent the inverse of 1:100 titer, the lowest sample dilution for which the positivity criteria optical is considered valid: density in the experimental wells > 0.1 optical density units and twofold greater than the optical density in the corresponding negative control wells. (C) Area under receiver operating characteristic curves for RV-specific IgA (AUROC = 0.59, CI 95%: 0.46–0.72, $p = 0.2$) and RV-specific SIg (AUROC = 0.64, CI 95%: 0.52–0.75, $p = 0.0478$) after D2. *(Source: Data were taken from vaccinated and placebo recipients described in publications Rojas et al., 2007; Herrera et al., 2013.)*

jointly and, unlike plasma RV-specific IgA, the frequency of protected children was significantly higher in children having RV-specific SIg (titers ≥ 1:100) than in those lacking these antibodies (titer < 1:100). Finally, the presence of RV-specific SIg conferred an almost 4 times increase in the probability of being protected against any RV gastroenteritis. However, analysis of ROC (receiver operating characteristic) curves for RV-specific IgA and RV-specific SIg (Fig. 2.9.5C) showed that both assays are inefficient at discriminating children that will or will not be protected by the vaccine, and more work is needed in the field. These results, and those reported in naturally infected children, indicate that RV-specific SIg might serve as a valuable correlate of protection for RV vaccines (Herrera et al., 2013), which needs to be assessed with current RV vaccines, especially because the formulation of RIX4414 we used was less immunogenic, although equally protective, than the final formulation of Rotarix (Rojas et al., 2007).

ACKNOWLEDGMENTS

This work was supported by funds from the Pontificia Universidad Javeriana.

REFERENCES

Angel, J., Franco, M.A., Greenberg, H.B., 2012. Rotavirus immune responses and correlates of protection. Curr. Opin. Virol. 2 (4), 419–425.

Angel, J., Steele, A.D., Franco, M.A., 2014. Correlates of protection for rotavirus vaccines: Possible alternative trial endpoints, opportunities, and challenges. Hum. Vaccin. Immunother. 10 (12), 3659–3671.

Azim, T., Ahmad, S.M., Sefat, E.K., et al., 1999. Immune response of children who develop persistent diarrhea following rotavirus infection. Clin. Diagn. Lab. Immunol. 6 (5), 690–695.

Banos, D.M., Lopez, S., Arias, C.F., Esquivel, F.R., 1997. Identification of a T-helper cell epitope on the rotavirus VP6 protein. J. Virol. 71 (1), 419–426.

Barreto, A., Rodriguez, L.-S., Lucia Rojas, O., et al., 2010. Membrane vesicles released by intestinal epithelial cells infected with rotavirus inhibit T-cell function. Viral Immunol. 23 (6), 595–608.

Cerutti, A., Cols, M., Puga, I., 2013. Marginal zone B cells: virtues of innate-like antibody-producing lymphocytes. Nat. Rev. Immunol. 13 (2), 118–132.

Cheuvart, B., Neuzil, K.M., Steele, A.D., et al., 2014. Association of serum anti-rotavirus immunoglobulin A antibody seropositivity and protection against severe rotavirus gastroenteritis: analysis of clinical trials of human rotavirus vaccine. Hum. Vaccin. Immunother. 10 (2), 505–511.

Deal, E.M., Jaimes, M.C., Crawford, S.E., Estes, M.K., Greenberg, H.B., 2010. Rotavirus structural proteins and dsRNA are required for the human primary plasmacytoid dendritic cell IFNalpha response. PLoS Pathog. **6** (6), e1000931.

Descatoire, M., Weller, S., Irtan, S., et al., 2014. Identification of a human splenic marginal zone B cell precursor with NOTCH2-dependent differentiation properties. J. Exp. Med. 211 (5), 987–1000.

Franco, M.A., Greenberg, H.B., 1997. Immunity to rotavirus in T cell deficient mice. Virology 238 (2), 169–179.

Franco, M., Greenberg, H., 2013. Rotavirus. Microbiol. Spectrum 1 (2), 1–10.

Franco, M.A., Tin, C., Greenberg, H.B., 1997. CD8+ T cells can mediate almost complete short-term and partial long-term immunity to rotavirus in mice. J. Virol. 71 (5), 4165–4170.

Franco, M.A., Angel, J., Greenberg, H.B., 2006. Immunity and correlates of protection for rotavirus vaccines. Vaccine 24 (15), 2718–2731.

Gladstone, B.P., Ramani, S., Mukhopadhya, I., et al., 2011. Protective effect of natural rotavirus infection in an Indian birth cohort. N. Engl. J. Med. 365 (4), 337–346.

Gonzalez, A.M., Jaimes, M.C., Cajiao, I., et al., 2003. Rotavirus-specific B cells induced by recent infection in adults and children predominantly express the intestinal homing receptor α4β7. Virology 305 (1), 93–105.

Grauballe, P.C., Hjelt, K., Krasilnikoff, P.A., Schiotz, P.O., 1981. ELISA for rotavirus-specific secretory IgA in human sera. Lancet 2 (8246), 588–589.

Herrera, D., Vásquez, C., Corthésy, B., Franco, M.A., Angel, J., 2013. Rotavirus specific plasma secretory immunoglobulin in children with acute gastroenteritis and children vaccinated with an attenuated human rotavirus vaccine. Hum. Vac. Immunotherap. 9 (11), 2409–2417.

Herrera, D., Rojas, O.L., Duarte-Rey, C., Mantilla, R.D., Ángel, J., Franco, M.A., 2014. Simultaneous assessment of rotavirus-specific memory B cells and serological memory after B cell depletion therapy with rituximab. PLoS One 9 (5), e97087.

Hjelt, K., Grauballe, P.C., Schiotz, P.O., Andersen, L., Krasilnikoff, P.A., 1985. Intestinal and serum immune response to a naturally acquired rotavirus gastroenteritis in children. J. Pediatr. Gastroenterol. Nutr. 4 (1), 60–66.

Hjelt, K., Grauballe, P.C., Andersen, L., Schiotz, P.O., Howitz, P., Krasilnikoff, P.A., 1986. Antibody response in serum and intestine in children up to six months after a naturally acquired rotavirus gastroenteritis. J. Pediatr. Gastroenterol. Nutr. 5 (1), 74–80.

Honeyman, M.C., Stone, N.L., Falk, B.A., Nepom, G., Harrison, L.C., 2010. Evidence for molecular mimicry between human T cell epitopes in rotavirus and pancreatic islet autoantigens. J. Immunol. 184 (4), 2204–2210.

Jaimes, M.C., Rojas, O.L., Gonzalez, A.M., et al., 2002. Frequencies of virus-specific CD4(+) and CD8(+) T lymphocytes secreting gamma interferon after acute natural rotavirus infection in children and adults. J. Virol. 76 (10), 4741–4749.

Jaimes, M.C., Rojas, O.L., Kunkel, E.J., et al., 2004. Maturation and trafficking markers on rotavirus-specific B cells during acute infection and convalescence in children. J. Virol. 78 (20), 10967–10976.

Jaimes, M.C., Feng, N., Greenberg, H.B., 2005. Characterization of homologous and heterologous rotavirus-specific T-cell responses in infant and adult mice. J. Virol. 79 (8), 4568–4579.

Jiang, B., Snipes-Magaldi, L., Dennehy, P., et al., 2003. Cytokines as mediators for or effectors against rotavirus disease in children. Clin. Diagn. Lab. Immunol. 10 (6), 995–1001.

Kallewaard, N.L., McKinney, B.A., Gu, Y., Chen, A., Prasad, B.V., Crowe, Jr., J.E., 2008. Functional maturation of the human antibody response to rotavirus. J. Immunol. 180 (6), 3980–3989.

Kaufhold, R.M., Field, J.A., Caulfield, M.J., et al., 2005. Memory T-cell response to rotavirus detected with a gamma interferon enzyme-linked immunospot assay. J. Virol. 79 (9), 5684–5694.

Lamichhane, A., Azegamia, T., Kiyonoa, H., 2014. The mucosal immune system for vaccine development. Vaccine 32 (49), 6711–6723.

Lee, M., Kiefel, H., LaJevic, M.D., et al., 2014. Transcriptional programs of lymphoid tissue capillary and high endothelium reveal control mechanisms for lymphocyte homing. Nat. Immunol. 15 (10), 982–995.

Mahnke, Y.D., Brodie, T.M., Sallusto, F., Roederer, M., Lugli, E., 2013. The who's who of T-cell differentiation: human memory T-cell subsets. Eur. J. Immunol. 43 (11), 2797–2809.

Mantis, N.J., Rol, N., Corthesy, B., 2011. Secretory IgA's complex roles in immunity and mucosal homeostasis in the gut. Mucosal Immunol. 4 (6), 603–611.

Mesa, M.C., Rodríguez, L.-S., Franco, M.A., Angel, J., 2007. Interaction of rotavirus with human peripheral blood mononuclear cells: Plasmacytoid dendritic cells play a role in stimulating memory rotavirus specific T cells in vitro. Virology 366 (1), 174–184.

Mesa, M.C., Gutiérrez, L., Duarte-Rey, C., Angel, J., Franco, M.A., 2010. A TGF-β mediated regulatory mechanism modulates the T cell immune response to rotavirus in adults but not in children. Virology 399 (1), 77–86.

Miller, A.D., Blutt, S.E., Conner, M.E., 2014. FoxP3+ regulatory T cells are not important for rotavirus clearance or the early antibody response to rotavirus. Microbes Infect. 16 (1), 67–72.

Narváez, C.F., Angel, J., Franco, M.A., 2005. Interaction of rotavirus with human myeloid dendritic cells. J. Virol. 79 (23), 14526–14535.

Narváez, C.F., Franco, M.A., Angel, J., Morton, J.M., Greenberg, H.B., 2010. Rotavirus differentially infects and polyclonally stimulates human B cells depending on their differentiation state and tissue of origin. J. Virol. 84 (9), 4543–4555.

Narváez, C.F., Feng, N., Vásquez, C., et al., 2012. Human rotavirus-specific IgM Memory B cells have differential cloning efficiencies and switch capacities and play a role in antiviral immunity in vivo. J. Virol. 86 (19), 10829–10840.

Parez, N., Garbarg-Chenon, A., Fourgeux, C., et al., 2004. The VP6 protein of rotavirus interacts with a large fraction of human naive B cells via surface immunoglobulins. J. Virol. 78 (22), 12489–12496.

Parra, M., Herrera, D., Calvo-Calle, J.M., et al., 2014a. Circulating human rotavirus specific CD4 T cells identified with a class II tetramer express the intestinal homing receptors $\alpha 4\beta 7$ and CCR9. Virology 452–453, 191–201.

Parra, M., Herrera, D., Jácome, M.F., et al., 2014b. Circulating rotavirus-specific T cells have a poor functional profile. Virology 468–470, 340–350.

Rodriguez, L.S., Narvaez, C.F., Rojas, O.L., Franco, M.A., Angel, J., 2012. Human myeloid dendritic cells treated with supernatants of rotavirus infected Caco-2 cells induce a poor Th1 response. Cell Immunol. 272 (2), 154–161.

Rodríguez, L.-S., Barreto, A., Franco, M.A., Angel, J., 2009. Immunomodulators released during rotavirus infection of polarized Caco-2 cells. Viral Immunol. 22 (3), 163–172.

Rojas, O.L., González, A.M., González, R., et al., 2003. Human rotavirus specific T cells: Quantification by ELISPOT and expression of homing receptors on CD4+ T cells. Virology 314 (2), 671–679.

Rojas, O.L., Caicedo, L., Guzmán, C., et al., 2007. Evaluation of circulating intestinally committed memory B cells in children vaccinated with attenuated human rotavirus vaccine. Viral Immunol. 20 (2), 300–311.

Rojas, O.L., Narváez, C.F., Greenberg, H.B., Angel, J., Franco, M.A., 2008. Characterization of rotavirus specific B cells and their relation with serological memory. Virology 380 (2), 234–242.

Rott, L.S., Rosé, J.R., Bass, D., Williams, M.B., Greenberg, H.B., Butcher, E.C., 1997. Expression of mucosal homing receptor alpha4beta7 by circulating CD4+ cells with memory for intestinal rotavirus. J. Clin. Invest. 100 (5), 1204–1208.

Seder, R.A., Darrah, P.A., Roederer, M., 2008. T-cell quality in memory and protection: implications for vaccine design. Nat. Rev. Immunol. 8 (4), 247–258.

Sigmundsdottir, H., Butcher, E.C., 2008. Environmental cues, dendritic cells and the programming of tissue-selective lymphocyte trafficking. Nat. Immunol. 9 (9), 981–987.

Ward, R.L., Kirkwood, C.D., Sander, D.S., et al., 2006. Reductions in cross-neutralizing antibody responses in infants after attenuation of the human rotavirus vaccine candidate 89-12. J. Infect. Dis. 194 (12), 1729–1736.

Yen, C., Tate, J.E., Hyde, T.B., et al., 2014. Rotavirus vaccines: current status and future considerations. Hum. Vaccin. Immunother. 10 (6), 1436–1448.

Zhao, W., Pahar, B., Sestak, K., 2008. Identification of rotavirus VP6-specific CD4+ T cell epitopes in a G1P[8] human rotavirus-infected rhesus macaque. Virol. Res. Treatment 1, 9–15.

Chapter 2.10

Molecular Epidemiology and Evolution of Rotaviruses

K. Bányai*, V.E. Pitzer**

**Institute for Veterinary Medical Research, Centre for Agricultural Research, Hungarian Academy of Sciences, Budapest, Hungary; **Department of Epidemiology of Microbial Diseases, Yale School of Public Health, New Haven, CT, United States*

1 CLASSIFICATION OF ROTAVIRUSES

Rotaviruses form a genus of the Reoviridae family and possess a genome of 11 segments of double-stranded (ds) RNA (Estes and Kapikian, 2007). Currently at least eight rotavirus species, designated *Rotavirus A* to *Rotavirus H* and a tentative ninth species, *Rotavirus I*, are known. *Rotavirus A* (RVA) occurs in birds and mammals, RVB, RVC, RVE, RVH, and RVI have been detected in one or more mammalian hosts, whereas RVD, RVF, and RVG have been detected only in birds (Matthijnssens et al., 2012b; Mihalov-Kovács et al., 2015).

Classification below the virus species level includes the differentiation of serotypes, defined by the surface antigens, VP7 and VP4. Using the two antigen specificities, a dual nomenclature has been introduced in order to designate rotavirus strains. Serotypes of VP7 are designated with the letter G followed by a number. The antigenic variants of VP4 are classified into P serotypes (Estes and Kapikian, 2007). Strains belonging to other RV species have not been classified into serotypes because they are difficult to culture, preventing serological cross neutralization studies.

The recognition that antigenic features of both serotypes correlate with deduced amino acid sequences allowed the introduction of a genotyping system. The dual nomenclature was extended to include genotype numbers, which in the case of VP7 were identical to those of the serotype specificity, whereas in the case of VP4 some discrepancies were seen. Thus, if both genotype and serotype specificities are known for the VP4, the P type is often labeled to show both type designations, for example, P1A[8], P2A[6], P3[9], where numbers outside and inside the bracket, respectively, denote the serotype and genotype specificities (Estes and Kapikian, 2007). The number of genotypes of the surface antigens varies across rotavirus species. For example, to date at least 27 G types and 37 P types

Viral Gastroenteritis

have been recognized within RVA, 21 G and 3 P types within RVB, and 10 G and 7 P types within RVC (Marthaler et al., 2012, 2013; Matthijnssens et al., 2011).

With the increasing availability of rotavirus sequence data, the genotyping system has been extended to all 11 genome segments permitting the empirical determination of cut-off values of genotypes within each gene (Matthijnssens et al., 2008b). In this new nomenclature the formula Gx-P[x]-Ix-Rx-Cx-Mx-Ax-Nx-Tx-Ex-Hx represents the genotypes of the VP7-VP4-VP6-VP1-VP2-VP3-NSP1-NSP2-NSP3-NSP4-NSP5-encoding gene segments, respectively, with "x' indicating the numbers of the corresponding genotypes. Within RVA, at least 27 G, 37 P, 16 I, 9 R, 9 C, 8 M, 16 A, 9 N, 12 T, 15 E, and 11 H genotypes have been identified, with many new genotypes identified in the past few years. This scheme was found helpful to describe common genotype constellations in various hosts as well as to understand the origin and evolution of unusual strains (Matthijnssens et al., 2011). Currently, a robust genotyping system is available for RVA only. Analogous systems for RVB and RVC are being developed.

2 MECHANISMS OF ROTAVIRUS EVOLUTION AND THEIR SIGNIFICANCE

Mechanisms of rotavirus evolution are driven by the interplay among viral genes, adaptation to the host species, and escaping the host immune response.

2.1 Point Mutation

Point mutations in rotavirus genomes occur due to the error prone activity of the viral RNA polymerase. The mutation rate has been estimated to range around the value of 5×10^{-5} per nucleotide, meaning that roughly one mutation arises for every new copy of the rotavirus genome synthesized (Blackhall et al., 1996). The rate of evolution due to accumulation of point mutations over time have been found to vary by genes and genotypes in molecular evolution studies, showing values in the range of 1.1×10^{-3} to 8.7×10^{-4} nucleotide substitutions per site per year (Donker and Kirkwood, 2012; Matthijnssens et al., 2010).

The majority of mutations do not cause changes in the amino acid sequence and there is evidence of negative selection in the evolution of most genes (Donker and Kirkwood, 2012; Song and Hao, 2009). Genomic regions that tend to accumulate missense mutations include those encoding key epitopes on the surface of the neutralization antigens. Such mutations have been implicated in changes to affinity to antibodies and likely contribute to the antigenic drift of field strains due to immune selection.

2.2 Reassortment

Reassortment of segmented RNA viruses is a mechanism by which cognate genome segments are exchanged in progeny viruses upon infection of a single

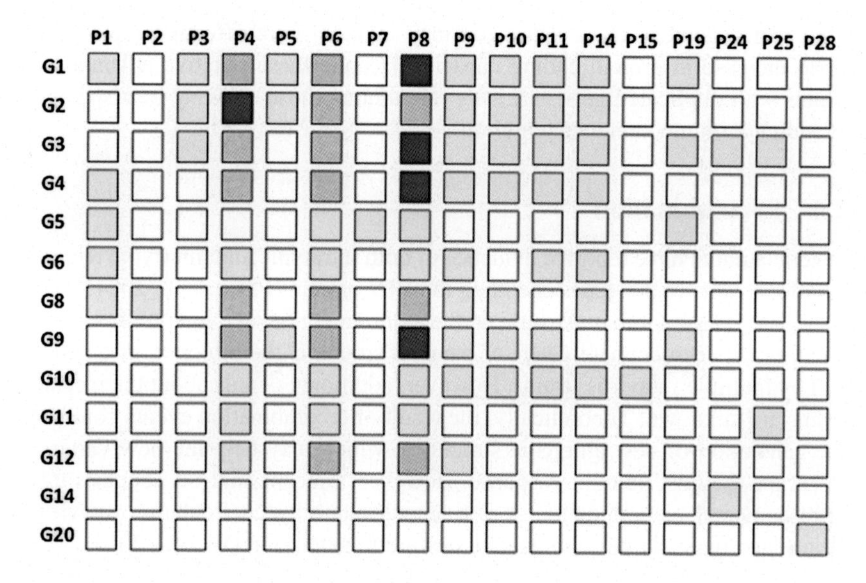

FIGURE 2.10.1 **Summary of human G/P genotype combinations found to date.** Color codes: red, globally common; orange, regionally common and unusual; yellow, rare.

cell by two or more closely related virus strains. When these viruses contain different cogent genes, new genotype constellations may emerge. Reassortment of the surface antigens could result in a large number of antigen combinations. However, while over 80 G–P antigen combinations have been identified so far, only six globally common G–P combinations occur in human, globally most common strains, even though many G and P types are expressed by cocirculating strains (Fig. 2.10.1) (Dóró et al., 2014).

The reason behind this phenomenon is unclear. One explanation operates with genetic or phenotypic incompatibility between some variants of the neutralizing antigen specificities (Iturriza-Gómara et al., 2001). An alternative scenario is that there is in vivo selection of combinations that result in masking of VP4 or VP7 neutralization epitopes. Strains hiding their neutralization epitopes after reassortment may evoke a weaker neutralizing antibody response, thus enhancing their fitness over less advantageous antigen combinations (Bányai et al., 2009).

2.3 Rearrangement

Rearrangement or intragenic recombination describes the insertions, deletions, and more commonly, gene duplications leading to sudden changes in the structure of genome segments. Normally, in gene duplications, the coding region of the template fragment remains unimpaired. Genome segments with duplicated fragments at the 3′ end are fairly common in rotavirus strains of some animal species as well as some atypical rotavirus strains shed by immuncompromised patients over prolonged periods (Desselberger, 1996). The evolutionary benefits from rearrangement are unclear, but may include more effective replication or an increase in protein coding capacity. Despite these putative advantages, strains with duplicated genes are very rare among those of many host species, including immunocompetent children.

2.4 Recombination

Several authors have reported evidence of both intratypic and intertypic recombination events in the genes encoding the VP7 protein (Parra et al., 2004; Phan et al., 2007). Similar to reassortment, the requirement beyond intergenic recombination between cognate genes of parental strains is the infection of a single cell by multiple rotavirus strains. However, additional details about the mechanisms are unknown. Theoretically, the result of recombination events could be the generation of new antigenic structures, which may help the new variants to escape the host immune response. However, a recent analysis suggests that intrasegmental recombination is rare among naturally cocirculating rotavirus strains and typically does not lead to onward transmission (Woods, 2015).

2.5 Interspecies Transmission

Initially, the distribution of rotaviruses was thought to be host-species restricted. More recent studies extending surveillance to many new geographic areas, including remote rural regions, and animal host species have revealed that animal strains play a key role in the genetic diversity of human rotavirus infections, primarily those caused by RVA strains.

There are two major routes by which heterologous rotaviruses may be acquired by a new host: (1) direct transmission of the rotavirus strain to a new host species and (2) reassortment between the rotavirus strains of any heterologous host species. Most interspecies direct transmission events lead to dead-end infection. Fully zoonotic strains are rarely detected in surveillance studies of human rotavirus infections (Bányai et al., 2012; Dóró et al., 2014). The low transmissibility in alternative hosts may be partially governed by host genetic factors or viral genes, for example, the one encoding NSP1 (see Chapter 2.8). Zoonotic transmission coupled by reassortment is a more efficient means for the introduction of new antigen specificities to which humans are immunologically naïve. An example is the introduction of G9 VP7 specificity during the 1990s from an animal, most likely the porcine, host (Iturriza-Gómara et al., 2000; Mijatovic-Rustempasic et al., 2011). Reassortants carrying a single or few animal-origin genes within a genetic background typical of human rotavirus strains may be more transmissible in the new host.

The zoonotic potential of RVAs has important implications for human rotavirus epidemiology. The diversity of rotaviruses in any host species, including humans, is readily refreshed by independent interspecies transmission events at any geographic location, coupled with global transmission of adapted strains.

Thus, eradication or even elimination of rotavirus infections through vaccination and other human-targeted interventions is unlikely due to the large number of host species whose homologous rotaviruses and gene pools are viable in the human host.

3 LABORATORY METHODS USED TO STUDY THE MOLECULAR EPIDEMIOLOGY OF ROTAVIRUSES

Understanding of the epidemiological features and evolutionary mechanisms of rotavirus infections has evolved over time along with improvements in laboratory methods. Early epidemiological studies relied on serotyping by MAb-EIA assays, RNA pattern analysis, and RNA-RNA hybridization, whereas today, most studies rely on multiplex genotyping reverse transcription-polymerase chain reactions (RT-PCRs) and more recently, on sequencing combined with phylogenetic analysis and other bioinformatics tools.

3.1 Serotyping and Subgrouping

Culturing rotaviruses has not become a routine laboratory method due to the varying isolation success of different rotavirus strains, although it was crucial in distinguishing serotypes (Wyatt et al., 1983). The findings that multiple serotype specificities exist prompted the realization that rotavirus vaccines would need to provide protection against an antigenically heterogeneous group of viruses. In routine laboratory procedures, serotype specificity has been determined by monoclonal antibody-based enzyme immunoassays (MAb-EIAs) targeting the VP7 (G serotypes) (Taniguchi et al., 1987). Large-scale surveillance studies conducted during the 1980s by MAb-EIAs uncovered the global distribution of four major VP7 serotypes of human rotaviruses, designated G1–G4 (Woods et al., 1992). In animal species, completely different serotype specificities were found to predominate, for example, G5 in pigs and serotypes G6 and G10 in cattle. In other cases, some overlap among the specificities of human and animal strains were observed, such as for G3 strains which prevailed in horses and dogs, but are also common in humans (Hoshino and Kapikian, 1994; Papp et al., 2013b,c).

The discovery of serotype-specific epitopes on the other surface antigen, VP4, was an important finding showing that the two surface antigens elicit neutralizing antibodies independently. However, this initially had little effect on routine epidemiological studies given that reliable MAb panels to distinguish VP4 (P) serotypes had not been developed. A limited number of studies using cross neutralization or cross-reactive MAbs in EIAs revealed that many human rotavirus strains carried the P1 (P1A and P1B) serotype specificity. Furthermore, some preferred combinations between G and P serotypes were observed: the G1, G3, and G4 VP7 specificities were more likely associated with P1A, whereas the G2 VP7 specificity showed a tendency to combine with the P1B VP4 gene (Estes and Kapikian, 2007). In animal rotavirus strains, a different distribution of VP4 serotypes (eg, P9 in pigs, and P6 to P8 in cattle) and surface antigen combinations were identified (Hoshino and Kapikian, 1994; Papp et al., 2013b,c).

Another set of MAbs developed against the VP6 protein of RVAs led to the recognition of antigenic variants of VP6 (Greenberg et al., 1983). Subgroup I strains were most commonly seen among animal RVAs and some human RVAs, whereas subgroup II strains were detected in most human and some porcine and lapine RVAs. A few other strains, mainly of animal origin, were characterized as [subgroup I + II] or [subgroup nonI, nonII] (Hoshino and Kapikian, 1994).

Altogether, studies using cross-neutralization assays and MAb-EIAs revealed differences in serotype and subgroup distribution among various host species.

3.2 Electropherotyping

From the 1980s onward, another arm of molecular epidemiology studies utilized RNA electropherotyping (also called RNA fingerprinting, RNA profiling, or RNA migration pattern analysis). The 11 segments of the rotavirus genome can be separated and visualized in polyacrylamide gels by silver staining (Herring et al., 1982). The relative migration of each segment in polyacrylamide gels has been found to be useful in diagnostic applications and can also be used to differentiate infections in local outbreak situations. Typically, RVA strains of mammalian hosts show a pattern of RNA bands in the 4-2-3-2 arrangement, while RVBs and RVCs are characterized by the 4-2-2-3 and 4-3-2-2 patterns, respectively (Saif and Jiang, 1994). Within RVAs, three major e-types were observed, including long, short, and super-short patterns. The difference lies in the relative migration of gene segment 11, which shows faster migration in long e-type strains and slower in short and super-short e-type strains. Within major e-types many additional variants occur; thus, in addition to the diagnostic benefits, the method has been found to be useful in molecular epidemiological studies demonstrating, for example, the changing prevalence of various e-types over time or permitting the tracking of viral spread in the community or during a hospital outbreak.

3.3 Whole Genome Hybridization

First insights into the genetic relationship among rotavirus strains were enabled by RNA–RNA hybridization studies. The number of hybridizing RNA bands between target RNAs and the isotope-labeled probe RNAs served as the basis to distinguish various genetic groups, referred to as genogroups. Among human rotaviruses, whole genome hybridization studies identified two major genogroups (Flores et al., 1982). The prototype strains were Wa (a serotype G1P1A[8] strain) and DS1 (a serotype G2P1B[4] strain). Molecular analysis of the medically important human strains revealed that the Wa-like genogroup mainly included long e-type strains that express the P1A VP4 and one of the common VP7 genes (G1, G3, and G4), whereas DS1-like strains included primarily short e-type strains that carry the P1B VP4 and G2 VP7 genes. Intergenogroup reassortants between Wa- and DS1-like strains have been identified, but are rare (Ward et al., 1990). Whole genome hybridization studies of animal strains revealed a limited relationship between strains detected in heterologous host species, including humans. On the other hand, RNA–RNA hybridization was the first method to firmly demonstrate that some unusual human strains (or parts of their genome) may originate from animals indicating the zoonotic potential of rotaviruses (Nakagomi and Nakagomi, 1996).

3.4 Multiplex Genotyping PCR and Sequencing

From the early 1990s, soon after the first rotavirus gene sequence data became available, a new RT-PCR-based method has been adopted for strain surveillance (Gentsch et al., 1992; Gouvea et al., 1990). This was corroborated by the finding that genotypes correspond closely to serotypes. Nested multiplex RT-PCR assays revolutionized routine surveillance and have become the method of choice recommended by the World Health Organization (WHO) in the generic protocol for hospital-based surveillance (WHO 2002; WHO 2009). Nested multiplex RT-PCR assays permitted genotyping of RVA strains based on the lengths of amplified DNA fragments that could be readily analyzed in agarose gels. This was the main method used for the genetic characterization of >150,000 human rotavirus strains in >100 countries worldwide (Bányai et al., 2012; Dóró et al., 2014), and has been adapted to both G and P type specificities. The flexibility of these RT-PCR-based methods permits rapid replacement of old typing primers or addition of new primers targeting emerging type specificities (Iturriza-Gómara et al., 2004).

A significant reduction in costs of traditional (ie, Sanger) nucleotide sequencing has opened new avenues in rotavirus strain typing. Over the past 30 years, >40,000 nucleotide sequences of rotaviruses have been deposited in GenBank, primarily for RVAs. Nearly 30,000 sequences were deposited within the past 5 years, showing a very strong motivation to understand the evolution of RV strains in the postvaccine marketing era. Sequencing and phylogenetic analysis have uncovered new type specificities and significant diversity within rotavirus genotypes. In large-scale sequencing and phylogenetic analyses, several genetic lineages and sublineages could be distinguished within most, if not all, G and P type specificities. Furthermore, classification of rotaviruses into lineages and sublineages typically permits the distinction of strains collected from various host species.

3.5 Whole Genome Sequencing

Although not routine yet, whole genome based strain characterization has become available worldwide to supplement traditional technologies. At present, over 1000 whole genome sequences are available for RVA, and a few dozens for RVB to RVI.

Assessment of rotavirus evolution and molecular epidemiology has been made possible by some large-scale whole genome based studies. These studies clearly demonstrated the relative conservation in the genotype constellations of backbone genes (ie, other than VP7 and VP4) of medically important RVA strains. Reassortment among cocirculating strains belonging to identical genotype constellations are fairly common (Iturriza-Gómara et al., 2001) whereas reassortment events among strains belonging to different genotype constellations are typically rare (Esona et al., 2013; Mijatovic-Rustempasic et al., 2011).

Many whole genome based studies have focused on the diversity and evolutionary origin of unusual human rotavirus strains. These studies, together with the increasing amount of genome sequence information collected from animal host species, have permitted new insights into the evolution of some unusual genotype specificities, for example, the canine or feline origin of many human G3P[3] and G3P[9] strains. In addition, bats in Africa and China have been recognized to carry rotaviruses with both usual and unusual human genes. More recently, pigs have been identified as a major source and reservoir of strains infecting humans, showing a complex multidirectional transmission pattern between these two host species (Matthijnssens et al., 2008a; Papp et al., 2013a; Theuns et al., 2015).

Further improvements and cost reduction of high-throughput sequencing technology will facilitate whole genome based comparisons, determination of preferred genotype constellations, identification of vaccine-derived genes in field strains, and tracking of evolution of vaccine strains and other rotavirus strains in the postvaccination surveillance era (Bányai and Gentsch, 2014).

4 TRENDS IN ROTAVIRUS STRAIN PREVALENCE IN HUMANS

Molecular epidemiology studies have revealed that the distribution of rotavirus strains varies over both space and time. Various RVA genotypes have been found to cocirculate in any given location and strong fluctuations in the distribution of the different genotypes can occur from year to year. The reasons for these fluctuations are not well understood, but can likely be attributed to slight differences in homotypic versus heterotypic host immunity, as well as stochastic effects acting on genetic variation produced through mutation and reassortment (Matthijnssens et al., 2012a).

4.1 Global Genotype Distribution

The majority of rotavirus gastroenteritis (RVGE) cases are caused by one of the five common genotypes—the Wa-like G1P[8], G3P[8], G4P[8], and G9P[8] strains or the DS-1-like G2P[4] strains. These five genotypes accounted for 75% of all strains genotyped between 1996 and 2007, prior to widespread rotavirus vaccination in most countries (Bányai et al., 2012). G8, found in combination with P[6], P[4] or P[8], is the sixth most common G-type, accounting for 0.9%

of all strains globally and 3.3–9.1% of strains in Africa (Bányai et al., 2012; Sanchez-Padilla et al., 2009). Recently, G12 strains have emerged and spread across the globe (Matthijnssens et al., 2010) and can cause a considerable proportion of RVGE cases during a given season. G12 strains accounted for 1.3% of genotyped strains between 2003 and 2007 (Bányai et al., 2012) and are typically found in combination with P[8] or P[6], with a Wa-like or reassortant backbone. Nontypeable and mixed infections (consisting of more than one G- or P-type) account for 9.5% and 5.6% of all typed infections, respectively (Bányai et al., 2012). There is no difference in the severity of RVGE caused by the different genotypes.

G1P[8] is the most common genotype, accounting for 38% of all genotyped strains (Bányai et al., 2012). However, the relative importance of G1P[8] strains has been declining in recent years. This genotype accounted for 65% of strains in studies published between 1989 and 2004 (Santos and Hoshino, 2005). The reasons for this decline are not well understood.

4.2 Regional Genotype Patterns

In the Americas, Europe, and the Western Pacific region, the five common RV G-types (G1–G4 and G9) accounted for 91–92% of all strains during the period from 1996 to 2007, whereas these five G-types accounted for 82% of strains in the Eastern Mediterranean Region and 71 and 75% of strains in Southeast Asia and Africa, respectively. G1 was the most common G-type in all six regions, but prevalence varied from 28% in Southeast Asia to 52% of strains in Europe. G2 was the second most common G-type in the Eastern Mediterranean (14%), Southeast Asia (20%) and Africa (10%), while G9 was more common in the Americas (18%) and Europe (13%), and G3 was more common in the Western Pacific (25%) (Fig. 2.10.2). G8 accounted for 9% of strains in Africa, while G12 accounted for 4% of strains in Southeast Asia; but these G-types accounted for <1% of strains in other regions. Nontypeable (NT) and mixed strain infections were more common in Africa, Southeast Asia, and the Eastern Mediterranean than in the Americas, Europe, and the Western Pacific (Fig. 2.10.2).

We can only speculate about the driving forces underlying these regional differences in rotavirus genotype distribution. Improved genotyping methods and increased surveillance efforts have led to the contribution of data from more and more countries in recent years, including greater representation from low income countries, which tend to exhibit a greater diversity of genotypes. Zoonotic transmission is likely responsible for the majority of rare and NT strains. Likewise, G8 strains appear to be closely related to bovine RVA strains, and therefore should be more common in countries where humans and animals live in close proximity, or where transmission can occur from unimproved water sources. Furthermore, mixed infections and infection with reassortant strains are expected to be more common in countries with a higher rate of rotavirus transmission.

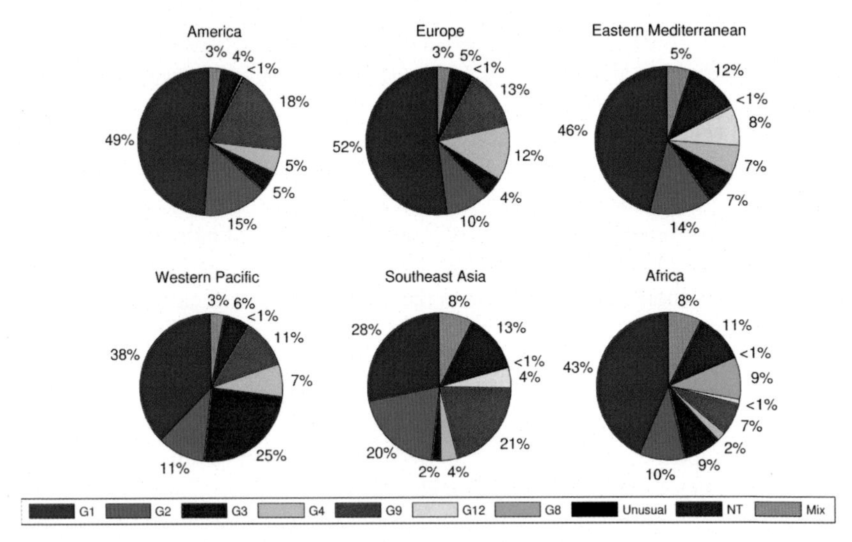

FIGURE 2.10.2 **Prevaccination genotype distributions by WHO region.** *(Source: Based on data from Bányai et al., 2012.)*

4.3 Local Genotype Patterns

Within a country, the distribution of RV genotypes in a given season can vary from one location to the next. For example, during Jul. 2003–Jun. 2004, in Australia, 86% of typed strains in Melbourne were G1, while G3 (56%) was the predominant G-type in Perth and G2 (77%) was the most common G-type in Alice Springs (Kirkwood et al., 2009). Similar geographic variation in the prevalence of different genotypes has been observed in countries, such as the United States (Gentsch et al., 2009), Brazil (Leite et al., 2008), and India (Kang et al., 2013).

4.4 Temporal Changes in Genotype Distributions

Rotavirus genotype distributions also change over time. While G1P[8] is the predominant genotype most years, particularly in high income countries, other genotypes can cause the majority of RVGE cases in any given season. For example, during the period between 1988 and 2006 in Budapest, Hungary, G1P[8] was the predominant genotype in 13 of the 18 seasons (Fig. 2.10.3) (Bányai et al., 2009). However, G2P[4] was the predominant genotype in 1996–97, while G9 strains were most common in 2002–03 and 2004–06 and G4P[8] was the predominant genotype in 2003–04. Furthermore, the predominant lineage within G1 changed in 6 out of 11 seasons with available sequence data in which G1P[8] was the predominant genotype (Bányai et al., 2009). Similar turnover of the multiple lineages within G1P[8], as well as of the other genotypes, has been observed (Arista et al., 2006; Giammanco et al., 2014; McDonald et al., 2009; Wu et al., 2014).

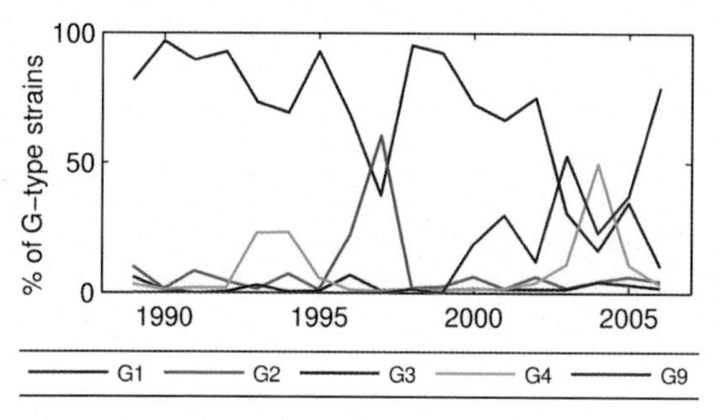

FIGURE 2.10.3 **Annual genotype distribution in Budapest, Hungary, 1989–2006.** *(Source: Adapted from Bányai et al., 2009; Pitzer et al., 2011a.)*

In settings with 10 or more years of continuous surveillance of RV genotypes, cycling of the predominant genotypes is observed. The period of oscillations in the prevalence of different genotypes has been found to vary between 3 and 11 (or more) years (Pitzer et al., 2011a). Therefore, at least 4 years of surveillance data are needed to reliably determine the underlying distribution of RV genotypes in any given location.

4.5 Local Persistence, Global Spread

Sequence data lend support for both local persistence of individual RV lineages and circulation of lineages on a global scale. In a given location, the same rotavirus lineage has been observed over multiple years, even in temperate countries where RVGE exhibits a distinct seasonality with very low incidence in the summer months (Arista et al., 2006; McDonald et al., 2012). This is suggestive of local persistence of the virus in the population, possibly through low levels of asymptomatic infection in adults and other partially immune subpopulations (Phillips et al., 2010).

Nevertheless, when RV strains collected from a single locale are compared to contemporary and archival sequences available for other regions of the world, the observed genetic diversity is similar (McDonald et al., 2012). The global spread of RV strains was particularly apparent following the recent emergence of G9 and G12 strains. In both cases, one particular lineage was responsible for the worldwide spread of these novel G-types (Matthijnssens et al., 2010). Within just over a decade, both of these strains had spread throughout the globe (Matthijnssens et al., 2009, 2010). However, larger than normal outbreaks of RVGE were not associated with the emergence of these new genotypes, which suggests that heterotypic immunity provides at least some protection against novel strains.

4.6 Impact of Vaccination on Genotype Distributions

Two rotavirus vaccines have been available since 2006, including the monovalent Rotarix® vaccine (GlaxoSmithKline Biologicals, Rixensart, and Belgium) and the pentavalent RotaTeq® vaccine (Merck and Co., Whitestation, NJ, and USA). These vaccines differ in their approach to eliciting immunity (Ward, 2009). The Rotarix vaccine consists of an attenuated Wa-like G1P[8] strain, whereas RotaTeq contains a mixture of 5 bovine-human monoreassortant RVA strains, where each strain contains either a human VP7 (G1, G2, G3, and G4) or VP4 (P1A[8]) gene, introduced by in vitro reassortment in the genetic background of the bovine WC3 rotavirus (G6P7[5]) strain. (See Chapter 2.11.)

Although there is a broad scientific consensus that both vaccines are effective in reducing the burden of severe RVGE, there is still controversy regarding the long-term effects of vaccination on the circulating RV genotype distribution (Matthijnssens et al., 2012a). Vaccine effectiveness has been found to be equally high against both homotypic and heterotypic strains (Leshem et al., 2014). Nevertheless, transient changes in the genotype distribution have been observed following large-scale vaccination with Rotarix and/or RotaTeq.

The introduction of RotaTeq was followed by an increase in the prevalence of the G3P[8] genotype in some places, including the United States and some Australian states (Hull et al., 2011; Kirkwood et al., 2011), raising concern that this could be linked to lower rates of seroconversion to the G3 component of the pentavalent vaccine (Matthijnssens et al., 2012a). However, similar observations have not persisted. Other genotypes, including G9P[8], G12P[8], and G2P[4], have predominated for a year or two in certain locations, but overall G1P[8] has remained the predominant genotype in countries using RotaTeq.

For Rotarix, a consistent and more prolonged increase in the relative prevalence of the fully heterotypic G2P[4] genotype has been observed following large-scale vaccination (Dóró et al., 2014), particularly in Brazil, Belgium, and some Australian states. However, the magnitude and duration of the increase varied between locations. In Australia, the increase G2P[4] was seen for two seasons following vaccine introduction (2007–09) and again in 2012, but G1P[8] and G3P[8] were the most common genotypes in 2009–11 and 2013, respectively, in states using Rotarix. A clear increase in G2P[4] after vaccine introduction has also been noted in Belgium, and has been sustained for at least seven seasons, 2006–13 (Zeller et al., 2010; Pitzer et al., 2015). In Brazil, G2P[4] has been the predominant genotype for at least six years since vaccine introduction (2006–11), with prevalence $\geq$50% in all years except 2009 (Gómez et al., 2014). Furthermore, G1P[8] has been relatively rare, having been detected in $\leq$20% of samples in all seasons since vaccine introduction in Brazil (Carvalho-Costa et al., 2011), while another fully heterotypic genotype (G8P[4]) emerged in Northeast Brazil in 2012 (Gurgel et al., 2014). The high prevalence of G2P[4] RVA strains in Brazil after introduction of the universal RV vaccination with Rotarix was not unique but also recorded in other South American countries where no RV vaccination program had been established (Matthijnssens et al., 2009).

Finally, a number of vaccine-derived reassortant strains have also been isolated from individuals both with and without a history of vaccination. In particular, vaccine-derived reassortant G1P[8] strains have been described in patients from the United States, Australia, and Finland following vaccination with RotaTeq. Reassortment between wild-type G1P[8] and the Rotarix vaccine strain has also occurred (Dóró et al., 2014). This has important implications for surveillance, as more advanced laboratory methods may be needed to differentiate between wild-type and vaccine-derived strains.

5 ROTAVIRUS TRANSMISSION DYNAMICS: DRIVING FORCES BEHIND EPIDEMIOLOGICAL TRENDS

Rotavirus is transmitted via the fecal-oral route. Improvements in sanitation have had only a minor impact on the level of rotavirus transmission, unlike for the bacterial causes of diarrhea. The overall incidence of RVGE and the age distribution of cases do not differ much between developed and developing countries (Parashar et al., 2003). The vast majority of reported RVGE cases occur in children <5 years of age, who have little or no immunity. However, infection with rotavirus continues to occur throughout life. Most infections in adults are mild or asymptomatic, but may contribute to maintenance of transmission in the population (Anderson and Weber, 2004).

In countries with temperate climate, most cases of RVGE occur during winter epidemics. In tropical countries, rotavirus tends to be more common during the dry season, but RVGE cases typically occur throughout the year. However, the strength of rotavirus seasonality is more closely linked to the level of country development than it is to geographic location or climate (Patel et al., 2013). For instance, high-income countries in the tropics, such as Hong Kong, tend to experience seasonal outbreaks, whereas lower-income countries in temperate regions, such as Uzbekistan, exhibit year-round disease (Fig. 2.10.4).

Nearly all children have been infected with rotaviruses by the age of 2 years. The likelihood and severity of RVGE is associated with the number of previous rotavirus infections. Infants who have not been previously exposed to the virus are more likely to experience RVGE, which may be moderate or severe. Maternal immunity provides some protection against symptoms, particularly during the first few months of life. Following two natural infections, moderate to severe RVGE is rare in most children (Velázquez et al., 1996). However, in poorer countries where immunity may be compromised by factors, such as malnutrition, concomitant infections of the gut, and tropical enteropathy, severe RVGE can continue to occur following three or more natural infections (Gladstone et al., 2011). Death due to rotavirus is infrequent in most developed countries, but worldwide rotavirus was estimated to cause nearly half a million deaths each year prior to the introduction of universal vaccination programs, with most deaths occurring in developing countries (Tate et al., 2012).

Infants appear to be major drivers of transmission of rotavirus in the community. Vaccination has resulted in both direct protection for infants who received the vaccine, as well as indirect protection resulting from a reduction in the transmission of rotavirus in the population and (possibly) contact immunity gained

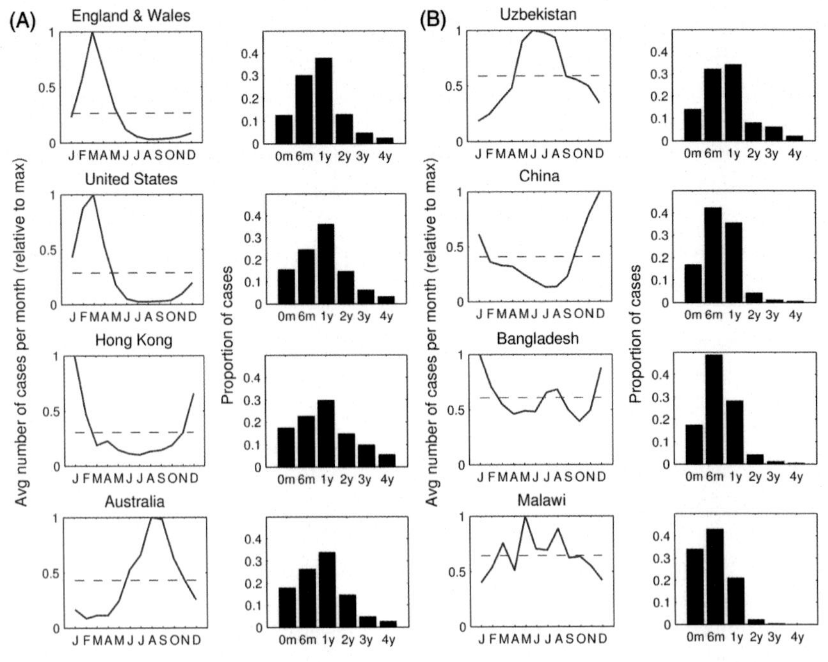

FIGURE 2.10.4 Age distribution and seasonality of rotavirus in different countries. The average number of RVGE cases per month (scaled relative to the maximum monthly number of RVGE cases) is plotted in blue; the average number of cases across all months is represented by the dashed black line. The proportion of cases in each age group (0–5 months, 6–11 months, 1–2 years, 2–3 years, 3–4 years, and 4–5 years old) is plotted according to the black bars for each country. Example countries are plotted by latitude (north to south) for (A) high-income countries and (B) developing countries. *(Source: Adapted from Pitzer et al., 2011b.)*

from shedding and circulation of vaccine strains. Decreases in the incidence of RVGE following vaccine introduction have been observed in age groups not eligible for the vaccines (Lopman et al., 2011; Payne et al., 2011). Furthermore, changes to the timing and seasonality of rotavirus epidemics have been observed following vaccine introduction in some temperate countries, including delayed epidemics and a shift from annual to biennial (every other year) epidemics in the United States (Tate et al., 2013). These observations, combined with insights from mathematical modeling, suggest that by preventing RVGE in infants, vaccination also serves to decrease the transmission of rotavirus in the community.

5.1 What can Mathematical Models Tell us About Rotavirus Transmission?

Mathematical models can provide insight into the underlying population dynamics of rotavirus infections and important aspects of transmission. By analyzing the natural history, immunity, and transmission feedbacks of rotaviruses and describing these processes using a series of equations or mathematical relationships, models were established that allow one to bridge the gap between the individual course of infection and the population-level spread of disease.

A variety of mathematical models for rotavirus infections have been developed. These models vary in their structure and parameters based on different data sources and interpretations of the literature. Most models were developed with the aim of predicting the direct and indirect effects of vaccination. Different models predict similar reductions in RVGE in the first 5 years following vaccine introduction, but predictions for the long-term impact of vaccination have varied depending on assumptions about the relationship between age at infection and the severity and reporting of RVGE, as well as the duration of vaccine-induced immunity (Pitzer et al., 2012).

Models are also useful for explaining observed patterns of data. For instance, one model was able to demonstrate that the southwest-to-northeast pattern of rotavirus epidemics in the United States prior to vaccine introduction could be linked to underlying variation in the birth rate (Pitzer et al., 2009). Higher birth rates typical of states in the southwest lead to a quicker build-up of numbers of fully susceptible infants, who serve as important drivers of transmission. Models also predicted that rotavirus epidemics could become biennial at high levels of vaccine coverage, which has aided in the interpretation of patterns that have emerged since vaccine introduction.

Models have also attempted to explain the interaction among different RV genotypes, for example, modifying this in light of a couple of recent publications (e.g. Pitzer et al., 2011a). This model was able to show that the observed cycling of genotypes could be explained by differences in homotypic versus heterotypic host immunity. If immunity is stronger against a second infection with the same genotype as that causing the first infection (ie, if homotypic immunity is stronger than heterotypic immunity), then over time there will be a build-up of population-level immunity to the prevailing strain. This provides a selection pressure in favor of the rarer genotypes, allowing them to infect more people. Thus, over time one of the rarer genotypes is expected to increase in prevalence, while the prevailing genotype decreases in prevalence, and then the cycle repeats (Fig. 2.10.5).

Models suggest that vaccination with a monovalent vaccine, such as Rotarix is expected to exert different selection pressures on the rotavirus population than vaccination with a multivalent vaccine, such as RotaTeq. This selection pressure can occur even in the absence of strain-specific variation in vaccine effectiveness against RVGE (Pitzer et al., 2015). If the relative risk of infection with a fully heterotypic strain (ie, G2P[4]) is greater than the risk of second infection with G1P[8] following vaccination with Rotarix, then this can explain the predominance of G2P[4] in the years following vaccine introduction in countries, such as Belgium. Nevertheless, both vaccines are estimated to provide a broadly heterotypic immunity in most fully vaccinated infants and to remain effective despite any possible changes to the genotype distributions (Pitzer et al., 2015). In order to understand and predict the impact of vaccination on the distribution of rotavirus genotypes, it is important to take into account differences in the risk of infection (and thus transmission), and not just differences in vaccine effectiveness against RVGE.

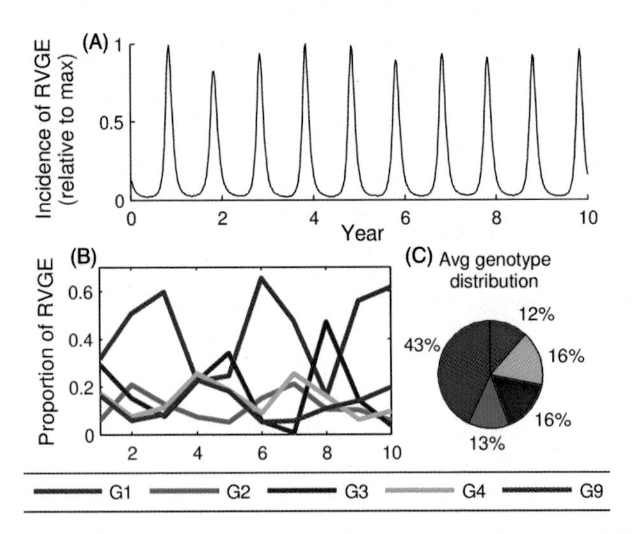

FIGURE 2.10.5 **Example output for strain-specific model of rotavirus transmission dynamics.** (A) The total number of severe RVGE cases per week (relative to the maximum number of weekly cases predicted by the model) is plotted for a 10-year period for model parameters typical of a high-income country (Pitzer et al., 2011a; Pitzer et al., 2015). (B) The mean annual proportion of severe RVGE cases caused by each strain, representing the five major G-types, is plotted for each year. (C) The average genotype distribution over the 10-year period.

Mathematical models are only as good as the data that are used to inform them. Often it is necessary to make simplifying assumptions in order for the models to be tractable. Individual rotavirus strains are continually evolving over time through evolutionary processes, such as point mutations and reassortment, which can be difficult to represent in a mathematical model. Nevertheless, models provide a "test-bed" for examining hypotheses regarding how individual-level characteristics of infection can affect the population-level spread of disease, and can help to highlight important unknown factors of outbreak developments.

6 CONCLUSIONS

The diversity of rotavirus strains infecting both human and animals has been recognized since soon after their discovery in the early 1970s. Recent advances in laboratory methods for the detection and classification of rotaviruses have aided in the study and interpretation of epidemiological and evolutionary patterns. The recent introduction of vaccines has made the continued surveillance and development of new approaches to interpret data on the molecular epidemiology of rotavirus an imperative. Our understanding of rotavirus epidemiology and evolution will no doubt be tested and continue to be refined in the near future.

ACKNOWLEDGMENTS

Financial support obtained from the Hungarian Scientific Research Fund (OTKA, T100727) (K.B.) and the US National Institutes of Health (R01 AI112970) (V.E.P.) is acknowledged. Fig. 2.10.1 was prepared and kindly shared by Renáta Dóró.

REFERENCES

Anderson, E.J., Weber, S.G., 2004. Rotavirus infection in adults. Lancet Infect. Dis. 4 (2), 91–99.

Arista, S., Giammanco, G.M., De Grazia, S., Ramirez, S., Lo Biundo, C., Colomba, C., et al., 2006. Heterogeneity and temporal dynamics of evolution of G1 human rotaviruses in a settled population. J. Virol. 80 (21), 10724–10733.

Bányai, K., Gentsch, J., 2014. Special issue on "Genetic diversity and evolution of rotavirus strains: Possible impact of global immunization programs.". Infect. Genet. Evol. 28, 375–376.

Bányai, K., Gentsch, J.R., Martella, V., Bogdán, A., Havasi, V., Kisfali, P., et al., 2009. Trends in the epidemiology of human G1P[8] rotaviruses: a Hungarian study. J. Infect. Dis. 200 (Suppl. (Suppl. 1)), S222–S227.

Bányai, K., László, B., Duque, J., Steele, A.D., Nelson, E.A.S., Gentsch, J.R., et al., 2012. Systematic review of regional and temporal trends in global rotavirus strain diversity in the pre rotavirus vaccine era: insights for understanding the impact of rotavirus vaccination programs. Vaccine 30 (Suppl. 1), A122–A130.

Blackhall, J., Fuentes a, Magnusson, G., 1996. Genetic stability of a porcine rotavirus RNA segment during repeated plaque isolation. Virology 225 (1), 181–190.

Carvalho-Costa, F.A., Volotão, E.D.M., de Assis, R.M.S., Fialho, A.M., de Andrade, J.D.S.R., Rocha, L.N., et al., 2011. Laboratory-based rotavirus surveillance during the introduction of a vaccination program, Brazil, 2005–2009. Pediatr. Infect. Dis. J. 30 (Suppl. 1), S35–S41.

Desselberger, U., 1996. Genome rearrangements of rotaviruses. Adv. Virus Res. 46, 69–95.

Donker, N.C., Kirkwood, C.D., 2012. Selection and evolutionary analysis in the nonstructural protein NSP2 of rotavirus A. Infect. Genet. Evol. 12 (7), 1355–1361.

Dóró, R., László, B., Martella, V., Leshem, E., Gentsch, J., Parashar, U., et al., 2014. Review of global rotavirus strain prevalence data from six years post vaccine licensure surveillance: is there evidence of strain selection from vaccine pressure? Infect. Genet. Evol. 28, 446–461.

Esona, M.D., Mijatovic-Rustempasic, S., Foytich, K., Roy, S., Banyai, K., Armah, G.E., et al., 2013. Human G9P[8] rotavirus strains circulating in Cameroon, 1999-2000: Genetic relationships with other G9 strains and detection of a new G9 subtype. Infect. Genet. Evol. 18, 315–324.

Estes, M.K., Kapikian, A.Z., 2007. Rotaviruses. In: Knipe, D.M., Howley, P.M., Griffin, D.E., Lamb, R.A., Martin, M.A., Roizman, B. et al., (Eds.), Fields Virology. 5th ed. Lippincott Williams & Wilkins, Philadelphia, PA USA, pp. 1917–1974.

Flores, J., Perez, I., White, L., Kalica a, R., Marquina, R., Wyatt, R.G., et al., 1982. Genetic relatedness among human rotaviruses as determined by RNA hybridization. Infect. Immun. 37 (2), 648–655.

Gentsch, J.R., Glass, R.I., Woods, P., Gouvea, V., Gorziglia, M., Flores, J., et al., 1992. Identification of group A rotavirus gene 4 types by polymerase chain reaction. J. Clin. Microbiol. 30 (6), 1365–1373.

Gentsch, J.R., Hull, J.J., Teel, E.N., Kerin, T.K., Freeman, M.M., Esona, M.D., et al., 2009 Nov 1. G and P types of circulating rotavirus strains in the United States during 1996-2005: nine years of prevaccine data. J. Infect. Dis. 200 (Suppl. (Suppl. 1)), S99–105.

Giammanco, G.M., Bonura, F., Zeller, M., Heylen, E., Van Ranst, M., Martella, V., et al., 2014. Evolution of DS-1-like human G2P[4] rotaviruses assessed by complete genome analyses. J. Gen. Virol. 95 (Pt 1), 91–109.

Gladstone, B.P., Ramani, S., Mukhopadhya, I., Muliyil, J., Sarkar, R., Rehman, A.M., et al., 2011. Protective effect of natural rotavirus infection in an Indian birth cohort. N. Engl. J. Med. 365 (4), 337–346.

Gómez, M.M., Carvalho-Costa, F.A., Volotão, E.D.M., Rose, T.L., da Silva, M.F.M., Fialho, A.M., et al., 2014. Prevalence and genomic characterization of G2P[4] group A rotavirus strains during monovalent vaccine introduction in Brazil. Infect. Genet. Evol. 28, 486–494.

Gouvea, V., Glass, R.I., Woods, P., Taniguchi, K., Clark, H.F., Forrester, B., et al., 1990. Polymerase chain reaction amplification and typing of rotavirus nucleic acid from stool polymerase chain reaction amplification and typing of rotavirus nucleic acid from stool specimens. J. Clin. Microbiol. 28 (2), 276–282.

Greenberg, H., McAuliffe, V., Valdesuso, J., Wyatt, R., Flores, J., Kalica, A., Hoshino, Y., Singh, N., 1983. Serological analysis of the subgroup protein of rotavirus, using monoclonal antibodies. Infect. Immun. 39 (1), 91–99.

Gurgel, R.Q., Alvarez, A.D.J., Rodrigues, A., Ribeiro, R.R., Dolabella, S.S., Da Mota, N.L., et al., 2014. Incidence of rotavirus and circulating genotypes in Northeast Brazil during 7 years of national rotavirus vaccination. PLoS One 9 (10), e110217.

Herring, A.J., Inglis, N.F., Ojeh, C.K., Snodgrass, D.R., Menzies, J.D., 1982. Rapid diagnosis of rotavirus infection by direct detection of viral nucleic acid in silver-stained polyacrylamide gels. J. Clin. Microbiol. 16 (3), 473–477.

Hoshino, Y., Kapikian, A.Z., 1994. Rotavirus antigens. Curr. Top. Microbiol. Immunology 185, 179–227.

Hull, J.J., Teel, E.N., Kerin, T.K., Freeman, M.M., Esona, M.D., Gentsch, J.R., et al., 2011. United States rotavirus strain surveillance from 2005 to 2008: genotype prevalence before and after vaccine introduction. Pediatr. Infect. Dis. J. 30 (1 Suppl.), S42–S47.

Iturriza-Gómara, M., Cubitt, D., Steele, D., Green, J., Brown, D., Kang, G., et al., 2000. Characterisation of rotavirus G9 strains isolated in the UK between 1995 and 1998. J. Med. Virol. 61 (4), 510–517.

Iturriza-Gómara, M., Isherwood, B., Desselberger, U., Gray, J., 2001. Reassortment in vivo: driving force for diversity of human rotavirus strains isolated in the United Kingdom between 1995 and 1999. J. Virol. 75 (8), 3696–3705.

Iturriza-Gómara, M., Kang, G., Gray, J., 2004. Rotavirus genotyping: keeping up with an evolving population of human rotaviruses. J. Clin. Virol. 31 (4), 259–265.

Kang, G., Desai, R., Arora, R., Chitamabar, S., Naik, T.N., Krishnan, T., et al., 2013. Diversity of circulating rotavirus strains in children hospitalized with diarrhea in India, 2005–2009. Vaccine 31 (27), 2879–2883.

Kirkwood, C.D., Boniface, K., Barnes, G.L., Bishop, R.F., 2011. Distribution of rotavirus genotypes after introduction of rotavirus vaccines, Rotarix® and RotaTeq®, into the National Immunization Program of Australia. Pediatr. Infect. Dis. J. 30 (1 Suppl.), S48–S53.

Kirkwood, C.D., Boniface, K., Bogdanovic-Sakran, N., Masendycz, P., Barnes, G.L., Bishop, R.F., 2009. Rotavirus strain surveillance—an Australian perspective of strains causing disease in hospitalised children from 1997 to 2007. Vaccine 27 (Suppl. 5), F102–F107.

Leite, J.P.G., Carvalho-Costa, F.A., Linhares, A.C., 2008. Group A rotavirus genotypes and the ongoing Brazilian experience: a review. Mem. Inst. Oswaldo Cruz. 103 (8), 745–753.

Leshem, E., Lopman, B., Glass, R., Gentsch, J., Bányai, K., Parashar, U., et al., 2014. Distribution of rotavirus strains and strain-specific effectiveness of the rotavirus vaccine after its introduction: a systematic review and meta-analysis. Lancet Infect. Dis. 14 (9), 847–856.

Lopman, B.A., Curns, A.T., Yen, C., Parashar, U.D., 2011. Infant rotavirus vaccination may provide indirect protection to older children and adults in the United States. J. Infect. Dis. 204 (7), 980–986.

Marthaler, D., Rossow, K., Culhane, M., Collins, J., Goyal, S., Ciarlet, M., et al., 2013. Identification, phylogenetic analysis and classification of porcine group C rotavirus VP7 sequences from the United States and Canada. Virology 446 (1–2), 189–198.

Marthaler, D., Rossow, K., Gramer, M., Collins, J., Goyal, S., Tsunemitsu, H., et al., 2012. Detection of substantial porcine group B rotavirus genetic diversity in the United States, resulting in a modified classification proposal for G genotypes. Virology 433 (1), 85–96.

Matthijnssens, J., Bilcke, J., Ciarlet, M., Martella, V., Bányai, K., Rahman, M., et al., 2009. Rotavirus disease and vaccination: impact on genotype diversity. Future Microbiol. 4 (10), 1303–1316.

Matthijnssens, J., Ciarlet, M., Heiman, E., Arijs, I., Delbeke, T., McDonald, S.M., Palombo, E.A., Iturriza-Gómara, M., Maes, P., Patton, J.T., Rahman, M., Van Ranst, M., 2008a. Full genome-based classification of rotaviruses reveals a common origin between human Wa-Like and porcine rotavirus strains and human DS-1-like and bovine rotavirus strains. J. Virol. 82 (7), 3204–3219.

Matthijnssens, J., Ciarlet, M., Rahman, M., Attoui, H., Bányai, K., Estes, M.K., Gentsch, J.R., Iturriza-Gómara, M., Kirkwood, C.D., Martella, V., Mertens, P.P., Nakagomi, O., Patton, J.T., Ruggeri, F.M., Saif, L.J., Santos, N., Steyer, A., Taniguchi, K., Desselberger, U., Van Ranst, M., 2008b. Recommendations for the classification of group A rotaviruses using all 11 genomic RNA segments. Arch. Virol. 153 (8), 1621–1629.

Matthijnssens, J., Ciarlet, M., McDonald, S.M., Attoui, H., Bányai, K., Brister, J.R., et al., 2011. Uniformity of rotavirus strain nomenclature proposed by the Rotavirus Classification Working Group (RCWG). Arch. Virol. 156 (8), 1397–1413.

Matthijnssens, J., Heylen, E., Zeller, M., Rahman, M., Lemey, P., Van Ranst, M., 2010. Phylodynamic analyses of rotavirus genotypes G9 and G12 underscore their potential for swift global spread. Mol. Biol. Evol. 27 (10), 2431–2436.

Matthijnssens, J., Nakagomi, O., Kirkwood, C.D., Ciarlet, M., Desselberger, U., Van Ranst, M., 2012a. Group A rotavirus universal mass vaccination: how and to what extent will selective pressure influence prevalence of rotavirus genotypes? Expert Rev. Vaccines 11 (11), 1347–1354.

Matthijnssens, J., Otto, P.H., Ciarlet, M., Desselberger, U., van Ranst, M., Johne, R., 2012b. VP6-sequence-based cutoff values as a criterion for rotavirus species demarcation. Arch. Virol. 157 (6), 1177–1182.

McDonald, S.M., Matthijnssens, J., McAllen, J.K., Hine, E., Overton, L., Wang, S., et al., 2009. Evolutionary dynamics of human rotaviruses: balancing reassortment with preferred genome constellations. PLoS Pathog. 5 (10), e1000634.

McDonald, S.M., McKell, A.O., Rippinger, C.M., McAllen, J.K., Akopov, A., Kirkness, E.F., et al., 2012. Diversity and relationships of cocirculating modern human rotaviruses revealed using large-scale comparative genomics. J. Virol. 86 (17), 9148–9162.

Mihalov-Kovács, E., Gellért, Á., Marton, S., Farkas, S.L., Fehér, E., Oldal, M., et al., 2015. Candidate New Rotavirus Species in Sheltered Dogs, Hungary. Emerg. Infect. Dis. 21 (4), 660–663.

Mijatovic-Rustempasic, S., Bányai, K., Esona, M.D., Foytich, K., Bowen, M.D., Gentsch, J.R., 2011. Genome sequence based molecular epidemiology of unusual US Rotavirus A G9 strains isolated from Omaha, USA between 1997 and 2000. Infect. Genet. Evol. 11 (2), 522–527.

Nakagomi, O., Nakagomi, T., 1996. Molecular epidemiology of human rotaviruses: genogrouping by RNA-RNA hybridization. Arch. Virol. Suppl. 12, 93–98.

Papp, H., Borzák, R., Farkas, S., Kisfali, P., Lengyel, G., Molnár, P., et al., 2013a. Zoonotic transmission of reassortant porcine G4P[6] rotaviruses in Hungarian pediatric patients identified sporadically over a 15 year period. Infect. Genet. Evol. 19, 71–80.

Papp, H., László, B., Jakab, F., Ganesh, B., De Grazia, S., Matthijnssens, J., et al., 2013b. Review of group A rotavirus strains reported in swine and cattle. Vet. Microbiol. 165 (3–4), 190–199.

Papp, H., Matthijnssens, J., Martella, V., Ciarlet, M., Bányai, K., 2013c. Global distribution of group A rotavirus strains in horses: A systematic review. Vaccine 31 (48), 5627–5633.

Parashar, U.D., Hummelman, E.G., Bresee, J.S., Miller, M.a, Glass, R.I., 2003. Global illness and deaths caused by rotavirus disease in children. Emerg. Infect. Dis. 9 (5), 565–572.

Parra, G.I., Bok, K., Martínez, M., Gomez, J.A, 2004. Evidence of rotavirus intragenic recombination between two sublineages of the same genotype. J. Gen. Virol. 85 (6), 1713–1716.

Patel, M.M., Pitzer, V.E., Alonso, W.J., Vera, D., Lopman, B., Tate, J., et al., 2013. Global seasonality of rotavirus disease. Pediatr. Infect. Dis. J. 32 (4), e134–e147.

Payne, D.C., Staat, M.A., Edwards, K.M., Szilagyi, P.G., Weinberg, G.A., Hall, C.B., et al., 2011. Direct and indirect effects of rotavirus vaccination upon childhood hospitalizations in 3 US Counties 2006–2009. Clin. Infect. Dis. 53 (3), 245–253.

Phan, T.G., Okitsu, S., Maneekarn, N., Ushijima, H., 2007. Evidence of intragenic recombination in G1 rotavirus VP7 genes. J. Virol. 81 (18), 10188–10194.

Phillips, G., Lopman, B., Rodrigues, L.C., Tam, C.C., 2010. Asymptomatic rotavirus infections in England: prevalence, characteristics, and risk factors. Am. J. Epidemiol. 171 (9), 1023–1030.

Pitzer, V.E., Atkins, K.E., de Blasio, B.F., Van Effelterre, T., Atchison, C.J., Harris, J.P., et al., 2012. Direct and indirect effects of rotavirus vaccination: comparing predictions from transmission dynamic models. PLoS One 7 (8), e42320.

Pitzer, V.E., Bilcke, J., Heylen, E., Crawford, F.W., Callens, M., De Smet, F., Van Ranst, M., Zeller, M., Matthijnssens, J., 2015. Did large-scale vaccination drive changes in the circulating rotavirus population in Belgium? Sci. Rep. 5, 18585.

Pitzer, V.E., Patel, M.M., Lopman, B.a, Viboud, C., Parashar, U.D., Grenfell, B.T., 2011a. Modeling rotavirus strain dynamics in developed countries to understand the potential impact of vaccination on genotype distributions. Proc. Natl. Acad. Sci. USA 108 (48), 19353–19358.

Pitzer, V.E., Viboud, C., Lopman, B a., Patel, M.M., Parashar, U.D., Grenfell, B.T., 2011b. Influence of birth rates and transmission rates on the global seasonality of rotavirus incidence. J. R. Soc. Interface 8 (64), 1584–1593.

Pitzer, V.E., Viboud, C., Simonsen, L., Steiner, C., Panozzo, C.A., Alonso, W.J., et al., 2009. Demographic variability, vaccination, and the spatiotemporal dynamics of rotavirus epidemics. Science 325 (5938), 290–294.

Saif, L.J., Jiang, B., 1994. Nongroup A rotaviruses of humans and animals. Curr. Top. Microbiol. Immunol. 185, 339–371.

Sanchez-Padilla, E., Grais, R.F., Guerin, P.J., Steele, A.D., Burny, M.-E., Luquero, F.J., 2009. Burden of disease and circulating serotypes of rotavirus infection in sub-Saharan Africa: systematic review and meta-analysis. Lancet Infect. Dis. 9 (9), 567–576.

Santos, N., Hoshino, Y., 2005. Global distribution of rotavirus serotypes/genotypes and its implication for the development and implementation of an effective rotavirus vaccine. Rev. Med. Virol. 15 (1), 29–56.

Song, X.F., Hao, Y., 2009. Adaptive evolution of rotavirus VP7 and NSP4 genes in different species. Comput. Biol. Chem. 33 (4), 344–349.

Taniguchi, K., Urasawa, T., Morita, Y., Greenberg, H.B., Urasawa, S., 1987. Direct serotyping of human rotavirus in stools by an enzyme-linked immunosorbent assay using serotype 1-, 2-, 3-, and 4-specific monoclonal antibodies to VP7. J. Infect. Dis. 155 (6), 1159–1166.

Tate, J.E., Burton, A.H., Boschi-Pinto, C., Steele, A.D., Duque, J., Parashar, U.D., 2012. 2008 Estimate of worldwide rotavirus-associated mortality in children younger than 5 years before the introduction of universal rotavirus vaccination programmes: a systematic review and meta-analysis. Lancet Infect. Dis. 12 (2), 136–141.

Tate, J.E., Haynes, A., Payne, D.C., Cortese, M.M., Lopman, B.A., Patel, M.M., et al., 2013. Trends in national rotavirus activity before and after introduction of rotavirus vaccine into the national immunization program in the United States, 2000 to 2012. Pediatr. Infect. Dis. J. 32 (7), 741–744.

Theuns, S., Heylen, E., Zeller, M., Roukaerts, I.D.M., Desmarets, L.M.B., van Ranst, M., Nauwynck, H.J., Matthijnssens, J., 2015. Complete genome characterization of recent and ancient Belgian pig group A rotaviruses and assessment of their evolutionary relationship with human rotaviruses. J. Virol. 89 (2), 1043–1057.

Velázquez, F., Matson, D., Calva, J.J., Guerrero, M.L., Morrow, A.L., Carter-Campbell, S., et al., 1996. Rotavirus infection in infants as protection against subsequent infections. N. Engl. J. Med. 335 (14), 1022–1028.

Ward, R., 2009. Mechanisms of protection against rotavirus infection and disease. Pediatr. Infect. Dis. J. 28 (Suppl. 3), S57–S59.

Ward, R., Nakagomi, O., Knowlton, D.R., McNeal, M.M., Nakagomi, T., Clemens, J.D., et al., 1990. Evidence for natural reassortants of human rotaviruses belonging to different genogroups. J. Virol. 64 (7), 3219–3225.

WHO. 2002. Generic protocols for (i) hospital-based surveillance to estimate the burden of rotavirus gastroenteritis in children and (ii) a community-based survey on utilization of health care services. WHO, Geneva.

WHO. 2009. Manual of rotavirus detection and characterization methods. Available from: http://whqlibdoc.who.int/hq/2008/who_ivb_08.17_eng.pdf

Woods, P.A., Gentsch, J., Gouvea, V., Mata, L., Simhon, A., Santosham, M., et al., 1992. Distribution of serotypes of human rotavirus in different populations. J. Clin. Microbiol. 30 (4), 781–785.

Woods, R.J., 2015. Intrasegmental recombination does not contribute to the long-term evolution of group A rotavirus. Infect. Genet. Evol. 32, 354–360.

Wu, F.-T., Bányai, K., Jiang, B., Wu, C.-Y., Chen, H.-C., Fehér, E., et al., 2014. Molecular epidemiology of human G2P[4] rotaviruses in Taiwan, 2004-2011. Infect. Genet. Evol. 28 (161), 530–536.

Wyatt, R.G., James, H.D., Pittman, A.L., Hoshino, Y., Greenberg, H.B., Kalica, A.R., et al., 1983. Direct Isolation in Cell Culture of Human Rotaviruses and Their Characterization into Four Serotypes. J. Clin. Microbiol. 18 (2), 310–317.

Zeller, M., Rahman, M., Heylen, E., De Coster, S., De Vos, S., Arijs, I., et al., 2010. Rotavirus incidence and genotype distribution before and after national rotavirus vaccine introduction in Belgium. Vaccine 28 (47), 7507–7513.

Chapter 2.11

Rotavirus Vaccines and Vaccination

T. Vesikari
Vaccine Research Center, University of Tampere Medical School, Tampere, Finland

1 INTRODUCTION

The discovery of RVs in 1973 by Bishop et al. (1973) and Flewett et al. (1973) was a major breakthrough because up to that point the etiology of acute gastroenteritis (GE) in young children was largely unknown. After this discovery it soon became apparent that RVs caused approximately one half of the cases of acute GE in industrialized countries and were also the most common causative agents of childhood diarrhea in developing countries (Steinhoff, 1980; Maki, 1981).

A recent estimate of the mortality attributable to RV disease puts the number at 453,000 deaths in 2008 in children less than 5 years of age (Tate et al., 2012). This may be an overestimate since total diarrheal disease mortality in children was estimated to be less than 0.6 million in 2012 (Liu et al., 2015), and RV disease-associated mortality of young children was at around 200,000 deaths in 2011 (Walker et al., 2013). Still, mortality from RVGE has declined less than the total mortality from diarrheal disease and other measures of diarrheal disease control, such as oral rehydration therapy, improvement of hygiene and sanitation, and promotion of breastfeeding have had proportionally less effect on RVGE associated deaths than mortality from other diarrhea. This emphasizes the global need for RV vaccination, and RVGE ranks high among vaccine-preventable diseases.

The incidence of RVGE in developed countries is not much different from that in developing countries, but the outcome is. In Europe, deaths from RVGE are rare and the main argument for RV vaccination is prevention of hospitalization (Soriano-Gabarro et al., 2006). Even severe RVGE can be successfully managed in appropriate facilities that are available in developed countries. Nevertheless, the episodes of severe RVGE cause distress to the infants and anxiety for the families, not to mention the costs of hospitalization. Thus, there are compelling medical and financial arguments for universal RV vaccination to be established also in developed countries.

Viral Gastroenteritis

2 ROTAVIRUS VACCINES

All past and current RV vaccines are live viruses given orally to multiply in the intestines and induce immunity mimicking that after natural RV infection. The vaccines can be divided into three categories: heterologous animal RVs, animal-human reassortant RV strains, and attenuated human RV strains (Table 2.11.1).

2.1 Heterologous (Animal) Rotavirus Vaccines

2.1.1 RIT4237 (NCDV) Bovine RV Vaccine

Bovine RV strain RIT4237 (G6P[5]) was the first RV vaccine tested in humans. The vaccine is highly attenuated with a history of 154 cell culture passages, much more than any other RV vaccine strain. The high level of attenuation was meant to minimize risk of RV disease in calves in case of transmission (and for possible use as veterinary vaccine) and was not for attenuation in humans as bovine RV would not cause any symptomatic disease in humans anyway. However, the adaptation to cell culture resulted in growth to a high titer over 10^8 $TCID_{50}$ /mL (Delem et al., 1984), which may explain the good efficacy of this vaccine in the early trials in Finland.

In its first efficacy trials in 1983, a single dose of RIT4237 vaccine induced 50% protection against any and 88% protection against severe (clinically significant) RV disease (Vesikari et al., 1984a), a result, which remains a benchmark for other vaccines and trials. Two doses of the same did not improve efficacy (Vesikari et al., 1985a). In Peru the efficacy was 40% against any and 75% against severe RVGE (Lanata et al., 1989) and in The Gambia the efficacy was 33% but severity of RVGE was not measured (Hanlon et al., 1987). Thus, a gradient of efficacy between developed and developing countries was observed, which has later been confirmed to be true for other RV vaccines as well.

The higher efficacy of RV vaccine against severe RV disease was also documented using a 20-point score, later often referred to as "Vesikari-score" (Ruuska and Vesikari, 1990). Figure 2.11.1 shows the efficacy of neonatal vaccination with bovine RV vaccine on RV disease of varying severity (Ruuska, 1991). The same score has later been applied to measure efficacy of the other RV vaccines.

The series of studies of RIT4237 bovine RV vaccine also established the following general principles that were largely confirmed later for other RV vaccines: (1) a higher titer of oral inoculum resulted in greater uptake (Vesikari et al., 1985b), (2) RV vaccine virus was sensitive to gastric acidity, and buffering such as milk feeding was needed for successful vaccine uptake (Vesikari et al., 1984b), (3) breast milk or breast feeding did not suppress the uptake to any significant degree, especially if the titer of vaccine was high (Vesikari et al., 1985b, 1986a). Earlier studies also showed that concurrent administration of live oral polio vaccine (OPV) suppressed the uptake of bovine RV vaccine, whereas RV vaccine had only minimal inhibitory effect (interference) on OPV (Vodopija et al., 1986; Giammanco et al., 1988).

The vaccine was withdrawn from further development for a combination of reasons: (1) the level of protection in developing countries was perceived

TABLE 2.11.1 Live Oral Rotavirus Vaccines Licensed or at Late Stage Development

Vaccine	Strain characteristic	Developer/manufacturer	Status in development
		Animal RV strains	
RIT4237	G6P[5]	SmithKline-RIT	Withdrawn
WC3	G6P[7]	Institut Merieux	Withdrawn
RRV	G3P[5]	NIH	Withdrawn
LLR	Lamb strain G10P[15]	Lanzhou Institute of Biologica products (China)	Licensed and in use in China
		Animal–human reassortant strains	
RotaShield®	Tetravalent human–rhesus reassortant G1–G4 + P7[5]	Wyeth (United States)	Withdrawn 1999, Phase II trials of neonatal administration in Ghana
RotaTeq®	Pentavalent human–bovine reassortant G1–G4 + P7[5]; G6 + P1A[8]	Merck (United States)	Licensed worldwide
BRV-TV/ BRV-PV	Bovine–human tetravalent/ pentavalent reassortants G1–G4 + P7[5]/ G1–G4 + G9–P7[5]	Biotecnics Instituto Butantan (Brazil) Shantha Biotecnics Limited (India) Serum Institute of India (India) Wuhan (China)	Phase I trial Phase II/III trials Phase III trial
		Human RV strains	
Rotarix®	Human strain G1P1A[8]	GlaxoSmithKline (Belgium)	Licensed worldwide
Rotavac®	Human neonatal strain 116E G9P[11]	Bharat Biotech International Limited (India)	Licensed in India
Rotavin-M1®	Human strain G1P1A[8]	Polyvan (Vietnam)	Licensed and in use in Vietnam
RV3-BB	Human neonatal strain G3P2A[6]	Murdoch Childrens Research Institute (Australia) and Biofarma (Indonesia)	Phase III trial

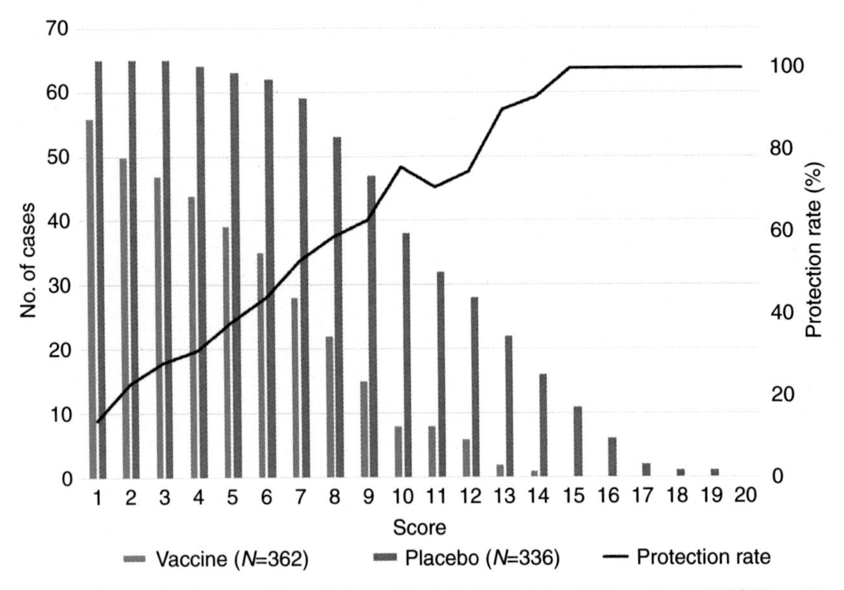

FIGURE 2.11.1 **Efficacy of neonatal vaccination with bovine RV vaccine RIT4237 on the severity of subsequent wild-type RV infections in a cohort of infants followed from birth to the age of 2–2½ years.** The line shows increasing vaccine efficacy with increasing severity of RV gastroenteritis. *(Source: Adapted from Ruuska, 1991.)*

as insufficient. (2) The concept of protection against severe disease only was new, and this end point was regarded as inadequate. (3) There was concern by the WHO that oral RV vaccine might compromise the success of the poliovirus eradication program, which was considered to be of the highest priority. The same issues remain relevant today for the current licensed vaccines, but are no longer regarded as obstacles for introduction of RV vaccination.

2.1.2 WC-3 Bovine RV Vaccine

WC-3 stands for Wistar Calf according to Wistar Institute where the vaccine was developed (Clark et al., 1986). This bovine G6P[7] RV strain is less adapted to cell culture (about 20 passages) and therefore grows only to a titer of approximately 10^7 PFU/mL. In the main efficacy trial in the United States the WC3 vaccine yielded a mere 20% protection against RVGE, and was not developed further (Bernstein et al., 1990). However, the strain is important because it is the backbone for bovine–human reassortants that are included in the current "pentavalent" (RV5) bovine–human reassortant RV vaccine (see later).

2.1.3 Rhesus RV (RRV) Vaccine

Rhesus (monkey) RV strain (G3P[5]) was grown 9 times in monkey kidney and 7 times in fetal rhesus lung cells and adopted as human vaccine

(Kapikian et al., 1986). Unlike bovine RV vaccine, RRV is clearly virulent in humans by causing fever, though not diarrhea (Vesikari et al., 1986b). At a high titer level of 10^5 PFU/dose the RRV vaccine was efficacious but reactogenic (Flores et al., 1987; Gothefors et al., 1989). Importantly, however, RRV served as a backbone for the development of the rhesus–human reassortant vaccines (Midthun et al., 1985), of which the tetravalent composition (RRV-TV) became the first licensed RV vaccine in 1998 (see later).

2.1.4 Lamb RV Vaccine

Lamb rotavirus strain LLR-37 (G10P[15]) was developed in Lanzhou Institute for Vaccines, China, and is a licensed RV vaccine in China (Li et al., 2015). No formal efficacy trial nor any head-to-head comparison of LLR-37 with other RV vaccines has been done, but there is post marketing evidence for efficacy (Fu et al., 2012). The vaccine is recommended for two doses at ages 2 months–2 years, and widely available in the private market in China. LLR has also served as a backbone for the development of a trivalent lamb-human reassortant vaccine, which is undergoing clinical trials in China.

2.2 Animal–Human Rotavirus Reassortant Vaccines

2.2.1 Rhesus–Human Reassortant Vaccine

Rhesus–human reassortant tetravalent (RRV-TV) vaccine (RotaShield, Wyeth) is the prototype of reassortant vaccines and contains four viruses expressing human G1, G2 and G4 VP7 antigens on the rhesus G3P[5] RV genetic backbone plus the RRV itself. RRV-TV retained the high reactogenicity (for fever) of RRV and accordingly, a relatively low dose of the vaccine (4×10^4 PFU) was used in the licensed formulation. RotaShield vaccine was in use in the United States in 1998–99, but was withdrawn because of association with intussusception (IS) (Murphy et al., 2001; Simonsen et al., 2005).

RRV-TV is a more efficacious vaccine than the parent rhesus RV vaccine (Rennels et al., 1996). In the pivotal efficacy trials before licensure of RotaShield the RRV-TV vaccine showed an efficacy of 100% against any hospitalization for RVGE and against any RVGE 68% in Finland (Joensuu et al., 1997) and 75 and 48%, respectively, in Venezuela (Pérez-Schael et al., 1997). These results were instrumental for wider acceptance of protection against severe RVGE as the primary end point of RV vaccine efficacy and for acknowledgment of the somewhat lower vaccine efficacy in developing countries as being satisfactory, thus considering the introduction of RV vaccination as worthwhile.

RotaShield vaccine was withdrawn in 1999 in the United States because of association with IS (see earlier), before it was launched in Europe or tested in developing countries. Recently, RRV-TV, given to neonates (to minimize risk of IS (Vesikari et al. 2006d)), was tested in Ghana with promising results, and the possibility of reintroduction in developing countries remains viable.

The vaccine efficacy was 63% for two doses (Armah et al., 2013). Whether transplacentally transmitted maternal RV-specific antibodies might have reduced the efficacy can only be speculated.

2.2.2 Bovine–Human Reassortant Vaccine

The "pentavalent" bovine–human reassortant RV vaccine (RotaTeq®, Merck, also termed RV5) is a combination of four G-type reassortants (for G1–G4) and one P-type (P[8]) reassortant on the WC-3 bovine RV genetic backbone (Clark et al., 1996). Since the WC-3 is a G6P[5] virus, these bovine G and P types are also present in the vaccine. The terms pentavalent or RV5 refer to the five mono-reassortants present in the vaccine. However, the term is not accurate as there are more RV types represented in the vaccine and more importantly because it is now well established that the protection induced by the vaccine is not limited to the G or P types contained in the product (see later).

The dose of the RotaTeq vaccine is approximately 10^6 PFU and was determined in a small scale dose-finding efficacy trial, which showed that a dose one log higher did not much improve efficacy (and would be more expensive to produce) whereas a dose one log lower was clearly less efficacious (Vesikari et al., 2006a). The vaccine is given in three doses. This was determined early on to accommodate the US childhood immunization program, but was also based on the demonstration of incremental immunogenicity up to the third dose (Clark et al., 2003).

The efficacy and safety of RotaTeq vaccine were established in a large (70,000 infants) Rotavirus Efficacy and Safety Trial (REST). The overall efficacy against health care utilization (combined end point of hospital admission and outpatient clinic treatment) was about 95% with a narrow confidence interval (Vesikari et al., 2006b). An extension study of REST in Finland of 21,000 children confirmed that RotaTeq was not only efficacious against G1, G3, and G4, all P[8], but also against G9P[8] and the fully heterologous G2 P[4], which are not among the G-types in the vaccine. The level of efficacy against various RV G-types was not significantly different (Vesikari et al., 2010). RotaTeq was licensed in 2006 and is now one of the two major RV vaccines used globally (the other one being Rotarix®, see later)

The G1 and P[8] reassortants included in RotaTeq vaccine may re-reassort with each other and form vaccine-derived (vd) double reassortants on the bovine RV VP6 core (Donato et al., 2012; Hemming and Vesikari, 2012). vdG1P[8] viruses may be more virulent than the original single reassortant vaccine viruses and they may be responsible for the low rate (about 1%) of diarrhea seen after vaccination, and may also be capable of transmission (Payne et al., 2010). If transmitted to immunocompromised subjects, vaccine viruses may cause prolonged infection (Patel et al., 2010). Shedding of live infectious virus after the first dose of vaccination was found in about 9% in the REST study, but using RT-PCR shedding may be found in about 50% of the recipients, with G1 reassortants being the most common ones (Markkula et al., 2015).

UK-bovine RV strain is another platform for reassortment. However, development of UK-bovine tetravalent (G1–G4) reassortant vaccine was discontinued by Wyeth at the same time as RotaShield vaccine was withdrawn. The technology has been licensed (by NIH) to several manufacturers in India, China and Brazil, but none of the vaccines have yet been licensed. The tetravalent UK-bovine reassortant was tested in Finland head to head with RRV-TV and found to have similar efficacy (Vesikari et al., 2006c).

2.3 Human Rotavirus Vaccines

2.3.1 Rotarix

Human RV vaccine RV1 (Rotarix®, GSK, also termed RV1) is the most extensively used RV vaccine today. It was derived from a G1P[8] RV isolate in Cincinnati, passaged 33 times in cell culture and designated 89-12 (Bernstein et al., 1999). The strain was acquired by GSK, cloned (by plaque purification) and passaged another 12 times in MRC-5 cells. In this process the virus lost its residual reactogenicity and is generally regarded as nonreactogenic for humans (Vesikari et al., 2004). The titer of the vaccine chosen for the final formulation is about 10^5 PFU/dose, although the virus could be grown to a one log higher titer. Rotarix multiplies effectively in humans, as characterized by a high rate of shedding (60% or even more) after the first dose, but does not cause diarrhea (Vesikari et al., 2004).

Rotarix is given in two doses. The uptake and immunogenicity are excellent (90%) even after the first dose when given in the presence of low level of maternal antibody, such as in Finland. The second dose under such circumstances does not add much but the uptake is prevented by the immunity induced after the first dose, as indicated by the lack of shedding and lack of booster response (Vesikari et al., 2004). The pivotal safety and efficacy trial for licensure was carried out in 60,000 children in Latin America (Ruiz-Palacios et al., 2006). In addition, before licensure in Europe, the vaccine was tested in five European countries and showed high efficacy (Vesikari et al., 2007). In 2004, Rotarix was the first new RV vaccine to be licensed after the withdrawal of RotaShield.

2.3.2 Other Human RV Vaccine Strains

Another human RV vaccine has been prepared from G1P[8] RV isolate in Vietnam and designated as "Rotavin" (Luan le et al., 2009). This vaccine is licensed in Vietnam, and is primarily intended for the local market only.

RV3 is based on a neonatal G3P[6] RV strain from Melbourne. A Vero cell adapted derivative of the strain, called RV3-BB, grows to a higher titer and is more immunogenic than the original RV3 (Danchin et al., 2013). The vaccine will be manufactured in Indonesia and is undergoing a phase III trial.

The RV 116E is an Indian neonatal strain of genotype G9P[11]. Natural attenuation of the neonatal strain by a modified P-antigen (VP4 gene of bovine origin) was regarded as an advantage in the selection of this strain as a candidate vaccine (Das et al., 1993). The efficacy of 116E against severe RVGE in a trial in India was about 55% for three doses (Bhandari et al., 2014), and the vaccine (Rotavac®, Bharat Biotech) is now licensed in India and is being introduced into the National Immunization Program (NIP) of India.

3 IMMUNE RESPONSE AND MECHANISM OF ACTION

While both single human strain vaccines (eg, Rotarix) and multiple reassortant strain vaccines (eg, RotaShield and RotaTeq) are highly efficacious it has been difficult to find, and agree on, immunological correlates of vaccine-induced protection against severe RVGE.

In practice, total RV serum IgA antibody has been the most widely used immunological marker for uptake of RV vaccine. From studies on consequent wildtype RV infections it is known that a high titer of serum RV IgA antibody correlates with protection (Ward et al., 1992; Velazquez et al., 1996). This can be applied to vaccine-induced immunity as well (Patel et al., 2013). High serum IgA responses are consistently induced by efficacious RV vaccines, such as Rotarix and RotaTeq, and it is logical to assume that serum IgA mediated immunity is not only a surrogate marker but also an actual mechanism of RV vaccine-induced protection against severe RV disease.

Wildtype RV infections induce neutralizing antibodies against the RV surface antigens VP7 and VP4, which protect against subsequent infections by the same serotype. Such antibodies are also induced after vaccinations, but the responses are inconsistent and the antibody levels often low (Vesikari, 2012). RV vaccines do not induce much protection against RV infection and it is unlikely that these neutralizing antibodies are a major mechanism of action that reduces the severity of RVGE, although they may play some role in vaccine-induced protection.

Importantly, the serum neutralizing antibodies may reflect also neutralizing antibodies in the gut. G-type specific immunity induced by the vaccines may be strong enough to "protect" against the homologous vaccine virus, that is, the subsequent dose of the vaccine itself. This is clearly seen by the lack of a booster effect to a monospecific human G1P[8] vaccine (Vesikari et al., 2004) and RRV (Rennels et al., 1996). The second dose of RV1 only "takes" if the first dose has not taken. Giving three or even five doses of monovalent human RV vaccine does not much improve the uptake of this vaccine in developing countries (Kompithra et al., 2014). The same is true for monotypic bovine RV vaccine, of which it is known that a second dose does not improve efficacy (Vesikari et al., 1985a). In contrast, the reassortant vaccines may have an advantage of multiple doses "taking," as it is unlikely that all reassortants succeed in taking upon the first dose. A second and third dose of the "multivalent" vaccine may have a better opportunity of infection and multiplication and inducing a booster effect, that is, an increase of RV IgA serum antibody level than a monotypic vaccine.

The target antigen of serum RV IgA is largely the RV inner core protein VP6. VP6 is a strong immunogen, and immune responses to it overshadow those to other structural RV antigens (Svensson et al., 1987a,b). Anti-VP6 antibodies are "nonneutralizing" in the sense that they do not prevent RV infection in vitro or in vivo (Feng et al., 2002; Burns et al., 1996). However, intracellularly acting anti-VP6 IgA antibodies block the infection at a later stage and conceivably, may protect against RV disease (Corthésy et al., 2006). A plausible mechanism of the intracellular action may be that the anti-VP6 antibodies stabilize the VP6 inner core particle and prevent the release of RV RNA from the inner core particle (Aiyegbo et al., 2013).

An intracellular mechanism of effect is also consistent with the concept of RV vaccines protecting against disease rather than infection.

4 COMPARATIVE EFFICACY

The two internationally licensed RV vaccines, human RV vaccine Rotarix and the bovine–human reassortant RV vaccine RotaTeq, have not been tested head-to-head for efficacy in any trial. On the other hand, both vaccines have been tested for efficacy in different environments, from developed to "intermediate" to developing countries, and the results of such studies can be compared with each other. Some examples are shown in Fig. 2.11.2. In general, the overall and serotype-specific efficacy against severe RVGE of the two vaccines are remarkably similar, but depend on the environment.

4.1 Developed Countries

The major European efficacy trial of Rotarix was conducted in five European countries in a total of about 3600 children. The follow-up was for two RV seasons and the primary end-point was severe RVGE as defined by score 11/20. Against this end-point, the efficacy for 2 years was 91%, with 96% efficacy in the first season and 86% in the second season, suggesting some decline over time. Against any RVGE the efficacy was 78% and 68% in the first year and in the second year, for a total of 72% over 2 years. Furthermore, against severe RVGE on 2 years of follow-up, the efficacy for various G-types was similar, ranging from 96% for G1P[8] to 86% for G2P[4], the differences were not statistically significant. For any RVGE, the efficacy point estimates were higher for G1, G3, G4, and G9 with P[8] than for with G2P[4] with 58% (Vesikari et al., 2007). The European efficacy trial was duplicated in Japan in a smaller population using a similar protocol. The point estimate for efficacy against severe RVGE over 2 years was the same as in Europe, that is, 91% (Kawamura et al., 2011).

Rotarix was also tested for efficacy in three affluent Asian countries, Taiwan, Hongkong, and Singapore. The end-point was severe RVGE as seen in hospital or outpatient clinics. The point estimates for efficacy in the 1st, 2nd

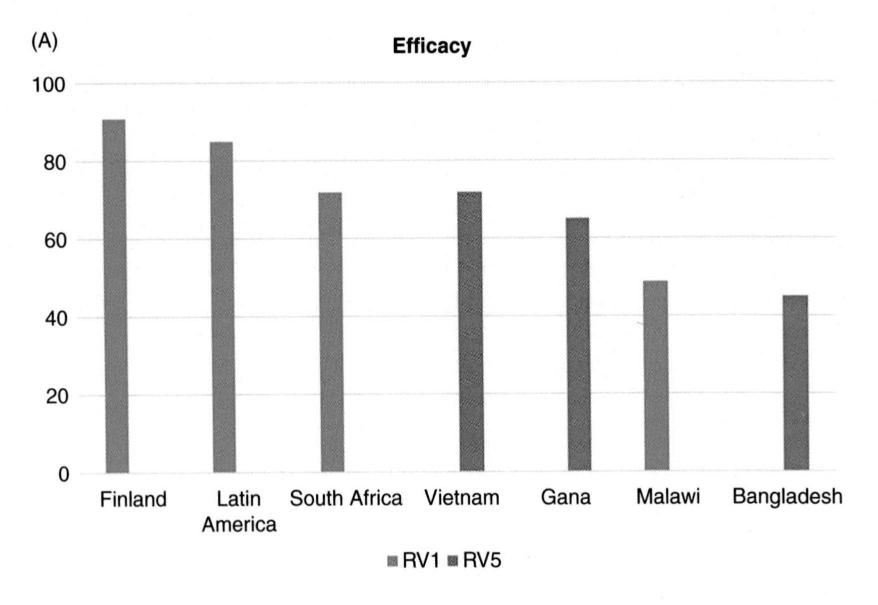

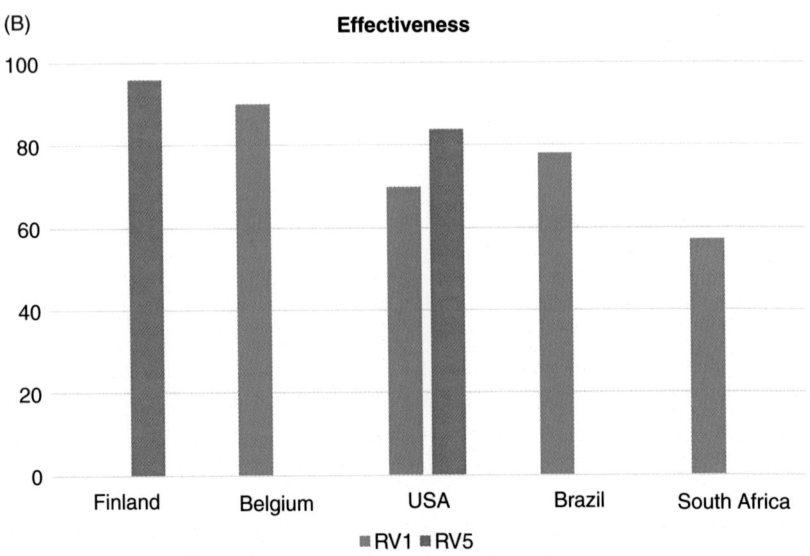

FIGURE 2.11.2 Representative efficacy point estimates against severe RVGE with score ≥11/20 of RV5 and RV1 in prelicensure trials (A) and postlicensure real life effectiveness studies for RVGE hospitalizations (B) in different geographical areas.

and 3rd year were 91%, 100%, and 90%, respectively (Phua et al., 2012). However, the numbers of cases remained small, and therefore this study is not equally conclusive as the European study. By comparison, RotaTeq was only tested in Japan in a relatively small trial with effectiveness of 90% over 2 years (Iwata et al., 2013).

The pivotal efficacy trial of RotaTeq was REST (Vesikari et al., 2006b). This trial included a detailed efficacy subset, the European part of which was about the same size as the European trial of Rotarix, and therefore the two would be comparable. However, the end-point was RVGE as scored by the Clark scale of 24 points (Clark et al., 2004). For severe RVGE defined as 16/24 points, the efficacy in the first year was 100% and in the second year 94%; for any RVGE the corresponding figures were 72% and 58% (Vesikari et al., 2009).

In the main REST, the primary end point was combined efficacy against hospitalization and outpatient clinic visits for RVGE. The efficacy in one-year follow-up was 95%. The follow-up was extended in Finland for up to 3 years (mean 2.3 years) in a cohort of 21,000 children. This Finnish extension study showed that the efficacy was sustained at the same level (95–96%) for up to the third year, and the larger follow-up made it possible to determine efficacy against five RV G-types with P[8], that is, G1, G3, G4, and G9 (not in the vaccine) as well as G2P[4]. The point estimate for G2P[4] was 83%, that is, slightly but not significantly lower than for the other G-types (Vesikari et al., 2010).

4.2 Latin America

Rotarix was extensively studied in Latin America whereas RotaTeq was not. In the pivotal trial for licensure of Rotarix about 20,000 children in 11 Latin American countries participated in the efficacy follow-up. Two end points were used: need for (oral) rehydration and disease score of 11/20. Both yielded approximately 85% efficacy in one-year follow-up (Ruiz-Palacios et al., 2006). An earlier, smaller, efficacy trial in three countries had yielded a similar overall efficacy, but suggested a significantly lower efficacy against G2P[4] (Salinas et al., 2005).

After licensure, another efficacy trial of Rotarix was conducted in Argentina and Chile. This study specifically looked at efficacy when Rotarix was coadministered with OPV; the resulting efficacy for severe RVGE was 86% (Tregnaghi et al., 2011).

REST trial included a minority of subjects enrolled in the Caribbean and Latin America, mainly Jamaica and Guatemala. In this subgroup the point estimate for efficacy of RotaTeq against health care resource utilization for RVGE was 90%, that is, almost as good as in the United States and Europe, but the confidence interval was wide (29–95%) (Vesikari et al., 2009).

Altogether, the efficacy of the currently licensed RV vaccines in Latin America appears to be about 10% lower than in Europe and affluent Asian countries.

4.3 Africa and South East Asia

Both Rotarix and RotaTeq were tested for efficacy in Africa, and RotaTeq also in less developed Asian countries Vietnam and Bangladesh. Apart from a generally lower efficacy, a characteristic finding in comparison with developed countries has been a sharp decline of efficacy from the first to second year of life. On the other hand, in developing countries the majority of cases of severe RVGE occur in the first year of life.

In South Africa and Malawi, the efficacy of two doses of Rotarix against severe RVGE (score 11/20) was 61.2 and 49.4%, respectively in 1 year follow-up (Madhi et al., 2010). Also in these African countries, there was no difference in vaccine efficacy against G1 and non-G1 (G2, G8, and G12) RVs (Madhi et al., 2012).

RotaTeq was tested in Kenya, Ghana, and Mali. The point estimates for three doses of RotaTeq were 83% in Kenya and 65% in Ghana for the first year of follow-up, with little or no efficacy in the second year. In Mali the results showed no efficacy at all (Armah et al., 2010).

RotaTeq was also tested in two developing countries in Asia, Vietnam, and Bangladesh. In Vietnam the efficacy against severe RVGE was 73% whereas in rural Bangladesh it was only 45%. Neither point estimate was significant because of small numbers, but the combined efficacy for "Asia" was 61%, which was statistically significant (Zaman et al., 2010).

To summarize, the results of efficacy trials of Rotarix and RotaTeq against severe RVGE show a general decline of efficacy from developed countries (Europe, United States, and Japan) to "intermediate" (Latin America) countries and further to developing countries in Africa and Asia (Fig. 2.11.2A). However, as the incidence of severe RVGE, and the number of resulting deaths, are higher in developing countries, it is generally held that even at a lower efficacy level these vaccines will prevent a substantial number of cases and deaths in Africa and Asia, and introduction of RV vaccination is justified (WHO, 2013).

In addition to Rotarix and RotaTeq, the Indian neonatal RV vaccine 116E (Rotavac, Bharat Biotech) has also been tested for efficacy in a proper controlled efficacy trial in India; when given in three doses this vaccine yielded a 55% efficacy (Bhandari et al., 2014). It is difficult to compare this result with those of the other licensed vaccines, since 116E has not been tested outside India, and no other RV vaccine has been tested for efficacy in India.

5 REAL LIFE EFFECTIVENESS

Studies on real life effectiveness of RV vaccines after introduction of immunization programs have been conducted in several countries and areas. As a whole there seems to be a similar gradient in vaccine effectiveness than in prelicensure efficacy trials between developed, "intermediate," and developing countries (Fig. 2.11.2), although the experience in Africa is still very limited. Since the Rotarix vaccine has also been used in more countries than RotaTeq, there are more data on this vaccine. In some countries the two vaccines have been used in parallel, allowing for limited head-to-head comparison.

A recent review has addressed both the difference between developed (high income) and middle income countries as well as the lack of serotype specificity in protection against severe RVGE in postlicensure effectiveness studies (Leshem et al., 2014). The gradient between developed and developing countries that was found in prelicensure studies is seen also in postlicensure real-life effectiveness studies. Some representative examples of the point estimates are shown in Fig. 2.11.2B.

In Europe, the examples of Finland and Belgium are well representative, showing over 90% effectiveness for hospitalization with RVGE over some period of time in countries that have reached a high coverage with RV5 (Finland) and RV1 (Belgium), respectively (Braeckman et al., 2012). In the United States, the 4-year follow-up of Payne et al reported 84% effectiveness for RotaTeq (Payne et al., 2013). In Brazil, representing Latin America, the point estimate of effectiveness is of the order of magnitude of 80% (Correia et al., 2010; Justino et al., 2011). In South Africa effectiveness of 57% was reported (Groome et al., 2014b).

5.1 Serotype (Genotype) Specific Effectiveness

The postintroduction studies have also collected a substantial amount of information on vaccine real life effectiveness against various RV genotypes (Leshem et al., 2014). Altogether, the postvaccination experience confirms the finding in prelicensure studies that RV vaccine effectiveness against severe RVGE is not strain-specific and that the terms "monovalent" for RV1 and "pentavalent" for RV5 are not functionally appropriate. Instead, these acronyms should only be used to refer to the number of constituents in each vaccine.

Predicted scenarios of serotype replacement have not materialized. New RV serotypes have not permanently replaced old ones, and RV vaccines have remained effective against all circulating RV strains. Soon after introduction of Rotarix vaccine in Latin America the G2P[4] RVs became dominant among the RVGE cases that remained in circulation and the shift was clearly in temporal relationship to widespread vaccinations. However, emergence of G2P[4] also happened in Latin American countries that did not introduce RV vaccine, and more recently, other G types, such as G1P[8] have again replaced G2P[4]. Even while G2P[4] was prevalent in Latin America, Rotarix G1P[8] vaccine remained effective against severe RVGE caused by this G type (Correia et al., 2010; Justino et al., 2011).

RV G type G3P[8] has possibly become more common after introduction of RV5 vaccine in Finland and the United States, but has never gained predominance (Hemming et al., 2013; Payne et al., 2013). G12P[8] has emerged in the United States and elsewhere, but has not remained predominant either. It is not possible to totally exclude the possibility that the emergence of these RV G types is related to introduction of RV vaccination, but such emergence has not been long lasting and of no clinical significance. In the United States, while G3P[8] and G12P[8] emerged as predominant G types, RotaTeq vaccine remained fully effective and specifically, showed the same effectiveness against G12P[8] as against the G-types included in the vaccine (Payne et al., 2013).

5.2 Impact of Rotavirus Vaccination—Direct and Indirect

In the first RV season after the introduction of RV vaccination in the United States, hospitals around the country reported that 2007–08 admissions due to

RV gastroenteritis were reduced by 90% or even more. As the coverage of RV vaccination was less than 50% this was too good to be true. Indeed, there was recurrence of RV activity in the following year, and since then RV epidemics have occurred biannually instead of every year (Payne et al., 2013). The US experience shows that RV GE cannot be eliminated at a vaccine coverage of 75–80%, as now is nationwide in the United States. More generally, the impact of universal RV vaccination cannot be reliably measured by short term follow-up only. First large scale introduction of RV vaccination will likely break the transmission chain of the wild type RV with resulting short term direct and indirect impact, but RV disease activity will resurge later.

The impact of RV vaccination in early introducing European countries was recently reviewed by Karafillakis et al. (2015). In Austria with coverage of 72–74% the reduction of RVGE hospitalizations in the target age group was 81–84% and remained sustained up to 3 years. However, while there was initially an indirect effect on unvaccinated children as well, after 3 years this was followed by an increase of RVGE hospitalizations in 5–14 year old children. In Belgium with a coverage rate of 90% the direct impact on RVGE hospitalization in the target group has been about 80% and the 3 year indirect effect in older children 20–64%. In Finland with RV vaccination coverage of 95–97%, the reduction in cases of RVGE seen in hospitals was 94% in a period of 4 years after vaccination, but specifically in age group 5–14 years no significant reduction has been seen over this period (Vesikari, unpublished). Actually, the direct impact in the target age group of RV vaccination has shifted the occurrence of RVGE to older unvaccinated children (Hemming et al., 2013).

It may be concluded that while a higher coverage results in greater impact of RV vaccination on severe RVGE, even a very high coverage does not lead to elimination of RVGE through herd protection. While RV vaccination has had a clear impact on RV disease in all age groups, as reported from the United States (Lopman et al., 2011; Gastañaduy et al., 2013) it remains to be seen if this effect is durable over time.

There is still only limited information on the impact of RV vaccination on deaths due to RVGE. An important milestone in this regard was the report of Richardson et al. (2011) from Mexico, showing that all diarrheal mortality in the RV vaccine eligible age group was decreased by 41% after vaccine introduction. This corresponds well to the share of RV in all severe gastroenteritis in Mexico. The decline of diarrheal mortality has also been confirmed in other Latin American countries. Further reports on RV mortality are awaited from Africa where many countries introduced RV vaccination in 2013–14 (Paternina-Caicedo et al., 2015).

Vaccine impact on all hospitalizations due to acute GE depends on the share of RV in all severe GE and the vaccine coverage. At best, the total reduction of hospitalizations from any GE may be as high as 70%, as observed in Finland over a period of 4 years (Vesikari, unpublished).

5.3 Introduction of Universal Rotavirus Vaccination

The first country to introduce universal RV vaccination was Brazil in March 2006, followed soon by Venezuela and Mexico, and about 20 more Latin American countries. With the exception of Nicaragua, which received RotaTeq as a donation from the manufacturer, all Latin American countries have chosen Rotarix vaccine for their programs. Universal recommendation for RotaTeq vaccine was issued in the United States soon after licensure in 2006 (Parashar et al., 2006). Other early introducers were Australia (both vaccines) and South Africa (Rotarix) since 2009.

In Europe, Austria (both vaccines) and Belgium (Rotarix) were the first introducers in 2007, followed by Finland (RotaTeq exclusively) in 2009. After that, there was a gap of a couple of years until Germany started vaccinations state by state. The most significant step forward is perhaps the UK introduction, and effective implementation, in 2014 (Marlow et al., 2015; Atchison et al., 2016).

Globally, the most significant recent progress has been introduction of RV vaccination in Africa since 2013, made possible by support from GAVI. As discussed earlier, the true impact of these programs can only be assessed after several years, although the first results may appear promising. With the exception of Rwanda, all the African countries have introduced Rotarix vaccine.

Eight of the 10 countries with highest mortality from RVGE are in Asia (Tate et al., 2012; Kawai et al., 2012) and none of these countries have implemented universal RV vaccination yet. India has announced a program using the locally produced Rotavac, and the Philippines has a government program for the poorest section of the population.

No country that has initiated a universal program has stopped it, but generally the programs are expanding in terms of coverage, and more countries are being added to the list. However, France has recently, in 2015, recalled the recommendation for RV vaccination for concerns of safety (intussusception), and Spain has withdrawn Rotarix vaccine for concern of porcine circovirus (PCV-1) contamination (see later).

5.4 Rotavirus Vaccine Recommendations

Before RV vaccine introduction and also apart from universal vaccination programs national recommendations for RV vaccination have been issued in many countries. A crucial one was the US Advisory Committee on Immunization Practices (ACIP) recommendation that was issued immediately after licensure of RotaTeq (Parashar et al., 2006). Globally, the most important one is the WHO position and universal recommendation (WHO, 2013). In Europe, there is no formal recommendation issuing body but the pediatric societies ESPID and ESPGHAN issued recommendations in 2008 that were updated as ESPID recommendations in 2015 (Vesikari et al., 2015).

All major recommendations have held that RV vaccination should be given to all children, simply because no special "risk groups" for RVGE can be reliably identified. Vaccination of special groups is a separate issue discussed later.

The scope and background of the recommendations differs and will have to consider programmatic and practical issues. For one thing, the recommendations for age of vaccination vary. The WHO recommendations start with the age of 6 weeks, but tend to allow the RV vaccine to be given without age restrictions, where the ESPID recommendations adhere to giving the first dose between 6 and 12 weeks of age, or, if possible, at 6–8 weeks of age. The implications of these differences are discussed under intussusception (IS) later.

5.5 Special Target Groups

Prelicensure efficacy trials of RV vaccine were conducted in healthy infants and had many exclusion criteria. Therefore, knowledge on safety and efficacy of RV vaccine in many special groups has accumulated slowly.

5.5.1 *Premature Infants*

Both RotaTeq and Rotarix vaccines can be given to prematurely born infants regardless of gestational age, but following the recommendations according to calendar age (Stumpf et al., 2013; Goveia et al., 2007; Omenaca et al., 2012). If the infant is still in hospital, a possible risk of transmission of vaccine virus must be considered locally. A much bigger issue is general neonatal immunization of full term and preterm infants which has not been explored with the currently licensed vaccines.

5.5.2 *HIV Infected Children*

Asymptomatic HIV-infected infants can be vaccinated normally according to calendar age without any safety issues using either Rotarix (Steele et al., 2011) or RotaTeq (Laserson et al., 2012). Screening for maternally acquired HIV-infection can often be done by the time of RV vaccination at 6–8 weeks of age, but the result is not needed for decision-making on RV vaccination (Vesikari et al., 2015)

5.5.3 *Immunodeficiency*

RV vaccine causes symptomatic disease (prolonged diarrhea and viral shedding) in children with severe combined immunodeficiency (SCID) (Werther et al., 2009; Bakare et al., 2010; Patel et al., 2010), and therefore vaccination is contraindicated and exposure to RV vaccine shedders should be avoided in such children. Other immunodeficiencies may be regarded similarly. Selective IgA deficiency may result in prolonged shedding of RV vaccine, but does not constitute a safety problem and, in any case, is usually not diagnosed by the time of RV vaccination (Vesikari et al., 2015).

5.5.4 *Short Gut Syndrome and Intestinal Failure*

Young children with intestinal failure are at risk for complications from RVGE and should, therefore be considered for RV vaccination. Two recent studies have shown that RV vaccine may cause substantial symptoms in such children, but given the severity of wildtype RV infection, they should nevertheless be vaccinated under close observation (Fang and Tingay, 2012; Javid et al., 2014).

5.6 Factors Associated With Reduced Uptake of Rotavirus Vaccine

5.6.1 Breast-Feeding

Early studies with bovine and rhesus RV vaccines showed little or no reduction of vaccine uptake in breast-fed infants compared with bottle-fed infants (Vesikari et al., 1986a, Ceyhan et al., 1993). Other studies have shown some inhibitory effect of breast-milk, particularly from developing countries, with high RV antibody content (Moon et al., 2013). The problem with the interpretation of the results is that women who have high levels of RV antibody in breast-milk are such that also have high levels of RV antibody in serum and serum antibody is more likely to prevent uptake of RV vaccine. The question of breast-feeding is largely theoretical, as there is no way recommending against breast-feeding for the uptake of RV vaccination. Specifically, it has been shown that a break of breast-feeding at the time of RV vaccine administration does not increase uptake (Rongsen-Chandola et al., 2014; Groome et al., 2014a).

5.6.2 Influence of Oral Polio Vaccine (OPV)

The early studies of bovine rotavirus vaccine administered with OPV showed that OPV suppressed the uptake of RV vaccine but not vice versa (Vodopija et al., 1986; Giammanco et al., 1988). The subject was recently reviewed by Patel et al. (2012) with the same conclusion. RV vaccines do not adversely affect OPV immunogenicity, but OPV does interfere with the uptake of (at least) the first dose of RV vaccine. Also the RV IgA antibody levels remain lower after a course of two doses of Rotarix or three doses of RotaTeq when administered concomitantly with OPV (Patel et al., 2012).

The suppressive effect by OPV on uptake of RV vaccine is probably greatest when the first dose of OPV is given concomitantly. However, in many studies the participating infants have already received a dose of OPV at birth and for the concomitant administration a second dose of OPV is given. This may multiply less efficiently than the first dose and therefore the suppressive effect may be less.

To avoid interference, spacing of OPV and RV vaccine (at least for the first dose) would be desirable (Vesikari et al., 2015). This must be balanced against programmatic needs. The time interval between the vaccines for no interference is not known but may be short.

If OPV suppresses/prevents the uptake of the first dose of Rotarix, the second dose 2 months later may become effectively the first dose. This could be the explanation why a small risk of IS after the second dose of Rotarix was identified in Brazil but not in Mexico (Patel et al., 2011).

5.6.3 Other Factors

Concomitant administration of OPV is likely one reason for lower uptake of RV vaccine in Africa. Similarly, it is probable that naturally occurring enterovirus infections interfere with RV vaccine virus in countries where enterovirus infections are highly prevalent at the time of EPI vaccination schedule.

Other intestinal factors, such as composition of microflora may also reduce the uptake of RV vaccine in developing countries (Harris, 2015). It would be important to find ways to modulate the microflora for more optimal conditions for RV vaccination, but no such intervention (eg, probiotics) is proven as yet. (See also Chapter 2.7.)

6 INTUSSUSCEPTION

Intussusception (IS) is the most important adverse effect of RV vaccination that has already caused the withdrawal of the first licensed RV vaccine, RotaShield, from the market. IS occurred mostly 3–7 days after the first dose of RotaShield, and the attributable risk was estimated at 1:10,000 (US consensus; Murphy et al., 2001). This rate was later challenged and a rate of 1:32,000 was proposed instead (Simonsen et al., 2005). Furthermore, the risk of IS was shown to be age-dependent, as most of the cases occurred in the catch-up vaccination programme in infants who were over 90 days of age at the time of the first dose (Simonsen et al., 2005).

It is now well established that both of the leading licensed RV vaccines, Rotarix and RotaTeq (and most likely other live attenuated RV vaccines as well), are also associated with IS. This could not be detected in prelicensure trials that were designed to rule out a risk of IS of similar magnitude as that of RotaShield. Later, in prospective surveillance studies after the launch, the risk estimates of IS for both vaccines are between 1:50,000 and 1:80,000, and the relative risks over background of both vaccines in the 7 day period after dose one are about 5 (GSK meta-analysis: http://www.gsk-clinicalstudyregister.com/study/ROTA-Meta-Analysis-201308#rs).

The small risk of IS is often weighed against the benefits of RV vaccination, and this comparison comes out in favor of vaccination in most settings (WHO, 2013). However, some consider IS preferentially as a risk only and for this reason France recently withdrew a recommendation for RV vaccination.

While it is not currently known whether the risk of IS associated with the currently licensed RV vaccines follows the same age related pattern as that of RotaShield it seems logical to assume that this is the case. It would seem prudent to follow the current ESPID recommendation and give the first dose of RV vaccine as early as possible, that is, at 6–8 weeks of age.

By the same logic, neonatal administration of live oral RV vaccines appears a better option than any current schedule. However, the currently licensed RV vaccines have not been studied for neonatal dose. RotaShield was actually studied, and a schedule including a dose at birth was not inferior for immunogenicity, but safer for (absence of) fever, than a schedule starting at 2 months of age (Vesikari et al., 2006d). RotaShield given according to a schedule beginning near birth (0–30 days of age) was also shown to be efficacious in Ghana (Armah et al., 2013). A general problem related to neonatal administration of live RV vaccines is possible suppression of uptake by transplacentally acquired maternal antibodies. Such antibodies are most likely targeted at common human RV G-types, specifically G1.

The Australian, Indonesian made, vaccine RV3-BB is based on a neonatal RV strain, and has from the beginning been called a "neonatal vaccine". The clinical trials are designed accordingly, and RV3-BB may well become the first RV vaccine licensed for neonatal administration.

7 PORCINE CIRCOVIRUS

In 2010 both licensed RV vaccines were found to have PCV as contaminants (Victoria et al., 2010; McClanahan et al., 2011). This created some concern, although PCV is not known to infect humans, but the WHO and EMA held that RV vaccines may continue to be used. Some European countries withdrew Rotarix temporarily, but the position is maintained only in Spain. In Rotarix, PCV contamination was traced to virus seed, but the manufacturer is committed to provide a PCV-free vaccine in the future. In RotaTeq the source of contamination was traced to batches of trypsin used in the manufacturing process (Ranucci et al., 2011), and with changes in the process PCV-free vaccine should be available.

8 NONLIVE ROTAVIRUS VACCINES

The need and rationale for the development of nonlive RV vaccines as alternatives for live oral RV vaccines are based on efficacy and safety concerns. As discussed earlier, IS remains a serious safety concern, although the magnitude of the problem is regarded as tolerable. This may, however, change. IS may be a class activity of all live attenuated RV vaccines. Also the possibility of contamination by adventitious agents like PCV is limited to live RV vaccines. As for efficacy, all live RV vaccines have shown a relatively (in comparison with developed countries) low efficacy in developing countries for reasons that may not be easily remedied, and repeated parenteral (or possibly intranasal) immunization might induce a high level of protection bypassing the intestinal obstacles.

Nonlive RV vaccines might be administered mucosally or parenterally. The types of vaccines under consideration include three main categories:

1. inactivated whole RVs (Jiang et al., 2008);
2. RV virus-like particles (VLPs) of double layered (VP2/VP6) or triple-layered (2/6/7 or 2/4/6/7) composition (Blutt et al., 2006; Istrate et al., 2008; El-Attar et al., 2009; Azevedo et al., 2010, 2013; Rodríguez-Limas et al., 2014);
3. recombinant RV proteins, such as VP8 or VP6.

The only nonlive RV vaccine that has reached human trials is the P-2 VP8* subunit vaccine in which VP8* from G1P[8] virus has been expressed in *Escherichia coli* and fused with P2 T-cell epitope of tetanus toxin to enhance immunogenicity (Wen et al., 2012).

Rotavirus VP6 alone forms large tubular structures under appropriate conditions (Lepault et al., 2001) that are strong immunogens. VP6 is also the simplest possible RV candidate vaccine consisting of only a single protein, which is a group antigen common to and largely conserved among all group A rotaviruses. RV VP6 vaccine has been shown to protect against live RV challenge in a mouse model (Lappalainen et al., 2014).

A possible future scenario is nonlive RV vaccine being used in parallel with live RV vaccines as a booster dose or to prime subjects before live RV vaccination. A whole new scenario might be a combined immunization against RV and norovirus GE using a RV VP6—norovirus VLP vaccine (Blazevic et al., 2011).

REFERENCES

Aiyegbo, M.S., Sapparapu, G., Spiller, B.W., Eli, I.M., Williams, D.R., Kim, R., Lee, D.E., Liu, T., Li, S., Woods, Jr., V.L., Nannemann, D.P., Meiler, J., Stewart, P.L., Crowe, Jr., J.E., 2013. Human rotavirus VP6-specific antibodies mediate intracellular neutralization by binding to a quaternary structure in the transcriptional pore. PLoS One 8, e61101.

Armah, G.E., Sow, S.O., Breiman, R.F., Dallas, M.J., Tapia, M.D., Feikin, D.R., et al., 2010. Efficacy of pentavalent rotavirus vaccine against severe rotavirus gastroenteritis in infants in developing countries in sub-Saharan Africa: a randomised, double-blind, placebo-controlled trial. Lancet 376 (9741), 606–614.

Armah, G.E., Kapikian, A.Z., Vesikari, T., Cunliffe, N., Jacobson, R.M., Burlington, D.B., Ruiz, Jr., L.P., 2013 Aug 1. Efficacy, immunogenicity, and safety of two doses of a tetravalent rotavirus vaccine RRV-TV in Ghana with the first dose administered during the neonatal period. J. Infect. Dis. 208 (3), 423–431.

Atchison, C.J., Stowe, J., Andrews, N., et al., 2016. Rapid declines in age group-specific rotavirus infection and acute gastroenteritis among vaccinated and unvaccinated individuals within 1 year of rotavirus vaccine introduction in England and Wales. J. Infect. Dis. 213 (2), 243–249.

Azevedo, M.S., Gonzalez, A.M., Yuan, L., Jeong, K.I., Iosef, C., Van Nguyen, T., Lovgren-Bengtsson, K., Morein, B., Saif, L.J., 2010. An oral versus intranasal prime/boost regimen using attenuated human rotavirus or VP2 and VP6 virus-like particles with immunostimulating complexes influences protection and antibody-secreting cell responses to rotavirus in a neonatal gnotobiotic pig model. Clin. Vaccine Immunol. 17, 420–428.

Azevedo, M.P., Vlasova, A.N., Saif, L.J., 2013. Human rotavirus virus-like particle vaccines evaluated in a neonatal gnotobiotic pig model of human rotavirus disease. Expert Rev. Vaccines 12, 169–181.

Bakare, N., Menschik, D., Tiernan, R., Hua, W., Martin, D., 2010. Severe combined immunodeficiency (SCID) and rotavirus vaccination: reports to the Vaccine Adverse Events Reporting System (VAERS). Vaccine 28 (40), 6609–6612.

Bernstein, D.I., Smith, V.E., Sander, D.S., Pax, K.A., Schiff, G.M., Ward, R.L., 1990. Evaluation of WC3 rotavirus vaccine and correlates of protection in healthy infants. J. Infect. Dis. 162 (5), 1055–1062.

Bernstein, D.I., Sack, D.A., Rothstein, E., Reisinger, K., Smith, V.E., O'Sullivan, D., Spriggs, D.R., Ward, R.L., 1999. Efficacy of live, attenuated, human rotavirus vaccine 89-12 in infants: a randomised placebo-controlled trial. Lancet 354 (9175), 287–290.

Bhandari, N., Rongsen-Chandola, T., Bavdekar, A., John, J., Antony, K., Taneja, S., et al., 2014. Efficacy of a monovalent human-bovine (116E) rotavirus vaccine in Indian infants: a randomised, double-blind, placebo-controlled trial. Lancet 383 (9935), 2136–2143.

Bishop, R.F., Davidson, G.P., Holmes, I.H., Ruck, B.J., 1973. Virus particles in epithelial cells of duodenal mucosa from children with acute non-bacterial gastroenteritis. Lancet 2 (7841), 1281–1283.

Blazevic, V., Lappalainen, S., Nurminen, K., Huhti, L., Vesikari, T., 2011. Norovirus VLPs and rotavirus VP6 protein as combined vaccine for childhood gastroenteritis. Vaccine 29, 8126–8133.

Blutt, S.E., Warfield, K.L., O'Neal, C.M., Estes, M.K., Conner, M.E., 2006. Host, viral, and vaccine factors that determine protective efficacy induced by rotavirus and virus-like particles (VLPs). Vaccine 24, 1170–1179.

Braeckman, T., Van Herck, K., Meyer, N., et al., 2012. RotaBel Study Group. Effectiveness of rotavirus vaccination in prevention of hospital admissions for rotavirus gastroenteritis among young children in Belgium: case-control study. BMJ 345, e4752.

Burns, J.W., Siadat-Pajouh, M., Krishnaney, A.A., Greenberg, H.B., 1996. Protective effect of rotavirus VP6-specific IgA monoclonal antibodies that lack neutralizing activity. Science 272, 104–107.

Ceyhan, M., Kanra, G., Seçmeer, G., Midthun, K., Davidson, B.L., Zito, E.T., Vesikari, T., 1993. Take of rhesus-human reassortant tetravalent rotavirus vaccine in breast-fed infants. Acta Paediatr. 82 (3), 223–227.

Clark, H.F., Furukawa, T., Bell, L.M., Offit, P.A., Perrella, P.A., Plotkin, S.A., 1986. Immune response of infants and children to low-passage bovine rotavirus (strain WC3). Am. J. Dis. Child 140, 350–356.

Clark, H.F., Offit, P.A., Ellis, R.W., Eiden, J.J., Krah, D., Shaw, A.R., Pichichero, M., Treanor, J.J., Borian, F.E., Bell, L.M., Plotkin, S.A., 1996. The development of multivalent bovine rotavirus (strain WC3) reassortant vaccine for infants. J. Infect. Dis. 174 (Suppl. 1), S73–S80.

Clark, H.F., Burke, C.J., Volkin, D.B., Offit, P., Ward, R.L., Bresee, J.S., Dennehy, P., Gooch, W.M., Malacaman, E., Matson, D., Walter, E., Watson, B., Krah, D.L., Dallas, M.J., Schödel, F., Kaplan kM, Heaton, P., 2003. Safety, immunogenicity and efficacy in healthy infants of G1 and G2 human reassortant rotavirus vaccine in a new stabilizer/buffer liquid formulation. Pediatr. Infect. Dis. J. 22, 914–920.

Clark, H.F., Bernstein, D.I., Dennehy, P.H., Offit, P., Pichichero, M., Treanor, J., Ward, R.L., Krah, D.L., Shaw, A., Dallas, M.J., Laura, D., Eiden, J.J., Ivanoff, N., Kaplan, K.M., Heaton, P., 2004. Safety, efficacy, and immunogenicity of a live, quadrivalent human-bovine reassortant rotavirus vaccine in healthy infants. J. Pediatr. 144, 184–190.

Correia, J.B., Patel, M.M., Nakagomi, O., et al., 2010. Effectiveness of monovalent rotavirus vaccine (Rotarix) against severe diarrhea caused by serotypically unrelated G2P[4] strains in Brazil. J. Infect. Dis. 20, 363–369.

Corthésy, B., Benureau, Y., Perrier, C., Fourgeux, C., Parez, N., Greenberg, H., Schwartz-Cornil, I., 2006. Rotavirus anti-VP6 secretory immunoglobulin A contributes to protection via intracellular neutralization but not via immune exclusion. J. Virol. 80, 10692–10699.

Danchin, M., Kirkwood, C.D., Lee, K.J., Bishop, R.F., Watts, E., Justice, F.A., et al., 2013. Phase I trial of RV3-BB rotavirus vaccine: a human neonatal rotavirus vaccine. Vaccine 31 (23), 2610–2616.

Das, B.K., Gentsch, J.R., Hoshino, Y., Ishida, S., Nakagomi, O., Bhan, M.K., et al., 1993. Characterization of the G serotype and genogroup of New Delhi newborn rotavirus strain 116E. Virology 197 (1), 99–107.

Delem, A., Lobmann, M., Zygraich, N., 1984. A bovine rotavirus developed as a candidate vaccine for use in humans. J. Biol. Stand. 12, 443–445.

Donato, C.M., Ch'ng, L.S., Boniface, K.F., Crawford, N.W., Buttery, J.P., Lyon, M., Bishop, R.F., Kirkwood, C.D., 2012. Identification of strains of RotaTeq rotavirus vaccine in infants with gastroenteritis following routine vaccination. J. Infect. Dis. 206 (3), 377–383.

El-Attar, L., Oliver, S.L., Mackie, A., Charpilienne, A., Poncet, D., Cohen, J., Bridger, J.C., 2009. Comparison of the efficacy of rotavirus VLP vaccines to a live homologous rotavirus vaccine in a pig model of rotavirus disease. Vaccine 27, 3201–3208.

Fang, A.Y., Tingay, D.G., 2012. Early observations in the use of oral rotavirus vaccination in infants with functional short gut syndrome. J. Paediatr. Child Health 48 (6), 512–516.

Feng, N., Lawton, J.A., Gilbert, J., Kuklin, N., Vo, P., Prasad, B.V., Greenberg, H.B., 2002. Inhibition of rotavirus replication by a non-neutralizing, rotavirus VP6-specific IgA mAb. J. Clin. Inves. 109, 1203–1213.

Flewett, T.H., Bryden, A.S., Davies, H., 1973. Letter. Virus particles in gastroenteritis. Lancet 2, 1497.

Flores, J., Perez-Schael, I., Gonzalez, M., Garcia, D., Perez, M., Daoud, N., Cunto, W., Chanock, R.M., Kapikian, A.Z., 1987. Protection against severe rotavirus diarrhoea by rhesus rotavirus vaccine in Venezuelan infants. Lancet 1 (8538), 882–884.

Fu, C., He, Q., Xu, J., Xie, H., Ding, P., Hu, W., et al., 2012. Effectiveness of the Lanzhou lamb rotavirus vaccine against gastroenteritis among children. Vaccine 31 (1), 154–158.

Gastañaduy, P.A., Curns, A.T., Parashar, U.D., Lopman, B.A., 2013. Gastroenteritis hospitalizations in older children and adults in the United States before and after implementation of infant rotavirus vaccination. JAMA 310, 851–853.

Giammanco, G., De Grandi, V., Lupo, L., Mistretta, A., Pignato, S., Teuween, D., Bogaerts, H., Andre, F.E., 1988. Interference of oral poliovirus vaccine on RIT 4237 oral rotavirus vaccine. Eur. J. Epidemiol. 4 (1), 121–123.

Gothefors, L., Wadell, G., Juto, P., Taniguchi, K., Kapikian, A.Z., Glass, R.I., 1989. Prolonged efficacy of rhesus rotavirus vaccine in Swedish children. J. Infect. Dis. 159 (4), 753–757.

Goveia, M.G., Rodriguez, Z.M., Dallas, M.J., Itzler, R.F., Boslego, J.W., Heaton, P.M., DiNubile, M.J., REST Study Team, 2007. Safety and efficacy of the pentavalent human-bovine (WC3) reassortant rotavirus vaccine in healthy premature infants. Pediatr. Infect. Dis. J. 26 (12), 1099–1104.

Groome, M.J., Moon, S.S., Velasquez, D., Jones, S., Koen, A., van Niekerk, N., Jiang, B., Parashar, U.D., Madhi, S.A., 2014a. Effect of breastfeeding on immunogenicity of oral live-attenuated human rotavirus vaccine: a randomized trial in HIV-uninfected infants in Soweto, South Africa. Bull. World Health Organ. 92 (4), 238–245.

Groome, M.J., Page, N., Cortese, M.M., et al., 2014b. Effectiveness of monovalent human rotavirus vaccine against admission to hospital for acute rotavirus diarrhoea in South African children : a case-control study. Lancet Infect. Dis. 14, 1096–1104.

Hanlon, P., Hanlon, L., Marsh, V., Byass, P., Shenton, F., Hassan-King, M., et al., 1987. Trial of an attenuated bovine rotavirus vaccine (RIT 4237) in Gambian infants. Lancet 1 (8546), 1342–1345.

Harris, V.C., 2015. Role of the intestinal microbiota in the uptake of rotavirus vaccines. Abstract, 4th European Rotavirus Expert Meeting on Rotavirus Vaccine. Santiago de Compostela, Spain, 23–25 March 2015.

Hemming, M., Räsänen, S., Huhti, L., Paloniemi, M., Salminen, M., Vesikari, T., 2013. Major reduction of rotavirus, but not norovirus, gastroenteritis in children seen in hospital after the introduction of RotaTeq vaccine into the National Immunization Programme in Finland. Eur. J. Pediatr. 172 (6), 739–746.

Hemming, M., Vesikari, T., 2012. Vaccine-derived human-bovine double reassortant rotavirus in infants with acute gastroenteritis. Pediatr. Infect. Dis. J. 31 (9), 992–994.

Istrate, C., Hinkula, J., Charpilienne, A., Poncet, D., Cohen, J., Svensson, L., Johansen, K., 2008. Parenteral administration of RF 8-2/6/7 rotavirus-like particles in a one-dose regimen induce protective immunity in mice. Vaccine 26, 4594–4601.

Iwata, S., Nakata, S., Ukae, S., Koizumi, Y., Morita, Y., Kuroki, H., Tanaka, Y., Shizuya, T., Schödel, F., Brown, M.L., Lawrence, J., 2013. Efficacy and safety of pentavalent rotavirus vaccine in Japan: a randomized, double-blind, placebo-controlled, multicenter trial. Hum. Vaccin. Immunother. 9 (8), 1626–1633.

Javid, P.J., Sanchez, S.E., Jacob, S., McNeal, M.M., Horslen, S.P., Englund, J.A., 2014. The safety and immunogenicity of rotavirus vaccination in infants with intestinal failure. J. Pediatric. Infect. Dis. Soc. 3 (1), 57–65.

Jiang, B., Wang, Y., Saluzzo, J.F., Bargeron, K., Frachette, M.J., Glass, R.I., 2008. Immunogenicity of a thermally inactivated rotavirus vaccine in mice. Hum. Vaccin. 4, 143–147.

Joensuu, J., Koskenniemi, E., Pang, X.L., Vesikari, T., 1997. Randomised placebo-controlled trial of rhesus-human reassortant rotavirus vaccine for prevention of severe rotavirus gastroenteritis. Lancet 350 (9086), 1205–1209.

Justino, M.C.A., Linhares, A.C., Lanzieri, T.M., et al., 2011. Effectiveness of the monovalent G1P[8] human rotavirus vaccine against hospitalization for severe G2P[4] rotavirus gastroenteritis in Belém, Brazil. Pediatr. Infect. Dis. J. 30, 396–401.

Kapikian, A.Z., Flores, J., Hoshino, Y., Glass, R.I., Midthun, K., Gorziglia, M., Chanock, R.M., 1986. Rotavirus: the major etiologic agent of severe infantile diarrhea may be controllable by a "Jennerian" approach to vaccination. J. Infect. Dis. 153 (5), 815–822.

Karafillakis, E., Hassounah, S., Atchison, C., 2015. Effectiveness and impact of rotavirus vaccines in Europe, 2006-2014. Vaccine 33 (18), 2097–2107.

Kawai, K., O'Brien, M.A., Goveia, M.G., Mast, T.C., El Khoury, A.C., 2012. Burden of rotavirus gastroenteritis and distribution of rotavirus strains in Asia: a systematic review. Vaccine 30, 1244–1254.

Kawamura, N., Tokoeda, Y., Oshima, M., Okahata, H., Tsutsumi, H., Van Doorn, L.J., Muto, H., Smolenov, I., Suryakiran, P.V., Han, H.H., 2011. Efficacy, safety and immunogenicity of RIX4414 in Japanese infants during the first two years of life. Vaccine 29 (37), 6335–6341.

Kompithra, R.Z., Paul, A., Manoharan, D., Babji, S., Sarkar, R., Mathew, L.G., et al., 2014. Immunogenicity of a three dose and five dose oral human rotavirus vaccine (RIX4414) schedule in south Indian infants. Vaccine 32 (Suppl. 1), A129–A133.

Lanata, C.F., Black, R.E., del Aguila, R., Gil, A., Verastegui, H., Gerna, G., et al., 1989. Protection of Peruvian children against rotavirus diarrhea of specific serotypes by one, two, or three doses of the RIT 4237 attenuated bovine rotavirus vaccine. J. Infect. Dis. 159 (3), 452–459.

Lappalainen, S., Pastor, A.R., Tamminen, K., López-Guerrero, V., Esquivel-Guadarrama, F., Palomares, L.A., Vesikari, T., Blazevic, V., 2014. Immune responses elicited against rotavirus middle layer protein VP6 inhibit viral replication in vitro and in vivo. Hum. Vaccin. Immunother. 10 (7), 2039–2047.

Laserson, K.F., Nyakundi, D., Feikin, D.R., Nyambane, G., Cook, E., Oyieko, J., Ojwando, J., Rivers, S.B., Ciarlet, M., Neuzil, K.M., Breiman, R.F., 2012. Safety of the pentavalent rotavirus vaccine (PRV), RotaTeq(®), in Kenya, including among HIV-infected and HIV-exposed infants. Vaccine 30 (Suppl. 1), A61–A70.

Lepault, J., Petitpas, I., Erk, I., Navaza, J., Bigot, D., Dona, M., Vachette, P., Cohen, J., Rey, F.A., 2001. Structural polymorphism of the major capsid protein of rotavirus. EMBO J. 20, 1498–1507.

Leshem, E., Lopman, B., Glass, R., et al., 2014. Distribution of rotavirus strains and strain-specific effectiveness of the rotavirus vaccine after its introduction: a systematic review and meta-analysis. Lancet Infect. Dis 14 (9), 847–856.

Li, D., Xu, Z., Xie, G., Wang, H., Zhang, Q., Sun, X., Guo, N., Pang, L., Duan, Z., 2015. Genotype of Rotavirus Vaccine Strain LLR in China is G10P[15]]. Bing Du Xue Bao 31 (2), 170–173.

Liu, L., Oza, S., Hogan, D., Perin, J., Rudan, I., Lawn, J.E., Cousens, S., Mathers, C., Black, R.E., 2015. Global, regional, and national causes of child mortality in 2000-13, with projections to inform post-2015 priorities: an updated systematic analysis. Lancet 385 (9966), 430–440.

Lopman, B.A., Curns, A.T., Yen, C., Parashar, U.D., 2011. Infant rotavirus vaccination may provide indirect protection to older children and adults in the United States. J. Infect. Dis. 204, 980–986.

Luan le, T., Trang, N.V., Phuong, N.M., Nguyen, H.T., Ngo, H.T., Nguyen, H.T., et al., 2009. Development and characterization of candidate rotavirus vaccine strains derived from children with diarrhoea in Vietnam. Vaccine 27 (Suppl. 5), F130–F138.

Madhi, S.A., Cunliffe, N.A., Steele, D., Witte, D., Kirsten, M., Louw, C., Ngwira, B., Victor, J.C., Gillard, P.H., Cheuvart, B.B., Han, H.H., Neuzil, K.M., 2010. Effect of human rotavirus vaccine on severe diarrhea in African infants. N. Engl. J. Med. 362 (4), 289–298.

Madhi, S.A., Kirsten, M., Louw, C., Bos, P., Aspinall, S., Bouckenooghe, A., Neuzil, K.M., Steele, A.D., 2012. Efficacy and immunogenicity of two or three dose rotavirus-vaccine regimen in South African children over two consecutive rotavirus-seasons: a randomized, double-blind, placebo-controlled trial. Vaccine 30 (Suppl. 1), A44–A51.

Maki, M., 1981. A prospective clinical study of rotavirus diarrhoea in young children. Acta Paediatr. Scand. 70 (1), 107–113.

Markkula, J., Hemming, M., Vesikari, T., 2015. Detection of vaccine-derived rotavirus strains in nonimmunocompromised children up to 3–6 months after RotaTeq vaccination. Pediatr. Infect. Dis. J. 34 (3), 296–298.

Marlow, R., Finn, A., Trotter, C., 2015. Quality of life impacts from rotavirus gastroenteritis on children and their families in the UK. Vaccine 33 (39), 5212–5216.

McClanahan, S.D., Krause, P.R., Uhlenhaut, C., 2011. Molecular and infectivity studies of porcine circovirus in vaccines. Vaccine 29, 4745–5453.

Midthun, K., Greenberg, H.B., Hoshino, Y., Kapikian, A.Z., Wyatt, R.G., Chanock, R.M., 1985. Reassortant rotaviruses as potential live rotavirus vaccine candidates. J. Virol. 53 (3), 949–954.

Moon, S.S., Tate, J.E., Ray, P., Dennehy, P.H., Archary, D., Coutsoudis, A., Bland, R., Newell, M.L., Glass, R.I., Parashar, U., Jiang, B., 2013. Differential profiles and inhibitory effect on rotavirus vaccines of nonantibody components in breast milk from mothers in developing and developed countries. Pediatr. Infect. Dis. J. 32 (8), 863–870.

Murphy, T.V., Gargiullo, P.M., Massoudi, M.S., Nelson, D.B., Jumaan, A.O., Okoro, C.A., Zanardi, L.R., Setia, S., Fair, E., LeBaron, C.W., Wharton, M., Livengood, J.R., Rotavirus Intussusception Investigation Team, 2001. Intussusception among infants given an oral rotavirus vaccine. N. Engl. J. Med. 344 (8), 564–572.

Omenaca, F., Sarlangue, J., Szenborn, L., Nogueira, M., Suryakiran, P.V., Smolenov, I.V., Han, H.H., ROTA-054 Study Group, 2012. Safety, reactogenicity and immunogenicity of the human rotavirus vaccine in preterm European Infants: a randomized phase IIIb study. Pediatr. Infect. Dis. J. 31, 487–493.

Parashar, U.D., Alexander, J.P., Glass, R.I., 2006. Prevention of rotavirus gastroenteritis among infants and children. Recommendations of the Advisory Committee on Immunization Practices (ACIP). MMWR Recomm. Rep. 55, 1–13.

Patel, N.C., Hertel, P.M., Estes, M.K., de la Morena, M., Petru, A.M., Noroski, L.M., Revell, P.A., Hanson, I.C., Paul, M.E., Rosenblatt, H.M., Abramson, S.L., 2010. Vaccine-acquired rotavirus in infants with severe combined immunodeficiency. N. Engl. J. Med. 362 (4), 314–319.

Patel, M.M., López-Collada, V.R., Bulhões, M.M., et al., 2011. Intussusception risk and health benefits of rotavirus vaccination in Mexico and Brazil. N. Engl. J. Med. 364 (24), 2283–2292.

Patel, M., Steele, A.D., Parashar, U.D., 2012. Influence of oral polio vaccines on performance of the monovalent and pentavalent rotavirus vaccines. Vaccine 30 (Suppl. 1), A30–A35.

Patel, M., Glass, R.I., Jiang, B., Santosham, M., Lopman, B., Parashar, U., 2013. A systematic review of anti-rotavirus serum IgA antibody titer as a potential correlate of rotavirus vaccine efficacy. J. Infect. Dis. 208 (2), 284–294.

Paternina-Caicedo, A., Parashar, U.D., Alvis-Guzmán, N., De Oliveira, L.H., Castaño-Zuluaga, A., Cotes-Cantillo, K., Gamboa-Garay, O., Coronell-Rodríguez, W., De la Hoz-Restrepo, F., 2015. Effect of rotavirus vaccine on childhood diarrhea mortality in five Latin American countries. Vaccine 33, 3923–3928.

Payne, D.C., Edwards, K.M., Bowen, M.D., Keckley, E., Peters, J., Esona, M.D., Teel, E.N., Kent, D., Parashar, U.D., Gentsch, J.R., 2010. Sibling transmission of vaccine-derived rotavirus (RotaTeq) associated with rotavirus gastroenteritis. Pediatrics 125 (2), e438–e441.

Payne, D.C., Boom, J.A., Staat, M.A., Edwards, K.M., Szilagyi, P.G., Klein, E.J., Selvarangan, R., Azimi, P.H., Harrison, C., Moffatt, M., Johnston, S.H., Sahni, L.C., Baker, C.J., Rench, M.A., Donauer, S., McNeal, M., Chappell, J., Weinberg, G.A., Tasslimi, A., Tate, J.E., Wikswo, M., Curns, A.T., Sulemana, I., Mijatovic-Rustempasic, S., Esona, M.D., Bowen, M.D., Gentsch, J.R., Parashar, U.D., 2013. Effectiveness of pentavalent and monovalent rotavirus vaccines in concurrent use among US children <5 years of age, 2009–2011. Clin. Infect. Dis. 57 (1), 13–20.

Pérez-Schael, I., Guntiñas, M.J., Pérez, M., Pagone, V., Rojas, A.M., González, R., Cunto, W., Hoshino, Y., Kapikian, A.Z., 1997. Efficacy of the rhesus rotavirus-based quadrivalent vaccine in infants and young children in Venezuela. N. Engl. J. Med. 337 (17), 1181–1187.

Phua, K.B., Lim, F.S., Lau, Y.L., Nelson, E.A., Huang, L.M., Quak, S.H., Lee, B.W., van Doorn, L.J., Teoh, Y.L., Tang, H., Suryakiran, P.V., Smolenov, I.V., Bock, H.L., Han, H.H., 2012. Rotavirus vaccine RIX4414 efficacy sustained during the third year of life: a randomized clinical trial in an Asian population. Vaccine 30 (30), 4552–4557.

Ranucci, C.S., Tagmyer, T., Duncan, P., 2011. Adventitious agent risk assessment case study: evaluation of RotaTeq for the presence of porcine circovirus. PDA J. Pharm. Sci. Technil. 65, 589–598.

Rennels, M.B., Glass, R.I., Dennehy, P.H., Bernstein, D.I., Pichichero, M.E., Zito, E.T., Mack, M.E., Davidson, B.L., Kapikian, A.Z., United States Rotavirus Vaccine Efficacy Group, 1996. Safety and efficacy of high-dose rhesus-human reassortant rotavirus vaccines--report of the National Multicenter Trial. Pediatrics 97 (1), 7–13.

Richardson, V., Parashar, U., Patel, M., 2011. Childhood diarrhea deaths after rotavirus vaccination in Mexico. N. Engl. J. Med. 365 (8), 772–773.

Rodríguez-Limas, W.A., Pastor, A.R., Esquivel-Soto, E., Esquivel-Guadarrama, F., Ramírez, O.T., Palomares, L.A., 2014. Immunogenicity and protective efficacy of yeast extracts containing rotavirus-like particles: a potential veterinary vaccine. Vaccine 32, 2794–2798.

Rongsen-Chandola, T., Strand, T.A., Goyal, N., et al., 2014. Effect of withholding breastfeeding on the immune response to a live oral rotavirus vaccine in North Indian infants. Vaccine 32 (Suppl. 1), A134–A139.

Ruiz-Palacios, G.M., Pérez-Schael, I., Velázquez, F.R., et al., 2006. Safety and efficacy of an attenuated vaccine against severe rotavirus gastroenteritis. N. Engl. J. Med. 354 (1), 11–22.

Ruuska, T., 1991. Childhood Diarrhoea. Thesis, University of Tampere.

Ruuska, T., Vesikari, T., 1990. Rotavirus disease in Finnish children: use of numerical scores for clinical severity of diarrhoeal episodes. Scand. J. Infect. Dis. 22 (3), 259–267.

Salinas, B., Pérez Schael, I., Linhares, A.C., Ruiz Palacios, G.M., Guerrero, M.L., Yarzábal, J.P., Cervantes, Y., Costa Clemens, S., Damaso, S., Hardt, K., De Vos, B., 2005. Evaluation of safety, immunogenicity and efficacy of an attenuated rotavirus vaccine, RIX4414: A randomized, placebo-controlled trial in Latin American infants. Pediatr. Infect. Dis. J. 24 (9), 807–816.

Simonsen, L., Viboud, C., Elixhauser, A., Taylor, R.J., Kapikian, A.Z., 2005. More on RotaShield and intussusception: the role of age at the time of vaccination. J. Infect. Dis. 192 (Suppl. 1), S36–S43.

Soriano-Gabarro, M., Mrukowicz, J., Vesikari, T., Verstraeten, T., 2006. Burden of rotavirus disease in European Union countries. Pediatr. Infect. Dis. J. 25 (Suppl.), 7–11.

Steele, A.D., Madhi, S.A., Louw, C.E., Bos, P., Tumbo, J.M., Werner, C.M., Bicer, C., De Vos, B., Delem, A., Han, H.H., 2011. Safety, reactogenicity, and immunogenicity of human rotavirus vaccine RIX4414 in human immunodeficiency virus-positive infants in South Africa. Pediatr. Infect. Dis. J. 30, 125–130.

Steinhoff, M.C., 1980. Rotavirus: the first five years. J. Pediatr. 96, 611–622.

Stumpf, K.A., Thompson, T., Sánchez, P.J., 2013. Rotavirus vaccination of very low birth weight infants at discharge from the NICU. Pediatrics 132, e662–e665.

Svensson, L., Sheshberadaran, H., Vene, S., Norrby, E., Grandien, M., Wadell, G., 1987a. Serum antibody responses to individual viral polypeptides in human rotavirus infections. J. Gen. Virol. 68 (Pt 3), 643–651.

Svensson, L., Sheshberadaran, H., Vesikari, T., Norrby, E., Wadell, G., 1987b. Immune response to rotavirus polypeptides after vaccination with heterologous rotavirus vaccines (RIT 4237, RRV-1). J. Gen. Virol. 68 (Pt 7), 1993–1999.

Tate, J.E., Burton, A.H., Boschi-Pinto, C., Steele, A.D., Duque, J., Parashar, U.D., et al., 2012. 2008 Estimate of worldwide rotavirus-associated mortality in children younger than 5 years before the introduction of universal rotavirus vaccination programmes: a systematic review and meta-analysis. Lancet Infect. Dis. 12 (2), 136–141.

Tregnaghi, M.W., Abate, H.J., Valencia, A., Lopez, P., Da Silveira, T.R., Rivera, L., Rivera Medina, D.M., Saez-Llorens, X., Gonzalez Ayala, S.E., De León, T., Van Doorn, L.J., Pilar Rubio, M.D., Suryakiran, P.V., Casellas, J.M., Ortega-Barria, E., Smolenov, I.V., Han, H.H., Rota-024 Study Group, 2011. Human rotavirus vaccine is highly efficacious when coadministered with routine expanded program of immunization vaccines including oral poliovirus vaccine in Latin America. Pediatr. Infect. Dis. J. 30 (6), e103–e108.

Velazquez, F.R., Matson, D.O., Calva, J.J., Guerrero, L., Morrow, A.L., Carter-Campbell, S., et al., 1996. Rotavirus infections in infants as protection against subsequent infections. N. Engl. J. Med. 335 (14), 1022–1028.

Vesikari, T., 2012. Rotavirus vaccination: a concise review. Clin. Microbiol. Infect. 18 (Suppl. 5), 57–63.

Vesikari, T., Isolauri, E., D'Hondt, E., Delem, A., André, F.E., Zissis, G., 1984a. Protection of infants against rotavirus diarrhoea by RIT 4237 attenuated bovine rotavirus strain vaccine. Lancet 1, 977–981.

Vesikari, T., Isolauri, E., D'Hondt, E., Delem, A., André, F.E., 1984b. Increased "take" rate of oral rotavirus vaccine in infants after milk feeding. Lancet 2, 700.

Vesikari, T., Isolauri, E., Delem, A., D'Hondt, E., Andre, F.E., Beards, G.M., et al., 1985a. Clinical efficacy of the RIT 4237 live attenuated bovine rotavirus vaccine in infants vaccinated before a rotavirus epidemic. J. Pediatr. 107 (2), 189–194.

Vesikari, T., Ruuska, T., Bogaerts, H., Delem, A., André, F., 1985b. Dose-response study of RIT 4237 oral rotavirus vaccine in breast-fed and formula-fed infants. Pediatr. Infect. Dis. 4, 622–625.

Vesikari, T., Ruuska, T., Delem, A., André, F.E., 1986a. Oral rotavirus vaccination in breast- and bottle-fed infants aged 6 to 12 months. Acta Paediatr. Scand. 75, 573–578.

Vesikari, T., Kapikian, A.Z., Delem, A., Zissis, G., 1986b. A comparative trial of rhesus monkey (RRV-1) and bovine (RIT 4237) oral rotavirus vaccines in young children. J. Infect. Dis. 153, 832–839.

Vesikari, T., Karvonen, A., Korhonen, T., Espo, M., Lebacq, E., Forster, J., et al., 2004. Safety and immunogenicity of RIX4414 live attenuated human rotavirus vaccine in adults, toddlers and previously uninfected infants. Vaccine 22, 2836–2842.

Vesikari, T., Clark, H.F., Offit, P.A., Dallas, M.J., DiStefano, D.J., Goveia, M.G., Ward, R.L., Schödel, F., Karvonen, A., Drummond, J.E., DiNubile, M.J., Heaton, P.M., 2006a. Effects of the potency and composition of the multivalent human-bovine (WC3) reassortant rotavirus vaccine on efficacy, safety and immunogenicity in healthy infants. Vaccine 24, 4821–4829.

Vesikari, T., Matson, D.O., Dennehy, P., Van Damme, P., Santosham, M., Rodriguez, Z., et al., 2006b. Safety and efficacy of a pentavalent human-bovine (WC3) reassortant rotavirus vaccine. N. Engl. J. Med. 354, 23–33.

Vesikari, T., Karvonen, A.V., Majuri, J., Zeng, S.Q., Pang, X.L., Kohberger, R., et al., 2006c. Safety, efficacy, and immunogenicity of 2 doses of bovine-human (UK) and rhesus-rhesus-human rotavirus reassortant tetravalent vaccines in Finnish children. J. Infect. Dis. 194, 370–376.

Vesikari, T., Karvonen, A., Forrest, B.D., Hoshino, Y., Chanock, R.M., Kapikian, A.Z., 2006d. Neonatal administration of rhesus rotavirus tetravalent vaccine. Pediatr. Infect. Dis. J. 25, 118–122.

Vesikari, T., Karvonen, A., Prymula, R., Schuster, V., Tejedor, J.C., Cohen, R., et al., 2007. Efficacy of human rotavirus vaccine against rotavirus gastroenteritis during the first 2 years of life in European infants: randomised, double-blind controlled study. Lancet 370, 1757–1763.

Vesikari, T., Itzler, R., Karvonen, A., Korhonen, T., Van Damme, P., Behre, U., Bona, G., Gothefors, L., Heaton, P.M., Dallas, M., Goveia, M.G., 2009. RotaTeq, a pentavalent rotavirus vaccine: efficacy and safety among infants in Europe. Vaccine 28 (2), 345–351.

Vesikari, T., Karvonen, A., Ferrante, S.A., Ciarlet, M., 2010. Efficacy of the pentavalent rotavirus vaccine, RotaTeq, in Finnish infants up to 3 years of age: the Finnish Extension Study. Eur. J. Pediatr. 169, 1379–1396.

Vesikari, T., Van Damme, P., Giaquinto, C., Dagan, R., Guarino, A., Szajewska, H., Usonis, V., 2015. European Society for Paediatric Infectious Diseases consensus recommendations for rotavirus vaccination in Europe: update 2014. Pediatr. Infect. Dis. J. 34, 635–643.

Victoria, J.G., Wang, C., Jones, M.S., Jaing, C., McLoughlin, K., Gardner, S., Delwart, E.L., 2010. Viral nucleic acids in live-attenuated vaccines: detection of minority variants and an adventitious virus. J. Virol. 84, 6033–6040.

Vodopija, I., Baklaic, Z., Vlatkovic, R., Bogaerts, H., Delem, A., Andre, F.E., 1986. Combined vaccination with live oral polio vaccine and the bovine rotavirus RIT 4237 strain. Vaccine 4, 233–236.

Walker, C.L., Rudan, I., Liu, L., Nair, H., Theodoratou, E., Bhutta, Z.A., O'Brien, K.L., Campbell, H., Black, R.E., 2013. Global burden of childhood pneumonia and diarrhoea. Lancet 381, 1405–1416.

Ward, R.L., Clemens, J.D., Knowlton, D.R., Rao, M.R., van Loon, F.P., Huda, N., et al., 1992. Evidence that protection against rotavirus diarrhea after natural infection is not dependent on serotype-specific neutralizing antibody. J. Infect. Dis. 166 (6), 1251–1257.

Wen, X., Cao, D., Jones, R.W., Li, J., Szu, S., Hoshino, Y., 2012. Construction and characterization of human recombinant VP8* subunit parenteral vaccine candidates. Vaccine 30, 6121–6126.

Werther, R.L., Crawford, N.W., Boniface, K., Kirkwood, C.D., Smart, J.M., 2009. Rotavirus vaccine induced diarrhea in a child with severe combined immune deficiency. J. Allergy Clin. Immunol. 124, 600.

WHO position paper, 2013. Rotavirus vaccines. Weekly Epid. Record 88, 49–64.

Zaman, K., Dang, D.A., Victor, J.C., Shin, S., Yunus, M., Dallas, M.J., et al., 2010. Efficacy of pentavalent rotavirus vaccine against severe rotavirus gastroenteritis in infants in developing countries in Asia: a randomised, double-blind, placebo-controlled trial. Lancet 376 (9741), 615–623.

Chapter 3.1

Structural Biology of Noroviruses

B.V. Venkataram Prasad*,, S. Shanker*, Z. Muhaxhiri*, J.-M. Choi*, R.L. Atmar**, M.K. Estes****

**Verna and Marrs McLean Department of Biochemistry and Molecular Biology, Baylor College of Medicine, Houston, TX, United States; **Departments of Molecular Virology and Microbiology, and Medicine, Baylor College of Medicine, Houston, TX, United States*

1 INTRODUCTION

Noroviruses (NoVs) constitute one of the five genera in the *Caliciviridae* family (Ramani et al., 2014; Green et al., 2000). They are the leading cause of epidemic acute gastroenteritis (Ahmed et al., 2014). It is estimated that these viruses are responsible for ~20 million total illnesses with a disease burden of ~2 billion dollars in the United States alone each year, and ~200,000 annual deaths of children under the age of 5 years worldwide (Patel et al., 2008). NoVs exhibit considerable genetic diversity. Based on phylogenetic analyses, NoVs are classified into six genogroups (GI-GVI), and they are further subdivided into genotypes (designated with an Arabic numeral) within each genogroup. While the genogroups GI, GII, and GIV predominantly contain human strains, the other genogroups only contain animal strains (Zheng et al., 2006). Epidemiological studies indicate that the NoVs belonging to genogroup II, genotype 4 (GII.4) are the most prevalent and account for up to 70-80% of the outbreaks worldwide (Ramani et al., 2014; Kroneman et al., 2008). These GII.4 NoVs undergo epochal evolution, similar to A/H3N2 influenza virus strains, with the emergence of new variants every 2 years coinciding with a new epidemic peak (Siebenga et al., 2007; Donaldson et al., 2008; Lindesmith et al., 2012). Recent epidemiological studies also show a considerable increase in the prevalence of GI outbreaks worldwide, with different genotypes, such as GI.4, GI.6, GI.3, and GI.7 predominating in different geographical regions (Vega et al., 2014; Grytdal et al., 2015). Several studies have demonstrated that susceptibility to many NoVs is determined by genetically controlled expression of histo-blood group antigens (HBGAs), which are also critical for NoV attachment to host cells (Ruvoen-Clouet et al., 2013) (see Chapter 3.3). Consistent with their high genetic diversity, these viruses exhibit extensive strain-dependent variation in the recognition of HBGAs, which

Viral Gastroenteritis

together with antigenic variations allow for their sustained evolution. The preponderance of global NoV outbreaks together with the recognition of new genogroups and rapid emergence of new variants within each genogroup signify a major health concern, particularly considering current lack of effective antiviral strategies either in terms of vaccines or in terms of small molecule drugs.

2 GENOME ORGANIZATION

Members of the *Caliciviridae*, including NoVs, are nonenveloped, icosahedral viruses typically 380–400 Å in diameter. The genome consists of a linear, positive-sense, single-stranded RNA of 7.4 to 8.3 kb in size with a covalently linked VPg at the 5′ end and a polyadenylated tail at the 3′ end (Green, 2007; Thorne and Goodfellow, 2014). Caliciviruses exhibit two distinct types of genome organization. In the members of the *Norovirus* and *Vesivirus* genera, the genome is organized into three open reading frames (ORFs), whereas in the *Sapovirus*, *Lagovirus,* and *Nebovirus* genera, the genome is organized into two ORFs (Thorne and Goodfellow, 2014; Smiley et al., 2002). In all cases, however, the calicivirus RNA encodes a large polyprotein, the major capsid protein VP1 (55–70 kDa), and a basic minor structural protein VP2 (Bertolotti-Ciarlet et al., 2003; Sosnovtsev et al., 2005). In the *Norovirus* and the *Vesivirus* genera, the large polyprotein, VP1 and VP2 are encoded separately by ORF1, ORF2, and ORF3, respectively. In contrast, in the *Sapovirus, Lagovirus,* and *Nebovirus* genera, the polyprotein and the major capsid protein VP1 are contiguously encoded by ORF1, and VP2 is encoded by the ORF2. In all caliciviruses, the polyprotein is posttranslationally processed by the viral protease, which itself is a component of the polyprotein, into nonstructural proteins (NSPs) that are essential for virus replication. In NoVs, these NSPs include p48, p41 (NTPase), p22, VPg, protease, and RNA-dependent RNA polymerase (RdRp) (Thorne and Goodfellow, 2014).

3 T=3 CAPSID ORGANIZATION

Capsid organization of NoVs and several other caliciviruses have been studied either by cryo-EM or by X-ray crystallographic techniques (Chen et al., 2004; Prasad et al., 1994a,b; Kumar et al., 2007; Katpally et al., 2008; Wang et al., 2013). The structures of recombinant NoV (rNoV) particles from different genogroups, murine NoV (MNV), and three animal caliciviruses are known. Since the human NoVs are so far resistant to growth in cell culture, recombinant virus-like particles (VLPs) have been produced by the coexpression of VP1 and VP2, preserving the morphological and antigenic features of the authentic virions, for use in structural studies. The first crystallographic structure of a calicivirus capsid was that of recombinant Norwalk virus (rNV), which is a GI.1 NoV (Prasad et al., 1999) (Fig. 3.1.1A). Since then, crystallographic structures of San Miguel Sealion virus (SMSV) (Chen et al., 2006) (Fig. 3.1.1B) and

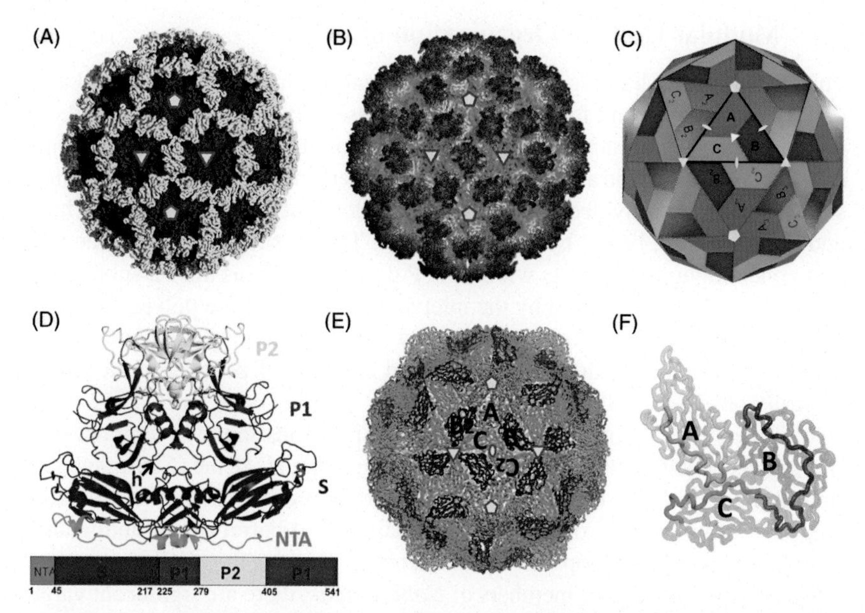

FIGURE 3.1.1 Structural organization of the NoV and animal calicivirus capsids. Crystal structures of (A) rNV capsid with S, P1 and P2 domains colored in deep blue, red, and yellow, respectively, viewed along the icosahedral twofold axis (PDB id: 1IHM); (B) SMSV capsid (PDB id: 2GH8), as an example of an animal calicivirus capsid structure is shown in the same orientation as rNV capsid with S, P1, and P2 colored in green, yellow, orange respectively. Note that both rNV and SMSV exhibit a similar structural organization; (C) model diagram of T=3 icosahedral lattice (in the same orientation as Fig. 3.1.1A) with 5-, 3-, 2-symmetry axes denoted by a pentagon, triangle, and an oval, respectively; locations of the quasi-equivalent A *(blue)*, B *(red)*, C *(green)* subunits are shown. The quasi-equivalent two- and threefold symmetry axes that relate the A and B_2 subunits *(oval)*, and A, B, and C subunits *(triangle)*, respectively, are also indicated; (D) the C/C dimer in the rNV capsid along with the domain organization (bar below) is shown. The NTA region is colored in green, hinge region denoted by "h" is shown by an arrow, and the S, P1, and P2 domains are colored as in Fig. 3.1.1A; (E) Tiling of the trapezoidal-shaped S domains with a jelly-roll fold, forming the icosahedral shell shown in same orientation as Fig. 3.1.1C with the S domains of the quasi-equivalent subunits A *(blue)*, B *(red)*, C *(green)* denoted. The "flat" C/C2 dimers are related by the icosahedral twofold axis, the bent A/B2 dimers are related by the quasi-twofold axis; and (F) the NTA interactions between the A, B, and C subunits in the rNV capsid (as viewed from the capsid interior). Only the S-domains are shown, using the same color scheme as Fig. 3.1.1E. The NTAs are shown in darker colors and the rest of the S domains in faded colors. Notice the NTA of the B subunit extends further to interact with the underside of the C subunit, whereas the equivalent regions of the NTAs in the A and C subunits are disordered and not seen in the capsid structure.

feline calicivirus (FCV) (Ossiboff et al., 2010) (*Vesivirus* genus), derived from authentic virions, and a GII.4 (HOV strain) recombinant NoV (rHOV) capsid (manuscript in preparation) have been determined. All these structural studies have consistently shown that calicivirus capsids, irrespective of the genera, have similar capsid architecture with a T=3 icosahedral symmetry (Fig. 3.1.1C), formed by 90 dimers of VP1 (Fig. 3.1.1D).

3.1 Modular Domain Organization of VP1

The capsid protein has a modular domain organization with an N-terminal arm (NTA) that is important for directing capsid assembly, followed by a shell (S) domain that is important for stabilizing the icosahedral scaffold (Bertolotti-Ciarlet et al., 2002), and a protruding (P) domain emanating from the icosahedral shell that is further divided into P1 and P2 subdomains (Figs. 3.1.1D–F). The S and P domains are linked by a flexible hinge. The P1 subdomain is formed by two noncontiguous segments within the P domain, whereas the P2 subdomain facing the exterior is formed by the intervening segment. The polypeptide fold in each of these domains is also essentially conserved among calicivirus structures. The S domain exhibits an 8-stranded antiparallel β-barrel motif that is typically observed in T=3 icosahedral viruses (Prasad and Schmid, 2012). The fold of the P1 subdomain, consisting of three β-strands in the N-terminal portion, a twisted antiparallel β-sheet formed by four strands in the C-terminal portion, and a well-defined α-helix, is novel and only seen in calicivirus structures (Fig. 3.1.2A). The fold of the P2 subdomain is a β-barrel of six antiparallel strands connected by loops of various lengths. Despite the similar structural characteristics among the members of *Caliciviridae*, there are significant variations within the capsid protein structure providing insight into how the unique modular organization of the capsid protein is conducive to the wide diversity and host specificity associated with this family of viruses. Comparisons of the calicivirus capsid protein sequences indicate that the S domain is highly

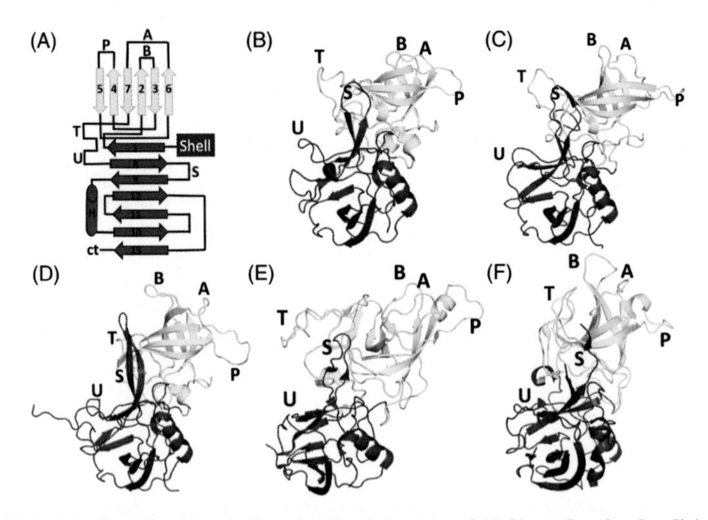

FIGURE 3.1.2 **Structural variations in the P domain of NoVs and animal caliciviruses.** (A) Topology diagram of the polypeptide folds in the P1 *(red)* and P2 *(yellow)* subdomains. P domain structures of (B) GI.1 NoV (PDB id: 2ZL5), (C) GII.4 NoV (PDB id: 3SEJ), (D) GV.1 MNV (PDB id: 3LQ6), (E) SMSV (PDB id: 2GH8), and (F) RHDV (PDB id: 4EGT). All the structures are shown in the same orientations to demonstrate the variations in the P2 subdomain following the same coloring scheme for P1 and P2 as in Fig. 3.1.2A.

conserved and the P1 subdomain is moderately conserved, whereas the distally located P2 subdomain is highly variable.

3.2 NTA Interactions

In the T=3 icosahedral lattice, the capsid protein is located in three quasi-equivalent positions, conventionally designated A, B, and C, which constitute the icosahedral asymmetric unit (Fig. 3.1.1C). The A subunits surround the icosahedral fivefold axis, whereas B and C subunits alternate around the icosahedral threefold axis giving rise to quasi sixfold symmetry. In the calicivirus capsid, as in other T=3 icosahedral capsids, A, B, and C subunits form two types of quasi-equivalent dimers, A/B dimer related by the quasi twofold axis (A/B_2) and C/C dimer related by the strict icosahedral twofold axis (C/C_2) (Fig. 3.1.2C). A common feature is that the A/B dimer has a "bent" conformation, whereas the C/C dimer has a "flat" conformation. Such a dual conformation imparts the necessary curvature for the formation of the T=3 icosahedral capsid. In many of the structurally characterized T=3 capsids, the NTA is implicated in providing a switch by undergoing an order to disorder transition to facilitate the bent A/B and the flat C/C conformations during T=3 capsid assembly (Harrison, 2001; Rossmann and Johnson, 1989). In these T=3 virus capsids, including rNV and rHOV, only one of the three NTAs of the quasi-equivalent subunits is ordered. In the case of rNoV structures, while the NTA of the B subunit is ordered to a larger extent, the equivalent regions in the A and C subunits are disordered (Fig. 3.1.1F). The ordered NTA portion of the B subunit interacts with the base of the S domain of the neighboring C subunit to stabilize the flat conformation of the C/C dimer (Fig. 3.1.1F). The equivalent interactions involving the NTA are not observed in the A/B dimer which adopts a bent conformation (Prasad et al., 1999). In contrast, although serving the same purpose, an ordered NTA of the C subunit provides a switch in the T=3 plant tombus- and sobemoviruses instead of the NTA of the B subunit as observed in rNoV capsids (Harrison, 2001; Rossmann and Johnson, 1989). In nodaviruses, which also exhibits a T=3 capsid organization, in addition to an ordered NTA of the C subunit, a piece of genomic RNA keeps the "flat" conformation of the C/C dimers (Fisher and Johnson, 1993). Interestingly, the SMSV and FCV capsid structures exhibit a novel and distinct variation from any of these viruses. In these structures, the NTAs of all three subunits are equally ordered, essentially maintaining the T=3 symmetry at this level. Instead of an order-to-disorder transition of the NTA, a distinct conformational change involving a Pro residue in the B subunit that leads to the formation of a ring-like structure around the fivefold axis appears to provide a switch (Chen et al., 2006; Ossiboff et al., 2010). Whether these unique NTA interactions found only in SMSV and FCV are influenced by the genome or the proteolytic processing of the capsid protein, a common feature in the members of *Vesivirus* genus, remains a question.

3.3 Hinge and Interdomain Flexibility

Like capsid proteins in other T=3 icosahedral viruses with distinct S and P domains, such as plant tombusviruses (Harrison, 2001), calicivirus VP1 has a flexible hinge between these two domains (Fig. 3.1.1D, denoted by h and an arrow). This hinge, which facilitates the interactions between the P1 subdomain and the upper portion of the S domain, is likely important for locking the A/B and C/C dimers in their appropriate conformations, as this interaction is seen only in the A/B dimers and not in the C/C dimers (Prasad et al., 1999; Chen et al., 2006; Ossiboff et al., 2010). Although such an interaction between the P1 and S domains is conserved in rNoV, SMSV, and FCV crystal structures as a structural requirement, the relative orientations between these two domains, likely because of the sequence changes, are noticeably altered. In the animal calicivirus structures including SMSV, FCV, and recently determined high resolution cryo-EM structure of rabbit hemorrhagic disease virus (RHDV) (Wang et al., 2013), this change in S–P1 orientation together with a compensatory change in the P1–P2 orientation causes only the distal P2 subdomain to participate in the dimeric interactions. This is in contrast to that observed in rNoV structures, in which both P1 and P2 subdomains participate in the dimeric interactions. A conserved glycine located at the junction of P1 and P2 is suggested to allow this compensatory change in the P1–P2 orientation. Thus, in addition to the flexibility between the S and P1 domains, there is an additional point of flexibility between the P1 and P2 subdomains that was not immediately apparent from the rNoV structures alone. The multiple points of flexibility could be an important factor in enhancing calicivirus diversity through structural variations within the context of a similar domain organization, somewhat akin to the interdomain flexibility seen in antibody structures with a hinge and an elbow.

In addition to structures of the calicivirus capsids, in recent years, structures of the recombinant P domain of different NoV genogroups and genotypes and of RHDV have been determined (Cao et al., 2007; Choi et al., 2008; Hansman et al., 2011; Kubota et al., 2012; Shanker et al., 2011, 2014; Taube et al., 2010). All these P domain structures show a similar dimeric conformation to that observed in the capsid structures (Choi et al., 2008). While all of the NoV P domain dimer structures determined to date consistently show both P1–P1 and P2–P2 dimeric interactions as observed in rNoV capsid (Prasad et al., 1999), the RHDV P domain dimer (Wang et al., 2013), as in SMSV (Chen et al., 2006) and the FCV capsid (Ossiboff et al., 2010) structures, only exhibits P2–P2 interactions. Thus, dimer-related P2–P2 interactions are a common feature in all of the caliciviruses for which structures have been determined. The underlying functional significance of this structural conservation across the caliciviruses is unclear.

3.4 Hypervariable P2 Subdomain—Antigenic Diversity and Receptor Binding

Comparison of the calicivirus sequences clearly indicates that the region that forms the P2 subdomain exhibits the most variability, consistent with its role in host specificity, antigenicity and receptor interactions. Accordingly, comparison of the animal calicivirus and norovirus capsid and P domain structures show that the P2 subdomain exhibits the most structural variability (Fig. 3.1.2B–D). Despite significant variation in the sequences, the basic polypeptide fold in the P2 subdomain, with six antiparallel β-strands forming a β-barrel structure, is conserved in all the calicivirus capsid structures so far determined (Fig. 3.1.2A). The main differences are in the orientations and the extension of the surface-exposed loops that connect these β-strands. These loops extend sideward from the dimeric surface with the conserved β-barrel structure at the dimeric interface. With shorter loops, the NV P2 subdomain exhibits the most compact structure, whereas the P2 subdomains of SMSV and FCV exhibit the most elaborate loop structures. The loop regions in the P2 subdomain of GII.4 are also more elaborate compared to GI.1 NV, exemplifying the intergenogroup differences in NoVs that play a role in differential glycan recognition as described in the next section. FCV is the only calicivirus for which a protein receptor has been identified (Makino et al., 2006). Cryo-EM structure of the FCV–fJAM-1 complex clearly shows that the receptor fJAM-1 (feline junctional adhesion molecule 1) binds to the P2 subdomain (Bhella et al., 2008). Known neutralization epitopes in the FCV capsid map to the loops in the P2 subdomain (Chen et al., 2006). Thus, crystallographic analyses of caliciviruses of different genera and genotypes demonstrate how the P2 subdomain in these viruses accommodates significant sequence alterations within the context of a conserved polypeptide fold to achieve antigenic diversity and strain-dependent receptor recognition.

3.5 Capsid Assembly

Based on the observation that dimeric interactions of the capsid protein particularly involving the P domain are a common feature in all the caliciviruses, a reasonable assumption is that the VP1 dimer is the building block for the assembly. The VP1 dimer may exist in a dynamic equilibrium between "bent" and "flat" conformations in solution prior to assembly, and during the assembly process may switch to appropriate conformations directed by the NTA arms. The role of the NTA arms and also that of dimeric interactions in the capsid assembly is substantiated by a systematic structure-directed mutational analysis of the NV VP1 (Bertolotti-Ciarlet et al., 2002). Based on the calicivirus capsid structures (Chen et al., 2006) and mass spectrometric analysis of rNV capsid and its pH-induced dissociation/association intermediates (Baclayon et al., 2011; Shoemaker et al., 2010), it is plausible that capsid assembly proceeds through the formation of trimers of dimers that are then brought together into an icosahedral structure.

3.6 Minor Structural Protein VP2

In the context of the capsid assembly and perhaps more particularly in the genome encapsidation, another important factor to be considered is the minor protein VP2. All calicivirus genomes encode this protein (Green et al., 2000), which is highly basic in nature. The association of VP2 with the infectious virus particles has been demonstrated in NV (Glass et al., 2000) as well as FCV (Sosnovtsev and Green, 2000) and is suspected to be present in all caliciviruses. The role of VP2, particularly in enhancing the capsid stability and size homogeneity, is clearly evident from the dynamic light scattering experiments of the NoV VLPs produced by the expression of VP1 alone and in comparison with those obtained from the coexpression of both VP1 and VP2 (Bertolotti-Ciarlet et al., 2003). Studies on FCV also suggest VP2 stabilizes the icosahedral capsid and is required for the production of infectious virus (Sosnovtsev et al., 2005). However, VP2 is not visualized in any of the calicivirus capsid structures, including the crystal structures of the rNoV particles produced by the coexpression of VP1 and VP2. This is likely because of the substoichiometric proportion of VP2 (suspected to be ~2–8 molecules per virion) which does not allow VP2 to interact with VP1 conforming to the icosahedral symmetry and causes it to be transparent in the structures in which the icosahedral symmetry is explicitly used for structure determination.

Recently, cotransfection studies of NoV VP1 and VP2 in mammalian cells accompanied by systematic mutational analysis demonstrated that VP2 directly interacts with the Ile52 residue of a highly conserved IPPWI motif in the NTA (Vongpunsawad et al., 2013). Although direct interaction with VP1 involves only a small region, this region inside the capsid is surrounded by a stretch of negatively charged residues, suggesting that highly basic VP2 may nonspecifically interact with the interior of the VP1 capsid spanning a larger area to influence both the stability and the size homogeneity of the capsid. Given that the calicivirus capsid lacks an abundance of basic residues in the interior surface, the bound VP2 may counteract the electrostatic repulsion between the RNA and capsid and help stabilize the encapsidated genome in infectious virions. By directly interacting with VP1, VP2 may also play a critical role in encapsidating the genome during capsid assembly, providing a rationale for the observation that VP2 is required for the production of infectious particles in the studies on FCV (Sosnovtsev et al., 2005).

4 GLYCAN RECOGNITION AND SPECIFICITY IN NOROVIRUSES

Susceptibility to some NoVs is associated with the expression of genetically determined histo-blood group antigens (HBGAs) (Ramani et al., 2014; Ruvoen-Clouet et al., 2013; Imbert-Marcille et al., 2014). These glycoconjugates, found in mucosal secretions and on epithelial cells (Hakomori, 1999), appear to function as initial receptors or coreceptors for human NoVs (Marionneau et al., 2002; Lindesmith et al., 2003; Hutson et al., 2004; Tan and Jiang, 2005). HBGAs are oligosaccharide epitopes with varying carbohydrate compositions and linkages between them (Marionneau et al., 2001). It has been proposed that human NoVs exploit the polymorphic nature of HBGAs in the host population to counter herd immunity during their evolution. (See also Chapter 3.3).

4.1 What are Histo-Blood Group Antigens (HBGAs)?

HBGAs are glycans that include the determinants of secretor-status and blood type of an individual (Ruvoen-Clouet et al., 2013). They are synthesized by the sequential addition of a monosaccharide to a terminal precursor disaccharide by genetically controlled expression of certain glycosyl-transferases. Depending upon the composition of the disaccharide backbone and the linkage, they are grouped in the four types: type 1 (Galβ1-3GlcNAcβ), type 2 (Galβ1-4GlcNAcβ), type 3 (Galβ1-3GalNAcα), and type 4 (Galβ1-3GalNAcβ). Although several studies have suggested that the secretor-positive status is a susceptibility factor for the majority of NoVs (Lindesmith et al., 2003; Hutson et al., 2005), recent epidemiological studies indicate that nonsecretors are susceptible to some GI and GII NoVs (Currier et al., 2015). The secretor-positive status of an individual is determined by expression of a functional FUT2 enzyme, a fucosyltransferase, that catalyzes the addition of an α fucose (SeFuc) to the β galactose (β Gal) of the disaccharide precursor to form the secretor epitope or the H-type HBGA. The H-type HBGA can be further modified by enzymes A or B by adding *N*-acetyl galactosamine (GalNAc) or Gal to the precursor β Gal to form A- or B-type HBGA, respectively (Ruvoen-Clouet et al., 2013). Similarly, the Lewis-positive status is determined by the activity of the fucosyl transferase 3 (FUT3) enzyme, which adds an α fucose (LeFuc) to the *N*-acetylglucosamine (GlcNAc) of the precursor disaccharide to form the Lewis epitope. Thus the sequential addition of carbohydrate moieties by the FUT2 and FUT3 along with enzymes A and B give rise to the secretor/nonsecretor Lewis and ABH families of HBGAs (Imbert-Marcille et al., 2014; Marionneau et al., 2005). Several studies using NoV VLPs with saliva, red blood cells, and synthetic carbohydrates demonstrate direct interaction between NoV VLPs and HBGAs (Tan and Jiang, 2005; Hutson et al., 2003; Huang et al., 2005; Shirato et al., 2008) and show that the specificity to HBGAs varies not only within a particular genogroup but also between the genogroups.

4.2 Structural Basis of Genogroup and Genotype-Dependent HBGA Specificity in Human Noroviruses

A typical strategy that is used to understand the structural basis of the HBGA interactions in NoVs is to determine the X-ray structure of the recombinantly expressed P domain of the NoV in complex with the HBGA (Cao et al., 2007; Choi et al., 2008; Tan et al., 2004). Following this strategy, in recent years there has been an explosion of crystallographic structures of the P domain of various NoVs in complex with a variety of HBGAs (Cao et al., 2007; Choi et al., 2008; Hansman et al., 2011; Kubota et al., 2012; Shanker et al., 2011, 2014; Prasad et al., 2014; Jin et al., 2015; Atmar et al., 2015). In addition to revealing that the HBGA interaction exclusively involves distally exposed regions of the hypervariable P2 subdomain, these studies have shown how the HBGA binding sites between the genogroups differ and how the sequence variations in the P2 subdomain within each genogroup affect the HBGA specificity.

4.3 HBGA Binding Sites in GI and GII are Differently Configured

A striking observation from the crystallographic studies is that the binding sites in GI and GII NoVs are distinctly different in their locations, structural characteristics, and in the modalities how the carbohydrate residues of the HBGA interact (Fig. 3.1.3A). In GI NoVs, while the majority of interactions with HBGA are localized within each subunit of the P domain dimer, in GII NoVs, they are shared between the opposing subunits of the P dimer (Fig. 3.1.3B). Another distinguishing feature is that in GI, the majority of the interactions with HBGA primarily involve a Gal moiety (Fig. 3.1.3C), whereas in GII, the interactions are centered on the Fuc residue (Fig. 3.1.3D). An exceptionally well-conserved feature in GI viruses is the hydrophobic interaction between the SeFuc moiety (as in the H-type) or the *N*-acetyl arm of *N*-acetylgalactosamine (as in the A-type) with a conserved Trp residue in the P2 subdomain (Fig. 3.1.3C). This combinatorial requirement of Gal and hydrophobic interactions appears to restrict the types of HBGAs that GI NoVs can bind. Several studies consistently show that most GI NoVs do not bind B-type HBGAs. A likely explanation is as follows: although B-type HBGA has a terminal Gal residue, it lacks an additional group, such as SeFuc present in the H-type or an *N*-acetyl arm present in the A-type that could engage in the hydrophobic interactions, resulting in lower affinity. HBGA binding in GII NoVs does not involve such a combinatorial requirement allowing them to bind all ABH HBGAs. This likely is one of the factors why GIIs, particularly GII.4 NoVs, are globally more prevalent.

4.4 Intragenogroup Differences in HBGA Binding

Sequence alterations in the P domain within the genogroup members also contribute to the variation of HBGA binding profiles (Prasad et al., 1999; Cao et al., 2007; Choi et al., 2008; Hansman et al., 2011; Kubota et al., 2012; Shanker et al., 2011, 2014; Bu et al., 2008; Chen et al., 2011). A generalizable concept from the crystallographic studies of the GI and GII P domains in complex with various HBGAs is that HBGA binding in human NoVs involves two

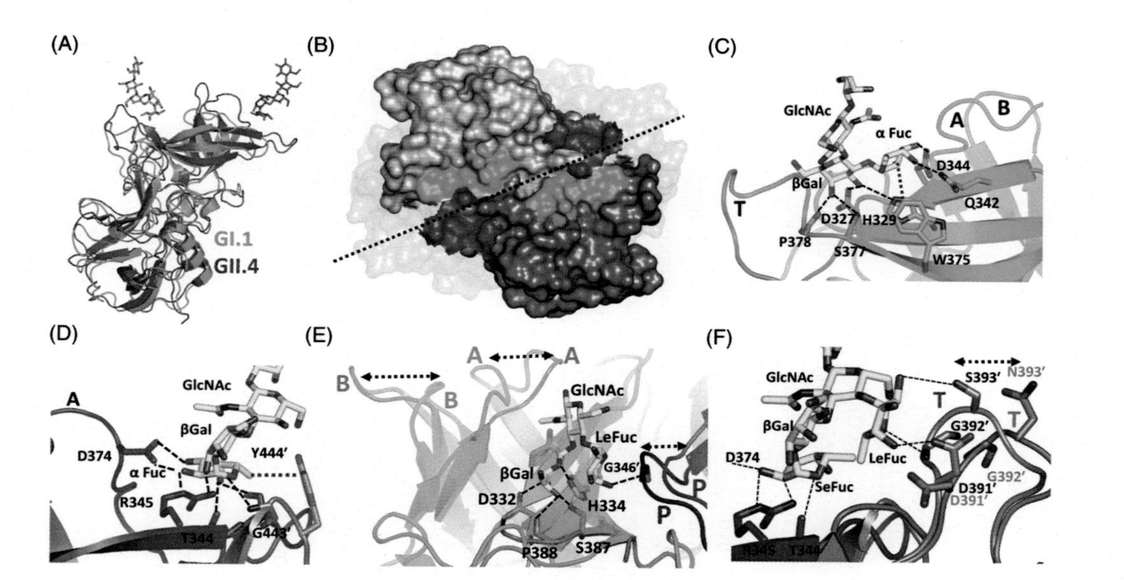

FIGURE 3.1.3 **Strain-dependent glycan interactions in human NoVs.** (A) Superposition of the P domain monomer in GI.1 (cyan, PDB id: 2ZL6) and GII.4 P (pink, PDB id: 3SEJ) with bound H-type HBGA shown to exemplify differences in the locations of the HBGA binding sites between GI and GII NoVs. (B) The HBGA binding sites in GI *(cyan)* and GII *(pink)* are shown in the context of the P dimer *(grey surface)*. The dotted line demarcates the two monomer subunits in the dimer. (C) Conserved "site 1" in GI NoVs (cyan) shows the highly conserved interactions involving Gal and Fuc residues of H-type HBGA. Hydrogen bond interactions are shown with black dotted lines, and hydrophobic interactions between the Fuc and His-Trp pairs are shown in red dotted lines. (D) Conserved "site 1" in GII NoVs *(pink)* shows the highly conserved interactions involving Fuc residues of the H-type HBGA. (E) Structural alterations in the loop regions *(double headed arrows)* due to genotypic sequence changes in GI NoVs that allow differential binding of nonsecretor Lewis HBGA as exemplified by the superposition of GI.1 *(cyan)* and GI.7 (PDB id: 4P12) shades of brown and red). Notice large movements in the A, B, and P loops. Because of the shorter P loop, G1.1 has low affinity for Lewis HBGA. (F) Structural alterations in the loop regions *(double headed arrow)* of GII.4 P domain, due to temporal sequence changes, that allow differential binding of secretor-positive Lewis HBGA as exemplified by the superposition of GII.4 1996 *(brown)* and 2004 *(grey)* variants. Due to the retracted T loop, the 1996 variant has low affinity for Lewis HBGA.

sites. The first site is formed by the highly conserved residues in the less flexible regions of the P2 subdomain, which preserves the Gal and Fuc dominant nature of interactions in GI and GII, respectively. Minor variations in this site could result in differences in the HBGA binding affinity. The second site is formed by the less conserved residues that are typically from the loop regions surrounding the first site. This site allows for differential binding to Lewis HBGAs in both GI and GII as discussed later.

In GI NoVs, sequence changes differentially alter their ability to bind nonsecretor monofucosyl Lewis HBGA ($Le^{a/x}$) as observed in GI.4, GI.6, GI.3, GI.2, and GI.7 NoVs (Lindesmith et al., 2012; Vega et al., 2014; Grytdal et al., 2015; Ruvoen-Clouet et al., 2013; Green, 2007). The GI.1, in contrast, does not bind $Le^{a/x}$. Crystallographic structures of GI.7 and GI.2 P domain with various HBGAs including Le^a have been determined (Shanker et al., 2014; Kubota et al., 2012). Comparative analysis of these crystal structures with GI.1 show while the Gal binding site remains invariant, genotypic sequence variations profoundly alter the loop structures to allow differential HBGA specificity and possibly antigenicity. Based on such comparative analyses, it is suggested that the threshold length and structure of one of the loops, the P loop, is the critical determinant for Le^a binding (Fig. 3.1.3E). The comparative analysis further showed significant differences in loops A and B. These two loops in GI.7 are significantly more separated in a distinctly "open" conformation in contrast to a "closed" conformation in GI.1 and GI.2 P domains. Interestingly, in the GI.1 NV, the B loop contains a residue critical for binding of HBGA blocking antibodies (Chen et al., 2013), and the corresponding loop in the P domains of murine NoV (genogroup V) (Katpally et al., 2008; Taube et al., 2010) and rabbit hemorrhagic disease virus (animal calicivirus) (Wang et al., 2013) contains the neutralization antigenic sites. Thus, this region is potentially a major site for differential antigenic presentations contributing to serotypic differences among the GI variants.

Similarly, in GII NoVs, including GII.4 variants that are suggested to undergo epochal evolution, while the first site involved in the interactions with Fuc is well conserved, the second site is susceptible to genotypic or temporal alterations and allows for differential binding of difucosyl Lewis HBGAs. A fascinating example is provided by the comparative analysis of the P domain structures of four GII.4 variants in which structural changes in the T-loop of the P2 subdomain, due to temporal sequence changes, modulate the binding strength of the difucosyl Lewis HBGAs between the variants (Shanker et al., 2014; Bu et al., 2008; Singh et al., 2015) (Fig. 3.1.3F). Interestingly, these crystallographic studies have also revealed a novel variation. In contrast to the three GII.4 variants (1996, 2004, and 2006) in which the first site interacts with the seFuc, in the 2012 variant, the first site is involved in anchoring the LeFuc residue, which is also observed in the GII.9 NoV. These studies suggest that the epochal evolution of GII.4 is driven by differentially (de Rougemont et al., 2011) targeting secretor-positive, Lewis-positive individuals.

Another important observation from the crystallographic studies of the P domain of GII.4 NoVs is that the temporal sequence changes contribute to distinct differences in the electrostatic landscape of the P2 subdomain, likely reflecting antigenic variations (Shanker et al., 2011). Some of these changes are in close proximity to the HBGA binding sites suggesting a coordinated interplay between antigenicity and HBGA binding in epochal evolution (Shanker et al., 2011). Despite the lack of a cell culture system and an efficient small animal model system for human NoVs, surrogate HBGA blockade assays with human antibodies, in lieu of neutralization assays, have shown how the variations within GII.4 variants affect antigenic profiles (Lindesmith et al., 2008, 2011, 2012). Circulating antibodies that block HBGA binding correlate with protection in chimpanzees (Bok et al., 2011). The importance of such surrogate neutralizing antibodies is further underscored by recent studies showing that circulating antibodies that block HBGA binding correlate with protection from NoV-associated illness (Atmar et al., 2011; Reeck et al., 2010). Although the effect of sequence changes on HBGA binding has been structurally well characterized, currently no structural studies have been reported on how HBGA-blocking antibodies interact with NoV strains.

5 NONSTRUCTURAL PROTEINS

The ORF1 in NoVs encodes a polyprotein that is proteolytically processed by virus-encoded protease into at least six NSPs (Thorne and Goodfellow, 2014). These NSPs (from N- to C-terminus of the polyprotein) include p48 of unknown function; p41, an NTPase with AAA+ sequence motifs similar to the picornavirus enzyme 2C; p22, which shares sequence similarities with picornavirus 3A, with a possible function as an antagonist of Golgi-dependent cellular protein secretion (Sharp et al., 2012); VPg that is covalently linked to the viral RNA; a protease that is analogous to the picornavirus 3C enzyme, and an RNA-dependent RNA polymerase (RdRp) orthologous to picornavirus $3D^{pol}$ (Pfister and Wimmer, 2001; Sosnovtsev et al., 2002; Hardy, 2005; Clarke and Lambden, 2000). In recent years, substantial progress has been made in understanding structure and function of three of these proteins including VPg, protease, and RdRp as discussed later. (See also Chapter 3.2).

5.1 VPg

Similar to picornaviruses and as demonstrated in animal caliciviruses, VPg (by inference in NoVs) is covalently linked to the genomic RNA (Thorne and Goodfellow, 2014). However, unlike picornavirus VPg, which is only about ~2 kDa in size, VPg in caliciviruses is significantly larger (12–15 kDa). The calicivirus genomes do not have a capped 5′-end, like cellular RNAs, or an internal ribosome entry site (IRES) as in picornaviral RNA, for RNA translation. Studies on animal caliciviruses and NoVs have attributed a dual role to this protein. First, VPg acts as a "cap substitute" and mediates translation initiation of viral RNA based on the observations that m^7-GTP cap can substitute for VPg to confer infectivity in vitro to synthesized FCV RNA (Sosnovtsev and Green, 1995), and that it can bind directly to initiation factor eIF4E (Daughenbaugh et al., 2003, 2006; Goodfellow et al., 2005). Second, analogous to picornavirus VPg, the NoV VPg has a priming function during RNA replication based on the observation that VPg is uridylylated at a conserved Tyr residue by the viral RdRp followed by elongation in the presence of RNA (Belliot et al., 2008; Han et al., 2010; Mitra et al., 2004; Royall et al., 2015; Chung et al., 2014).

Currently, there is no structural information on full-length VPg for any calicivirus. However, NMR structures of the central core of VPg, consisting of about ~55 residues, from FCV (Leen et al., 2013), porcine sapovirus (PSV) (Hwang et al., 2015), and murine NoV (MNV) (Leen et al., 2013) have been determined. The N-terminal and the C-terminal regions flanking the central core are considered mostly disordered. The VPg core structures of FCV and PSV are very similar with a well-defined three-helical bundle, whereas that of MNV consists of only two of these helical segments (Fig. 3.1.4A). The Tyr residue within a conserved DDEYDEW motif is suggested to function as a nucleotide acceptor for viral replication and translation (Belliot et al., 2008; Han et al., 2010; Mitra et al., 2004). In all the three structures, the location of this residue, which is fully solvent exposed, is conserved. Although currently there are no structural studies on calicivirus VPg-RdRp, crystallographic structures of VPg-RdRp complex of foot-and-mouth disease virus (a picornavirus) both in the presence and absence of oligoadenylate substrate have been determined, and provide mechanistic details of how VPg interacts with RdRp in carrying out its priming function (Ferrer-Orta et al., 2006). Given that calicivirus VPg is significantly larger than picornavirus VPg, it remains to be seen how much of the structural details of the interactions between VPg and RdRp in caliciviruses remain similar.

5.2 Protease

The role of the NoV protease in the polyprotein processing of the polyprotein, the primary cleavage sites, and the order in which it cleaves the polyprotein have been firmly established. The proteolytic processing of the polyprotein occurs in a sequential manner as a mechanism to regulate viral genome expression and replication (Belliot et al., 2003; Hardy et al., 2002). Protease cleaves the polyprotein at five sites with three different cleavage junctions—Gln/Gly, Glu/Gly, and Glu/Ala, first cleaving the two Gln/Gly junctions between p48/p41 and p42/p22 followed Glu/Gly (VPg/Pro) and Glu/Ala (Pro/RdRp) junctions (Sosnovtsev and Green, 2000; Hardy et al., 2002; Sosnovtsev et al., 2006). These sites exhibit significant variations in the amino acid composition in both the N- (P5–P2) and C-terminal (P2′–P5′) sides flanking the scissile bond (P1/P1′)

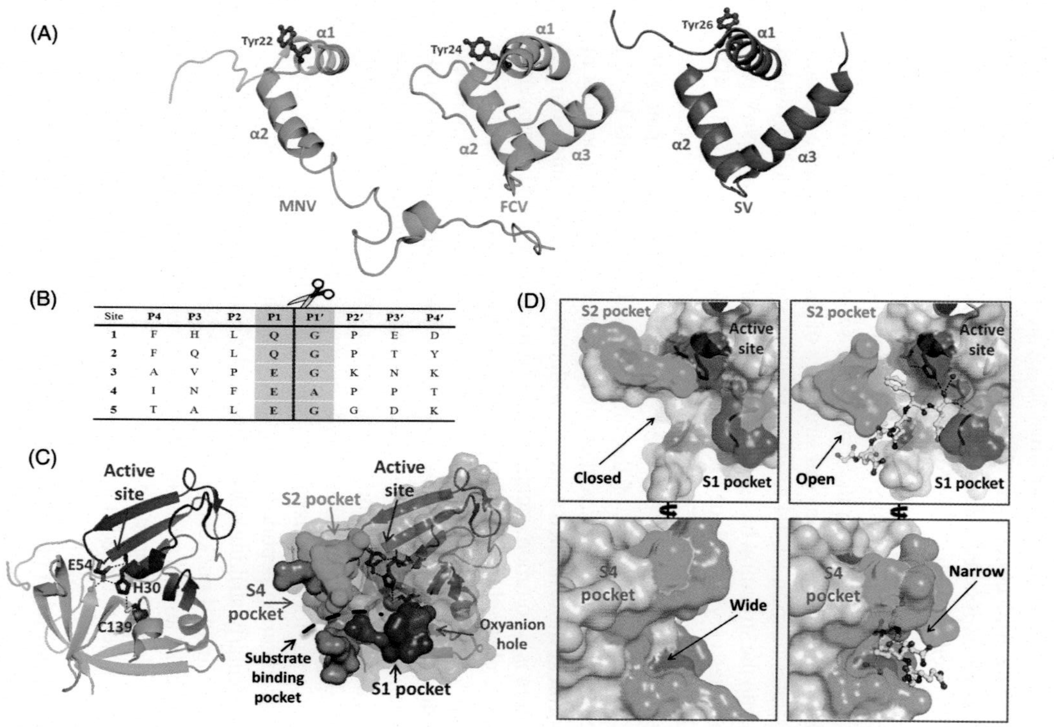

Site	P4	P3	P2	P1	P1′	P2′	P3′	P4′
1	F	H	L	Q	G	P	E	D
2	F	Q	L	Q	G	P	T	Y
3	A	V	P	E	G	K	N	K
4	I	N	F	E	A	P	P	T
5	T	A	L	E	G	G	D	K

FIGURE 3.1.4 **Calicivirus VPg and protease structures.** (A) NMR structures of the VPg core of MNV (left) (PDB id: 2M4G), FCV (middle) (PDB id: 2M4H), and porcine Sapovirus (right) (PDB id: 2MXD). The exposed Tyr residue that is be uridylylated by the RdRp in each structure is shown in stick representation. (B) Cleavage sites in the NV polyprotein, from the N- to C-terminal, along with the surrounding N-terminal P1-P4 and C-terminal P1′-P4′ residues. (C) *Left*. Cartoon representation of the NV protease structure (PDB id: 4IN1), the N-terminal and C terminal domains are shown in blue and cyan, respectively. The catalytic triad is shown in red. *Right*. Surface representation of the protease structure showing the locations of the S1 *(blue)*, S2 *(green)*, and S4 *(gold)* pockets, and the oxyanion hole *(pink)*, with respect to the active site *(red)* (see Zeitler et al., 2006; Muhaxhiri et al., 2013). The substrate binding cleft between the two β-barrel domains is shown by a black dashed line. (D) Coordinated structural changes in the S2 *(top)* and S4 *(bottom)* pockets before *(left)* (PDB id: 2FYQ) and after *(right)* (PDB id: 4IN2) substrate binding. The substrate binding pockets are depicted in the same color as in Fig. 3.1.4C, right. Substrate is shown as yellow sticks.

(Fig. 3.1.4B). Mutational analysis has shown that residues surrounding the cleavage sites contribute to proteolytic efficiency (Hardy et al., 2002). An interesting question is how the protease recognizes such nonhomologous sites within the polyprotein with differential affinities.

Crystallographic structures of proteases from different NoVs show that NoV protease, similar to the picornavirus $3C_{Pro}$, is a cysteine protease with a chymotrypsin-like fold comprised of two domains separated by a groove where the active site is located (Hussey et al., 2011; Leen et al., 2012; Nakamura et al., 2005; Zeitler et al., 2006; Muhaxhiri et al., 2013) (Fig. 3.1.4C). The active site consists of a catalytic triad with Cys as a nucleophile, His as the general base catalyst, and Glu as the anion to orient the imidazole ring of His, similar to the Ser-His-Asp triad in the trypsin-like serine proteases (Bazan and Fletterick, 1988). More recently, crystal structures of the NoV protease in complex with substrates bearing P1–P4 residues or substrate-mimics have provided novel insights into the structural basis of substrate recognition and how NoV protease accommodates varying residue compositions of the substrate (Fig. 3.1.4D) (Leen et al., 2013; Hussey et al., 2011; Muhaxhiri et al., 2013). These studies show that the substrate adopts an extended β-strand conformation to pair with a β-strand in the active site cleft of the protease, and the side chains P1–P4 optimally interact with the S1–S4 pockets of the protease. The P1 positions in the NoV polyprotein show limited variation with only Glu or Gln. Interestingly, Gln in P1 position is a common occurrence in the picornavirus and coronavirus substrates that are cleaved by their respective proteases, which are structurally similar to the NoV protease. The S1 pocket in the NoV protease is ideally suited for optimal hydrophobic and hydrogen bond interactions with either Glu or Gln and remains unaltered with variations in P2–P4 positions. A novel observation from these protease-substrate structures is the conformational change induced by the substrate binding in the main chain amide group of a conserved Gly adjacent to the catalytic Cys to form the oxyanion hole required for stabilizing the tetrahedral intermediate during peptide hydrolysis (Muhaxhiri et al., 2013).

Another particularly striking observation is the coordinated conformational alterations that S2 and S4 pockets undergo in response to variations in the residue composition at P2 and P4 positions of the substrate (Muhaxhiri et al., 2013). The S2 pocket undergoes transition from an "open" state as observed in the apo-structure to a gradual closed state depending upon the bulkiness of the sidechain in the P2 position, whereas the S4 pocket shows a reverse trend from a closed (apo-structure) to an open state in response to P4 sidechain interactions Fig. 3.1.4D). In contrast to P1, P2, and P4 positions, which show extensive interactions with well-defined S1, S2, and S4 pockets, respectively, the interactions between the P3 residue and the S3 pocket are minimal suggesting that this position can tolerate variations.

Similar to proteases of other RNA viruses, the NoV protease, particularly of the G1.1 NV, has been targeted for structure-assisted design and development of small molecule inhibitors (Muhaxhiri et al., 2013; Deng et al., 2013;

Prior et al., 2013). Currently, these studies have focused on the characterization of substrate-based peptido-mimetics containing aldehydes, ketones, esters, or bisulfite adducts as electrophilic warheads attached to the P1 residue that form a covalent bond with the sulfhydryl group of the catalytic cysteine. Structures of the NV protease with three of the inhibitors (Muhaxhiri et al., 2013), containing a terminal aldehyde and selected based on their potency of inhibition, have been determined. These studies show that the mechanism of inhibition, in addition to the formation of the covalent adducts, involves prevention of the conformational change necessary for the formation of the oxyanion hole. They further suggest that peptido-mimetics with suitable warheads, a Glu-like chemical entity at P1 for optimal interactions with S1, and an appropriate combination of hydrophobic residues at P2 and P4 that maximizes the interactions with S2 and S4, have a direct impact on the potency of the inhibitors. Further studies should be anticipated that are directed at optimizing the design strategies and improving the pharmacokinetic properties of such inhibitors as antivirals using structural analysis and cell-based assays (Qu et al., 2014; Chang et al., 2006).

5.3 RdRp

In caliciviruses, this obligatory enzyme that is similar to the picornavirus $3D^{pol}$, is responsible for synthesizing both negative-sense RNA as well as newly made positive-sense genomic RNA. X-ray structures of RdRps from several caliciviruses including RHDV (Ng et al., 2002), GII NoV (Ng et al., 2004; Zamyatkin et al., 2008), sapovirus (Fullerton et al., 2007), and MNV (Lee et al., 2011; Hogbom et al., 2009) have been determined. Calicivirus RdRp exhibits a typical "right hand" configuration of palm, finger, and thumb domains as observed in all RNA/DNA polymerases (Ng et al., 2008). In addition to these domains, like in other RdRps, it has a distinct N-terminal domain bridging the fingers and the thumb domains. The active site of the RdRp is located in the thumb domain, which consists of three conserved Asp residues critical for mediating catalysis through a two metal-ion mechanism (Fig. 3.1.5C) and other key residues such as Arg, Asn, and Ser required for substrate binding and catalysis.

Comparative analysis of the various calicivirus RdRps show the following: While the conformations of the individual domains, despite only marginal sequence similarities, are highly conserved, the conformations of the loop regions, N-terminal domain, C-terminal region, and interdomain orientations are susceptible to significant variations. Norovirus RdRp can exist in two principal conformations: an "open" active site conformation that represents the inactive state of the RdRp, as observed generally in the apo RdRp structures (Ng et al., 2002; Ng et al., 2004; Fullerton et al., 2007) (Fig. 3.1.5A); and a "closed" active site conformation that is primed for catalyzing the nucleotidyl transfer reaction, as observed in the RdRp structure in complex with divalent metal ions, the primer-template RNA duplex, and the NTP that is required for template elongation (Zamyatkin et al., 2008) (Fig. 3.1.5B). The "closed" conformation allows optimal positioning of the NTP, RNA and the metal ions for the catalytic reaction to occur (Fig. 3.1.5C). Transition from the "open" to "closed" conformation involves displacement of the C-terminal tail, which triggers the central helix in the palm domain to rotate by about 20 degree (Fig. 3.1.5D). The C-terminal tail appears to function as a lid to regulate the access of the active site. In the inactive "open" conformation, it is located in the active site restricting the access for the RNA, whereas in the active "closed" conformation, it moves away and essentially becomes unstructured. This region, which is relatively well conserved in all NoVs, is suggested to play a role not only in regulating the initiation of RNA synthesis but also in mediating interactions with accessory proteins during replication.

In addition to its known interaction with VPg during VPg-primed RNA synthesis, recent studies of human and murine NoVs have shown that the major capsid protein VP1 can interact with RdRp through its S domain in a species-specific and concentration-dependent manner to modulate the rate and kinetics of RdRp activity (Subba-Reddy et al., 2012). Such an interaction is suggested to play a significant role in the temporal regulation of RdRp activity during genome replication, capsid assembly, and genome encapsidation. Previous to these studies, the possibility of VP1–RdRp interaction in replication complexes was also suggested in the case of FCV based on a yeast two-hybrid assay (Kaiser et al., 2006). Further studies are required to understand the mechanistic basis of how the S domain of VP1 influences the RdRp activity and whether VPg and VP1 are the only interacting partners of RdRp or whether other viral proteins, such as p41 with its NTPase activity or protease with its recently discovered property to bind RNA (Viswanathan et al., 2013), can also exert influence on the RdRp activity. This is feasible, considering their likely close proximity in the membranous replication complexes.

Structural studies on calicivirus RdRps have provided a rational basis to embark on structure-based in silico screening for small molecule inhibitors (Mastrangelo et al., 2012; Croci et al., 2014). These studies identified two molecules, suramin and its analogue NF023, consisting of naphthalene-trisulfonic acid moiety, that inhibit NoV RdRp with IC_{50}s of 24.6 and 71.5 nM, respectively. Crystallographic analysis show that both inhibitors interact with the RdRp at a similar region along the access pathway of the NTPs between the fingers and thumb domains. Future studies using similar structure-based techniques as well as high throughput screening (Eltahla et al., 2014) are likely to provide more potent small molecule RdRp inhibitors with desirable pharmacokinetic properties.

6 CONCLUSIONS AND FUTURE DIRECTIONS

In the last two decades, starting from cryo-EM and X-ray crystallographic analyses of the recombinant NV capsid, there have been a significant number of structural studies that have led to a better understanding of the structure and function of NoVs and their encoded proteins. These studies have provided insights into how the elements required for capsid assembly, strain diversity, and

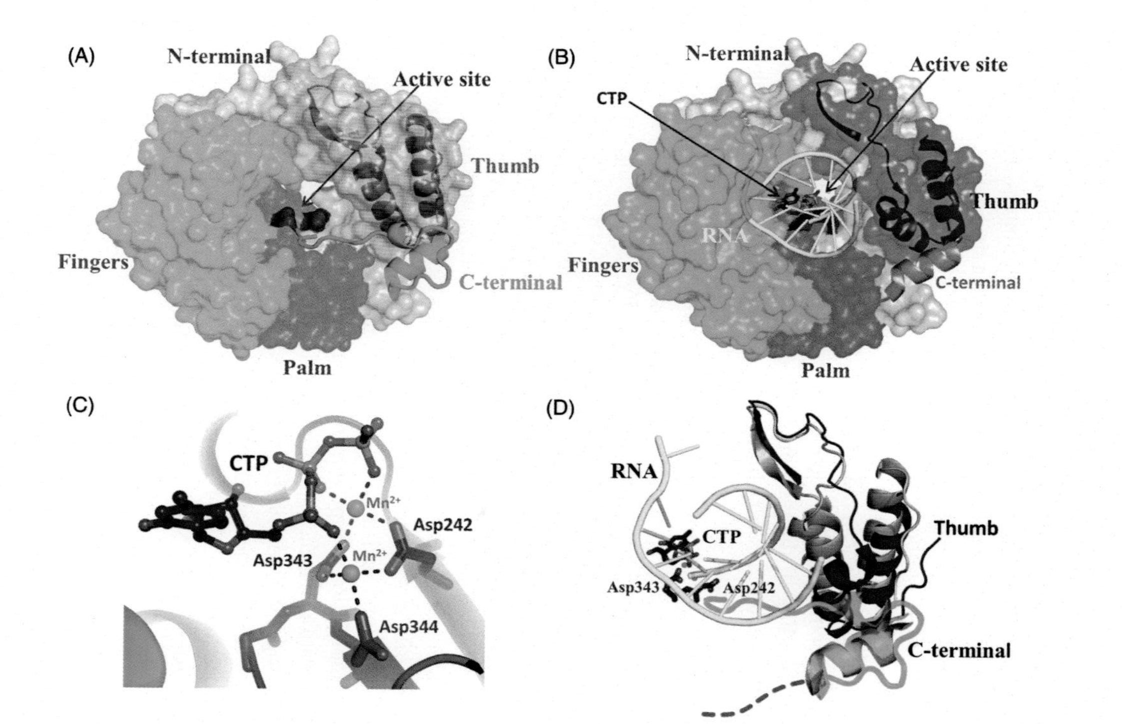

FIGURE 3.1.5 NoV RdRp structure. (A) Open conformation (PDB id: 1SH0) of the apo-RdRp. The thumb *(yellow)*, fingers *(blue)*, palm *(pink)* domains, active site location *(red)*, along with the N- *(light yellow)* and C-terminal *(green)* regions are indicated. Notice the C-terminal tail located close to the active site. (B) "Closed" conformation (PDB id: 3BSO) with each of the structural elements shown using the same coloring scheme as in Fig. 3.1.5A with the exception of the moving parts, C-terminal tail and thumb domain, which are shown in blue and gold, respectively. Bound primer-template RNA duplex and CTP are shown in yellow and dark blue, respectively. (C) Conserved catalytic Asp residues *(gold sticks)* coordinating the two Mn^{2+} ions *(cyan balls)* along with the bound CTP *(blue)*. (D) Movement of the central helix in the thumb domain *(in yellow and blue)* and the C-terminal tail *(in green and gold)* from an inactive "open" conformation to a catalytically active "closed" conformation shown along with the bound RNA *(yellow)*, CTP *(blue)*, and Asp residues *(gold sticks)*.

immunogenicity are integrated into a single capsid protein through an elegant modular domain organization and how the distally located P2 subdomain with a unique fold provides an efficient platform for genotype-dependent variations in glycan recognition and antigenicity to facilitate virus evolution. Structural studies on nonstructural proteins, such as protease and RdRp have uncovered fascinating novel mechanistic details that underlie their enzymatic functions, and allowed design and development of small molecule inhibitors. However, there are still several significant questions that merit further studies.

For the capsid proteins, these questions include: (1) what is the structural basis of how 'neutralizing' antibodies block HBGA binding in human NoVs? The results may inform the design of immunotherapeutic agents. (2) What is the structure and function of the minor protein VP2? Such data may provide insights into capsid assembly and genome encapsidation. For the nonstructural proteins, our understanding of the proteins, such as p48, p41, and p22 both in terms of structure and function is very limited. Sequence analysis and available experimental data suggest that these proteins are likely membrane-associated and have a role in initiating and structuring vesicular replication compartments. 2C-like p41 with NTPase activity is particularly enigmatic with distinct AAA^+ motifs—raising the question: how is the NTPase activity used during replication? VPg is largely disordered except for the central core. As it has to interact with multiple partners during replication, such a flexible state prior to interacting with its partners could be a necessity. Further structural studies may provide insight into how this protein interacts with eIF4E and RdRp. The recent discovery that protease can interact with RNA opens further structural and functional studies to understand how this interaction interferes with the enzymatic activity, and whether it has a role in replication. Many of these proteins including RdRp are likely to have temporal and transient interactions with each other within the confines of the replication compartments to regulate and coordinate various stages of virus replication. Further structural studies of virus infected cells including electron tomographic approaches may provide mechanistic insights into these processes. Considering the recent advances in producing human NoVs in cultured cells (Katayama et al., 2014; Jones et al., 2014), exciting progress in furthering our understanding of NoV structure-function is to be expected.

ACKNOWLEDGMENTS

We acknowledge support from NIH grant PO1 AI057788 (MKE, RLA, BVVP), and a grant (Q1279) from the Robert Welch foundation (BVVP).

REFERENCES

Ahmed, S.M., Hall, A.J., Robinson, A.E., Verhoef, L., Premkumar, P., Parashar, U.D., et al., 2014. Global prevalence of norovirus in cases of gastroenteritis: a systematic review and meta-analysis. Lancet Infect. Dis. 14, 725–730.

Atmar, R.L., Bernstein, D.I., Harro, C.D., Al-Ibrahim, M.S., Chen, W.H., Ferreira, J., et al., 2011. Norovirus vaccine against experimental human Norwalk Virus illness. N. Engl. J. Med. 365, 2178–2187.

Atmar, R.L., Bernstein, D.I., Lyon, G.M., Treanor, J.J., Al-Ibrahim, M.S., Graham, D.Y., et al., 2015. Serological correlates of protection against a GII.4 Norovirus. Clin. Vaccine Immunol. 22, 923–929.

Baclayon, M., Shoemaker, G.K., Uetrecht, C., Crawford, S.E., Estes, M.K., Prasad, B.V., et al., 2011. Prestress strengthens the shell of Norwalk virus nanoparticles. Nano Lett. 11, 4865–4869.

Bazan, J.F., Fletterick, R.J., 1988. Viral cysteine proteases are homologous to the trypsin-like family of serine proteases: structural and functional implications. Proc. Natl. Acad. Sci. USA 85, 7872–7876.

Belliot, G., Sosnovtsev, S.V., Mitra, T., Hammer, C., Garfield, M., Green, K.Y., 2003. In vitro proteolytic processing of the MD145 norovirus ORF1 nonstructural polyprotein yields stable precursors and products similar to those detected in calicivirus-infected cells. J. Virol. 77, 10957–10974.

Belliot, G., Sosnovtsev, S.V., Chang, K.O., McPhie, P., Green, K.Y., 2008. Nucleotidylylation of the VPg protein of a human norovirus by its proteinase-polymerase precursor protein. Virology 374, 33–49.

Bertolotti-Ciarlet, A., White, L.J., Chen, R., Prasad, B.V., Estes, M.K., 2002. Structural requirements for the assembly of Norwalk virus-like particles. J. Virol. 76, 4044–4055.

Bertolotti-Ciarlet, A., Crawford, S.E., Hutson, A.M., Estes, M.K., 2003. The 3' end of Norwalk virus mRNA contains determinants that regulate the expression and stability of the viral capsid protein VP1: a novel function for the VP2 protein. J. Virol. 77, 11603–11615.

Bhella, D., Gatherer, D., Chaudhry, Y., Pink, R., Goodfellow, I.G., 2008. Structural insights into calicivirus attachment and uncoating. J. Virol. 82, 8051–8058.

Bok, K., Parra, G.I., Mitra, T., Abente, E., Shaver, C.K., Boon, D., et al., 2011. Chimpanzees as an animal model for human norovirus infection and vaccine development. Proc. Natl. Acad. Sci. USA 108, 325–330.

Bu, W., Mamedova, A., Tan, M., Xia, M., Jiang, X., Hegde, R.S., 2008. Structural basis for the receptor binding specificity of Norwalk virus. J. Virol. 82, 5340–5347.

Cao, S., Lou, Z., Tan, M., Chen, Y., Liu, Y., Zhang, Z., et al., 2007. Structural basis for the recognition of blood group trisaccharides by norovirus. J. Virol. 81, 5949–5957.

Chang, K.O., Sosnovtsev, S.V., Belliot, G., King, A.D., Green, K.Y., 2006. Stable expression of a Norwalk virus RNA replicon in a human hepatoma cell line. Virology 353, 463–473.

Chen, R., Neill, J.D., Noel, J.S., Hutson, A.M., Glass, R.I., Estes, M.K., et al., 2004. Inter- and intragenus structural variations in caliciviruses and their functional implications. J. Virol. 78, 6469–6479.

Chen, R., Neill, J.D., Estes, M.K., Prasad, B.V., 2006. X-ray structure of a native calicivirus: structural insights into antigenic diversity and host specificity. Proc. Natl. Acad. Sci. USA 103, 8048–8053.

Chen, Y., Tan, M., Xia, M., Hao, N., Zhang, X.C., Huang, P., et al., 2011. Crystallography of a Lewis-binding norovirus, elucidation of strain-specificity to the polymorphic human histo-blood group antigens. PLoS Pathog. 7, e1002152.

Chen, Z., Sosnovtsev, S.V., Bok, K., Parra, G.I., Makiya, M., Agulto, L., et al., 2013. Development of Norwalk virus-specific monoclonal antibodies with therapeutic potential for the treatment of Norwalk virus gastroenteritis. J. Virol. 87, 9547–9557.

Choi, J.M., Hutson, A.M., Estes, M.K., Prasad, B.V., 2008. Atomic resolution structural characterization of recognition of histo-blood group antigens by Norwalk virus. Proc. Natl. Acad. Sci. USA 105, 9175–9180.

Chung, L., Bailey, D., Leen, E.N., Emmott, E.P., Chaudhry, Y., Roberts, L.O., et al., 2014. Norovirus translation requires an interaction between the C Terminus of the genome-linked viral protein VPg and eukaryotic translation initiation factor 4G. J. Biol. Chem. 289, 21738–21750.

Clarke, I.N., Lambden, P.R., 2000. Organization and expression of calicivirus genes. J. Infect. Dis. 181 (Suppl 2), S309–S316.

Croci, R., Pezzullo, M., Tarantino, D., Milani, M., Tsay, S.C., Sureshbabu, R., et al., 2014. Structural bases of norovirus RNA dependent RNA polymerase inhibition by novel suramin-related compounds. PLoS One 9, e91765.

Currier, R.L., Payne, D.C., Staat, M.A., Selvarangan, R., Shirley, S.H., Halasa, N., et al., 2015. Innate susceptibility to norovirus infections influenced by FUT2 genotype in a United States pediatric population. Clin. Infect. Dis. 60, 1631–1638.

Daughenbaugh, K.F., Fraser, C.S., Hershey, J.W., Hardy, M.E., 2003. The genome-linked protein VPg of the Norwalk virus binds eIF3, suggesting its role in translation initiation complex recruitment. EMBO J. 22, 2852–2859.

Daughenbaugh, K.F., Wobus, C.E., Hardy, M.E., 2006. VPg of murine norovirus binds translation initiation factors in infected cells. Virol. J. 3, 33.

Deng, L., Muhaxhiri, Z., Estes, M.K., Palzkill, T., Prasad, B.V., Song, Y., 2013. Synthesis, activity and structure-activity relationship of noroviral protease inhibitors. Medchemcomm, 4(10). doi: 10.1039/C3MD00219E.

de Rougemont, A., Ruvoen-Clouet, N., Simon, B., Estienney, M., Elie-Caille, C., Aho, S., Pothier, P., Le Pendu, J., Boireau, W., Belliott, G., 2011. Qualitative and quantitative analysis of the binding of GII.4 norovirus variants onto human blood group antigens. J. Virol. 85, 4057–4070.

Donaldson, E.F., Lindesmith, L.C., Lobue, A.D., Baric, R.S., 2008. Norovirus pathogenesis: mechanisms of persistence and immune evasion in human populations. Immunol. Rev. 225, 190–211.

Eltahla, A.A., Lim, K.L., Eden, J.S., Kelly, A.G., Mackenzie, J.M., White, P.A., 2014. Nonnucleoside inhibitors of norovirus RNA polymerase: scaffolds for rational drug design. Antimicrob. Agents Chemother. 58, 3115–3123.

Ferrer-Orta, C., Arias, A., Agudo, R., Perez-Luque, R., Escarmis, C., Domingo, E., et al., 2006. The structure of a protein primer-polymerase complex in the initiation of genome replication. EMBO J. 25, 880–888.

Fisher, A.J., Johnson, J.E., 1993. Ordered duplex RNA controls capsid architecture in an icosahedral animal virus. Nature 361, 176–179.

Fullerton, S.W., Blaschke, M., Coutard, B., Gebhardt, J., Gorbalenya, A., Canard, B., et al., 2007. Structural and functional characterization of sapovirus RNA-dependent RNA polymerase. J. Virol. 81, 1858–1871.

Glass, P.J., White, L.J., Ball, J.M., Leparc-Goffart, I., Hardy, M.E., Estes, M.K., 2000. Norwalk virus open reading frame 3 encodes a minor structural protein. J. Virol. 74, 6581–6591.

Goodfellow, I., Chaudhry, Y., Gioldasi, I., Gerondopoulos, A., Natoni, A., Labrie, L., et al., 2005. Calicivirus translation initiation requires an interaction between VPg and eIF 4 E. EMBO Rep. 6, 968–972.

Green, K.Y., 2007. Caliciviridae: The Noroviruses. Lippincoot Williams & Wilkins, Philadelphia.

Green, K.Y., Ando, T., Balayan, M.S., Berke, T., Clarke, I.N., Estes, M.K., et al., 2000. Taxonomy of the caliciviruses. J. Infect. Dis. 181 (Suppl 2), S322–S330.

Grytdal, S.P., Rimland, D., Shirley, S.H., Rodriguez-Barradas, M.C., Goetz, M.B., Brown, S.T., et al., 2015. Incidence of medically-attended norovirus-associated acute gastroenteritis in four veteran's affairs medical center populations in the United States, 2011-2012. PLoS One 10, e0126733.

Hakomori, S., 1999. Antigen structure and genetic basis of histo-blood groups A, B and O: their changes associated with human cancer. Biochim. Biophys. Acta 1473, 247–266.

Han, K.R., Choi, Y., Min, B.S., Jeong, H., Cheon, D., Kim, J., et al., 2010. Murine norovirus-1 3Dpol exhibits RNA-dependent RNA polymerase activity and nucleotidylylates on Tyr of the VPg. J. Gen. Virol. 91, 1713–1722.

Hansman, G.S., Biertumpfel, C., Georgiev, I., McLellan, J.S., Chen, L., Zhou, T., et al., 2011. Crystal structures of GII.10 and GII.12 norovirus protruding domains in complex with histo-blood group antigens reveal details for a potential site of vulnerability. J. Virol. 85, 6687–6701.

Hardy, M.E., 2005. Norovirus protein structure and function. FEMS Microbiol. Lett. 253, 1–8.

Hardy, M.E., Crone, T.J., Brower, J.E., Ettayebi, K., 2002. Substrate specificity of the Norwalk virus 3C-like proteinase. Virus Res. 89, 29–39.

Harrison, S.C., 2001. The familiar and the unexpected in structures of icosahedral viruses. Curr. Opin. Struct. Biol. 11, 195–199.

Hogbom, M., Jager, K., Robel, I., Unge, T., Rohayem, J., 2009. The active form of the norovirus RNA-dependent RNA polymerase is a homodimer with cooperative activity. J. Gen. Virol. 90, 281–291.

Huang, P., Farkas, T., Zhong, W., Tan, M., Thornton, S., Morrow, A.L., et al., 2005. Norovirus and histo-blood group antigens: demonstration of a wide spectrum of strain specificities and classification of two major binding groups among multiple binding patterns. J. Virol. 79, 6714–6722.

Hussey, R.J., Coates, L., Gill, R.S., Erskine, P.T., Coker, S.F., Mitchell, E., et al., 2011. A structural study of norovirus 3C protease specificity: binding of a designed active site-directed peptide inhibitor. Biochemistry 50, 240–249.

Hutson, A.M., Atmar, R.L., Marcus, D.M., Estes, M.K., 2003. Norwalk virus-like particle hemagglutination by binding to H histo-blood group antigens. J. Virol. 77, 405–415.

Hutson, A.M., Atmar, R.L., Estes, M.K., 2004. Norovirus disease: changing epidemiology and host susceptibility factors. Trends Microbiol. 12, 279–287.

Hutson, A.M., Airaud, F., LePendu, J., Estes, M.K., Atmar, R.L., 2005. Norwalk virus infection associates with secretor status genotyped from sera. J. Med. Virol. 77, 116–120.

Hwang, H.J., Min, H.J., Yun, H., Pelton, J.G., Wemmer, D.E., Cho, K.O., et al., 2015. Solution structure of the porcine sapovirus VPg core reveals a stable three-helical bundle with a conserved surface patch. Biochem. Biophys. Res. Commun. 459, 610–616.

Imbert-Marcille, B.M., Barbe, L., Dupe, M., Le Moullac-Vaidye, B., Besse, B., Peltier, C., et al., 2014. A FUT2 gene common polymorphism determines resistance to rotavirus A of the P(Donaldson et al., 2008) genotype. J. Infect. Dis. 209, 1227–1230.

Jin, M., Tan, M., Xia, M., Wei, C., Huang, P., Wang, L., et al., 2015. Strain-specific interaction of a GII.10 Norovirus with HBGAs. Virology 476, 386–394.

Jones, M.K., Watanabe, M., Zhu, S., Graves, C.L., Keyes, L.R., Grau, K.R., et al., 2014. Enteric bacteria promote human and mouse norovirus infection of B cells. Science 346, 755–759.

Kaiser, W.J., Chaudhry, Y., Sosnovtsev, S.V., Goodfellow, I.G., 2006. Analysis of protein-protein interactions in the feline calicivirus replication complex. J. Gen. Virol. 87, 363–368.

Katayama, K., Murakami, K., Sharp, T.M., Guix, S., Oka, T., Takai-Todaka, R., et al., 2014. Plasmid-based human norovirus reverse genetics system produces reporter-tagged progeny virus containing infectious genomic RNA. Proc. Natl. Acad. Sci. USA 111, E4043–E4052.

Katpally, U., Wobus, C.E., Dryden, K., Virgin, H.W., 4th, Smith, T.J., 2008. Structure of antibody-neutralized murine norovirus and unexpected differences from viruslike particles. J. Virol. 82, 2079–2088.

Kroneman, A., Verhoef, L., Harris, J., Vennema, H., Duizer, E., van Duynhoven, Y., et al., 2008. Analysis of integrated virological and epidemiological reports of norovirus outbreaks collected within the Foodborne Viruses in Europe network from 1 July 2001 to 30 June 2006. J. Clin. Microbiol. 46, 2959–2965.

Kubota, T., Kumagai, A., Ito, H., Furukawa, S., Someya, Y., Takeda, N., et al., 2012. Structural basis for the recognition of Lewis antigens by genogroup I norovirus. J. Virol. 86, 11138–11150.

Kumar, S., Ochoa, W., Kobayashi, S., Reddy, V.S., 2007. Presence of a surface-exposed loop facilitates trypsinization of particles of Sinsiro virus, a genogroup II.3 norovirus. J. Virol. 81, 1119–1128.

Lee, J.H., Alam, I., Han, K.R., Cho, S., Shin, S., Kang, S., et al., 2011. Crystal structures of murine norovirus-1 RNA-dependent RNA polymerase. J. Gen. Virol. 92, 1607–1616.

Leen, E.N., Baeza, G., Curry, S., 2012. Structure of a murine norovirus NS6 protease-product complex revealed by adventitious crystallisation. PLoS One 7, e38723.

Leen, E.N., Kwok, K.Y., Birtley, J.R., Simpson, P.J., Subba-Reddy, C.V., Chaudhry, Y., et al., 2013. Structures of the compact helical core domains of feline calicivirus and murine norovirus VPg proteins. J. Virol. 87, 5318–5330.

Lindesmith, L., Moe, C., Marionneau, S., Ruvoen, N., Jiang, X., Lindblad, L., et al., 2003. Human susceptibility and resistance to Norwalk virus infection. Nat. Med. 9, 548–553.

Lindesmith, L.C., Donaldson, E.F., Lobue, A.D., Cannon, J.L., Zheng, D.P., Vinje, J., et al., 2008. Mechanisms of GII.4 norovirus persistence in human populations. PLoS Med. 5, e31.

Lindesmith, L.C., Donaldson, E.F., Baric, R.S., 2011. Norovirus GII.4 strain antigenic variation. J. Virol. 85, 231–242.

Lindesmith, L.C., Beltramello, M., Donaldson, E.F., Corti, D., Swanstrom, J., Debbink, K., et al., 2012. Immunogenetic mechanisms driving norovirus GII.4 antigenic variation. PLoS Pathog. 8, e1002705.

Makino, A., Shimojima, M., Miyazawa, T., Kato, K., Tohya, Y., Akashi, H., 2006. Junctional adhesion molecule 1 is a functional receptor for feline calicivirus. J. Virol. 80, 4482–4490.

Marionneau, S., Cailleau-Thomas, A., Rocher, J., Le Moullac-Vaidye, B., Ruvoen, N., Clement, M., et al., 2001. ABH and Lewis histo-blood group antigens, a model for the meaning of oligosaccharide diversity in the face of a changing world. Biochimie 83, 565–573.

Marionneau, S., Ruvoen, N., Le Moullac-Vaidye, B., Clement, M., Cailleau-Thomas, A., Ruiz-Palacois, G., et al., 2002. Norwalk virus binds to histo-blood group antigens present on gastroduodenal epithelial cells of secretor individuals. Gastroenterology 122, 1967–1977.

Marionneau, S., Airaud, F., Bovin, N.V., Le Pendu, J., Ruvoen-Clouet, N., 2005. Influence of the combined ABO, FUT2, and FUT3 polymorphism on susceptibility to Norwalk virus attachment. J. Infect. Dis. 192, 1071–1077.

Mastrangelo, E., Pezzullo, M., Tarantino, D., Petazzi, R., Germani, F., Kramer, D., et al., 2012. Structure-based inhibition of norovirus RNA-dependent RNA polymerases. J. Mol. Biol. 419, 198–210.

Mitra, T., Sosnovtsev, S.V., Green, K.Y., 2004. Mutagenesis of tyrosine 24 in the VPg protein is lethal for feline calicivirus. J. Virol. 78, 4931–4935.

Muhaxhiri, Z., Deng, L., Shanker, S., Sankaran, B., Estes, M.K., Palzkill, T., et al., 2013. Structural basis of substrate specificity and protease inhibition in Norwalk virus. J. Virol. 87, 4281–4292.

Nakamura, K., Someya, Y., Kumasaka, T., Ueno, G., Yamamoto, M., Sato, T., et al., 2005. A norovirus protease structure provides insights into active and substrate binding site integrity. J. Virol. 79, 13685–13693.

Ng, K.K., Cherney, M.M., Vazquez, A.L., Machin, A., Alonso, J.M., Parra, F., et al., 2002. Crystal structures of active and inactive conformations of a caliciviral RNA-dependent RNA polymerase. J. Biol. Chem. 277, 1381–1387.

Ng, K.K., Pendas-Franco, N., Rojo, J., Boga, J.A., Machin, A., Alonso, J.M., et al., 2004. Crystal structure of norwalk virus polymerase reveals the carboxyl terminus in the active site cleft. J. Biol. Chem. 279, 16638–16645.

Ng, K.K., Arnold, J.J., Cameron, C.E., 2008. Structure-function relationships among RNA-dependent RNA polymerases. Curr. Top. Microbiol. Immunol. 320, 137–156.

Ossiboff, R.J., Zhou, Y., Lightfoot, P.J., Prasad, B.V., Parker, J.S., 2010. Conformational changes in the capsid of a calicivirus upon interaction with its functional receptor. J. Virol. 84, 5550–5564.

Patel, M.M., Widdowson, M.A., Glass, R.I., Akazawa, K., Vinje, J., Parashar, U.D., 2008. Systematic literature review of role of noroviruses in sporadic gastroenteritis. Emerg. Infect. Dis. 14, 1224–1231.

Pfister, T., Wimmer, E., 2001. Polypeptide p41 of a Norwalk-like virus is a nucleic acid-independent nucleoside triphosphatase. J. Virol. 75, 1611–1619.

Prasad, B.V., Schmid, M.F., 2012. Principles of virus structural organization. Adv. Exp. Med. Biol. 726, 17–47.

Prasad, B.V., Rothnagel, R., Jiang, X., Estes, M.K., 1994a. Three-dimensional structure of baculovirus-expressed Norwalk virus capsids. J. Virol. 68, 5117–5125.

Prasad, B.V., Matson, D.O., Smith, A.W., 1994b. Three-dimensional structure of calicivirus. J. Mol. Biol. 240, 256–264.

Prasad, B.V., Hardy, M.E., Dokland, T., Bella, J., Rossmann, M.G., Estes, M.K., 1999. X-ray crystallographic structure of the Norwalk virus capsid. Science 286, 287–290.

Prasad, B.V.V., Shanker, S., Hu, L., Choi, J.-M., Sue, E.C., Sasirekha Ramani, et al., 2014. Structural basis of glycan interaction in gastroenteric viral pathogens. Curr. Opin. Virol. 7, 119–127.

Prior, A.M., Kim, Y., Weerasekara, S., Moroze, M., Alliston, K.R., Uy, R.A., et al., 2013. Design, synthesis, and bioevaluation of viral 3C and 3C-like protease inhibitors. Bioorg. Med. Chem. Lett. 23, 6317–6320.

Qu, L., Vongpunsawad, S., Atmar, R.L., Prasad, B.V., Estes, M.K., 2014. Development of a Gaussia luciferase-based human norovirus protease reporter system: cell type-specific profile of Norwalk virus protease precursors and evaluation of inhibitors. J. Virol. 88, 10312–10326.

Ramani, S., Atmar, R.L., Estes, M.K., 2014. Epidemiology of human noroviruses and updates on vaccine development. Curr. Opin. Gastroenterol. 30, 25–33.

Reeck, A., Kavanagh, O., Estes, M.K., Opekun, A.R., Gilger, M.A., Graham, D.Y., et al., 2010. Serological correlate of protection against norovirus-induced gastroenteritis. J. Infect. Dis. 202, 1212–1218.

Rossmann, M.G., Johnson, J.E., 1989. Icosahedral RNA virus structure. Annu. Rev. Biochem. 58, 533–573.

Royall, E., Doyle, N., Abdul-Wahab, A., Emmott, E., Morley, S.J., Goodfellow, I., et al., 2015. Murine norovirus 1 (MNV1) replication induces translational control of the host by regulating eIF4E activity during infection. J. Biol. Chem. 290, 4748–4758.

Ruvoen-Clouet, N., Belliot, G., Le Pendu, J., 2013. Noroviruses and histo-blood groups: the impact of common host genetic polymorphisms on virus transmission and evolution. Rev. Med. Virol. 23, 355–366.

Shanker, S., Choi, J.M., Sankaran, B., Atmar, R.L., Estes, M.K., Prasad, B.V., 2011. Structural analysis of histo-blood group antigen binding specificity in a norovirus GII.4 epidemic variant: implications for epochal evolution. J. Virol. 85, 8635–8645.

Shanker, S., Czako, R., Sankaran, B., Atmar, R.L., Estes, M.K., Prasad, B.V., 2014. Structural analysis of determinants to HBGA binding specificity in GI noroviruses. J. Virol. 88, 6168–6180.

Sharp, T.M., Crawford, S.E., Ajami, N.J., Neill, F.H., Atmar, R.L., Katayama, K., et al., 2012. Secretory pathway antagonism by calicivirus homologues of Norwalk virus nonstructural protein p22 is restricted to noroviruses. Virol. J. 9, 181.

Shirato, H., Ogawa, S., Ito, H., Sato, T., Kameyama, A., Narimatsu, H., et al., 2008. Noroviruses distinguish between type 1 and type 2 histo-blood group antigens for binding. J. Virol. 82, 10756–10767.

Shoemaker, G.K., van Duijn, E., Crawford, S.E., Uetrecht, C., Baclayon, M., Roos, W.H., et al., 2010. Norwalk virus assembly and stability monitored by mass spectrometry. Mol. Cell Proteomics. 9, 1742–1751.

Siebenga, J.J., Vennema, H., Renckens, B., de Bruin, E., van der Veer, B., Siezen, R.J., et al., 2007. Epochal evolution of GGII.4 norovirus capsid proteins from 1995 to 2006. J. Virol. 81, 9932–9941.

Singh, B.K., Leuthold, M.M., Hansman, G.S., 2015. Human noroviruses' fondness for histo-blood group antigens. J. Virol. 89, 2024–2040.

Smiley, J.R., Chang, K.O., Hayes, J., Vinje, J., Saif, L.J., 2002. Characterization of an enteropathogenic bovine calicivirus representing a potentially new calicivirus genus. J. Virol. 76, 10089–10098.

Sosnovtsev, S., Green, K.Y., 1995. RNA transcripts derived from a cloned full-length copy of the feline calicivirus genome do not require VpG for infectivity. Virology 210, 383–390.

Sosnovtsev, S.V., Green, K.Y., 2000. Identification and genomic mapping of the ORF3 and VPg proteins in feline calicivirus virions. Virology 277, 193–203.

Sosnovtsev, S.V., Garfield, M., Green, K.Y., 2002. Processing map and essential cleavage sites of the nonstructural polyprotein encoded by ORF1 of the feline calicivirus genome. J. Virol. 76, 7060–7072.

Sosnovtsev, S.V., Belliot, G., Chang, K.O., Onwudiwe, O., Green, K.Y., 2005. Feline calicivirus VP2 is essential for the production of infectious virions. J. Virol. 79, 4012–4024.

Sosnovtsev, S.V., Belliot, G., Chang, K.O., Prikhodko, V.G., Thackray, L.B., Wobus, C.E., et al., 2006. Cleavage map and proteolytic processing of the murine norovirus nonstructural polyprotein in infected cells. J. Virol. 80, 7816–7831.

Subba-Reddy, C.V., Yunus, M.A., Goodfellow, I.G., Kao, C.C., 2012. Norovirus RNA synthesis is modulated by an interaction between the viral RNA-dependent RNA polymerase and the major capsid protein, VP1. J. Virol. 86, 10138–10149.

Tan, M., Jiang, X., 2005. Norovirus and its histo-blood group antigen receptors: an answer to a historical puzzle. Trends Microbiol. 13, 285–293.

Tan, M., Hegde, R.S., Jiang, X., 2004. The P domain of norovirus capsid protein forms dimer and binds to histo-blood group antigen receptors. J. Virol. 78, 6233–6242.

Taube, S., Rubin, J.R., Katpally, U., Smith, T.J., Kendall, A., Stuckey, J.A., et al., 2010. High-resolution x-ray structure and functional analysis of the murine norovirus 1 capsid protein protruding domain. J. Virol. 84, 5695–5705.

Thorne, L.G., Goodfellow, I.G., 2014. Norovirus gene expression and replication. J. Gen. Virol. 95, 278–291.

Vega, E., Barclay, L., Gregoricus, N., Shirley, S.H., Lee, D., Vinje, J., 2014. Genotypic and epidemiologic trends of norovirus outbreaks in the United States, 2009 to 2013. J. Clin. Microbiol. 52, 147–155.

Viswanathan, P., May, J., Uhm, S., Yon, C., Korba, B., 2013. RNA binding by human Norovirus 3C-like proteases inhibits protease activity. Virology 438, 20–27.

Vongpunsawad, S., Venkataram Prasad, B.V., Estes, M.K., 2013. Norwalk Virus Minor Capsid Protein VP2 Associates within the VP1 Shell Domain. J. Virol. 87, 4818–4825.

Wang, X., Xu, F., Liu, J., Gao, B., Liu, Y., Zhai, Y., et al., 2013. Atomic model of rabbit hemorrhagic disease virus by cryo-electron microscopy and crystallography. PLoS Pathog. 9, e1003132.

Zamyatkin, D.F., Parra, F., Alonso, J.M., Harki, D.A., Peterson, B.R., Grochulski, P., et al., 2008. Structural insights into mechanisms of catalysis and inhibition in Norwalk virus polymerase. J. Biol. Chem. 283, 7705–7712.

Zeitler, C.E., Estes, M.K., Venkataram Prasad, B.V., 2006. X-ray crystallographic structure of the Norwalk virus protease at 1.5-A resolution. J. Virol. 80, 5050–5058.

Zheng, D.P., Ando, T., Fankhauser, R.L., Beard, R.S., Glass, R.I., Monroe, S.S., 2006. Norovirus classification and proposed strain nomenclature. Virology 346, 312–323.

Chapter 3.2

Calicivirus Replication and Reverse Genetics

I. Goodfellow*, S. Taube**

**Division of Virology, Department of Pathology, University of Cambridge, Addenbrooke's Hospital, Cambridge, United Kingdom; **Institute of Virology and Cell Biology, University of Lübeck, Lübeck, Germany*

1 CALICIVIRUS GENOME ORGANIZATION

Members of the *Caliciviridae* infect a variety of animals, including birds and reptiles, as well as humans (reviewed in Green, 2013). Based on phylogenetic differences, the family is currently divided into five genera *Lagovirus*, *Nebovirus*, *Norovirus*, *Sapovirus*, and *Vesivirus* (Fig. 3.2.1), with several additional genera pending approval. So far only noroviruses and sapoviruses have been shown to infect man, although there is also increasing evidence for vesivirus infection in the human population (Lee et al., 2012; Smith et al., 2006). Noroviruses are further subdivided into six genogroups (GI-GVI) and several genotypes (reviewed in Green, 2013). Human noroviruses (HuNoV) include GI, GII, and GIV and are the predominant cause for acute gastroenteritis. GIII has only been found in bovines, GV in rodents, and GVI in canines. All members of the *Caliciviridae* family have a single-stranded, positive-sense RNA genome ranging from 7.3 to 8.5 kb in size (Fig. 3.2.2). The genome is flanked by short, but highly conserved untranslated regions (UTR): 5′ UTR (4–19 nt) and 3′ UTR (46–108 nt). The 5′ UTR invariably starts with a pGpU sequence, and is covalently linked to a small virus-encoded protein (VPg, viral protein genome-linked) (Alhatlani et al., 2015). The polycistronic genome contains two or more open reading frames (ORFs) and terminates with a 3′ UTR (46–108 nt) followed by a polyadenosine sequence (Herbert et al., 1997). A shorter than genome or subgenomic RNA (sgRNA) is produced during replication which is 3′ coterminal to the genome RNA (gRNA) (Meyers et al., 1991). The sgRNA contains many of the features of gRNA, including a short, conserved 5′UTR and is also covalently linked to VPg at the 5′ end (Alhatlani et al., 2015). The overall genome organization is conserved among all known members of the *Caliciviridae* family. The nonstructural genes (NS1–NS7) are

Viral Gastroenteritis

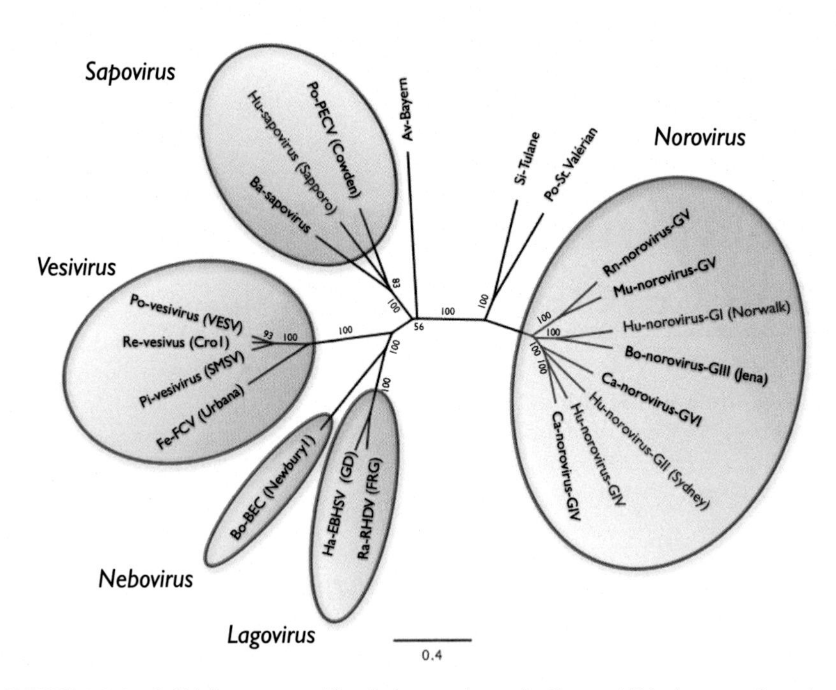

FIGURE 3.2.1 **Calicivirus genera.** The phylogram shows the five established genera: *Lagovirus (LaV), Nebovirus (NeV), Norovirus (NoV), Sapovirus (SaV),* and *Vesivirus (VeV)* based on the 2014 classification of the International Committee on Taxonomy of Viruses (ICTV). Representative strains include LaV: Rabbit Haemorrhagic Disease Virus (RHDV), NeV: Newbury1-Virus (NBV), NoV: Norwalk Virus (NV), SaV: Sapporo Virus (SV), VeV: Feline Calicivirus (FCV). Additional genera have been proposed and include the simian *Recovirus* (ReV) (ie, Tulane virus), the porcine *Valovirus* (VaV) (ie, St. Valérian virus), and the avian *Balovirus (BaV)* (ie, Bayern virus). Phylogenetic analysis was performed using Geneious 9 (Kearse et al., 2012). Construction of the neighbor-joining tree was based on maximum likelihood using the Jukes-Cantor model of nucleotide substitution. Bootstrap values (1000 replicates) are shown for confidence levels higher than 50%. The scale bar represents the number of fixed mutations per nucleotide position. Species abbreviations: avian *(Av)*; bat *(Ba)*; bovine *(Bo)*; canine *(Ca)*; feline *(Fe)*; hare *(Ha)*; human *(Hu, highlighted in red)*; pinneped *(Pi)*; porcine *(Po)*; rabbit *(Ra)*; rat *(Rn)*; reptile *(Re)*; simian *(Si)*.

located towards the 5′ end of the genome and the two structural genes, encoding the major structural protein VP1 and the minor structural protein VP2, towards the 3′ end (Fig. 3.2.2). The first ORF encodes a large polyprotein that is co- and posttranslationally processed into precursors and final proteins primarily by the virus-encoded protease (NS6), but at least in the case of murine noroviruses host caspases may also contribute (Sosnovtsev et al., 2006). There are a number of significant differences in the processing of the large polyproteins among genera; while noroviruses and vesiviruses separate nonstructural and structural proteins between ORF1 and ORF2, lagoviruses, recoviruses, and sapoviruses express the major structural protein (VP1) from the polyprotein encoded by ORF1 (Fig. 3.2.2B,C). Importantly however, in the cases where VP1 is encoded as a component of the large polyprotein of ORF1, it is also encoded by the sgRNA, a process that may serve to increase the abundance of

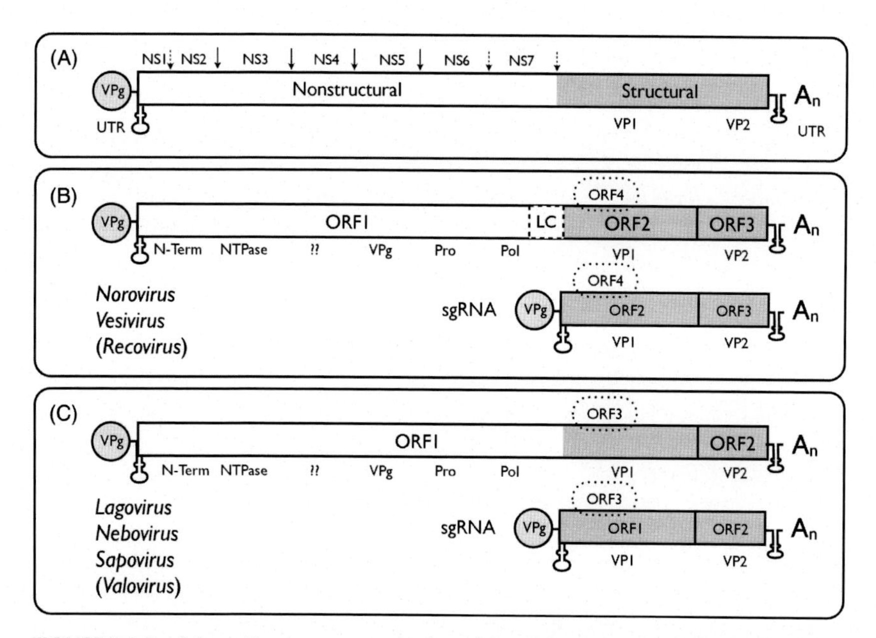

FIGURE 3.2.2 Schematic genome organization of the major genera in the family Caliciviridae. (A) General genome organization depicts the untranslated regions (UTR) and shows genes encoding nonstructural proteins NS1-7 *(white)* and genes encoding structural proteins VP1 and VP2 *(grey)*. Arrows depict protease cleavage sites. Genome linked VPg is shown as a grey circle at the 5′ end. A_n marks the polyadenosine sequence at the 3′ end. (B) Genome organization of the *Norovirus*, *Vesivirus*, and the proposed *Recovirus* genera. Subgenomic RNA *(sgRNA)* includes ORF2 and ORF3 (Lin et al., 2015). ORF4 encodes for the virulence factor VF1 and has only been observed for MNV *(dashed oval)*. (C) Genome organization of the *Nebovirus*, *Sapovirus*, *Lagovirus* and the proposed *Valovirus* genera. ORF3 has been proposed for sapoviruses *(dashed oval)*.

VP1 during the later stages of the viral life cycle. Other notable differences include the observation that the vesivirus VP1 is expressed as a precursor protein containing an N-terminal leader protein (LC) and is also processed by the viral protease (Sosnovtsev et al., 1998) and host caspases (Al-Molawi et al., 2003) (Fig. 3.2.2B). In addition, the NS6 and NS7 proteins are expressed as a single NS6-7 fusion protein in the case of vesiviruses and sapoviruses. The individual functions of the seven nonstructural proteins have yet to be fully elucidated but functional activities have been predicted based on sequence conservation or determined by biochemical studies. These include NTPase/helicase activity for NS3, protein priming and ribosome recruitment for NS5 (VPg), chymotrypsin-like protease for NS6, and RNA-dependent RNA polymerase (RdRp) activity for NS7 (reviewed in Green, 2013). Cleavage between NS6 (protease) and NS7 (RdRp) is essential for murine norovirus (MNV) (Ward et al., 2007) but is dispensable for feline calicivirus (FCV) (Sosnovtsev et al., 2002). The minor structural protein VP2 is always encoded by the last ORF of the calicivirus genome and expressed from the sgRNA (Meyers et al., 1991). All calicivirus sgRNAs are polycistronic, and both structural proteins, VP1 and VP2, are

TABLE 3.2.1 Productive Calicivirus Cell Culture Systems

Genus	Virus (strain)	Cell type (prototype cell lines)	References
Lagovirus	RHDV (CHA/JX/97)	Rabbit kidney cell line (RK13)	Liu et al. (2008)
Norovirus	MNV (all tested strains)	Murine macrophages and dendritic cells (ie, RAW264,7,BV-2, WEHI)	Cox et al. (2009), Wobus et al. (2004)
Norovirus	MNV (MNV-1, MNV-3)	Primary B-cells	Jones et al. (2014)
Norovirus	HuNoV (Sydney, GII.4)	Human B-cells (BJAB)	Jones et al. (2014)
Vesivirus	FCV (all tested strains)	Crandell-Rees feline kidney cells (CRFK)	Crandell et al. (1973)
Recovirus	ReV (TV)	Monkey kidney cell line (LLC-MK2)	Farkas et al. (2008)
Sapovirus	PEC (Cowden)	Porcine kidney cells (LLC-PK, requires filtered intestinal content from gnotobiotic pigs or bile)	Chang et al. (2004), Flynn and Saif (1988)

expressed from distinct ORFs (Fig. 3.2.2B), even in cases where the VP1 protein is also produced as a component of the ORF1 polyprotein translated from the gRNA. An upstream element, denoted as a termination upstream ribosomal binding site (TURBS), facilitates the reinitiation of VP2 translation after VP1 termination (Meyers, 2007). Murine noroviruses further contain a fourth ORF, which overlaps with ORF2 and encodes a protein referred to as virulence factor (VF1) linked to immune modulation (McFadden et al., 2011). The presence of an additional ORF overlapping with ORF2 has also been proposed for some sapovirus strains, the function of which has yet to be determined.

2 CALICIVIRUS MODEL SYSTEMS

The study of the calicivirus life cycle has been significantly hampered by the lack of a robust cell culture system for many members of the family. Susceptible cell culture systems are also critical for efficient virus recovery by reverse genetics, described in more detail later, and functional analysis of recombinant viruses. However, only a few members of the *Caliciviridae* can be propagated in cell culture (summarized in Table 3.2.1). Despite their first discovery in 1972 (Kapikian et al., 1972), HuNoV have long resisted being cultivated. Animal models to study HuNoV infection, described in more detail in Chapter 3.4, include gnotobiotic pigs and calves (Cheetham et al., 2006) but these systems are elaborate and costly and not easily available. A mouse model for HuNoV, which relies on an immunocompromised mouse, has only recently been developed (Taube et al., 2013). As a consequence, our understanding of HuNoV biology and pathogenesis lags behind that of other RNA viruses and is largely based on studies using cultivatable animal caliciviruses, such as MNV and FCV (Vashist et al., 2009).

Animal caliciviruses that can be efficiently cultivated include FCV strains, MNV strains, and recoviruses (ReV, ie, Tulane virus, TV), summarised in Table 3.2.1 (reviewed in Bridgen, 2012). Porcine enteric calicivirus (PEC) is the only member of the *Sapovirus* genus that can be cultivated, but replication requires the presence of bile acids which appear to function in the uncoating process (Parwani et al., 1991; Shivanna et al., 2015). A cell culture system for HuNoV has very recently been described (Jones et al., 2014) which uses an immortalized B cell line (BJAB) but also requires the presence of histo-blood group antigen (HBGA)-expressing bacteria or free HBGA. While this is a major breakthrough in the field, the system is still under development and the yields of replicated virus (~20-fold over the inoculum) are still too low to enable the utility of the system for deciphering the finer detail of the norovirus life cycle.

3 THE CALICIVIRUS LIFE CYCLE

A diagrammatic representation of the calicivirus life cycle is shown in Fig. 3.2.3 and has been recently reviewed in detail (Thorne and Goodfellow, 2014). The calicivirus life cycle is initiated following the interaction of the infectious virus particle with attachment and entry receptors present on the surface of permissive cells. A number of molecules have been identified as functioning receptors for caliciviruses including proteins, carbohydrates and glycolipids (Tan and Jiang, 2010). A more detailed description of the interaction of noroviruses with their carbohydrate receptors can be found in Chapter 3.3.

Following the release of the viral genome into the cytoplasm of the permissive cell, the first process to occur is the translation of the incoming parental viral RNA into proteins. The process of calicivirus translation differs markedly from that of host cell mRNAs, which require a 5′ cap structure that functions to recruit the eIF4F cap-binding complex, leading to ribosome recruitment and protein synthesis. In contrast, caliciviruses use the VPg protein, covalently linked to the 5′ end of the viral gRNA and sgRNA, as a proteinaceous cap-substitute. While VPg-linked RNA has been observed in some plant viruses, only two other vertebrate viruses, astroviruses and picornaviruses, also produce viral RNA that is covalently linked to a VPg protein at the 5′ end (Fuentes et al., 2012; Paul and Wimmer, 2015). The functions of calicivirus VPg differ significantly from those of picornaviruses; based on size (13–15 kDa) and sequence homology, the calicivirus VPg proteins appear more closely related to that expressed by astroviruses (~11 kDa) than the picornavirus VPg (2–6 kDa) (Fuentes et al., 2012; Goodfellow, 2011). While broadly speaking, all calicivirus VPg proteins function in a similar way in that they recruit translation initiation factors to the 5′ end of the viral genome, the translation initiation factors that bind directly to VPg differ between viruses. FCV and PEC VPg proteins appear to function essentially identically to the cellular cap structure

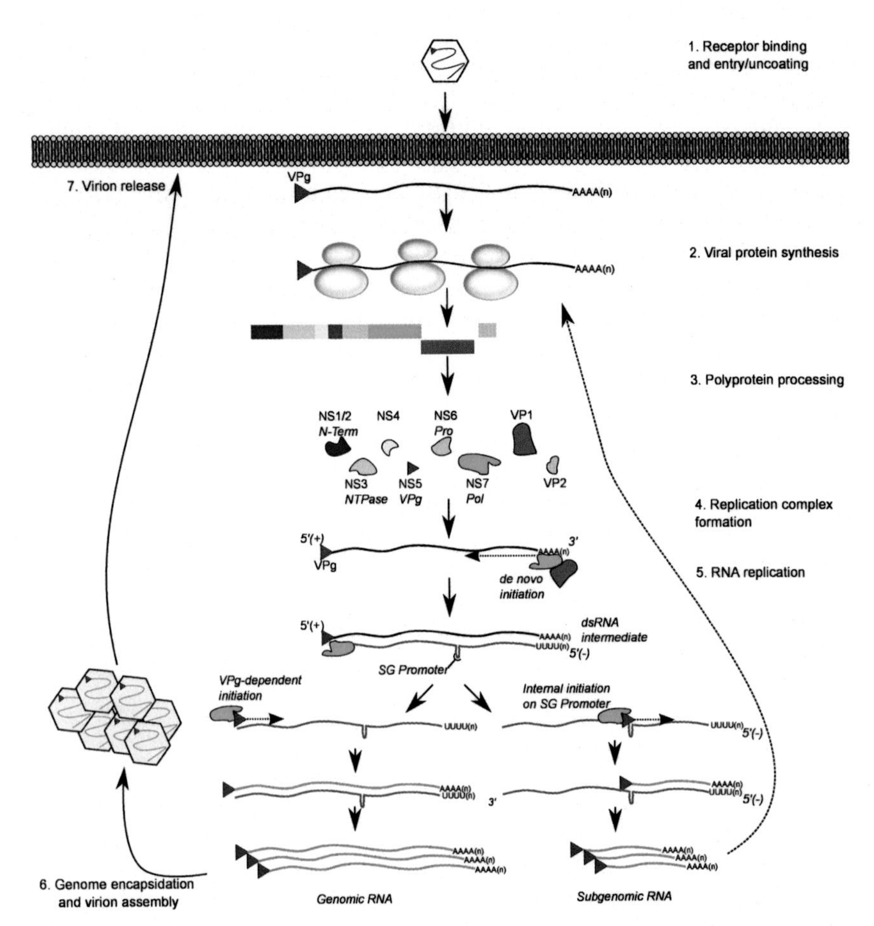

FIGURE 3.2.3 **Schematic overview of the life cycle of members of the *Caliciviridae* family.** (1) Binding, entry and uncoating of caliciviruses occurs via a direct interaction with cellular receptors that can include carbohydrates, glycolipids and proteins on the surface of the target cells. (2) Following entry, ribosome recruitment occurs via the interaction of translation initiation factors with VPg. (3) Viral protein translation follows and is accompanied by polyprotein processing. (4) Newly synthesized viral proteins then function to assemble a membrane-bound replication complex in which RNA replication occurs (5). (6) Genome encapsidation is then followed by virion release (7).

by interacting with the eIF4E component of the eIF4F complex (Goodfellow et al., 2005; Hosmillo et al., 2014). In contrast, the norovirus VPg interacts with several components of the translation initiation factor complex. An interaction between the norovirus VPg protein and eIF4E has been described (Chaudhry et al., 2006), yet this interaction appears not to be required for viral translation but may contribute to the regulation of the host response to infection (Royall et al., 2015). The norovirus VPg is also reported to interact with components of the eIF3 complex but a functional role for this interaction has yet to be determined (Daughenbaugh et al., 2003). In contrast, an interaction between the highly conserved C-terminus of the norovirus VPg and the central domain of eIF4G appears to be essential for translation of the norovirus genome (Chung et al., 2014). In addition to the initiation factors that bind to VPg, caliciviruses also recruit numerous cellular factors that may contribute to viral translation via RNA-protein interactions with RNA structures located at the 5′ and 3′ termini of the viral gRNA and sgRNA (reviewed in Alhatlani et al., 2015).

The first translation products to be generated following genome release into the cytoplasm are the products of ORF1, namely the large polyprotein that encodes the viral replicase components. As described earlier, the polyprotein is co- and posttranslationaly cleaved by the viral encoded protease NS6 to produce up to 7 nonstructural proteins. In the case of MNV, caspases are also thought to contribute to additional cleavage of the NS1 protein (Sosnovtsev et al., 2006). The function of the replicase components is to manipulate the cellular environment to generate specific subcellular compartments for viral replication and to subsequently synthesise new viral RNA genomes, producing infectious viral particles.

Like all positive sense RNA viruses, replication of the calicivirus RNA genome occurs via a double-stranded RNA intermediate and takes place in membrane-associated replication complexes. In the case of MNV, the replication complexes are thought to be localized proximal to the microtubule organizing center (Hyde et al., 2012) and contain membranes considered to originate from the secretory pathway (Hyde et al., 2009). The association of the nonstructural proteins with components of the secretory pathway is also observed in HuNoV; the p22 protein (NS4) inhibits protein secretion and leads to Golgi disassembly (Sharp et al., 2010), and the p48 protein (NS1/2) interacts with the VAP-A protein (Ettayebi and Hardy, 2003). In contrast, the FCV, the NS2, 3 and 4 proteins localize to the endoplasmic reticulum (ER) with NS3 or NS4 expression alone leading to reorganization of the ER (Bailey et al., 2010a; Ettayebi and Hardy, 2003). Therefore the source of the membranes may differ markedly between members of the *Caliciviridae*, and this is an area that clearly requires further study.

Following the "pioneer round" of translation whereby the incoming parental RNA genome is translated, the RNA genome is then replicated to produce a double-stranded RNA replication intermediate (Fig. 3.2.3). This process most likely occurs via a de novo mechanism of RNA synthesis performed by the viral RdRp (NS7) following binding to the 3′ end of the viral genome (Subba-Reddy et al., 2012). The dsRNA intermediate then functions as a template for VPg-primed RNA synthesis to produce VPg-linked genomic and subgenomic RNAs. VPg-linked gRNA is synthesised by priming from the 3′ end of the negative sense RNA, whereas VPg-linked sgRNA is produced by priming at the sgRNA promoter, a conserved stem-loop that is positioned six nucleotides 3′ of the start site of the sgRNA RNA in all caliciviruses (Yunus et al., 2015). Efficient viral genome replication is thought to require the circularization of the calicivirus genome, which, at least in the case of MNV, occurs via the interaction of the cellular proteins PCBP2 and hnRNP A1 with the 5′ and 3′ ends of the genome (López-Manríquez et al., 2013).

Recent work has also highlighted how the activity of the NS7 RdRp can be regulated by the nonstructural protein NS1/2 and the structural proteins VP1 and VP2 (Subba-Reddy et al., 2011). Loop sequences within the shell domain of the VP1 capsid protein interact directly with the NS7 RdRp in a species-specific manner and function to stimulate the de novo mechanism of RNA synthesis (Subba-Reddy et al., 2012). This observation suggests that the incoming VP1 protein forming the parental capsid and/or the newly synthesized VP1 molecules produced early during the viral life cycle, function to promote negative strand RNA synthesis (Fig. 3.2.3). As VP1 levels increase, concentration dependent dimerization of VP1 occurs, preventing the association with the viral RdRp, therefore favouring a VPg-primed mechanism of RNA synthesis. The interaction between the RdRp and the major capsid protein VP1 is also seen in FCV (Kaiser et al., 2006). This process provides caliciviruses with a mechanism to regulate aspects of gene expression and replication; however, whether or not this mechanism is conserved between all caliciviruses has yet to be determined.

As new gRNA and sgRNAs accumulate, they function as templates for additional rounds of translation, leading to the accumulation of high levels of all the viral proteins. Importantly, the sgRNA appears to be preferentially replicated over the gRNA, which may at least in part be due to the shorter size. This serves to increase the levels of the major and minor structural proteins relative to the nonstructural proteins. As a single infectious calicivirus capsid consists of 180 copies of the VP1 protein (Prasad et al., 1999) and at least 1–2 copies of VP2, the increased levels of the subgenomic RNA may function simply to provide the addition translation-competent templates required to increase VP1 and VP2 levels (Vongpunsawad et al., 2013).

Following the production of sufficient viral structural proteins and viral RNA, genome encapsidation and virion release occur. The process of calicivirus genome encapsidation has yet to be studied in any detail, however an interaction between the VPg protein and the major capsid protein VP1 has been reported in FCV (Kaiser et al., 2006) and observed in noroviruses (Alexis de Rougemont and Ian Goodfellow, unpublished data). This may indicate a process of selection of VPg-linked RNAs from the cytoplasm of infected cells via a direct interaction between VPg and VP1. Recent work has highlighted a role for the cellular molecular chaperone Hsp90 in the stabilization of the norovirus VP1 protein and a potential role in norovirus capsid formation (Vashist et al., 2015). RNAi or small molecule-mediated modulation of Hsp90 activity and/or levels resulted in a >100-fold reduction in infectious MNV production with only a minor impact on viral RNA synthesis. Surprisingly, Hsp90 was initially identified as a component of a ribonucleoprotein complex that forms on the 5′ and 3′ extremities of the MNV genome, indicating direct genome binding activity (Vashist et al., 2012). The RNA binding activity of Hsp90 had previously been described in bamboo mosaic virus (Huang et al., 2012). The RNA binding activity of Hsp90, combined with its critical role in determining the stability of the VP1 protein, would provide an additional mechanism by which viral RNA could be recruited to VP1 to enable encapsidation.

Encapsidation of viral RNA is followed by release of the newly synthesized capsids by an as yet unknown mechanism. All caliciviruses that replicate in cell culture induce cell death, and at least for FCV and MNV this process is known to involve the activation of caspases (Bok et al., 2009; Natoni et al., 2006; Roberts et al., 2003; Sosnovtsev et al., 2003), which for MNV involved the down regulation of survivin, an inhibitor of apoptosis (Bok et al., 2009). Whether other nonlytic mechanisms of virus release are involved in calicivirus exit from infected cells is not known, but persistent infection of B-cells has recently been described (Jones et al., 2014).

4 CALICIVIRUS REVERSE GENETICS

Reverse genetics allows the generation of recombinant viruses from a genomic cDNA clone, and by permitting the analysis of the linkage of genotypes to phenotypes, it constitutes one of the most powerful tools available to dissect aspects of a virus life cycle. Using site-directed mutagenesis, genomic cDNA can be deliberately modified, and if mutations are tolerated, genetically modified viruses can be rescued. Studying the impact of targeted mutations on phenotypic traits (reverse genetics) has greatly advanced our knowledge of the replication and pathogenesis of many virus families and is extensively used in drug development and vaccine generation (reviewed in Bridgen, 2012). Most human and animal viruses have reverse genetics systems available and significant progress has been made for members of the *Caliciviridae* (summarized in Table 3.2.2).

4.1 Infectious Calicivirus RNA

Positive-sense RNA viruses are particularly suitable for reverse genetics because their genomes are typically infectious in permissive cells and can be immediately translated by the host's protein-synthesis machinery. Importantly, for this to occur, in vitro transcribed viral genomic RNA needs to be "translation-ready". For most positive-sense RNA viruses this is facilitated by the generation of a capped 5′ end or the natural presence of an internal ribosome entry site (IRES) (Picard-Jean et al., 2013), both of which will direct ribosome recruitment and translation initiation. However, as described earlier, caliciviruses use a different strategy for protein synthesis whereby the VPg protein functions as a cap-substitute (Goodfellow, 2011). The calicivirus VPg protein is covalently linked to the viral genomic and subgenomic RNAs as it is used as a protein primer during the production of the viral genomic RNA. During protein synthesis, the calicivirus VPg recruits cellular translation initiation factors to the 5′ end of the genomic and subgenomic RNA and thereby mediates the formation of the translation initiation complex (Thorne and Goodfellow, 2014). Since there is still no method available to efficiently generate VPg-capped RNA in

TABLE 3.2.2 Recovery of Infectious Calicivirus by Reverse Genetics

Date	Genus	Virus (strain)	Genome (expression)	Helper virus	Cell	References
1995	*Vesivirus*	FCV (Urbana)	Capped RNA (IVT, T7)	No	CRFK[a]	Sosnovtsev and Green (1995)
2004	*Vesivirus*	FCV (2024)	DNA (T7) or Capped RNA (IVT, T7)	Modified vaccinia (MVA-T7)	CRFK[a]	Thumfart and Meyers (2002)
2005	*Sapovirus*	PEC (Cowden)	Capped RNA (IVT, T7)	No	LLC-PK[a]	Chang et al. (2005b)
2007	*Norovirus*	MNV (MNV-1)	DNA (CMV)	No	293T, RAW269.7[a]	Ward et al. (2007)
2007	*Norovirus*	MNV (MNV-1)	DNA (T7)	Modified baculovirus (Bac-T7)	293T/BHK/COS7//HepG2 → RAW269.7[a]	Ward et al. (2007)
2007	*Norovirus*	MNV (MNV-1)	DNA (T7)	Modified fowlpox, (FWPV-T7)	BHK → RAW269.7[a]	Chaudhry et al. (2007)
2008	*Lagovirus*	RHDV (JX/CHA/97)	RNA (IVT, T7)	No	BHK and in vivo	Liu et al. (2006)
2008	*Recovirus*	ReV (TV)	Capped RNA (IVT, T7)	No	LLC-MK2[a]	Wei et al. (2008)
2010	*Norovirus*	MNV (MNV-1)	Capped RNA (IVT, T7)	No	293T/ BSR-T7 → RAW269.7[a]	Yunus et al. (2010)
2012	*Norovirus*	MNV (CR6)	DNA (CMV)	No	293T → RAW269.7[a]	Strong et al. (2012)
2012	*Norovirus*	MNV (MNV-3)	Capped RNA (IVT, T7)	No	RAW269.7[a]	Arias et al. (2012)
2014	*Norovirus*	HuNoV (U201 and others)	DNA (EF-1α)	No	293T/ COS-7	Katayama et al. (2014)
2014	*Norovirus*	MNV (S7)	DNA (EF-1α)	No	293T/ COS-7→ RAW269.7[a]	Katayama et al. (2014)
2014	*Vesivirus*	FCV (F4)	DNA (EF-1α)	No	CRFK[a]	(Oka et al., 2014)
2015	*Norovirus*, *Vesivirus*	MNV (MNV-1), FCV (Urbana)	DNA (T7) or RNA (IVT) with 5′ IRES (EMCV)	No	293T/ BHK-21 → RAW269.7[a]	Sandoval-Jaime et al. (2015)

→ Indicates passaging in susceptible cells for amplification
[a]*Susceptible cell line*

vitro, this rare mechanism of translation initiation adds additional challenges to the generation of recombinant caliciviruses by reverse genetics. A major breakthrough in the field was the observation that the addition of a synthetic cap structure [$m^7G(5')ppp(5')G$] to the 5′-end of in vitro transcribed FCV RNA can substitute for VPg and produce infectious virus (Sosnovtsev and Green, 1995). Although the yield of recombinant virus from transfecting capped RNA was lower than that of purified VPg-linked viral RNA (Chaudhry et al., 2007), it circumvented the problem of generating VPg-linked viral RNA. Capped RNAs still constitute the basis of most calicivirus reverse genetics systems today (reviewed in Bridgen, 2012), although a recent alternative described later uses an EMCV IRES to initiate protein synthesis overcoming the need of capped RNA (Sandoval-Jaime et al., 2015).

4.2 Strategies for Recombinant Calicivirus Recovery

Many different strategies have been developed to generate viral RNA from a genomic cDNA clone, including RNA and DNA-based delivery systems, helper virus dependent and independent approaches, and complete vector-based cDNA delivery systems. Each approach has its own strengths and limitations and because members of the calicivirus family differ in aspects of their biology, compatibility with distinct reverse genetics systems may vary. A summary of available calicivirus reverse genetics strategies is summarized in Fig. 3.2.4 and Table 3.2.2. Strategies to rescue individual genera are discussed in more detail later.

The most common approach to generate genomic RNA from a cDNA clone is to use bacteriophage T7 RNA polymerase because it allows cytoplasmic transcription without posttranscriptional modifications. A truncated T7 promotor permits the generation of a precise 5′ end, without the addition of nonviral sequences. Transcription starts with GTP, the first base of all calicivirus 5′ untranslated regions (Ikeda, 1992). While T7 works well in vitro, it does not cap RNA, therefore RNA produced in vitro is uncapped unless a cap analogue is included in the transcription reaction or a posttranscriptional capping system is subsequently employed. Evidence suggests that incompletely capped RNA stimulates immune sensors, which in turn hinders calicivirus replication (Yunus et al., 2010). This hypothesis was further supported by the observation that a significant increase in PEC recovery can be achieved by transfecting capped RNA into cells that are engineered to lack aspects of the innate immune response (Hosmillo et al., 2015). In cell culture, the T7 RNA polymerase can be expressed from a transfected plasmid or from a helper-virus, such as modified vaccinia Ankara (MVA-T7), or modified fowlpox virus (FPV-T7) (Britton et al., 1996; Sutter and Moss, 1995). Both helper-viruses express their own cytoplasmic capping machinery therefore cap at least a subset of the transcribed RNA and generate translation ready viral RNA (Binns et al., 1990; Fuerst and Moss, 1989).

The 3′ end of the genomic RNA of caliciviruses contains two important regions, the 3′ untranslated region (UTR) and a poly(A) sequence. An authentic

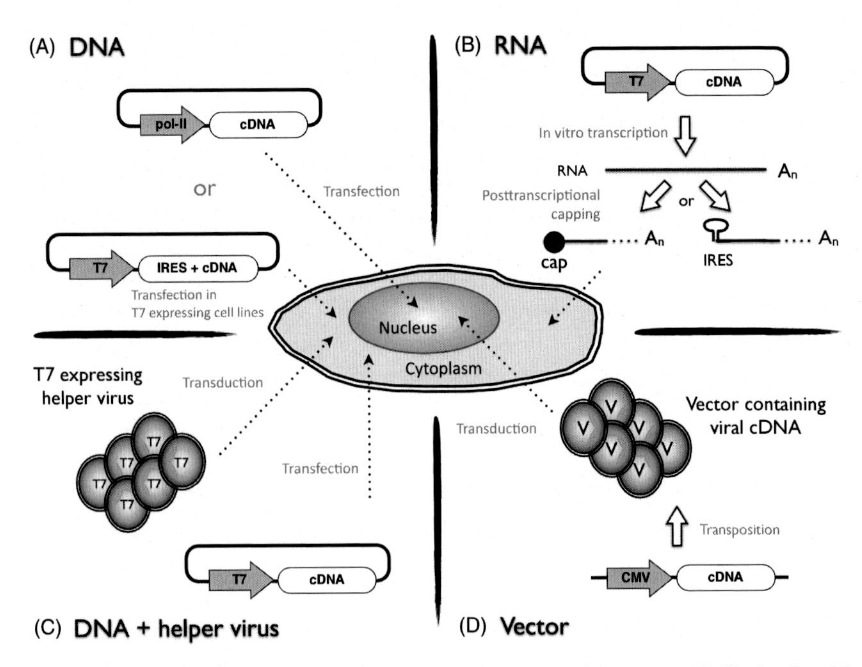

FIGURE 3.2.4 **Principles of available calicivirus reverse genetics systems.** (A) Single-plasmid based approach, transfecting viral cDNA under the control of a polymerase II (pol II) or T7 promoter. (B) RNA-based approach, transfecting in vitro transcribed (IVT) RNA that is either capped or containing a heterologous internal ribosome entry site (IRES). (C) Helper virus-dependent approach, transfecting viral cDNA under the control of a T7 promoter into cells transduced with T7-expressing helper virus. (D) Vector-based transduction of a cDNA clone under the control of a pol II promoter.

3′ end appears to be critical for efficient virus recovery; in case of MNV, a single incorrect nucleotide immediately upstream of the poly(A) tail was shown to completely abrogate virus recovery (Chaudhry et al., 2007). Two main strategies are used to generate an authentic 3′ terminus: first by linearizing the cDNA clone during in vitro transcription (IVT), second, if transcription is carried out inside the cell, by incorporating a self-cleaving RNA sequence, such as the hepatitis delta virus ribozyme, immediately downstream of the 3′ end, which cleaves off any nonvirus specific sequences (Perrotta and Been, 1998).

While RNA-based reverse genetics systems are well established and provide the most efficient method for recovery of infectious caliciviruses, they elaborate and require the use of costly cap analogs or posttranscriptional capping kits. Alternative approaches are DNA-based and make use of the mammalian or bacteriophage promotors, such as polymerase II (pol II) and the T7 promoters, respectively. Since the pol II promoters lead to nuclear RNA transcription of cDNA, the RNA product is subject to posttranscriptional modifications, including capping, 3′ polyadenylation, and eventually splicing. Despite these limitations, multiple pol II-based reverse genetics systems have been shown to generate recombinant virus (Katayama et al., 2014; Ward et al., 2007). While single plasmid-based system approaches represent the simplest and most cost-effective strategies, they come at the cost of lower recovery rates; ~200–5,000 infectious units per transfection (35 mm dish), compared to >500,000 infectious units for transfection of capped IVT RNA (Yunus et al., 2010). As a result, capped in vitro transcribed RNA often provides a method with which to recover replication-debilitated viruses that would otherwise appear nonviable by any other method of recovery. The single-plasmid based approach was recently adapted by initiating protein synthesis using the encephalomyocarditis virus (EMCV) IRES (Sandoval-Jaime et al., 2015). This allowed translation of uncapped RNA and is helper virus-independent. This is the first single plasmid reverse genetics system that allows RNA synthesis in the cytoplasm, avoiding the restrictions of previous pol II promotor systems and has been successfully applied for MNV and FCV (Sandoval-Jaime et al., 2015).

Another strategy makes use of a two-component baculovirus-based system to deliver the MNV genome into various mammalian cell lines (Ward et al., 2007). Here, one baculovirus contains the viral cDNA clone under the control of a tetracycline-inducible promotor, while the other baculovirus produces the respective transactivator. Transduction of both vectors in multiple permissive cell lines led to the recovery of recombinant MNV-1.

The cellular tropism of the virus has great implications on the reverse genetics strategy. The choice of the target cell depends foremost on susceptibility to allow for multiple rounds of infection. This works well for FCV and TV, which replicate in epithelial cell lines but is challenging for MNV, which infects monocyte-derived phagocytic cells; these are intrinsically difficult to transfect and sensitive to immune stimulation. Consequently, particularly MNV reverse genetics systems follow a two-step approach. First, a permissive cell line that is easily transfected is used for primary virus recovery (ie, 293T, BSR-T7, or COS-7), and a second susceptible cell line (ie, RAW 264.7 or BV-2) is used for amplifying the recovered virus. Some cell lines, in particular 293T cells can be transfected at very high efficiencies, and the large T antigen facilitates the amplification of plasmids containing the SV40 origin of replication (Van Craenenbroeck et al., 2000). This has been applied for multiple single plasmid based approaches (Katayama et al., 2014; Ward et al., 2007). Other cell lines such as Vero and BHK cells are deficient in interferon production, which in turn may be beneficial for virus recovery. In the case of PEC, engineering porcine kidney cells to express innate immune system antagonists greatly enhance recovery following RNA transfection (Hosmillo et al., 2015).

4.2.1 Feline Calicivirus (FCV)

The first calicivirus reverse genetics system was established for FCV in 1995 by Sosnovtsev and Green following transfection of IVT-RNA of the FCV Urbana strain into the susceptible feline kidney cell line CRFK

(Sosnovtsev and Green, 1995). Transcription was performed in the presence of the cap analog m7G(5′)ppp(5′)G to render the RNA translation-ready. In fact, viral RNA produced in this manner was infectious in many cell types but only susceptible cells expressing the feline Jam-1 receptor allowed for multiple rounds of infection (Makino et al., 2006; Sosnovtsev et al., 1997). To bypass the need of capped IVT-RNA, a helper virus dependent approach was later developed using the Modified Vaccinia Virus Ankara virus expressing T7 RNA polymerase (MVA-T7) as a helper virus (Thumfart and Meyers, 2002). Here, recombinant virus was recovered from transfecting the plasmid containing FCV (2024) cDNA under the control of a T7 promotor into MVA-T7 transduced CRFK cells. Since the development of the first FCV recovery system in 1995, many questions about FCV biology have been addressed, including the identification of essential cleavage sites of the FCV polyprotein, the necessity of VP2 for efficient virus replication, and the role of host factors in calicivirus infection (reviewed in Bridgen, 2012). This system also led to the first chimeric FCV strains (Neill et al., 2000) and fluorescent marker expressing variants to visualize FCV infection in real time (Abente et al., 2010). Further single-plasmid based helper virus independent FCV recovery systems have since been developed including the first IRES-driven expression system (Katayama et al., 2014; Oka et al., 2014; Sandoval-Jaime et al., 2015).

4.2.2 Porcine Enteric Calicivirus (PEC)

Porcine enteric calicivirus (PEC) Cowden strain still remains the only sapovirus that can be efficiently grown in cell culture (Flynn and Saif, 1988). Replication of PEC in primary porcine kidney cell cultures and the LLC-PK cell line requires the addition of intestinal content of gnotobiotic piglets or bile acids (Chang et al., 2004; Flynn and Saif, 1988), the effect of which is to enable virus entry and release from endosome via the inhibition of acid sphingomyelinase (Shivanna et al., 2015). The first PEC reverse genetics system was described in 2005 transfecting capped IVT-RNA into susceptible LLC-PK cells but also required the addition of intestinal content of gnotobiotic piglets or bile (Chang et al., 2005a). Recombinant virus was infectious in piglets and recapitulated the same attenuated phenotype as the parental cell culture adapted strain (Farkas et al., 2008). A significant enhancement ($>$ 100-fold) in virus recovery was obtained by engineering permissive cells to express antagonists of the innate immune response, effectively inactivating aspects of the innate immune response (Hosmillo et al., 2015).

4.2.3 Rabbit Hemorrhagic Disease Virus (RHDV)

RHDV, a member of the *Lagovirus* genus is highly lethal in rabbits causing liver necrosis and haemorrhage. First isolated in China, RHDV has spread worldwide and is now established in wild rabbit populations throughout Western Europe, Australia, and New Zealand (Cooke, 2002). Here, uncapped in vitro

RNA transcribed RNA was shown to be infectious, when mixed with Lipofectin and injected into the liver of rabbits (Liu et al., 2006, 2008). This is in stark contrast to other calicivirus systems where capped RNA is essential for efficient virus recovery (Chaudhry et al., 2007; Sosnovtsev et al., 2005). However, a similar approach, injecting capped IVT-RNA from a hepatitis E cDNA clone into nude rats was recently shown to be successful for viral recovery of infectious hepatitis E virus (Li et al., 2015), suggesting viral recovery from RNA injection into a host species constitutes a broadly applicable strategy. A DNA-based reverse genetics system under the control of a mammalian CMV promotor also led to robust recovery of infectious virions after transfecting the plasmid into the rabbit kidney cell line (RK13) (Liu et al., 2006). Surprisingly, deletion of the poly(A) sequence or the deletion of the minor structural protein VP2, did not abolish infectivity. This is also in contrast to FCV where VP2 is essential for RNA replication (Sosnovtsev et al., 2005). While these unique features of RHDV warrant further investigations, they highlight considerable differences between caliciviruses and their requirements for virus recovery by reverse genetics systems.

4.2.4 Tulane Virus (TV)

TV is the prototype recovirus and was first isolated from stool samples from juvenile rhesus macaques in 2008 (Farkas et al., 2008). Similar to PEC, which infects cultured porcine kidney epithelial cells (LLC-PK), TV grows in rhesus macaque kidney epithelial cells (LLC-MK2). Recoviruses are evolutionarily and biologically related to HuNoV and recapitulate key features of HuNoV biology, including gastrointestinal disease and HBGA-binding (Zhang et al., 2015). The presence of recovirus-neutralizing antibodies in human serum samples and detection of the TV genome in human stools samples suggest a possible zoonotic capacity of this genus. Thus, TV constitutes a valuable surrogate to study human norovirus pathogenesis and replication (Smits et al., 2012). A TV reverse genetics system was developed in 2008 by means of transfecting capped RNA into susceptible LLC-MK2 cells (Wei et al., 2008).

4.2.5 Murine Noroviruses (MNV)

Despite great advances in cultivating HuNoV (Jones et al., 2014), MNV remains the only norovirus that replicates efficiently in cell culture. The prototype strain (MNV-1), was first described in 2003 and shown to cause fatal disease in STAT1$^{-/-}$ mice (Karst et al., 2003). MNV shares a similar genome organization with HuNoV, as well as comparable biological and molecular characteristics (Wobus et al., 2006). Contrary to HuNoV infection, MNV infection is mostly chronic and clinically silent in the immunocompetent host. MNV displays a tropism for monocyte-derived cells and B cells in vitro and in vivo and readily infects various cell lines (Cox et al., 2009; Wobus et al., 2004). MNV also comes with a tractable, low-cost small animal model, and because it can be studied in its natural host, it has become a widely used model to study norovirus biology and pathogenesis (Wobus et al., 2006).

MNV was the first norovirus for which a reverse genetics system has been developed (Chaudhry et al., 2007; Ward et al., 2007). Since then multiple reverse genetics systems have been developed for various MNV strains including MNV-1, CR6, MNV-3, and S9 (Arias et al., 2012; Katayama et al., 2014; Strong et al., 2012). Reverse genetics strategies include IVT and capped RNA-based systems, helper virus-dependent and -independent DNA-based systems, as well as a two-component vector-mediated transduction system (Arias et al., 2012; Chaudhry et al., 2007; Sandoval-Jaime et al., 2015; Ward et al., 2007; Yunus et al., 2010). Initial attempts to reproduce the capped IVT RNA or MVA-T7 based systems developed for FCV failed because transfection efficiencies of available susceptible cells (RAW264.7) were insufficient and because the MVA-T7 helper virus impaired MNV-1 replication (Chaudhry et al., 2007). These obstacles were overcome by replacing the MVA-T7 with FPV-T7 and RAW264.7 with alternative cell lines, such as BHK (Chaudhry et al., 2007). Interestingly, cells that endogenously express T7 RNA polymerase, such as BSR-T7, were not sufficient to rescue MNV, presumably because they lack the cytoplasmic capping activities provided by the helper viruses that are required to enhance translation of the viral RNA. An alternative approach was initially developed as a two-component baculovirus-based transduction system relying on one baculovirus containing the cDNA clone under the control of a tetracycline-inducible pol II promotor and the other expressing the tetracycline-transactivator (Ward et al., 2007). This system was further simplified by directly transfecting 293T cells with a single plasmid containing the cDNA clone under the control of a minimal CMV promotor (Ward et al., 2007). While this system avoided IVT of capped RNA, it came at the expense of misplacing the viral RNA into the nucleus, exposing it to post-transcriptional modifications, and reducing yields of recovered virus. Reported recovery of the single plasmid recovery system are approximately 200–5,000 infectious units per transfection (35 mm dish), which is at least ~10-fold less than the FPV-T7 based system (>50,000 infectious units) (Chaudhry et al., 2007). Initial attempts to rescue MNV directly from transfected RAW264.7 cells failed, possibly because transfection-rates in these cells were insufficient. This obstacle was overcome using the capillary-based Neon transfection system (Invitrogen), which greatly improved transfection efficiencies and facilitated electroporation of in vitro transcribed capped RNA directly into RAW264.7 cells (Yunus et al., 2015). At 24 h post transfection (hpt) $\sim 5 \times 10^5$ infectious units can be obtained from 1 μg of IVT capped RNA, compared to 1×10^8 infectious units obtained from purified VPg-linked RNA. The overall infectious units converged over time, with $\sim 10^9$ infectious units obtained by IVT capped RNA at 72 hpt. Transfection of capped IVT MNV RNA into BHK or 293T cells also produces a highly robust and efficient reverse genetics approach for MNV with $>5 \times 10^5$ infectious units being produced per 35 mm dish. Detailed protocols for MNV recovery systems have been described (Hwang et al., 2014).

MNV reverse genetics has been used to unravel many features of norovirus replication and pathogenesis. At a molecular level it was shown that the ablation of the cleavage site between NS6 (protease) and NS7 (polymerase) of MNV abolished recovery (Ward et al., 2007); this was in stark contrast to FCV, where NS6/7 cleavage was not essential (Sosnovtsev et al., 2002). The last nucleotide of the MNV genome, immediately upstream of the poly(A) tail was shown to be essential for virus recovery (Chaudhry et al., 2007; Ward et al., 2007). Furthermore, the influence of genome-scale RNA structures on calicivirus replication was functionally characterized in vitro and in vivo (McFadden et al., 2013; Simmonds et al., 2008). The first mouse-human chimeric virus was constructed, and by exchanging a antibody neutralization epitope in MNV-1 with the corresponding amino acids from a HuNoV, the recombinant virus lost its sensitivity to the neutralizing antibody (Taube et al., 2010). The availability of a small animal model paired with reverse genetics is a powerful tool to study viral pathogenesis. Locations in ORF1 (nt 2151, MNV-1) and ORF2 (nt. 5941, MNV-1) associated with pathogenicity were mapped and functionally characterized (Bailey et al., 2008; Bailey et al., 2010b). Chimeric viruses were constructed between an acute strain MNV-1 (CW3) and a chronic strain (CR6). Systematically exchanging each gene in MNV-1 (CW3) background with its CR6 counterpart revealed that the viral capsid is a major determinant for lethal infection (Strong et al., 2012) and that NS1/2 is involved in persistence (Nice et al., 2013). It was further shown that MNV-1 CW3 containing the CR6 capsid was more sensitive to interferon β and λ than the parental strain (Nice et al., 2015).

The discovery of MNV has been a great milestone for studies into norovirus molecular biology. In its own right, MNV is an important pathogen highly prevalent in laboratory mice. MNV reverse genetics systems and the availability of numerous genetic mouse models make this platform a great tool to study viral pathogenesis in general. This has been exemplified by showing that MNV acts as a trigger for inflammation in susceptible mice (Cadwell et al., 2010) and by demonstrating the ability of the virus to complement beneficial effects of the intestinal microbiome (Kernbauer et al., 2014).

4.2.6 Human Norovirus (HuNoV)

The major obstacle for the development of a HuNoV reverse genetics system has been the inability to grow the virus efficiently in cell culture. Numerous attempts to grow HuNoV in gastrointestinal epithelial cells and other human and animal tissues have been unsuccessful (Duizer et al., 2004; Lay et al., 2010). A 3-dimensional organoid culture system (Straub et al., 2007; Straub et al., 2011) could not be validated in other laboratories (Herbst-Kralovetz et al., 2013; Takanashi et al., 2014). Despite these setbacks, there have been great advances to replicate the HuNoV genome in cells, particularly by the generation of stable replicon containing cell lines (Chang et al., 2006, 2008; Chang, 2009), transient replicons (Katayama et al., 2014), and the observation that HuNoV RNA isolated from stool is infectious (Guix et al., 2007). These observations strongly suggest that the bottleneck for achieving multiple rounds of HuNoV infection in cell culture is due to events prior to replication, namely entry and uncoating.

Two HuNoV recovery systems have been developed in the absence of a susceptible cell culture platform. The first system by Asanaka et al. (2005) was an adaptation of the MVA-T7 helper-virus driven approach successfully applied for FCV (Thumfart and Meyers, 2002). Transfection of NV cDNA under the control of a T7 promotor into cells transduced with MVA-T7 resulted in the recovery of empty particles only despite the presence of viral replication components and sgRNA (Asanaka et al., 2005). Similar results were later obtained with a GII.3 strain (U201) also showing the production of viral negative-strand RNA (Katayama et al., 2006). The second system by Katayama and colleagues removed the helper virus dependence by exchanging the T7 with a eukaryotic E1α promotor (Katayama et al., 2014; reviewed in Taube and Wobus, 2014). While this system allowed for the recovery of genome packed particles, it was still restricted to a single round of infection. The authors then conducted surrogate passaging experiments, transfecting viral RNA purified from the de novo generated particles into COS7 cells, showing that the packed RNA led to the production of nonstructural proteins. Embedding a GFP gene in ORF1 between NS3 and NS4 generated a tagged genome that could also be packed and produced fluorescent cells during surrogate passaging.

5 SUMMARY AND OUTLOOK

A number of approaches have now been developed for calicivirus reverse genetics to generate infectious RNA genomes including RNA- or DNA-based systems, helper virus-dependent systems, and vector-based approaches (summarized in Fig. 3.2.4 and Table 3.2.2). With the development of alternative reverse genetics strategies for animal caliciviruses and the recent development of a basic human norovirus cell culture system and a small animal model, investigations into human norovirus replication and pathogenicity have become feasible in the foreseeable future. A number of important questions remain to be addressed including the identity of the target cell during authentic HuNoV infection in man, the duration of protective immune responses to noroviruses and the identification of genetic determinants of HuNoV pathogenesis. With the availability of reverse genetics systems and potential animal models, the ability to address these questions has been greatly improved, and these efforts will undoubtedly provide new insights into the biology of these important animal and human pathogens.

REFERENCES

Abente, E.J., Sosnovtsev, S.V., Bok, K., Green, K.Y., 2010. Visualization of feline calicivirus replication in real-time with recombinant viruses engineered to express fluorescent reporter proteins. Virology 400 (1), 18–31.

Al-Molawi, N., Beardmore, V.A., Carter, M.J., Kass, G.E., Roberts, L.O., 2003. Caspase-mediated cleavage of the feline calicivirus capsid protein. J. Gen. Virol. 84 (Pt 5), 1237–1244.

Alhatlani, B., Vashist, S., Goodfellow, I., 2015. Functions of the 5′ and 3′ ends of calicivirus genomes. Virus Res. 206, 134–143.

Arias, A., Bailey, D., Chaudhry, Y., Goodfellow, I., 2012. Development of a reverse-genetics system for murine norovirus 3: long-term persistence occurs in the caecum and colon. J. Gen. Virol. 93 (Pt 7), 1432–1441.

Asanaka, M., Atmar, R.L., Ruvolo, V., Crawford, S.E., Neill, F.H., Estes, M.K., 2005. Replication and packaging of Norwalk virus RNA in cultured mammalian cells. Proc. Natl. Acad. Sci. USA 102 (29), 10327–10332.

Bailey, D., Thackray, L.B., Goodfellow, I.G., 2008. A single amino acid substitution in the murine norovirus capsid protein is sufficient for attenuation in vivo. J. Virol. 82 (15), 7725–7728.

Bailey, D., Kaiser, W.J., Hollinshead, M., et al., 2010a. Feline calicivirus p32, p39 and p30 proteins localize to the endoplasmic reticulum to initiate replication complex formation. J. Gen. Virol. 91 (Pt 3), 739–749.

Bailey, D., Karakasiliotis, I., Vashist, S., et al., 2010b. Functional analysis of RNA structures present at the 3′ extremity of the murine norovirus genome: the variable polypyrimidine tract plays a role in viral virulence. J. Virol. 84 (6), 2859–2870.

Binns, M.M., Britton, B.S., Mason, C., Boursnell, M.E., 1990. Analysis of the fowlpox virus genome region corresponding to the vaccinia virus D6 to A1 region: location of, and variation in, non-essential genes in poxviruses. J. Gen. Virol. 71 (Pt 12), 2873–2881.

Bok, K., Prikhodko, V.G., Green, K.Y., Sosnovtsev, S.V., 2009. Apoptosis in murine norovirus-infected RAW264.7 cells is associated with downregulation of survivin. J. Virol. 83 (8), 3647–3656.

Bridgen, A. (Ed.), 2012. Reverse Genetics of RNA Viruses. John Wiley & Sons, Chichester, UK.

Britton, P., Green, P., Kottier, S., et al., 1996. Expression of bacteriophage T7 RNA polymerase in avian and mammalian cells by a recombinant fowlpox virus. J. Gen. Virol. 77 (Pt 5), 963–967.

Cadwell, K., Patel, K.K., Maloney, N.S., et al., 2010. Virus-plus-susceptibility gene interaction determines Crohn's disease gene Atg16L1 phenotypes in intestine. Cell 141 (7), 1135–1145.

Chang, K.-O., Sosnovtsev, S.V., Belliot, G., Wang, Q., Saif, L.J., Green, K.Y., 2005a. Reverse genetics system for porcine enteric calicivirus, a prototype sapovirus in the Caliciviridae. J. Virol. 79 (3), 1409–1416.

Chang, K.-O., Belliot, G., King, A.D., Green, K.Y., 2006. Stable expression of a Norwalk virus RNA replicon in a human hepatoma cell line. Virology 353 (2), 463–473.

Chang, K.-O., George, D.W., Patton, J.B., Green, K.Y., 2008. Leader of the capsid protein in feline calicivirus promotes replication of Norwalk virus in cell culture. J. Virol. 82 (19), 9306–9317.

Chang, K.O., Sosnovtsev, S.V., Belliot, G., Kim, Y., Saif, L.J., Green, K.Y., 2004. Bile acids are essential for porcine enteric calicivirus replication in association with down-regulation of signal transducer and activator of transcription 1. Proc. Natl. Acad. Sci. USA 101 (23), 8733–8738.

Chang, K.O., Sosnovtsev, S.V., Belliot, G., Wang, Q., Saif, L.J., Green, K.Y., 2005b. Reverse genetics system for porcine enteric calicivirus, a prototype sapovirus in the Caliciviridae. J. Virol. 79 (3), 1409–1416.

Chang, K.O., 2009. Role of cholesterol pathways in norovirus replication. J. Virol. 83 (17), 8587–8595.

Chaudhry, Y., Nayak, A., Bordeleau, M.E., et al., 2006. Caliciviruses differ in their functional requirements for eIF4F components. J. Biol. Chem. 281 (35), 25315–25325.

Chaudhry, Y., Skinner, M.A., Goodfellow, I.G., 2007. Recovery of genetically defined murine norovirus in tissue culture by using a fowlpox virus expressing T7 RNA polymerase. J. Gen. Virol. 88 (Pt 8), 2091–2100.

Cheetham, S., Souza, M., Meulia, T., Grimes, S., Han, M.G., Saif, L.J., 2006. Pathogenesis of a genogroup II human norovirus in gnotobiotic pigs. J. Virol. 80 (21), 10372–10381.

Chung, L., Bailey, D., Leen, E.N., et al., 2014. Norovirus translation requires an interaction between the C Terminus of the genome-linked viral protein VPg and eukaryotic translation initiation factor 4G. J. Biol. Chem. 289 (31), 21738–21750.

Cooke, B.D., 2002. Rabbit haemorrhagic disease: field epidemiology and the management of wild rabbit populations. Rev. Sci. Tech. 21 (2), 347–358.

Cox, C., Cao, S., Lu, Y., 2009. Enhanced detection and study of murine norovirus-1 using a more efficient microglial cell line. Virol. J. 6, 196.

Crandell, R.A., Fabricant, C.G., Nelson-Rees, W.A., 1973. Development, characterization, and viral susceptibility of a feline (Felis catus) renal cell line (CRFK). In Vitro 9 (3), 176–185.

Daughenbaugh, K.F., Fraser, C.S., Hershey, J.W., Hardy, M.E., 2003. The genome-linked protein VPg of the Norwalk virus binds eIF3, suggesting its role in translation initiation complex recruitment. EMBO J. 22 (11), 2852–2859.

Duizer, E., Schwab, K.J., Neill, F.H., Atmar, R.L., Koopmans, M.P., Estes, M.K., 2004. Laboratory efforts to cultivate noroviruses. J. Gen. Virol. 85 (Pt 1), 79–87.

Ettayebi, K., Hardy, M.E., 2003. Norwalk virus nonstructural protein p48 forms a complex with the SNARE regulator VAP-A and prevents cell surface expression of vesicular stomatitis virus G protein. J. Virol. 77 (21), 11790–11797.

Farkas, T., Sestak, K., Wei, C., Jiang, X., 2008. Characterization of a rhesus monkey calicivirus representing a new genus of Caliciviridae. J. Virol. 82 (11), 5408–5416.

Flynn, W.T., Saif, L.J., 1988. Serial propagation of porcine enteric calicivirus-like virus in primary porcine kidney cell cultures. J. Clin. Microbiol. 26 (2), 206–212.

Fuentes, C., Bosch, A., Pinto, R.M., Guix, S., 2012. Identification of human astrovirus genome-linked protein (VPg) essential for virus infectivity. J. Virol. 86 (18), 10070–10078.

Fuerst, T.R., Moss, B., 1989. Structure and stability of mRNA synthesized by vaccinia virus-encoded bacteriophage T7 RNA polymerase in mammalian cells. Importance of the 5' untranslated leader. J. Mol. Biol. 206 (2), 333–348.

Goodfellow, I., Chaudhry, Y., Gioldasi, I., et al., 2005. Calicivirus translation initiation requires an interaction between VPg and eIF 4 E. EMBO Rep. 6 (10), 968–972.

Goodfellow, I., 2011. The genome-linked protein VPg of vertebrate viruses - a multifaceted protein. Curr. Opin. Virol. 1 (5), 355–362.

Green, K.Y., 2013. Caliciviridae: the noroviruses. In: Knipe, D.M., Howley, P.M. et al., (Eds.), Fields Virology, sixth ed. Wolters Kluwer Health/Lippincott Williams & Wilkins, Philadelphia, PA, pp. 949–979.

Guix, S., Asanaka, M., Katayama, K., et al., 2007. Norwalk virus RNA is infectious in mammalian cells. J. Virol. 81 (22), 12238–12248.

Herbert, T.P., Brierley, I., Brown, T.D., 1997. Identification of a protein linked to the genomic and subgenomic mRNAs of feline calicivirus and its role in translation. J. Gen. Virol. 78 (Pt 5), 1033–1040.

Herbst-Kralovetz, M.M., Radtke, A.L., Lay, M.K., et al., 2013. Lack of norovirus replication and histo-blood group antigen expression in 3-dimensional intestinal epithelial cells. Emerg. Infect. Dis. 19 (3), 431–438.

Hosmillo, M., Chaudhry, Y., Kim, D.S., Goodfellow, I., Cho, K.O., 2014. Sapovirus translation requires an interaction between VPg and the cap binding protein eIF4E. J. Virol. 88 (21), 12213–12221.

Hosmillo, M., Sorgeloos, F., Hiraide, R., Lu, J., Goodfellow, I., Cho, K.O., 2015. Porcine sapovirus replication is restricted by the type I interferon response in cell culture. J. Gen. Virol. 96 (Pt 1), 74–84.

Huang, Y.W., Hu, C.C., Liou, M.R., et al., 2012. Hsp90 interacts specifically with viral RNA and differentially regulates replication initiation of Bamboo mosaic virus and associated satellite RNA. PLoS Pathog. 8 (5), e1002726.

Hwang, S., Alhatlani, B., Arias, A., et al., 2014. Murine norovirus: propagation, quantification, and genetic manipulation. Curr. Protoc. Microbiol. 33, 15K.2.1–K.2.61.

Hyde, J.L., Green, K.Y., Wobus, C.E., Virgin IV, H.W., Mackenzie, J.M., 2009. Mouse norovirus replication is associated with virus-induced vesicle clusters originating from membranes derived from the secretory pathway. J. Virol. 83 (19), 9709–9719.

Hyde, J.L., Gillespie, L.K., Mackenzie, J.M., 2012. Mouse norovirus 1 utilizes the cytoskeleton network to establish localization of the replication complex proximal to the microtubule organizing center. J. Virol. 86 (8), 4110–4122.

Ikeda, R.A., 1992. The efficiency of promoter clearance distinguishes T7 class II and class III promoters. J. Biol. Chem. 267 (16), 11322–11328.

Jones, M.K., Watanabe, M., Zhu, S., et al., 2014. Enteric bacteria promote human and mouse norovirus infection of B cells. Science 346 (6210), 755–759.

Kaiser, W.J., Chaudhry, Y., Goodfellow, I.G., 2006. Analysis of protein-protein interactions in the feline calicivirus replication complex. J. Gen. Virol. 87 (Pt 2), 363–368.

Kapikian, A.Z., Wyatt, R.G., Dolin, R., Thornhill, T.S., Kalica, A.R., Chanock, R.M., 1972. Visualization by immune electron microscopy of a 27-nm particle associated with acute infectious nonbacterial gastroenteritis. J. Virol. 10 (5), 1075–1081.

Karst, S.M., Wobus, C.E., Lay, M., Davidson, J., Virgin, H.W., 2003. STAT1-dependent innate immunity to a Norwalk-like virus. Science 299 (5612), 1575–1578.

Katayama, K., Hansman, G.S., Oka, T., Ogawa, S., Takeda, N., 2006. Investigation of norovirus replication in a human cell line. Arch. Virol. 151 (7), 1291–1308.

Katayama, K., Murakami, K., Sharp, T.M., et al., 2014. Plasmid-based human norovirus reverse genetics system produces reporter-tagged progeny virus containing infectious genomic RNA. Proc. Natl. Acad. Sci. USA 111 (38), E4043–E4052.

Kearse, M., Moir, R., Wilson, A., Stones-Havas, S., Cheung, M., Sturrock, S., Buxton, S., Cooper, A., Markowitz, S., Duran, C., Thierer, T., Ashton, B., Mentjies, P., Drummond, A., 2012. Geneious basic: an integrated and extendable desktop software platform for the organization and analysis of sequence data. Bioinformatics 28 (12), 1647–1649.

Kernbauer, E., Ding, Y., Cadwell, K., 2014. An enteric virus can replace the beneficial function of commensal bacteria. Nature 516 (7529), 94–98.

Lay, M.K., Atmar, R.L., Guix, S., et al., 2010. Norwalk virus does not replicate in human macrophages or dendritic cells derived from the peripheral blood of susceptible humans. Virology 406 (1), 1–11.

Lee, H., Cho, Y.H., Park, J.S., Kim, E.C., Smith, A.W., Ko, G., 2012. Elevated post-transfusion serum transaminase values associated with a highly significant trend for increasing prevalence of anti-Vesivirus antibody in Korean patients. J. Med. Virol. 84 (12), 1943–1952.

Li, T.C., Yang, T., Yoshizaki, S., et al., 2015. Construction and characterization of an infectious cDNA clone of rat hepatitis E virus. J. Gen. Virol. 96 (Pt 6), 1320–1327.

Lin, X., Thorne, L., Jin, Z., et al., 2015. Subgenomic promoter recognition by the norovirus RNA-dependent RNA polymerases. Nucleic Acids Res. 43 (1), 446–460.

Liu, G., Zhang, Y., Ni, Z., et al., 2006. Recovery of infectious rabbit hemorrhagic disease virus from rabbits after direct inoculation with in vitro-transcribed RNA. J. Virol. 80 (13), 6597–6602.

Liu, G.Q., Ni, Z., Yun, T., et al., 2008. Rabbit hemorrhagic disease virus poly(A) tail is not essential for the infectivity of the virus and can be restored in vivo. Arch. Virol. 153 (5), 939–944.

López-Manríquez, E., Vashist, S., Ureña, L., et al., 2013. Norovirus genome circularization and efficient replication are facilitated by binding of PCBP2 and hnRNP A1. J. Virol. 87 (21), 11371–11387.

Makino, A., Shimojima, M., Miyazawa, T., Kato, K., Tohya, Y., Akashi, H., 2006. Junctional adhesion molecule 1 is a functional receptor for feline calicivirus. J. Virol. 80 (9), 4482–4490.

McFadden, N., Bailey, D., Carrara, G., et al., 2011. Norovirus regulation of the innate immune response and apoptosis occurs via the product of the alternative open reading frame 4. PLoS Pathog. 7 (12), e1002413.

McFadden, N., Arias, A., Dry, I., et al., 2013. Influence of genome-scale RNA structure disruption on the replication of murine norovirus--similar replication kinetics in cell culture but attenuation of viral fitness in vivo. Nucleic Acids Res. 41 (12), 6316–6331.

Meyers, G., Wirblich, C., Thiel, H.J., 1991. Genomic and subgenomic RNAs of rabbit hemorrhagic disease virus are both protein-linked and packaged into particles. Virology 184 (2), 677–686.

Meyers, G., 2007. Characterization of the sequence element directing translation reinitiation in RNA of the calicivirus rabbit hemorrhagic disease virus. J. Virol. 81 (18), 9623–9632.

Natoni, A., Kass, G.E., Carter, M.J., Roberts, L.O., 2006. The mitochondrial pathway of apoptosis is triggered during feline calicivirus infection. J. Gen. Virol. 87 (Pt 2), 357–361.

Neill, J.D., Sosnovtsev, S.V., Green, K.Y., 2000. Recovery and altered neutralization specificities of chimeric viruses containing capsid protein domain exchanges from antigenically distinct strains of feline calicivirus. J. Virol. 74 (3), 1079–1084.

Nice, T.J., Strong, D.W., McCune, B.T., Pohl, C.S., Virgin IV, H.W., 2013. A single-amino-acid change in murine norovirus NS1/2 is sufficient for colonic tropism and persistence. J. Virol. 87 (1), 327–334.

Nice, T.J., Baldridge, M.T., McCune, B.T., et al., 2015. Interferon-lambda cures persistent murine norovirus infection in the absence of adaptive immunity. Science 347 (6219), 269–273.

Oka, T., Takagi, H., Tohya, Y., 2014. Development of a novel single step reverse genetics system for feline calicivirus. J. Virol. Methods 207, 178–181.

Parwani, A.V., Flynn, W.T., Gadfield, K.L., Saif, L.J., 1991. Serial propagation of porcine enteric calicivirus in a continuous cell line. Effect of medium supplementation with intestinal contents or enzymes. Arch. Virol. 120 (1-2), 115–122.

Paul, A.V., Wimmer, E., 2015. Initiation of protein-primed picornavirus RNA synthesis. Virus Res. 206, 12–26.

Perrotta, A.T., Been, M.D., 1998. A toggle duplex in hepatitis delta virus self-cleaving RNA that stabilizes an inactive and a salt-dependent pro-active ribozyme conformation. J. Mol. Biol. 279 (2), 361–373.

Picard-Jean, F., Tremblay-Létourneau, M., Serra, E., et al., 2013. RNA 5′-end maturation: A crucial step in the replication of viral genomes. In: Romanowski, V. (Ed.), Current Issues in Molecular Virology—Viral Genetics, Biotechnological Applications, 27–56.

Prasad, B.V., Hardy, M.E., Dokland, T., Bella, J., Rossmann, M.G., Estes, M.K., 1999. X-ray crystallographic structure of the Norwalk virus capsid. Science 286 (5438), 287–290.

Roberts, L.O., Al-Molawi, N., Carter, M.J., Kass, G.E., 2003. Apoptosis in cultured cells infected with feline calicivirus. Ann. NY Acad. Sci. 1010, 587–590.

Royall, E., Doyle, N., Abdul-Wahab, A., et al., 2015. Murine norovirus 1 (MNV1) replication induces translational control of the host by regulating eIF4E activity during infection. J. Biol. Chem. 290 (8), 4748–4758.

Sandoval-Jaime, C., Green, K.Y., Sosnovtsev, S.V., 2015. Recovery of murine norovirus and feline calicivirus from plasmids encoding EMCV IRES in stable cell lines expressing T7 polymerase. J. Virol. Methods 217, 1–7.

Sharp, T.M., Guix, S., Katayama, K., Crawford, S.E., Estes, M.K., 2010. Inhibition of cellular protein secretion by norwalk virus nonstructural protein p22 requires a mimic of an endoplasmic reticulum export signal. PLoS One 5 (10), e13130.

Shivanna, V., Kim, Y., Chang, K.O., 2015. Ceramide formation mediated by acid sphingomyelinase facilitates endosomal escape of caliciviruses. Virology 483, 218–228.

Simmonds, P., Karakasiliotis, I., Bailey, D., Chaudhry, Y., Evans, D.J., Goodfellow, I.G., 2008. Bioinformatic and functional analysis of RNA secondary structure elements among different genera of human and animal caliciviruses. Nucleic Acids Res. 36 (8), 2530–2546.

Smith, A.W., Iversen, P.L., Skilling, D.E., Stein, D.A., Bok, K., Matson, D.O., 2006. Vesivirus viremia and seroprevalence in humans. J. Med. Virol. 78 (5), 693–701.

Smits, S.L., Rahman, M., Schapendonk, C.M.E., et al., 2012. Calicivirus from novel Recovirus genogroup in human diarrhea, Bangladesh. Emerg. Infect. Dis. 18 (7), 1192–1195.

Sosnovtsev, S.V., Green, K.Y., 1995. RNA transcripts derived from a cloned full-length copy of the feline calicivirus genome do not require VpG for infectivity. Virology 210 (2), 383–390.

Sosnovtsev, S.V., Sosnovtseva, S.A., Green, K.Y., Virology, ESoV., 1997. Recovery of feline calicivirus from plasmid DNA containing a full-length copy of the genome. In: Chasey, D., Gaskell, R., Clarke, I.N. (Eds.), First International Symposium on Calciviruses; 1997. European Society for Veterinary Virology and Central Veterinary Laboratory, Reading, UK.

Sosnovtsev, S.V., Sosnovtseva, S.A., Green, K.Y., 1998. Cleavage of the feline calicivirus capsid precursor is mediated by a virus-encoded proteinase. J. Virol. 72 (4), 3051–3059.

Sosnovtsev, S.V., Garfield, M., Green, K.Y., 2002. Processing map and essential cleavage sites of the nonstructural polyprotein encoded by ORF1 of the feline calicivirus genome. J. Virol. 76 (14), 7060–7072.

Sosnovtsev, S.V., Prikhod'ko, E.A., Belliot, G., Cohen, J.I., Green, K.Y., 2003. Feline calicivirus replication induces apoptosis in cultured cells. Virus Res. 94 (1), 1–10.

Sosnovtsev, S.V., Belliot, G., Chang, K.O., Onwudiwe, O., Green, K.Y., 2005. Feline calicivirus VP2 is essential for the production of infectious virions. J. Virol. 79 (7), 4012–4024.

Sosnovtsev, S.V., Belliot, G., Chang, K.O., et al., 2006. Cleavage map and proteolytic processing of the murine norovirus nonstructural polyprotein in infected cells. J. Virol. 80 (16), 7816–7831.

Straub, T.M., Honer zu Bentrup, K., Orosz-Coghlan, P., et al., 2007. In vitro cell culture infectivity assay for human noroviruses. Emerg. Infect. Dis. 13 (3), 396–403.

Straub, T.M., Bartholomew, R.A., Valdez, C.O., et al., 2011. Human norovirus infection of caco-2 cells grown as a three-dimensional tissue structure. J. Water Health 9 (2), 225–240.

Strong, D.W., Thackray, L.B., Smith, T.J., Virgin IV, H.W., 2012. Protruding domain of capsid protein is necessary and sufficient to determine murine norovirus replication and pathogenesis in vivo. J. Virol. 86 (6), 2950–2958.

Subba-Reddy, C.V., Goodfellow, I., Kao, C.C., 2011. VPg-primed RNA synthesis of norovirus RNA-dependent RNA polymerases by using a novel cell-based assay. J. Virol. 85 (24), 13027–13037.

Subba-Reddy, C.V., Yunus, M.A., Goodfellow, I.G., Kao, C.C., 2012. Norovirus RNA synthesis is modulated by an interaction between the viral RNA-dependent RNA polymerase and the major capsid protein. VP1. J. Virol. 86 (18), 10138–10149.

Sutter, G., Moss, B., 1995. Novel vaccinia vector derived from the host range restricted and highly attenuated MVA strain of vaccinia virus. Dev. Biol. Stand. 84, 195–200.

Takanashi, S., Saif, L.J., Hughes, J.H., et al., 2014. Failure of propagation of human norovirus in intestinal epithelial cells with microvilli grown in three-dimensional cultures. Arch. Virol. 159 (2), 257–266.

Tan, M., Jiang, X., 2010. Virus-host interaction and cellular receptors of caliciviruses. In: Hansman, G.S., Jiang, X., Green, K.Y. (Eds.), Caliciviruses, Molecular and Cellular Virology. first ed. Caister Academic Press, Norfolk, UK, pp. 111–130.

Taube, S., Rubin, J.R., Katpally, U., et al., 2010. High-resolution x-ray structure and functional analysis of the murine norovirus 1 capsid protein protruding domain. J. Virol. 84 (11), 5695–5705.

Taube, S., Kolawole, A.O., Höhne, M., et al., 2013. A mouse model for human norovirus. mBio 4 (4), , e00450-13-e-13.

Taube, S., Wobus, C.E., 2014. A novel reverse genetics system for human norovirus. Trends Microbiol. 22 (11), 604–606.

Thorne, L.G., Goodfellow, I.G., 2014. Norovirus gene expression and replication. J. Gen. Virol. 95 (Pt 2), 278–291.

Thumfart, J.O., Meyers, G., 2002. Feline calicivirus: recovery of wild-type and recombinant viruses after transfection of cRNA or cDNA constructs. J. Virol. 76 (12), 6398–6407.

Van Craenenbroeck, K., Vanhoenacker, P., Haegeman, G., 2000. Episomal vectors for gene expression in mammalian cells. Eur. J. Biochem. 267 (18), 5665–5678.

Vashist, S., Bailey, D., Putics, A., Goodfellow, I., 2009. Model systems for the study of human norovirus biology. Future Virol. 4 (4), 353–367.

Vashist, S., Ureña, L., Chaudhry, Y., Goodfellow, I.G., 2012. Identification of RNA-protein interaction networks involved in the norovirus life cycle. J. Virol. 86 (22), 11977–11990.

Vashist, S., Ureña, L., Gonzalez-Hernandez, M.B., et al., 2015. The molecular chaperone Hsp90 is a therapeutic target for noroviruses. J. Virol. 89 (12), 6352–6363.

Vongpunsawad, S., Prasad, B.V., Estes, M.K., 2013. Norwalk Virus Minor Capsid Protein VP2 Associates within the VP1 Shell Domain. J. Virol. 87 (9), 4818–4825.

Ward, V.K., McCormick, C.J., Clarke, I.N., et al., 2007. Recovery of infectious murine norovirus using pol II-driven expression of full-length cDNA. Proc. Natl. Acad. Sci. USA 104 (26), 11050–11055.

Wei, C., Farkas, T., Sestak, K., Jiang, X., 2008. Recovery of infectious virus by transfection of in vitro-generated RNA from tulane calicivirus cDNA. J. Virol. 82 (22), 11429–11436.

Wobus, C.E., Karst, S.M., Thackray, L.B., et al., 2004. Replication of Norovirus in cell culture reveals a tropism for dendritic cells and macrophages. PLoS Biol. 2 (12), e432.

Wobus, C.E., Thackray, L.B., Virgin IV, H.W., 2006. Murine norovirus: a model system to study norovirus biology and pathogenesis. J. Virol. 80 (11), 5104–5112.

Yunus, M.A., Chung, L.M., Chaudhry, Y., Bailey, D., Goodfellow, I., 2010. Development of an optimized RNA-based murine norovirus reverse genetics system. J. Virol. Methods 169 (1), 112–118.

Yunus, M.A., Lin, X., Bailey, D., et al., 2015. The murine norovirus core subgenomic RNA promoter consists of a stable stem-loop that can direct accurate initiation of RNA synthesis. J. Virol. 89 (2), 1218–1229.

Zhang, D., Huang, P., Zou, L., Lowary, T.L., Tan, M., Jiang, X., 2015. Tulane virus recognizes the A type 3 and B histo-blood group antigens. J. Virol. 89 (2), 1419–1427.

Chapter 3.3

Human Norovirus Receptors

J. Le Pendu*, G.E. Rydell**, W. Nasir†, G. Larson†

*Inserm, CNRS, Nantes University, IRS UN, Nantes, France; **Department of Infectious Diseases, Sahlgrenska Academy, University of Gothenburg, Gothenburg, Sweden; †Department of Clinical Chemistry and Transfusion Medicine, Sahlgrenska Academy, University of Gothenburg, Gothenburg, Sweden

1 INTRODUCTION

Human noroviruses (NoVs) are a leading cause of viral gastroenteritis, infecting all age groups worldwide and as such represent a major public health problem. Despite being mainly transmitted through person-to-person contact these viruses are the principal cause of toxic food-borne viral infections in Europe and in the United States (Ahmed et al., 2014; Hall et al., 2012; Patel et al., 2009). Although they generally cause a relatively mild disease of short duration, NoVs can be responsible for severe dehydration that may lead to hospitalization and even death of at-risk patients, notably young children, elderly people in poor health and immunocompromised patients (Bok and Green, 2012; Gustavsson et al., 2011). Thus, these small nonenveloped RNA viruses represent the main cause of acute gastroenteritis leading to hospitalization in adults and the second leading cause after rotavirus in young children in the United States (Hall, 2012). In countries with established universal rotavirus vaccination programs, NoVs are the most frequent cause of acute gastroenteritis also in children (Payne et al., 2013; Koo et al., 2013; Hemming et al., 2013; Bucardo et al., 2014). NoV present with a very large genetic diversity. By comparing sequences that code for the polymerase and the capsid protein of a large number of human and animal NoV strains, they have been classified into distinct groups (genogroups GI–GVI). GI and GII NoVs include strains that are the most frequently implicated in human infections. GIII exclusively comprises bovine strains. The GIV genogroup is rarely found in humans and mainly includes strains isolated in lions and dogs. GV and GVI include murine and canine strains, respectively (Mesquita et al., 2010). Finally, the existence of a seventh genogroup has been proposed after the discovery of new canine strains (Vinjé, 2015). In each genogroup, the analysis of complete ORF2 sequences that encode the capsid protein has revealed huge genetic diversity, making it possible to distinguish between eight genotypes in GI (GI.1–GI.8) and 21 genotypes in GII (GII.1–GII.21). GII NoVs are primarily found in humans. The three exceptions are GII.11, GII.18, and GII.19, which are exclusively found in pigs (Hall et al., 2011; Zheng et al., 2006). From an epidemiological point of view, GII NoVs are currently predominant. Among these, the GII.4 genotype alone is responsible for over 70% of all cases of NoV gastroenteritis (Glass et al., 2009; Hall et al., 2011). (See also Chapter 3.5).

The first step of a viral infection is characterized by attachment onto a susceptible cell through surface components that will contribute to virus entry. Those cell surface components involved in the process of binding and entry function as viral receptors. Yet, in many instances, binding to a single attachment or adhesion factor is not sufficient for infection since additional molecules, or coreceptors are required for entry to take place (Verdaguer et al., 2014). In the case of human NoVs, recognition and binding to several glycans has been demonstrated. However, their exact role in the infection process remains ill-defined since until very recently it has not been possible to cultivate human NoVs, hampering studies of the entry process.

Following the discovery that Rabbit Hemorrhagic Disease Virus (RHDV), another member of the *Caliciviridae* family, could attach to glycans of the histo-blood group family (HBGAs) (Ruvoën-Clouet et al., 2000), it was observed that human NoVs recognize similar carbohydrates (Marionneau et al., 2002). It was then suggested that human NoVs could additionally bind to heparan sulfate (Tamura et al., 2004) and sialylated glycans such as sialyl Le^x (SLe^x) and sialyl-type 2 precursor (Rydell et al., 2009b). Very recently, specific NoV recognition of gangliosides, that is, glycosphingolipids carrying one (or more) sialic acids, was demonstrated in addition to that of HBGAs (Han et al., 2014), an intriguing observation reminiscent of the recognition of HBGAs and gangliosides by various strains of rotavirus (Hu et al., 2012; Huang et al., 2012; Martinez et al., 2013).

2 HUMAN NOROVIRUS BINDS TO HISTO-BLOOD GROUP ANTIGENS

HBGAs, which include the ABH, Lewis, I/i, and P antigens, are complex sugars typically present on the erythrocytes of humans and some of the great apes (Storry and Olsson, 2004). However, HBGAs are also found on the surface of epithelial cells of different tissues in a wide range of vertebrate species and can be secreted in free or complex forms in biological fluids such as saliva and milk (Marionneau et al., 2001). Since most human NoVs recognize ABH and/or Lewis antigens we will concentrate on these carbohydrate motifs. Their synthesis is under the control of several genes that code for glycosyltransferases and display common genetic polymorphisms (Lowe, 1993) (Fig. 3.3.1).

The α1,2-fucosyltransferase encoded by the *FUT2* gene is key to the process because it contributes to the synthesis of the A, B, H, Le^b, and Le^y antigens. Accordingly, the presence of at least one of these antigens characterizes

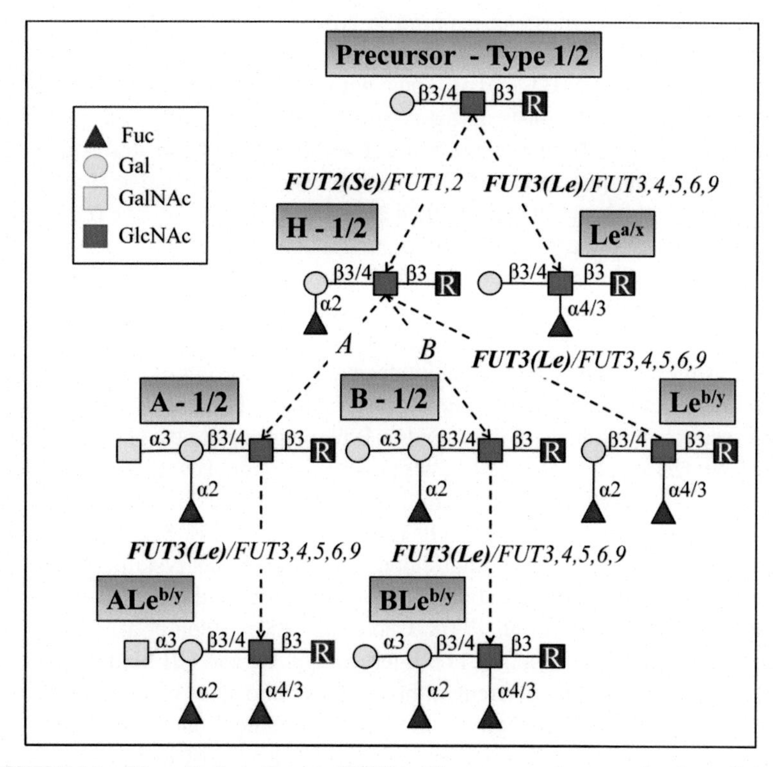

FIGURE 3.3.1 Biosynthetic pathways of HBGAs. The precursors from type-1 and type-2 chains glycans are considered. The names of antigens *(blue boxes)*, genetic loci of glycosyltransferases *(italic)* and the glycosidic linkages *(solid lines)* are separated by forward slash *(/)* to specify the pathways for either chain type. The *FUT2* and the *FUT3* genetic loci driving the biosynthesis of the secretor *(Se)* and the Lewis *(Le)* antigens, respectively, are also highlighted *(bold)*.

the so-called secretor phenotype, whereas its absence, due to mutations in the enzyme's gene coding sequence, generates the so-called nonsecretor phenotype. The secretory ABH and Lewis phenotypes of each individual depend on the combined polymorphisms at the three loci *ABO*, *FUT2*, and *FUT3*. Using virus-like particles (VLPs), spontaneously formed from the recombinantly expressed capsid protein of NoVs, specific recognition patterns of HBGAs have been described. This was performed by analyzing the attachment of VLPs to tissue sections of digestive tissues and to saliva mucins from HBGA phenotyped individuals, to immobilized synthetic oligosaccharides or to cells transfected in order to express the appropriate glycosyltransferases. Through these methods, it was initially noticed that A, H, and Le^b antigens are ligands of the Norwalk Virus (NV), a prototype virus belonging to the GI.1 genotype (Marionneau et al., 2002). In contrast, B and Le^a antigens are not recognized by this viral strain (Huang et al., 2003). Following these initial studies the glycan specificity of many other strains has been studied using VLPs belonging to all genotypes.

It was observed that most strains bind to HBGAs, albeit important variations in specificities were found within each of the main human genogroups (GI and GII) (Tan and Jiang, 2011, 2014). Thus, within GI two major types of binding patterns could be distinguished, that is, strains that are similar to the previously described NV GI.1 strain, recognize the A and H epitopes and strains that bind to the Lewis epitopes, targeting the α1,3/4-linked fucose residue (Fig. 3.3.1). The latter strains show variation in their ability to accommodate the α1,2-linked fucose residue and the A or B epitopes, and a dominant binding to the α1,3/4-linked fucose, allowing them to recognize nonsecretor individuals as observed by their binding to saliva from such individuals. Among GII strains three types of binding patterns have been discerned. The first one has been called ABH binder as these strains attach to saliva from all secretor individuals, regardless of the ABO and Lewis status. The second, called A/B binder, attach to saliva from secretor individuals of the A and/or B blood groups only. The third, called Lewis binder, attach to saliva of Lewis positive individuals with more or less influence of the secretor phenotype, indicating strong involvement of the α1,3/4-linked fucose (Tan and Jiang, 2011; Tan and Jiang, 2014). Structural analysis of the capsid protein's protruding region, or P-domain, cocrystallized with oligosaccharides allowed characterization of the exact binding sites for various GI and GII strains (Bu et al., 2008; Cao et al., 2007; Chen et al., 2011; Choi et al., 2008; Hansman et al., 2011). Interestingly, GI and GII binding sites appeared to be completely different, although the amino acids involved are highly conserved within each genogroup. In addition binding modes between the two genogroups are very different (Cao et al., 2007; Ruvoën-Clouet et al., 2013). The GI binding sites involve contact mainly with galactose or *N*-acetylgalactosamine residues, while the GII sites involved essentially fucose recognition (Cao et al., 2007; Koppisetty et al., 2010; Nasir et al., 2012). Nevertheless, in both instances slight amino acid changes generate subtle modifications in the range of preferred oligosaccharides, allowing recognition of distinct subgroups of individuals or populations. The different ways by which GI and GII strains cope with their host's glycan diversity are strongly suggestive of a convergent evolution and underline the important role of HBGA binding in the process of infection. These structural aspects are presented in detail in Chapter 3.1.

3 BINDING OF NOROVIRUS TO HBGAS IS INVOLVED IN INFECTION

The importance of HBGA binding in infection was first assessed through studies on volunteers. Two independent studies involving ingestion of the virus by healthy volunteers showed that nonsecretors are resistant to infection by NV (GI.1) (Hutson et al., 2005; Lindesmith et al., 2003). Indeed, the challenged nonsecretor individuals presented neither clinical signs nor immune response to the infection, and the virus was absent from their feces. It was additionally observed that among individuals with the secretor phenotype, those with blood group B were not infected or remained asymptomatic, consistent with the lack of binding to the B blood group antigen by the NV strain. Thus, a combination of *FUT2* and *ABO* alleles determines susceptibility or resistance to NV infection. More recently a new study conducted in volunteers clearly demonstrated that one GII.4 strain infected secretors almost exclusively, and regardless of their ABO phenotype (Frenck et al., 2012). Interestingly, it was recently shown in concordant studies that both secretor and Lewis status mediate susceptibility to rotavirus infection (Imbert-Marcille et al., 2014; Nordgren et al., 2014; Kambhampati et al., 2016).

The influence of the secretor status on susceptibility to NoV infection in authentic outbreaks was first investigated in Sweden (Thorven et al., 2005). Thirty-eight symptomatic individuals from three hospital outbreaks caused by GII.4 strains were genotyped for secretor status. In addition, the same analysis was performed on 15 symptomatic individuals from three community outbreaks caused by GI.6 and GII.6 strains. Thus, in total, 53 symptomatic and 62 asymptomatic individuals were genotyped for polymorphisms of *FUT2* at nucleotides 385, 428, and 571. Strikingly, no nonsecretors were identified among the symptomatic individuals. As the secretor status was determined by genotyping, the influence of homozygosity and heterozygosity could be determined. However, no difference was identified between heterozygous and homozygous secretors. Subsequently, the resistance of nonsecretors to GII.4 strains in authentic outbreaks was confirmed by a similar study in Denmark (Kindberg et al., 2007). Similar results were also reported from a GII.4 outbreak in China (Tan et al., 2008). In that study, secretor status was determined by phenotyping of saliva samples and the weak-secretors were grouped with the secretors. In addition, two studies have investigated the influence of secretor status on susceptibility to infection with recently emerged GII.4 strains. Currier et al. (2015) investigated sporadic pediatric NoV infections and could not identify any nonsecretors among 155 patients with GII.4 infections, even though 24% of the NoV negative control individuals were nonsecretors. The GII.4 viruses identified belonged to the Den Haag 2006b (18%), New Orleans 2009 (46%), and Sydney 2012 (36%) clusters. In another pediatric material from Ecuador all 27 GII.4 single infections and all 4 coinfections that were positive for GII.4 strains were found in secretors (Lopman et al., 2015). In addition, a number of studies of NoV outbreaks caused by strains from other genogroups have reported that only secretors were identified among the symptomatically infected patients, even though the materials were not large enough to prove any statistical significance (Bucardo et al., 2009; Tan et al., 2008). Furthermore, a study of Swedish blood donors showed that secretor positive individuals had significantly higher sero-prevalence and IgG antibody titers to NoV GII.4 virus than nonsecretors (Larsson et al., 2006). The remarkable match between susceptibility to infection, the presence of either A, B, or H antigens in the digestive tract and the carbohydrate specificity of the virus strongly suggested that the interaction between the virus and these glycans is an essential step of infection.

4 THE COMPLEX AND GENOTYPE-DEPENDENT HBGA SPECIFICITIES PRECLUDE A PERFECT MATCH BETWEEN INFECTION AND THE ABO AND SECRETOR CHARACTERS

However, not all strains show a clear-cut specificity for HBGA epitopes expressed by secretors as discussed earlier. Indeed, a recent study indicated that a GII.2 strain weakly binds to saliva from either A and O blood group secretors or nonsecretors Lewis positive individuals, in addition to their strong binding to saliva from B and AB secretors (Yazawa et al., 2014).

Likewise, GI.3 strains have been reported to attach to saliva from secretors as well as nonsecretors of the A, AB and O blood groups and of the Lewis positive phenotype (Shirato et al., 2008; Yazawa et al., 2014). GII.7 strains bind to saliva from nearly all individuals, except Lewis negative nonsecretors (Shirato et al., 2008; Yazawa et al., 2014). Besides strong binding to saliva from secretors, some binding to saliva of Lewis positive nonsecretors of a GII.6 strain was also reported (Yazawa et al., 2014). Regarding old GII.4 strains from the 1990s (US95/96), all studies indicate a strong preference for recognition of HBGAs present in secretors (Huang et al., 2005; Lindesmith et al., 2008; Ruvoën-Clouet et al., 2014; Rydell et al., 2009b). Nonetheless, in some studies, weak binding to saliva from nonsecretor Lewis positive individuals have been observed (Ruvoën-Clouet et al., 2014). In addition, more recent strains of the GII.4 den Haag and Osaka subtypes readily attach to saliva from nonsecretors, but only if they are Lewis positive, in accordance with their ability to recognize Lewis antigens (de Rougemont et al., 2011). From these data it is to be expected that some nonsecretor individuals should be infected, albeit less frequently than secretors, and indeed this has been observed on several occasions.

Thus, a volunteers' study performed using the GII.2 Snow Mountain Virus (SMV) strain failed to reveal an association between infection and HBGA phenotypes (Lindesmith et al., 2005), despite the fact that SMV reportedly showed strong preference for the B blood group (Harrington et al., 2002). However, that volunteers' study was conducted on a small number of subjects, some of whom had received very high doses of virus. The volunteers' study performed using a GII.4 strain, showed one subject of the nonsecretor phenotype that was infected, albeit presenting very mild symptoms (Frenck et al., 2012). Furthermore, a limited number of outbreaks studies also showed that nonsecretor individuals can be infected by NoV (Carlsson et al., 2009; Lindesmith et al., 2005; Nordgren et al., 2010; Rockx et al., 2005). Both Rockx et al. (2005) and Nordgren et al. (2010) observed that GI.3 NoV can infect secretors as well as nonsecretors, in agreement with GI.3 VLP binding studies demonstrating binding of nonsecretor saliva and Le^a glycoconjugates (Shirato et al., 2008; Yazawa et al., 2014). Rockx et al. (2005) investigated a water-borne GI.3 NoV outbreak and found that 20/22 (91%) of the secretors and four of seven (57%) of the nonsecretors exposed to the virus became infected. Nordgren et al. (2010) investigated a food-borne GI.3 NoV outbreak, in which symptoms suggestive of NoV infection were found in seven of 15 (47%) of the nonsecretors and 26/68 (38%) of the secretors. In another study, Carlsson et al. (2009) showed that even though nonsecretor individuals had a significantly lower risk of getting infected in a GII.4 NoV outbreak in Spain, the virus also infected one individual, genotyped as a nonsecretor, suggesting that nonsecretors are not totally resistant to infections caused by the dominating NoV genotype. A similar result was obtained in a study in Burkina Faso, in which one nonsecretor and Lewis negative individual was infected with a GII.4 strain (Nordgren et al., 2013).

Although these occasional infections of nonsecretor individuals by strains such as GII.4 that show a clear preference for secretors may be explained by their ability to recognize the α1,3/4-linked fucose residue of the Lewis type present in Lewis positive (*FUT3* +) individuals even in absence of α1,2-linked fucose as in nonsecretors (*FUT2*−), another explanation for these cases has recently been raised. It was shown that chronic infection by virulent strains of the bacterium *Helicobacter pylori* aberrantly induces expression of α1,2-fucosylated motifs in the stomach mucosa of nonsecretors, which may contribute to facilitate infection of these individuals by NoV strains that otherwise show a strong preference for secretors (Ruvoën-Clouet et al., 2014).

In addition to the occasional infections of nonsecretors that seemed to contradict the early reports on the genetic susceptibility to NoV, several reports have presented apparent contradictory results concerning the association between the ABO phenotype and infection (Fretz et al., 2005; Halperin et al., 2008; Meyer et al., 2004; Miyoshi et al., 2005). It must be noted that these studies did not take into account the strains involved and their binding specificities for HBGAs. Their interpretations are therefore dubious since, as discussed earlier, NoV strain specificities for HBGAs are variable. It is highly likely that the differences observed in NoV epidemics stem from differences in specificity for ABH antigens of the strains involved. Finally, the existence of NoVs VLPs that do not appear to attach to HBGAs has been reported (Huang et al., 2005; Shirato et al., 2008; Takanashi et al., 2011; Yazawa et al., 2014), suggesting either the existence of human NoVs that can infect independently of HBGAs attachment or that the methods of VLPs binding characterization may not always reflect the behavior of true viruses.

5 ADDITIONAL BINDING SPECIFICITIES OF HUMAN NOROVIRUS

In addition to ABO(H) histo-blood group glycans, human NoV have been shown to have a number of other carbohydrate binding specificities. An early report showed that VLPs representing three different GII strains bind to a number of cell types in a heparan sulfate-dependent manner (Tamura et al., 2004). In contrast, VLPs from two GI strains only bound weakly via heparan sulfate. In addition, GII VLP have been shown to recognize sialylated glycans (Rydell et al., 2009b). In this study, VLPs from GII.4 and GII.3, but not from GI.1, were shown to recognize neoglycoproteins conjugated with SLe^x, $SdiLe^x$, and sialyl type-2 precursor glycans. The binding was specific as the VLPs did not bind to structural analogues including Le^x and SLe^a. A subsequent X-ray crystal structure showed that a GII.9 NoV VLP bound SLe^x in the HBGA binding site (Hansman et al., 2011). In this structure, the α1,3-linked fucose was tightly bound in the fucose binding site whereas no direct contacts were identified between the protein and the sialic acid. However, for the GII.3 and GII.4 strains the sialic acid is important for the interaction as the nonfucosylated structure sialyl type-2 precursor also binds to these VLPs (de Rougemont et al., 2011; Rydell et al., 2009b). Recently, binding to several additional sialylated glycans including the ganglioside GM3 was reported for GII.4 and GI.3 VLPs (Han et al., 2014). This ganglioside is widely expressed on many cell types and unlikely to confer tissue specificity. In addition, the short glycosphingolipid galactosylceramide (GalCer), organized in membrane domains and also widely expressed, has been shown to bind to human NoV (Bally et al., 2012b). Collectively, these studies indicate that charged glycans and GalCer could be involved in the NoV infection process. However, this remains to be proven, which may become possible since a culture system for human NoVs was very recently described (Jones et al., 2014) along with a novel plasmid based human NoV reverse genetic system which produces reporter virions with infectious RNA (Katayama et al., 2014).

6 GLYCAN-BINDING PROPERTIES OF ANIMAL NOROVIRUSES

NoV strains infecting animals also bind to glycans. Glycan binding thus appears to be a shared property within the NoV species. Thus, a recent report indicated that canine strains of the GIV and GVI genogroups recognized HBGAs expressed in the canine gut with specificities quite similar to those of the human NV strain (Caddy et al., 2014). The presence of shared ligands between humans and dogs recognized by these strains may contribute to cross-species transmission. It had previously been shown that the bovine specific GIII strains bind to the so-called alpha-Gal epitope, which is a glycan epitope synthesized by an α1,3-galactosyltransferase of the same enzyme family as the A and B blood group enzymes (Zakhour et al., 2009, 2010). Humans are not able to synthesize that epitope since the corresponding *GGTA* glycosyltransferase gene has become a pseudogene during evolution of the apes lineage (Macher and Galili, 2008). By contrast, the alpha-Gal epitope is present in the bovine gut mucosa where it may serve as a ligand, akin to the ABH and Lewis antigens for human strains as described earlier. In this instance, the lack of expression in the human gut mucosa may confer protection from cross-species transmission.

NoV GV strains exclusively infect mice and do not appear to recognize HBGAs. Yet, they bind to glycosphingolipids of the ganglioside type. Since these strains can be cultured, it has been possible to show that the ganglioside GD1a functions as receptor for these strains (Taube et al., 2009). This is particularly interesting in view to the recent discovery that human GI and GII strains could also bind to gangliosides with affinities in the same range as those involved in HBGAs binding (Han et al., 2014).

7 GLYCOSPHINGOLIPIDS AS RECEPTORS FOR NOROVIRUS

HBGAs are found both on glycoproteins in saliva, milk, and other body fluids, but also on membrane bound glycoproteins and glycosphingolipids (GSLs). Since glycosphingolipids, in contrast to most natural glycoproteins, only contain one glycan per molecule, they may be used to determine the precise binding specificities of glycan-binding viruses. The glycosphingolipids also, due to their amphipathic character, offer the unique possibilities to study VLP-glycan binding in dynamic membranes and not only in solution or on solid surfaces. Furthermore, such studies are motivated since glycosphingolipids have indeed been shown to function as true cellular receptors for a number of viruses (Ewers et al., 2010; Neu et al., 2009; Schmidt and Chiorini, 2006), including murine NoV (see earlier).

Thin-layer chromatogram binding assay (CBA) was used to show that the Norwalk virus VLP recognizes both type-1 and type-2 chain glycosphingolipids terminated with blood group A and H but not B epitopes (Fig. 3.3.2) (Nilsson et al., 2009). Subsequently, the binding of a Norwalk and Dijon (GII.4 US95/96) VLP to glycosphingolipids incorporated in fluid solid-supported lipid bilayers was studied using quartz crystal microbalance with dissipation monitoring (QCM-D) (Fig. 3.3.2) (Rydell et al., 2009a). Both VLPs recognized bilayers containing H type-1, whereas no binding was observed to bilayers containing Le^a. For both VLPs, the concentration of H type-1 glycosphingolipid in the bilayer had to be above a threshold value for binding to occur, suggesting that a multivalent interaction is needed to stably attach the VLPs to the bilayer since the binding affinity of single ligands is expected to be low. Interestingly, the threshold concentration was one order of magnitude higher for the Dijon strain VLP, possibly suggesting the Norwalk strain to have a higher affinity for the H type-1 glycosphingolipid compared to the Dijon strain.

It may be calculated that the secretor α1,2 linked fucose of the H type 1 glycan cocrystallized with GII.4 VA387 (Cao et al., 2007) contributes to more than half of the glycan-protein binding energy (Koppisetty et al., 2010). Through molecular dynamics simulations, it has been suggested that human GII NoV may recognize fucosylated glycans either through secretor (α1,2 linked) or through the Lewis (α1,3/4 linked) fucose residues in the binding site (Nasir et al., 2012). The hypothesis was confirmed by crystal structures of Lewis antigens (SLe^x and Le^y) in complex with GII NoV (Hansman et al., 2011). These so called "secretor" and "Lewis" poses (Nasir et al., 2012) could have implications in binding complex branched structures of glycoproteins or glycosphingolipids in membranes (Fig. 3.3.3).

To further quantitatively characterize the virus membrane interactions and reveal the complex mathematical details of capsid protein to glycan interactions, Total Internal Reflection Fluorescent Microscopy (TIRF-M) based binding

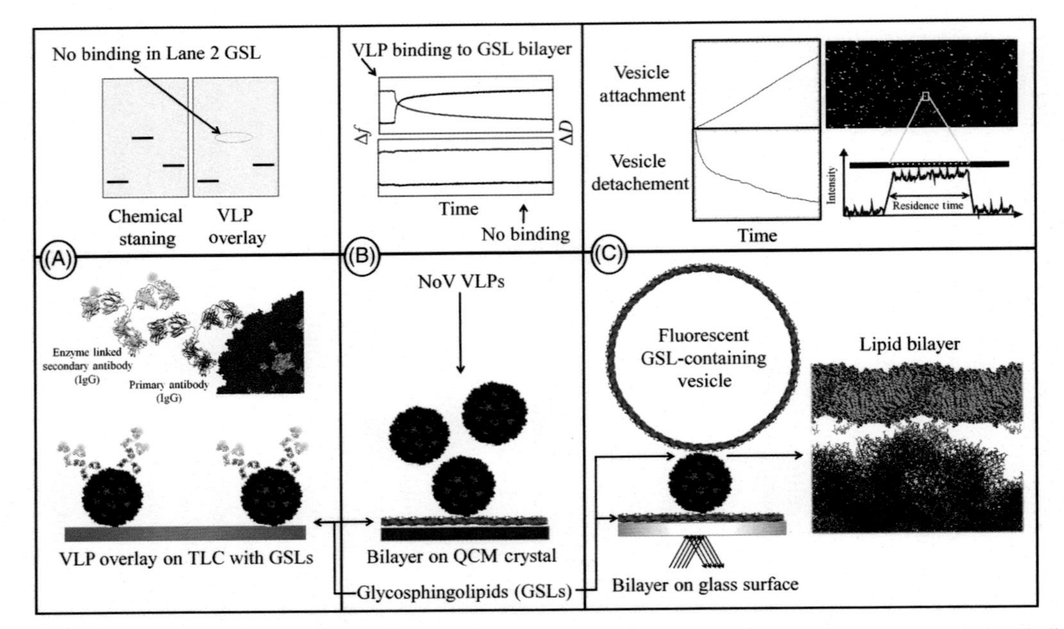

FIGURE 3.3.2 **Detection principles of surface based glycosphingolipid binding assays with nonlabeled VLPs.** (A) Chromatogram binding assay: qualitative assay detecting VLPs bound to glycosphingolipids, separated on thin-layer chromatograms *(TLC),* after incubations with primary antivirus antibody and secondary enzyme-linked antibodies for visualization by substrate conversion. (B) Quartz crystal microbalance with dissipation monitoring *(QCM-D).* The binding kinetics of attachment to membrane associated glycosphingolipids is probed in real time based on measurements of the bound mass (Δf) and the physicochemical properties of the attached object (ΔD) (Rydell et al., 2009a). (C) Total-internal-reflection fluorescence microscopy (TIRF-M). Attachment and detachment events of single vesicles, containing glycosphingolipids and a fluorescent lipid tag, interacting with VLPs bound to glycosphingolipids in a solid supported lipid bilayer. Fluorescence is registered in real time into a depth of <200 nm from the membrane due to the weak evanescent field (Bally et al., 2011). The method allows determination of the association and dissociation rates, providing a description of the complex binding kinetics (Nasir et al., 2015).

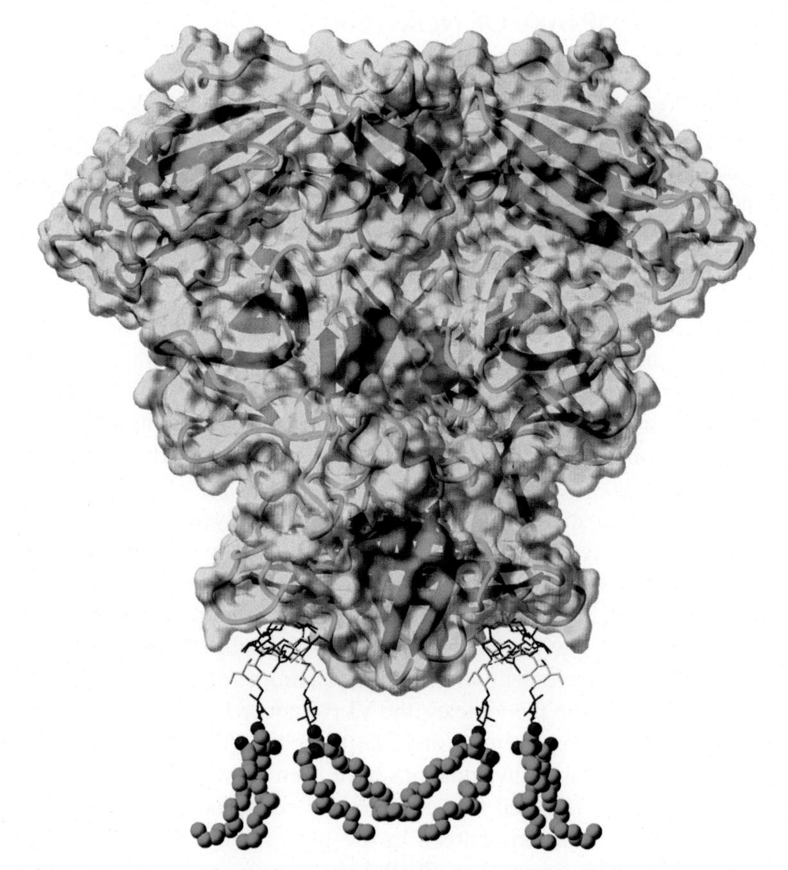

FIGURE 3.3.3 **Close up view of possible interactions of the NoV VA387 dimer with Lewis b glycosphingolipids.** The dimer and its ligand were obtained from the protein database id 2OBT (Cao et al., 2007) and the Lewis b glycosphingolipids were fitted, by fucose superimposition, into both the secretor and the Lewis poses at the two binding sites (Nasir et al., 2012). Although only two glycosphingolipids are expected to bind simultaneously to one dimer, the caption illustrates how different poses may affect the packing of glycosphingolipids in the membrane. In the secretor pose, with the secretor gene dependent α1,2-linked fucose *(red)* in the VLP binding site, the distance between the two glucose residues *(purple)*, glycosidically linked to the ceramide moiety of glycosphingolipids is ~ 40 Å, while in the Lewis pose, with the Lewis gene dependent α1,4-linked fucose *(red)* in the binding site, the distance is only ~27 Å. A virtual membrane bilayer simulation was not available at the time of generating the illustration.

assay was recently established (Fig. 3.3.2) (Bally et al., 2011, 2012a). This assay makes it possible to record the attachment and detachment events of single glycosphingolipid-containing vesicles to VLPs bound to fluid solid-supported lipid bilayers in real time (Fig. 3.3.2). Quantitative binding data describing the vesicle attachment to and detachment from Dijon (Bally et al., 2011) and Ast6139 GII.4 VLPs (Nasir et al., 2015) were obtained at both transient and steady state conditions using TIRF-M methodology.

8 CELLULAR UPTAKE OF NOROVIRUS

The lack of cell culture methods has made it very difficult to study NoV cell entry. However, a recent study showed replication of human NoV in cultured B lymphocytes in the presence of bacteria from the microbiota (Jones et al., 2014). The study suggested that the virus binds to HBGA expressing bacteria, which transport the virus into the cells. The significance of this mechanism for the pathogenesis of the infection remains to be determined. The low level of replication observed in cultured B cells is inconsistent with the high viral loads shed in feces of infected individuals. In addition, the mechanism does not explain the clinical resistance of nonsecretors. Thus, although they may be infected, B lymphocytes are unlikely the major cellular targets of human NoVs. In support of primary infection of epithelial cells, bovine NoV capsid protein is in the early period of the infection detected exclusively in enterocytes (Otto et al., 2011). At later time points the capsid protein is progressively detected in leucocytes of the lamina propria.

In an attempt to address the receptor function of glycosphingolipids, NoV interactions with glycosphingolipids embedded in giant unilamellar vesicles (GUVs) was studied using fluorescently labelled GII.4 VLPs and phospholipid vesicles containing glycosphingolipids and a fluorescent lipid (Rydell et al., 2013). The binding pattern to glycosphingolipids incorporated into GUVs was in full agreement with thin-layer chromatography (CBA) experiments. Upon binding to the vesicles, the VLPs induced the formation of membrane invaginations that were positive both for the fluorescent lipid and the VLP (Fig. 3.3.4). Similar invaginations have been shown to correspond to endocytosis intermediates used for cell entry for endogenous lectins, microbial toxins, viruses, and bacteria known to use glycosphingolipids for cell entry (Eierhoff et al., 2014; Ewers et al., 2010; Lakshminarayan et al., 2014; Roemer et al., 2007). The formation of the membrane invaginations does not require clathrin or any other cytosolic coat proteins. Instead, the multivalent binding of receptor glycosphingolipids induces the formation of lipid-nanodomains that

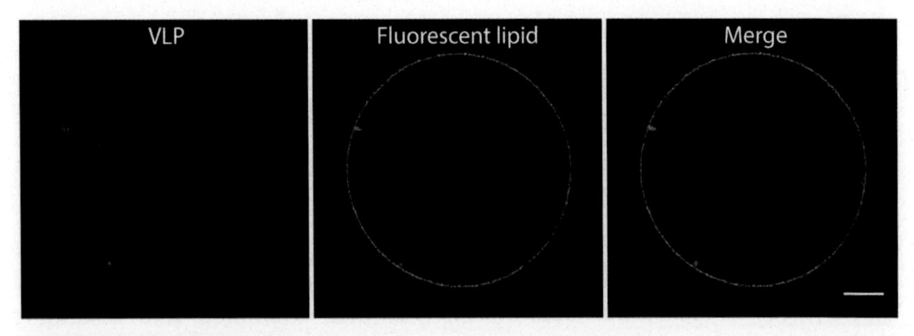

FIGURE 3.3.4 **Human Norovirus induces membrane invaginations upon binding to HBGA active glycosphingolipids.** Confocal image of a GII.4 VLP binding to ALe^b glycosphingolipids incorporated into a giant unilamellar vesicle (Rydell et al., 2013). Scale bar, 10 μm

spontaneously invaginate to form tubular structures. The mechanisms by which the membrane is bent is still not clear, but likely involves line tension, asymmetric compressive stress, a specific protein–lipid geometry, and the organization of lipids in specific orientation fields in the membrane (Aigal et al., 2015; Arnaud et al., 2014; Johannes et al., 2014). The observation that NoV VLP has an intrinsic property to induce membrane invagination on model membranes suggests that glycosphingolipids may be functional receptors also for human NoV. Some strains of murine NoV has indeed been shown to use gangliosides as receptors on macrophages (Taube et al., 2009). When complex glycosphingolipids were depleted from the cells the infectivity of the virus was markedly decreased. Furthermore, the infectivity could be rescued by addition of the ganglioside GD1a to the cells. In addition, the entry of murine NoV has been described to be independent of caveolin and clathrin, but dependent on cholesterol (Gerondopoulos et al., 2010; Perry and Wobus, 2010). These properties are characteristic also for the bacterial toxins and viruses that use the glycosphingolipid-dependent membrane invagination pathway to enter cells (Ewers et al., 2010; Roemer et al., 2007).

The apical surface of epithelial cells presents a large glycocalyx, up to 500 nm long in the duodenum that is decorated with HBGAs. It is tempting to speculate that human NoVs first attach to these motifs of the glycocalyx in order to reach their glycolipid targets that protrude a couple of nanometers only above the lipid bilayer (Fig. 3.3.2C). The HBGA-mediated glycocalyx binding would be a primary attachment and tethering step, whilst the glycosphingolipid binding would directly contribute to the entry process, as observed with the GUV experiments.

In conclusion, detailed understanding of the genetic coevolution and the cell-surface molecular interactions between enteric viruses and host receptors will hopefully contribute to the development of novel antiviral therapies and prevention of gastrointestinal infections.

REFERENCES

Ahmed, S.M., Hall, A.J., Robinson, A.E., et al., 2014. Global prevalence of norovirus in cases of gastroenteritis: a systematic review and meta-analysis. Lancet Infect Dis. 14, 725–730.

Aigal, S., Claudinon, J., Roemer, W., 2015. Plasma membrane reorganization: A glycolipid gateway for microbes. Biochim. Biophys. Acta 1853, 858–871.

Arnaud, J., Tröndle, K., Claudinon, J., et al., 2014. Membrane deformation by neolectins with engineered glycolipid binding sites. Angew. Chem. Int. Ed. 53, 9267–9270.

Bally, M., Dimitrievski, K., Larson, G., et al., 2012a. Interaction of virions with membrane glycolipids. Phys. Biol. 9, 026011.

Bally, M., Gunnarsson, A., Svensson, L., et al., 2011. Interaction of single viruslike particles with vesicles containing glycosphingolipids. Phys. Rev. Lett. 107, 188103.

Bally, M., Rydell, G.E., Zahn, R., et al., 2012b. Norovirus GII.4 virus-like particles recognize galactosylceramides in domains of planar supported lipid bilayers. Angew. Chem. Int. Ed. Engl. 51, 12020–12024.

Bok, K., Green, K.Y., 2012. Norovirus gastroenteritis in immunocompromised patients. N. Engl. J. Med. 367, 2126–2132.

Bu, W., Mamedova, A., Tan, M., et al., 2008. Structural basis for the receptor binding specificity of Norwalk virus. J. Virol. 82, 5340–5347.

Bucardo, F., Kindberg, E., Paniagua, M., et al., 2009. Genetic susceptibility to symptomatic norovirus infection in Nicaragua. J. Med. Virol. 81, 728–735.

Bucardo, F., et al., 2014. Predominance of norovirus and sapovirus in Nicaragua after implementation of universal rotavirus vaccination. PLoS One 9 (5), e98201.

Caddy, S., Breiman, A., le Pendu, J., et al., 2014. Genogroup IV and VI canine noroviruses interact with histo-blood group antigens. J. Virol. 88, 10377–10391.

Cao, S., Lou, Z., Tan, M., et al., 2007. Structural basis for the recognition of blood group trisaccharides by norovirus. J. Virol. 81, 5949–5957.

Carlsson, B., Kindberg, E., Buesa, J., et al., 2009. The G428A nonsense mutation in FUT2 provides strong but not absolute protection against symptomatic GII.4 Norovirus infection. PLoS One 4, e5593.

Chen, Y., Tan, M., Xia, M., et al., 2011. Crystallography of a Lewis-binding norovirus, elucidation of strain-specificity to the polymorphic human histo-blood group antigens. PLoS Pathog. 7, e1002152.

Choi, J.-M., Hutson, A.M., Estes, M.K., et al., 2008. Atomic resolution structural characterisation of recognition of histo-blood group antigens by Norwalk virus. Proc. Natl. Acad. Sci. USA 105, 9175–9180.

Currier, R.L., Payne, D.C., Staat, M.A., et al., 2015. Innate susceptibility to norovirus infections influenced by FUT2 genotype in a United States pediatric population. Clin. Infect. Dis. 60, 1631–1638.

de Rougemont, A., Ruvoën-Clouet, N., Simon, B., et al., 2011. Qualitative and quantitative analysis of the binding of GII.4 norovirus variants onto human blood group antigens. J. Virol. 85, 4057–4070.

Eierhoff, T., Bastian, B., Thuenauer, R., et al., 2014. A lipid zipper triggers bacterial invasion. Proc. Natl. Acad. Sci. U S A 111, 12895–12900.

Ewers, H., Roemer, W., Smith, A.E., et al., 2010. GM1 structure determines SV40-induced membrane invagination and infection. Nat. Cell. Biol. 12, 11–18.

Frenck, R., Bernstein, D.I., Xia, M., et al., 2012. Predicting susceptibility to norovirus GII.4 by use of a challenge model involving humans. J. Infect. Dis. 206, 1386–1393.

Fretz, R., Svoboda, P., Schorr, D., et al., 2005. Risk factors for infections with norovirus gastrointestinal illness in Switzerland. Eur. J. Clin. Microbiol. Infect. Dis. 24, 256–261.

Gerondopoulos, A., Jackson, T., Monaghan, P., et al., 2010. Murine norovirus-1 cell entry is mediated through a non-clathrin-, non-caveolae-, dynamin- and cholesterol-dependent pathway. J. Gen. Virol. 91, 1428–1438.

Glass, R.I., Parashar, U.D., Estes, M.K., 2009. Norovirus gastroenteritis. N. Engl. J. Med. 361, 1776–1785.

Gustavsson, L., Andersson, L.M., Lindh, M., et al., 2011. Excess mortality following community-onset norovirus enteritis in the elderly. J. Hosp. Infect. 79, 27–31.

Hall, A., 2012. Noroviruses: the perfect human pathogens? J. Infect. Dis. 205, 1622–1626.

Hall, A.J., Eisenbart, V.G., Etingue, A.L., et al., 2012. Epidemiology of foodborne norovirus outbreaks, United States, 2001-2008. Emerg. Infect. Dis. 18, 1566–1573.

Hall, A.J., Vinjé, J., Lopman, B., et al., 2011. Center for Disease Control and Prevention. Updated norovirus outbreak management and disease prevention guidelines. MMWR 60, 1–15.

Halperin, T., Vennema, H., Koopmans, M., et al., 2008. No association between histo-blood group antigens and susceptibility to clinical infections with genogroup II norovirus. J. Infect. Dis. 197, 63–65.

Han, L., Tan, M., Xia, M., et al., 2014. Gangliosides are ligands for human noroviruses. J. Am. Chem. Soc. 136, 12631–12637.

Hansman, G.S., Biertümpfel, C., Georgiev, I., et al., 2011. Crystal structures of GII.10 and GII. 12 norovirus protruding domains in complex with histo-blood group antigens reveal details for a potential site of vulnerability. J. Virol. 85, 6687–6701.

Harrington, P.R., Lindensmith, L., Yount, B., et al., 2002. Binding of Norwalk virus-like particles to ABH histo-blood group antigens is blocked by antisera from infected human volunteers or experimentally vaccinated mice. J. Virol. 76, 12325–12343.

Hemming, M., et al., 2013. Major reduction of rotavirus, but not norovirus, gastroenteritis in children seen in hospital after the introduction of RotaTeq vaccine into the National Immunization Programme in Finland. Eur. J. Pediatr. 172, 739–746.

Hu, L., Crawford, S.E., Czako, R., et al., 2012. Cell attachment protein VP8* of a human rotavirus specifically interacts with A-type histo-blood group antigen. Nature 485, 256–259.

Huang, P., Farkas, T., Zhong, W., et al., 2005. Norovirus and histo-blood group antigens: demonstration of a wide spectrum of strain specificities and classification of two major binding groups among multiple binding patterns. J. Virol. 79, 6714–6722.

Huang, P., Xia, M., Tan, M., et al., 2012. Spike protein VP8* of human rotavirus recognizes histo-blood group antigens in a type-specific manner. J. Virol. 86, 4833–4843.

Huang, P.W., Farkas, T., Marionneau, S., et al., 2003. Norwalk-like viruses bind to ABO, Lewis and secretor histo-blood group antigens but different strains bind to distinct antigens. J. Infect. Dis. 188, 19–31.

Hutson, A.M., Airaud, F., Le Pendu, J., et al., 2005. Norwalk virus infection associates with secretor status genotyped from sera. J. Med. Virol. 77, 116–120.

Imbert-Marcille, B.-M., Barbé, L., Dupé, M., et al., 2014. A FUT2 gene common polymorphism determines resistance to rotavirus A of the P [8] genotype. J. Infect. Dis. 209, 1227–1230.

Johannes, L., Wunder, C., Bassereau, P., 2014. Bending "on the rocks"—a cocktail of biophysical modules to build endocytic pathways. Cold Spring Harbor Perspect. Biol. 6, a016741.

Jones, M.K., Watanabe, M., Zhu, S., et al., 2014. Enteric bacteria promote human and mouse norovirus infection of B cells. Science 346, 755–759.

Kambhampati, A., Payne, D.C., Costantini, V., Lopman, B., 2016. Host genetic susceptibility to enteric viruses: a systematic review and metaanalysis. Clin. Infect. Dis. 62, 11–18.

Katayama, K., Murakami, K., Sharp, T.M., et al., 2014. Plasmid-based human norovirus reverse genetics system produces reporter-tagged progeny virus containing infectious genomic RNA. Proc. Natl. Acad. Sci. USA 111, E4043–E4052.

Kindberg, E., Akerlind, B., Johnsen, C., et al., 2007. Host genetic resistance to symptomatic norovirus (GGII.4) infections in Denmark. J. Clin. Microbiol. 45, 2720–2722.

Koo, H.L., et al., 2013. Noroviruses: the most common pediatric viral enteric pathogen at a large unviersity hospital after introduction of rotavirus vaccination. J. Pediatric Infect. Dis. Soc. 2 (1), 57–60.

Koppisetty, C.A., Nasir, W., Strino, F., et al., 2010. Computational studies on the interaction of ABO-active saccharides with the norovirus VA387 capsid protein can explain experimental binding data. J. Comput. Aided Mol. Des. 24, 423–431.

Lakshminarayan, R., Wunder, C., Becken, U., et al., 2014. Galectin-3 drives glycosphingolipid-dependent biogenesis of clathrin-independent carriers. Nat. Cell. Biol. 16, 595–606.

Larsson, M.M., Rydell, G.E., Grahn, A., et al., 2006. Antibody prevalence and titer to norovirus (genogroup II) correlate with secretor (FUT2) but not with ABO phenotype or Lewis (FUT3) genotype. J. Infect. Dis. 194, 1422–1427.

Lindesmith, L., Moe, C., Lependu, J., et al., 2005. Cellular and humoral immunity following Snow Mountain virus challenge. J. Virol. 79, 2900–2909.

Lindesmith, L., Moe, C.L., Marionneau, S., et al., 2003. Human susceptibility and resistance to Norwalk virus infection. Nat. Med. 9, 548–553.

Lindesmith, L.C., Donaldson, E.F., Lobue, A.D., et al., 2008. Mechanisms of GII. 4 norovirus persistence in human populations. PLoS Med. 5, e31.

Lopman, B.A., Trivedi, T., Vicuna, Y., et al., 2015. Norovirus infection and disease in an Ecuadorian birth cohort: Association of certain norovirus genotypes with host FUT2 secretor status. J. Infect. Dis. 211, 1813–1821.

Lowe, J.B., 1993. The blood group-specific human glycosyltransferases. Baillieres Clin. Haematol. 6, 465–492.

Macher, B.A., Galili, U., 2008. The Gal⟨1,3Gal®1,4GlcNAc-R (⟨-Gal) epitope: A carbohydrate of unique evolution and clinical relevance. Biochem. Biophys. Acta 1780, 75–88.

Marionneau, S., Cailleau-Thomas, A., Rocher, J., et al., 2001. ABH and Lewis histo-blood group antigens, a model for the meaning of oligosaccharide diversity in the face of a changing world. Biochimie 83, 565–573.

Marionneau, S., Ruvoen, N., Le Moullac-Vaidye, B., et al., 2002. Norwalk virus binds to histo-blood group antigens present on gastroduodenal epithelial cells of secretor individuals. Gastroenterology 122, 1967–1977.

Martinez, M.A., Lopez, S., Arias, C.F., et al., 2013. Gangliosides have a functional role during rotavirus cell entry. J. Virol. 87, 1115–1122.

Mesquita, J.R., Barclay, L., Nascimento, M.S., et al., 2010. Novel norovirus in dogs with diarrhea. Emerg. Infect. Dis. 16, 980–982.

Meyer, E., Ebner, W., Scholz, R., et al., 2004. Nosocomial outbreak of norovirus gastroenteritis and investigation of ABO histo-blood group type in infected staff and patients. J. Hosp. Infect. 56, 64–66.

Miyoshi, M., Yoshizumi, S., Sato, C., et al., 2005. Relationship between ABO histo-blood group type and an outbreak of norovirus gastroenteritis among primary and junior high school students: results of questionnaire-based study. Kansenshogaku Zasshi 79, 664–671.

Nasir, W., Bally, M., Zhdanov, V.P., et al., 2015. Interaction of virus-like particles with vesicles containing glycolipids: Kinetics of detachment. J. Phys. Chem. B 119, 11466–11472.

Nasir, W., Frank, M., Koppisetty, C.A., et al., 2012. Lewis histo-blood group alpha1,3/alpha1,4 fucose residues may both mediate binding to GII.4 noroviruses. Glycobiology 22, 1163–1172.

Neu, U., Stehle, T., Atwood, W.J., The Polyomaviridae:, 2009. Contributions of virus structure to our understanding of virus receptors and infectious entry. Virology 384, 389–399.

Nilsson, J., Rydell, G.E., Le Pendu, J., et al., 2009. Norwalk virus-like particles bind specifically to A, H and difucosylated Lewis but not to B histo-blood group active glycosphingolipids. Glycoconj. J. 26 (9), 1171–1180.

Nordgren, J., Kindberg, E., Lindgren, P.E., et al., 2010. Norovirus gastroenteritis outbreak with a secretor-independent susceptibility pattern, Sweden. Emerg. Infect. Dis. 16, 81–87.

Nordgren, J., Nitiema, L.W., Ouermi, D., et al., 2013. Host genetic factors affect susceptibility to norovirus infections in Burkina Faso. PLoS One 8, e69557.

Nordgren, J., Sharma, S., Bucardo, F., et al., 2014. Both Lewis and secretor status mediate susceptibility to rotavirus infections in a rotavirus genotype dependent manner. Clin. Infect. Dis. 59, 1567–1573.

Otto, P.H., Clarke, I.N., Lambden, P.R., et al., 2011. Infection of calves with bovine norovirus GIII.1 strain Jena virus: an experimental model to study the pathogenesis of norovirus infection. J. Virol. 85, 12013–12021.

Patel, M.M., Hall, A.J., Vinje, J., et al., 2009. Noroviruses: a comprehensive review. J. Clin. Virol. 44, 1–8.

Payne, D.C., et al., 2013. Norovirus and medically attended gastroenteritis in U.S. children. N Engl. J. Med. 368 (12), 1121–1130.

Perry, J.W., Wobus, C.E., 2010. Endocytosis of murine norovirus 1 into murine macrophages is dependent on dynamin II and cholesterol. J. Virol. 84, 6163–6176.

Rockx, B.H., Vennema, H., Hoebe, C.J., et al., 2005. Association of histo-blood group antigens and susceptibility to norovirus infections. J. Infect. Dis. 191, 749–754.

Roemer, W., Berland, L., Chambon, V., et al., 2007. Shiga toxin induces tubular membrane invaginations for its uptake into cells. Nature 450, 670–675.

Ruvoën-Clouet, N., Belliot, G., Le Pendu, J., 2013. Noroviruses and histo-blood groups: the impact of common host genetic polymorphisms on virus transmission and evolution. Rev. Med. Virol. 23, 355–366.

Ruvoën-Clouet, N., Ganière, J.P., André-Fontaine, G., et al., 2000. Binding of Rabbit Hemorrhagic Disease Virus to antigens of the ABH histo-blood group family. J. Virol. 74, 11950–11954.

Ruvoën-Clouet, N., Magalhaes, A., Marcos-Silva, L., et al., 2014. Increase in genogroup II.4 norovirus host spectrum by CagA-positive Helicobacter pylori infection. J. Infect. Dis. 210, 183–191.

Rydell, G.E., Dahlin, A.B., Hook, F., et al., 2009a. QCM-D studies of human norovirus VLPs binding to glycosphingolipids in supported lipid bilayers reveal strain-specific characteristics. Glycobiology 19, 1176–1184.

Rydell, G.E., Nilsson, J., Rodriguez-Diaz, J., et al., 2009b. Human noroviruses recognize sialyl Lewis x neoglycoprotein. Glycobiology 19, 309–320.

Rydell, G.E., Svensson, L., Larson, G., et al., 2013. Human GII.4 norovirus VLP induces membrane invaginations on giant unilamellar vesicles containing secretor gene dependent alpha1,2-fucosylated glycosphingolipids. Biochim. Biophys. Acta 1828, 1840–1845.

Schmidt, M., Chiorini, J.A., 2006. Gangliosides are essential for bovine adeno-associated virus entry. J. Virol. 80, 5516–5522.

Shirato, H., Ogawa, S., Ito, H., et al., 2008. Noroviruses distinguish between type 1 and type 2 histo-blood group antigens for binding. J. Virol. 82, 10756–10767.

Storry, J.R., Olsson, M.L., 2004. Genetic basis of blood group diversity. Br. J. Haematol. 126, 759–771.

Takanashi, S., Wang, Q., Chen, N., et al., 2011. Characterization of emerging GII.g/GII.12 noroviruses from a gastroenteritis outbreak in the United States in 2010. J. Clin. Microbiol. 49, 3234–3244.

Tamura, M., Natori, K., Kobayashi, M., et al., 2004. Genogroup II noroviruses efficiently bind to heparan sulfate proteoglycan associated with the cellular membrane. J. Virol. 78, 3817–3826.

Tan, M., Jiang, X., 2011. Norovirus-host interaction: Multi-selections by human histo-blood group antigens. Trends Microbiol. 19, 382–388.

Tan, M., Jiang, X., 2014. Histo-blood group antigens: a common niche for norovirus and rotavirus. Expert Rev. Mol. Med. 16, e5.

Tan, M., Jin, M., Xie, H., et al., 2008. Outbreak studies of a GII-3 and a GII-4 norovirus revealed an association between HBGA phenotypes and viral infection. J. Med. Virol. 80, 1296–1301.

Taube, S., Perry, J.W., Yetming, K., et al., 2009. Ganglioside-linked terminal sialic acid moieties on murine macrophages function as attachment receptors for murine noroviruses. J. Virol. 83, 4092–4101.

Thorven, M., Grahn, A., Hedlund, K.O., et al., 2005. A homozygous nonsense mutation (428G– > A) in the human secretor (FUT2) gene provides resistance to symptomatic norovirus (GGII) infections. J. Virol. 79, 15351–15355.

Verdaguer, N., Ferrero, D., Murthy, M.R., 2014. Viruses and viral proteins. IUCrJ 1, 492–504.
Vinjé, J., 2015. Advances in laboratory methods for detection and typing of norovirus. J. Clin. Microbiol. 53, 373–381.
Yazawa, S., Yokobori, T., Ueta, G., et al., 2014. Blood group substances as potential therapeutic agents for the prevention and treatment of infection with noroviruses proving novel binding patterns in human tissues. PLoS One 9, e89071.
Zakhour, M., Maalouf, H., Di Bartolo, I., et al., 2010. Bovine norovirus: carbohydrate ligand, environmental contamination, and potential cross-species transmission via oysters. Appl. Environ. Microbiol. 76, 6404–6411.
Zakhour, M., Ruvoën-Clouet, N., Charpilienne, A., et al., 2009. The alphaGal epitope of the histo-blood group antigen family is a ligand for bovine norovirus Newbury2 expected to prevent cross-species transmission. PLoS Pathog. 5, e1000504.
Zheng, D.P., Ando, T., Fankhauser, R.L., et al., 2006. Norovirus classification and proposed strain nomenclature. Virology 346, 312–323.

Chapter 3.4

Animal Models of Norovirus Infection

C.E. Wobus*, J.B. Cunha*, M.D. Elftman, A.O. Kolawole***
**Department of Microbiology and Immunology, University of Michigan Medical School, Ann Arbor, MI, United States; **Department of Biomedical and Diagnostic Sciences, University of Detroit Mercy, School of Dentistry, Detroit, MI, United States*

1 INTRODUCTION

Advances in the understanding of human noroviruses (HuNoVs) are challenging because, despite being such ubiquitous pathogens, these viruses cause species-specific infections and, until very recently (Jones et al., 2014), have not been successfully cultured in vitro. The use of animal models to address questions that are difficult or impossible to answer in human studies is a common approach to complement in vitro and epidemiological data for human viruses in general. Animal models are based either on studies of human viruses in animals, often requiring the use of immunocompromised hosts and/or host-adapted viruses, or the use of related viruses in their natural hosts. As outlined in this chapter, both approaches have been used to study HuNoVs.

Human volunteer-based studies and epidemiologic analyses have been invaluable in determining features of HuNoV infection and disease (Vashist et al., 2009; Kaufman et al., 2014). These are: (1) short incubation time (average 1 day) (Lee et al., 2013); (2) quick (1–4 days) resolution of disease symptoms, characterized mainly by diarrhea and/or vomiting (Kaufman et al., 2014; Atmar et al., 2008); (3) viral shedding for weeks to months with or without disease symptoms (Kaufman et al., 2014; Teunis et al., 2014); (4) viral spread mainly by the fecal–oral route and person-to-person transmission (Bitler et al., 2013), (5) histo-blood group antigen (HBGA) expression and secretor status as genetic susceptibility factors (Lindesmith et al., 2003; Hutson et al., 2002, 2005; Tan and Jiang, 2014). However, studies relying on human subjects have ethical limitations and require vast resources, therefore preventing them from being routinely conducted (Vashist et al., 2009). Therefore, animal models for HuNoV infection, as well as HuNoV surrogates, are valuable and necessary tools that can elucidate basic aspects of HuNoV infection and pathogenesis (Karst et al., 2014; Kniel, 2014).

Viral Gastroenteritis

Ideally, the best HuNoV animal model would use HuNoV or a virus with high genetic similarity that closely mirrors the biology and clinical features of HuNoV infection in an animal host. However, HuNoVs are members of the genetically diverse *Norovirus* genus within the *Caliciviridae* family that exhibit very narrow species specificity. Thus, HuNoV infection in a nonnative host is limited to some closely related primates (chimpanzees, newborn pigtail macaques), immunocompromised animals (ie, mice), and gnotobiotic (Gn) pigs and calves, albeit with varying levels of clinical signs (Bok et al., 2011; Cheetham et al., 2006; Souza et al., 2008; Taube et al., 2013). Alternatively, many studies have been carried out using related caliciviruses as HuNoV surrogate models. Historically, feline calicivirus (FCV) was the first surrogate for human caliciviruses due to the availability of an established cell culture system, and its study provided an early understanding of calicivirus biology (Radford et al., 2004). However, FCV causes upper respiratory tract infections in cats (Radford et al., 2007), and therefore has limited utility as a model for gastroenteritis and will not be discussed further. In contrast, porcine sapovirus (PoSaV) causes diarrhea in swine and can be propagated in culture in the presence of bile acids (Flynn and Saif, 1988), providing an animal model of calicivirus gastroenteritis. However, it was not until the discovery of murine norovirus 1 (MNV-1), a norovirus that infects a small animal species (Karst et al., 2003) and the subsequent development of a cell culture model (Wobus et al., 2004), that a HuNoV surrogate became widely available for both in vitro and in vivo studies (Wobus et al., 2006). MNV has been instrumental in elucidating many basic aspects of NoV infection and pathogenesis (Karst et al., 2003; Gonzalez-Hernandez et al., 2013, 2014; Taube et al., 2012, 2009; Chachu et al., 2008). More recently, Tulane virus (TV) was discovered as a causative agent of diarrhea in rhesus macaques (Sestak et al., 2012). Combined with the ability to readily grow in cell culture (Farkas et al., 2008) and availability of an infectious clone (Wei et al., 2008), TV is another important HuNoV surrogate (Farkas, 2015). In this chapter, the characteristics of HuNoV animal and surrogate models under investigation in the field will be discussed, including their advantages and disadvantages. We will also point out contributions of each model to our collective understanding of NoV biology and therapeutic approaches for HuNoVs.

2 ANIMAL MODELS OF HuNoV INFECTION

To date, a limited number of HuNoV strains was shown to infect mice deficient in the recombination activating gene (Rag) and the common gamma chain (γc) (Taube et al., 2013), gnotobiotic pigs (Souza et al., 2007a), gnotobiotic calves (Souza et al., 2008), macaques (Subekti et al., 2002), and chimpanzees (Bok et al., 2011). So far, no single animal model recapitulates all aspects of HuNoV infection. As outlined later, each model has strengths to allow the study of specific aspects of HuNoV biology (Table 3.4.1).

TABLE 3.4.1 Summary of the Discussed HuNoV Animal Models With Their Strengths and Limitations

Host	Virus	Strengths	Limitations
Balb/c Rag2/$\gamma c^{-/-}$ mouse (Taube et al., 2013)	HuNoV (GII.4)	Genetically tractable host Relatively low cost	Immune defects in host Transient infection Not orally susceptible Asymptomatic infection Lack of fecal shedding Not natural host
Chimpanzee (Bok et al., 2011)	HuNoV (GI.1, untyped strain)	Genetically similar to human Orally susceptible Transmission between animals Similar shedding and antibody responses to humans	Asymptomatic infection High cost Not natural host Moratorium on use
Newborn pigtail macaques (Subekti et al., 2002)	HuNoV (GII.3)	Orally susceptible Symptomatic infection Fecal shedding Similar serum IgG responses to humans Transmission between animals	High cost Unknown IgA responses Not natural host
Gnotobiotic pigs (Cheetham et al., 2006; Souza et al., 2007a; Bui et al., 2013; Takanashi et al., 2011; Cheetham et al., 2007)	HuNoV (GII.4, GII.12)	Symptomatic and asymptomatic infections Orally susceptible Fecal shedding Susceptibility partially mediated by HBGAs Similar incubation period, duration of symptoms to humans Mild villous atrophy Similar cytokine, antibody responses to humans	High cost Limited ability to passage Lack of microbiota Underdeveloped immune system Not natural host
Gnotobiotic calves (Souza et al., 2008)	HuNoV (GII.4)	Symptomatic infection Orally susceptible Fecal shedding Villous atrophy	High cost Lack of microbiota Underdeveloped immune system Not natural host

TABLE 3.4.2 Summary of the Discussed HuNoV Surrogates With Their Strengths and Limitations

Host	Virus	Strengths	Limitations
Wild-type mouse (Wobus et al., 2006; Chang et al., 2005; Farkas et al., 2012)	MNV (all strains)	Natural host Fecal shedding Orally susceptible Genetically tractable host Relatively low cost Similar histopathology to humans (villous blunting, mild intestinal inflammation) Available cell culture system Available reverse genetics system	Asymptomatic infection Does not bind HBGAs Relatively limited genetic diversity of MNV strains
STAT1-/-AG129 mouse (Karst et al., 2003; Hsu et al., 2006)	MNV (eg, MNV-1, MNV-3, CR3)	Orally susceptible Symptomatic infection Fecal shedding Relatively low cost Available cell culture system Available reverse genetics system	Immune deficiency Systemic infection Exacerbated histopathology Does not bind HBGAs Relatively limited genetic diversity
Rhesus macaques (Sestak et al., 2012; Farkas, 2015)	Recoviruses/ Tulane virus	Natural host Orally susceptible Symptomatic infection Virus binds HBGAs Large genetic diversity Genomic organization similar to HuNoVs Fecal shedding Similar histopathology to humans (villous blunting, mild intestinal inflammation) Transient neutralizing antibody response like humans Available cell culture system Available reverse genetics system	Not a norovirus High cost
Gnotobiotic pigs (Rydell et al., 2009; Scheuer et al., 2013)	PoSaV	Natural host Orally susceptible Symptomatic infection Fecal shedding Villous atrophy and blunting Available cell culture system Available reverse genetics system	Not a norovirus High cost Lack of microbiota Underdeveloped immune system Does not bind HBGAs

2.1 Mice as an Animal Model

The first small animal model to study HuNoV infection was described in 2013 (Taube et al., 2013). Balb/c mice deficient in recombination activating genes (Rag$^{-/-}$) 1 or 2 and common gamma chain ($\gamma c^{-/-}$) (Rag$^{-/-}$/$\gamma c^{-/-}$) engrafted with human CD34+ hematopoietic stem cells (humanized mice), nonengrafted siblings, and immunocompetent wild type controls were challenged with a pool of HuNoV strains (Fig. 3.4.1). Viral genome titers were measured by quantitative polymerase chain reaction preceded by reverse transcription (RT-qPCR) in intestinal and nonintestinal tissues. Infection was measured by detecting increased genome titers over input in both humanized and nonhumanized Balb/c Rag$^{-/-}$/$\gamma c^{-/-}$ mice. Viral genomes were present in all tissues analyzed, except the brain. However, infection was asymptomatic and transient (lasting 2 days). Immunocompetent wild type Balb/c or B6/B10 Rag$^{-/-}$/$\gamma c^{-/-}$ mice were not infected, suggesting both the immune deficiency and genetic background of the mouse as determinants of susceptibility in this model. Unlike natural HuNoV infection, mice were susceptible only via the intraperitoneal, not the oral route,

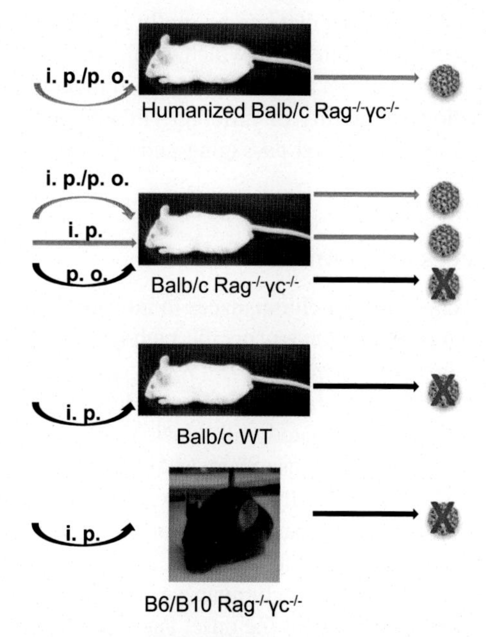

FIGURE 3.4.1 Human Norovirus Mouse Model. Balb/c mice lacking the recombination activation genes (Rag) 1 or 2 and common gamma chain (γc) (Rag$^{-/-}\gamma c^{-/-}$) were humanized with CD34+ hematopoietic stem cells. Humanized and nonhumanized Balb/c Rag$^{-/-}\gamma c^{-/-}$, wild-type (WT) Balb/c controls and B6/B10 Rag$^{-/-}\gamma c^{-/-}$ mice were challenged with a pool of HuNoV strains by intraperitoneal *(i.p.)* and/or oral *(p.o.)* routes. Viral infection was determined by detecting increased viral genome titers over input and viral protein expression by immunohistochemistry. Productive infection routes are indicated by green arrows and presence of a virion, while nonproductive infections are indicated by black arrows and crossed-out virions. *(Source: Summary of data from Taube et al., 2013.)*

and no shedding of virus in the stool was detected. The former is likely a result of the genetic defects of this mouse strain, which lacks Peyer's patches and associated M cells, a known gateway for NoV entry into the host (Gonzalez-Hernandez et al., 2014). Expression of structural and/or nonstructural proteins in cells with macrophage-like morphology in the spleen, small intestinal lamina propria, and Kupffer cells in the liver of Balb/c Rag$^{-/-}$/$\gamma c^{-/-}$, but not of mock-infected mice, verified replication of HuNoV.

A mouse model of HuNoV infection can facilitate translational and basic science applications, such as drug development and identification of viral and host factors mediating infection. The advantages of a mouse model as opposed to large animal models are manifold, including reduced cost, shorter gestation length, and the ability to genetically manipulate the host. However, the necessity for deleting components of the immune system to permit HuNoV to cross the species barrier represents an inherent caveat that limits the use of this model for certain applications. For example, vaccine testing is not possible due to the absence of T and B cells in Rag-deficient animals. Therefore, the identification of mouse strains with other immune deficiencies that permit HuNoV infection could expand the utility of the mouse model. Furthermore, the discovery of different permissive mouse strains that can be orally infected, shed virus in feces and have a longer duration of infection, which would improve this model. Some of the open questions to address are the minimum number of particles required to initiate an infection, the ability of different HuNoV genotypes to cause an infection, and the ability for serial passaging and adaptation of HuNoV in the murine host.

2.2 Chimpanzees as an Animal Model

The close genetic relatedness of chimpanzees to humans led to the use of chimpanzees as animal model for human-specific pathogens. HuNoV infections of chimpanzees were first performed in 1978 (Wyatt et al., 1978). Although none of the chimpanzees challenged orally with filtered HuNoV-containing fecal samples developed symptoms associated with HuNoV infection, infected animals developed NoV-specific antibody responses, and viral antigen was detected in feces. Importantly, HuNoV was serially passaged via feces to additional uninfected animals, demonstrating that chimpanzees support asymptomatic HuNoV infection. This finding was confirmed a few decades later when Bok et al. (2011) intravenously inoculated seronegative chimpanzees with HuNoV. No clinical signs of gastroenteritis were observed but the onset and duration of virus shedding in stool and serum antibody responses were similar to those observed in humans (Fig. 3.4.2). HuNoV capsid protein was detected in intestinal lamina propria cells from the B cell and dendritic cell lineages, and NoV RNA was found in intestinal and liver biopsies. Two infected chimpanzees were protected against re-infection of homologous HuNoV when challenged either at 4 and 24 months, or at 10 months after the first infection. In addition, chimpanzees vaccinated

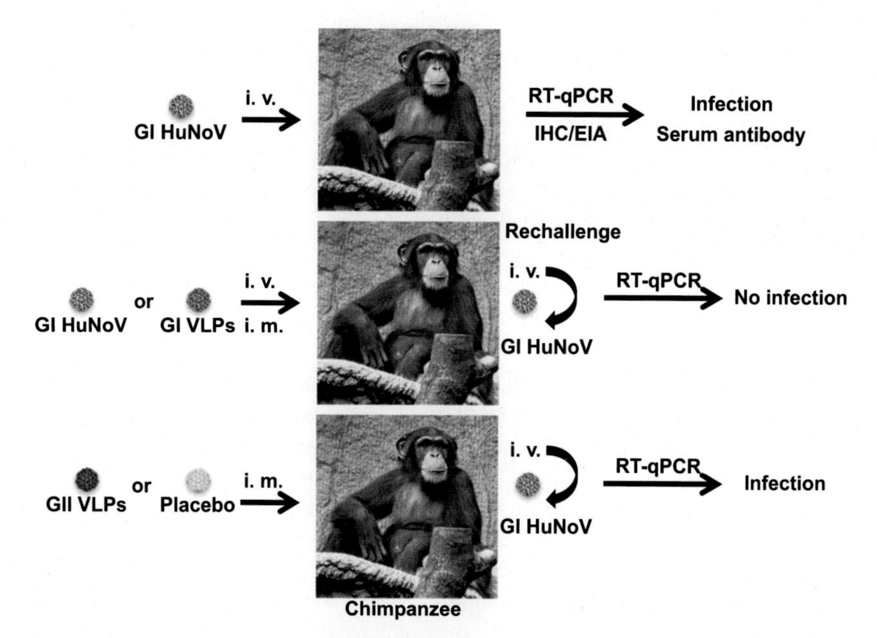

FIGURE 3.4.2 Human Norovirus Chimpanzee Model. Seronegative chimpanzees were intravenously *(i.v.)* inoculated with genogroup I (GI) HuNoV (Norwalk virus). Virus replication was measured by RT-qPCR, virus protein by IHC and serum antibody by EIA. Infected chimpanzees were protected against re-infection with homologous HuNoV when rechallenged after the first infection or vaccinated intramuscularly *(i.m.)* with GI (Norwalk virus) VLPs. GII VLPs and placebo did not protect from challenge with GI HuNoV, thus providing important information regarding the ability of VLP-based vaccination to protect from homologous but not heterologous, cross-genotype challenge. *(Source: Schematic summary of data from Bok et al., 2011.)*

intramuscularly with Norwalk virus (GI) virus-like particles (VLPs), but not those derived from GII VLPs, were protected from homologous Norwalk virus infection 18 months after vaccination and developed antibodies that blocked VLP binding to HBGAs in a surrogate neutralization assay. Although this study did not use the natural route of infection, it provided important information regarding the ability of VLP-based vaccination to protect from homologous but not heterologous, cross-genogroup challenge. Protection from homologous challenge is also observed in human clinical trials (Atmar et al., 2011), while vaccine efficacy after heterologous challenge remains to be studied in humans. These data demonstrate the utility of chimpanzees as a valuable infection model for HuNoV but the NIH moratorium on the use of chimpanzees for biomedical research has halted any future NIH-funded studies in this HuNoV animal model.

2.3 Macaques as an Animal Model

Nonhuman primates other than chimpanzees may provide an alternative HuNoV animal model but only limited infections in a few species have been performed to date. The susceptibility of common marmosets, cotton top tamarins, cynomolgus, and rhesus macaques to oral HuNoV infection was tested in one study (Rockx et al., 2005). Only rhesus macaques were found susceptible to Norwalk virus infection, albeit asymptomatically, with virus shedding in some animals and development of NoV-specific IgG and IgM, but not IgA, antibodies. While the asymptomatic infection status and duration of shedding resembles that found in humans (Graham et al., 1994; Rockx et al., 2002), the lack of IgA responses limits the use of this species as a vaccine model.

More promising results were observed in a study which showed newborn pigtail macaques (*Macaca nemestrina*) to be symptomatically infected with Toronto virus P2-A (a GII.3 HuNoV strain) (Subekti et al., 2002). Two of three monkeys developed diarrhea and dehydration. All animals developed IgG antibody responses and shed virus in stool for at least 3 weeks. Similar development of disease and shedding results were obtained following passage into additional newborn macaques. In addition, the mothers of the infected infant monkeys became symptomatically infected via natural infection, demonstrating the ability of HuNoV to be transmitted and passaged in this species of monkeys. This promising first study warrants further development of *Macaca nemestrina* as a HuNoV disease and transmission model.

2.4 Gnotobiotic Pigs and Calves as Animal Models

Gnotobiotic (Gn) pigs are models for enteric diseases because their gastrointestinal anatomy and physiology are similar to that of humans (Saif et al., 1996). Gn pigs support symptomatic HuNoV infection characterized by mild diarrhea, transient viremia, viral shedding in feces, and the presence of HuNoV structural and nonstructural proteins in enterocytes (Cheetham et al., 2006) (Fig. 3.4.3). In one study (Cheetham et al., 2006), Gn pigs were orally inoculated with fecal filtrates of the NoV/GII/4/HS66/2001/US strain, or with pig-passaged intestinal

FIGURE 3.4.3 **Human Norovirus Gnotobiotic Pig Model.** Gnotobiotic pigs support symptomatic HuNoV infection characterized by mild diarrhea, transient viremia, viral shedding in feces, and detection of HuNoV structural and nonstructural proteins in enterocytes by immunofluorescence microscopy. *(Source: Schematic representation of findings from Cheetham et al., 2006 and Souza et al., 2007a.)*

contents. Two thirds of the inoculated animals developed mild diarrhea, 57% seroconverted, but only 44% of pigs shed low levels of virus in their feces, suggesting disease symptoms can develop without fecal shedding or seroconversion. Using different virus strains [a GII.4 2006b variant (Bui et al., 2013) or the GII.g/GII.12 HS206 strain (Takanashi et al., 2011)], more infected animals shed virus than developed diarrhea, demonstrating asymptomatic infections can occur in Gn pigs. Increases in genome titers over input were detected in most (18/22) pigs (Bui et al., 2013), confirming productive infection. The short incubation period (24–48 h) and duration of disease symptoms (1–3 days) or asymptomatic infection resemble clinical features of HuNoV infection in humans (Lee et al., 2013; Lindesmith et al., 2003; Lopman et al., 2004). To date, only GII.4 and GII.12 strains have been reported to infect pigs, but further studies are needed to test susceptibility of Gn pigs to additional genogroups/genotypes.

Susceptibility factors of HuNoV infection in the human population are secretor status and HBGA type (Lindesmith et al., 2003; Hutson et al., 2002). A similar correlation between infection outcome and HBGA expression was seen in Gn pigs, wherein increased incidence of diarrhea, fecal shedding, and seroconversion rate were observed in pigs with the A+/H+ genotype compared to the A−/H− genotype (Cheetham et al., 2007).

Histopathologic examination of HuNoV infected Gn pigs detected mild villous atrophy in the duodenum of one out of seven animals and an increased number of apoptotic cells in intestinal villi of all infected animals (Cheetham et al., 2006), features also reported during HuNoV infection (Troeger et al., 2009). Structural or nonstructural HuNoV antigens were detected in enterocytes of the proximal small intestine, suggesting HuNoV replication occurs in these cells, but whether additional cell types also support infection in this model remains unknown. The tropism for enterocytes differs from the recently described tropism of HuNoV for B cells in vitro (Jones et al., 2014), but the cellular tropism of HuNoV in infected human intestines remains undefined (Lay et al., 2010).

The power of animals models for pathogenesis studies is highlighted by a study where naïve Gn pigs were infected with the HuNoV HS66 GII.4 strain, and serum and intestinal contents were analyzed for intestinal and systemic humoral and cellular immune responses (Souza et al., 2007a). Such studies are difficult to perform in human volunteers with unknown and variable NoV-exposure history. Infection induced low antibody titers and antibody-secreting cell numbers, which may explain the incomplete seroconversion rate observed in pigs. However, the kinetics of IgA and IgG antibody-secreting cell responses and the greater induction of a Th1 compared to Th2 and proinflammatory IL-6 responses are also observed in HuNoV-infected volunteers (Lindesmith et al., 2003; Ramani et al., 2015).

Animal models of disease are important for the development of therapeutics and Gn pigs have been evaluated as test models for both vaccines and antivirals. Similar to vaccine trials in humans (Atmar et al., 2011; Bernstein et al., 2015; Treanor et al., 2014), HuNoV VLPs are immunogenic in Gn pigs and induce homologous protection (Souza et al., 2007b). In an additional study, Gn pigs were vaccinated with VLPs and P particles (globular structures formed by multimerization of 12–36 copies of the P domain (Bereszczak et al., 2012)) and challenged with a closely related GII HuNoV. Primary NoV infection protected 83% of challenged animals against diarrhea and 57% against virus shedding. In contrast, VLP or P particles provided protection against diarrhea in 60% or 47%, respectively, and against virus shedding in 0% or 11%, respectively (Kocher et al., 2014). These studies suggest that a replicating vaccine (eg, attenuated virus) should be pursued as an alternative vaccination approach as it will likely confer superior protection to the recombinant subunit vaccine of VLPs or P particles.

The use of Gn pigs as an antiviral efficacy test model for HuNoV infection is only just beginning but will increase as more compounds move through the drug development pipeline (Arias et al., 2013; Rocha-Pereira et al., 2014). One study (Jung et al., 2012) looked at interferon (IFN) α an important innate immune mediator with proven antiviral activity in humans (Feld and Hoofnagle, 2005). Treatment of Gn pigs with IFNα decreased viral shedding with a delayed onset and shorter duration of shedding, but viral shedding increased after treatment was discontinued (Jung et al., 2012). The inability of type I IFNs (IFNα/β) to clear NoV infection has also been seen in mice (Nice et al., 2014). Taking into account the known side effects of IFNα treatment in humans (Gota and Calabrese, 2003), IFNα will unlikely become a widely used treatment option for HuNoV infection in humans, and more specific anti-NoV drugs will need to be developed. Another area of future development will be studies regarding the utility of Gn pigs as a transmission model. Although the HS66 strain was serially passaged three times through Gn pigs with evidence of virus shedding at each passage and evidence of HuNoV-infected cells, viral titers and duration of fecal shedding decreased with each passage and no evidence of mutations was observed in the region analyzed (ie, ORF1/2 junction) (Cheetham et al., 2006). Thus, whether adaptive mutations occur in the HuNoV genome during infection in pigs remains to be addressed.

Additional Gn pig studies demonstrated that susceptibility to HuNoV infection changes with age, with lower median infectious dose (ID_{50}) in neonates (4–5 days old) compared to older (33–34 days old) pigs (Bui et al., 2013). Whether susceptibility similarly changes with age in humans remains unknown. However, Gn pig studies (Bui et al., 2013; Jung et al., 2012) did confirm earlier observations in humans that the use of statins (cholesterol lowering drugs) exacerbates HuNoV disease severity (Rondy et al., 2011). Simvastatin use in Gn pigs increased the incidence of diarrhea, and led to earlier onset and longer virus shedding (Bui et al., 2013; Jung et al., 2012). Treated animals showed impaired Toll-like receptor (TLR) 3-dependent innate immune responses and lowered IFNα expression by alveolar macrophages and intestinal dendritic cells (Jung et al., 2012), but future studies are needed to determine whether similar mechanisms operate during NoV infection of humans.

In addition to Gn pigs, Gn calves were also investigated as an alternative large animal model (Souza et al., 2008), although to a limited extent. HuNoV infection of Gn calves with the GII.4 HS66 strain resulted in intestinal lesions (villous atrophy and loss of epithelial cells) in the small intestine, diarrhea and virus shedding in all analyzed animals. One of five calves also developed viremia. Viral capsid antigen was detected in small intestinal enterocytes and in lamina propria macrophage-like cells. HuNoV infection induced low levels of antibodies with only 67% (2/3) seroconversion but 100% (3/3) coproconversion (ie, antibodies in feces) in calves. Furthermore, early proinflammatory TNFα responses during diarrhea, high antiinflammatory IL-10 responses, and low to moderate Th1 (IL-12, IFNγ) and Th2 (IL-4) responses were observed. While more pronounced intestinal lesions were induced in Gn calves compared to Gn pigs, coinciding with increased numbers of IgA and IgG antibody secreting cells, most features of HuNoV infection in Gn calves are similar to those in Gn pigs.

In summary, both species of Gn animals are suitable models for the study of disease mechanisms, and thus represent important large animal models for preclinical, IND (investigational new drug)-enabling studies. However, the high cost of maintenance and required specialized care limit the broader use of this model in the scientific community. In addition, the lack of commensal bacteria in Gn animals, which promote development of the immune system (Sommer and Backhed, 2013) and of enteric viral infections (Kuss et al., 2011; Uchiyama et al., 2014; Kane et al., 2011), including NoV (Jones et al., 2014; Miura et al., 2013), represent caveats for studies of viral pathogenesis and immune responses. However, recent studies of Gn animals colonized with human commensal bacteria (Kandasamy et al., 2014; Vlasova et al., 2013) promise to overcome this limitation in future studies of HuNoV infection.

3 HUMAN NOROVIRUS SURROGATES AS ANIMAL MODELS

The alternative approach to studying human viruses in a nonnative animal host is the study of surrogate viruses in their native hosts. The most widely used surrogate viruses for modeling HuNoV infection in vivo were PoSaV and are now MNV and recoviruses. All these viruses share the ability to readily grow in cell culture while sharing varying degrees of genetic relatedness and biological features with HuNoVs. Although no surrogate animal model recapitulates all aspects of HuNoV infection, each model has contributed important knowledge to our understanding of calicivirus biology as detailed later.

3.1 The Porcine Sapovirus Model

Sapoviruses (SaVs) are gastrointestinal pathogens transmitted by the fecal–oral route (Oka et al., 2015). Like HuNoVs, human SaVs (HuSaVs) are enteric viruses that cause outbreaks and sporadic cases of acute gastroenteritis in people of all ages around the globe (Oka et al., 2015). The clinical features of HuSaV disease are indistinguishable from HuNoV infections and laboratory diagnosis is required for accurate identification of these viruses (Oka et al., 2015). Unlike

HuNoVs, HuSaVs do not bind HBGAs (Shirato-Horikoshi et al., 2007) but other cell attachment factors like sialic acid might be shared (Kim et al., 2014; Rydell et al., 2009). No host genetic factor(s) of susceptibility has been identified for HuSaVs to date (Oka et al., 2015). PoSaVs are genetically closely related to HuSaVs (Scheuer et al., 2013), and the PoSaV Cowden strain is currently the only member of the *Sapovirus* genus with a tissue culture system that requires the addition of intestinal contents or bile acids during infection (Flynn and Saif, 1988; Chang et al., 2004). This combined with the ability to genetically manipulate the viral genome (Chang et al., 2005) and development of acute gastroenteritis in PoSaV-infected pigs (Guo et al., 2001; Flynn et al., 1988) led to the use of PoSaV as a surrogate for HuNoV and HuSaV infections.

Gnotobiotic pigs orally infected with the PoSaV prototype strain Cowden develop mild to severe diarrhea, shed virus in the feces, and some develop viremia (Guo et al., 2001). Histological analysis of small intestinal segments from PoSaV-infected pigs showed villous atrophy, villous shortening and blunting, villous fusion, crypt hyperplasia, and cytosolic vacuolization of enterocytes. PoSaV antigen detection indicates that the major site of PoSaV replication is in villous enterocytes in the proximal small intestine. Therefore, the clinical and pathological features of PoSaV infection resemble those found in HuSaV (Oka et al., 2015) and HuNoV (Green, 2013) infections in humans. However, whether the cellular tropism of PoSaV mirrors the in vivo tropism of HuNoVs and HuSaVs remains to be determined. Importantly, this model demonstrated that limited amino acid changes in a virus strain following serial passaging in cell culture resulted in an attenuated PoSaV phenotype in vivo (Guo et al., 2001). Given the large genetic variability of HuNoVs and HuSaVs, this suggests a wide range of phenotypes for these human viruses.

In summary, the PoSaV experimental infection model using Gn pigs is a valuable tool that has and will undoubtedly contribute to a greater understanding of SaV pathogenesis and disease. Future studies using the PoSaV model also have the potential to advance translational studies towards vaccine and antiviral development for HuSaVs. However, the use of PoSaV as a surrogate for HuNoV is less ideal given the availability of MNV and TV, which are more closely related to HuNoV and for which the molecular tools and infection or disease models are available that are needed to perform detailed mechanistic and translational studies.

3.2 The Mouse Norovirus Model

Murine norovirus (MNV) is a natural pathogen of wild rodents and laboratory mice (Henderson, 2008; Farkas et al., 2012; Tsunesumi et al., 2012; Smith et al., 2012). To this date, MNV is the only NoV that has a small animal model and can be efficiently cultivated in tissue culture (Wobus et al., 2006). Since its discovery in 2003 (Karst et al., 2003), the MNV model has enabled significant advances in the understanding of norovirus-host interactions on the molecular, cellular, and organismal levels and has provided valuable perspectives on host–pathogen interactions in general (reviewed, eg, in Karst et al., 2014; Wobus et al., 2006). The utility of MNV as a model system for other NoVs is based on the following main features: (1) MNV can readily be cultivated in primary cells and stable cell lines derived from its natural host (Wobus et al., 2004), enabling a thorough analysis of the entire virus life cycle from attachment to release at the cellular and molecular levels. (2) The genetic tractability of mice has allowed researchers to identify diverse host factors that are important for survival and clearance of MNV, providing insights into host defenses that are likely to be important for protection during HuNoV infection (reviewed in Karst et al., 2014, 2015; Karst, 2010). (3) The availability of several reverse-genetics systems (Ward et al., 2007; Yunus et al., 2010; Chaudhry et al., 2007; Sandoval-Jaime et al., 2015) permits a detailed understanding of viral determinants of NoV infections, including receptor utilization, tissue tropism and viral persistence (eg, Taube et al., 2012; Nice et al., 2013; Strong et al., 2012). While MNV does not recapitulate clinical symptoms of HuNoV infection in immunocompetent hosts, it provides novel perspectives toward understanding interactions between intestinal viruses and their natural, coevolved host (Jones et al., 2014; Cadwell, 2015; Kernbauer et al., 2014; Cadwell et al., 2010).

Sequence analysis of the MNV genome shows that these viruses are homologous to other viruses in the NoV genus, including HuNoV, but form a separate genogroup (GV) (Karst et al., 2003; Smith et al., 2012). Three of MNV's four open reading frames (ORFs) encoding the nonstructural proteins (ORF1), major and minor capsid proteins (ORF2 and ORF3, respectively) are shared with all other NoVs, while an additional fourth ORF is only present within GV (McFadden et al., 2011; Thackray et al., 2007). Studies of NoV proteins have suggested that products of the ORF1 polyprotein of MNV and HuNoV have common functional properties, such as the rearrangement of cellular membranes during virus replication or during translation initiation (reviewed in Thorne and Goodfellow, 2014). In addition, MNV shares with HuNoVs the use of carbohydrates for cell attachment, although the specific carbohydrates only partially overlap (Tan and Jiang, 2014; Taube et al., 2009). Many strains of HuNoV bind to HBGA (reviewed in Tan and Jiang, 2014) but some HuNoV strains also bind sialic acids (Rydell et al., 2009; Maalouf et al., 2010; Esseili et al., 2012). MNV binds to terminal sialic acids on gangliosides and N- or O-linked glycoproteins in a strain-dependent manner (Taube et al., 2009, 2012), although no strains of MNV are known to date that use HBGA (Taube et al., 2010). Thus, while the MNV model does not allow studies into the function of HBGA during NoV pathogenesis, it does provide a system to understand the role of carbohydrates during NoV infection in general.

MNVs and HuNoVs also share the physical properties of the capsid and route of transmission. Like HuNoV, MNV virions are highly stable in the environment and are relatively resistant to heat, dessication, and low pH

(Cannon et al., 2006; Arthur and Gibson, 2015; Kotwal and Cannon, 2014; Tuladhar et al., 2012). In addition, viral particles are also highly infectious. The estimated 50% HuNoV infectious dose is between 18 and 2800 particles depending on the study (Teunis et al., 2008; Atmar et al., 2014), and MNV infection can be initiated with as few as 10 particles (Liu et al., 2009). These similarities are consistent with their common fecal–oral route of transmission making MNV a widely used surrogate for food safety studies (reviewed in Kniel, 2014).

Given the shared route of transmission, viral shedding in the stool is characteristic for all NoVs. While the infectious potential of shed HuNoV virions remains unknown, genomes are detected in the stool for weeks to months in immunocompetent individuals, long after symptomatic disease has resolved, and for years in immunocompromised individuals (Kaufman et al., 2014; Teunis et al., 2014; Green, 2014). Infectious MNV is shed for months as most strains cause a persistent infection in wild-type mice (Thackray et al., 2007; Hsu et al., 2006). In addition, the MNV-1 strain, which causes a self-resolving, acute infection in wild-type mice with transient shedding, causes long-term shedding in mice lacking components of the adaptive immune system (ie, Rag-deficient mice) (Karst et al., 2003). Thus, different durations of viral shedding can be modeled depending on the combinations of MNV and mouse strains. Multiple factors influencing shedding have been identified; viral gene products (ie, NS1/2 [N-terminal] and VP1 [viral capsid]) (Nice et al., 2013; Strong et al., 2012), host factors (eg, adaptive immune responses, interferon λ) (Karst et al., 2003; Nice et al., 2014), and the microbiota (Baldridge et al., 2015). While transmission occurs readily between mice in a cage (Manuel et al., 2008) (Fig. 3.4.4), the parameters that influence transmission remain largely unexplored. Nevertheless, a recent study successfully demonstrated the ability of an antiviral compound to disrupt transmission of MNV (Rocha-Pereira et al., 2015), highlighting the utility of the MNV system as a transmission model.

Despite many similarities, important differences exist between MNVs and HuNoVs. One of the greatest differences is that MNV does not cause overt clinical disease in an immunocompetent host. HuNoV infections frequently result in vomiting and diarrhea but mice are unable to vomit and MNV-infected wild-type mice do not develop overt diarrhea. Nevertheless, MNV-infected mice show histopathologic changes in the intestine (mild inflammation, limited villous atrophy) similar to those observed in HuNoV-infected volunteers (Karst et al., 2015; Mumphrey et al., 2007). It is also important to note that MNV causes diarrhea in mouse strains with impaired innate immunity, including strains lacking signal transducer and activator of transcription (Stat) 1 or the type I and II interferon receptors (Karst et al., 2003; Mumphrey et al., 2007; Rocha-Pereira et al., 2013). While the immunodeficient status of the host limits the use of those strains for studies of disease mechanisms, it does provide a small animal disease model for antiviral efficacy testing to develop compounds that not only reduce viral loads but also disease scores, as one recent study elegantly demonstrated (Rocha-Pereira et al., 2013).

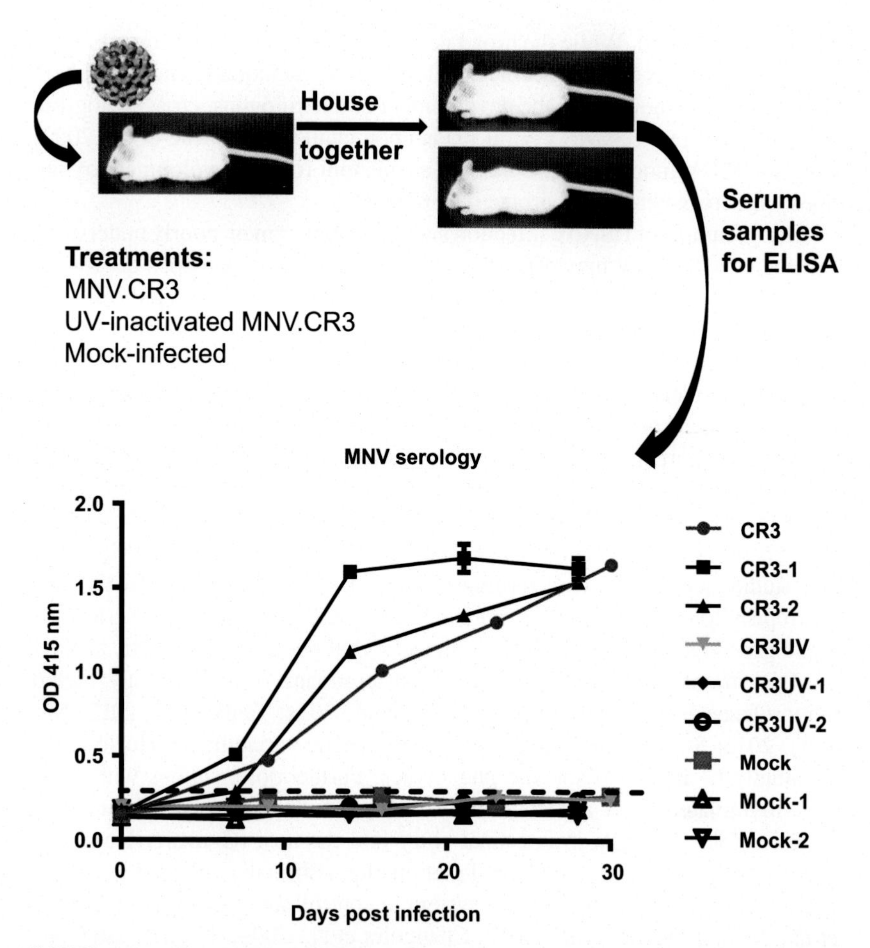

FIGURE 3.4.4 Transmission of MNV between mice in a cage. One Balb/c mouse each was orally infected with 1×10^6 pfu of MNV.CR3, the same dose of UV-inactivated MNV.CR3, or mock-infected. Two days later, each infected mouse was cohoused with 2 naïve mice for 28 days. Serum was collected from all animals on days 0, 7, 14, 21, and 28 of cohousing. ELISA was performed on a 1:100 dilution of serum as described (Wobus et al., 2004) and absorbance values were graphed. Naïve mice housed with a MNV.CR3-inoculated, but not mock or UV inactivated virus-inoculated cage mate seroconverted, indicating viral infection.

Differences exist also at the level of genetic diversity. HuNoV strains identified to date exhibit a broader genetic diversity (up to 45% divergence at the nucleotide level), spanning 3 genogroups (GI, GII, and GIV) with at least 29 genotypes (Zheng et al., 2006) than known MNV strains. MNVs identified to date separate into two genotypes with ~23% nucleotide divergence (Smith et al., 2012). However, a recently identified rat NoV that clusters with MNVs in GV diverges at the nucleotide level by 31% (Tsunesumi et al., 2012). Thus, it is likely that the diversity of rodent NoVs will increase in the future as more strains are identified. While the broad genetic diversity of HuNoV requires specific considerations during vaccine development, vaccination studies of HuNoV VLPs in mice demonstrated the development of heterologous, cross-genogroup immunity that reduces MNV viral loads upon challenge (LoBue et al., 2009). Thus, the MNV model can also be used as one tool for the development of vaccines with cross-genogroup protective efficacy.

Many features of HuNoV infection and replication remain poorly understood, but studies with MNV may offer clues to these processes. One open question in the field is the cell tropism of NoV. The identity of HuNoV-infected cell types in vivo remains unknown. However, viral antigen from both MNV and HuNoV has been detected within the intestinal lamina propria of infected mice and humans, respectively (Lay et al., 2010; Mumphrey et al., 2007), suggesting a tropism for immune cells (Karst and Wobus, 2015). This is consistent with the cell types that support HuNoV and MNV replication in vitro, specifically B cells for HuNoV (Jones et al., 2014) and macrophages, dendritic cells, microglia, and B cells (Jones et al., 2014; Wobus et al., 2004; Cox et al., 2009) for MNV. Thus, to date a partial overlap in cellular tropism exists between HuNoV and MNV. Future studies are needed to determine whether human and murine NoVs share the tropism for all these different immune cells. Another open question is how the broad genetic diversity of HuNoV influences viral phenotypes. Genetic analysis of MNV strains has already revealed viral determinants of NoV biology, including virulence and glycan utilization (Karst et al., 2014; Taube et al., 2012; Nice et al., 2013; Strong et al., 2012). Similarly, genetic variability in HuNoV may eventually be linked to specific phenotypes. Furthermore, studies with MNV have illuminated many aspects of NoV biology, for example, identifying roles for microfold (M) cells in intestinal entry into the host (Gonzalez-Hernandez et al., 2014), for dendritic cells in dissemination (Elftman et al., 2013), and for components of the innate immune system in controlling infections (McCartney et al., 2008; Thackray et al., 2012; Changotra et al., 2009). Future studies will need to determine the translatability of these findings to HuNoV infections.

One of the most promising aspects of the MNV model is the opportunity to study an enteric pathogen in its natural host and elucidate how the host's genetic background and microbial environment influence an encounter with a pathogen. For example, studies using MNV demonstrate that the interaction between enteric viruses, the microbiome, and the host's genetic background can influence the development of complex conditions such as inflammatory bowel disease (Cadwell et al., 2010). More recent studies have shown how concurrent helminth infections create a type 2 cytokine environment resulting in poor control of NoV replication (Osborne et al., 2014), and that the microbiome as a whole or specific members promote MNV infection and persistence (Jones et al., 2014; Baldridge et al., 2015), at least in the latter case via modulating the host immune response.

Overall, MNV provides a unique opportunity to mechanistically dissect, not only the effects of host factors and environment on a naturally coevolved host–pathogen relationship, but also many aspects of NoV biology. Despite the limitations of MNV as a model for clinical disease (ie, need for an immunodeficient host), MNV infection of wild-type mice represents a genetically tractable infection and transmission model.

3.3 The Recovirus Model

Rhesus enteric caliciviruses represent the most recently established HuNoV model. Recoviruses were first described in 2008, after the detection and isolation of a novel calicivirus from fecal samples of rhesus macaques (Farkas et al., 2008). These viruses are phylogenetically distinct from the genus *Norovirus* and from the proposed genus *Recovirus*. However, they share many properties with HuNoV, including their genomic organization, large genetic diversity, the ability to bind to HBGA, transmission by the fecal–oral route, and clinical disease presentations. Recovirus genetic diversity encompasses three genogroups and five genotypes (Farkas, 2015). GI and GII strains are found in rhesus macaques (Farkas et al., 2014), while GIII strains were discovered in samples collected from six humans with gastroenteritis in Bangladesh (Smits et al., 2012). The prototypic Tulane virus (TV), a GI.1 strain, replicates efficiently in cell culture in the rhesus macaque kidney epithelial cell line LLC-MK2 (Farkas et al., 2008), has a reverse genetics system available (Wei et al., 2008), and is associated with diarrheal disease in rhesus macaques after experimental infection (Sestak et al., 2012), providing crucial tools for mechanistic studies.

One of the most important advantages of the recovirus model is that the clinical disease presentation, histopathology, and transient serological response to TV are reminiscent of volunteer studies of HuNoV infection (Sestak et al., 2012). The similarities are multi-fold. First, intragastric infection of juvenile macaques with TV causes fever, a semiliquid diarrhea, and viral shedding in feces, although neither vomiting nor viremia were observed (Sestak et al., 2012). Second, intestinal biopsies from the duodenum of these monkeys revealed slightly blunted villi and mild mononuclear infiltration of the lamina propria. Third, experimentally infected macaques developed a robust but transient neutralizing antibody response against TV. However, future studies are needed to determine the ability of these antibodies to protect from infection or disease. Fourth, TV has a tropism for B cells. TV antigen-positive B cells were observed in the intestinal lamina propria, and infection of peripheral blood mononuclear cells obtained from healthy macaques led to an increase in viral RNA and antigen expression within B cells (Sestak et al., 2012). This tropism is similar to at least one HuNoV GII.4 strain (Jones et al., 2014).

Another important similarity of HuNoV and recoviruses is their ability to bind HBGAs. Diverse HuNoV strains have different patterns of HBGA-utilization and dependency (Huang et al., 2005) (see Chapter 3.3). Three HBGA binding patterns have been identified to date for recoviruses (Farkas, 2015). In case of TV, binding occurs with the type 3 form of the A antigen and all forms of the B antigen (Zhang et al., 2015). HBGAs are a known susceptibility factor for HuNoV infection (reviewed in Tan and Jiang, 2014) but the primate colonies used for recovirus studies exhibit a fairly homogeneous HBGA phenotype with greater than 95% of monkeys producing B antigen (Farkas et al., 2010). Thus, future experiments in a more phenotypically diverse population of macaques are needed to investigate whether specific HBGAs determine susceptibility of macaques to recovirus infection. Multiple roles have been proposed for HBGAs and its different forms (ie, free, bacterially associated, cell associated) during NoV infection (Karst and Wobus, 2015). Thus, the ability of recoviruses to bind HBGA and cause disease in a conventionally housed (ie, bacterially colonized) immunocompetent host provides a unique opportunity to understand the role of HBGA during enteric calicivirus infection and disease in the future.

The relationship between genetic and serologic diversity of HuNoVs remains unknown. Study of recoviruses with a similarly broad genetic diversity and available cell culture system has yielded initial insights; namely that genotypic classifications correlate generally with serotypes, although exceptions exist (Farkas, 2015; Farkas et al., 2014). These data suggest that the great HuNoV genetic diversity is likely accompanied by broad serological diversity. While this may present challenges in vaccine design, the recently described HuNoV culture system (Jones et al., 2014) will now enable future serotyping studies.

In summary, the recovirus model combines many beneficial features of an ideal HuNoV model. Most importantly, recoviruses can be genetically manipulated in vitro (Wei et al., 2008), replicate efficiently in tissue culture (Farkas et al., 2008), use the same HBGA attachment receptors as HuNoV (Sestak, 2014), and cause similar disease in vivo (Sestak et al., 2012). While the financial cost of in vivo studies and the inability to genetically manipulate the host are drawbacks of this model, the limitations are outweighed by many benefits, foremost the availability of an immunologically and microbially competent disease model. This model might be particularly useful in elucidating various roles of HBGAs in calicivirus pathogenesis, including how changing HBGA binding affinities and escape from neutralizing immune responses influence individual and herd immunity, an important driver of HuNoV evolution (Debbink et al., 2013).

4 CONCLUSIONS

Animal models are an integral part of our studies of human viruses. Large animal models are particularly valuable for translational studies and preclinical efficacy testing of antivirals and vaccines, while small animal models, especially knock-out and transgenic mouse models, are invaluable to gain a greater mechanistic understanding of all aspects of virus—host interactions at the organismal and cellular levels.

Although the past two decades have seen much progress in our understanding of HuNoVs, our understanding of these viruses lags behind that of other human viruses. The study of HuNoV has been particularly difficult due to the long-term lack of a cell culture system (Duizer et al., 2004; Herbst-Kralovetz et al., 2013; Papafragkou et al., 2013; Takanashi et al., 2014) and thus independent validation of the recently described B-cell culture model (Jones et al., 2014) is critical. The field also struggles to identify an animal model that closely reproduces HuNoV disease and is widely accepted. No "perfect" model has been developed to date but multiple different animal models for the study of HuNoV are available either by overcoming the species barrier of HuNoV or using HuNoV surrogates. Each of these models has its strengths and limitations (Tables 3.4.1 and 3.4.2) and overcoming some of these limitations remains an important goal for future research. One potentially fruitful avenue could be the generation of host-adapated HuNoVs, which has not been reported for any HuNoV animal model. Improvements might also come from the study of little explored (eg, newborn pigtail macaques) or new animal hosts, the identification of new surrogate viruses, or discovery of additional strains of HuNoV that more readily adapt or transmit to other species. While data obtained from any of the existing or future models are critical for our improved understanding of HuNoVs, validation of findings in the human host will ultimately determine how appropriate each model is to study a given aspect of HuNoV biology.

ACKNOWLEDGMENTS

We apologize to all colleagues whose work could not be cited due to lengths restrictions. C.E.W. is a recipient of the Friedrich Wilhelm Bessel award from the Alexander von Humboldt Foundation. Work in her laboratory was supported by NIH grants R21/R33 AI102106, R01 AI080611, R21 AI103961, and DARPA grant HR0011-11-C-0093. J.B.C. was funded by Coordenação de Aperfeiçoamento de Pessoal de Nível Superior (CAPES), Brasília, Brazil.

REFERENCES

Arias, A., Emmott, E., Vashist, S., Goodfellow, I., 2013. Progress towards the prevention and treatment of norovirus infections. Future Microbiol. 8, 1475–1487.

Arthur, S.E., Gibson, K.E., 2015. Comparison of methods for evaluating the thermal stability of human enteric viruses. Food Environ. Virol. 7 (1), 14–26.

Atmar, R.L., Opekun, A.R., Gilger, M.A., et al., 2008. Norwalk virus shedding after experimental human infection. Emerg. Infect. Dis. 14 (10), 1553–1557.

Atmar, R.L., Bernstein, D.I., Harro, C.D., et al., 2011. Norovirus vaccine against experimental human Norwalk Virus illness. N. Engl. J. Med. 365 (23), 2178–2187.

Atmar, R.L., Opekun, A.R., Gilger, M.A., et al., 2014. Determination of the 50% human infectious dose for Norwalk virus. J. Infect. Dis. 209 (7), 1016–1022.

Baldridge, M.T., Nice, T.J., McCune, B.T., et al., 2015. Commensal microbes and interferon-lambda determine persistence of enteric murine norovirus infection. Science 347 (6219), 266–269.

Bereszczak, J.Z., Barbu, I.M., Tan, M., et al., 2012. Structure, stability and dynamics of norovirus P domain derived protein complexes studied by native mass spectrometry. J. Struct. Biol. 177 (2), 273–282.

Bernstein, D.I., Atmar, R.L., Lyon, G.M., et al., 2015. Norovirus vaccine against experimental human GII.4 virus illness: a challenge study in healthy adults. J. Infect. Dis. 211 (6), 870–878.

Bitler, E.J., Matthews, J.E., Dickey, B.W., Eisenberg, J.N., Leon, J.S., 2013. Norovirus outbreaks: a systematic review of commonly implicated transmission routes and vehicles. Epidemiol. Infect. 141 (8), 1563–1571.

Bok, K., Parra, G.I., Mitra, T., et al., 2011. Chimpanzees as an animal model for human norovirus infection and vaccine development. Proc. Natl. Acad. Sci. USA 108 (1), 325–330.

Bui, T., Kocher, J., Li, Y., et al., 2013. Median infectious dose of human norovirus GII.4 in gnotobiotic pigs is decreased by simvastatin treatment and increased by age. J. Gen. Virol. 94 (Pt 9), 2005–2016.

Cadwell, K., 2015. Expanding the Role of the Virome: Commensalism in the Gut. J. Virol. 89 (4), 1951–1953.

Cadwell, K., Patel, K.K., Maloney, N.S., et al., 2010. Virus-plus-susceptibility gene interaction determines Crohn's disease gene Atg16L1 phenotypes in intestine. Cell 141 (7), 1135–1145.

Cannon, J.L., Papafragkou, E., Park, G.W., Osborne, J., Jaykus, L.A., Vinje, J., 2006. Surrogates for the study of norovirus stability and inactivation in the environment: A comparison of murine norovirus and feline calicivirus. J. Food Protect. 69 (11), 2761–2765.

Chachu, K.A., Strong, D.W., LoBue, A.D., Wobus, C.E., Baric, R.S., Virgin IV, H.W., 2008. Antibody is critical for the clearance of murine norovirus infection. J. Virol. 82 (13), 6610–6617.

Chang, K.O., Sosnovtsev, S.V., Belliot, G., Kim, Y., Saif, L.J., Green, K.Y., 2004. Bile acids are essential for porcine enteric calicivirus replication in association with down-regulation of signal transducer and activator of transcription 1. Proc. Natl. Acad. Sci. USA 101 (23), 8733–8738.

Chang, K.O., Sosnovtsev, S.V., Belliot, G., Wang, Q., Saif, L.J., Green, K.Y., 2005. Reverse genetics system for porcine enteric calicivirus, a prototype sapovirus in the Caliciviridae. J. Virol. 79 (3), 1409–1416.

Changotra, H., Jia, Y., Moore, T.N., et al., 2009. Type I and type II interferon inhibit the translation of murine norovirus proteins. J. Virol. 83 (11), 5683–5692.

Chaudhry, Y., Skinner, M.A., Goodfellow, I.G., 2007. Recovery of genetically defined murine norovirus in tissue culture by using a fowlpox virus expressing T7 RNA polymerase. J. Gen. Virol. 88 (Pt 8), 2091–2100.

Cheetham, S., Souza, M., Meulia, T., Grimes, S., Han, M.G., Saif, L.J., 2006. Pathogenesis of a genogroup II human norovirus in gnotobiotic pigs. J. Virol. 80 (21), 10372–10381.

Cheetham, S., Souza, M., McGregor, R., Meulia, T., Wang, Q., Saif, L.J., 2007. Binding patterns of human norovirus-like particles to buccal and intestinal tissues of gnotobiotic pigs in relation to A/H histo-blood group antigen expression. J. Virol. 81 (7), 3535–3544.

Cox, C., Cao, S., Lu, Y., 2009. Enhanced detection and study of murine norovirus-1 using a more efficient microglial cell line. Virol. J. 6, 196.

Debbink, K., Lindesmith, L.C., Donaldson, E.F., et al., 2013. Emergence of new pandemic GII.4 Sydney Norovirus strain correlates with escape from herd immunity. J. Infect. Dis. 208 (11), 1877–1887.

Duizer, E., Schwab, K.J., Neill, F.H., Atmar, R.L., Koopmans, M.P., Estes, M.K., 2004. Laboratory efforts to cultivate noroviruses. J. Gen. Virol. 85 (Pt 1), 79–87.

Elftman, M.D., Gonzalez-Hernandez, M.B., Kamada, N., et al., 2013. Multiple effects of dendritic cell depletion on murine norovirus infection. J. Gen. Virol. 94 (Pt 8), 1761–1768.

Esseili, M.A., Wang, Q., Saif, L.J., 2012. Binding of human GII.4 norovirus virus-like particles to carbohydrates of romaine lettuce leaf cell wall materials. Appl. Environ. Microbiol. 78 (3), 786–794.

Farkas, T., 2015. Rhesus enteric calicivirus surrogate model for human norovirus gastroenteritis. J. Gen. Virol. 96 (7), 1504–1514.

Farkas, T., Sestak, K., Wei, C., Jiang, X., 2008. Characterization of a rhesus monkey calicivirus representing a new genus of Caliciviridae. J. Virol. 82 (11), 5408–5416.

Farkas, T., Cross, R.W., Hargitt, 3rd., E., Lerche, N.W., Morrow, A.L., Sestak, K., 2010. Genetic diversity and histo-blood group antigen interactions of rhesus enteric caliciviruses. J. Virol. 84 (17), 8617–8625.

Farkas, T., Fey, B., Keller, G., Martella, V., Egyed, L., 2012. Molecular detection of murine noroviruses in laboratory and wild mice. Vet. Microbiol. 160 (3–4), 463–467.

Farkas, T., Lun, C.W., Fey, B., 2014. Relationship between genotypes and serotypes of genogroup 1 recoviruses: a model for human norovirus antigenic diversity. J. Gen. Virol. 95 (Pt 7), 1469–1478.

Feld, J.J., Hoofnagle, J.H., 2005. Mechanism of action of interferon and ribavirin in treatment of hepatitis C. Nature 436 (7053), 967–972.

Flynn, W.T., Saif, L.J., 1988. Serial propagation of porcine enteric calicivirus-like virus in primary porcine kidney cell cultures. J. Clin. Microbiol. 26 (2), 206–212.

Flynn, W.T., Saif, L.J., Moorhead, P.D., 1988. Pathogenesis of porcine enteric calicivirus-like virus in four-day-old gnotobiotic pigs. Am. J. Vet. Res. 49 (6), 819–825.

Gonzalez-Hernandez, M.B., Liu, T., Blanco, L.P., Auble, H., Payne, H.C., Wobus, C.E., 2013. Murine norovirus transcytosis across an in vitro polarized murine intestinal epithelial monolayer is mediated by M-like cells. J. Virol. 87 (23), 12685–12693.

Gonzalez-Hernandez, M.B., Liu, T., Payne, H.C., et al., 2014. Efficient Norovirus and Reovirus Replication in the Mouse Intestine Requires Microfold (M) Cells. J. Virol. 88 (12), 6934–6943.

Gota, C., Calabrese, L., 2003. Induction of clinical autoimmune disease by therapeutic interferon-alpha. Autoimmunity 36 (8), 511–518.

Graham, D.Y., Jiang, X., Tanaka, T., Opekun, A.R., Madore, H.P., Estes, M.K., 1994. Norwalk virus infection of volunteers: new insights based on improved assays. J. Infect. Dis. 170 (1), 34–43.

Green, K.Y., 2013. Caliciviridae: the noroviruses. In: Knipe, D.M., Howley, P.M., Cohen, J.I., Griffin, D.I., Lamb, R.A., Martin, M.A., Racaniello, V.R., Roizman, B. (Eds.), Fields Virology. Lippincott Williams & Wilkins, a Wolters Kluwer Business, Philadelphia, pp. 582–608.

Green, K.Y., 2014. Norovirus infection in immunocompromised hosts. Clin. Microbiol. Infect. 20 (8), 717–723.

Guo, M., Hayes, J., Cho, K.O., Parwani, A.V., Lucas, L.M., Saif, L.J., 2001. Comparative pathogenesis of tissue culture-adapted and wild-type Cowden porcine enteric calicivirus (PEC) in gnotobiotic pigs and induction of diarrhea by intravenous inoculation of wild-type PEC. J. Virol. 75 (19), 9239–9251.

Henderson, K.S., 2008. Murine norovirus, a recently discovered and highly prevalent viral agent of mice. Lab. Animal 37 (7), 314–320.

Herbst-Kralovetz, M.M., Radtke, A.L., Lay, M.K., et al., 2013. Lack of norovirus replication and histo-blood group antigen expression in 3-dimensional intestinal epithelial cells. Emerg. Infect. Dis. 19 (3), 431–438.

Hsu, C.C., Riley, L.K., Wills, H.M., Livingston, R.S., 2006. Persistent infection with and serologic cross-reactivity of three novel murine noroviruses. Comp. Med. 56 (4), 247–251.

Huang, P., Farkas, T., Zhong, W., et al., 2005. Norovirus and histo-blood group antigens: demonstration of a wide spectrum of strain specificities and classification of two major binding groups among multiple binding patterns. J. Virol. 79 (11), 6714–6722.

Hutson, A.M., Atmar, R.L., Graham, D.Y., Estes, M.K., 2002. Norwalk virus infection and disease is associated with ABO histo-blood group type. J. Infect. Dis. 185 (9), 1335–1337.

Hutson, A.M., Airaud, F., LePendu, J., Estes, M.K., Atmar, R.L., 2005. Norwalk virus infection associates with secretor status genotyped from sera. J. Med. Virol. 77 (1), 116–120.

Jones, M.K., Watanabe, M., Zhu, S., et al., 2014. Enteric bacteria promote human and mouse norovirus infection of B cells. Science 346 (6210), 755–759.

Jung, K., Wang, Q., Kim, Y., et al., 2012. The effects of simvastatin or interferon-alpha on infectivity of human norovirus using a gnotobiotic pig model for the study of antivirals. PLoS One 7 (7), e41619.

Kandasamy, S., Chattha, K.S., Vlasova, A.N., Rajashekara, G., Saif, L.J., 2014. Lactobacilli and Bifidobacteria enhance mucosal B cell responses and differentially modulate systemic antibody responses to an oral human rotavirus vaccine in a neonatal gnotobiotic pig disease model. Gut Microbes 5 (5), 639–651.

Kane, M., Case, L.K., Kopaskie, K., et al., 2011. Successful transmission of a retrovirus depends on the commensal microbiota. Science 334 (6053), 245–249.

Karst, S.M., 2010. Pathogenesis of noroviruses, Emerging RNA viruses. Viruses 2, 748–781.

Karst, S.M., Wobus, C.E., 2015. A working model of how noroviruses infect the intestine. PLoS Pathog. 11 (2), e1004626.

Karst, S.M., Wobus, C.E., Lay, M., Davidson, J., Virgin IV, .H.W., 2003. STAT1-dependent innate immunity to a Norwalk-like virus. Science 299 (5612), 1575–1578.

Karst, S.M., Wobus, C.E., Goodfellow, I.G., Green, K.Y., Virgin, H.W., 2014. Advances in norovirus biology. Cell Host Microbe 15 (6), 668–680.

Karst, S.M., Zhu, S., Goodfellow, I.G., 2015. The molecular pathology of noroviruses. J. Pathol. 235 (2), 206–216.

Kaufman, S.S., Green, K.Y., Korba, B.E., 2014. Treatment of norovirus infections: moving antivirals from the bench to the bedside. Antiviral. Res. 105, 80–91.

Kernbauer, E., Ding, Y., Cadwell, K., 2014. An enteric virus can replace the beneficial function of commensal bacteria. Nature 516 (7529), 94–98.

Kim, D.S., Hosmillo, M., Alfajaro, M.M., et al., 2014. Both alpha2,3- and alpha2,6-Linked Sialic Acids on O-Linked Glycoproteins Act as Functional Receptors for Porcine Sapovirus. PLoS Pathog. 10 (6), e1004172.

Kniel, K.E., 2014. The makings of a good human norovirus surrogate. Curr. Opin. Virol. 4, 85–90.

Kocher, J., Bui, T., Giri-Rachman, E., et al., 2014. Intranasal P particle vaccine provided partial cross-variant protection against human GII.4 norovirus diarrhea in gnotobiotic pigs. J. Virol. 88 (17), 9728–9743.

Kotwal, G., Cannon, J.L., 2014. Environmental persistence and transfer of enteric viruses. Curr. Opin. Virol. 4, 37–43.

Kuss, S.K., Best, G.T., Etheredge, C.A., et al., 2011. Intestinal microbiota promote enteric virus replication and systemic pathogenesis. Science 334 (6053), 249–252.

Lay, M.K., Atmar, R.L., Guix, S., et al., 2010. Norwalk virus does not replicate in human macrophages or dendritic cells derived from the peripheral blood of susceptible humans. Virology 406 (1), 1–11.

Lee, R.M., Lessler, J., Lee, R.A., et al., 2013. Incubation periods of viral gastroenteritis: a systematic review. BMC Infect. Dis. 13, 446.

Lindesmith, L., Moe, C., Marionneau, S., et al., 2003. Human susceptibility and resistance to Norwalk virus infection. Nature Med. 9 (5), 548–553.

Liu, G., Kahan, S.M., Jia, Y., Karst, S.M., 2009. Primary high-dose murine norovirus 1 infection fails to protect from secondary challenge with homologous virus. J. Virol. 83 (13), 6963–6968.

LoBue, A.D., Thompson, J.M., Lindesmith, L., Johnston, R.E., Baric, R.S., 2009. Alphavirus-adjuvanted norovirus-like particle vaccines: heterologous, humoral, and mucosal immune responses protect against murine norovirus challenge. J. Virol. 83 (7), 3212–3227.

Lopman, B.A., Reacher, M.H., Vipond, I.B., Sarangi, J., Brown, D.W., 2004. Clinical manifestation of norovirus gastroenteritis in health care settings. Clin. Infect. Dis. 39 (3), 318–324.

Maalouf, H., Zakhour, M., Le Pendu, J., Le Saux, J.C., Atmar, R.L., Le Guyader, F.S., 2010. Distribution in tissue and seasonal variation of norovirus genogroup I and II ligands in oysters. Appl. Environ. Microbiol. 76 (16), 5621–5630.

Manuel, C.A., Hsu, C.C., Riley, L.K., Livingston, R.S., 2008. Soiled-bedding sentinel detection of murine norovirus 4. J. Am. Assoc. Lab. Anim. Sci. 47 (3), 31–36.

McCartney, S.A., Thackray, L.B., Gitlin, L., Gilfillan, S., Virgin IV, H.W., Colonna, M., 2008. MDA-5 recognition of a murine norovirus. PLoS Pathog. 4 (7), e1000108.

McFadden, N., Bailey, D., Carrara, G., et al., 2011. Norovirus regulation of the innate immune response and apoptosis occurs via the product of the alternative open reading frame 4. PLoS Pathog. 7 (12), e1002413.

Miura, T., Sano, D., Suenaga, A., et al., 2013. Histo-blood group antigen-like substances of human enteric bacteria as specific adsorbents for human noroviruses. J. Virol. 87 (17), 9441–9451.

Mumphrey, S.M., Changotra, H., Moore, T.N., et al., 2007. Murine norovirus 1 infection is associated with histopathological changes in immunocompetent hosts, but clinical disease is prevented by STAT1-dependent interferon responses. J. Virol. 81 (7), 3251–3263.

Nice, T.J., Strong, D.W., McCune, B.T., Pohl, C.S., Virgin, H.W., 2013. A single-amino-acid change in murine norovirus NS1/2 is sufficient for colonic tropism and persistence. J. Virol. 87 (1), 327–334.

Nice, T.J., Baldridge, M.T., McCune, B.T., et al., 2014. Interferon lambda cures persistent murine norovirus infection in the absence of adaptive immunity. Science 347 (6219), 269–273.

Oka, T., Wang, Q., Katayama, K., Saif, L.J., 2015. Comprehensive review of human sapoviruses. Clin. Microbiol. Rev. 28 (1), 32–53.

Osborne, L.C., Monticelli, L.A., Nice, T.J., et al., 2014. Coinfection. Virus-helminth coinfection reveals a microbiota-independent mechanism of immunomodulation. Science 345 (6196), 578–582.

Papafragkou, E., Hewitt, J., Park, G.W., Greening, G., Vinje, J., 2013. Challenges of culturing human norovirus in three-dimensional organoid intestinal cell culture models. PLoS One 8 (6), e63485.

Radford, A.D., Gaskell, R.M., Hart, C.A., 2004. Human norovirus infection and the lessons from animal caliciviruses. Curr. Opin. Infect. Dis. 17 (5), 471–478.

Radford, A.D., Coyne, K.P., Dawson, S., Porter, C.J., Gaskell, R.M., 2007. Feline calicivirus. Vet. Res. 38 (2), 319–335.

Ramani, S., Neill, F.H., Opekun, A.R., et al., 2015. Mucosal and cellular immune responses to Norwalk virus. J. Infect. Dis. 212 (3), 397–405.

Rocha-Pereira, J., Jochmans, D., Debing, Y., Verbeken, E., Nascimento, M.S., Neyts, J., 2013. The viral polymerase Inhibitor 2'-C-methylcytidine inhibits Norwalk virus replication and protects against Norovirus-induced diarrhea and mortality in a mouse model. J. Virol. 87 (21), 11798–11805.

Rocha-Pereira, J., Neyts, J., Jochmans, D., 2014. Norovirus: targets and tools in antiviral drug discovery. Biochem. Pharmacol. 91 (1), 1–11.

Rocha-Pereira, J., Jochmans, D., Neyts, J., 2015. Prophylactic treatment with the nucleoside analogue 2'-C-methylcytidine completely prevents transmission of norovirus. J. Antimicrob. Chemother. 70 (1), 190–197.

Rockx, B., De Wit, M., Vennema, H., et al., 2002. Natural history of human calicivirus infection: a prospective cohort study. Clin. Infect. Dis. 35 (3), 246–253.

Rockx, B.H., Bogers, W.M., Heeney, J.L., van Amerongen, G., Koopmans, M.P., 2005. Experimental norovirus infections in non-human primates. J. Med. Virol. 75 (2), 313–320.

Rondy, M., Koopmans, M., Rotsaert, C., et al., 2011. Norovirus disease associated with excess mortality and use of statins: a retrospective cohort study of an outbreak following a pilgrimage to Lourdes. Epidemiol. Infect. 139 (3), 453–463.

Rydell, G.E., Nilsson, J., Rodriguez-Diaz, J., et al., 2009. Human noroviruses recognize sialyl Lewis x neoglycoprotein. Glycobiology 19 (3), 309–320.

Saif, L.J., Ward, L.A., Yuan, L., Rosen, B.I., To, T.L., 1996. The gnotobiotic piglet as a model for studies of disease pathogenesis and immunity to human rotaviruses. Arch. Virol. Suppl. 12, 153–161.

Sandoval-Jaime, C., Green, K.Y., Sosnovtsev, S.V., 2015. Recovery of murine norovirus and feline calicivirus from plasmids encoding EMCV IRES in stable cell lines expressing T7 polymerase. J. Virol. Methods 217, 1–7.

Scheuer, K.A., Oka, T., Hoet, A.E., et al., 2013. Prevalence of porcine noroviruses, molecular characterization of emerging porcine sapoviruses from finisher swine in the United States, and unified classification scheme for sapoviruses. J. Clin. Microbiol. 51 (7), 2344–2353.

Sestak, K., 2014. Role of histo-blood group antigens in primate enteric calicivirus infections. World J. Virol. 3 (3), 18–21.

Sestak, K., Feely, S., Fey, B., et al., 2012. Experimental inoculation of juvenile rhesus macaques with primate enteric caliciviruses. PLoS One 7 (5), e37973.

Shirato-Horikoshi, H., Ogawa, S., Wakita, T., Takeda, N., Hansman, G.S., 2007. Binding activity of norovirus and sapovirus to histo-blood group antigens. Arch. Virol. 152 (3), 457–461.

Smith, D.B., McFadden, N., Blundell, R.J., Meredith, A., Simmonds, P., 2012. Diversity of murine norovirus in wild-rodent populations: species-specific associations suggest an ancient divergence. J. Gen. Virol. 93 (Pt 2), 259–266.

Smits, S.L., Rahman, M., Schapendonk, C.M., et al., 2012. Calicivirus from novel Recovirus genogroup in human diarrhea. Emerg. Infect. Dis. 18 (7), 1192–1195.

Sommer, F., Backhed, F., 2013. The gut microbiota--masters of host development and physiology. Nature Rev. 11 (4), 227–238.

Souza, M., Cheetham, S.M., Azevedo, M.S., Costantini, V., Saif, L.J., 2007a. Cytokine and antibody responses in gnotobiotic pigs after infection with human norovirus genogroup II.4 (HS66 strain). J. Virol. 81 (17), 9183–9192.

Souza, M., Costantini, V., Azevedo, M.S., Saif, L.J., 2007b. A human norovirus-like particle vaccine adjuvanted with ISCOM or mLT induces cytokine and antibody responses and protection to the homologous GII.4 human norovirus in a gnotobiotic pig disease model. Vaccine 25 (50), 8448–8459.

Souza, M., Azevedo, M.S., Jung, K., Cheetham, S., Saif, L.J., 2008. Pathogenesis and immune responses in gnotobiotic calves after infection with the genogroup II.4-HS66 strain of human norovirus. J. Virol. 82 (4), 1777–1786.

Strong, D.W., Thackray, L.B., Smith, T.J., Virgin, H.W., 2012. Protruding domain of capsid protein is necessary and sufficient to determine murine norovirus replication and pathogenesis in vivo. J. Virol. 86 (6), 2950–2958.

Subekti, D.S., Tjaniadi, P., Lesmana, M., et al., 2002. Experimental infection of Macaca nemestrina with a Toronto Norwalk-like virus of epidemic viral gastroenteritis. J. Med. Virol. 66 (3), 400–406.

Takanashi, S., Wang, Q., Chen, N., et al., 2011. Characterization of emerging GII.g/GII.12 noroviruses from a gastroenteritis outbreak in the United States in 2010. J. Clin. Microbiol. 49 (9), 3234–3244.

Takanashi, S., Saif, L.J., Hughes, J.H., et al., 2014. Failure of propagation of human norovirus in intestinal epithelial cells with microvilli grown in three-dimensional cultures. Arch. Virol. 159 (2), 257–266.

Tan, M., Jiang, X., 2014. Histo-blood group antigens: a common niche for norovirus and rotavirus. Exp. Rev. Mol. Med. 16, e5.

Taube, S., Perry, J.W., Yetming, K., et al., 2009. Ganglioside-linked terminal sialic acid moieties on murine macrophages function as attachment receptors for murine noroviruses. J. Virol. 83 (9), 4092–4101.

Taube, S., Jiang, M., Wobus, C.E., 2010. Glycosphingolipids as receptors for non-enveloped viruses. Viruses 2 (4), 1011–1049.

Taube, S., Perry, J.W., McGreevy, E., et al., 2012. Murine noroviruses bind glycolipid and glycoprotein attachment receptors in a strain-dependent manner. J. Virol. 86 (10), 5584–5593.

Taube, S., Kolawole, A.O., Hohne, M., et al., 2013. A mouse model for human norovirus. MBio 4 (4), e00450–e00513.

Teunis, P.F., Moe, C.L., Liu, P., et al., 2008. Norwalk virus: how infectious is it? J. Med. Virol. 80 (8), 1468–1476.

Teunis, P.F., Sukhrie, F.H., Vennema, H., Bogerman, J., Beersma, M.F., Koopmans, M.P., 2014. Shedding of norovirus in symptomatic and asymptomatic infections. Epidemiol. Infect., 1–8.

Thackray, L.B., Wobus, C.E., Chachu, K.A., et al., 2007. Murine noroviruses comprising a single genogroup exhibit biological diversity despite limited sequence divergence. J. Virol. 81 (19), 10460–10473.

Thackray, L.B., Duan, E., Lazear, H.M., et al., 2012. Critical role for interferon regulatory factor 3 (IRF-3) and IRF-7 in type I interferon-mediated control of murine norovirus replication. J. Virol. 86 (24), 13515–13523.

Thorne, L.G., Goodfellow, I.G., 2014. Norovirus gene expression and replication. J. Gen. Virol. 95 (Pt 2), 278–291.

Treanor, J.J., Atmar, R.L., Frey, S.E., et al., 2014. A Novel Intramuscular Bivalent Norovirus Virus-Like Particle Vaccine Candidate-Reactogenicity, Safety, and Immunogenicity in a Phase 1 Trial in Healthy Adults. J. Infect. Dis. 210 (11), 1763–1771.

Troeger, H., Loddenkemper, C., Schneider, T., et al., 2009. Structural and functional changes of the duodenum in human norovirus infection. Gut 58 (8), 1070–1077.

Tsunesumi, N., Sato, G., Iwasa, M., Kabeya, H., Maruyama, S., Tohya, Y., 2012. Novel Murine Norovirus-Like Genes in Wild Rodents in Japan. J. Vet. Med. Sci. 74, 1221–1224.

Tuladhar, E., Bouwknegt, M., Zwietering, M.H., Koopmans, M., Duizer, E., 2012. Thermal stability of structurally different viruses with proven or potential relevance to food safety. J. Appl. Microbiol. 112 (5), 1050–1057.

Uchiyama, R., Chassaing, B., Zhang, B., Gewirtz, A.T., 2014. Antibiotic treatment suppresses rotavirus infection and enhances specific humoral immunity. J. Infect. Dis. 210 (2), 171–182.

Vashist, S., Bailey, D., Putics, A., Goodfellow, I., 2009. Model systems for the study of human norovirus Biology. Future Virol. 4 (4), 353–367.

Vlasova, A.N., Chattha, K.S., Kandasamy, S., et al., 2013. Lactobacilli and bifidobacteria promote immune homeostasis by modulating innate immune responses to human rotavirus in neonatal gnotobiotic pigs. PLoS One 8 (10), e76962.

Ward, V.K., McCormick, C.J., Clarke, I.N., et al., 2007. Recovery of infectious murine norovirus using pol II-driven expression of full-length cDNA. Proc. Natl. Acad. Sci. USA 104 (26), 11050–11055.

Wei, C., Farkas, T., Sestak, K., Jiang, X., 2008. Recovery of infectious virus by transfection of in vitro-generated RNA from tulane calicivirus cDNA. J. Virol. 82 (22), 11429–11436.

Wobus, C.E., Karst, S.M., Thackray, L.B., et al., 2004. Replication of Norovirus in cell culture reveals a tropism for dendritic cells and macrophages. PLoS Biol. 2 (12), e432.

Wobus, C.E., Thackray, L.B., Virgin IV, H.W., 2006. Murine norovirus: a model system to study norovirus biology and pathogenesis. J. Virol. 80 (11), 5104–5112.

Wyatt, R.G., Greenberg, H.B., Dalgard, D.W., et al., 1978. Experimental infection of chimpanzees with the Norwalk agent of epidemic viral gastroenteritis. J. Med. Virol. 2 (2), 89–96.

Yunus, M.A., Chung, L.M., Chaudhry, Y., Bailey, D., Goodfellow, I., 2010. Development of an optimized RNA-based murine norovirus reverse genetics system. J. Virol. Methods 169 (1), 112–118.

Zhang, D., Huang, P., Zou, L., Lowary, T.L., Tan, M., Jiang, X., 2015. Tulane virus recognizes the A type 3 and B histo-blood group antigens. J. Virol. 89 (2), 1419–1427.

Zheng, D.P., Ando, T., Fankhauser, R.L., Beard, R.S., Glass, R.I., Monroe, S.S., 2006. Norovirus classification and proposed strain nomenclature. Virology 346 (2), 312–323.

Chapter 3.5

Molecular Epidemiology and Evolution of Noroviruses

K.Y. Green
Caliciviruses Section, Laboratory of Infectious Diseases, National Institute of Allergy and Infectious Diseases, National Institutes of Health, Bethesda, MD, United States

1 INTRODUCTION

Noroviruses are enteric pathogens associated with acute gastroenteritis. The consequences of infection can range from asymptomatic to life-threatening (Hall et al., 2013). The virus is highly infectious and can spread easily in a population, causing outbreaks in settings where individuals are in close proximity or exposed to a common source of virus in water or food (Glass et al., 2009). The cost of outbreak management and medical visits for treatment is high; it is estimated that a vaccine of only 50% efficacy could save the US economy one billion dollars per year (Bartsch et al., 2012). Noroviruses are the leading cause of severe pediatric diarrhea following the successful implementation of rotavirus vaccines, prompting an estimated one million health care visits per year in US infants and young children (Payne et al., 2013). Vaccines to prevent norovirus disease are in clinical trials (reviewed in Ramani et al., 2014 and in Chapter 3.6), and there is an acknowledged need for antiviral drugs to treat immunocompromised patients, who are at high risk for morbidity from norovirus diarrhea (Kaufman et al., 2014).

Understanding of norovirus epidemiology and evolution is advancing rapidly with the availability of state-of-the-art diagnostic assays and new nucleotide sequencing technologies (reviewed in Vinje, 2015). Furthermore, progress has been made in the development of cell culture and animal model systems for the study of some noroviruses (reviewed in Karst et al., 2014). This chapter will provide an overview of the features of norovirus genetic diversity and adaptation responsible for the epidemiologic success of these abundant pathogens and how this diversity challenges vaccine design.

Viral Gastroenteritis

2 CLASSIFICATION AND NOMENCLATURE OF THE NOROVIRUSES

2.1 Genus *Norovirus* in the Family *Caliciviridae*

Noroviruses belong to the virus family *Caliciviridae*, a group of small icosahedral, nonenveloped viruses that possess a polyadenylated single-stranded positive-sense RNA genome (Jiang et al., 1990). Caliciviruses are diverse in their associated disease syndromes, host specificity, and genetic characteristics. However, a striking property shared by nearly all caliciviruses is their global and ubiquitous distribution in their respective hosts. The human noroviruses are highly successful in this characteristic, with evidence for multiple infections throughout life in the majority of individuals (Jing et al., 2000; Smit et al., 1997; O'Ryan et al., 1998).

The current taxonomic status of the virus family *Caliciviridae* as defined by the International Committee on Taxonomy of Viruses (ICTV) includes five distinct genera: *Norovirus*, *Sapovirus*, *Lagovirus*, *Nebovirus*, and *Vesivirus* (Table 3.5.1) (Clarke et al., 2012). The current calicivirus genera are further divided into one or more species (Table 3.5.1), with each species represented by a prototype strain. The genus *Norovirus* is represented by Norwalk virus, which was the first norovirus discovered (Kapikian et al., 1972), and the first to be characterized at the genome level as a calicivirus (Jiang et al., 1990). Classification schemes below the species level are not addressed by the ICTV, including the genetic typing (genotyping) systems implemented by norovirus researchers to track the molecular epidemiology of norovirus strains (Kroneman et al., 2013) (see later).

2.2 Norovirus Genogroups and Genotypes

Human noroviruses are genetically diverse, and early studies noted the presence of two major genetic groups, now called genogroups (G) (Ando et al., 1994; Wang et al., 1994). The norovirus RNA genome (approx. 7.5–7.6 kb in size) is organized into three major open reading frames (ORFs 1–3). The length of the RNA genome can vary among strains, and the nucleotide boundaries of the ORFs are shown for representative GI (Norwalk virus) and GII (Lordsdale virus) human norovirus strains in Fig. 3.5.1A.

2.2.1 ORF1

The norovirus ORF1 encodes the nonstructural proteins (designated NS1 through NS7) that are cleaved from the newly-synthesized polyprotein precursor by a virus-encoded protease (NS6) (reviewed in Sosnovtsev, 2010; Thorne and Goodfellow, 2014) (Fig. 3.5.1B). The nonstructural proteins play a major role in viral replication, and enzymatic activity has been identified for the NS3 (NTPase), NS6 (protease) and NS7 (RNA-dependent RNA polymerase) proteins. Nonstructural proteins NS1-2 (or Nterm) and NS4 (p22 or "3A-like") contain hydrophobic

TABLE 3.5.1 Taxonomic Structure of the *Caliciviridae*

Genus	Species	Representative strain
Norovirus (NoV)	*Norwalk virus* (NV)	NoV/GI/Hu/US/1968/ GI.P1-GI.1/Norwalk
Sapovirus (SaV)	*Sapporo virus* (SV)	SaV/GI/Hu/JP/1982/GI.1/ Sapporo
Lagovirus (LaV)	*Rabbit hemorrhagic disease virus* (RHDV) *European brown hare syndrome virus* (EBHSV)	LaV/RHDV/Ra/DE/1988/GH LaV/EBHSV/Ha/FR/1989/GD
Vesivirus (VeV)	*Vesicular exanthema of swine virus* (VESV) *Feline calicivirus* (FCV)	VeV/VESV/Po/US/1948/ VESV-A48 VeV/FCV/Fe/US/1958/F9
Nebovirus (NeV)	*Newbury-1 virus* (NBV)	NeV/NBV/Bo/UK/1976/ Newbury-1

The cryptograms in this table are organized as follows: genus/genogroup or virus species/host of origin/country of origin/year of occurrence/Pol genotype (if known)/Capsid genotype (if known)/ strain name. Abbreviations for the host species are: *Bo*, bovine; *Fe*, feline; *Ha*, Hare; *Hu*, Human; *Po*, Porcine; *Ra*, Rabbit. Country abbreviations are: *DE*, Germany; *FR*, France; *JP*, Japan; *UK*, United Kingdom; *US*, United States. GenBank Accession numbers of representative viruses: NV, M87661; SV, U65427; RHDV, M67473, EBHSV, Z69620; VESV, AF181082; FCV, M86379; NBV, DQ013304.

domains and interact with host cell membranes to establish the sites of viral replication (Hyde et al., 2009). The VPg protein (NS5) is covalently linked to the 5′-end of the viral RNA genome and plays an essential role in translation and RNA replication. The RNA-dependent RNA polymerase (abbreviated as RdRp, NS7, or Pol), mapping to the C-terminal end of ORF1, is the enzyme involved in viral RNA replication and one of the most conserved proteins.

2.2.2 ORF2

The structural proteins of the virion, VP1, and VP2, are encoded by ORFs 2 and 3, respectively. The VP1 (approximately 58 kDa) is the most abundant protein in the virion, and 180 copies of VP1 form the icosahedral capsid of the virus (Prasad et al., 1996). Expression of the norovirus VP1 by recombinant baculovirus yields virus-like particles (VLPs) that resemble native virions, and VLPs have become an essential tool in norovirus research and vaccine development (Jiang et al., 1992). The VP1 is organized into two major domains, designated Shell (S) and Protruding (P) (Prasad et al., 1996). In dimeric form with another VP1 protein, the P domain forms an arch-like structure in which two arms (the P1 subdomains) present a highly variable P2 subdomain to the surface of the virion (Prasad et al., 1996). Structural studies have demonstrated that both antibodies and histo-blood group antigen (HBGA) carbohydrates (proposed attachment factors in the gut) bind within the norovirus P2 domain, confirming its importance in host interactions and immunity (Cao et al., 2007; Bu et al., 2008;

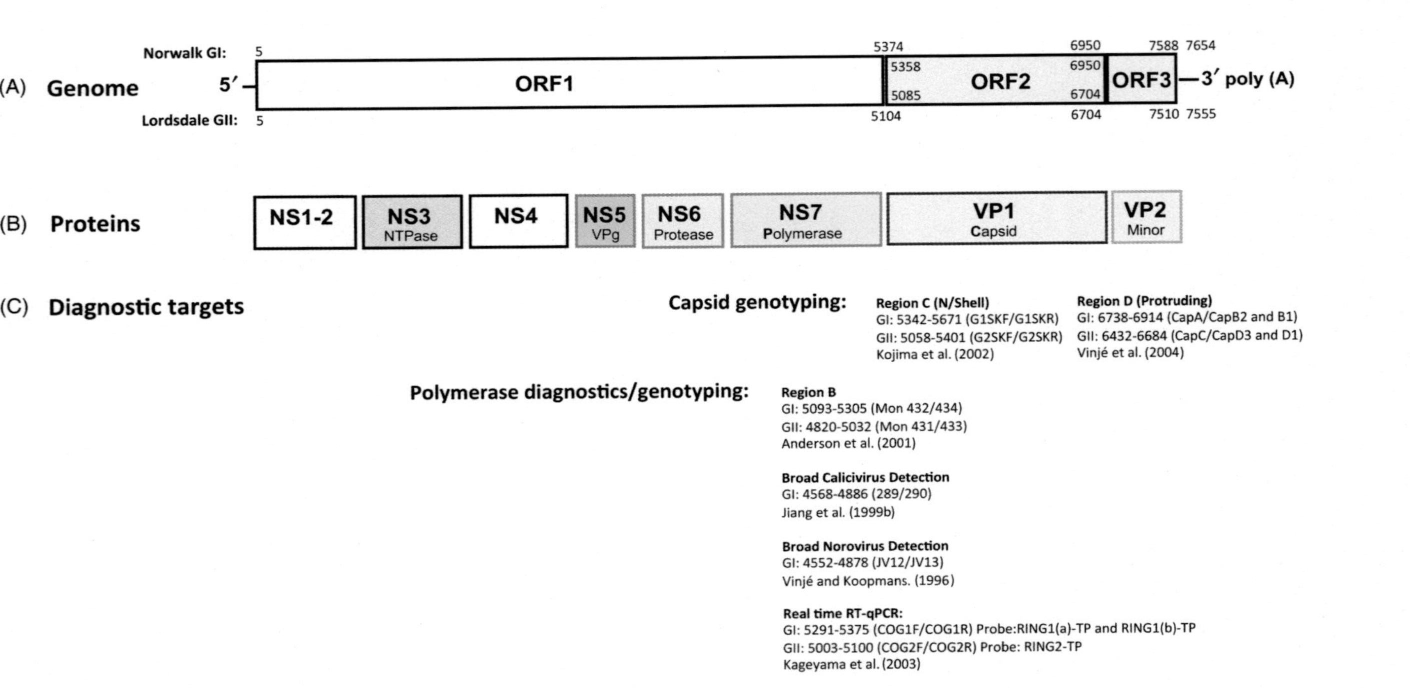

FIGURE 3.5.1 **Norovirus genome organization, proteins, and diagnostic primers.** (A) Genome organization and nucleotide boundaries of the ORFs of representative human norovirus strains Genogroup I Norwalk virus (M87661) (top) and Genogroup II Lordsdale virus (X86557) (bottom). (B) Proteins encoded by the norovirus genome. ORF1 encodes the nonstructural proteins of the virus, NS1-NS7. ORF2 encodes the major capsid protein, VP1, and ORF3 encodes the minor capsid protein, VP2. (C) Genomic regions targeted for the norovirus dual genotyping system involve RNA-dependent RNA polymerase (P) and capsid (C) sequences. The regions of interest are amplified in a reverse transcription-polymerase chain reaction (RT-PCR) assay, sequenced, and then interrogated against a sequence database using the norovirus typing tool (Kroneman et al., 2013). References for representative primer sets are: (Anderson et al., 2001; Jiang et al., 1999b; Kojima et al., 2002; Newman and Leon, 2015; Vinje and Koopmans, 1996; Vinje et al., 2004). A widely-used real time RT-qPCR assay for norovirus detection is based on the primers of Kageyama et al. (Kageyama et al., 2003). Many additional primer pairs have been developed, but periodic optimization may be needed to encompass ever-evolving norovirus strains (Kong et al., 2015).

Katpally et al., 2008). Furthermore, because multiple antigenic epitopes and HBGA binding sites overlap (Parra et al., 2012a), the measurement of antibodies that block HBGA binding to virus particles has been taken as a surrogate marker of virus neutralization and been accepted widely as a correlate of protection from norovirus infection and disease (Bok et al., 2011; Reeck et al., 2010).

2.2.3 ORF3 (and ORF4)

The ORF3 encodes a small, basic protein, VP2, considered a minor structural protein because of its low abundance in virions (Sosnovtsev and Green, 2000). The VP2 plays a role in particle stability and infectivity (Bertolotti-Ciarlet et al., 2003; Sosnovtsev et al., 2005), and it interacts directly with a highly conserved region of the inner shell domain in VP1 (Vongpunsawad et al., 2013). The norovirus VP2 may exert host effects: a murine norovirus VP2 was recently shown to dampen the maturation of antigen presenting cells (Zhu et al., 2013).

An ORF4 encoding a host regulatory protein (designated virulence factor-1) has been described for murine norovirus (McFadden et al., 2011), a member of the genus that replicates efficiently in cell culture (Wobus et al., 2004), but an ORF4 (or functional homolog) has not yet been identified in human strains.

2.2.4 Genotyping Methods

Diagnostic RT-PCR primers for the analysis of human noroviruses are targeted to conserved regions of the genome, and references describing representative primer sets for the detection and genotyping of noroviruses are diagrammed in Fig. 3.5.1C. The RNA-dependent RNA polymerase and capsid encoding regions are important targets for virus genotyping (see later).

2.2.5 Genotyping System Based on the Viral Capsid

Comparative phylogenetic analysis of the complete capsid VP1-encoding gene (ORF2) has led to the division of noroviruses into seven genogroups (G) designated with Roman numerals as GI through GVII (Vinje, 2015; Kroneman et al., 2013) (Fig. 3.5.2). The genogroups are sufficiently divergent to require the use of different diagnostic primer sets for identification (Fig. 3.5.1C). The genogroups are further subdivided into capsid (C) genotypes that are numbered sequentially as they are identified (Table 3.5.2). The two genogroups with the majority of human pathogens, GI and GII, currently contain 9 and 22 capsid genotypes, respectively. The reference strain for a genotype, by convention, is the first full-length capsid sequence entered into a public sequence database. Each capsid genotype is further segregated into "clusters" or "variants," and this diversity is especially noteworthy for genotype GII.4 because of its major role in epidemic disease (reviewed in White, 2014). Eight pandemic variants of norovirus GII.4 have emerged since 1995, when molecular diagnostic assays began to gain widespread use (Kroneman et al., 2013). The GII.4 variants and

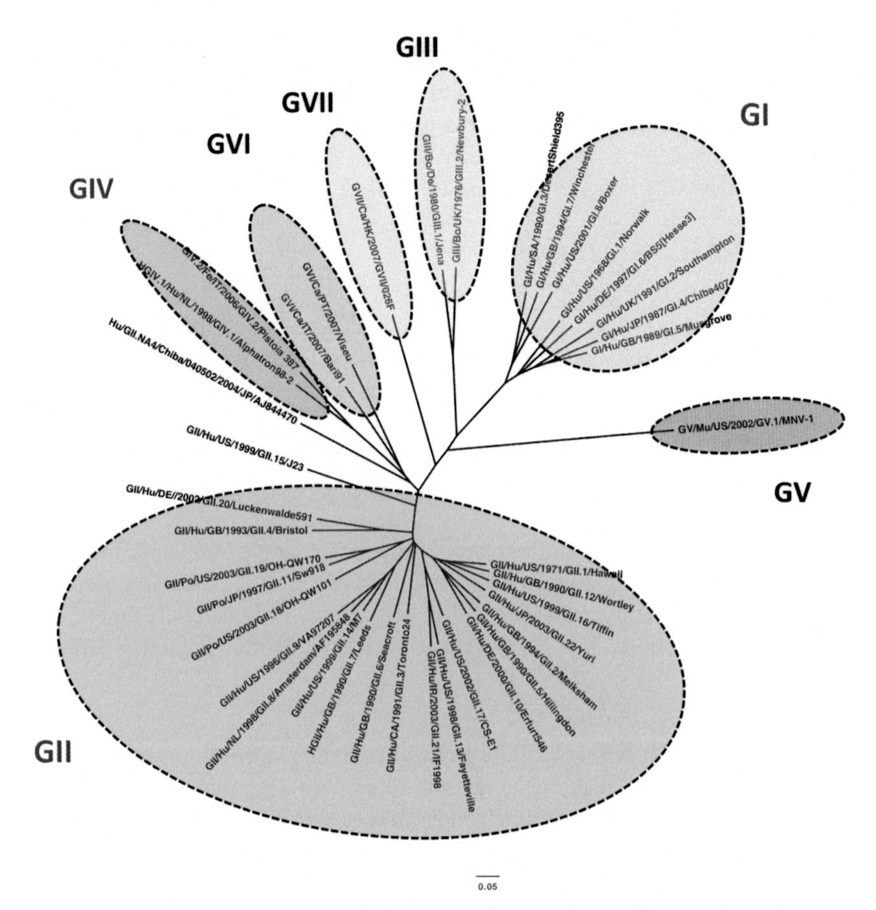

FIGURE 3.5.2 **Phylogenetic relationships among the noroviruses.** There are presently seven major genogroups within the genus *Norovirus*, with human pathogens found in GI, GII, and GIV. Porcine norovirus strains have been found also in GII, and canine and feline strains have been detected in GIV. Strains belonging to certain genogroups have not been detected in humans thus far and include GIII (bovine), GV (murine), and GVII (canine). Phylogenetic analyses were carried out with MEGA v6 using Neighbor-Joining as the algorithm for reconstruction and amino acid sequences from the entire capsid protein VP1. Scale bar represents the number of amino acids substitutions per site. *(Source: Analysis and image courtesy of Gabriel I. Parra.)*

their GenBank accession numbers are listed in the footnote of Table 3.5.2. The variants are named according to one of the early sites where the emergence of a new pandemic strain was first recognized.

2.2.6 Genotyping System Based on the Viral RNA-Dependent RNA Polymerase

The RdRp (Pol) gene can vary among strains of the same capsid genotype (Hardy et al., 1997; Jiang et al., 1999a), which has led to the development of

TABLE 3.5.2 Norovirus Genogroups and Genotypes as Determined by Capsid (C) Gene Relatedness

Reference virus	Genogroup. C genotype	GenBank accession number
GI/Hu/US/1968/GI.1/Norwalk	GI.1	M87661
GI/Hu/UK/1991/GI.2/Southampton	GI.2	L07418
GI/Hu/SA/1990/GI.3/DesertShield 395	GI.3	U04469
GI/Hu/JP/1987/GI.4/Chiba 407	GI.4	AB042808
GI/Hu/GB/1989/GI.5/Musgrove	GI.5	AJ277614
GI/Hu/DE/1997/GI.6/BS5(Hesse3)	GI.6	AF093797
GI/Hu/GB/1994/GI.7/Winchester	GI.7	AJ277609
GI/Hu/US/2001/GI.8/Boxer	GI.8	AF538679
GI/Hu/CA/2004/GI.9/Vancouver730	GI.9	HQ637267
GII/Hu/US/1971/GII.1/Hawaii	GII.1	U07611
GII/Hu/GB/1994/GII.2/Melksham	GII.2	X81879
GII/Hu/CA/1991/GII.3/Toronto 24	GII.3	U02030
GII/Hu/GB/1993/GII.4/Bristol[a]	GII.4	X76716
GII/Hu/GB/1990/GII.5/Hillingdon	GII.5	AJ277607
GII/Hu/GB/1990/GII.6/Seacroft	GII.6	AJ277620
GII/Hu/GB/1990/GII.7/Leeds	GII.7	AJ277608
GII/Hu/NL/1998/GII.8/Amsterdam	GII.8	AF195848
GII/Hu/US/1996/GII.9/VA97207	GII.9	AY038599
GII/Hu/DE/2000/GII.10/Erfurt546	GII.10	AF427118
GII/Po/JP/1997/GII.11/Sw918	GII.11	AB074893
GII/Hu/GB/1990/GII.12/Wortley	GII.12	AJ277618
GII/Hu/US/1998/GII.13/Fayetteville	GII.13	AY113106
GII/Hu/US/1999/GII.14/M7	GII.14	AY130761
GII/Hu/US/1999/GII.15/J23	GII.15	AY130762
GII/Hu/US/1999/GII.16/Tiffin	GII.16	AY502010
GII/Hu/US/2002/GII.17/CS-E1	GII.17	AY502009
GII/Po/US/2003/GII.18/OH-QW101	GII.18	AY823304
GII/Po/US/2003/GII.19/OH-QW170	GII.19	AY823306
GII/Hu/DE/2002/GII.20/Luckenwalde591	GII.20	EU373815
GII/Hu/IR/2003/GII.21/IF1998	GII.21	AY675554
GII/Hu/JP/2003/GII.22/Yuri	GII.22	AB083780
GIII/Bo/De/1980/GIII.1/Jena	GIII.1	AJ011099
GIII/Bo/GB/1976/GIII.2/Newbury-2	GIII.2	AF097917

(Continued)

TABLE 3.5.2 Norovirus Genogroups and Genotypes as Determined by Capsid (C) Gene Relatedness (*cont.*)

Reference virus	Genogroup. C genotype	GenBank accession number
GIII/Ov/NZ/2007/GIII.3/Norsewood30	GIII.3	EU193658
GIV.1/Hu/NL/1998/GIV.1/Alphatron 98-2	GIV.1	AF195847
GIV.2/Fe/IT/2006/GIV.2/Pistoia 387	GIV.2	EF450827
GV/Mu/US/2002/GV.1/MNV-1	GV.1	AY228235
GV/Rn/HK/2011/GV.2/HKU_CT2	GV.2	JX486101
GVI/Ca/IT/2007/Bari 91	GVI.1	FJ875027
GVI/Ca/PT/2007/Viseu	GVI.2	GQ443611
GVII/Ca/HK/2007/GVII/026F	GVII	FJ692500

Note: According to classification system of the online norovirus typing tool at http://www.rivm.nl/mpf/norovirus/typingtool (Kroneman et al., 2011)
The cryptogram is organized as follows: Genogroup/host species of origin /country of origin/year of occurrence/Capsid genotype/strain name
Host species abbreviations are: *Bo*, bovine; *Ca*, canine, *Fe*, feline; *Hu*, human; *Mu*, murine; *Ov*, ovine; *Po*, porcine; *Rn*, rat
Country abbreviations are: *CA*, Canada; *DE*, Germany; *IR*, Iraq; *IT*, Italy; *JP*, Japan; *NL*, Netherlands; *NZ*, New Zealand; *PT*, Portugal; *SA*, Saudi Arabia; *UK*, United Kingdom; *US*, United States
[a]*The pandemic GII.4 variants and their GenBank accession numbers are: US95_96 (AJ004864), Farmington_Hills_2002 (AY485642), Asia_2003 (AB220921), Hunter_2004 (AY883096), Yerseke_2006a (EF126963), Den Haag_2006b (EF126965), NewOrleans_2009 (GU445325), and Sydney_2012 (JX459908) (Kroneman et al., 2013).*

a polymerase (P) genotyping system based on phylogenetic analysis of the 3′-terminal 1300 nucleotides of the norovirus ORF1 (that encodes a large part of the RdRp protein) (Kroneman et al., 2013). In an effort to unify the disparate nomenclature in the literature, the known P genotypes were recently assigned number or letter designations (Kroneman et al., 2013). The P genotyping system currently focuses on GI and GII noroviruses, each containing 14 and 27 unique P types, respectively (Table 3.5.3). A number of reference viruses share the same P and C genotyping designations (Table 3.5.3), but recombination events (Jiang et al., 1999a) can lead to varying combinations of ORF1 and ORF2. Some C genotypes are found in combination with two or more P genotypes, as illustrated by the GII.4 Bristol (GII/Hu/GB/1993/GII.P4-GII.4/Bristol), Sakai (GII/Hu/JP/2005/GII.P12-GII.4/Sakai/04-179, and OC07138 (GII/Hu/JP/2007/GII.Pe-GII.4/OC07138) noroviruses (Table 3.5.3).

2.2.7 Norovirus Online Genotyping Tool

An online typing tool has been developed for identification of norovirus genogroup, genotype (P or C), and GII.4 variant cluster (Kroneman et al., 2011; Verhoef et al., 2011) (http://www.rivm.nl/mpf/norovirus/typingtool). Full-length norovirus genomic or partial nucleotide sequences can be submitted

TABLE 3.5.3 Norovirus Genogroups and Genotypes as Determined by Polymerase (P) Gene Relatedness

Reference Virus	Genogroup. P Genotype	GenBank Accession Number
GI/Hu/US/1968/GI.P1-GI.1/Norwalk	**GI.P1**	**M87661**
GI/Hu/GB/1991/GI.P2-GI.2/Southampton	**GI.P2**	**L07418**
GI/Hu/US/1998/GI.P3-GI.3/VA98115	**GI.P3**	**AY038598**
GI/Hu/JP/1987/GI.P4-GI.4/Chiba407	**GI.P4**	**AB042808**
GI/Hu/SE/2005/GI.P5-unknown/07_1	GI.P5	**EU007765**
GI/Hu/DE/1997/GI.P6-GI.6/BS5(Hesse)	**GI.P6**	**AF093797**
GI/Hu/SE/2008/GI.P7-GI.7/Lilla Edet	**GI.P7**	**JN603251**
GI/Hu/US/2008/GI.P8-GI.8/890321	**GI.P8**	**GU299761**
GI/Hu/FR/2004/GI.P9-GI.9/ Chatellerault709	**GI.P9**	**EF529737**
GI/Hu/SA/1990/GI.Pa-GI.3/DesertShield	GI.Pa	**U04469**
GI/Hu/JP/2002/GI.Pb-GI.6/WUG1	GI.Pb	**AB081723**
GI/Hu/JP/2000/GI.Pc-GI.5/SzUG1	GI.Pc	**AB039774**
GI/Hu/FR/2003/GI.Pd-GI.3/Vesoul576	GI.Pd	**EF529738**
GI/Hu/JP/1979/GI.Pf-GI.3/Otofuke	GI.Pf	**AB187514**
GII/Hu/US/1971/GII.P1-GII.1/Hawaii	**GII.P1**	**U07611**
GII/Hu/GB/1994/GII.P2-GII.2/Melksham	**GII.P2**	**X81879**
GII/Hu/CA/1991/GII.P3-GII.3/Toronto	**GII.P3**	**U02030**
GII/Hu/GB/1993/GII.P4-GII.4/Bristol	**GII.P4**	**X76716**
GII/Hu/HU/1999/GII.P5-GII.5/MOH	**GII.P5**	**AF397156**
GII/Hu/JP/2002/GII.P6-GII.6/Saitama U16	**GII.P6**	**AB039778**
GII/Hu/JP/2002/GII.P7-GII.6/Saitama U4	GII.P7	**AB039777**
GII/Hu/JP/2002/GII.P8-GII.8/Saitama U25	**GII.P8**	**AB039780**
GII/Po/US/1997/GII.P11-GII.11/Sw918	**GII.P11**	**AB074893**
GII/Hu/JP/2005/GII.P12-GII.4/ Sakai/04-179	GII.P12	**AB220922**
GII/Hu/FR/2004/GII.P13-GII.17/ Briancon870	GII.P13	**EF529741**
GII/Hu/JP/2006/GII.P15-GII.15/ Hiroshima66	**GII.P15**	**AB360387**
GII/Hu/DE/2000/GII.P16-GII.16/ Neustrelitz260	**GII.P16**	**AY772730**
GII/Po/US/2003/GII.P18-GII.18/OH-QW101	**GII.P18**	**AY823304**

(Continued)

TABLE 3.5.3 Norovirus Genogroups and Genotypes as Determined by Polymerase (P) Gene Relatedness (*cont.*)

Reference Virus	Genogroup. P Genotype	GenBank Accession Number
GII/Hu/DE/2005/GII.P20-GII.20/Leverkusen267	**GII.P20**	**EU424333**
GII/Hu/FR/2004/GII.P21-GII.2/Pont de Roide673	GII.P21	**AY682549**
GII/Hu/JP/2003/GII.P22-GII.5/Hokkaido133	GII.P22	**AB212306**
GII/Hu/JP/2004/GII.Pa-GII.3/SN2000JA	GII.Pa	**AB190457**
GII/Hu/US/1976/GII.Pc-GII.2/SnowMountain	GII.Pc	**AY134748**
GII/Hu/JP/2007/GII.Pe-GII.4/OC07138	GII.Pe	**AB434770**
GII/Hu/FR/1999/GII.Pf-GII.5/S63	GII.Pf	**AY682550**
GII/Hu/AU/1983/GII.Pg-GII.13/Goulburn Valley	GII.Pg	**DQ379714**
GII/Hu/JP/1997/GII.Ph-GII.2/OC97007	GII.Ph	**AB089882**
GII/Hu/GR/1997/GII.Pj-GII.2/E3	GII.Pj	**AY682552**
GII/Hu/JP/1996/GII.Pk-Unknown/OC96065	GII.Pk	**AF315813**
GII/Hu/IN/2006/GII.Pm-GII.12/PunePC24	GII.Pm	**EU921353**
GII/Hu/CN/2007/GII.Pn-GII.22/Beijing53931	GII.Pn	**GQ856469**

Note: According to classification system of the online norovirus typing tool at http://www.rivm.nl/mpf/norovirus/typingtool (Kroneman et al., 2011)
Cryptogram is organized as follows: Genogroup/host species of origin/country of origin/year of occurrence/Pol (P) genotype-Capsid (C) genotype/strain name
Host species abbreviations: *Hu*, human; *Po*, porcine
Country abbreviations are: *AU*, Australia; *CA*, Canada; *CN*, China; *DE*, Germany; *GB*, Great Britain; *GR*, Greece; *FR*, France; *HU*, Hungary; *IN*, India; *IT*, Italy; *JP*, Japan; *SA*, Saudi Arabia; *SE*, Sweden; *UK*, United Kingdom; *US*, United States
Bold lettering represents P genotypes with matched C genotype numbering

and quickly matched to a database of known reference strains for genotype assignment. New genotype and variant designations are assigned by an international norovirus working group (reviewed in Kroneman et al., 2013). The development of a unified nomenclature system should facilitate surveillance efforts to track the emergence and spread of epidemic norovirus strains. Data sharing networks such as the international NoroNet group (Siebenga et al., 2009) and CaliciNet in the United States (Vega et al., 2011) provide platforms that allow communication of molecular epidemiological findings in real time.

3 MECHANISMS FOR THE GENERATION OF NOROVIRUS DIVERSITY

Noroviruses are diverse and constantly changing. A number of mechanisms have been identified that allow the generation of diversity and the emergence of new norovirus strains (White, 2014; Bull and White, 2011). Genetic drift, transmission bottlenecks, and recombination are among the factors driving norovirus evolution, and are briefly described later.

3.1 Genetic Drift

The RdRp of the *Caliciviridae*, like that of other positive strand RNA viruses, does not have an editing function, resulting in a comparably higher error rate per genome per replication cycle than DNA viruses (Duffy et al., 2008). The subsequent nucleotide mutations can lead to fixed substitutions at the viral population level, a key component in adaptation. Genetic drift may best be illustrated by the evolutionary patterns of the predominant GII.4 noroviruses that show the emergence of a new pandemic variant every few years (Fig. 3.5.3). The genetic changes of an emerging variant are often clustered in epitopes in the surface-exposed P2 region of the capsid, suggesting a role for herd immunity and selective antibody pressure (antigenic drift) in its evolution (reviewed in Debbink et al., 2012). In immunocompromised individuals, the population of RNA molecules is even more diverse, presumably because of the absence of selective pressure to drive escape from antibodies and other immune responses (Bull et al., 2012).

One study compared the replication and mutation rates of circulating norovirus strains that varied in global prevalence to assess whether the RdRp error rate might influence strain success. Three pandemic GII.4 strains, a nonpandemic recombinant GII.4/GII.10 strain, a prevalent GII.b/GII.3 virus, and the infrequently detected GII.7 strain were evaluated (Bull et al., 2010). Evidence was found for a higher error rate (decreased fidelity) for the pandemic GII.4 RdRp, leading to the suggestion that a more error-prone polymerase might facilitate the rapid evolution of GII.4 viruses by introducing a higher rate of adaptive substitutions, especially in the VP1 (Bull et al., 2010).

Estimates of the evolutionary rates of two prevalent norovirus genotypes, GII.3 and GII.4, were determined by comparative phylogenetic analysis of viruses occurring over a span of approximately 40 years (Boon et al., 2011; Bok et al., 2009). The GII.3 and GII.4 norovirus genotypes were calculated to evolve at similar rates of $4.16 \times 10(-3)$ and $4.3 \times 10(-3)$ nucleotide substitutions/site/year (strict clock), respectively, even though differences in their evolutionary patterns within the VP1 sequence were noted (Boon et al., 2011; Bok et al., 2009). This evolutionary rate is comparable to that of other positive strand RNA viruses with rates of within one order of magnitude of $1 \times 10(-3)$ nucleotide substitutions/site/year (Duffy et al., 2008). These data suggest that

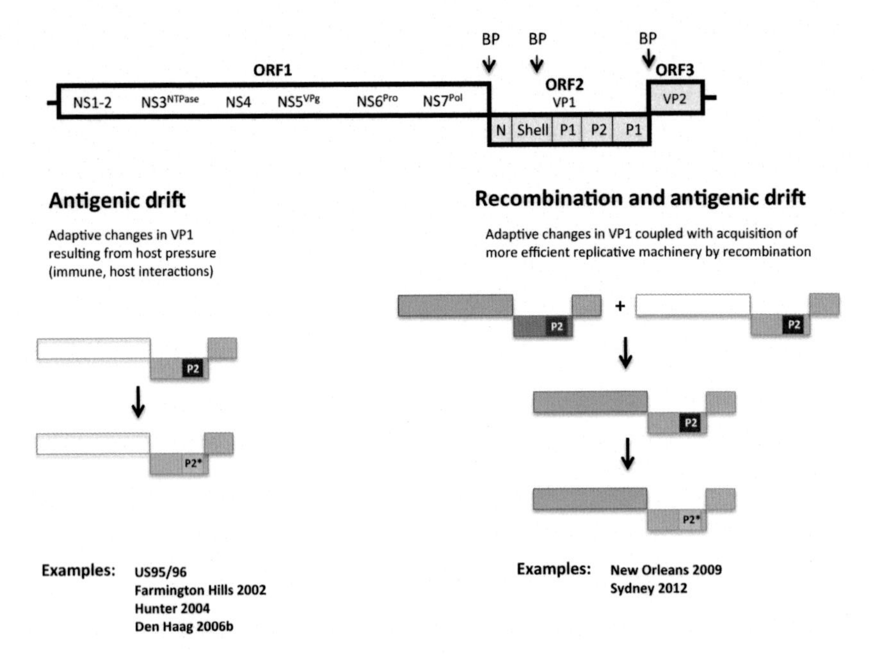

FIGURE 3.5.3 **Mechanisms by which GII.4 noroviruses evolve to become pandemic.** Evidence for two distinct mechanisms in the emergence of pandemic GII.4 noroviruses has been documented in molecular epidemiologic studies: antigenic drift and recombination (White, 2014). The VP1 capsid protein is organized into domains (Shell and Protruding) and subdomains (P1 and P2) (Prasad et al., 1999). A short N-terminal region (N) precedes the highly conserved S domain. The P2 domain, highly variable among strains, bears key antigenic and host carbohydrate ligand binding sites and is a major site of host selective immune pressure (Debbink et al., 2012). The asterisk represents a GII.4 P2 domain that has sustained amino acid substitutions that escape herd immunity. Proposed recombination breakpoints (BP) in GII.4 recombinants are indicated, with BPs identified at the ORF1/ORF2 junction, within ORF2, and near the ORF2/ORF3 junction (Eden et al., 2013). Examples of pandemic GII.4 variants and the mechanism by which they may have emerged are shown.

noroviruses, although diverse and subject to genetic drift, are not equipped with a replication machinery that drives them to evolve at higher rates than most other positive strand RNA viruses.

3.2 Transmission Bottlenecks

Deep sequencing of clinical specimens has shown that norovirus strains exist as a mixed population of genetically closely related RNA variants in the host, with a predominant RNA sequence that can comprise >60% of the total population (Bull et al., 2012). However, transmission of this mixed population into a new host may create a fitness bottleneck, and the emergence of a minor population can give rise to new population diversity (Bull et al., 2012). The highly infectious nature of norovirus and its efficient spread by person-to-person transmission (Koopmans, 2008) creates ample opportunity for the emergence of new variants via this mechanism. A recent study of the norovirus genomic RNA population in an immunocompromised host showed evidence for the presence of novel GII.4 variants with the potential for transmission into a susceptible population (Vega et al., 2014a). The impact of transmission bottlenecks in norovirus epidemiology will require additional viral metagenomics studies. (See Chapter 5.1.)

3.3 Recombination

During calicivirus replication, two major positive-sense RNA species are generated, with one corresponding to the full-length genome (approx. 7.6 kb in size) and the other to a 3′-end coterminal subgenomic region (approx. 2.6 kb in size) (Neill and Mengeling, 1988). The abundant subgenomic RNA serves as a template for translation of the VP1 and VP2 capsid proteins. There is evidence that during coinfection of a cell with two caliciviruses, the two genomic RNAs replicate in proximity so that recombination events can readily occur (Abente et al., 2013), likely due to template switching by the RdRp (Bull et al., 2005). A recent in vitro study found that after coinfection of cells with two parental feline calicivirus strains, recombination was detected as early as 6 h after infection at a rate of $6.8 \times 10(-6)$ single direction recombinant genomes generated per parental virus genome (Symes et al., 2015). Although the recombination rate is not known for the human noroviruses, epidemiological studies suggest that recombination is not a rare event (Lu et al., 2015). A hotspot for recombination in the noroviruses exists at the junction between ORF1 and ORF2 as a breakpoint (BP) for template switching, yielding new combinations of VP1 and polymerase that may confer adaptive advantages to the new strain (Bull et al., 2005; Mans et al., 2014; Shen et al., 2012) (Fig. 3.5.3). For example, the "GII.Pb" polymerase has been detected globally for nearly a decade and has been associated with at least five different GII capsid proteins (GII.1, GII.2, GII.3, GII.4, and GII.13) (Lim et al., 2013; Reuter et al., 2005). The GII.4 noroviruses may have additional recombination BPs within ORF2 and between ORF2 and 3, leading to new combinations of VP1 and VP2 (Kamel et al., 2009; Eden et al., 2013). Evidence for additional BPs in ORF1 was found in the murine norovirus (MNV) VPg, RdRp, and protease genes following coinfection of mice with two distinct MNV strains (Zhang et al., 2015). A common feature of the MNV BP sites was high sequence identity between the two parental strains, strongly suggesting that any highly conserved region may, in theory, be a site for template switching and recombination.

Tracking the molecular epidemiology of the Pol gene (ORF1) in the natural history of norovirus infection may give insight into the factors driving recombination and why different capsid and RdRp combinations occur in nature (Hasing et al., 2014). Recombination occurs predominantly among strains of the same genogroup (intragenogroup), although rare intergenogroup recombinants have been reported (Nayak et al., 2008; Nataraju et al., 2011). There may be specific structural constraints for capsid and polymerase interactions as evidence has been reported for a direct interaction between the two proteins during RNA replication (Subba-Reddy et al., 2012).

4 MOLECULAR EPIDEMIOLOGY AND TRANSMISSION

4.1 Distribution of Norovirus Genotypes

Norovirus genotypes vary among geographical locations, seasons, and years. Outbreaks occur more often in the winter months of temperate climates (Ahmed et al., 2013). A link has been proposed between higher particle stability and low absolute humidity (below 0.007 kg/kg air), which could explain, in part, the increase of norovirus transmission and illness during cold months when these conditions predominate (Colas de la Noue et al., 2014). The marked diversity and unpredictable pattern of norovirus genotypes is reflected in epidemiologic surveys conducted across the globe. The diversity is especially striking in environmental studies, reflecting the ubiquitous distribution of these viruses (Fernandez et al., 2012; van den Berg et al., 2005). Two consistently observed features in most clinical surveys are the predominance of GII compared to GI viruses, and the higher prevalence of the GII.4 genotype (Matthews et al., 2012; Vega et al., 2014b). In contrast, studies of environmental samples, waterborne outbreaks, or shellfish may find a predominance of GI strains (Matthews et al., 2012; Lowther et al., 2012). The molecular basis for this variation in genotypic distribution is not yet known, but the emergence of a pathogenic strain likely involves properties of the virus (eg, physicochemical features, replicative fitness) and the susceptible host (eg, immune status, HBGA profile).

Most human norovirus capsid proteins interact with HBGA carbohydrate ligands displayed on intestinal tissue (Marionneau et al., 2002), and there is a diverse array of HBGA molecules and other glycan-containing molecules such as galactosylceramide in the gut that could potentially serve as attachment factors (Bally et al., 2012). The precise mechanism of viral entry into host cells is not known. The HBGA phenotype of an individual, as well as the presentation of these HBGA antigens on intestinal epithelial cells (ie, secretor status) is determined by inherited alleles that affect norovirus binding to host cells and thus, susceptibility to infection (reviewed in Le Pendu, 2004 and in Chapter 3.3). Ancestry-based differences in HBGA genetics can influence norovirus susceptibility in different populations (Currier et al., 2015). Norovirus strains vary in their recognition of HBGA carbohydrates, and these interactions have been studied extensively at the structural level with recombinant norovirus P domains and synthetic saccharides (Cao et al., 2007; Huang et al., 2003). The GI and GII noroviruses exhibit sequence differences in their HBGA binding interfaces, although an overall structural similarity in the HBGA recognition sites of the P2 domain has remained conserved during evolution of the two distinct phylogenetic lineages (Tan et al., 2009). An association between susceptibility to GII.4 noroviruses and the presence of a functional FUT2 enzyme has been noted in epidemiologic surveys, suggesting that the higher prevalence of secretors (80%) versus nonsecretors (20%) in many populations insures a large pool of susceptible individuals for this predominant genotype (Currier et al., 2015; Frenck et al., 2012). The specific amino acid sequences for HBGA binding are often highly conserved in VP1 proteins of the same genotype or cluster, but subtle changes in sequence can affect the binding of a norovirus strain to a specific HBGA ligand that consequently might influence host cell interactions (Tan et al., 2003; de Rougemont et al., 2011).

4.2 Diversity and Norovirus Vaccine Development

Serologic surveys have shown evidence for the acquisition of both GI and GII norovirus-specific antibodies early in life in most populations, although regional differences in rates of antibody acquisition and specificity have been reported (Jiang et al., 2000; Menon et al., 2013; Gray et al., 1993). Susceptibility to norovirus disease spans all age groups, and this lifetime susceptibility enables the occurrence of sharp gastroenteritis outbreaks in a wide range of settings involving otherwise healthy individuals. Severe, life-threatening norovirus illness occurs most often in the young and old, suggesting that both, immunological immaturity in the young and immunosenescence in the elderly, hamper virus clearance. This is consistent with the link between immunocompromised patients and chronic norovirus infection, in which impaired immune function affects virus clearance (Vega et al., 2014a, reviewed in Bok and Green, 2012).

A key goal in norovirus vaccine development is the formulation of a vaccine that will provide broad protection against the myriad of circulating genotypes. Early cross-challenge studies had shown that Norwalk virus (Genogroup I) and Hawaii virus (Genogroup II) were serotypically distinct (Wyatt et al., 1974), consistent with a proposed minimal need for representative strains from these two genogroups in a vaccine (Bok et al., 2011; Malm et al., 2015). Presently, a bivalent norovirus VLP-based candidate is under investigation containing both GI and GII components (Treanor et al., 2014), with the GII component consisting of an engineered consensus GII.4 VLP (Parra et al., 2012b). Efficacy data will be important to assess whether additional antigenic components are needed. Immunity to noroviruses is complex, and infection with a virus of one genotype may not induce immunity to a second genotype in young individuals (Parra and Green, 2014). This observation is reflected in natural history studies as well, in which a high rate of sequential reinfection occurs in children with noroviruses belonging to different genotypes within the same genogroup (Ayukekbong et al., 2014; Saito et al., 2014).

Efforts are underway to identify cross-reactive antigenic sites that are shared among the diverse genotypes (Kitamoto et al., 2002; Parker et al., 2005; Crawford et al., 2015; Parra et al., 2013). An understanding of heterotypic immunity will be important in the development of broadly-protective vaccines or immunotherapy (see Chapter 3.6).

4.3 Transmission and Site of Replication

Noroviruses are transmitted by the fecal–oral route, and shed in the feces. It is estimated that approximately 14% of norovirus outbreaks are foodborne (Verhoef et al., 2015). The virus is apparently most infectious in the early phase of the acute illness, when the symptoms of vomiting and/or diarrhea are most severe and norovirus shedding is at its peak (Zelner et al., 2013). The infectious dose has been estimated in volunteer studies. In one such study, aggregation of Norwalk virus particles was detected and the ID_{50} dose of disaggregated virus was estimated to be approximately between 18 and 1015 genome equivalents (Teunis et al., 2008). In a second dose-response study of Norwalk virus, a higher ID_{50} dose of 1320–2800 genome equivalents was reported, with dose depending on secretor and blood group status (Atmar et al., 2014). Norovirus is shed to similar levels (as measured by RT-qPCR) in symptomatic and asymptomatic individuals, making it problematic to use virus genome copy numbers as a marker of disease (Teunis et al., 2014).

Viral replication is thought to occur predominantly in the small intestine, and studies in adult volunteers (Agus et al., 1973) and a calf model (Otto et al., 2011) show a striking pathology of blunted villi. Extra-intestinal sites of replication have not been confirmed in humans, but murine norovirus can be detected in multiple organs in the immunocompromised mouse host (Karst et al., 2003). A recent study found no evidence of viremia in healthy adults challenged with norovirus, suggesting local replication in the gut (Newman et al., 2015). Norovirus RNA has been detected in the serum of immunocompromised children (Frange et al., 2012), but the presence of the virus in organs outside the gut has not been confirmed. Norovirus strains may vary in their pathogenicity, as has been noted for the GII.4 noroviruses (Desai et al., 2012), but there is presently no marker to track or predict the virulence of circulating viruses. Identification of the mechanisms of norovirus pathogenesis among all members of this diverse genus will be an important area of future research.

5 SUMMARY AND FUTURE DIRECTIONS

Noroviruses are major enteric pathogens, and efforts are underway to develop vaccines and antiviral drugs to control morbidity and mortality. Molecular epidemiologic studies will continue to play a key role in the development of effective strategies. The implementation of a recently developed unified genotyping nomenclature system will facilitate the tracking (and possible prediction) of emerging pandemic strains and inform vaccine design. Investigations will undoubtedly continue to develop tractable and efficient cell culture systems as well as "humanized" animal models to elucidate the mechanisms of norovirus pathogenesis and immunity. These model systems, in concert with clinical investigations and vaccine efficacy trials, will be empowered by ground-breaking new technologies to understand norovirus evolution and immunity at the molecular level. Elucidation of the complex interactions between virus and host that result in acute and chronic norovirus infection will undoubtedly remain a compelling area of enteric virus research.

ACKNOWLEDGMENTS

This work was supported by the Division of Intramural Research (DIR), NIAID, NIH, Bethesda, Maryland. The author would like to thank Gabriel I. Parra, LID, NIAID, for data analysis and critical review. In addition, the author thanks Jordan Johnson, Stanislav V. Sosnovtsev, Eric Levenson, and Allison Behrle of LID, NIAID, for reviewing the chapter and providing constructive comments.

REFERENCES

Abente, E.J., Sosnovtsev, S.V., Sandoval-Jaime, C., Parra, G.I., Bok, K., Green, K.Y., 2013. The feline calicivirus leader of the capsid protein is associated with cytopathic effect. J. Virol. 87 (6), 3003–3017.

Agus, S.G., Dolin, R., Wyatt, R.G., Tousimis, A.J., Northrup, R.S., 1973. Acute infectious nonbacterial gastroenteritis: intestinal histopathology. Histologic and enzymatic alterations during illness produced by the Norwalk agent in man. Ann. Intern. Med. 79 (1), 18–25.

Ahmed, S.M., Lopman, B.A., Levy, K., 2013. A systematic review and meta-analysis of the global seasonality of norovirus. PLoS One 8 (10), e75922.

Anderson, A.D., Garrett, V.D., Sobel, J., et al., 2001. Multistate outbreak of Norwalk-like virus gastroenteritis associated with a common caterer. Am. J. Epidemiol. 154 (11), 1013–1019.

Ando, T., Mulders, M.N., Lewis, D.C., Estes, M.K., Monroe, S.S., Glass, R.I., 1994. Comparison of the polymerase region of small round structured virus strains previously classified in three antigenic types by solid-phase immune electron microscopy. Arch. Virol. 135 (1-2), 217–226.

Atmar, R.L., Opekun, A.R., Gilger, M.A., et al., 2014. Determination of the 50% human infectious dose for Norwalk virus. J. Infect. Dis. 209 (7), 1016–1022.

Ayukekbong, J.A., Fobisong, C., Tah, F., Lindh, M., Nkuo-Akenji, T., Bergstrom, T., 2014. Pattern of circulation of norovirus GII strains during natural infection. J. Clin. Microbiol. 52 (12), 4253–4259.

Bally, M., Rydell, G.E., Zahn, R., et al., 2012. Norovirus GII.4 virus-like particles recognize galactosylceramides in domains of planar supported lipid bilayers. Angew. Chem. Int. Ed. Engl. 51 (48), 12020–12024.

Bartsch, S.M., Lopman, B.A., Hall, A.J., Parashar, U.D., Lee, B.Y., 2012. The potential economic value of a human norovirus vaccine for the United States. Vaccine 30 (49), 7097–7104.

Bertolotti-Ciarlet, A., Crawford, S.E., Hutson, A.M., Estes, M.K., 2003. The 3' end of Norwalk virus mRNA contains determinants that regulate the expression and stability of the viral capsid protein VP1: a novel function for the VP2 protein. J. Virol. 77 (21), 11603–11615.

Bok, K., Green, K.Y., 2012. Norovirus gastroenteritis in immunocompromised patients. N. Engl. J. Med. 367 (22), 2126–2132.

Bok, K., Abente, E.J., Realpe-Quintero, M., et al., 2009. Evolutionary dynamics of GII.4 noroviruses over a 34-year period. J. Virol. 83 (22), 11890–11901.

Bok, K., Parra, G.I., Mitra, T., et al., 2011. Chimpanzees as an animal model for human norovirus infection and vaccine development. Proc. Natl. Acad. Sci. USA 108 (1), 325–330.

Boon, D., Mahar, J.E., Abente, E.J., et al., 2011. Comparative evolution of GII.3 and GII.4 norovirus over a 31-year period. J. Virol. 85 (17), 8656–8666.

Bu, W., Mamedova, A., Tan, M., Xia, M., Jiang, X., Hegde, R.S., 2008. Structural basis for the receptor binding specificity of Norwalk virus. J. Virol. 82 (11), 5340–5347.

Bull, R.A., White, P.A., 2011. Mechanisms of GII.4 norovirus evolution. Trends Microbiol. 19 (5), 233–240.

Bull, R.A., Hansman, G.S., Clancy, L.E., Tanaka, M.M., Rawlinson, W.D., White, P.A., 2005. Norovirus recombination in ORF1/ORF2 overlap. Emerg. Infect. Dis. 11 (7), 1079–1085.

Bull, R.A., Eden, J.S., Rawlinson, W.D., White, P.A., 2010. Rapid evolution of pandemic noroviruses of the GII.4 lineage. PLoS Pathog. 6 (3), e1000831.

Bull, R.A., Eden, J.S., Luciani, F., McElroy, K., Rawlinson, W.D., White, P.A., 2012. Contribution of intra- and interhost dynamics to norovirus evolution. J. Virol. 86 (6), 3219–3229.

Cao, S., Lou, Z., Tan, M., et al., 2007. Structural basis for the recognition of blood group trisaccharides by norovirus. J. Virol. 81 (11), 5949–5957.

Clarke, I.N., Estes, M.K., Green, K.Y., Hansman, G.S., Knowles, N.J., Koopmans, M.K., Matson, D.O., Meyers, G., Neill, J.D., Radford, A., Smith, A.W., Studdert, M.J., Thiel, H.-J., Vinje, J., 2012. Caliciviridae. In: King, A.M.Q., Adams, M.J., Carstens, E.B., Lefkowitz, E.J. (Eds.), Virus Taxonomy: Ninth Report of the International Committee on Taxonomy of Viruses. Elsevier Academic Press, London, pp. 977–986.

Colas de la Noue, A., Estienney, M., Aho, S., et al., 2014. Absolute humidity influences the seasonal persistence and infectivity of human norovirus. Appl. Environ. Microbiol. 80 (23), 7196–7205.

Crawford, S.E., Ajami, N., Parker, T.D., et al., 2015. Mapping broadly reactive norovirus genogroup I and II monoclonal antibodies. Clin. Vaccine. Immunol. 22 (2), 168–177.

Currier, R.L., Payne, D.C., Staat, M.A., et al., 2015. Innate susceptibility to norovirus infections influenced by FUT2 genotype in a United States pediatric population. Clin. Infect. Dis. 60 (11), 1631–1638.

de Rougemont, A., Ruvoen-Clouet, N., Simon, B., et al., 2011. Qualitative and quantitative analysis of the binding of GII.4 norovirus variants onto human blood group antigens. J. Virol. 85 (9), 4057–4070.

Debbink, K., Lindesmith, L.C., Donaldson, E.F., Baric, R.S., 2012. Norovirus immunity and the great escape. PLoS Pathog. 8 (10), e1002921.

Desai, R., Hembree, C.D., Handel, A., et al., 2012. Severe outcomes are associated with genogroup 2 genotype 4 norovirus outbreaks: a systematic literature review. Clin. Infect. Dis. 55 (2), 189–193.

Duffy, S., Shackelton, L.A., Holmes, E.C., 2008. Rates of evolutionary change in viruses: patterns and determinants. Nat. Rev. Genet. 9 (4), 267–276.

Eden, J.S., Tanaka, M.M., Boni, M.F., Rawlinson, W.D., White, P.A., 2013. Recombination within the pandemic norovirus GII.4 lineage. J. Virol. 87 (11), 6270–6282.

Fernandez, M.D., Torres, C., Poma, H.R., et al., 2012. Environmental surveillance of norovirus in Argentina revealed distinct viral diversity patterns, seasonality and spatio-temporal diffusion processes. Sci. Total Environ. 437, 262–269.

Frange, P., Touzot, F., Debre, M., et al., 2012. Prevalence and clinical impact of norovirus fecal shedding in children with inherited immune deficiencies. J. Infect. Dis. 206 (8), 1269–1274.

Frenck, R., Bernstein, D.I., Xia, M., et al., 2012. Predicting susceptibility to norovirus GII.4 by use of a challenge model involving humans. J. Infect. Dis. 206 (9), 1386–1393.

Glass, R.I., Parashar, U.D., Estes, M.K., 2009. Norovirus gastroenteritis. N. Engl. J. Med. 361 (18), 1776–1785.

Gray, J.J., Jiang, X., Morgan-Capner, P., Desselberger, U., Estes, M.K., 1993. Prevalence of antibodies to Norwalk virus in England: detection by enzyme-linked immunosorbent assay using baculovirus-expressed Norwalk virus capsid antigen. J. Clin. Microbiol. 31, 1022–1025.

Hall, A.J., Lopman, B.A., Payne, D.C., et al., 2013. Norovirus disease in the United States. Emerg. Infect. Dis. 19 (8), 1198–1205.

Hardy, M.E., Kramer, S.F., Treanor, J.J., Estes, M.K., 1997. Human calicivirus genogroup II capsid sequence diversity revealed by analyses of the prototype Snow Mountain agent. Arch. Virol. 142 (7), 1469–1479.

Hasing, M.E., Hazes, B., Lee, B.E., Preiksaitis, J.K., Pang, X.L., 2014. Detection and analysis of recombination in GII.4 norovirus strains causing gastroenteritis outbreaks in Alberta. Infect Genet Evol 27, 181–192.

Huang, P., Farkas, T., Marionneau, S., et al., 2003. Noroviruses bind to human ABO, Lewis, and secretor histo-blood group antigens: identification of 4 distinct strain-specific patterns. J. Infect. Dis. 188 (1), 19–31.

Hyde, J.L., Sosnovtsev, S.V., Green, K.Y., Wobus, C., Virgin, H.W., Mackenzie, J.M., 2009. Mouse norovirus replication is associated with virus-induced vesicle clusters originating from membranes derived from the secretory pathway. J. Virol. 83 (19), 9709–9719.

Jiang, X., Graham, D.Y., Wang, K., Estes, M.K., 1990. Norwalk virus genome cloning and characterization. Science 250, 1580–1583.

Jiang, X., Wang, M., Graham, D.Y., Estes, M.K., 1992. Expression, self-assembly, and antigenicity of the Norwalk virus capsid protein. J. Virol. 66 (11), 6527–6532.

Jiang, X., Espul, C., Zhong, W.M., Cuello, H., Matson, D.O., 1999a. Characterization of a novel human calicivirus that may be a naturally occurring recombinant. Arch. Virol. 144 (12), 2377–2387.

Jiang, X., Huang, P.W., Zhong, W.M., Farkas, T., Cubitt, D.W., Matson, D.O., 1999b. Design and evaluation of a primer pair that detects both Norwalk- and Sapporo-like caliciviruses by RT-PCR. J. Virol. Methods 83 (1–2), 145–154.

Jiang, X., Wilton, N., Zhong, W.M., et al., 2000. Diagnosis of human caliciviruses by use of enzyme immunoassays. J. Infect. Dis. 181 (S2), S349–S359.

Jing, Y., Qian, Y., Huo, Y., Wang, L.P., Jiang, X., 2000. Seroprevalence against Norwalk-like human caliciviruses in Beijing. China. J Med Virol 60 (1), 97–101.

Kageyama, T., Kojima, S., Shinohara, M., et al., 2003. Broadly reactive and highly sensitive assay for Norwalk-like viruses based on real-time quantitative reverse transcription-PCR. J Clin Microbiol 41 (4), 1548–1557.

Kamel, A.H., Ali, M.A., El-Nady, H.G., de Rougemont, A., Pothier, P., Belliot, G., 2009. Predominance and circulation of enteric viruses in the region of Greater Cairo. Egypt. J Clin Microbiol 47 (4), 1037–1045.

Kapikian, A.Z., Wyatt, R.G., Dolin, R., Thornhill, T.S., Kalica, A.R., Chanock, R.M., 1972. Visualization by immune electron microscopy of a 27-nm particle associated with acute infectious nonbacterial gastroenteritis. J. Virol. 10 (5), 1075–1081.

Karst, S.M., Wobus, C.E., Lay, M., Davidson, J., Virgin, H.W. 4th, 2003. STAT1-dependent innate immunity to a Norwalk-like virus. Science 299 (5612), 1575–1578.

Karst, S.M., Wobus, C.E., Goodfellow, I.G., Green, K.Y., Virgin, H.W., 2014. Advances in norovirus biology. Cell Host Microbe 15 (6), 668–680.

Katpally, U., Wobus, C.E., Dryden, K., Virgin, H.W., 4th, Smith, T.J., 2008. Structure of antibody-neutralized murine norovirus and unexpected differences from viruslike particles. J. Virol. 82 (5), 2079–2088.

Kaufman, S.S., Green, K.Y., Korba, B.E., 2014. Treatment of norovirus infections: Moving antivirals from the bench to the bedside. Antiviral Res 105C, 80–91.

Kitamoto, N., Tanaka, T., Natori, K., et al., 2002. Cross-reactivity among several recombinant calicivirus virus-like particles (VLPs) with monoclonal antibodies obtained from mice immunized orally with one type of VLP. J Clin Microbiol. 40 (7), 2459–2465.

Kojima, S., Kageyama, T., Fukushi, S., Genogroup-specific PCR, et al., 2002. Primers for detection of Norwalk-like viruses. J. Virol. Methods 100 (1–2), 107–114.

Kong, B.H., Lee, S.G., Han, S.H., Jin, J.Y., Jheong, W.H., Paik, S.Y., 2015. Development of enhanced primer sets for detection of norovirus. Biomed Res Int 2015, 103052.

Koopmans, M., 2008. Progress in understanding norovirus epidemiology. Curr Opin Infect Dis 21 (5), 544–552.

Kroneman, A., Vennema, H., Deforche, K., et al., 2011. An automated genotyping tool for enteroviruses and noroviruses. J Clin Virol 51 (2), 121–125.

Kroneman, A., Vega, E., Vennema, H., et al., 2013. Proposal for a unified norovirus nomenclature and genotyping. Arch. Virol. 158 (10), 2059–2068.

Le Pendu, J., 2004. Histo-blood group antigen and human milk oligosaccharides: genetic polymorphism and risk of infectious diseases. Adv. Exp. Med. Biol. 554, 135–143.

Lim, K.L., Eden, J.S., Oon, L.L., White, P.A., 2013. Molecular epidemiology of norovirus in Singapore, 2004-2011. J. Med. Virol. 85 (10), 1842–1851.

Lowther, J.A., Gustar, N.E., Powell, A.L., Hartnell, R.E., Lees, D.N., 2012. Two-year systematic study to assess norovirus contamination in oysters from commercial harvesting areas in the United Kingdom. Appl. Environ. Microbiol. 78 (16), 5812–5817.

Lu, Q.B., Huang, D.D., Zhao, J., et al., 2015. An increasing prevalence of recombinant GII norovirus in pediatric patients with diarrhea during 2010-2013 in China. Infect. Genet. Evol. 31, 48–52.

Malm, M., Tamminen, K., Lappalainen, S., Uusi-Kerttula, H., Vesikari, T., Blazevic, V., 2015. Genotype considerations for virus-like particle-based bivalent norovirus vaccine composition. Clin. Vaccine Immunol. 22 (6), 656–663.

Mans, J., Murray, T.Y., Taylor, M.B., 2014. Novel norovirus recombinants detected in South Africa. Virol. J. 11, 168.

Marionneau, S., Ruvoen, N., Le Moullac-Vaidye, B., et al., 2002. Norwalk virus binds to histo-blood group antigens present on gastroduodenal epithelial cells of secretor individuals. Gastroenterology 122 (7), 1967–1977.

Matthews, J.E., Dickey, B.W., Miller, R.D., et al., 2012. The epidemiology of published norovirus outbreaks: a review of risk factors associated with attack rate and genogroup. Epidemiol. Infect. 140 (7), 1161–1172.

McFadden, N., Bailey, D., Carrara, G., et al., 2011. Norovirus regulation of the innate immune response and apoptosis occurs via the product of the alternative open reading frame 4. PLoS Pathog. 7 (12), e1002413.

Menon, V.K., George, S., Aladin, F., et al., 2013. Comparison of age-stratified seroprevalence of antibodies against norovirus GII in India and the United Kingdom. PLoS One 8 (2), e56239.

Nataraju, S.M., Pativada, M.S., Kumar, R., et al., 2011. Emergence of novel Norovirus recombinants with NVGII.1/NVGII.5 RdRp gene and NVGII.13 capsid gene among children and adults in Kolkata, India. Int. J. Mol. Epidemiol. Genet. 2 (2), 130–137.

Nayak, M.K., Balasubramanian, G., Sahoo, G.C., et al., 2008. Detection of a novel intergenogroup recombinant Norovirus from Kolkata. India Virol. 377 (1), 117–123.

Neill, J.D., Mengeling, W.L., 1988. Further characterization of the virus-specific RNAs in feline calicivirus infected cells. Virus Res. 11 (1), 59–72.

Newman, K.L., Leon, J.S., 2015. Norovirus immunology: Of mice and mechanisms. Eur. J. Immunol. 45 (10), 2742–2757.

Newman, K.L., Marsh, Z., Kirby, A.E., Moe, C.L., Leon, J.S., 2015. Immunocompetent adults from human norovirus challenge studies do not exhibit norovirus viremia. J. Virol. 89 (13), 6968–6969.

O'Ryan, M.L., Vial, P.A., Mamani, N., et al., 1998. Seroprevalence of Norwalk virus and Mexico virus in Chilean individuals: assessment of independent risk factors for antibody acquisition. Clin. Infect. Dis. 27 (4), 789–795.

Otto, P.H., Clarke, I.N., Lambden, P.R., Salim, O., Reetz, J., Liebler-Tenorio, E.M., 2011. Infection of calves with bovine norovirus GIII.1 strain Jena virus: an experimental model to study the pathogenesis of norovirus infection. J. Virol. 85 (22), 12013–12021.

Parker, T.D., Kitamoto, N., Tanaka, T., Hutson, A.M., Estes, M.K., 2005. Identification of Genogroup I and Genogroup II broadly reactive epitopes on the norovirus capsid. J. Virol. 79 (12), 7402–7409.

Parra, G.I., Green, K.Y., 2014. Sequential gastroenteritis episodes caused by 2 norovirus genotypes. Emerg. Infect. Dis. 20 (6), 1016–1018.

Parra, G.I., Abente, E.J., Sandoval-Jaime, C., Sosnovtsev, S.V., Bok, K., Green, K.Y., 2012a. Multiple antigenic sites are involved in blocking the interaction of GII.4 norovirus capsid with ABH histo-blood group antigens. J. Virol. 86 (13), 7414–7426.

Parra, G.I., Bok, K., Taylor, R., et al., 2012b. Immunogenicity and specificity of norovirus Consensus GII.4 virus-like particles in monovalent and bivalent vaccine formulations. Vaccine 30 (24), 3580–3586.

Parra, G.I., Azure, J., Fischer, R., et al., 2013. Identification of a broadly cross-reactive epitope in the inner shell of the norovirus capsid. PLoS One 8 (6), e67592.

Payne, D.C., Vinje, J., Szilagyi, P.G., et al., 2013. Norovirus and medically attended gastroenteritis in U.S. children. N. Engl. J. Med. 368 (12), 1121–1130.

Prasad, B.V., Hardy, M.E., Jiang, X., Estes, M.K., 1996. Structure of Norwalk virus. Arch. Virol. Suppl 12, 237–242.

Prasad, B.V., Hardy, M.E., Dokland, T., Bella, J., Rossmann, M.G., Estes, M.K., 1999. X-ray crystallographic structure of the Norwalk virus capsid. Science 286 (5438), 287–290.

Ramani, S., Atmar, R.L., Estes, M.K., 2014. Epidemiology of human noroviruses and updates on vaccine development. Curr. Opin. Gastroenterol. 30 (1), 25–33.

Reeck, A., Kavanagh, O., Estes, M.K., et al., 2010. Serological correlate of protection against norovirus-induced gastroenteritis. J. Infect. Dis. 202 (8), 1212–1218.

Reuter, G., Krisztalovics, K., Vennema, H., Koopmans, M., Szucs, G., 2005. Evidence of the etiological predominance of norovirus in gastroenteritis outbreaks—emerging new-variant and recombinant noroviruses in Hungary. J. Med. Virol. 76 (4), 598–607.

Saito, M., Goel-Apaza, S., Espetia, S., et al., 2014. Multiple norovirus infections in a birth cohort in a Peruvian Periurban community. Clin. Infect. Dis. 58 (4), 483–491.

Shen, Q., Zhang, W., Yang, S., et al., 2012. Genomic organization and recombination analysis of human norovirus identified from China. Mol. Biol. Rep. 39 (2), 2–1275.

Siebenga, J.J., Vennema, H., Zheng, D.P., et al., 2009. Norovirus illness is a global problem: emergence and spread of norovirus GII.4 variants, 2001-2007. J. Infect. Dis. 200 (5), 802–812.

Smit, T.K., Steele, A.D., Peenze, I., Jiang, X., Estes, M.K., 1997. Study of Norwalk virus and Mexico virus infections at Ga-Rankuwa Hospital, Ga-Rankuwa, South Africa. J. Clin. Microbiol. 35 (9), 2381–2385.

Sosnovtsev, S.V., 2010. Proteolytic cleavage and viral proteins. In: Hansman, G.S., Jiang, X., Green, K.Y. (Eds.), Caliciviruses: Molecular and Cellular Virology. Caister Academic Press, Norfolk, UK, pp. 65–94.

Sosnovtsev, S.V., Green, K.Y., 2000. Identification and genomic mapping of the ORF3 and VPg proteins in feline calicivirus virions. Virology 277 (1), 193–203.

Sosnovtsev, S.V., Belliot, G., Chang, K.O., Onwudiwe, O., Green, K.Y., 2005. Feline calicivirus VP2 Is essential for the production of infectious virions. J. Virol. 79 (7), 4012–4024.

Subba-Reddy, C.V., Yunus, M.A., Goodfellow, I.G., Kao, C.C., 2012. Norovirus RNA synthesis is modulated by an interaction between the viral RNA-dependent RNA polymerase and the major capsid protein VP1. J. Virol. 86 (18), 10138–10149.

Symes, S.J., Job, N., Ficorilli, N., Hartley, C.A., Browning, G.F., Gilkerson, J.R., 2015. Novel assay to quantify recombination in a calicivirus. Vet. Microbiol. 177 (1–2), 25–31.

Tan, M., Huang, P., Meller, J., Zhong, W., Farkas, T., Jiang, X., 2003. Mutations within the P2 domain of norovirus capsid affect binding to human histo-blood group antigens: evidence for a binding pocket. J. Virol. 77 (23), 12562–12571.

Tan, M., Xia, M., Chen, Y., et al., 2009. Conservation of carbohydrate binding interfaces: evidence of human HBGA selection in norovirus evolution. PLoS One 4 (4), e5058.

Teunis, P.F., Moe, C.L., Liu, P., et al., 2008. Norwalk virus: how infectious is it? J. Med. Virol. 80 (8), 1468–1476.

Teunis, P.F., Sukhrie, F.H., Vennema, H., Bogerman, J., Beersma, M.F., Koopmans, M.P., 2014. Shedding of norovirus in symptomatic and asymptomatic infections. Epidemiol. Infect., 1–8.

Thorne, L.G., Goodfellow, I.G., 2014. Norovirus gene expression and replication. J. Gen. Virol. 95 (Pt 2), 278–291.

Treanor, J.J., Atmar, R.L., Frey, S.E., et al., 2014. A novel intramuscular bivalent norovirus virus-like particle vaccine candidate—reactogenicity, safety, and immunogenicity in a phase 1 trial in healthy adults. J. Infect. Dis. 210 (11), 1763–1771.

van den Berg, H., Lodder, W., van der Poel, W., Vennema, H., de Roda Husman, A.M., 2005. Genetic diversity of noroviruses in raw and treated sewage water. Res. Microbiol. 156 (4), 532–540.

Vega, E., Barclay, L., Gregoricus, N., Williams, K., Lee, D., Vinje, J., 2011. Novel surveillance network for norovirus gastroenteritis outbreaks United States. Emerg. Infect. Dis. 17 (8), 1389–1395.

Vega, E., Donaldson, E., Huynh, J., et al., 2014a. RNA populations in immunocompromised patients as reservoirs for novel norovirus variants. J. Virol. 88 (24), 14184–14196.

Vega, E., Barclay, L., Gregoricus, N., Shirley, S.H., Lee, D., Vinje, J., 2014b. Genotypic and epidemiologic trends of norovirus outbreaks in the United States, 2009 to 2013. J. Clin. Microbiol. 52 (1), 147–155.

Verhoef, L., Kouyos, R.D., Vennema, H., et al., 2011. An integrated approach to identifying international foodborne norovirus outbreaks. Emerg. Infect. Dis. 17 (3), 412–418.

Verhoef, L., Hewitt, J., Barclay, L., et al., 2015. Norovirus genotype profiles associated with foodborne transmission, 1999-2012. Emerg. Infect. Dis. 21 (4), 592–599.

Vinje, J., 2015. Advances in laboratory methods for detection and typing of norovirus. J. Clin. Microbiol. 53 (2), 373–381.

Vinje, J., Koopmans, M.P., 1996. Molecular detection and epidemiology of small round-structured viruses in outbreaks of gastroenteritis in the Netherlands. J. Infect. Dis. 174 (3), 610–615.

Vinje, J., Hamidjaja, R.A., Sobsey, M.D., 2004. Development and application of a capsid VP1 (region D) based reverse transcription PCR assay for genotyping of genogroup I and II noroviruses. J. Virol. Methods 116 (2), 109–117.

Vongpunsawad, S., Venkataram Prasad, B.V., Estes, M.K., 2013. Norwalk virus minor capsid protein VP2 associates within the VP1 shell domain. J. Virol. 87 (9), 4818–4825.

Wang, J., Jiang, X., Madore, H.P., et al., 1994. Sequence diversity of small, round-structured viruses in the Norwalk virus group. J. Virol. 68 (9), 5982–5990.

White, P.A., 2014. Evolution of norovirus. Clin. Microbiol. Infect. 20 (8), 741–745.

Wobus, C.E., Karst, S.M., Thackray, L.B., et al., 2004. Replication of Norovirus in cell culture reveals a tropism for dendritic cells and macrophages. PLoS Biol. 2 (12), e432.

Wyatt, R.G., Dolin, R., Blacklow, N.R., et al., 1974. Comparison of three agents of acute infectious nonbacterial gastroenteritis by cross-challenge in volunteers. J. Infect. Dis. 129 (6), 709–714.

Zelner, J.L., Lopman, B.A., Hall, A.J., Ballesteros, S., Grenfell, B.T., 2013. Linking time-varying symptomatology and intensity of infectiousness to patterns of norovirus transmission. PLoS One 8 (7), e68413.

Zhang, H., Cockrell, S.K., Kolawole, A.O., et al., 2015. Isolation and analysis of rare norovirus recombinants from co-infected mice using drop-based microfluidics. J. Virol. 89 (15), 7722–7734.

Zhu, S., Regev, D., Watanabe, M., et al., 2013. Identification of immune and viral correlates of norovirus protective immunity through comparative study of intracluster norovirus strains. PLoS Pathog. 9 (9), e1003592.

Chapter 3.6

Norovirus Vaccine Development

S. Ramani*, M.K. Estes*,, R.L. Atmar*,****
**Department of Molecular Virology and Microbiology, Baylor College of Medicine, Houston, TX, United States; **Department of Medicine, Baylor College of Medicine, Houston, TX, United States*

1 BACKGROUND

Human noroviruses (NoV) are a leading cause of sporadic and epidemic gastroenteritis across all age groups and are associated with nearly one-fifth of all cases of acute gastroenteritis worldwide (Ahmed et al., 2014). A systematic review of global data showed that NoVs were detected in approximately 24% of acute gastroenteritis cases in the community, 20% of cases in outpatient settings and 17% among inpatients. In the United States, NoV infections result in 19–21 million total illnesses annually, leading to 1.7–1.9 million outpatient visits, 400,000 emergency department visits, 56,000–71,000 hospitalizations and 570–800 deaths (Hall et al., 2013). The economic burden of NoV disease is also high with NoV hospitalizations costing an estimated $500 million annually in the US while foodborne NoV infections result in $2 billion in healthcare and loss of productivity costs (Bartsch et al., 2012). While NoV infections occur in individuals of all age groups, severe outcomes are seen in the elderly and in immunocompromised populations (Bok and Green, 2012; Trivedi et al., 2013). NoVs are also important pediatric pathogens, accounting for nearly 12% of diarrheal hospitalizations in children under 5 years of age (Ramani et al., 2014). In countries where rotavirus vaccines are effective, NoVs are rapidly replacing rotavirus as the most common cause of viral gastroenteritis in this age group (Koo et al., 2013; Payne et al., 2013). The high clinical and economic impact of NoV disease warrants the need for safe and effective vaccines. While there are currently no licensed NoV vaccines, many candidate vaccines are under development or are in clinical trials.

NoVs are classified into six genogroups (GI–GVI) of which genogroups I, II and IV are known to infect humans (Ramani et al., 2014). The prototype virus Norwalk virus (NV) belongs to genogroup I (GI) while most cases of NoV infections described worldwide are caused by genogroup II (GII) NoVs. In particular, genotype GII.4 is the predominant genotype detected worldwide. The epidemiology of human NoVs is complex and new variants of the GII.4 genotype are known to emerge every 2–3 years, replacing the previously dominant variant. (see Chapter 3.5.) Histo-blood group antigens (HBGAs) are cell attachment factors for many human NoVs (Hutson et al., 2003; Lindesmith et al., 2003; Huang et al., 2005), and immune pressure-driven antigenic variation in epitopes surrounding the HBGA binding domain of the capsid protein VP1 may contribute to the evolution and emergence of new variants, especially among GII.4 viruses (Lindesmith et al., 2012). This chapter briefly summarizes data on immune responses to NoVs, details the developments and progress in the field of NoV vaccines and highlights the challenges remaining.

2 IMMUNE CORRELATES

A correlate of protection (CoP) can be defined as an immune marker that statistically correlates with protection from infection or illness, be it induced by natural infection or in response to vaccination. Specifically in the case of vaccines, the CoP marker correlates with vaccine efficacy or protection from disease after vaccination. (Plotkin and Gilbert, 2012). The identification of such a marker (or markers) is often based on studies of immune responses to natural infection with the pathogen. Seroprevalence studies in many populations indicate that most adults have high levels of antibodies to NoVs (Greenberg et al., 1979; O'Ryan et al., 1998; Menon et al., 2013). However, the absence of a fully permissive cell culture system or a small animal models for human NoVs has limited the study of immunity to NoVs. In particular, this poses a problem for performing virus neutralization assays to assess protective immunity.

Much of our understanding of immune responses to natural infection therefore comes from human volunteer challenge studies (Parrino et al., 1977; Ryder et al., 1985; Johnson et al., 1990; Atmar et al., 2014). In one study using NV as the challenge virus, serum IgA levels peaked at day 14 post infection while serum IgG peaked at day 28 (Kavanagh et al., 2011). IgM antibodies were also present in most infected adults. While serological responses to infection were observed in many studies, the data on correlation between protection from disease and levels of antibodies as measured by enzyme linked immunosorbent assays (ELISA) remained inconsistent. The first CoP against NoV disease was identified when it was demonstrated that functional antibodies in serum that block the binding of VLPs to HBGAs reduces the risk of gastroenteritis following challenge with NV (Reeck et al., 2010). This finding was further validated by similar results obtained when the blocking activity was measured using a hemagglutination inhibition assay (Czako et al., 2012). Interestingly, HBGA-blocking antibody responses following NV infection were found to include heterotypic responses (Lindesmith et al., 2010; Czako et al., 2015). Although the peak titers of heterotypic antibodies and fold increases in blocking antibody levels were modest in comparison to the response seen with the homologous challenge GI.1 virus, blocking activity was seen against other GI viruses as well as GII.4 variants (Fig. 3.6.1A–C). A particularly remarkable finding was the detection of HBGA-blocking antibodies to GII.4 variants that were not circulating at the time of challenge or sample collection (Fig. 3.6.1B,C). These findings support the possibility of development of broadly cross-protective NoV vaccines using a limited number of NoV strains as immunogens. It is important to note that while data from early volunteer challenge studies suggest that natural infection may not confer long-term protection, a recent mathematical model estimated that immunity following natural NoV infection lasts 4–9 years (Simmons et al., 2013).

Apart from serum HBGA blocking antibodies, mucosal and cellular immunity also appears to play an important role in protection from NoV disease. In a NV volunteer challenge study, NV-specific IgA levels in saliva and stool peaked on day 14 after challenge in the infected participants (Ramani et al., 2015). Antibody-secreting cell (ASC) responses peaked on day 7 and were biased toward IgA. Memory B-cell responses peaked on day 14 and were biased toward IgG. In addition, NV-specific memory B cells but not ASCs persisted 180 days after infection. The study of mucosal and cellular immune responses resulted in the identification of two new potential correlates of protection against NV gastroenteritis. Prechallenge levels of NV-specific salivary IgA and NV-specific IgG memory B cells correlated with protection against NV-gastroenteritis (Ramani et al., 2015). NV-specific salivary IgA levels before challenge correlated with reduced severity of gastroenteritis. In addition, NV-specific fecal IgA levels before challenge were associated with a reduction in peak viral load, whereas fecal IgA measured on day 7 after infection correlated with a shorter duration of virus shedding. A rapid salivary IgA response was previously demonstrated to be associated with protection from infection following challenge with NV in a genetically-susceptible population (Lindesmith et al., 2003). A summary of immune correlates of protection is given in Table 3.6.1.

The efficacy of a vaccine can be measured using clinical end-points or by measuring a CoP that acts as a surrogate marker predicting clinical outcome. Often, the measurement of clinical end-points requires more expensive and effort-intensive study designs. The measurement of a CoP can thus facilitate studies on vaccine efficacy and contribute greatly to vaccine development. The identification of many immune markers or effector molecules that correlate with protection from gastroenteritis in the challenge studies, however, raises the important question on whether there are multiple mechanisms that contribute to protection from NV gastroenteritis. It also raises questions on whether the effector molecules identified are directly responsible for mediating protection or are a reflection of changes in levels of other effector molecules that covary and actually result in protection (mechanistic and nonmechanistic correlates of protection, respectively). These are critical questions in the context

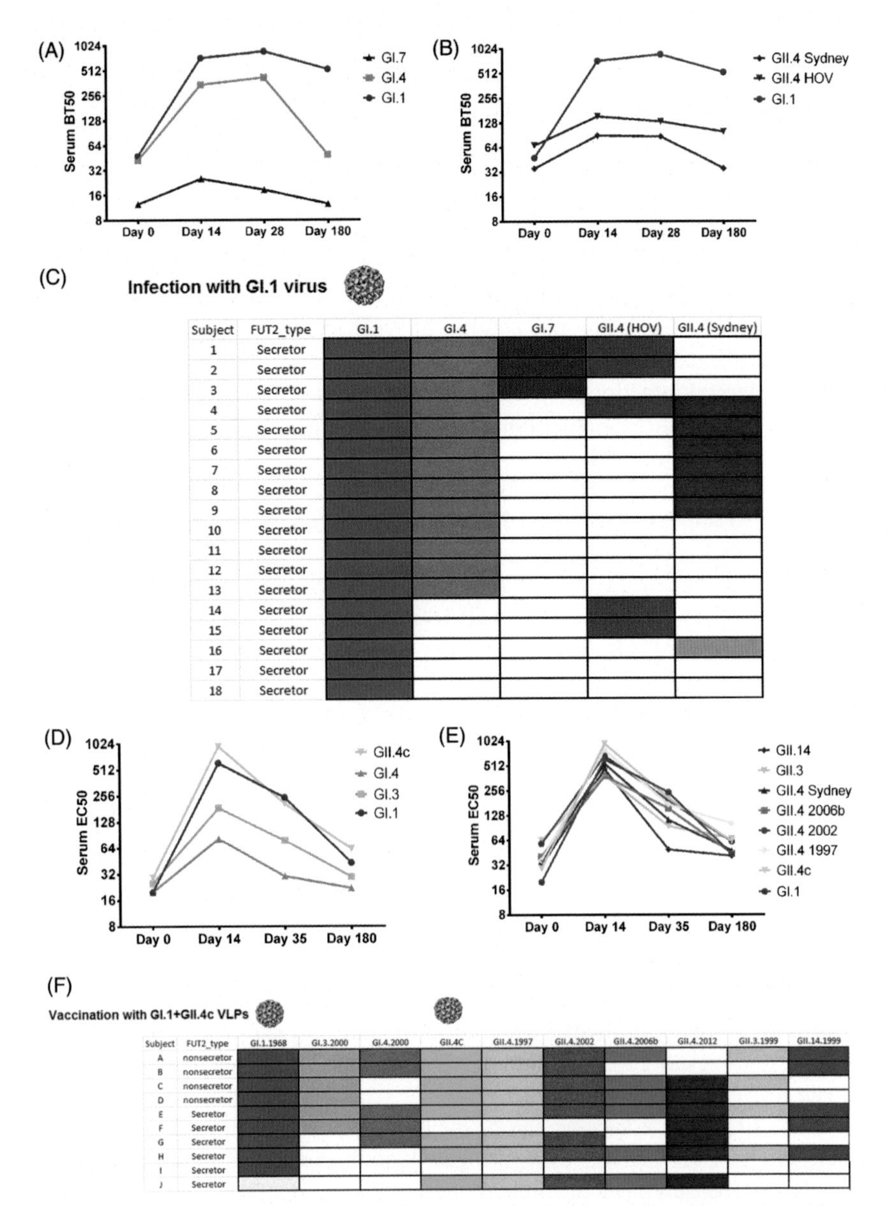

FIGURE 3.6.1 Heterotypic HBGA blocking antibody responses after norovirus infection and NoV VLP vaccination. Panels A–C show blocking antibody responses following infection with GI.1 virus while panels D–F show blocking antibody responses following vaccination with GI.1.and GII.4c VLPs. Y-axis in panels A, B, D, E represents 50% percent blocking titers (BT50 for infection, EC50 for vaccination as defined in the original studies describing these results) and is defined as the serum titer at which 50% of the VLP-carbohydrate interaction is blocked when compared to the positive control. The geometric mean titers (GMT) of blocking antibodies to GI VLPs following infection and vaccination are seen in Panels A, D respectively. Panels B, E show geometric mean titers (GMT) of blocking antibodies to GII VLPs following infection and vaccination. Shaded cells in panels C and F represent individuals who showed a >fourfold change in blocking antibody response post infection and vaccination, respectively. The response to each NoV genotype or variant is represented by a different color.

TABLE 3.6.1 Immune Correlates of Protection

Correlate	Outcome	References
Preexposure		
Serum histo-blood group antigen (HBGA)-blocking antibody	Illness	Reeck et al. (2010), Atmar et al. (2011), Atmar et al. (2015)
	Infection	Atmar et al. (2011), Atmar et al. (2015)
Serum hemagglutination inhibition (HAI) antibody	Illness	Czako et al. (2012)
Salivary IgA antibody	Illness	Ramani et al. (2015)
Memory B cells	Illness	Ramani et al. (2015)
Fecal IgA antibody	Peak virus shedding	Ramani et al. (2015)
Postexposure		
Rapid salivary IgA response	Infection	Lindesmith et al. (2003)
Day 7 fecal IgA response	Duration of virus shedding	Ramani et al. (2015)

of measuring immune response to vaccines and add complexity to study designs for NoV vaccine trials.

3 PRECLINICAL STUDIES WITH NOROVIRUS VACCINE CANDIDATES

In the absence of efficient cell culture systems to grow human NoVs, the possibility of developing live-attenuated or killed NoV vaccines is limited. However, a number of vaccine candidate for NoVs have been evaluated (Table 3.6.2). The expression of NoV capsid proteins in vitro results in self-assembly of virus-like particles (VLPs) that are morphologically and antigenically identical to particles of the infectious virus (Jiang et al., 1992; Green et al., 1993). The production of VLPs was first demonstrated when the NV capsid was produced by expressing the second and third open reading frames of the viral genome in insect cells infected with a recombinant baculovirus (Jiang et al., 1992). VLPs contain 180 copies of the VP1 capsid protein, and their particulate nature may enhance immune activation and uptake by antigen presenting cells in the Peyer's patches of the gastrointestinal tract. NoV VLPs have been proposed as candidate vaccines since their first description. Preclinical studies evaluated the immunogenicity of VLPs in mice, rabbits, gnotobiotic pigs and chimpanzees using various routes of administration (Table 3.6.3). The effect of different adjuvants on immune response was also tested in several studies.

Early preclinical studies evaluated the immunogenicity of VLPs in mice. Dose-response and kinetics of serum and intestinal immune responses were

TABLE 3.6.2 Norovirus Vaccine Candidates

Antigen	Expression system	References
Virus-like particles	Baculovirus	Jiang et al. (1992)
	Pichia (yeast)	Xia et al. (2007)
	Venezuelan equine encephalitis virus replicon	Harrington et al. (2002)
	Plants (tobacco, potato, tomato)	Mason et al. (1996), Zhang et al. (2006)
	Recombinant adenovirus	Guo et al. (2008)
P particles	*E. coli*	Tan and Jiang (2005)

evaluated in inbred and outbred mice that were administered NV VLPs orally (Ball et al., 1998). The VLPs were found to be immunogenic when administered in the presence or absence of cholera toxin (CT) as a mucosal adjuvant. The use of CT was associated with the induction of higher levels of serum IgG. Subsequent studies showed that the VLPs were highly immunogenic when administered intranasally, both in the presence and absence of heat-labile toxin of *Escherichia coli* (LT) as a mucosal adjuvant (Guerrero et al., 2001). In these studies, the intranasal route of delivery was found to be more effective than oral immunization, inducing responses to lower doses of VLPs. Intranasal immunization was also found to induce a stronger immune response compared to oral administration when a modified CT was used as adjuvant (Periwal et al., 2003). The use of CT as an adjuvant resulted in a stronger Th2 type response. Higher numbers of antigen-specific IL-4-producing cells as well as antigen-specific IgA-secreting cells were observed in the Peyer's patches. Similar results were demonstrated with GII VLPs administered orally or by the intranasal route, with LT or a nontoxic mutant LT as mucosal adjuvant. Intranasal immunization resulted in high serum and fecal IgA responses that were enhanced in the presence of adjuvant (Nicollier-Jamot et al., 2004). A mixed Th1/Th2-like response was observed in cervical lymph nodes and Peyer's patch cultures with either adjuvant. A dry powder formulation of NV VLPs along with 3-O-desacyl-4′-monophosphoryl lipid A (MPL) as adjuvant and chitosan as a muco adherent induced NV-specific antibody responses in serum and cellular immunity including NV-specific IgA memory B cells in the mesenteric lymph nodes and Peyer's patches in rabbits (Richardson et al., 2013). In preclinical studies with a bivalent VLP formulation containing NV and GII.4 VLPs in rabbits, higher homologous and heterologous antibody responses were seen following immunization of animals by the intramuscular route when compared to the intranasal route (Parra et al., 2012). In another study, gnotobiotic pigs were vaccinated with human GII.4 VLPs with immune-stimulating complexes or mutant LT as

TABLE 3.6.3 Select Preclinical Studies of NoV Vaccine Candidates

Immunogen	Animal	Route	Adjuvant	Findings	References
Baculovirus-expressed GI.1/Norwalk VLPs	Mice	Oral gavage	+/– cholera toxin	Oral administration of unadjuvanted vaccine induced systemic and mucosal immune responses, and use of adjuvant augmented responses	Ball et al. (1998)
Baculovirus-expressed GI.1/Norwalk VLPs	Mice	Intranasal (IN) versus oral gavage	+/– *E. coli* mutant (R192G) labile toxin (LT)	Intranasal administration of unadjuvanted vaccine induced systemic and mucosal immune responses, and use of adjuvant augmented responses at lower dosages than following oral gavage	Guerrero et al. (2001)
Baculovirus-expressed GI.1/Norwalk VLPs	Mice	Intranasal versus oral gavage	+/– mutant (E29H) cholera toxin	Adjuvant enhanced cellular and serological immune responses, and responses were higher following IN administration compared to the oral route	Periwal et al. (2003)
Baculovirus-expressed GII.4/Dijon VLPs	Mice	Intranasal versus oral gavage	+/– *E. coli* wild type (wt) or mutant (R192G) labile toxin	Adjuvant enhanced cellular and serological immune responses, and responses were higher following IN administration compared to the oral route. Responses were similar when wt LT or mutant LT were used as adjuvant	Nicollier-Jamot et al. (2004)
Baculovirus-expressed GII.4/HS66 VLPs	Gnotobiotic pigs	Oral and intranasal	*E. coli* mutant (R192G) labile toxin or ISCOM matrix	Both adjuvanted vaccines induced serological and cellular immune responses and were associated with decreased viral shedding and diarrhea after virus challenge compared to control animals	Souza et al. (2007)

(*Continued*)

TABLE 3.6.3 Select Preclinical Studies of NoV Vaccine Candidates (*cont.*)

Immunogen	Animal	Route	Adjuvant	Findings	References
Baculovirus-expressed GI.1/Norwalk versus GII.4/MD145 VLPs	Chimpanzees	Intramuscular (IM)	Aluminum hydroxide	Seroresponses occurred after vaccination; only immunization with the homologous antigen protected against infection following intravenous challenge with Norwalk virus	Bok et al. (2011)
GI.1/Norwalk VLPs—tobacco plant extracts and potato expressing VLPs	Mice	Oral gavage	+/– cholera toxin	Plant-expressed VLPs were immunogenic	Mason et al. (1996)
GII.4/VA387 VLPs—*Pichia pastoris*-expressed raw extract	Mice	Oral gavage or intramuscular	Ribi adjuvant with IM injection	Yeast-expressed VLPs immunogenic after IM delivery and when orally-delivered, induced serum and fecal antibody responses, including production of serum HBGA-blocking antibody	Xia et al. (2007)
Venezuelan equine encephalitis (VEE) replicons and VEE replicon-expressed GI.1/Norwalk VLPs	Mice	Footpad inoculation or oral gavage	None	Serum antibody responses were induced by both routes of immunization, but the replicon induced higher antibody levels than obtained following oral VLP administration	Harrington et al. (2002)
VEE replicons expressing GI.1/Norwalk, GII.1/Hawaii, GII.2/Snow Mountain, GII.4/Lordsdale VLPs	Mice	Footpad inoculation	None	Monovalent preparations induced only homotypic HBGA-blocking antibody responses, while a trivalent (GI.1, GII.1, GII.2) preparation induced a heterotypic response to a fourth genotype (GII.4), although to a lesser degree than when the GII.4 strain was included in a vaccination regimen (as monovalent or quadrivalent)	LoBue et al. (2006)

GII.4/VA387 P particles expressed in *E. coli* and yeast	Mice	Intranasal	None	P particle vaccine had immunogenicity similar to that achieved using VLPs as the immunogen, including induction of strain-specific HBGA-blocking antibody	Tan et al. (2008)
E. coli-expressed GII.4/VA387 P particles and chimeric P particles with rotavirus VP8	Mice	Intranasal, subcutaneous	Freund's adjuvant (subcutaneous only)	The chimera enhanced responses to VP8 compared to free VP8 delivered IN; it induced immune responses that provided protection against rotavirus challenge in a mouse model; and it induced homotypic norovirus HBGA-blocking antibodies	Tan et al. (2011)
GII.4/VA387 baculovirus-expressed VLPs and *E. coli*-expressed P particles and P dimers	Mice	Intranasal	None	VLPs and P particles induced higher levels of antibody and CD4 T cellular immune responses compared to the P dimers	Fang et al. (2013)
Baculovirus-expressed GII.4/1999 VLPs; *E. coli*-expressed GII.4/1999 P particles	Mice	Intramuscular, intradermal	None	VLPs induced higher serum antibody levels, heterotypic ELISA antibody responses, more balanced Th1/Th2 antibody response, and IFN-gamma-expressing cellular responses compared to P particles	Tamminen et al. (2012)
Baculovirus-expressed GI.3, GII.4/1999, rotavirus rVP6	Mice	Intramuscular	None	Trivalent vaccine induced homotypic and heterotypic serum antibody responses, mucosal antibody responses and T cell responses	Tamminen et al. (2013)
E. coli-expressed GII.4/VA387 P dimers or fusion proteins with hepatitis E virus P domain dimers	Mice	Intranasal	None	Immunization with the fusion protein induced hepatitis E virus neutralizing antibodies and homotypic norovirus HBGA-blocking antibodies	Wang et al. (2014)

adjuvant (Souza et al., 2007). Both formulations showed high rates of seroconversion and fecal IgA responses as well as protection from live virus infection 28 days after vaccination. VLPs administered with LT induced Th1/Th2 serum cytokines and cytokine-secreting cells, while animals immunized with VLPs and immune-stimulating complexes showed Th2-biased responses. In a chimpanzee model of NV infection, the presence of NV-specific serum antibodies correlated with protection against re-infection at 4, 10, and 24 months following primary infection (Bok et al., 2011). In addition, chimpanzees vaccinated intramuscularly with GI VLPs were protected from NV infection when challenged 2 and 18 months after vaccination.

Apart from baculovirus-expression systems, VLPs produced in a number of other expression systems, such as transgenic plants (tobacco and potatoes), yeast, bacterial expression systems, and alphavirus replicon systems, have also been evaluated in preclinical studies. Recombinant NV (rNV) VLPs expressed in transgenic plants were immunogenic when delivered orally in mice (Mason et al., 1996). Serum IgG antibody was induced in response to the tobacco leaf extracts and potato tubers while NV-specific secretory IgA responses was seen with the plant expressed rNV. Oral administration of raw material from *Pichia pastoris* yeast lysates that contained GII.4 VLPs resulted in systemic and mucosal immune responses in mice without the use of adjuvants (Xia et al., 2007). In this study, both serum and fecal antibodies blocked the binding of homologous NoV VLPs to HBGA antigens. Venezuelan equine encephalitis (VEE) virus and other alphaviruses have been used as vectors to express heterologous proteins. In this approach, mice subcutaneously inoculated with VEE replicon particles expressing the NV capsid protein developed systemic and mucosal immune responses to the homologous NV VLPs as well as heterotypic antibody responses to another genotype of GI virus (Harrington et al., 2002). The administration of multivalent VEE replicon particles encoding various NoV VLPs induced heterotypic blocking antibody responses, including to a genotype not included in the cocktail (LoBue et al., 2006). With the exception of the Hawaii virus, the monovalent vaccines induced the highest titers of serum IgG response.

A number of other approaches including NoV P-particles, P-dimers, polyvalent norovirus P domain–glutathione-S-transferase (GST) complexes and combination vaccines produced in bacteria have also been evaluated as candidate vaccines (Table 3.6.3). P-particles are 24-mer particles produced in *Escherichia coli* that comprises the P domain of the NoV capsid. Intranasal immunization of mice with the P particles without an adjuvant or subcutaneous immunization with Freund's adjuvant resulted in the stimulation of humoral immune response to NoV VLPs (Tan et al., 2008; Tan et al., 2011). In addition, P-particles were shown to elicit cellular immunity after intranasal immunization in mice (Fang et al., 2013). Immune responses generated after immunization with VLPs and P-particles were compared in one study (Tamminen et al., 2012). Higher antibody responses were achieved after immunization with a single dose of VLPs whereas a booster dose was required for P-particle immunization. In addition, high avidity antibodies were raised only on immunization with VLPs. VLPs also resulted in a balanced Th1/Th2 type response whereas P-particles induced a Th2-biased response.

Combination vaccines targeting NoV and other viral pathogens are also being developed. In one study, a trivalent combination vaccine was developed consisting of VP1 from a GI NoV and a GII.4 NoV strain along with a tubular form of VP6, the most abundant rotavirus (RV) protein (Tamminen et al., 2013). Each component was produced by a recombinant baculovirus expression system and combined in vitro. High levels of NoV- and RV-specific serum and intestinal IgG antibodies were detected following intramuscular immunization in mice. Serum antibodies blocked binding of homologous and heterologous VLPs to HBGA antigens. A polyvalent complex platform was used to develop a bivalent vaccine against NoV and Hepatitis E virus (HEV) (Wang et al., 2014). Dimeric P domains of NoV and HEV were fused together, linked with dimeric GST and expressed in *E. coli*. The fusion protein assembled into polyvalent complexes and was found to be immunogenic in mice.

4 CLINICAL STUDIES

Following preclinical studies that demonstrated the immunogenicity of VLPs, clinical studies were initiated to assess the safety, immunogenicity and efficacy of VLP vaccines in humans (Table 3.6.4). The earliest phase I clinical studies evaluated the safety and immunogenicity of oral immunization with increasing doses of rNV VLPs in healthy adults. In one study, two doses of 100 or 250 μg of the rNV VLP vaccine were administered without adjuvant and a dose-dependent increase in serum IgG responses to rNV VLPs was observed (Ball et al., 1999). All vaccinees who received 250ug of rNV VLPs had a > fourfold increase in serum IgG titers, although most responses were seen after the first dose of vaccine. The vaccine was well tolerated, and no serious adverse events were reported. In another study, healthy adult volunteers received increasing doses of rNV VLP vaccines ranging from 250 to 2000 μg and humoral, mucosal and cellular immune responses were assessed (Tacket et al., 2003). The vaccine was well tolerated, and no volunteer experienced fever, diarrhea or vomiting in the 3 days following vaccination. Three vaccinees reported mild cramps while one person each in the vaccine and placebo arms experienced nausea. Headache and malaise were reported more commonly in the placebo arm of the study. Serum antibody responses were observed in 90% of vaccinees who received two doses of 250 μg of VLPs administered three weeks apart, and there was no further increase in response in the higher dosage groups. All vaccinees developed significant rises in IgA ASCs. Mucosal IgA responses were seen in less than 50% of vaccinees while transient lymphoproliferative and interferon gamma (IFNγ) responses were observed in the two lower dosage groups. Immune responses

TABLE 3.6.4 Summary of Immunogenicity From Clinical Studies of Norovirus Vaccine Candidates

Vaccine (Manufacturer)	Route	Dosage (Adjuvant)	Schedule	Number of Subjects (Age range)	Peak Serum IgG GMFR Reported (95% CI)	Peak Serum IgA GMFR Reported (95% CI)	Serum HAI or HBGA-blocking antibody GMFR Reported (95% CI)	References
GI.1/Norwalk (Baylor College of Medicine)	Oral	100 mcg (none)	2 doses 3 wks apart	5 (18–46 yrs)	5.3 (1.3–22.3)	3.0 (1.4–6.5)	ND	Ball et al. (1999)
		250 mcg (none)	2 doses 3 wks apart	15 (18–46 yrs)	8.8 (5.4–14.2)	4.4 (2.1–9.2)	ND	
GI.1/Norwalk (Baylor College of Medicine)	Oral	250 mcg (none)	2 doses 3 wks apart	10 (18–40 yrs)	~13[a] (NR)	NR	ND	Tacket et al. (2003)
		500 mcg (none)	2 doses 3 wks apart	10 (18–40 yrs)	~13[a] (NR)	NR	ND	
		2000 mcg (none)	2 doses 3 wks apart	10 (18–40 yrs)	~18[a] (NR)	NR	ND	
GI.1/Norwalk in potato (Boyce Thompson Institute for Plant Research)	Oral	150 g potato containing 215–751 mcg VLP	2 doses 3 wks apart	10 (adult)	8[a] (not calculable)	ND	ND	Tacket et al. (2000)
		150 g potato containing 215–751 mcg VLP	3 doses on days 0, 7, and 21	10 (adult)	13.3[a] (NR)	ND	ND	

GI.1/Norwalk (Ligocyte/ Takeda)	Intranasal	5 mcg (chitosan, MPL)	2 doses 3 wks apart	5 (18–49 yrs)	0.9 (NR)	1.2 (NR)	ND	El-Kamary et al. (2010)
		15 mcg (chitosan, MPL)	2 doses 3 wks apart	5 (18–49 yrs)	1.9 (NR)	2.5 (NR)	ND	
		50 mcg (chitosan, MPL)	2 doses 3 wks apart	9 (18–49 yrs)	4.7 (NR)	4.5 (NR)	ND	
		50 mcg (chitosan, MPL)	2 doses 3 wks apart	18 (18–49 yrs)	4.6 (2.5–8.6)	7.6 (4.2–13.8)	4.0 (2.0–7.9)—HAI	
		100 mcg (chitosan, MPL)	2 doses 3 wks apart	19 (18–49 yrs)	4.8 (3.2–7.1)	9.1 (4.7–17.6)	9.1 (4.0–20.7)—HAI	
GI.1/Norwalk (Ligocyte/ Takeda)	Intranasal	100 mcg (chitosan, MPL)	2 doses 3 wks apart	37 (18–50 yrs)	4.5 (3.1–6.5)	7.5 (4.6–12.2)	3.6 (2.8–4.7) HBGA-blocking	Atmar et al. (2011)
GI.1/Norwalk and GII.4/ consensus (Ligocyte/ Takeda)	Intramuscular	5 mcg each (MPL, aluminum hydroxide)	2 doses 4 wks apart	10 (18–49 yrs)	NR	NR	NR	Treanor et al. (2014)
		15 mcg each (MPL, aluminum hydroxide)	2 doses 4 wks apart	10 (18–49 yrs)	NR	NR	NR	
		150 mcg each (MPL, aluminum hydroxide)	2 doses 4 wks apart	9 (18–49 yrs)	NR	NR	NR	

(*Continued*)

TABLE 3.6.3 Summary of Immunogenicity From Clinical Studies of Norovirus Vaccine Candidates (*cont.*)

Vaccine (Manufacturer)	Route	Dosage (Adjuvant)	Schedule	Number of Subjects (Age range)	Peak Serum IgG GMFR Reported (95% CI)	Peak Serum IgA GMFR Reported (95% CI)	Serum HAI or HBGA-blocking antibody GMFR Reported (95% CI)	References
		50 mcg each (MPL, aluminum hydroxide)	2 doses 4 wks apart	18 (18–49 yrs)	GI.1: ~29 (NR) GII.4: ~11 (NR)	GI.1: ~39 (NR) GII.4: ~14 (NR)	GI.1: 16 (NR) GII.4: 10 (NR) HBGA-blocking	
		50 mcg each (MPL, aluminum hydroxide)	2 doses 4 wks apart	9 (50–64 yrs)	GI.1: ~57 (NR) GII.4: ~13 (NR)	GI.1: ~45 (NR) GII.4: ~10 (NR)	GI.1: 16 (NR) GII.4: 11 (NR) HBGA-blocking	
		50 mcg each (MPL, aluminum hydroxide)	2 doses 4 wks apart	10 (65–85 yrs)	GI.1: (~13 (NR) GII.4: ~6 (NR)	GI.1: ~42 (NR) GII.4: ~5 (NR)	GI.1: 16 (NR) GII.4: 6 (NR) HBGA-blocking	
GI.1/Norwalk and GII.4/ consensus (Ligocyte/ Takeda)	Intramuscular	50 mcg each (MPL, aluminum hydroxide)	2 doses 4 wks apart	49 (18–50 yrs)	GI.1: 27.8 (19.2, 40.2) GII.4: 8.8 (6.5, 11.9)	GI.1: 17.5 (11.9, 25.8) GII.4: 9.3 (6.7, 12.7)	GI.1: 31.6 (24.6, 40.5) GII.4: 7.8 (5.9, 10.2) HBGA-blocking	Bernstein et al. (2015), Atmar et al. (2015)

Abbreviations: *HAI*, hemagglutination inhibition; *HBGA*, histoblood group antigen; *MPL*, monophosphoryl lipid A; *ND*, not done; *NR*, not reported; *wks*, weeks; *yrs*, years; *GMFR*, geometric mean fold rise; *CI*, confidence interval; *mcg*, microgram.

[a]*Among responders only.*

were also measured in persons who consumed transgenic potatoes expressing the NV capsid protein (Tacket et al., 2000). Although IgG and IgA antibody responses were modest, 95% of vaccinees had a significant increase in virus-specific IgA-ASCs.

Intranasal administration of adjuvanted NV VLP vaccines was the next approach evaluated. A dry powder formulation of the vaccine consisting of NV VLPs produced using a recombinant baculovirus expression system, MPL as adjuvant, chitosan as a mucoadherent, sucrose and mannitol excipients as bulking agents and preservatives was tested in healthy adults (El-Kamary et al., 2010). No vaccine-related serious adverse events were reported, and the vaccine was found to be immunogenic. Mild local symptoms were reported in the vaccinees including nasal stuffiness, discharge and sneezing. Headache and malaise were the most commonly reported systemic symptoms and were not significantly different between vaccine and placebo recipients. Dose-dependent increases in serum IgA and IgG antibodies were observed, and functional antibodies were induced as measured by hemagglutination inhibition assay. Importantly, intranasal immunization was shown to elicit NV-specific IgA and IgG ASCs with different homing potentials. IgA-specific ASCs exhibited homing potential to the gut mucosa and peripheral lymphoid tissues while IgG-ASCs expressed homing receptors to peripheral lymphoid tissues. Another study also demonstrated the production of dose-dependent, functional memory B-cell responses to an enteric pathogen following intranasal immunization with VLP vaccines (Ramirez et al., 2012). The frequency of antigen-specific memory B cells correlated with serum antibody responses and mucosally-primed ASCs. This vaccine was then tested in a proof-of-concept efficacy trial in healthy, susceptible adults who received two doses of vaccine or placebo 3 weeks apart, followed by challenge with NV given at approximately 10 times the human infectious dose 50% (HID50) (Atmar et al., 2011). No vaccine-related serious adverse events were reported in this study. Similar to the previous study, the most common local symptoms after vaccination were nasal stuffiness, discharge, and sneezing. Local and systemic reactogenicity occurred with similar frequency in the vaccine and placebo groups after the first dose of the vaccine while local symptoms were reported more frequently in the vaccine group after the second dose of the vaccine. Although the level of protection was only modest, vaccinated individuals were less likely to develop infection or gastroenteritis compared with placebo recipients. The incidence of NV infection was lower in the vaccine arm compared to the placebo group (61 vs 82% respectively, $P = 0.05$). Importantly, the incidence of protocol-defined gastroenteritis in the vaccine arm was 37% compared to 69% in the placebo arm ($P = 0.006$). Delayed onset of illness and overall reduction in disease severity also were observed in vaccinees who developed gastroenteritis. Higher prechallenge levels of HBGA blocking antibodies correlated with protection against infection and illness. In particular, HBGA blocking titers of 200 or greater were associated with more than 50% reduction in the frequency of illness and infection.

Early cross-challenge studies in human volunteers had demonstrated that protective immunity did not extend to heterotypic virus strains (Wyatt et al., 1974). Since the majority of human infections are caused by GII.4 viruses and clinical trials had previously been carried out with GI.1 NV VLP vaccines, a new bivalent vaccine formulation was developed to provide broader protection extending across GI and GII virus strains. The new formulation included GI.1 NV VLPs and a GII.4 "consensus" VLP designed by aligning the VP1 sequences from three human GII.4 viruses (Parra et al., 2012). The three strains included for designing the consensus VLP sequence were TCH186 (Houston/TCH186/2002/US/ABY27560), 2006a (Yerseke38/2006/NL/ABL74391), and 2006b (DenHaag89/2006/NL/ABL74395). At amino acid positions there were differences between the three sequences, the residue in the 2006a strain was chosen since fewer substitutions were needed to achieve consensus (Parra et al., 2012). The amino acid residues in the two HBGA binding sites (Shanker et al., 2011) and the putative epitopes for HBGA blocking (Lindesmith et al., 2012) for the GII.4 consensus VLP and the three parental strains are shown in Fig. 3.6.2. The vaccine was adjuvanted with MPL and aluminum hydroxide and tested by intramuscular administration since preclinical studies of the new formulation in rabbits showed that intramuscular delivery of

Putative epitopes for HBGA blocking antibody

	(A)						(B)		(C)		(D)			(E)		
	294	296	297	298	368	372	333	382	340	376	393	394	395	407	412	413
TCH186	A	T	H	D	T	N	V	K	G	E	N	S	A	N	T	G
2006a	A	T	Q	E	S	S	V	R	R	E	S	T	T	D	D	S
2006b	A	S	R	N	S	E	V	K	G	E	S	T	T	S	N	V
GII.4c	A	T	Q	E	S	S	V	K	G	E	S	T	T	D	D	S

HBGA binding sites

	Site 1												Site 2									
	342	343	344	345	346	347	374	440'	441'	442'	443'	444'	387	388	389	390	391	382	393	394	395	396
TCH186	G	S	T	R	G	H	D	G	C	S	G	Y	G	V	V	Q	D	G	N	S	A	H
2006a	G	S	T	R	G	H	D	G	C	S	G	Y	G	V	V	Q	D	G	S	T	T	H
2006b	G	S	T	R	G	H	D	G	C	S	G	Y	G	V	I	Q	D	G	S	T	T	H
GII.4c	G	S	T	R	G	H	D	G	C	S	G	Y	G	V	V	Q	D	G	S	T	T	H

FIGURE 3.6.2 **Putative HBGA-blocking antibody epitopes and HBGA-binding site on the GII.4 capsid.** Five putative HBGA blocking epitopes (Lindesmith et al., 2012) and two sites involved in HBGA binding (Shanker et al., 2011) are shown. Amino acid residues are shown based upon numbering of the TCH186/2002 VP1 capsid sequence. Yellow highlighted residues are present in the TCH186 strain, blue highlighted residues are changes present in the 2006a strain, and orange residues are unique to the 2006b strain. GII.4c refers to the GII.4 "consensus" VLP that was designed by aligning the VP1 sequence from three human GII.4 viruses and is part of the bivalent VLP vaccine.

this vaccine induced higher antibody levels than the intranasal route. A dose escalation study was conducted in 48 adults aged 18–49 years, and each person received either 2 doses of vaccine (5, 15, 50, or 150 μg of each VLP) or 2 doses of placebo administered 4 weeks apart (Treanor et al., 2014). No serious adverse events related to vaccination were reported during the 1-year safety surveillance for the whole study. Local and systemic reactogenicity (such as pain or tenderness at injection site and mild heachache, respectively) were more common in vaccine recipients than in those receiving placebo. A greater magnitude of seroresponse was observed to the GI.1 VLP component compared to that of GII.4 VLPs. All subjects (100%) displayed a ≥fourfold increase in titer from baseline after the first dose for GI.1 VLPs, regardless of vaccine dosage. On the other hand, only 56% had a seroresponse to the GII.4 component in the 5/5-μg group, and only 75–88% responded in the higher-dosage groups. There were no significant increases in antibody responses following the second dose. In contrast to previous studies using oral or nasal administration of NoV VLPs, serum antibody response to the intramuscular vaccine occurred earlier with peak responses seen at day 7 post vaccination. PBMCs were analyzed for B cell activation and mucosal homing markers (Sundararajan et al., 2015). High frequencies of ASCs specific for both GI and GII.4 VLPs were detected 7 days after the first dose of vaccine. IgA ASC responses predominated over IgG ASC responses. The plasmablasts exhibited a mucosal-homing phenotype, with more than 98% of ASCs also expressing CCR10, a chemokine receptor with a broad role in mucosal homing (Sundararajan et al., 2015). The highest immune responses for the GI.1 VLP were observed in the 150/150-μg dose, whereas the highest immune responses to the GII.4 VLP antigen were observed with the 50/50-μg dose. Based on these findings, safety and immunogenicity studies were carried out with the 50/50-μg vaccine formulation in different age groups including 18–49, 50–64, and 65–85 years. Although lower fold changes in antibody response were observed in the higher age groups, most subjects in all 3 groups developed serum HBGA-blocking titers of ≥200, a level that was associated with significant protection from NoV disease in the prior NV intranasal vaccine challenge study. An efficacy evaluation of the bivalent formulation was carried out using a challenge dose of 4400 RT-PCR units of a GII.4/2002-like NoV heterovariant (ie, distinct from the consensus GII.4 VLP) following vaccination (Bernstein et al., 2015). Only 63% of placebo recipients were infected, and only 1/3 developed norovirus-associated gastroenteritis. There were no vaccine-related severe adverse events in the study. Local reactions such as tenderness at the injection site were more frequent in vaccine recipients compared to the placebo group. Ninety percent of vaccine recipients had a seroresponse to the vaccine, and the vaccine appeared to reduce acute gastroenteritis induced by the challenge virus, although vaccination did not meet the primary endpoints of the study using predefined criteria for illness and infections. Self-reported cases of severe gastroenteritis were lower in the vaccine group when compared to placebo (0 vs 8.3% respectively, $P = .054$). Similar results were also observed for moderate or greater and any gastroenteritis, although the differences did not achieve statistical significance. Vaccination significantly reduced the severity of illness as measured by a modified Vesikari score (from 7.3 to 4.5, $P = 0.002$) and vomiting or diarrhea of any severity (42 vs 20%, $P = 0.028$). Higher levels of prechallenge HBGA-blocking antibody were associated with a lower frequency of infection and illness (Atmar et al., 2015). However, there were apparent differences in the threshold of this CoP since HBGA-blocking antibody levels prechallenge were significantly higher among the vaccinees than among the placebo recipients, while significant differences in illness frequency were still evident based upon prechallenge antibody levels.

To further explore heterotypic protection from the bivalent vaccine, serum samples from ten persons who participated in the dose escalation evaluation of the bivalent formulation were analyzed for IgG and HBGA blocking antibodies to additional VLPs, including GI and GII genotypes not present in the vaccine as well as other GII.4 variants (Lindesmith et al., 2015) (Fig. 3.6.1 D–F). Rapid rises in IgG and HBGA-blocking antibodies to diverse strains were observed, and, importantly, vaccination induced HBGA-blocking antibodies to GII.4 strains that were not circulating at the time of vaccination.

5 SUMMARY AND CHALLENGES

Data from epidemiological studies and surveillance networks across the world highlight the public health burden of NoV disease. Depending on the cost of NoV vaccines, level of protective efficacy and duration of protection, modelling studies show that vaccination can offer important healthcare and economic benefits (Bartsch et al., 2012). Significant progress has been made in the development and evaluation of NoV vaccines. Several vaccine candidates are in development, of which the VLP vaccines have progressed farthest. VLPs appear to be attractive vaccine candidates as they have been shown to be safe and immunogenic in clinical studies, have demonstrated efficacy in proof-of-concept human experimental infection models and can be manufactured in a consistent and reproducible manner using cost-effective production systems.

Significant challenges remain in the development of effective NoV vaccines (summarized in Table 3.6.5). New discoveries on immune response to NoVs and CoP continue to be made from the volunteer challenge studies and vaccine trials. Aspects of immunity, such as T-cell and cytokine responses, remain to be characterized. Findings in these areas will likely influence the design of newer NoV vaccines. The epidemiology of NoVs is complex with new variants emerging every few years. The rapid evolution of virus strains through antigenic drift and recombination as well as new HBGA binding specificities pose challenges to vaccine development. The antigenic and genetic diversity of NoVs lead to questions about the number of different strains that will need to be included in a broadly effective vaccine. In addition, the apparent continuum of emerging strains raises questions on whether NoV vaccines will need to be

TABLE 3.6.5 Hurdles and Challenges to Development of an Effective Norovirus Vaccine

Demonstration of efficacy in field trials
Determination of breadth of protection (across genotypes and variants within a genotype)
Determination of duration of vaccine protection
Influence of antigenic drift (epochal evolution) on vaccine efficacy
Evaluation in children
Importance of immunological priming (in children) on immune response
Influence of increased age on vaccine responses and efficacy

updated periodically like influenza vaccines. Breadth of immune response to the vaccine VLPs is an important consideration in this context. New data from the challenge studies and vaccine trials in this area are promising as they indicate that heterotypic immunity is induced after natural infection and vaccination (Fig. 3.6.1). The duration of protective immunity needs to be determined since vaccines with short duration of protection may not have broad applicability, although they may be useful in specific settings. These are important questions to be addressed in field efficacy (and later effectiveness) studies that should provide an accurate estimate of protection in the natural environment where multiple NoV genotypes and variants cocirculate.

Challenge studies and vaccine trials so far have been carried out on healthy adults in fairly controlled settings. It remains to be seen whether findings from these studies will hold true in the general population. All adults in the vaccine trials had preexisting antibodies to NoVs, and it remains to be seen if immunological priming from previous infections is important for generating a robust immune response to the VLP vaccines. This is particularly relevant in the context of pediatric populations where the burden of NoV illness is high and where an effective vaccine will be highly useful. The other population where NoV infections result in high morbidity and mortality is the elderly. It remains to be seen whether the vaccines will be effective in the elderly and in those with chronic diseases where immune responses may not be achieved to the same extent as seen in healthy adults.

Data from the present studies with NoV vaccines are highly promising. There are many on-going studies with the VLP vaccines and other NoV vaccine candidates. Findings from these studies will influence the further development of NoV vaccines. As exciting progress in efforts to grow human NoVs continues to be made, more fundamental questions regarding the biology of these viruses will be answered/amenable to exploration. These findings will influence the development of NoV vaccines and our understanding of whether neutralization antibodies correlate with HBGA-blocking antibodies and protection from infection/disease.

REFERENCES

Ahmed, S.M., Hall, A.J., Robinson, A.E., et al., 2014. Global prevalence of norovirus in cases of gastroenteritis: a systematic review and meta-analysis. Lancet Infect. Dis. 14 (8), 725–730.

Atmar, R.L., Bernstein, D.I., Harro, C.D., et al., 2011. Norovirus vaccine against experimental human Norwalk Virus illness. N. Engl. J. Med. 365 (23), 2178–2187.

Atmar, R.L., Opekun, A.R., Gilger, M.A., et al., 2014. Determination of the 50% human infectious dose for Norwalk virus. J. Infect. Dis. 209 (7), 1016–1022.

Atmar, R.L., Bernstein, D.I., Lyon, G.M., et al., 2015. Serological correlates of protection against a GII.4 norovirus. Clin. Vaccine Immunol. 22 (8), 923–929.

Ball, J.M., Hardy, M.E., Atmar, R.L., et al., 1998. Oral immunization with recombinant Norwalk virus-like particles induces a systemic and mucosal immune response in mice. J. Virol. 72 (2), 1345–1353.

Ball, J.M., Graham, D.Y., Opekun, A.R., et al., 1999. Recombinant Norwalk virus-like particles given orally to volunteers: phase I study. Gastroenterology 117 (1), 40–48.

Bartsch, S.M., Lopman, B.A., Hall, A.J., et al., 2012. The potential economic value of a human norovirus vaccine for the United States. Vaccine 30 (49), 7097–7104.

Bernstein, D.I., Atmar, R.L., Lyon, G.M., et al., 2015. Norovirus vaccine against experimental human GII.4 virus illness: A challenge study in healthy adults. J. Infect. Dis. 211 (6), 870–878.

Bok, K., Green, K.Y., 2012. Norovirus gastroenteritis in immunocompromised patients. N. Engl. J. Med. 367 (22), 2126–2132.

Bok, K., Parra, G.I., Mitra, T., et al., 2011. Chimpanzees as an animal model for human norovirus infection and vaccine development. Proc. Natl. Acad. Sci. USA 108 (1), 325–330.

Czako, R., Atmar, R.L., Opekun, A.R., et al., 2012. Serum hemagglutination inhibition activity correlates with protection from gastroenteritis in persons infected with Norwalk virus. Clin. Vaccine Immunol. 19 (2), 284–287.

Czako, R., Atmar, R.L., Opekun, A.R., et al., 2015. Experimental human infection with Norwalk virus elicits a surrogate neutralizing antibody response with cross-genogroup activity. Clin. Vaccine Immunol. 22 (2), 221–228.

El-Kamary, S.S., Pasetti, M.F., Mendelman, P.M., et al., 2010. Adjuvanted intranasal Norwalk virus-like particle vaccine elicits antibodies and antibody-secreting cells that express homing receptors for mucosal and peripheral lymphoid tissues. J. Infect. Dis. 202 (11), 1649–1658.

Fang, H., Tan, M., Xia, M., et al., 2013. Norovirus P particle efficiently elicits innate, humoral and cellular immunity. PLoS One 8 (4), e63269.

Green, K.Y., Lew, J.F., Jiang, X., et al., 1993. Comparison of the reactivities of baculovirus-expressed recombinant Norwalk virus capsid antigen with those of the native Norwalk virus antigen in serologic assays and some epidemiologic observations. J. Clin. Microbiol. 31 (8), 2185–2191.

Greenberg, H.B., Valdesuso, J., Kapikian, A.Z., et al., 1979. Prevalence of antibody to the Norwalk virus in various countries. Infect. Immun. 26 (1), 270–273.

Guerrero, R.A., Ball, J.M., Krater, S.S., et al., 2001. Recombinant Norwalk virus-like particles administered intranasally to mice induce systemic and mucosal (fecal and vaginal) immune responses. J. Virol. 75 (20), 9713–9722.

Guo, L., Wang, J., Zhou, H., et al., 2008. Intranasal administration of a recombinant adenovirus expressing the norovirus capsid protein stimulates specific humoral, mucosal, and cellular immune responses in mice. Vaccine 26 (4), 460–468.

Hall, A.J., Lopman, B.A., Payne, D.C., et al., 2013. Norovirus disease in the United States. Emerg. Infect. Dis. 19 (8), 1198–1205.

Harrington, P.R., Yount, B., Johnston, R.E., et al., 2002. Systemic, mucosal, and heterotypic immune induction in mice inoculated with Venezuelan equine encephalitis replicons expressing Norwalk virus-like particles. J. Virol. 76 (2), 730–742.

Huang, P., Farkas, T., Zhong, W., et al., 2005. Norovirus and histo-blood group antigens: demonstration of a wide spectrum of strain specificities and classification of two major binding groups among multiple binding patterns. J. Virol. 79 (11), 6714–6722.

Hutson, A.M., Atmar, R.L., Marcus, D.M., et al., 2003. Norwalk virus-like particle hemagglutination by binding to h histo-blood group antigens. J. Virol. 77 (1), 405–415.

Jiang, X., Wang, M., Graham, D.Y., et al., 1992. Expression, self-assembly, and antigenicity of the Norwalk virus capsid protein. J. Virol. 66 (11), 6527–6532.

Johnson, P.C., Mathewson, J.J., DuPont, H.L., et al., 1990. Multiple-challenge study of host susceptibility to Norwalk gastroenteritis in US adults. J. Infect. Dis. 161 (1), 18–21.

Kavanagh, O., Estes, M.K., Reeck, A., et al., 2011. Serological responses to experimental Norwalk virus infection measured using a quantitative duplex time-resolved fluorescence immunoassay. Clin. Vaccine Immunol. 18 (7), 1187–1190.

Koo, H.L., Neill, F.H., Estes, M.K., et al., 2013. Noroviruses: The most common pediatric viral enteric pathogen at a large university hospital after introduction of rotavirus vaccination. J. Pediatric. Infect. Dis. Soc. 2 (1), 57–60.

Lindesmith, L., Moe, C., Marionneau, S., et al., 2003. Human susceptibility and resistance to Norwalk virus infection. Nat. Med. 9 (5), 548–553.

Lindesmith, L.C., Donaldson, E., Leon, J., et al., 2010. Heterotypic humoral and cellular immune responses following Norwalk virus infection. J. Virol. 84 (4), 1800–1815.

Lindesmith, L.C., Beltramello, M., Donaldson, E.F., et al., 2012. Immunogenetic mechanisms driving norovirus GII.4 antigenic variation. PLoS Pathog. 8 (5), e1002705.

Lindesmith, L.C., Ferris, M.T., Mullan, C.W., et al., 2015. Broad blockade antibody responses in human volunteers after immunization with a multivalent norovirus VLP candidate vaccine: Immunological analyses from a phase I clinical trial. PLoS Med. 12 (3), e1001807.

LoBue, A.D., Lindesmith, L., Yount, B., et al., 2006. Multivalent norovirus vaccines induce strong mucosal and systemic blocking antibodies against multiple strains. Vaccine 24 (24), 5220–5234.

Mason, H.S., Ball, J.M., Shi, J.J., et al., 1996. Expression of Norwalk virus capsid protein in transgenic tobacco and potato and its oral immunogenicity in mice. Proc. Natl. Acad. Sci. USA 93 (11), 5335–5340.

Menon, V.K., George, S., Aladin, F., et al., 2013. Comparison of age-stratified seroprevalence of antibodies against norovirus GII in India and the United Kingdom. PLoS One 8 (2), e56239.

Nicollier-Jamot, B., Ogier, A., Piroth, L., et al., 2004. Recombinant virus-like particles of a norovirus (genogroup II strain) administered intranasally and orally with mucosal adjuvants LT and LT(R192G) in BALB/c mice induce specific humoral and cellular Th1/Th2-like immune responses. Vaccine 22 (9–10), 1079–1086.

O'Ryan, M.L., Vial, P.A., Mamani, N., et al., 1998. Seroprevalence of Norwalk virus and Mexico virus in Chilean individuals: assessment of independent risk factors for antibody acquisition. Clin. Infect. Dis. 27 (4), 789–795.

Parra, G.I., Bok, K., Taylor, R., et al., 2012. Immunogenicity and specificity of norovirus Consensus GII.4 virus-like particles in monovalent and bivalent vaccine formulations. Vaccine 30 (24), 3580–3586.

Parrino, T.A., Schreiber, D.S., Trier, J.S., et al., 1977. Clinical immunity in acute gastroenteritis caused by Norwalk agent. N. Engl. J. Med. 297 (2), 86–89.

Payne, D.C., Vinje, J., Szilagyi, P.G., et al., 2013. Norovirus and medically attended gastroenteritis in U.S. children. N. Engl. J. Med. 368 (12), 1121–1130.

Periwal, S.B., Kourie, K.R., Ramachandaran, N., et al., 2003. A modified cholera holotoxin CT-E29H enhances systemic and mucosal immune responses to recombinant Norwalk virus-virus like particle vaccine. Vaccine 21 (5–6), 376–385.

Plotkin, S.A., Gilbert, P.B., 2012. Nomenclature for immune correlates of protection after vaccination. Clin. Infect. Dis. 54 (11), 1615–1617.

Ramani, S., Atmar, R.L., Estes, M.K., 2014. Epidemiology of human noroviruses and updates on vaccine development. Curr. Opin. Gastroenterol. 30 (1), 25–33.

Ramani, S., Neill, F.H., Opekun, A.R., et al., 2015. Mucosal and cellular immune responses to Norwalk virus. J. Infect. Dis. 212 (3), 397–405.

Ramirez, K., Wahid, R., Richardson, C., et al., 2012. Intranasal vaccination with an adjuvanted Norwalk virus-like particle vaccine elicits antigen-specific B memory responses in human adult volunteers. Clin. Immunol. 144 (2), 98–108.

Reeck, A., Kavanagh, O., Estes, M.K., et al., 2010. Serological correlate of protection against norovirus-induced gastroenteritis. J. Infect. Dis. 202 (8), 1212–1218.

Richardson, C., Bargatze, R.F., Goodwin, R., et al., 2013. Norovirus virus-like particle vaccines for the prevention of acute gastroenteritis. Expert. Rev. Vaccines 12 (2), 155–167.

Ryder, R.W., Singh, N., Reeves, W.C., et al., 1985. Evidence of immunity induced by naturally acquired rotavirus and Norwalk virus infection on two remote Panamanian islands. J. Infect. Dis. 151 (1), 99–105.

Shanker, S., Choi, J.M., Sankaran, B., et al., 2011. Structural analysis of histo-blood group antigen binding specificity in a norovirus GII.4 epidemic variant: implications for epochal evolution. J. Virol. 85 (17), 8635–8645.

Simmons, K., Gambhir, M., Leon, J., et al., 2013. Duration of immunity to norovirus gastroenteritis. Emerg. Infect. Dis. 19 (8), 1260–1267.

Souza, M., Costantini, V., Azevedo, M.S., et al., 2007. A human norovirus-like particle vaccine adjuvanted with ISCOM or mLT induces cytokine and antibody responses and protection to the homologous GII.4 human norovirus in a gnotobiotic pig disease model. Vaccine 25 (50), 8448–8459.

Sundararajan, A., Sangster, M.Y., Frey, S., et al., 2015. Robust mucosal-homing antibody-secreting B cell responses induced by intramuscular administration of adjuvanted bivalent human norovirus-like particle vaccine. Vaccine 33 (4), 568–576.

Tacket, C.O., Mason, H.S., Losonsky, G., et al., 2000. Human immune responses to a novel norwalk virus vaccine delivered in transgenic potatoes. J. Infect. Dis. 182 (1), 302–305.

Tacket, C.O., Sztein, M.B., Losonsky, G.A., et al., 2003. Humoral, mucosal, and cellular immune responses to oral Norwalk virus-like particles in volunteers. Clin. Immunol. 108 (3), 241–247.

Tamminen, K., Huhti, L., Koho, T., et al., 2012. A comparison of immunogenicity of norovirus GII-4 virus-like particles and P-particles. Immunology 135 (1), 89–99.

Tamminen, K., Lappalainen, S., Huhti, L., et al., 2013. Trivalent combination vaccine induces broad heterologous immune responses to norovirus and rotavirus in mice. PLoS One 8 (7), e70409.

Tan, M., Jiang, X., 2005. The P domain of norovirus capsid protein forms a subviral particle that binds to histo-blood group antigen receptors. J. Virol. 79 (22), 14017–14030.

Tan, M., Fang, P., Chachiyo, T., et al., 2008. Noroviral P particle: structure, function and applications in virus-host interaction. Virology 382 (1), 115–123.

Tan, M., Huang, P., Xia, M., et al., 2011. Norovirus P particle, a novel platform for vaccine development and antibody production. J. Virol. 85 (2), 753–764.

Treanor, J.J., Atmar, R.L., Frey, S.E., et al., 2014. A novel intramuscular bivalent norovirus virus-like particle vaccine candidate--reactogenicity, safety, and immunogenicity in a phase 1 trial in healthy adults. J. Infect. Dis. 210 (11), 1763–1771.

Trivedi, T.K., Desai, R., Hall, A.J., et al., 2013. Clinical characteristics of norovirus-associated deaths: a systematic literature review. Am. J. Infect. Control. 41 (7), 654–657.

Wang, L., Cao, D., Wei, C., et al., 2014. A dual vaccine candidate against norovirus and hepatitis E virus. Vaccine 32 (4), 445–452.

Wyatt, R.G., Dolin, R., Blacklow, N.R., et al., 1974. Comparison of three agents of acute infectious nonbacterial gastroenteritis by cross-challenge in volunteers. J. Infect. Dis. 129 (6), 709–714.

Xia, M., Farkas, T., Jiang, X., 2007. Norovirus capsid protein expressed in yeast forms virus-like particles and stimulates systemic and mucosal immunity in mice following an oral administration of raw yeast extracts. J. Med. Virol. 79 (1), 74–83.

Zhang, X., Buehner, N.A., Hutson, A.M., et al., 2006. Tomato is a highly effective vehicle for expression and oral immunization with Norwalk virus capsid protein. Plant Biotechnol. J. 4 (4), 419–432.

Chapter 4.1

Studies of Astrovirus Structure–Function Relationships

S. Marvin, V.A. Meliopoulos, S. Schultz-Cherry
Department of Infectious Diseases, St. Jude Children's Research Hospital, Memphis, TN, United States

1 INTRODUCTION

Though still in its infancy as compared to other enteric viruses, astrovirus structural studies have provided important new information that will aid in our understanding of the viral lifecycle and disease pathogenesis as well as assist in development of antiviral therapies and vaccines. Cryo-electron microscopy studies have yielded information on the conformational changes in the virion as a result of proteolytic maturation. Additionally, high resolution X-ray crystallography studies of the astrovirus capsid spike domain and the protease have contributed insights into receptor binding and polyprotein events. The mechanism(s) underlying astrovirus-induced diarrhea remains to be elucidated, but data from in vitro and in vivo systems suggests that the astrovirus capsid alone may play an important role independent of productive viral replication. The goal of this chapter is to summarize the current knowledge on astrovirus structure–function relationships in terms of viral replication and disease pathogenesis.

2 ASTROVIRUS VIRION STRUCTURE AND MATURATION

Astroviruses (AstV) are single stranded, positive-sense, nonenveloped RNA viruses of icosahedral particle symmetry that cause a variety of clinical diseases ranging from diarrhea to encephalitis, or asymptomatic infection in a wide range of mammals and birds (Bosch et al., 2014; De Benedictis et al., 2011). Human astrovirus (HAstV) was first observed in 1975 associated with outbreaks of gastroenteritis in infants (Appleton and Higgins, 1975; Madeley and Cosgrove, 1975). HAstV are one of the leading causes of diarrhea in children under 2 years of age worldwide, represent a high disease burden in developing countries, and can be especially dangerous to immunocompromised children (Goodgame, 1999, 2001; Maldonado et al., 1998; Bjorkholm et al., 1995;

Viral Gastroenteritis

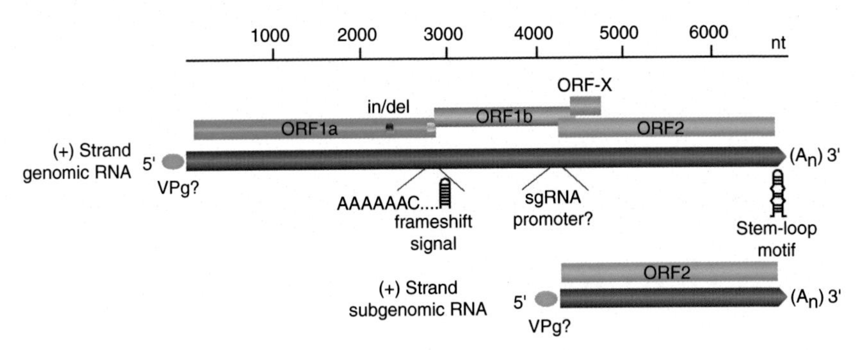

FIGURE 4.1.1 **Genome organization of human astrovirus.** The genomic RNA, of approximately 6.8 kb, contains three recognized open reading frames (ORF1a, ORF1b, and ORF2). ORF-X has been proposed as functional, given its conservation among astroviruses. The genome contains three elements conserved among all members of the family: the frameshift signal *(yellow square)*, the sequence upstream of ORF2 that putatively acts as promoter for synthesis of the subgenomic *(sg)* RNA, and the stem-loop at the 3′-end. Also shown are the insertion/deletion *(in/del)* region *(red square)* and the presence of a poly(A) tail at the 3′-end. A putative VPg protein is depicted as bound to the 5′-end of the genomic and subgenomic RNAs. *(Source: From Méndez et al. 2013 and used with kind permission from Springer Science and Business Media.)*

Coppo et al., 2000; Cox et al., 1994; Grohmann et al., 1993; Liste et al., 2000; Wood et al., 1988; Yuen et al., 1998). Until 2008, human astrovirus infections were thought to be due solely to HAstV 1–8 serotypes. Through advances in viral surveillance and next generation sequencing, several genetically distinct astroviruses including MLB1-3 and VA/HMO genotypes were identified in diarrheal patients (Bosch et al., 2014; Schultz-Cherry, 2013, 2015). (See also Chapter 4.3.) The AstV genome contains three open reading frames: ORF1a, ORF1b, and ORF2 (Fig. 4.1.1). ORF1a and 1b encode two nonstructural polyproteins that are cleaved by viral and cellular proteases to form several proteins that likely function in viral replication. ORF2 encodes the viral capsid protein (Fuentes et al., 2012).

Astroviruses were named for their star-like appearance in negative-stain electron microscopy (EM) studies (Caul and Appleton, 1982). Infectious HAstV propagated in cell culture showed spherical particles studded with spikes on the surface (Geigenmuller et al., 2002a,b) that are assembled from the VP90 capsid structural protein, an 87–90 kDa precursor protein that possesses a highly conserved N-terminal domain and a variable C-terminal domain (Royuela, 2010). The VP90 capsid protein goes through a series of proteolytic cleavages to obtain the infectious virion (Fig. 4.1.2). Cleavage by cellular caspase at the C-terminus leaves a 70–79 kDa protein, referred to as VP70, or the immature particle, which appears to be required for particle release from cells (Mendez et al., 2004; Banos-Lara and Mendez, 2010). Viral infectivity in vitro depends upon trypsin cleavage of VP70 to yield the viral proteins VP34, VP27, and VP25. VP34 contains the N-terminus of the capsid

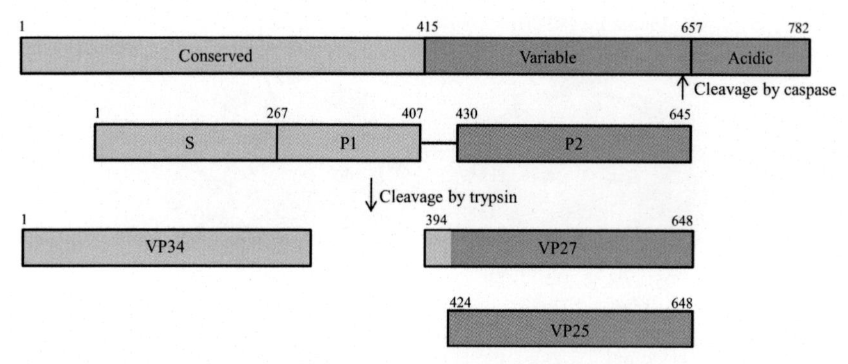

FIGURE 4.1.2 Astrovirus capsid cleavage steps and products. The VP90 domain structure of human astrovirus serotype 8 contains a conserved region, a variable region and an acidic region. The S and P1 domains are mapped to the conserved regions and the P2 domain is mapped to the variable region. The line from amino acids 408–429 represents an extended (P1–P2) linker. Cleavage by cellular proteases (caspase) removes the acidic domain to generate VP70. Cleavage in vitro by trypsin yields infectious particles containing VP34, VP27, and VP25.

protein while VP27 and VP25 contain the variable C-terminal domain (reviewed in Krishna, 2005). It is the C-terminal domain that is referred to as the spike receptor binding domain, although no astrovirus receptors have been identified to date (see Section 2.3). It is important to note that our current knowledge on astrovirus structure comes from studies with the HAstV 1–8 serotypes, due primarily to the lack of in vitro culture systems for the recently identified human strains.

The structures of immature and mature particles have been calculated to 25Å-resolution by electron cryomicroscopy and averaged image analysis (Fig. 4.1.3) (Dryden et al., 2012). Both immature and mature particles display two layers of density. The inner layer is continuous and the outer layer is composed of globular densities with poorly defined connections to the underlying continuous capsid layer. The capsid shell is organized as $T = 3$ icosahedral symmetry composed of 180 subunits, which are most likely VP34 subunits.

The key difference between immature and mature particles is the loss of the spike domain after cleavage by trypsin or cellular proteases. Immature particles have 90 spikes whereas the mature particles contain only 30 spikes on the surface. It was first thought that trypsin cleavage of VP70 forms VP34, VP27, and VP25 and would remove two-thirds of the spikes from the particle. However, previously published SDS-PAGE gels of trypsin-digested virus revealed that the amount of VP34 was comparable to levels of VP27 and VP25 (Sanchez-Fauquier et al., 1994; Mendez et al., 2007; Bass and Qiu, 2000), indicating that most of the spikes remained on the capsid. Alternatively, the number of spikes on the capsid surface may remain the same, but cleavage by trypsin allows for flexibility, leading to a loss of resolution upon averaging.

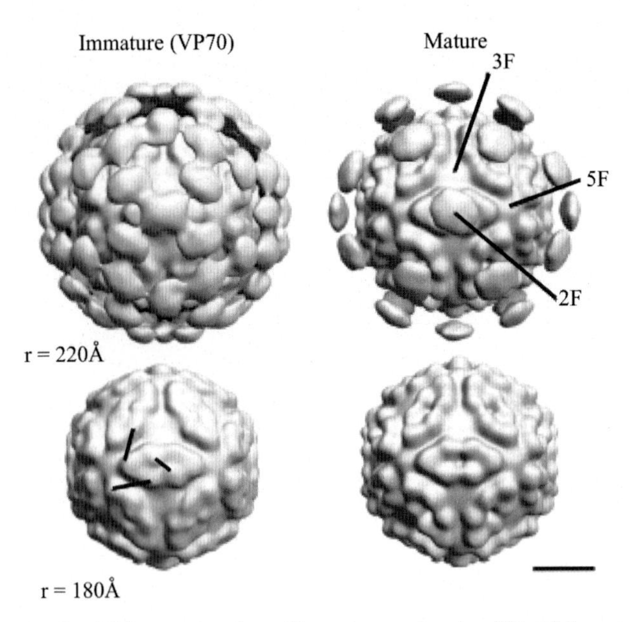

FIGURE 4.1.3 **CryoEM reconstructions of immature and mature HAstV.** Immature (left) and mature (right) particles have very similar capsids as displayed by truncating the maps at radius 180Å (lower), while the mature particles lack 60 of the 90 spikes visible on immature virions (upper). Symmetry axes (2F, 3F, 5F) are indicated on mature virion. One set of probable subunit contributors to the dimer spikes is displayed on the immature capsid. Scale Bar = 100Å. *(Source: From Dubois et al. (2013b) and used with kind permission from Springer Science and Business Media.)*

2.1 Crystal Structure of the Human Astrovirus Spike

To date only the HAstV-8 spike variable domain (P2) has been crystalized (Dong et al., 2011). Dong et al. produced soluble capsid spike protein in *Escherichia coli* containing $P2^{415-646}$ residues, which is longer than VP25 but shorter than VP27 at the N-terminus (Dong et al., 2011). $P2^{415-646}$ had a molecular mass of 60 kDa by gel filtration chromatography, suggesting that the protein exists as a dimer in solution. The crystal structure of $P2^{415-646}$ dimer was determined to 1.8Å resolution. The final model consists of a β-structure containing 11 β-strands. The overall structure of each spike molecule contains a core composed of a six-stranded β-barrel with a tightly packed hydrophobic core. A three-strand β-sheet is packed against the outside of the β-barrel away from the dimer interface. Two long loops, L1 (residues 439–466) and L4 (residues 597–632) span across the top of the β-barrel and a shorter loop, L2 (residues 491–506) runs across the bottom. A β-hairpin motif, consisting of 2 β-strands and loop L3 (residues 554–574), sits on top of L1 and L4, which is thought to reinforce the structural conformation of the two long loops.

Structural analysis using the PISA server confirmed the dimeric configuration of $P2^{415-646}$ and suggested a dimer dissociation constant within the low nanomolar range. Therefore, the dimer interaction likely plays an important role in stabilizing the viral capsid. The overall morphology of the P2$^{415–646}$ dimer matches that of the dimeric surface spikes of a cryo-EM reconstitution image of an astrovirus.

2.2 Crystal Structure of the Turkey Astrovirus Spike

A study by DuBois et al. determined the crystal structure of the turkey AstV-2 (TAstV-2) spike domain to 1.5Å resolution (DuBois et al., 2013a). The C-terminal region (residues 421–724; 35 kDa) was produced in *E. coli* and purified. Pepsin digestion of the C-terminal region yielded a 26 kDa domain (residues 423–630) that eluted from a size-exclusion chromatography column at a molecular weight of approximately 60 kDA similar to the HAstV-8 P2$^{415–646}$ domain, suggesting that the TAstV-2 spike domain also exists as a dimer in solution.

The structure of the TAstV-2 spike yielded differences from the HAstV-8 spike structure (Fig. 4.1.4). While the HAstV-8 spike domain consisted solely of loops and β-strands, the TAstV-2 spike contained 3 alpha helices. The structural scaffold comprised an antiparallel β-barrel composed of 8 strands with a tightly packed hydrophobic core. The β-barrel is capped at one end by a β-strand and an alpha helix and is flanked by two additional alpha helices, a β-hairpin and four, 3/10 η-helices. The TAstV-2 spike resembles a tightly packed heart compared to the bowl-like structure of the HAstV-8 spike. However, both predict a role for carbohydrates in astrovirus binding.

2.3 Putative Receptor Binding Sites

To date, no receptor for any astrovirus genotype has been identified. The crystal structure analysis of both the HAstV-8 and TAstV-2 spike domains yielded a putative carbohydrate binding site. Putative receptor-binding sites located at the top of the HAstV-8 and TAstV-2 spike domains contain conserved charged residues typically found in carbohydrate-binding pockets. Addition of heparin, heparin sulfate, or dextran sulfate to HAstV-8 prior to infecting Caco-2 cells, which support HAstV infection and replication, partially blocked HAstV-8 infection, with heparin having the most pronounced effect (Dong et al., 2011). Addition of glycophorin (a protein rich in sialic acid) to the virus prior to infection or sialidase treatment of Caco-2 cells had no effect on HAstV-8 infectivity.

Since there is no cell culture system for TAstV-2, DuBois et al. used HAstV-1 to determine the role of carbohydrates on HAstV-1 binding and infectivity by removing heparin moieties by heparinase treatment of Caco-2 cells. Heparinase treatment did not reduce HAstV-1 binding or replication in Caco-2 cells. Additionally, chondroitinase ABC treatment of Caco-2 cells, to remove chondroitin sulfates, did not decrease HAstV-1 binding, but did produce a small, but consistently significant decrease in HAstV-1 infectivity in a dose-dependent manner

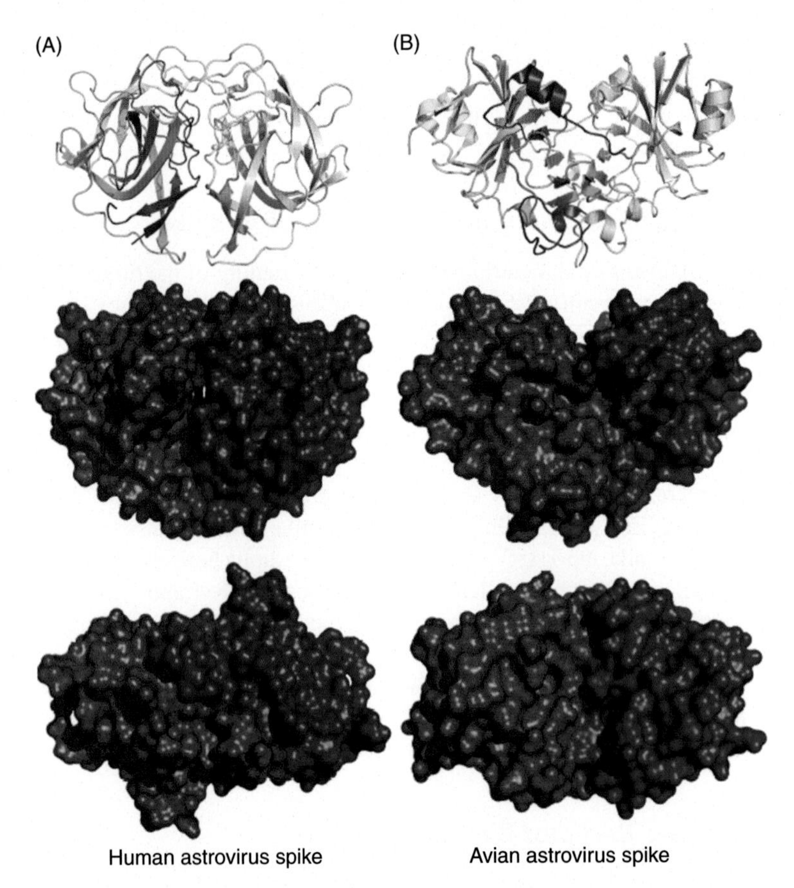

FIGURE 4.1.4 **Comparison of human and avian astrovirus capsid spike structures.** (A) Dimeric structure of the human astrovirus 8 capsid spike shown as cartoon (top) or surface (bottom) representation. (PDB entry 3QSQ) (B) Dimeric structure of the turkey astrovirus 2 spike shown as cartoon (top) or surface (bottom) representation. (PDB entry 3TS3). In the cartoon representations, one protomer is colored in a blue-to-red rainbow from N- to C-termini whereas the other protomer is colored gray. In the surface representations, one protomer is colored red and the other is colored blue. *(Source: From Dubois et al. (2013b) and used with kind permission from Springer Science and Business Media.)*

(DuBois et al., 2013a). Future studies are needed to further define the role of carbohydrates in astrovirus binding and infection.

3 ASTROVIRUS CAPSID STUDIES AND DISEASE

3.1 Astrovirus Disease

Although the spike crystal structures of HAstV-8 and TAstV-2 spike domain have been defined (Dong et al., 2011; DuBois et al., 2013a), little is known about astrovirus pathogenesis, and how the capsid structure relates to the mechanism of pathogenesis. However, there is evidence to support that the astrovirus capsid structure plays a critical role in disease pathogenesis.

Astrovirus infection is characterized by fever, headache, malaise, and most notably loose watery stool. Intestinal pathogens can cause diarrhea in a variety of ways. Common mechanisms include extensive inflammation (Goodgame, 2001) or destruction of the intestinal lining itself due to cell death and atrophy of the intestinal villi (Lundgren et al., 2000). However, several studies have shown that this is not the case with astroviruses. In a small animal (turkey) model, although TAstV-2 infection caused distention and thinning of the intestines as well as fluid accumulation, only mild histopathological changes were noted, and there was an absence of inflammation and cellular infiltrates (Koci et al., 2003). This finding is similar to that of the first identified viral enterotoxin, the rotavirus NSP4 protein, which induces diarrhea in neonatal mice in the absence of histological damage to the intestinal mucosa (Zhang et al., 2000) through a number of mechanisms including impaired glucose absorption and increased chloride reabsorption (reviewed in Lorrot and Vasseur, 2007) (see Chapter 2.4). However, in the TAstV-2 model, elevated levels of transforming growth factor-β (TGF-β), an immunosuppressive cytokine, were found both in the serum of infected turkey poults, and increased levels of active TGF-β were found in embryonated turkey eggs that had been inoculated with TAstV-2 (Koci et al., 2003). Although this may partially explain the lack of inflammation, more extensive studies need to be performed, preferably in a mammalian model, to understand the lack of inflammation seen during astrovirus infection.

TAstV-2 was also shown by RT-PCR to spread extra-intestinally to sites including the bursa, thymus, spleen, kidney, liver, muscle, and pancreas and the animals developed viremia (Behling-Kelly et al., 2002). Although the virus did not replicate outside the intestines, TAstV-2-infected poults had decreased thymus size suggesting an alteration of immune response, and experienced growth depression, which might be due to nutrient malabsorption from damaged intestine. Intriguingly, several recent studies demonstrated astrovirus in the brains of immunocompromised children and an adult with encephalitis (Quan et al., 2010; Wunderli et al., 2011; Naccache et al., 2015), and viremia in an MLB2-infected febrile child (Holtz et al., 2011), implying that human astroviruses also have the ability to leave the intestine and go systemic. In order to get out of the intestinal lumen and access the vasculature, the virus must transverse the epithelial layer. How does this occur in the absence of cell death or basolateral release? The capsid protein may be crucial for this ability.

3.2 Astrovirus-Induced Barrier Permeability by Disruption of Tight Junctions is Mediated by the Capsid Protein

The intestine is lined with specialized epithelial cells tightly joined to each other by cell-cell contacts that include tight junctions (Coskun, 2014). These cellular junctions are crucial for protecting the body from substances in the intestinal lumen and are tightly regulated. Human astroviruses increase the permeability of differentiated human intestinal epithelial cells in culture by disrupting these cellular junctions (Moser et al., 2007). This also translates to an in vivo model, as TAstV-2-infected turkey poults also show increased intestinal permeability during infection (Nighot et al., 2010). Mechanistically, it was shown that HAstV-1 increased epithelial permeability by inducing redistribution of tight junction proteins and the actin cytoskeleton independent of viral replication (Moser et al., 2007). Addition of purified HAstV-1 capsid protein was sufficient to increase epithelial permeability in vitro (Moser et al., 2007), and TAstV-2 capsid protein was sufficient in vivo (manuscript in preparation) suggesting that it may be a novel viral enterotoxin. Studies are underway to determine whether the astrovirus capsid can affect permeability of other epithelial and endothelial cells and to explore the role of this in systemic spread during infection.

3.3 Astrovirus Infection and Cell Death

Another contribution to intestinal disease is pathogen-induced cell death. Several studies have investigated whether astrovirus infection causes cell death including HAstV-4 and HAstV-8 infection in Caco-2 cells (Mendez et al., 2004; Guix et al., 2004). Nuclear morphological changes, including chromatin condensation and the formation of apoptotic bodies were observed in 28% of HAstV-4 infected cells at 48 h postinfection (hpi) compared to 6% of mock infected cells. The TUNEL assay was also performed, and 71% of HAstV-4 infected cells were apoptotic at 72 hpi compared to 0.2% of mock infected cells. This was associated with caspase-8 activation. Inhibition of caspase 8 partially inhibited HAstV-4-induced cell death. Mendez et al. (2004) also found morphological changes of nuclear DNA and TUNEL staining in HAstV-8 infected cells. The study further went on to find that caspase-8 activation during HAstV-4 infection increases infectious progeny virus yield. The viral proteins responsible for HAstV-4/8-induced apoptosis are unknown, but transient transfection and expression of the ORF1a gene induced apoptosis suggesting that an ORF1a protein product may have been responsible (Guix et al., 2004).

In contrast, no apoptosis was observed in HAstV-1 infected Caco-2 cells as measured by trypan blue exclusion and caspase activation (Moser et al., 2007). Further, TUNEL staining of TAstV-2 infected turkey poults suggested a lack of cell death during TAstV-2 infection in spite of extensive diarrhea (Koci et al., 2003). The studies between HAstV-1 and HAstV-4 and -8 used different methods to determine cell death. Additionally, the apoptotic pathway can be activated, but be blocked at a later step to prevent death of the cell. No study had directly compared astrovirus types in the same experiment. Further studies are needed to determine if HAstV-induced cell death is type dependent.

3.4 Astrovirus Suppression of Complement

The complement system is a fundamental component of the innate immune response in vertebrates that functions as an immune surveillance system that can rapidly respond to pathogens (Stoermer and Morrison, 2011). It is comprised of soluble and cell surface proteins that detect bacteria and viruses through one of three pathways, the classical, the mannose-binding lectin (MBL), and alternative pathways. Once activated, a protease cascade is activated that eliminates pathogens, regulates inflammatory responses, and helps to shape the adaptive immune responses. Intriguingly, the HAstV-1 capsid protein inhibits activation of the classical and lectin complement pathways by binding C1q, disrupting its interaction with the cognate serine protease complex (C1r–C1s–C1s–C1r), and preventing C1s cleavage (Bonaparte et al., 2008; Gronemus et al., 2010; Hair et al., 2010). HAstV-1 capsid was also shown to bind to MBL and possibly interact with the MASP2 binding site on MBL to inhibit the lectin pathway (Hair et al., 2010). Two HAstV-1 capsid peptides, E23A (PAICQRATATLGTVG-SNTSGTTEIAACILL, 30 aa) and d8–22 (PAICQRAEIEACILL, 15 aa), were identified that bound C1q and inhibited classical pathway activation in vitro and in vivo (Gronemus et al., 2010). Whether this region is also involved in disruption of tight junctions remains unknown. Instead, the ability of the HAstV-1 capsid to bind to complement might be a mechanism of evading the immune system. Innate and adaptive immunity play an important role in controlling astrovirus infection in mice (Yokoyama et al., 2012), and antibody therapy has been used in immunocompromised humans to treat persistent astrovirus infections (reviewed in Koci, 2005).

4 SUMMARY AND FUTURE STUDIES

Novel astrovirus strains are continually being identified in new mammalian and avian hosts through increased viral surveillance and pathogen discovery, highlighting the extensive/expansive tropism of this viral family. Yet our understanding of the basic virology and pathogenesis of this medically and agriculturally important viral family remains understudied. Which cells support viral replication, what are the receptor(s), and how these viruses induce disease are some of the many questions that need to be addressed. To date only two astrovirus spike proteins have been crystallized. Analysis of additional genotypes may provide invaluable insight into structural similarities including possible receptor binding sites. Further work also needs to be performed on the capsid protein alone, given its ability to inhibit complement and increase epithelial permeability through disruption of cellular junctions. Are similar regions of the capsid involved in these distinct activities? Do all genotypes possess these abilities? Hopefully, these and many other questions can be answered in the near future.

REFERENCES

Appleton, H., Higgins, P.G., 1975. Letter: viruses and gastroenteritis in infants. Lancet 1 (7919), 1297.

Banos-Lara, M.D., Mendez, E., 2010. Role of individual caspases induced by astrovirus on the processing of its structural protein and its release from the cell through a non-lytic mechanism. Virology 401 (2), 322–332.

Bass, D.M., Qiu, S., 2000. Proteolytic processing of the astrovirus capsid. J. Virol. 74 (4), 1810–1814.

Behling-Kelly, E., Schultz-Cherry, S., Koci, M., Kelley, L., Larsen, D., Brown, C., 2002. Localization of astrovirus in experimentally infected turkeys as determined by in situ hybridization. Vet. Pathol. 39 (5), 595–598.

Bjorkholm, M., Celsing, F., Runarsson, G., Waldenstrom, J., 1995. Successful intravenous immunoglobulin therapy for severe and persistent astrovirus gastroenteritis after fludarabine treatment in a patient with Waldenstrom's macroglobulinemia. Int. J. Hematol. 62 (2), 117–120.

Bonaparte, R.S., Hair, P.S., Banthia, D., Marshall, D.M., Cunnion, K.M., Krishna, N.K., 2008. Human astrovirus coat protein inhibits serum complement activation via C1, the first component of the classical pathway. J. Virol. 82 (2), 817–827.

Bosch, A., Pinto, R.M., Guix, S., 2014. Human astroviruses. Clin. Microbiol. Rev. 27 (4), 1048–1074.

Caul, E.O., Appleton, H., 1982. The electron microscopical and physical characteristics of small round human fecal viruses: an interim scheme for classification. J. Med. Virol. 9 (4), 257–265.

Coppo, P., Scieux, C., Ferchal, F., Clauvel, J., Lassoued, K., 2000. Astrovirus enteritis in a chronic lymphocytic leukemia patient treated with fludarabine monophosphate. Ann. Hematol. 79 (1), 43–45.

Coskun, M., 2014. Intestinal epithelium in inflammatory bowel disease. Front. Med. 1, 24.

Cox, G.J., Matsui, S.M., Lo, R.S., Hinds, M., Bowden, R.A., Hackman, R.C., et al., 1994. Etiology and outcome of diarrhea after marrow transplantation: a prospective study. Gastroenterology 107 (5), 1398–1407.

De Benedictis, P., Schultz-Cherry, S., Burnham, A., Cattoli, G., 2011. Astrovirus infections in humans and animals - molecular biology, genetic diversity, and interspecies transmissions. Infect. Genet. Evol. 11 (7), 1529–1544.

Dong, J., Dong, L., Mendez, E., Tao, Y., 2011. Crystal structure of the human astrovirus capsid spike. Proc. Natl. Acad. Sci. U S A 108 (31), 12681–12686.

Dryden, K.A., Tihova, M., Nowotny, N., Matsui, S.M., Mendez, E., Yeager, M., 2012. Immature and mature human astrovirus: structure, conformational changes, and similarities to hepatitis E virus. J. Mol. Biol. 422 (5), 650–658.

DuBois, R.M., Freiden, P., Marvin, S., Reddivari, M., Heath, R.J., White, S.W., et al., 2013a. Crystal structure of the avian astrovirus capsid spike. J. Virol. 87 (14), 7853–7863.

Dubois, R., Dryden, K., Yeager, M., Tao, Y., 2013b. Astrovirus structure and assembly. In: Schultz-Cherry, S. (Ed.), Astrovirus Research. Springer, New York, pp. 47–64.

Fuentes, C., Bosch, A., Pinto, R.M., Guix, S., 2012. Identification of human astrovirus genome-linked protein (VPg) essential for virus infectivity. J. Virol. 86 (18), 10070–10078.

Geigenmuller, U., Chew, T., Ginzton, N., Matsui, S.M., 2002a. Processing of nonstructural protein 1a of human astrovirus. J. Virol. 76 (4), 2003–2008.

Geigenmuller, U., Ginzton, N.H., Matsui, S.M., 2002b. Studies on intracellular processing of the capsid protein of human astrovirus serotype 1 in infected cells. J. Gen. Virol. 83, 1691–1695.

Goodgame, R.W., 1999. Viral infections of the gastrointestinal tract. Curr. Gastroenterol. Rep. 1 (4), 292–300.

Goodgame, R.W., 2001. Viral causes of diarrhea. Gastroenterol. Clin. North Am. 30 (3), 779–795.

Grohmann, G.S., Glass, R.I., Pereira, H.G., Monroe, S.S., Hightower, A.W., Weber, R., et al., 1993. Enteric viruses and diarrhea in HIV-infected patients. Enteric Opportunistic Infections Working Group. N. Engl. J. Med. 329 (1), 14–20.

Gronemus, J.Q., Hair, P.S., Crawford, K.B., Nyalwidhe, J.O., Cunnion, K.M., Krishna, N.K., 2010. Potent inhibition of the classical pathway of complement by a novel C1q-binding peptide derived from the human astrovirus coat protein. Mol. Immunol. 48 (1–3), 305–313.

Guix, S., Bosch, A., Ribes, E., Dora Martinez, L., Pinto, R.M., 2004. Apoptosis in astrovirus-infected CaCo-2 cells. Virology 319 (2), 249–261.

Hair, P.S., Gronemus, J.Q., Crawford, K.B., Salvi, V.P., Cunnion, K.M., Thielens, N.M., et al., 2010. Human astrovirus coat protein binds C1q and MBL and inhibits the classical and lectin pathways of complement activation. Mol. Immunol. 47 (4), 792–798.

Holtz, L.R., Wylie, K.M., Sodergren, E., Jiang, Y., Franz, C.J., Weinstock, G.M., et al., 2011. Astrovirus MLB2 viremia in febrile child. Emerg. Infect. Dis. 17 (11), 2050–2052.

Koci, M.D., 2005. Immunity and resistance to astrovirus infection. Viral Immunol. 18 (1), 11–16.

Koci, M.D., Moser, L.A., Kelley, L.A., Larsen, D., Brown, C.C., Schultz-Cherry, S., 2003. Astrovirus induces diarrhea in the absence of inflammation and cell death. J. Virol. 77 (21), 11798–11808.

Krishna, N.K., 2005. Identification of structural domains involved in astrovirus capsid biology. Viral Immunol. 18 (1), 17–26.

Liste, M.B., Natera, I., Suarez, J.A., Pujol, F.H., Liprandi, F., Ludert, J.E., 2000. Enteric virus infections and diarrhea in healthy and human immunodeficiency virus-infected children. J. Clin. Microbiol. 38 (8), 2873–2877.

Lorrot, M., Vasseur, M., 2007. How do the rotavirus NSP4 and bacterial enterotoxins lead differently to diarrhea? Virol. J. 4, 31.

Lundgren, O., Peregrin, A.T., Persson, K., Kordasti, S., Uhnoo, I., Svensson, L., 2000. Role of the enteric nervous system in the fluid and electrolyte secretion of rotavirus diarrhea. Science 287 (5452), 491–495.

Madeley, C.R., Cosgrove, B.P., 1975. Letter: 28 nm particles in faeces in infantile gastroenteritis. Lancet 2 (7932), 451–452.

Maldonado, Y., Cantwell, M., Old, M., Hill, D., Sanchez, M.L., Logan, L., et al., 1998. Population-based prevalence of symptomatic and asymptomatic astrovirus infection in rural Mayan infants. J. Infect. Dis. 178 (2), 334–339.

Mendez, E., Salas-Ocampo, E., Arias, C.F., 2004. Caspases mediate processing of the capsid precursor and cell release of human astroviruses. J. Virol. 78 (16), 8601–8608.

Mendez, E., Aguirre-Crespo, G., Zavala, G., Arias, C.F., 2007. Association of the astrovirus structural protein VP90 with membranes plays a role in virus morphogenesis. J. Virol. 81 (19), 10649–10658.

Méndez, E., Murillo, A., Velázquez, R., Burnham, A., Arias, C., 2013. Replication cycle of astroviruses. In: Schultz-Cherry, S. (Ed.), Astrovirus Research. Springer, New York, pp. 19–45.

Moser, L.A., Carter, M., Schultz-Cherry, S., 2007. Astrovirus increases epithelial barrier permeability independently of viral replication. J. Virol. 81 (21), 11937–11945.

Naccache, S.N., Peggs, K.S., Mattes, F.M., Phadke, R., Garson, J.A., Grant, P., et al., 2015. Diagnosis of neuroinvasive astrovirus infection in an immunocompromised adult with encephalitis by unbiased next-generation sequencing. Clin. Infect. Dis. 60 (6), 919–923.

Nighot, P.K., Moeser, A., Ali, R.A., Blikslager, A.T., Koci, M.D., 2010. Astrovirus infection induces sodium malabsorption and redistributes sodium hydrogen exchanger expression. Virology 401 (2), 146–154.

Quan, P.L., Wagner, T.A., Briese, T., Torgerson, T.R., Hornig, M., Tashmukhamedova, A., et al., 2010. Astrovirus encephalitis in boy with X-linked agammaglobulinemia. Emerg. Infect. Dis. 16 (6), 918–925.

Royuela, E., 2010. Molecular cloning, expression and first antigenic characterization of human astrovirus VP26 structural protein and a C-terminal deleted form. Comp. Immunol. Microbiol. Infect. Dis. 33 (1), 1–14.

Sanchez-Fauquier, A., Carrascosa, A.L., Carrascosa, J.L., Otero, A., Glass, R.I., Lopez, J.A., et al., 1994. Characterization of a human astrovirus serotype 2 structural protein (VP26) that contains an epitope involved in virus neutralization. Virology 201 (2), 312–320.

Schultz-Cherry, S. (Ed.), 2013. Astrovirus Research: Essential Ideas, Everyday Impacts, Future Directions. Springer Verlag, Berlin-Heidelberg.

Schultz-Cherry, S., 2015. Astroviruses. John Wiley & Sons, Ltd, Chichester.

Stoermer, K.A., Morrison, T.E., 2011. Complement and viral pathogenesis. Virology 411 (2), 362–373.

Wood, D.J., David, T.J., Chrystie, I.L., Totterdell, B., 1988. Chronic enteric virus infection in two T-cell immunodeficient children. J. Med. Virol. 24 (4), 435–444.

Wunderli, W., Meerbach, A., Gungor, T., Berger, C., Greiner, O., Caduff, R., et al., 2011. Astrovirus infection in hospitalized infants with severe combined immunodeficiency after allogeneic hematopoietic stem cell transplantation. PLoS One 6 (11), e27483.

Yokoyama, C.C., Loh, J., Zhao, G., Stappenbeck, T.S., Wang, D., Huang, H.V., et al., 2012. Adaptive immunity restricts replication of novel murine astroviruses. J. Virol. 86 (22), 12262–12270.

Yuen, K.Y., Woo, P.C., Liang, R.H., Chiu, E.K., Chen, F.F., Wong, S.S., et al., 1998. Clinical significance of alimentary tract microbes in bone marrow transplant recipients. Diagn. Microbiol. Infect. Dis. 30 (2), 75–81.

Zhang, M., Zeng, C.Q., Morris, A.P., Estes, M.K., 2000. A functional NSP4 enterotoxin peptide secreted from rotavirus-infected cells. J. Virol. 74 (24), 11663–11670.

Chapter 4.2

Astrovirus Replication and Reverse Genetics

C.F. Arias, T. López, A. Murillo
Departamento de Genética del Desarrollo y Fisiología Molecular, Instituto de Biotecnología, Universidad Nacional Autónoma de México, Cuernavaca, Morelos, México

1 INTRODUCTION

Astroviruses are small, nonenveloped viruses with a single-stranded positive-sense RNA genome. Astroviruses are members of the *Astroviridae* family, which is divided into two genera: *Mamastrovirus* including viruses infecting mammals and *Avastrovirus* including viruses that infect avian species. Human astroviruses (HAstV) were originally classified into 8 serotypes, however, genetically diverse astrovirus strains have been recently detected, and the astrovirus taxonomy was modified (Guix et al., 2013). At least four genotype species of astrovirus that infect humans are now recognized: the classical 8 original serotypes are classified now as Mamastrovirus genotype 1, and the novel HAstVs that include the MLB, VA, and HMO virus lineages are classified into three additional genotypes (Guix et al., 2013). The novel HAstVs are more closely related to animal astroviruses. Most of the information described in this chapter, unless otherwise stated, has been obtained from characterization of the classical HAstV-1 and -8 serotypes.

Astroviruses have been long known to be a significant cause of gastroenteritis in children and in the young of many animal species (Méndez and Arias, 2013). The novel HAstVs have been isolated from stools, although a clear correlation with gastroenteritis is not established yet. They have been implicated, however, as agents of encephalitis in immunocompromised patients. In a fatal case of progressive encephalitis an astrovirus belonging to a recently discovered VA/HMO clade was identified in the brain of a 42-year-old man with chronic lymphocytic leukemia (Naccache et al., 2015), and an astrovirus phylogenetically related to ovine, mink, and bat astroviruses was reported to be the etiological agent of encephalitis in a 15-year-old boy with agammaglobulinemia (Quan et al., 2015). In another case report, the astrovirus HAstV-VA1/HMO-C-UK1 was identified in the brain and cerebrospinal fluid of an 18-month-old boy

Viral Gastroenteritis

with encephalopathy, indicating that VA1/HMO-C viruses, unlike the classic astrovirus HAstV-1-8 serotypes, can be neuropathic (Brown et al., 2015). In an additional report, a child with severe primary combined immunodeficiency died with a disseminated viral infection, and astroviral RNA was detected in the brain and different other organs, while immunochemistry confirmed infection of gastrointestinal tissues (Wunderli et al., 2011). Similarly, astroviruses have been detected in brain tissue from minks with the neurological shaking syndrome (Blomstrom et al., 2010) and from a young adult crossbreed steer with acute onset of neurologic disease (Li et al., 2013). In avian species, astroviruses produce not only enteritis, but also more diverse pathologies such as hepatitis, poult enteritis mortality syndrome, and nephritis (Méndez and Arias, 2013; Liu et al., 2014).

2 GENOME STRUCTURE AND ORGANIZATION

The genomes of astroviruses vary from 6.1 to 6.8 kb in size for mammalian viruses and from 6.9 to 7.7 kb for bird strains, including the 5′ and 3′ untranslated regions (UTRs). The 5′ UTR is short, ranging from 11 to 85 nt, while the 3′ UTR ranges from 80 to 85 in HAstVs and can be longer (130–305 nt) in avian astroviruses (Méndez and Arias, 2013). Regardless of its size, the astrovirus genome is organized into tree open reading frames, named ORF1a, 1b, and 2, is polyadenylated at the 3′ end and at the 5′ is linked covalently to VPg, a viral protein that is essential for viral replication (Fuentes et al., 2012; Méndez et al., 2013) (Fig. 4.2.1). A highly conserved sequence motif that forms a stem-loop structure at the 3′ end of the astroviral genome has been suggested

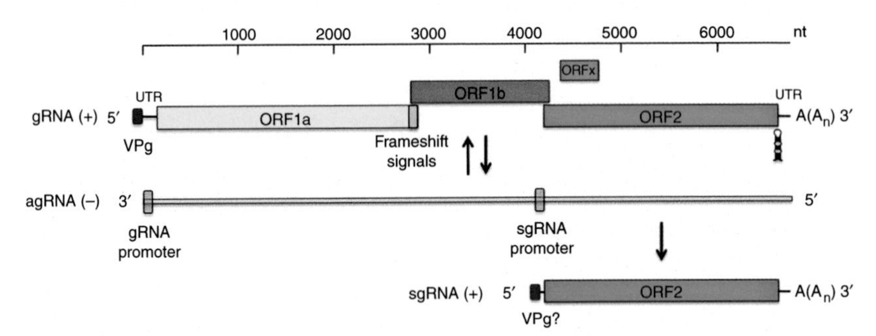

FIGURE 4.2.1 Genome organization of human astrovirus. The genome of astrovirus (shown here for HAstV-8) is organized in three ORFs: 1a, 1b, and 2. A fourth hypothetical ORFx is shown. Three well conserved sequence elements are present in the genome: the ribosomal frame shift signals, the sequence in the overlap region between ORF1b and OFR2, predicted to be the promoter for the subgenomic RNA (sgRNA), and a structural sequence motif at the 3′ UTR. The VPg protein is covalently bound to the 5′ end of the genomic RNA (gRNA), and could be also present in the sgRNA. The genomic RNA is around 6.8 kb and the sgRNA about 2.4 kb in size; both positive-sense RNAs have a polyA tail at the 3′ end. The negative-sense antigenomic RNA (agRNA), which serves as template to synthesize the full length positive-sense gRNA and sgRNA is shown.

to be relevant for the virus genome replication. Although some strains lack the conserved sequence, a stem-loop in this region is also predicted (De Benedictis et al., 2011). Similar sequence and predicted structures have been observed in other virus families, such as the *Caliciviridae, Picornaviridae, and Coronaviridae* (Kofstad and Jonassen, 2011).

The 5′-most ORFs 1a and 1b code for two nonstructural polyproteins (nsp's). Nsp1a includes sequences encoded only in ORF1a, while nsp1ab includes sequences derived from both ORF1a and 1b. Protein nsp1ab is produced by a – 1 translational frameshift mechanism. There is an overlap of 10–148 encoding nt in the genome of mammalian viruses, and between 10 and 45 nt in avian viruses between ORFs 1a and 1b. This overlapping region contains a conserved heptameric sequence (AAAAAAC) followed by a potential stem-loop structure, motifs that are both essential for the frameshift and the synthesis of the viral polymerase encoded in ORF1b (Jiang et al., 1993; Marczinke et al., 1994). ORF2 encodes the precursor of the virus capsid polypeptides. A short predicted ORF X of 91–122 codons, conserved in all HAstV and some other mammalian astroviruses, is present in a + 1 reading frame within ORF 2, although there is no evidence of a protein derived from this ORF (Firth and Atkins, 2010). The capsid polyprotein precursor is synthesized from a subgenomic RNA (sgRNA) of about 2.4 kb in size (Monroe et al., 1993) that might be transcribed by the recognition of a *cis*-element that acts as a promoter on the antigenomic RNA (agRNA) (Méndez and Arias, 2013).

3 VIRUS ENTRY

A critical step in virus infection is the interaction of the virus with receptors on the cell surface. Astrovirus receptors have not been described, however, the observation that the susceptibility of different cell lines to infection is dependent on the HAstV serotype suggests that more than one receptor for the virus might exist (Brinker et al., 2000). The crystal structures of HAstV-8 and turkey astrovirus type 2 viral spikes have been recently determined. Their overall structures showed only distant structural similarities, but putative polysaccharide receptor binding sites were described in both capsid spike structures (Dong et al., 2011; DuBois et al., 2013). For cell entry, the infectivity of HAstVs requires to be activated by treatment of the virus with trypsin, which processes the structural polyprotein precursor into the final capsid products and induces a change in the structure of the immature viral VP70 particles (Méndez and Arias, 2013; Dryden et al., 2012).

HAstV-1 and -8 are internalized into the cell by clathrin-dependent endocytosis, and the half-time for release of the genomic RNA (gRNA) into the cytoplasm for HAstV-8 is around 130 min (Donelli et al., 1992; Méndez et al., 2014). Drugs that disrupt endosome acidification and actin filament polymerization, as well as others that reduce the presence of cholesterol in the cell membrane decrease the infectivity of HAstV-8. Furthermore, in cells where the expression of Rab 7 is down regulated, the infectivity of HAstV-8 is also reduced. Altogether, these data support the notion that during cell entry astroviruses arrive to late endosomes, where the viral genome could exit into the cytosol (Méndez et al., 2014). During virus entry cellular signaling pathways are activated; in particular, it has been reported that astrovirus induces the transient phosphorylation of ERK1/2, independently of replication. Inhibitors of ERK affect viral protein and RNA synthesis with the consequent reduction of viral progeny (Moser and Schultz-Cherry, 2008). PI3K was also shown to reduce HAstV-1 infectivity, independently of MAPK, probably during cell entry (Tange et al., 2013).

4 GENOME TRANSCRIPTION AND REPLICATION

Upon infection of the host cell, the gRNA is translated into the nonstructural proteins. The viral RdRp uses the gRNA as template to synthesize a full-length negative-sense, antigenomic RNA (agRNA), which in turn serves as template to produce more copies of the full-length gRNA and many copies of an sgRNA (Méndez and Arias, 2013) (Fig. 4.2.1). The agRNA abundance ranges from 0.7% to 4% of that of the gRNA (Jang et al., 2010), while the sgRNA can reach up to 5- to 10-fold higher molar abundance than the full-length gRNA (Monroe et al., 1991). The viral and/or cellular factors that regulate the synthesis of all three species of RNA are not known, however, viruses that differ only in the hypervariable, carboxy-terminal region of nsp1a (see later) display differences in their RNA replication and growth properties (Guix et al., 2005). On the other hand, two predicted stem-loop structures conserved in the 3′ UTR of HAstV serotypes were recently found to bind in vitro to the polypyrimidine tract binding protein. This protein redistributes from nucleus to cytoplasm in HAstV8-infected Caco-2 cells and down regulation of its expression reduces the synthesis of gRNA and virus yield, suggesting that it might have a role in viral RNA replication (Espinosa-Hernandez et al., 2014).

Based on the transcription initiation site determined for the sgRNA in HAstV-1 and HAstV-2, ORF1b and ORF2 overlap in 8 nt; however, the length of this overlapping may vary and is not present in some viruses. The highly conserved sequence of around 40 nt upstream of the ORF2 start codon has been suggested to be part of the promoter for synthesis of the sgRNA. This conserved sequence shows partial identity with the 5′ end of the gRNA, suggesting that it has an important role for the synthesis of both gRNA and sgRNA (Méndez and Arias, 2013).

5 SYNTHESIS OF VIRAL PROTEINS

The gRNA of astrovirus is infectious when transfected into cells, implicating that it has the capacity to be translated and to initiate a virus replication cycle. A VPg protein covalently attached to the 5′ end of the viral gRNA seems to be

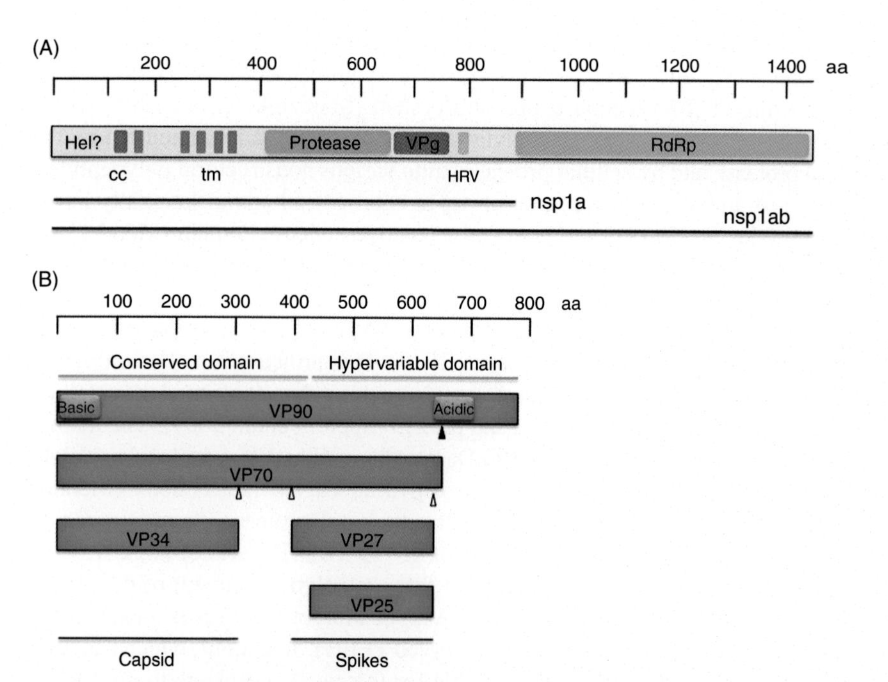

FIGURE 4.2.2 **Structural and functional features of viral proteins.** (A) Three nonstructural proteins have been positively identified and their function determined: the viral protease, a VPg protein, and the RNA-dependent RNA polymerase. In addition, other conserved structural motifs in the nsp1a polyprotein have been identified: a putative helicase domain *(Hel)*, a coiled-coil domain *(cc)*, several transmembrane domains *(tm)*, and a hypervariable region *(HVR)*. The intermediate protease (viral and cellular) cleavage products of polyproteins nsp1a and nsp1b, as well as the final produced polypeptides are not shown here, but can be identified for HAstV-8 from recent reviews (Méndez et al., 2003; Méndez and Arias, 2013). (B) The structural polyprotein precursor VP90 contains an N-terminal conserved domain and a C-terminal hypervariable domain. The conserved domain has a basic amino acid region that is thought to interact with the viral RNA, and a region that has been predicted to be the shell of the viral capsid. The hypervariable domain encodes the amino acid sequences that conform the viral spikes, and an acidic region toward the carboxy terminus of VP90 that is processed by caspases to yield VP70, present in the extracellular virions. Three final proteins generated by trypsin cleavage, VP25, VP27, and VP34, form the capsid of the fully infectious viral particles. The closed triangle represents a site for caspase cleavage and the open triangles sites for trypsin cleavage. The intermediate caspases and trypsin cleavage products are not shown, but have been described in recent reviews (Méndez et al., 2003; Méndez and Arias, 2013).

required for the efficient translation and/or replication of the viral RNA (Fuentes et al., 2012; Velázquez-Moctezuma et al., 2012). The gRNA directs the synthesis of the two nonstructural precursor polyproteins, nsp1a and nsp1ab, which are proteolytically processed by viral and cellular proteases (Fig. 4.2.2A) (Méndez and Arias, 2013). ORF1a codes for nsp1a, of between 787 aa for the HAstV sequence to 1240 aa for the avian astroviruses (Méndez et al., 2013). ORF1b is translated by a -1 ribosomal frameshift mechanism as fusion with ORF1a to produce the nsp1ab protein of about 160 kDa (Méndez and Arias, 2013). The efficiency of the signal that modulates the frameshift has been evaluated with reporter genes and varies from 6 to 7% in in vitro translation to 25–28% in cells using de T7-vaccinia expression system (Marczinke et al., 1994; Lewis and Matsui, 1995, 1996). Proteolytic processing of nsp1a is carried out by the viral protease and by cellular proteases, into various nonstructural polypeptides. The most amino-terminal region of nsp1a is processed cotranslationally into a final product of 20 kDa and contains a putative helicase domain (Méndez and Arias, 2013; Méndez et al., 2003), although a virus-encoded helicase activity has not been identified. After this domain there is a predicted coiled-coil structure, a predicted region of transmembrane domains, and a region coding for the viral protease (Méndez and Arias, 2013). This protease is a classic trypsin-like enzyme, and biochemical and crystal structural data confirmed the catalytic triad and the catalytic activity of the protein (Speroni et al., 2009). The mature product has been detected as a 26-kDa protein in HAstV8, a molecular weight that is consistent with processing of nsp1a at Val409 and Glu654 (Méndez et al., 2003; Speroni et al., 2009). After the viral protease, there is a region in nsp1a that encodes a VPg protein of 13–15 kDa (Fuentes et al., 2012). Its functional role is supported by the fact that proteolytic treatment of the gRNA leads to loss of virus infectivity and the replication of the virus is abolished by mutagenesis of the Tyr-693 at the conserved TEEEY-like motif, which has been postulated to be the residue responsible for the covalent linkage to viral RNA (Fuentes et al., 2012). The VPg sequence is followed by an in/del hypervariable region, which has been proposed to be relevant for viral replication (Guix et al., 2005). In addition, some polypeptides derived from the carboxy-terminal region of nsp1a, including the hypervariable region, are phosphorylated and colocalize with the gRNA and the endoplasmic reticulum (Guix et al., 2004b). Finally, processing of the nsp1ab precursor polyprotein yields, in addition to the proteins encoded in ORF1a, the viral RNA-dependent RNA polymerase (RdRp) of 57–59 kDa size (Méndez et al., 2003; Willcocks et al., 1999).

The structural proteins of the virus are translated from the sgRNA as a primary product of 87–90 kDa, named VP90 (Fig. 4.2.2B). The mechanism of translation of the sgRNA is not known, but it could depend on VPg (Fuentes et al., 2012). VP90 contains two domains classified by their sequence identity: a highly conserved amino-terminal domain (residues 1–415) and a highly divergent domain (residues 416 to the carboxy terminus) (Méndez and Arias, 2013; Mendez-Toss et al., 2000). The amino-terminal region contains a region of basic amino acids predicted to interact with the viral genome and a region proposed to form the capsid core (Méndez and Arias, 2013). The region between amino acids 416 and 647 are thought to form the spikes of the virion (Dong et al., 2011; Krishna, 2005), while the carboxy-terminal region of the hypervariable domain contains an acidic region that includes several potential sites for processing by cellular caspases (Méndez and Arias, 2013; Dong et al., 2011). Three proteins of 25, 27, and 34 kDa represent the final trypsin cleavage products present in the capsid of the mature HAstV-8 virion (Méndez and Arias, 2013).

6 VIRUS REPLICATION SITES

Positive-strand RNA viruses are known to replicate in the cytoplasm of the host cell in association with membranes. There is evidence suggesting that astrovirus replication is also carried out in association with membranes. Structural (VP90) and nonstructural (protease, RNA polymerase) proteins, as well as genomic and antigenomic RNA and infectious virus particles have been reported to associate with membranes (Méndez et al., 2007; Murillo et al., 2015). Additionally, a protein corresponding to the carboxy-terminus of nsp1a was shown to colocalize in cells with virus gRNA and calnexin, and to interact with the RNA polymerase (Guix et al., 2004b; Fuentes et al., 2011). The architecture of membrane rearrangement in astrovirus infected cells is not yet known, however, ultrastructural analysis by electron microscopy of astrovirus-infected Caco-2 cells showed large groups of viral particles surrounding "O-ring" vesicles, probably corresponding to the double-membrane vesicles reported by Guix et al. (2004b). Structural proteins and the viral RNA polymerase were detected inside these vesicles by immunoelectron microscopy (Guix et al., 2004b; Méndez et al., 2007). Additional structures, similar to those found in HeLa cells infected with poliovirus, have been observed in Caco-2 cells infected with HAstV-8, around which astroviral particles are found (Méndez et al., 2007) (Fig 4.2.3).

As part of their replication cycle, viruses subvert intracellular membranes where viral and host factors cooperatively generate distinct organelle-like structures that serve as platform to form the replication complex of the virus (Paul and Bartenschlager, 2013). Recently, the cellular proteins present in membranes to which astroviral proteins and RNA associate were determined by LC-MS/MS. Functional analysis of the protein–protein interaction network showed some biological processes that were enriched in these membranes, such as gluconeogenesis, fatty acid beta-oxidation, fatty acid synthesis, long chain fatty acid synthesis and tricarboxylic acid cycle (Murillo et al., 2015). These findings are consistent with the fact that modification of the lipid metabolism is emerging as a landmark of the infection of positive-sense-viruses (Paul and Bartenschlager, 2013). Although it is not known if the ubiquitin/proteasome system contributes to cell membrane remodelling, it is also required for astrovirus replication; proteasomal inhibitors and knockdown of ubiquitin expression produce a reduction in viral protein synthesis, and gRNA and viral progeny production (Casorla et al., in preparation).

7 ASSEMBLY AND EXIT OF VIRAL PARTICLES

Astrovirus particles assemble into virus-like particles when the ORF2 is expressed using different expression systems. In LLCMK2 cells, a recombinant vaccinia virus expressing the structural polyprotein of HAstV-2 induced the formation of VLPs with size and morphology similar to native HAstV-2 particles, and these VLPs remained stable during the purification procedure using sucrose

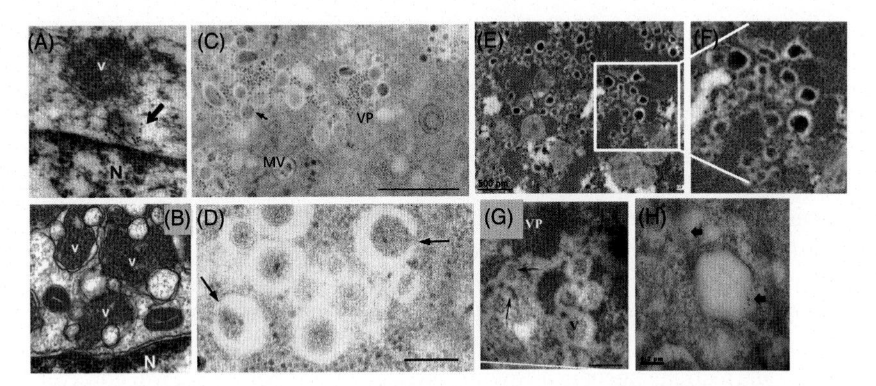

FIGURE 4.2.3 **Electron microscopy localization of astrovirus particles in intracellular vesicles.** (A) Immunodetection of the C-terminal region of the nsp1a protein of HAstV-4 in Caco-2 cells at 48 hpi; N, nucleus; V, viral aggregates. Arrow indicates area positive for immunostaining. (B) Astrovirus particles observed in clusters inside double membrane vacuoles in Caco-2 cells infected with of HAstV-4 (24 hpi); V, viral clusters; N, nucleus. (C) Morphology of poliovirus-induced membranes in HeLa cells (5 hpi). The arrow indicates a poliovirus-induced structure with viral particles in the enclosed region; VP, viral particles; MV, multivesicular body. The calibration bar represents 1 μm. (D) Selected horseshoe-shaped membranes induced by poliovirus at 5 hpi. Arrows indicate the opening between the enclosed region and the cytoplasm. The calibration bar represents 0.2 μm. (E) Morphology of HAstV-8 infected Caco-2 cells (18 hpi). (F) Enlargement of the indicated area showing some horseshow-like structures, similar to those observed in poliovirus infected cells. (G) HAstV-8 particles observed in clusters in Caco-2 cells (24 hpi). The calibration bar represents 0.2 μm. Particles that look partially assembled inside or at the edges of vesicles induced during infection are marked with arrows; VP, virus clusters; V, vesicles (H) Immunodetection of astrovirus RNA polymerase in HAstV-8 infected Caco-2 cells (18 hpi) using the primary antibody 1b2, followed by incubation with goat antirabbit IgG labeled with 10-nm gold particles as detection antibody. Arrows indicate positive gold signal. *(Source: Images A and B were reproduced from Guix et al., 2004b, C and D from Schlegel et al., 1996, and G from Méndez et al., 2007, with permission. Images E, F, and H were provided by Murillo et al., unpublished data.)*

gradient ultracentrifugation (Dalton et al., 2003). The ORF2 of HAstV-1 also produced VLPs of around 38 nm when expressed in insect cells using a recombinant baculovirus (Caballero et al., 2004). In both cases, the assembled VLP were antigenically, biochemically, and structurally similar to native astrovirus particles. These features make VLPs an interesting candidate to be tested for the development of an astrovirus vaccine. Of interest, structurally and antigenically correct VLPs were also obtained when a truncated form of HAstV-1 ORF2 lacking the amino-terminal 70 aa was expressed (Caballero et al., 2004).

In astrovirus-infected Caco-2 cells, VP90 assembles intracellularly into viral particles, and is then processed by cellular caspases to yield viruses containing VP70 in a process that might involve at least three intermediate cleavage products of 82, 78, and 75 kDa (Méndez et al., 2004). Astrovirus infection induces apoptosis (Méndez et al., 2004; Guix et al., 2004a), and down regulation of the expression of caspases 3 and 9 blocks the processing of HAstV-8 VP90 to VP70 at the sequence motif TYVD657 (Baños-Lara and Méndez, 2010). The cleavage of VP90–VP70 is associated with release of the virus from the cell through a mechanism that does not involve cell lysis (Méndez et al., 2004; Baños-Lara and Méndez, 2010). All extracellular virions grown in cells in the absence of trypsin are composed by VP70, and to be fully infective these viral particles have to be proteolytically processed to yield the mature virions composed of three capsid proteins (Bass and Qiu, 2000; Méndez et al., 2002).

8 REVERSE GENETICS

The availability of a reverse genetics system, that is, the ability to recover infectious virus from a cDNA copy of the RNA genome, represents an invaluable tool to study the biology of RNA viruses. To this end, Geigenmüller et al. used a full-length cDNA clone of the HAstV1 gRNA to develop a reverse genetics system that was used to analyze specific mutations in the structural (Geigenmuller et al., 1997; Geigenmuller et al., 2002) and nonstructural proteins (Guix et al., 2005) of the virus. In this system, the in vitro transcribed RNA was introduced into the highly transfectable, but poorly permissive BHK-21 cells and the virus produced was propagated in the permissive Caco-2 cell line (Geigenmuller et al., 1997). The disadvantage of this system was the low permissiveness of BHK-21 cells, and the subsequent requirement to propagate the virus in Caco-2 cells to reach acceptable titers. This implied harvesting the virus after several rounds of replication, what might cause the accumulation of mutations in the viral progeny. In fact, viruses with nucleotide changes in the nonstructural region, not related to the original introduced mutations, were found when this method was used to recover viruses (Guix et al., 2005).

To overcome this problem, Velázquez-Moctezuma et al. (2012) developed a reverse genetic system using the highly transfectable hepatoma Huh7.5.1 cells to pack the virus. Although these cells were infected with a 100-fold lower efficiency than Caco-2 cells, the yield of infectious virus in both cells was similar, and in Huh7.5.1 cells the VP90 precursor of the astrovirus capsid polypeptides was found to be efficiently processed. Importantly, virus titers near to 10^8 infectious units per ml were obtained in Huh7.5.1 at 16–20 h after transfection with total RNA isolated from astrovirus-infected Caco-2 cells. On the other hand, virus titers close to 10^6 were recovered by using in vitro transcribed RNA from a full-length cDNA HAstV-1 clone; this virus yield was about two log steps higher than that obtained in BHK-21 cells (Geigenmuller et al., 1997). The lower efficiency of virus production in Huh7.5.1 cells when in vitro synthesized RNA was compared to gRNA purified from infected cells, suggested that a factor was missing for the efficient translation or replication of the transfected, synthetic viral RNA (Velázquez-Moctezuma et al., 2012). The 5′-end modification of the viral RNA seemed to determine the specific infectivity of the RNA, since virus recovery was abolished when the total RNA isolated from infected cells was treated with a protease, while it increased when the in vitro transcribed RNA was capped. These observations are consistent with the more recently description of the existence of a VPg protein covalently bound to the 5′-end of the astrovirus gRNA, essential for virus replication (Fuentes et al., 2012). Using this reverse genetic system, amino acids relevant for the function of VPg were recently explored (Fuentes et al., 2012).

9 PERSPECTIVES

The availability of a reverse genetic system for human astroviruses provides a invaluable tool to close knowledge gaps of the various replication stages of astrovirus, and it can also contribute to define virulence determinants, and eventually to develop safe vaccine candidates. Advances in the structural characterization of different astrovirus strains and their components, as well as the identification of cellular factors relevant for virus replication, represent important avenues to better understanding of the biology of this genetically diverse family of viruses. It will also be important to develop reverse genetics systems, adaptation to cell culture, and animal models for some of the novel viruses that have been recently discovered, which have been associated with diseases other than enteritis both in mammalian and avian species.

ABBREVIATIONS USED

agRNA antigenomic RNA
ERK1/2 extracellular signal-regulated kinases 1 and 2
gRNA genomic RNA
HAstV human astrovirus
in/del insertion/deletion
LC-MS/MS liquid chromatography–tandem mass spectrometry
MAPK mitogen-activated protein kinases
Nsp nonstructural protein
ORF open reading frame
PI3K phosphatidylinositol-4,5-bisphosphate 3-kinase
Rab 7 ras-related protein Rab-7
RdRp RNA-dependent RNA polymerase
sgRNA subgenomic RNA
UTR untranslated region
VLP virus-like particle

REFERENCES

Baños-Lara M del, R., Méndez, E., 2010. Role of individual caspases induced by astrovirus on the processing of its structural protein and its release from the cell through a non-lytic mechanism. Virology 401, 322–332.

Bass, D.M., Qiu, S., 2000. Proteolytic processing of the astrovirus capsid. J. Virol. 74, 1810–1814.

Blomstrom, A.L., Widen, F., Hammer, A.S., et al., 2010. Detection of a novel astrovirus in brain tissue of mink suffering from shaking mink syndrome by use of viral metagenomics. J. Clin. Microbiol. 48, 4392–4396.

Brinker, J.P., Blacklow, N.R., Herrmann, J.E., 2000. Human astrovirus isolation and propagation in multiple cell lines. Arch. Virol. 145, 1847–1856.

Brown, J.R., Morfopoulou, S., Hubb, J., et al., 2015. Astrovirus VA1/HMO-C: An increasingly recognized neurotropic pathogen in immunocompromised patients. Clin. Infect. Dis. 60, 881–888.

Caballero, S., Guix, S., Ribes, E., Bosch, A., et al., 2004. Structural requirements of astrovirus virus-like particles assembled in insect cells. J. Virol. 78, 13285-13292.

Dalton, R.M., Pastrana, E.P., Sanchez-Fauquier, A., 2003. Vaccinia virus recombinant expressing an 87-kilodalton polyprotein that is sufficient to form astrovirus-like particles. J. Virol. 77, 9094–9098.

De Benedictis, P., Schultz-Cherry, S., Burnham, A., et al., 2011. Astrovirus infections in humans and animals—molecular biology, genetic diversity, and interspecies transmissions. Infect. Genet. Evol. 11, 1529–1544.

Donelli, G., Superti, F., Tinari, A., et al., 1992. Mechanism of astrovirus entry into Graham 293 cells. J. Med. Virol. 38, 271–277.

Dong, J., Dong, L., Méndez, E., et al., 2011. Crystal structure of the human astrovirus capsid spike. Proc. Natl. Acad. Sci. U S A 108, 12681–12686.

Dryden, K.A., Tihova, M., Nowotny, N., et al., 2012. Immature and mature human astrovirus: structure, conformational changes, and similarities to hepatitis E virus. J. Mol. Biol. 422, 650–658.

DuBois, R.M., Freiden, P., Marvin, S., et al., 2013. Crystal structure of the avian astrovirus capsid spike. J. Virol. 87, 7853–7863.

Espinosa-Hernandez, W., Velez-Uriza, D., Valdes, J., et al., 2014. PTB binds to the 3′ untranslated region of the human astrovirus type: 8 a possible role in viral replication. PLoS One 9, e113113.

Firth, A.E., Atkins, J.F., 2010. Candidates in Astroviruses, Seadornaviruses, Cytorhabdoviruses and Coronaviruses for +1 frame overlapping genes accessed by leaky scanning. Virol. J. 7, 17.

Fuentes, C., Guix, S., Bosch, A., et al., 2011. The C-terminal nsP1a protein of human astrovirus is a phosphoprotein that interacts with the viral polymerase. J. Virol. 85, 4470–4479.

Fuentes, C., Bosch, A., Pinto, R.M., et al., 2012. Identification of human astrovirus genome-linked protein (VPg) essential for virus infectivity. J. Virol. 86, 10070–10078.

Geigenmuller, U., Ginzton, N.H., Matsui, S.M., 1997. Construction of a genome-length cDNA clone for human astrovirus serotype 1 and synthesis of infectious RNA transcripts. J. Virol. 71, 1713–1717.

Geigenmuller, U., Chew, T., Ginzton, N., et al., 2002. Processing of nonstructural protein 1a of human astrovirus. J. Virol. 76, 2003–2008.

Guix, S., Bosch, A., Ribes, E., et al., 2004a. Apoptosis in astrovirus-infected CaCo-2 cells. Virology 319, 249–261.

Guix, S., Caballero, S., Bosch, A., et al., 2004b. C-terminal nsP1a protein of human astrovirus colocalizes with the endoplasmic reticulum and viral RNA. J. Virol. 78, 13627–13636.

Guix, S., Caballero, S., Bosch, A., et al., 2005. Human astrovirus C-terminal nsP1a protein is involved in RNA replication. Virology 333, 124–131.

Guix, S., Bosch, A., Pintó, R.M., 2013. Astrovirus taxonomy. In: Schultz-Cherry, S. (Ed.), Astrovirus Research: Essentials Ideas, Everyday Impacts. Springer, New York, pp. 97–118.

Jang, S.Y., Jeong, W.H., Kim, M.S., et al., 2010. Detection of replicating negative-sense RNAs in CaCo-2 cells infected with human astrovirus. Arch. Virol. 155, 1383–1389.

Jiang, B., Monroe, S.S., Koonin, E.V., et al., 1993. RNA sequence of astrovirus: distinctive genomic organization and a putative retrovirus-like ribosomal frameshifting signal that directs the viral replicase synthesis. Proc. Natl. Acad. Sci. USA 90, 10539–10543.

Kofstad, T., Jonassen, C.M., 2011. Screening of feral and wood pigeons for viruses harbouring a conserved mobile viral element: characterization of novel Astroviruses and Picornaviruses. PLoS One 6, e25964.

Krishna, N.K., 2005. Identification of structural domains involved in astrovirus capsid biology. Viral Immunol. 18, 17–26.

Lewis, T.L., Matsui, S.M., 1995. An astrovirus frameshift signal induces ribosomal frameshifting in vitro. Arch. Virol. 140, 1127–1135.

Lewis, T.L., Matsui, S.M., 1996. Astrovirus ribosomal frameshifting in an infection-transfection transient expression system. J. Virol. 70 (5), 2869–2875.

Li, L., Diab, S., McGraw, S., et al., 2013. Divergent astrovirus associated with neurologic disease in cattle. Emerg. Infect. Dis. 19, 1385–1392.

Liu, N., Wang, F., Shi, J., et al., 2014. Molecular characterization of a duck hepatitis virus 3-like astrovirus. Vet. Microbiol. 170, 39–47.

Marczinke, B., Bloys, A.J., Brown, T.D., et al., 1994. The human astrovirus RNA-dependent RNA polymerase coding region is expressed by ribosomal frameshifting. J. Virol. 68, 5588–5595.

Méndez, E., Aguirre-Crespo, G., Zavala, G., et al., 2007. Association of the astrovirus structural protein VP90 with membranes plays a role in virus morphogenesis. J. Virol. 81, 10649–10658.

Méndez, E., Arias, C.F., 2013. Astroviruses. In: Knipe, D.M., Howley, P.M. et al., (Eds.), Fields Virology, sixth ed. Lippincott Williams & Wilkins, Philadelphia, pp. 609–628.

Méndez, E., Fernández-Luna, T., López, S., et al., 2002. Proteolytic processing of a serotype 8 human astrovirus ORF2 polyprotein. J. Virol. 76, 7996–8002.

Méndez, E., Salas-Ocampo, M.P., Munguía, M.E., et al., 2003. Protein products of the open reading frames encoding nonstructural proteins of human astrovirus serotype 8. J. Virol. 77, 11378–11384.

Méndez, E., Salas-Ocampo, E., Arias, C.F., 2004. Caspases mediate processing of the capsid precursor and cell release of human astroviruses. J. Virol. 78, 8601–8608.

Méndez, E., Murillo, A., Velázquez, R., et al., 2013. Replication cycle of astroviruses. In: Schultz-Cherry, S. (Ed.), Astrovirus Research: Essentials Ideas, Everyday Impacts. Springer, New York, pp. 19–45.

Méndez, E., Muñoz-Yañez, C., Sánchez-San Martín, C., et al., 2014. Characterization of human astrovirus cell entry. J. Virol. 88, 2452–2460.

Mendez-Toss, M., Romero-Guido, P., Munguia, M.E., et al., 2000. Molecular analysis of a serotype 8 human astrovirus genome. J. Gen. Virol. 81, 2891–2897.

Monroe, S.S., Stine, S.E., Gorelkin, L., et al., 1991. Temporal synthesis of proteins and RNAs during human astrovirus infection of cultured cells. J. Virol.. 65, 641–648.

Monroe, S.S., Jiang, B., Stine, S.E., Koopmans, M, Glass, R.I., 1993. Subgenomic RNA sequence of human astrovirus supports classification of Astroviridae as a new family of RNA viruses. J. Virol. 67, 3611–3614.

Moser, L.A., Schultz-Cherry, S., 2008. Suppression of astrovirus replication by an ERK1/2 inhibitor. J. Virol. 82, 7475–7482.

Murillo, A., Vera-Estrella, R., Barkla, B.J., et al., 2015. Identification of host cell factors associated with astrovirus replication in Caco-2 cells. J. Virol. 89, 10359–10370.

Naccache, S.N., Peggs, K.S., Mattes, F.M., et al., 2015. Diagnosis of neuroinvasive astrovirus infection in an immunocompromised adult with encephalitis by unbiased next-generation sequencing. Clin. Infect. Dis. 60, 919–923.

Paul, D., Bartenschlager, R., 2013. Architecture and biogenesis of plus-strand RNA virus replication factories. World J. Virol. 2, 32–48.

Quan, P.L., Wagner, T.A., Briese, T., et al., 2015. Astrovirus encephalitis in boy with X-linked agammaglobulinemia. Emerg. Infect. Dis. 16, 918–925.

Schlegel, A., Giddings, Jr., T.H., Ladinsky, M.S., et al., 1996. Cellular origin and ultrastructure of membranes induced during poliovirus infection. J. Virol. 70, 6576–6588.

Speroni, S., Rohayem, J., Nenci, S., et al., 2009. Structural and biochemical analysis of human pathogenic astrovirus serine protease at 2.0 A resolution. J. Mol. Biol. 387, 1137–1152.

Tange, S., Zhou, Y., Nagakui-Noguchi, Y., et al., 2013. Initiation of human astrovirus type 1 infection was blocked by inhibitors of phosphoinositide 3-kinase. Virol. J. 10, 153.

Velázquez-Moctezuma, R., Baños-Lara, M., del, R., Acevedo, Y., et al., 2012. Alternative cell lines to improve the rescue of infectious human astrovirus from a cDNA clone. J. Virol. Methods 179, 295–302.

Willcocks, M.M., Boxall, A.S., Carter, M.J., 1999. Processing and intracellular location of human astrovirus non-structural proteins. J. Gen. Virol. 80, 2607–2611.

Wunderli, W., Meerbach, A., Gungor, T., et al., 2011. Astrovirus infection in hospitalized infants with severe combined immunodeficiency after allogeneic hematopoietic stem cell transplantation. PLoS One 6, e27483.

Chapter 4.3

Molecular Epidemiology of Astroviruses

P. Khamrin*, N. Maneekarn*, H. Ushijima**
**Department of Microbiology, Faculty of Medicine, Chiang Mai University, Chiang Mai, Thailand; **Division of Microbiology, Department of Pathology and Microbiology, Nihon University School of Medicine, Tokyo, Japan*

1 INTRODUCTION

Astroviruses are enteric viruses associated with acute gastroenteritis in humans and a number of animal species (Mendez and Arias, 2013). Since the first human astrovirus was discovered using electron microscope in 1975, extensive epidemiological studies on astrovirus infection and disease have been carried out and these viruses are considered to be among the most common viral agents that causes gastroenteritis in young children worldwide (Appleton and Higgins, 1975; Bosch et al., 2014; De Benedictis et al., 2011; Madeley and Cosgrove, 1975). Soon after the first description in humans, astrovirus infections have been reported from a wide variety of mammals and avian species, including lambs, sheep, calves, pigs, dogs, cats, deer, mice, minks, bats, cheetahs, sea lions, dolphins, rats, rabbits, chickens, ducks, turkeys, and pigeons (Bosch et al., 2014; De Benedictis et al., 2011; Madeley and Cosgrove, 1975). The chronology of astrovirus discoveries is shown in Table 4.3.1.

Astroviruses belong to the family *Astroviridae*, which are small, nonenveloped, single-stranded, positive-sense RNA viruses. The genomes range from 6.1 to 7.9 kb in size and contain three open reading frames (ORFs). The ORF1a and ORF1b encode nonstructural proteins, whereas ORF2 encodes the capsid protein precursor (Mendez and Arias, 2013) (see Chapter 4.1). Although astroviruses have been known to be major causative agents for gastroenteritis in humans worldwide, there is relatively little information on their association with gastroenteritis diseases in other animal species. Surveillance investigations conducted previously in several countries worldwide indicated that human astroviruses are involved in 2–30% of diarrhea in children (De Benedictis et al., 2011).

Viral Gastroenteritis

TABLE 4.3.1 Chronology of Astrovirus Discoveries From Human and Animal Species

Host	Year of detection	Disease associated	References
Human (Classic HAstV)	1975	Gastroenteritis in children	Madeley and Cosgrove (1975)
Ovine	1977	Diarrhea in lambs	Snodgrass and Gray (1977)
Bovine	1978	Diarrhea in calves, asymptomatic	Woode and Bridger (1978)
Chicken	1979	Interstitial nephritis in young chicks, enteritis	Yamaguchi et al. (1979)
Turkey	1980	Poultry enteritis and mortality	McNulty et al. (1980)
Pig	1980	Diarrhea in piglets, asymptomatic	Bridger (1980)
Dog	1980	Diarrhea in pups, asymptomatic	Williams (1980)
Cat	1981	Pyrexia and mild diarrhea, asymptomatic	Hoshino et al. (1981)
Red deer	1981	Diarrhea	Tzipori et al. (1981)
Duck	1984	Acute hepatitis	Gough et al. (1984)
Mouse	1985	Diarrhea, asymptomatic	Kjeldsberg and Hem (1985)
Mink	2002	Preweaning diarrhea, shaking mink syndrome	Englund et al. (2002)
Guinea fowl	2005	Enteritis	Cattoli et al. (2005)
Insectivorous bat	2008	—	Chu et al. (2008)
Human (MLB1)	2008	Acute diarrhea	Finkbeiner et al. (2008)
Human (VA1)	2009	Gastroenteritis	Finkbeiner et al. (2009b)
Cheetah	2009	Lethargy and anorexia, watery diarrhea	Atkins et al. (2009)
California sea lion	2010	Diarrhea (pups), asymptomatic (adults)	Rivera et al. (2010)
Steller sea lion	2010	Asymptomatic	Rivera et al. (2010)
Bottlenose dolphin	2010	Asymptomatic	Rivera et al. (2010)
Brown rat	2010	—	Chu et al. (2010)

(*Continued*)

TABLE 4.3.1 Chronology of Astrovirus Discoveries From Human and Animal Species (*cont.*)

Host	Year of detection	Disease associated	References
Roe deer	2010	Diarrhea	Smits et al. (2010)
Rabbit	2011	Rabbit enteritis	Martella et al. (2011b)
Feral pigeon	2011	—	Kofstad and Jonassen (2011)
Wood pigeon	2011	—	Kofstad and Jonassen (2011)
Shorebird	2014	Asymptomatic	Honkavuori et al. (2014)

2 CLASSIFICATION

Astroviruses have been found to be morphologically unique and typically appear as a star-like structures by electron microscopy, with the icosahedral capsid particles measuring approximately 28–30 nm in diameter (Appleton and Higgins, 1975; Madeley and Cosgrove, 1975). The term "astron" (greek) refers to the appearance of the icosahedral capsid which is similar to a star shape. The *Astroviridae* family contains several important human and animal pathogens and is further differentiated into two genera: *Mamastrovirus* (MAstV) and *Avastrovirus* (AAstV). MAstV and AAstV have been known to infect many mammalian and avian species, respectively. Astroviruses identified thus far have been classified on the basis of their host tropism and viral genomic variation by the International Committee on Taxonomy of Viruses (ICTV). Currently, 19 species within the *Mamastrovirus* genus (MAstV1-19) and 3 species within *Avastrovirus* (AAstV1-3) have been officially approved (Fig. 4.3.1). Recently, several novel astroviruses from different animal host species have been discovered, mostly by metagenomic pyrosequencing.

The prototype astrovirus species was originally isolated from humans, called the classic human astrovirus (HAstV) and now belongs to MAstV1 species. The classic human astroviruses are genetically highly diversed and have been classified into eight genotypes (HAstV1-HAstV8). Beside MAstV1, three additional astrovirus species, MAstV6, MAstV8, and MAstV9 have been recently identified in human stool samples. The MAstV6 or MLB1 was detected during an outbreak of diarrhea in Australia (Finkbeiner et al., 2008). MAstV8 is comprised of the human astrovirus strains VA2, VA4, and HMOAstV-A, while MAstV9 includes the VA1, VA3, HMOAstV-B, and HMOAstV-C strains (Finkbeiner et al., 2009a,b; Jiang et al., 2013; Kapoor et al., 2009). Therefore,

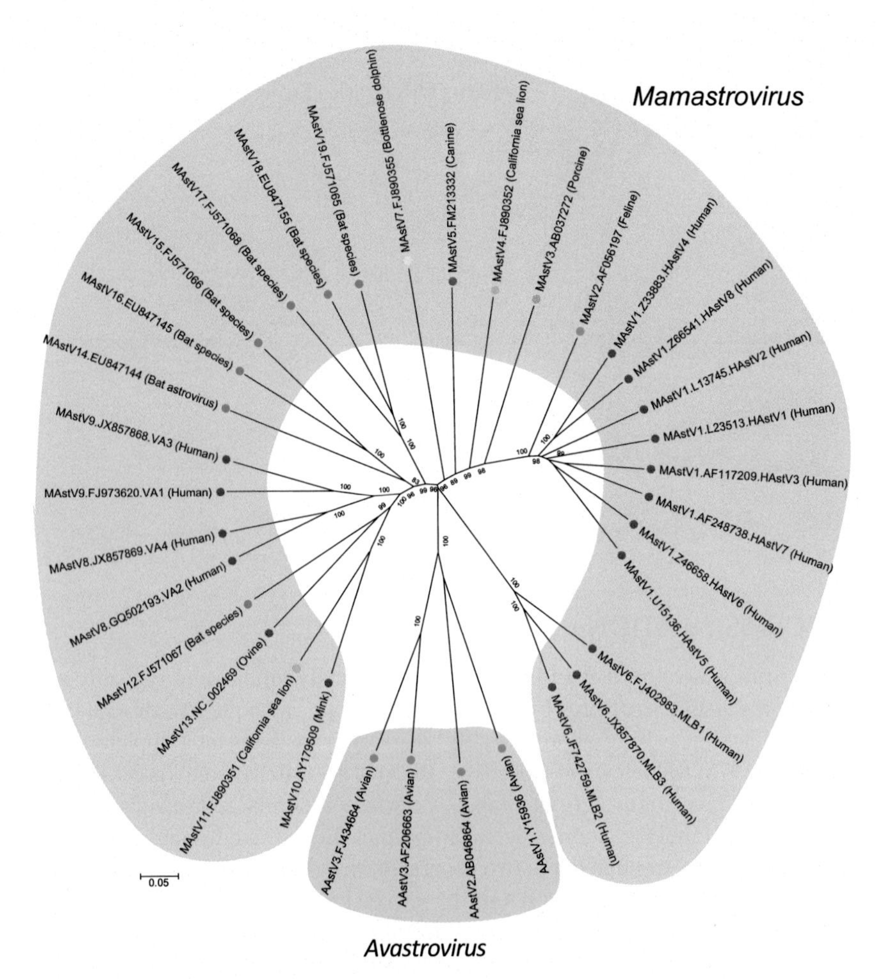

FIGURE 4.3.1 **Phylogenetic relationship among the members in *Astroviridae* family.** The phylogenetic tree was constructed by neighbor-joining clustering method using MEGA5.2. Scale bar indicates nucleotide substitutions per site and bootstrap values (>80) are indicated at the corresponding nodes. The representative reference strains from each species were obtained from GenBank. *MAstV, Mamastrovirus*; *AAstv, Avastrovirus.*

it is evident that there are at least three clades or genetic clusters of astroviruses cocirculating in humans: MAstV1 (classic human astrovirus: HAstV), MAstV6 (MLB clade), and MAstV8 together with MAstV9 (VA clade). In addition to humans, a wide range of animals has been found to be infected by different astrovirus species, such as cats, pigs, dogs, dolphin, mink, and sheep, infected by MAstV2, 3, 5, 7, 10, and 13, respectively. Moreover, California sea lions have been discovered to be infected by MAstV4 and 11, while several bat species are reported to be positive for MAstV12 and MAstV14–19 (Bosch et al., 2014; De Benedictis et al., 2011). The avian astroviruses are clustered exclusively within the *Avastrovirus* genus (AAstV); most recently a mammalian-like astrovirus has been reported in birds in 2011 (Pankovics et al., 2015).

3 DETECTION AND DIAGNOSIS

By electron microscopy, the astrovirus particle is a small, round virus with a distinctive 5–6 pointed star-like structure (Appleton and Higgins, 1975; Madeley and Cosgrove, 1975). In 1981, the virus was successfully propagated in a primary cell culture, initially using human kidney epithelial cells (HEK) and later a human colon carcinoma (CaCo-2) cell line, which has been found to be more sensitive for the propagation of most human astrovirus strains (Willcocks et al., 1990). In addition, astroviruses could be propagated in several types of adenocarcinoma cell lines (SK-CO-1, T-84, and HT-29) and monkey kidney cell lines (Vero, MA104, and Cos-1) (Brinker et al., 2000). The success of growing human astroviruses in cell cultures was a big step forward in the characterization of these viruses. For the diagnosis of astrovirus infection, several different methods are now available. These include enzyme immunoassay (EIA), an immunochromatography (IC) test, reverse transcription-polymerase chain reaction (RT-PCR), real-time PCR, loop-mediated isothermal amplification, and nucleic acid sequence analysis (Jonassen et al., 1995; Khamrin et al., 2010; Wei et al., 2013). Of all of these methods, RT-PCR and nucleic acid sequence analysis are the most widely used methods for the detection and genotype identification of astroviruses. These techniques have replaced the traditional culture assays and immunological tests and become the gold standard for diagnosis of astrovirus infections in both humans and animal species.

4 MOLECULAR EPIDEMIOLOGY

4.1 Human Astroviruses

4.1.1 Classic Human Astroviruses (HAstV)

Since the human astrovirus was discovered, epidemiological studies of classic human astroviruses (HAstVs) have been extensively performed and reported from several countries around the world (Bosch et al., 2014; De Benedictis et al., 2011). The viruses have been identified in sporadic nonbacterial diarrhea cases and have been reported to be associated with a wide range of clinical illnesses including diarrhea, vomiting, fever, abdominal pain, bronchiolitis, and otitis (Mendez and Arias, 2013).

The HAstV positivity rates in acute gastroenteritis patients have been reviewed and summarized in Table 4.3.2. Surveillance of HAstV conducted in several countries worldwide indicate that the prevalence of HAstVs is variable from <1–40% of diarrhea cases. The variations of positivity rates are dependent on geographical area, study periods, age groups, and screening methods. However, in most studies the prevalence rate of HAstVs is approximately 10%.

TABLE 4.3.2 Human Astrovirus Positivity Rates in Patients With Acute Gastroenteritis (Reviewed From 60 Studies Worldwide)

Region/Country		Year of sample collection	Detection method	No. of specimens tested	No. astrovirus positive (%)	Age	Coinfecting viruses	Site	References
Asia	Japan	2006–07	RT-PCR	628	15 (2.4%)	≤5 years		O	Dey et al. (2010)
	Japan	2008–09	RT-PCR	662	11 (1.7%)	≤5 years		O	Chan-it et al. (2010)
	Japan	2009–14	RT-PCR	2,908	86 (3%)	≤5 years	NoV	O	Thongprachum et al. (2015)
	Thailand	2000–03, 2005, 2007–08, 2010–11	RT-PCR	1,022	14 (1.4%)	≤5 years	RVA	H	Malasao et al. (2012)
	China	1998–2005	EIA/RT-PCR	1,668	91 (5.5%)	≤5 years		H	Fang et al. (2006)
	China	2005–07	RT-PCR	664	52 (7.8%)	≤13 years		O	Guo et al. (2010)
	China	2007–08	RT-PCR	Children 361 Adult 301	Children 49 (13.6%) Adult 7 (2.3%)	Children ≤3 years Adult ≥50 years		—	Wang et al. (2011)
	China	2008	RT-PCR	279	23 (8.2%)	≤6 years		H	Shan et al. (2009)
	Russia	2009	Real-time PCR	495	7 (1.4%)	≤5 years		H	Chhabra et al. (2014)
	Korea	2002–07	EIA/RT-PCR	106,027	2,057 (1.3%)			—	Jeong et al. (2011)
	Korea	2010–11	RT-PCR	186	3 (1.6%)	≤11 years		H	So et al. (2013)
	Korea	2008–12	RT-PCR	9,597	94 (1%)	≤78 years		O	Ham et al. (2014)
	Taiwan	2009	RT-PCR	989	16 (1.6%)	≤5 years	RVA	H	Tseng et al. (2012)
	Vietnam	2005–06	RT-PCR	502	70 (13.9%)	≤35 years	RVA	B	Nguyen et al. (2008)

	Pakistan	1990–94	RT-PCR	517	58 (11.2%)	≤5 years		H	Phan et al. (2004)
	Pakistan	2009–10	RT-PCR	563	48 (8.5%)	≤31 years		H	Alam et al. (2015)
	India	1999–2004	RT-PCR	857	50 (5.8%)	All ages	RVA	H	Bhattacharya et al. (2006)
	India	2007–09	RT-PCR	2,535	60 (2.4%)	All ages	RVA	H	Pativada et al. (2012)
	Bangladesh	2005–06	RT-PCR	138	13 (9.4%)	≤5 years	NoV, BoV, AdV	H	Mitui et al. (2014)
	Bangladesh	2010–12	RT-PCR	826	26 (3.1%)	≤3 years	RVA, NoV, AdV	H	Afrad et al. (2013)
	Turkey	2004–05	RT-PCR	150	4 (2.7%)	≤5 years	AdV	H	Mitui et al. (2014)
	Saudi Arabia	2002–03	EIA/RT-PCR	1,000	19 (1.9%)	≤6 years		B	Tayeb et al. (2010)
	Qatar	2009	Real-time PCR	288	1 (0.3%)	All ages		—	Al-Thani et al. (2013)
Africa	Egypt	2005–07	EIA	2,112	56 (2.7%)	≤5 years		O	El-Mohammady et al. (2012)
	Egypt	2006–07	EIA/RT-PCR	230	5 (2.2%)	≤18 years	RVA	O	Kamel et al. (2009)
	Egypt	2006–07	RT-PCR	364	23 (6.3%)	≤5 years		—	Ahmed et al. (2011)
	Tunisia	2003–07	EIA/RT-PCR	788	28 (3.6%)	14 days–12 years	RVA	B	Sdiri-Loulizi et al. (2009)
	Nigeria	2002	EIA	134	7 (5.2%)	≤5 years		O	Aminu et al. (2008)
	Nigeria	2007–08	EIA	161	65 (40.4%)	≤5 years		H	Ayolabi et al. (2012)
	Malawi	1997–99	EIA/RT-PCR	In patients 786 Out patients 400	In patients 15 (1.9%) Out patients 9 (2.3%)	≤5 years	RVA	B	Cunliffe et al. (2002)
	Kenya	1999–2005	EIA	476	30 (6.3%)	≤10 years		O	Kiulia et al. (2007)

(Continued)

TABLE 4.3.2 Human Astrovirus Positivity Rates in Patients With Acute Gastroenteritis (Reviewed From 60 Studies Worldwide) (*cont.*)

Region	Country	Year of sample collection	Detection method	No. of specimens tested	No. astrovirus positive (%)	Age	Coinfecting viruses	Site	References
	Madagascar	2004–05	RT-PCR	237	5 (2.1%)	≤16 years	RVA	—	Papaventsis et al. (2008)
	Ghana	2005–06	RT-PCR	367	12 (3.3%)	≤11 years		O	Silva et al. (2008)
Europe	Italy	1999–2000	EIA/RT-PCR	157	5 (3.2%)	≤2 years		H	De Grazia et al. (2004)
Europe	Italy	2002–05	EIA/RT-PCR	708	28 (4.0%)	≤5 years	RVA	H	De Grazia et al. (2011)
Europe	Italy	2008–09	RT-PCR	1,321	49 (3.7%)	≤5 years		H	De Grazia et al. (2013)
Europe	Spain	1997–2000	Southern Blot Hybridization/RT-PCR	2,341	116 (4.9%)	≤82 years		B	Guix et al. (2002)
Europe	Hungary	2002	RT-PCR	607	10 (1.6%)	49–60 months		H	Jakab et al. (2004)
Europe	Hungary	2003–05	RT-PCR	449	12 (2.7%)			H	Jakab et al. (2009)
Europe	France	2001–04	EIA/RT-PCR	457	7 (1.5%)	≤15 years		H	Lorrot et al. (2011)
Europe	France	2007	EIA	973	18 (1.8%)	≤6 years		H	Tran et al. (2010)
Europe	UK	2000–03	RT-PCR	685	22 (3.2%)	≤6 years	RVA	B	Iturriza Gomara et al. (2008)
Europe	UK	2006–07	RT-PCR	576	28 (4.9%)	≤16 years	RVA	B	Cunliffe et al. (2010)
Europe	UK	2012–13	RT-PCR	Children 200 Adult 195	Children 35 (17.5%) Adult 1 (0.5%)	Children ≤5 years Adult ≥65 years		H	Borrows and Turner (2014)
Europe	Netherlands	2010–13	Real-time PCR	1802	131 (7.3%)	≤47 years		O	Enserink et al. (2015)
Europe	Bulgaria	2009	RT-PCR	115	7 (6.0%)	40 days–3 years		H	Mladenova et al. (2015)

North America	USA	2008–09	RT-PCR/Real-time PCR	782	38 (4.9%)	≤5 years	NoV	B	Chhabra et al. (2013)
	Mexico	1988–91	EIA/RT-PCR/Cell culture	365	23 (6.3%)	≤18 months		—	Walter et al. (2001)
	Mexico	1994–95	EIA/IEM/RT-PCR	522	24 (4.6%)	≤5 years		B	Mendez-Toss et al. (2004)
	Guatemala	1987–89	EIA	321	124 (38.6%)	0–3 years		—	Cruz et al. (1992)
South America	Brazil	1994–96, 1998–2003	RT-PCR	1,588	57 (4.1%)	≤5 years		H	Silva et al. (2009)
	Brazil	2000–04	RT-PCR	354	11 (3.1%)	≤3 years	RVA	H	Andreasi et al. (2008)
	Brazil	1994–2008	Real-time PCR	539	19 (6.3%)	≤5 years		O	Ferreira et al. (2012)
	Argentina	1995–98	EIA/RT-PCR	1,070	40 (3.7%)	≤3 years		B	Espul et al. (2004)
	Argentina	1997–98	EIA/RT-PCR	66	5 (7.6%)	≤3 years		O	Bereciartu et al. (2002)
	Argentina	2001–02	EIA	97	12 (12.4%)	≤33 months		O	Giordano et al. (2004)
	Colombia	1997–99	EIA	251	12 (5%)	≤4 years		—	Medina et al. (2000)
	Venezuela	1994–95	RT-PCR	29	3 (10%)	≤4 years		—	Medina et al. (2000)
Australia	Australia	1995	Northern blot hybridization/RT-PCR	378	16 (4.2%)	≤5 years		—	Palombo and Bishop (1996)
	Australia	1995–98	RT-PCR	1,327	40 (3.0%)	≤5 years		—	Mustafa et al. (2000)
	Australia	1997–98	EIA/RT-PCR	414	19 (4.6%)	≤6 years		B	McIver et al. (2000)
	Australia	1995–98	Southern-blot hybridization/RT-PCR	774	33 (4.3%)	3 weeks–5 years		—	Schnagl et al. (2002)

EIA, Enzyme immunoassay; *IEM*, immune electron microscopy; *RT-PCR*, reverse transcription-polymerase chain reaction; *RVA*, rotavirus of species A; *NoV*, norovirus; *BoV*, bocavirus; *AdV*, adenovirus; *H*, hospitalized; *O*, outpatient; *B*, both hospitalized and outpatient.

The virus is detected at a lower prevalence in comparison with group A rotavirus (RVA) and norovirus GII (NoV GII) (Chaimongkol et al., 2012; Chhabra et al., 2013). HAstVs predominantly infect young children with the highest rate of infection in children under 2 years of age; mixed infections with RVA are frequently observed (Iturriza Gomara et al., 2008). Although different seasonal peaks have been recorded in HAstV studies, the highest rates of infection have been reported for the cool season in the USA, Netherlands, Spain, China, and Japan (Chan-it et al., 2010; Chhabra et al., 2013; Enserink et al., 2015; Fang et al., 2006; Guix et al., 2002).

Among HAstV circulating in the human population, at least eight genotypes have been described (HAstV1–HAstV8). Based on the genome sequences and phylogenetic analyses, the genotypes of HAstV are tightly correlated with the serotypes (Mendez and Arias, 2013). The HAstV1 has been reported to be the most predominant genotype worldwide in children with diarrhea. However, the detection rate varies by geographical region (Fig. 4.3.2). HAstV1 is a major genotype which represents 30–84% of all HAstV genotypes in each continent. It should be noted that the HAstV1 alone represents over 80% of the strains circulating in Asia, while other genotypes (HAstV2–HAstV8) are found to be uncommon and responsible for only 1–6% of astrovirus infections. Lower detection rates of HAstV1 are observed in North America, South America, and Africa at 30, 43, and 46%, respectively. The HAstV1 together with HAstV2-HAstV4 are found regularly in Europe, North America, and South America, accounting for 89–98% of all astrovirus strains, while HAstV5–HAstV8 are less common and represent only 2–11% of astrovirus infections. In contrast, the distribution of HAstV genotypes in Africa is different from that of other regions since the uncommon genotypes, HAstV8 and HAstV5, are found to be the third and fourth common types after HAstV1 and HAstV3.

Within each HAstV genotype, the strains can be further subdivided into lineages or subtypes. Based on the capsid sequences, the new subtype must have a nucleotide sequence divergence of $\geq$7% from the existing lineages (Guix et al., 2002). Thus far, HAstV1 has been subdivided into six lineages (1a–1f), HAstV2 into four lineages (2a–2d), HAstV3 into two lineages (3a–3b), and HAstV4 into three lineages (4a–4c) (Martella et al., 2013). In Japan, molecular characterization of HAstV1 lineages from the year 2006–15 demonstrated that during the first half of the surveillance studies from 2006–09 the detection rate of HAstV1d reached a peak of 96% prevalence, while HAstV1a was detected with a low prevalence of 4%. In contrast, from 2010 to 2015 HAstV1a reemerged with the prevalence of up to 98% as a predominant genotype (Chan-it et al., 2010; Dey et al., 2010; Thongprachum et al., 2015). Moreover, a surveillance study in Vietnam carried out in 2005-2006 revealed that HAstV1d was the most predominated astrovirus subtype (Nguyen et al., 2008). In Italy, during the period 2002–05 all HAstV1 were classified as HAstV1d, while during 2008 and 2009 both HAstV1a and HAstV1d subtypes were found to be equally predominant (De Grazia et al., 2013). These data imply that HAstVs

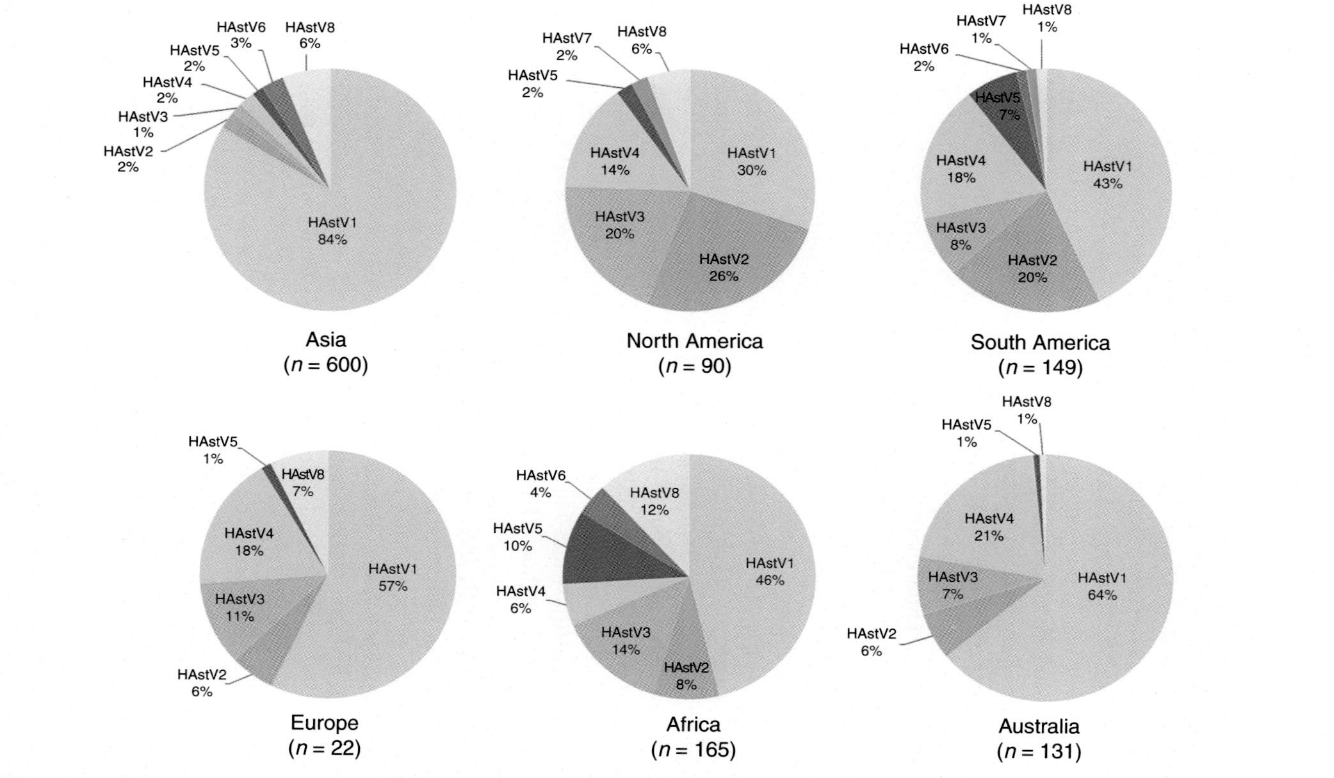

FIGURE 4.3.2 **The global distribution of human astrovirus genotypes ascertained by analysis of strains collected between 1998–2014.** *Asia*: Afrad et al. (2013), Chan-it et al. (2010), Chhabra et al. (2014), Dey et al. (2010), Fang et al. (2006), Guo et al. (2010), Jeong et al. (2011), Malasao et al. (2012), Mitui et al. (2014), Nguyen et al. (2008), Pativada et al. (2012), Phan et al. (2004), Shan et al. (2009), Tayeb et al. (2010), Thongprachum et al. (2015), Tseng et al. (2012), Wang et al. (2011) *North America*: Chhabra et al. (2013), Mendez-Toss et al. (2004), Walter et al. (2001) *South America*: Andreasi et al. (2008), Espul et al. (2004), Ferreira et al. (2012), Medina et al. (2000), Silva et al. (2009), Ulloa et al. (2005) *Europe*: De Grazia et al. (2004), De Grazia et al. (2013), Guix et al. (2002), Jakab et al. (2004, 2009), Mladenova et al. (2015) *Africa*: Ahmed et al. (2011), Cunliffe et al. (2002), Naficy et al. (2000), Papaventsis et al. (2008), Sdiri-Loulizi et al. (2009), Silva et al. (2008) *Australia*: McIver et al. (2000), Mustafa et al. (2000), Palombo and Bishop (1996), Schnagl et al. (2002). n, the total number of human astrovirus genotypes reported from each continent.

cocirculating in children with diarrhea worldwide are genetically highly diverse pathogens.

4.1.2 MLB Astroviruses

The first MLB astrovirus prototype strain was identified in 2008 using the mass genome sequencing approach (Finkbeiner et al., 2008). The virus was detected from archived stool samples collected in 1999 from a 3-year-old boy with acute gastroenteritis in Melbourne, Australia, and later the virus was named as "MLB1." The complete nucleotide sequence and genetic organization of MLB1 were characterized and demonstrated that MLB1 was considerably distinct from other known classic human astroviruses (HAstV1–HAstV8) and should be classified as a new astrovirus species in the genus *Mamastrovirus* (Fig. 4.3.1). One year later, the same research group reported the discovery of MLB2 in stool specimens collected from children with diarrhea in the United States and India (Finkbeiner et al., 2009a). In 2013, MLB3 astrovirus strains were identified in five stool specimens which had been collected in India. Within these, four of the MLB3 positive samples were obtained from patients with diarrhea, while 1 positive sample was from an asymptomatic case (Jiang et al., 2013).

The role of astroviruses of the MLB clade as causative agents of gastroenteritis has not been clearly established. Although most of the positive stool samples for this new clade were found in patients with diarrhea, a cohort study with a control population, carried out in Indian children, found no association of MLB1 with diarrhea (Holtz et al., 2011a). In addition, it was reported that MLB1 astrovirus infections were common in an American population since a greater than 60% seropositivity rate was found in children younger than 1-year old and reached 100% by adulthood (Holtz et al., 2014). However, the potential role of a systemic infection being related to febrile illness has been recently described in patient with MLB2 viremia (Holtz et al., 2011b).

After the discovery of the novel MLB astrovirus clade, these viruses have been reported in acute gastroenteritis pediatric patients in the USA, Nigeria, India, Japan, China, Hong Kong, Bhutan, Russia, Italy, Egypt, and Turkey (Bosch et al., 2014). In Japan, during 2013 and 2014, MLB astroviruses were found in 35 out of 330 (10.6%) fecal specimens collected from children with acute diarrhea. Among these, 2 types of MLB were identified (MLB1: $n = 32$ and MLB2: $n = 3$). It is interesting to note that the detection rate of MLB astroviruses was higher than that of classic human astroviruses (5.2%); the peak incidence was found in March and April (Khamrin et al., 2016). Since the first MLB astrovirus was identified in 2008, research into MLB astrovirus molecular biology, pathogenicity, and molecular epidemiology remained scarce.

4.1.3 VA Astroviruses

The VA astrovirus prototype strain was initially isolated during a sporadic outbreak of acute diarrhea in a child care center in Virginia, USA, in 2008. Using the pyrosequencing approach, the viral genome was found to be highly divergent from those of all previously known astroviruses, and initially named as "VA1" after the original place of viral discovery (Finkbeiner et al., 2009b). At the same time, another research group also reported the discovery of a novel astrovirus in fecal specimens collected from Nigeria, Pakistan, and Nepal, and the characterization of their genomes revealed that these viruses were phylogentically closely related to mink and ovine astroviruses. Hence, these viruses are provisionally named Human, Mink, and Ovine-like astrovirus (HMOAstV) and subdivided into the three types of HMOAstV-A, HMOAstV-B, and HMOAstV-C (Kapoor et al., 2009). Later, it was found that HMOAstV-C shared a close genetic relationship with VA1 astrovirus. Subsequently, VA2 and VA3 were identified in a cohort study of children with diarrhea in India and the USA (Finkbeiner et al., 2009a). Most recently, VA4 astrovirus strains were identified in two stool specimens collected from children with diarrhea in Nepal (Jiang et al., 2013). Taken together, VA astroviruses are classified into two phylogenetic clusters as VA1 and VA3, together with HMOAstV-B, HMOAstV-C, and are genetically more closely related with *Mamastrovirus 9*, while VA2, VA4, and HMOAstV-A are grouped together within *Manastrovirus 8* (Fig. 4.3.1) (Finkbeiner et al., 2009a; Jiang et al., 2013; Kapoor et al., 2009).

Since the discovery of VA astrovirus was initially published in 2009, the role of VA astrovirus as a cause of gastroenteritis remains unclear (Finkbeiner et al., 2009a; Jiang et al., 2013; Kapoor et al., 2009). Although VA astrovirus infections are thought to be limited to the gastrointestinal tract, some studies have recently described the association of VA1 astrovirus with neurologic disorders in immunocompromised individuals. The VA1 astrovirus has been reported to cause encephalitis in a 15-year-old boy with agammaglobulinemia (Quan et al., 2010). The VA1 astrovirus can escape from the gastrointestinal tract into the circulatory system and is capable of invading the brain, with viral RNA being detected in brain biopsies and cerebrospinal fluid (CSF) (Naccache et al., 2015). In recent epidemiological studies, the virus has been detected at low prevalence rates in sporadic acute gastroenteritis cases in the USA, India, Pakistan, Nepal, Japan, China, and Nigeria (Bosch et al., 2014; Finkbeiner et al., 2009a; Jiang et al., 2013; Kapoor et al., 2009; Khamrin et al., 2016).

4.2 Animal Astroviruses

4.2.1 Ovine Astroviruses (OAstV)

The OAstV was first identified in 1977 in Scotland in lambs suffering from diarrhea shortly after the description of astroviruses in human. The OAstV was identified by electron microscopy (Snodgrass and Gray, 1977) and was purified and characterized structurally in 1981. Subsequently, the nucleotide sequences of the capsid region (ORF2) and the complete genome sequence were determined in the period from 2001 to 2003, and the virus was named OAstV-1 (Y15937) (Jonassen et al., 2003). Recently another OAstV strain was identified in samples taken from healthy domestic sheep in Hungary (Reuter et al., 2012). Sequence and phylogenetic analyses of ORF1b/ORF2/3′UTR of this isolate showed genetic divergence from OAstV-1, and it was therefore named OAstV-2. So far, only two OAstVs have been identified, one from lamb with diarrhea and the other one from a healthy sheep, suggesting that OAstV may not play a crucial causative role of diarrhea in sheep.

4.2.2 Bovine Astroviruses (BoAstV)

The first BoAstV was reported in England in 1978 in samples taken from calves with acute enteritis (Woode and Bridger, 1978). Several years later, another BoAstV, antigenically related to the first BoAstV, was isolated from a calf with diarrhea in Florida, USA. Although, these BoAstVs were isolated from calves with acute diarrhea, they could not induce diarrhea in calves infected experimentally. It was, therefore concluded that BoAstV is not associated with diarrhea in calves under natural conditions. Most recently, three bovine astroviruses associated with neurologic symptoms, so called BoAstV-Neuro S1 (KF233994), were detected in the brain tissue of cattle with histologically confirmed encephalomyelitis and ganglioneuritis, suggesting the virus to be a potential cause of neurologic disease in cattle (Li et al., 2013). Additionally, another bovine neurotropic astrovirus strain (BoAstV-CH13) was identified in cattle with nonsuppurative encephalitis in Switzerland (Bouzalas et al., 2014). Phylogenetically, BoAstV-CH13 was closely related to BoAstV-Neuro S1. These findings support the notion that infection with BoAstV is a common cause of encephalitis in cattle.

4.2.3 Porcine Astroviruses (PoAstV)

The PoAstV was first detected in piglets with diarrhea in association with calicivirus and rotavirus infections (Bridger, 1980). The virus was isolated later in 1990 in an established cell line derived from porcine embryonic kidney tissue. Molecular characterization of the capsid sequence (ORF2 gene) of this isolate was successfully carried out in 2001 and this PoAstV was named as PoAstV-1 (Jonassen et al., 2001). Later, PoAstVs were reported from several countries around the world, suggesting a wide geographical distribution. To date, five different PoAstV types, PoAstV-1–PoAstV-5, have been identified from different countries, including the Czech Republic, Columbia, Canada, the USA, China, and Hungary (De Benedictis et al., 2011). Although PoAstVs were described as a common virus detected in piglets with diarrhea, they were also detected in apparently asymptomatic healthy pigs (Luo et al., 2011). A prevalence of PoAstV infection in pigs of 62% has been reported recently from the United States (Mor et al., 2012). However, examination of the fecal contents of healthy slaughtered finisher pigs revealed a 79% positivity rate for PoAstV (Luo et al., 2011). Taken together, the clinical significance of PoAstV infection in pigs has not been completely elucidated.

4.2.4 Canine Astroviruses (CaAstV)

The CaAstV was first detected by electron microscopy in stool samples of beaglepuppies with gastroenteritis (Williams, 1980). The presumptive identification made by electron microscopy was confirmed by the genetic analysis of viral RNA extracted from clinical specimens. Recently, CaAstV has been propagated in MDCK cell cultures and virus growth induced a clear cytopathic effects (CPE) (Martella et al., 2011a). To date, CaAstVs have been reported from the USA, the UK, Australia, Italy, Germany, France, Brazil, China, Korea, and Japan (De Benedictis et al., 2011; Caddy and Goodfellow, 2015; Takano et al., 2015).

The prevalence of CaAstV infection associated with gastroenteritis in puppies was reported from Italy at 25%, France at 21%, the United Kingdom at 6%, China at 12%, Korea at 2%, and Japan at 10% (De Benedictis et al., 2011; Caddy and Goodfellow, 2015; Takano et al., 2015). It should be noted that the prevalence of CaAstV infection in puppies in Europe is relatively higher than that in Asia. CaAstV was also detected in asymptomatic puppies at an incidence of 9% (Martella et al., 2011a). The information relating to the clinical significance of CaAstV in dogs is limited and requires further investigation.

4.2.5 Feline Astroviruses (FeAstV)

The FeAstV was first detected using electron microscopy in the feces of domestic kittens with diarrhea (Hoshino et al., 1981). Further investigation demonstrated that FeAstV was commonly found in the stools of cats with and without diarrhea (Rice et al., 1993). So far, FeAstV in the feces of domestic cats has been reported from the USA, Australia, Germany, England, New Zealand, and Italy (De Benedictis et al., 2011). The clinical impact of FeAstV infection in cats seems to be low since kittens experimentally infected with FeAstV showed only pyrexia and mild diarrhea (Harbour et al., 1987).

4.2.6 Mink Astroviruses (MiAstV)

The MiAstV has been shown in a case-control study as a significant risk factor for the preweaning diarrhea syndrome in mink (Englund et al., 2002). The full-length positive-stranded RNA genome of MiAstV has been characterized, and phylogenetic analysis revealed that MiAstV is distantly related to ovine astrovirus (OAstV) with a nucleotide sequence similarity of less than 67%. Recently, MiAstV was detected in the brain of three mink kids suffering from shaking mink syndrome, and the complete coding region of this virus was closely related to that of the previously reported preweaning diarrhea MiAstV (Blomstrom et al., 2010).

4.2.7 Bat Astroviruses (BatAstV)

The BatAstVs were first detected in the insectivorous bats in Hong Kong in 2008 (Chu et al., 2008). The virus detection rates in this study were variable from 36% to 100% depending on bat species. At almost the same time, BatAstV was also searched for in 20 bat species in 11 provinces in mainland China (Zhu et al., 2009) and detected in 12 of the 20 species, with significant differences in the detection rates between bat species. The highest positivity rate was 93%, and some bat species were positive even though the sampling number was small. BatAstVs detected in Hong Kong and in Mainland China showed a remarkably high genetic diversity. Phylogenetic analysis of the RdRp gene sequence revealed that all BatAstVs cluster within the genus *Mamastrovirus* and can be further subdivided into seven species (MAstV12, MAstV14-19) (Zhu et al., 2009). Most recently, novel European lineages of BatAstVs were identified in Hungary from different bat species (Kemenesi et al., 2014). The highest detection rate of BatAstV in Hungary was 42.8%. The phylogenetic analysis of the RdRp gene showed that the Hungarian BatAstV strains clustered together with BatAstVs from China, but also displayed a notable genetic variability. These findings indicate that various bat species are important reservoirs for astroviruses with high prevalence rates and broad genetic diversity (Chu et al., 2008; Kemenesi et al., 2014; Zhu et al., 2009).

4.2.8 Rabbit Astroviruses (RaAstV)

The RaAstVs were detected in fecal samples from domestic rabbits with enteritis in Italy during the period 2005–08 (Martella et al., 2011b). The virus detection rates were 43% in rabbits with enteric disease while only 18% in asymptomatic rabbits. However, the pathogenicity of astroviruses in rabbits remains unclear. There was a report of an enterocolitis outbreak with high mortality (~90%) in a domestic rabbit colony in Tennessee, USA, and astrovirus-like particles were demonstrated in stool samples. Metagenomic sequencing enabled the recovery of the astrovirus genome, and the ORF1a/protease, ORF1b/RdRp, and ORF2/capsid proteins of this virus were found to share 30%, 59%, and 25% pairwise amino acid identity, respectively, with the corresponding human astrovirus MLB1 protein sequences (Stenglein et al., 2012). In conclusion, although astroviruses cause gastroenteritis in other mammals, the pathogenicity of RaAstV and the relationship to disease in rabbits remain unclear.

4.2.9 Duck Astroviruses (DAstV)

The hepatitis-associated astrovirus in ducks was discovered by electron microscopy in 1984 (Gough et al., 1984). The virus was reported to cause an outbreak of hepatitis in England in commercial ducks with a mortality rate of 25% in 4–6 week-old and up to 50% in 6–14 days old animals. An outbreak of astrovirus-associated hepatitis in ducks was also reported in a commercial duck flock in China in 2008 (Fu et al., 2009). The mortality rate of this outbreak was about 50% in 1–2-week old ducklings. The complete genome sequence and phylogenetic analyses revealed that DAstV was more closely related to turkey astrovirus type-2 (TAstV-2), TAstV-3, and TAstV/MN/01 than to TAstV-1 or other astrovirus types (Fu et al., 2009). The findings suggest that astroviruses may be transmitted between ducks and turkeys.

4.2.10 Turkey Astroviruses (TAstV)

The TAstV was first reported as a cause of gastroenteritis and mortality in turkeys of 6 to 11 days of age in the United Kingdom in 1980 and the astrovirus particles were demonstrated by electron microscopy (McNulty et al., 1980). Since then, astrovirus outbreaks in turkeys have been reported in the United States. These virus isolates were classified later as TAstV-1. Subsequently, TAstV-2 was detected and characterized in association with poult enteritis mortality syndrome (PEMS) in the United States in 1996. Experimentally, PEMS-associated TAstV-2 has been shown to cause enteritis and mortality in turkeys (Behling-Kelly et al., 2002). The TAstV-2 is one of the most commonly detected viruses in turkey poults suffering from diarrhea. The PEMS-associated TAstV-2 in the feces of turkeys has been reported from Brazil with a prevalence of 100%, while a 38.9% prevalence was found in Poland (De Benedictis et al., 2011). Generally, the prevalence of TAstV in symptomatic poults varies from 40% to 100%. In the field, observations indicate that TAstV-2 is more prevalent in turkeys than TAstV-1 (De Benedictis et al., 2011).

4.2.11 Avian Nephritis Viruses (ANV)

The ANV was first isolated from the feces of healthy broiler chickens in 1976, by culturing in chicken kidney cells (Yamaguchi et al., 1979). The ANV was initially classified as a picornavirus based on its virion morphology and it was reclassified later as a new member of the *Astroviridae* family based on complete genome sequence analysis. The ANV is widely distributed throughout chickens worldwide and causes a wide range of symptoms in chickens from subclinical infection to severe clinical manifestation and mortality. Currently, at least two serotypes, ANV-1 and ANV-2, have been reported (Pantin-Jackwood et al., 2011). Additionally, ANV has also been detected in pigeons suffering from diarrhea and gastrointestinal illness, and its complete genome sequence has been characterized (Zhao et al., 2011).

4.2.12 Chicken Astroviruses (CAstV)

The CAstV was first detected by electron microscopy in the feces of chickens in 1990 (McNeilly et al., 1994; McNulty et al., 1990). The characterization of CAstVs revealed that they were antigenically and genetically distinct from the ANV (Baxendale and Mebatsion, 2004). Analysis of ORF1 of CAstV revealed a distant relationship to the nonstructural proteins of TAstV-1, TAstV-2, and ANV at 58%, 62%, and 55% similarity, respectively. The CAstVs are distributed across the world, and multiple serotypes cocirculate (Baxendale and Mebatsion, 2004; Pantin-Jackwood et al., 2011).

5 RECOMBINATION AND INTERSPECIES TRANSMISSION

Astroviruses, like other RNA viruses, express their own RdRp. Due to lack of proofreading activity, the AstV RdRp has a high error rate, leading to significant genetic diversity of these viruses, as demonstrated in both, the *Mamastrovirus* and *Avastrovirus* genera (Fu et al., 2009; Luo et al., 2011; Pantin-Jackwood et al., 2011; Zhu et al., 2009). One of the evolutionary mechanisms of astrovirus is genome recombination. There is evidence of recombination events among human astrovirus strains which have been reported from several countries. Increasing numbers of inter- and intragenotype recombinant astrovirus strains are detected. Various types of intergenotype recombinations have been documented, such as among HAstV1 and HAstV2, HAstV1 and HAstV3, HAstV1 and HAstV4, HAstV3 and HAstV2 strains (De Grazia et al., 2012; Martella et al., 2013; Medici et al., 2015). Recombination of human astroviruses HAstV1 and HAstV4 has also been detected in Russia (Babkin et al., 2014). Genetic analysis of these naturally occurring astrovirus recombinants revealed that the sites of recombinantion breakpoints are located mainly at the junction of ORF1b and ORF2. Recently, analysis of whole-genome sequences indicates that the HAstV RNA can be a mosaic genome as multiple recombination events among HAstV6, HAstV3, and HAstV2 were detected within one virus strain (Babkin et al., 2014). Moreover, intragenotype recombinations within HAstV1 (1b/1d) and HAstV2 (2c/2b) have been identified (De Grazia et al., 2012; Martella et al., 2014).

For animal astroviruses, there is growing evidence that astroviruses can cross species barriers (Chu et al., 2008; Kapoor et al., 2009; Zhu et al., 2009). Interspecies transmission has been demonstrated among avian astroviruses causing acute hepatitis in ducks and for turkey astrovirus type 2 (TAstV-2), TAstV-3, and TAstV/MN/01 (Fu et al., 2009). Additionaly, TAstV-2 has been detected in guinea fowl, suggesting either direct or indirect interspecies transmission of astroviruses is likely occurred between turkey and guinea fowl (De Battisti et al., 2012). Furthermore, interspecies transmission of astroviruses has also been demonstrated between mammalian and avian astroviruses (Atkins et al., 2009; van Hemert et al., 2007). Although no direct evidence of zoonotic transmission of animal astroviruses to humans has been documented, transmission of an astrovirus from pigs to cats and then to humans has been described, with unidentified intermediate hosts potentially being involved (Lukashov and Goudsmit, 2002). Recently, antibodies against turkey astrovirus (TAstV-2) were detected in 26% of poultry abattoir workers, suggesting that people in contact with turkeys can develop serological response to turkey astrovirus. However, further work is needed to determine whether the exposure results in virus replication and/or clinical disease in humans (Meliopoulos et al., 2014). Taken together, the data emphasize that recombination is a major mechanism underlying genetic variation that enables the viruses to change their properties rapidly. The findings also highlight the importance of continued surveillance of the astrovirus genotypes globally to obtain more information about the potential of interspecies transmission.

6 WATERBORNE AND FOODBORNE DISEASE OUTBREAKS

Transmission of astroviruses is primarily via the fecal–oral route, either by direct contact with infected persons or by ingestion of contaminated food and water. Astroviruses are considered to be an important cause of acute gastroenteritis in both children and adults in community-based gastroenteritis, and to be responsible for sporadic cases and several outbreaks in various epidemiological settings, including schools, day-care centers, military recruits, and hospitals (Bosch et al., 2014). Astroviruses have been isolated directly from water, food, and environmental sources. There are several reports confirming that water is an important source for astrovirus contamination and pathway of transmission. Several outbreaks of waterborne astrovirus infection have been related to the contamination of drinking water, municipal water supplies, raw sewage, and river water (Lizasoain et al., 2015; Sezen et al., 2015). In addition, shellfish and oysters have also been reported as possible important sources of astrovirus infection (Le Guyader et al., 2008). A large outbreak of acute gastroenteritis in Osaka, Japan, in 1991 was caused by an astrovirus and has been linked to contaminated food from a common supplier/in the outbreak illness was recorded in >4700 persons (Oishi et al., 1994).

7 CONCLUSIONS AND FUTURE PERSPECTIVES

Astroviruses are increasingly being recognized as significant gastrointestinal tract pathogens in both, humans and a large variety of animal species. However, the full clinical impact of astrovirus-associated gastroenteritis remains unclear, especially in the cases of novel, recently discovered astrovirus types (MLB and VA). Detection of astroviruses at low prevalence rates in diarrhea cases and high seroprevalence in certain populations suggest that the virus might be responsible for mild or asymptomatic infections. Most recently, astrovirus infections have been documented in association with neurologic disorders in humans, cattle, and mink. Therefore, further study is necessary to determine the full spectrum of clinical symptoms in infected groups. Molecular detection of classic human astroviruses, screening for novel MLB/VA astroviruses in humans, as well as detection of the viruses in other animal species, are important tools to enable a better understanding of the overall epidemiological characteristics, genetic diversity, and potential of zoonotic transmission of these viruses.

REFERENCES

Afrad, M.H., Karmakar, P.C., Das, S.K., et al., 2013. Epidemiology and genetic diversity of human astrovirus infection among hospitalized patients with acute diarrhea in Bangladesh from 2010 to 2012. J. Clin. Virol. 58, 612–618.

Ahmed, S.F., Sebeny, P.J., Klena, J.D., et al., 2011. Novel astroviruses in children, Egypt. Emerg. Infect. Dis. 17, 2391–2393.

Alam, M.M., Khurshid, A., Shaukat, S., et al., 2015. Viral etiologies of acute dehydrating gastroenteritis in Pakistani children: confounding role of parechoviruses. Viruses 7, 378–393.

Al-Thani, A., Baris, M., Al-Lawati, N., et al., 2013. Characterising the aetiology of severe acute gastroenteritis among patients visiting a hospital in Qatar using real-time polymerase chain reaction. BMC Infect. Dis. 13, 329.

Aminu, M., Esona, M.D., Geyer, A., et al., 2008. Epidemiology of rotavirus and astrovirus infections in children in northwestern Nigeria. Ann. Afr. Med. 7, 168–174.

Andreasi, M.S., Cardoso, D., Fernandes, S.M., et al., 2008. Adenovirus, calicivirus and astrovirus detection in fecal samples of hospitalized children with acute gastroenteritis from Campo Grande, MS, Brazil. Mem. Inst. Oswaldo Cruz. 103, 741–744.

Appleton, H., Higgins, P.G., 1975. Letter: Viruses and gastroenteritis in infants. Lancet 1, 1297.

Atkins, A., Wellehan, Jr., J.F., Childress, A.L., et al., 2009. Characterization of an outbreak of astroviral diarrhea in a group of cheetahs (Acinonyx jubatus). Vet. Microbiol. 136, 160–165.

Ayolabi, C., Ojo, D., Akpan, I., 2012. Astrovirus infection in children in Lagos, Nigeria. Afr. J. Infect. Dis. 6, 1–4.

Babkin, I.V., Tikunov, A.Y., Sedelnikova, D.A., et al., 2014. Recombination analysis based on the HAstV-2 and HAstV-4 complete genomes. Infect. Genet. Evol. 22, 94–102.

Baxendale, W., Mebatsion, T., 2004. The isolation and characterisation of astroviruses from chickens. Avian. Pathol. 33, 364–370.

Behling-Kelly, E., Schultz-Cherry, S., Koci, M., et al., 2002. Localization of astrovirus in experimentally infected turkeys as determined by in situ hybridization. Vet. Pathol. 39, 595–598.

Bereciartu, A., Bok, K., Gomez, J., 2002. Identification of viral agents causing gastroenteritis among children in Buenos Aires, Argentina. J. Clin. Virol. 25, 197–203.

Bhattacharya, R., Sahoo, G.C., Nayak, M.K., et al., 2006. Molecular epidemiology of human astrovirus infections in Kolkata, India. Infect. Genet. Evol. 6, 425–435.

Blomstrom, A.L., Widen, F., Hammer, A.S., et al., 2010. Detection of a novel astrovirus in brain tissue of mink suffering from shaking mink syndrome by use of viral metagenomics. J. Clin. Microbiol. 48, 4392–4396.

Borrows, C.L., Turner, P.C., 2014. Seasonal screening for viral gastroenteritis in young children and elderly hospitalized patients: Is it worthwhile? J. Hosp. Infect. 87, 98–102.

Bosch, A., Pinto, R.M., Guix, S., 2014. Human astroviruses. Clin. Microbiol. Rev. 27, 1048–1074.

Bouzalas, I.G., Wuthrich, D., Walland, J., et al., 2014. Neurotropic astrovirus in cattle with nonsuppurative encephalitis in Europe. J. Clin. Microbiol. 52, 3318–3324.

Bridger, J.C., 1980. Detection by electron microscopy of caliciviruses, astroviruses and rotavirus-like particles in the faeces of piglets with diarrhoea. Vet. Rec. 107, 532–533.

Brinker, J.P., Blacklow, N.R., Herrmann, J.E., 2000. Human astrovirus isolation and propagation in multiple cell lines. Arch. Virol. 145, 1847–1856.

Caddy, S.L., Goodfellow, I., 2015. Complete genome sequence of canine astrovirus with molecular and epidemiological characterisation of UK strains. Vet. Microbiol. 177, 206–213.

Cattoli, G., Toffan, A., De Battisti, C., et al., 2005. Astroviruses found in the intestinal contents of guinea fowl suffering from enteritis. Vet. Rec. 156, 220.

Chaimongkol, N., Khamrin, P., Suantai, B., et al., 2012. A wide variety of diarrhea viruses circulating in pediatric patients in Thailand. Clin. Lab. 58, 117–123.

Chan-it, W., Thongprachum, A., Okitsu, S., et al., 2010. Epidemiology and molecular characterization of sapovirus and astrovirus in Japan, 2008-2009. Jpn. J. Infect. Dis. 63, 302–303.

Chhabra, P., Payne, D.C., Szilagyi, P.G., et al., 2013. Etiology of viral gastroenteritis in children <5 years of age in the United States, 2008-2009. J. Infect. Dis. 208, 790–800.

Chhabra, P., Samoilovich, E., Yermalovich, M., et al., 2014. Viral gastroenteritis in rotavirus negative hospitalized children <5 years of age from the independent states of the former Soviet Union. Infect. Genet. Evol. 28, 283–288.

Chu, D.K., Poon, L.L., Guan, Y., et al., 2008. Novel astroviruses in insectivorous bats. J. Virol. 82, 9107–9114.

Chu, D.K., Chin, A.W., Smith, G.J., et al., 2010. Detection of novel astroviruses in urban brown rats and previously known astroviruses in humans. J. Gen. Virol. 91, 2457–2462.

Cruz, J.R., Bartlett, A.V., Herrmann, J.E., et al., 1992. Astrovirus-associated diarrhea among Guatemalan ambulatory rural children. J. Clin. Microbiol. 30, 1140–1144.

Cunliffe, N.A., Dove, W., Gondwe, J.S., et al., 2002. Detection and characterisation of human astroviruses in children with acute gastroenteritis in Blantyre, Malawi. J. Med. Virol. 67, 563–566.

Cunliffe, N.A., Booth, J.A., Elliot, C., et al., 2010. Healthcare-associated viral gastroenteritis among children in a large pediatric hospital, United Kingdom. Emerg. Infect. Dis. 16, 55–62.

De Battisti, C., Salviato, A., Jonassen, C.M., et al., 2012. Genetic characterization of astroviruses detected in guinea fowl (Numida meleagris) reveals a distinct genotype and suggests cross-species transmission between turkey and guinea fowl. Arch. Virol. 157, 1329–1337.

De Benedictis, P., Schultz-Cherry, S., Burnham, A., et al., 2011. Astrovirus infections in humans and animals - molecular biology, genetic diversity, and interspecies transmissions. Infect. Genet. Evol. 11, 1529–1544.

De Grazia, S., Giammanco, G.M., Colomba, C., et al., 2004. Molecular epidemiology of astrovirus infection in Italian children with gastroenteritis. Clin. Microbiol. Infect. 10, 1025–1029.

De Grazia, S., Platia, M.A., Rotolo, V., et al., 2011. Surveillance of human astrovirus circulation in Italy 2002-2005: emergence of lineage 2c strains. Clin. Microbiol. Infect. 17, 97–101.

De Grazia, S., Medici, M.C., Pinto, P., et al., 2012. Genetic heterogeneity and recombination in human type 2 astroviruses. J. Clin. Microbiol. 50, 3760–3764.

De Grazia, S., Martella, V., Chironna, M., et al., 2013. Nationwide surveillance study of human astrovirus infections in an Italian paediatric population. Epidemiol. Infect. 141, 524–528.

Dey, S.K., Phan, G.T., Nishimura, S., et al., 2010. Molecular and epidemiological trend of sapovirus, and astrovirus infection in Japan. J. Trop. Pediatr. 56, 205–207.

El-Mohammady, H., Mansour, A., Shaheen, H.I., et al., 2012. Increase in the detection rate of viral and parasitic enteric pathogens among Egyptian children with acute diarrhea. J. Infect. Dev. Ctries. 6, 774–781.

Englund, L., Chriel, M., Dietz, H.H., et al., 2002. Astrovirus epidemiologically linked to preweaning diarrhoea in mink. Vet. Microbiol. 85, 1–11.

Enserink, R., van den Wijngaard, C., Bruijning-Verhagen, P., et al., 2015. Gastroenteritis attributable to 16 enteropathogens in children attending day care: significant effects of rotavirus, norovirus, astrovirus, cryptosporidium and giardia. Pediatr. Infect. Dis. J. 34, 5–10.

Espul, C., Martinez, N., Noel, J.S., et al., 2004. Prevalence and characterization of astroviruses in Argentinean children with acute gastroenteritis. J. Med. Virol. 72, 75–82.

Fang, Z.Y., Sun, Y.P., Ye, X.H., et al., 2006. Astrovirus infection among hospitalized children with acute diarrhea in seven regions of China, 1998-2005. Zhonghua Liu Xing Bing Xue Za Zhi 27, 673–676.

Ferreira, M.S., Xavier Mda, P., Tinga, A.C., et al., 2012. Assessment of gastroenteric viruses frequency in a children's day care center in Rio De Janeiro, Brazil: a fifteen year study (1994-2008). PLoS One 7, e33754.

Finkbeiner, S.R., Kirkwood, C.D., Wang, D., 2008. Complete genome sequence of a highly divergent astrovirus isolated from a child with acute diarrhea. Virol. J. 5, 117.

Finkbeiner, S.R., Holtz, L.R., Jiang, Y., et al., 2009a. Human stool contains a previously unrecognized diversity of novel astroviruses. Virol. J. 6, 161.

Finkbeiner, S.R., Li, Y., Ruone, S., et al., 2009b. Identification of a novel astrovirus (astrovirus VA1) associated with an outbreak of acute gastroenteritis. J. Virol. 83, 10836–10839.

Fu, Y., Pan, M., Wang, X., et al., 2009. Complete sequence of a duck astrovirus associated with fatal hepatitis in ducklings. J. Gen. Virol. 90, 1104–1108.

Giordano, M.O., Martinez, L.C., Isa, M.B., et al., 2004. Childhood astrovirus-associated diarrhea in the ambulatory setting in a Public Hospital in Cordoba city, Argentina. Rev. Inst. Med. Trop. Sao Paulo 46, 93–96.

Gough, R.E., Collins, M.S., Borland, E., et al., 1984. Astrovirus-like particles associated with hepatitis in ducklings. Vet. Rec. 114, 279.

Guix, S., Caballero, S., Villena, C., et al., 2002. Molecular epidemiology of astrovirus infection in Barcelona, Spain. J. Clin. Microbiol. 40, 133–139.

Guo, L., Xu, X., Song, J., et al., 2010. Molecular characterization of astrovirus infection in children with diarrhea in Beijing, 2005-2007. J. Med. Virol. 82, 415–423.

Ham, H., Oh, S., Jang, J., et al., 2014. Prevalence of human astrovirus in patients with acute gastroenteritis. Ann. Lab. Med. 34, 145–147.

Harbour, D.A., Ashley, C.R., Williams, P.D., et al., 1987. Natural and experimental astrovirus infection of cats. Vet. Rec. 120, 555–557.

Holtz, L.R., Bauer, I.K., Rajendran, P., et al., 2011a. Astrovirus MLB1 is not associated with diarrhea in a cohort of Indian children. PLoS One 6, e28647.

Holtz, L.R., Wylie, K.M., Sodergren, E., et al., 2011b. Astrovirus MLB2 viremia in febrile child. Emerg. Infect. Dis. 17, 2050–2052.

Holtz, L.R., Bauer, I.K., Jiang, H., et al., 2014. Seroepidemiology of astrovirus MLB1. Clin. Vaccine. Immunol. 21, 908–911.

Honkavuori, K.S., Briese, T., Krauss, S., et al., 2014. Novel coronavirus and astrovirus in Delaware Bay shorebirds. PLoS One 9, e93395.

Hoshino, Y., Zimmer, J.F., Moise, N.S., et al., 1981. Detection of astroviruses in feces of a cat with diarrhea. Brief report. Arch. Virol. 70, 373–376.

Iturriza Gomara, M., Simpson, R., Perault, A.M., et al., 2008. Structured surveillance of infantile gastroenteritis in East Anglia, UK: incidence of infection with common viral gastroenteric pathogens. Epidemiol. Infect. 136, 23–33.

Jakab, F., Meleg, E., Banyai, K., et al., 2004. One-year survey of astrovirus infection in children with gastroenteritis in a large hospital in Hungary: occurrence and genetic analysis of astroviruses. J. Med. Virol. 74, 71–77.

Jakab, F., Varga, L., Nyul, Z., et al., 2009. Acute viral gastroenteritis among hospitalized children between 2003 and 2005 in Baranya County, Hungary. J. Clin. Virol. 44, 340–341.

Jeong, A.Y., Jeong, H.S., Jo, M.Y., et al., 2011. Molecular epidemiology and genetic diversity of human astrovirus in South Korea from 2002 to 2007. Clin. Microbiol. Infect. 17, 404–408.

Jiang, H., Holtz, L.R., Bauer, I., et al., 2013. Comparison of novel MLB-clade, VA-clade and classic human astroviruses highlights constrained evolution of the classic human astrovirus nonstructural genes. Virology 436, 8–14.

Jonassen, T.O., Monceyron, C., Lee, T.W., et al., 1995. Detection of all serotypes of human astrovirus by the polymerase chain reaction. J. Virol. Methods 52, 327–334.

Jonassen, C.M., Jonassen, T.O., Saif, Y.M., et al., 2001. Comparison of capsid sequences from human and animal astroviruses. J. Gen. Virol. 82, 1061–1067.

Jonassen, C.M., Jonassen, T.T., Sveen, T.M., et al., 2003. Complete genomic sequences of astroviruses from sheep and turkey: comparison with related viruses. Virus Res. 91, 195–201.

Kamel, A.H., Ali, M.A., El-Nady, H.G., et al., 2009. Predominance and circulation of enteric viruses in the region of Greater Cairo, Egypt. J. Clin. Microbiol. 47, 1037–1045.

Kapoor, A., Li, L., Victoria, J., et al., 2009. Multiple novel astrovirus species in human stool. J. Gen. Virol. 90, 2965–2972.

Kemenesi, G., Dallos, B., Gorfol, T., et al., 2014. Novel European lineages of bat astroviruses identified in Hungary. Acta Virol. 58, 95–98.

Khamrin, P., Dey, S.K., Chan-it, W., et al., 2010. Evaluation of a rapid immunochromatography strip test for detection of astrovirus in stool specimens. J. Trop. Pediatr. 56, 129–131.

Khamrin, P., Thongprachum, A., Okitsu, S., et al., 2016. Multiple astrovirus MLB1, MLB2, VA2 clades and classic human astrovirus in children with acute gastroenteritis in Japan. J. Med. Virol. 88, 356–360.

Kiulia, N.M., Mwenda, J.M., Nyachieo, A., et al., 2007. Astrovirus infection in young Kenyan children with diarrhoea. J. Trop. Pediatr. 53, 206–209.

Kjeldsberg, E., Hem, A., 1985. Detection of astroviruses in gut contents of nude and normal mice. Arch. Virol. 84, 135–140.

Kofstad, T., Jonassen, C.M., 2011. Screening of feral and wood pigeons for viruses harbouring a conserved mobile viral element: characterization of novel Astroviruses and Picornaviruses. PLoS One 6, e25964.

Le Guyader, F.S., Le Saux, J.C., Ambert-Balay, K., et al., 2008. Aichi virus, norovirus, astrovirus, enterovirus, and rotavirus involved in clinical cases from a French oyster-related gastroenteritis outbreak. J. Clin. Microbiol. 46, 4011–4017.

Li, L., Diab, S., McGraw, S., et al., 2013. Divergent astrovirus associated with neurologic disease in cattle. Emerg. Infect. Dis. 19, 1385–1392.

Lizasoain, A., Tort, L.F., Garcia, M., et al., 2015. Environmental assessment of classical human astrovirus in Uruguay. Food Environ. Virol. 7, 142–148.

Lorrot, M., Bon, F., El Hajje, M.J., et al., 2011. Epidemiology and clinical features of gastroenteritis in hospitalised children: prospective survey during a 2-year period in a Parisian hospital, France. Eur. J. Clin. Microbiol. Infect. Dis. 30, 361–368.

Lukashov, V.V., Goudsmit, J., 2002. Evolutionary relationships among Astroviridae. J. Gen. Virol. 83, 1397–1405.

Luo, Z., Roi, S., Dastor, M., et al., 2011. Multiple novel and prevalent astroviruses in pigs. Vet. Microbiol. 149, 316–323.

Madeley, C.R., Cosgrove, B.P., 1975. Viruses in infantile gastroenteritis. Lancet 2, 124.

Malasao, R., Khamrin, P., Chaimongkol, N., et al., 2012. Diversity of human astrovirus genotypes circulating in children with acute gastroenteritis in Thailand during 2000-2011. J. Med. Virol. 84, 1751–1756.

Martella, V., Moschidou, P., Lorusso, E., et al., 2011a. Detection and characterization of canine astroviruses. J. Gen. Virol. 92, 1880–1887.

Martella, V., Moschidou, P., Pinto, P., et al., 2011b. Astroviruses in rabbits. Emerg. Infect. Dis. 17, 2287–2293.

Martella, V., Medici, M.C., Terio, V., et al., 2013. Lineage diversification and recombination in type-4 human astroviruses. Infect. Genet. Evol. 20, 330–335.

Martella, V., Pinto, P., Tummolo, F., et al., 2014. Analysis of the ORF2 of human astroviruses reveals lineage diversification, recombination and rearrangement and provides the basis for a novel sub-classification system. Arch. Virol. 159, 3185–3196.

McIver, C.J., Palombo, E.A., Doultree, J.C., et al., 2000. Detection of astrovirus gastroenteritis in children. J. Virol. Methods 84, 99–105.

McNeilly, F., Connor, T.J., Calvert, V.M., et al., 1994. Studies on a new enterovirus-like virus isolated from chickens. Avian. Pathol. 23, 313–327.

McNulty, M.S., Curran, W.L., McFerran, J.B., 1980. Detection of astroviruses in turkey faeces by direct electron microscopy. Vet. Rec. 106, 561.

McNulty, M.S., Connor, T.J., McNeilly, F., et al., 1990. Biological characterisation of avian enteroviruses and enterovirus-like viruses. Avian. Pathol. 19, 75–87.

Medici, M.C., Tummolo, F., Martella, V., et al., 2015. Genetic heterogeneity and recombination in type-3 human astroviruses. Infect. Genet. Evol. 32, 156–160.

Medina, S.M., Gutierrez, M.F., Liprandi, F., et al., 2000. Identification and type distribution of astroviruses among children with gastroenteritis in Colombia and Venezuela. J. Clin. Microbiol. 38, 3481–3483.

Meliopoulos, V.A., Kayali, G., Burnham, A., et al., 2014. Detection of antibodies against Turkey astrovirus in humans. PLoS One 9, e96934.

Mendez, E., Arias, C.F., 2013. Astroviruses. In: Knipe, D.M., Howley, P.M. et al., (Eds.), Fields Virology. sixth ed. Lippincott Williams and Wilkins, Philadelphia, PA, pp. 609–628.

Mendez-Toss, M., Griffin, D.D., Calva, J., et al., 2004. Prevalence and genetic diversity of human astroviruses in Mexican children with symptomatic and asymptomatic infections. J. Clin. Microbiol. 42, 151–157.

Mitui, M.T., Bozdayi, G., Ahmed, S., et al., 2014. Detection and molecular characterization of diarrhea causing viruses in single and mixed infections in children: a comparative study between Bangladesh and Turkey. J. Med. Virol. 86, 1159–1168.

Mladenova, Z., Steyer, A., Steyer, A.F., et al., 2015. Aetiology of acute paediatric gastroenteritis in Bulgaria during summer months: prevalence of viral infections. J. Med. Microbiol. 64, 272–282.

Mor, S.K., Chander, Y., Marthaler, D., et al., 2012. Detection and molecular characterization of Porcine astrovirus strains associated with swine diarrhea. J. Vet. Diagn. Invest. 24, 1064–1067.

Mustafa, H., Palombo, E.A., Bishop, R.F., 2000. Epidemiology of astrovirus infection in young children hospitalized with acute gastroenteritis in Melbourne, Australia, over a period of four consecutive years, 1995 to 1998. J. Clin. Microbiol. 38, 1058–1062.

Naccache, S.N., Peggs, K.S., Mattes, F.M., et al., 2015. Diagnosis of neuroinvasive astrovirus infection in an immunocompromised adult with encephalitis by unbiased next-generation sequencing. Clin. Infect. Dis. 60, 919–923.

Naficy, A.B., Rao, M.R., Holmes, J.L., et al., 2000. Astrovirus diarrhea in Egyptian children. J. Infect. Dis. 182, 685–690.

Nguyen, T.A., Hoang, L., Pham le, D., et al., 2008. Identification of human astrovirus infections among children with acute gastroenteritis in the Southern Part of Vietnam during 2005-2006. J. Med. Virol. 80, 298–305.

Oishi, I., Yamazaki, K., Kimoto, T., et al., 1994. A large outbreak of acute gastroenteritis associated with astrovirus among students and teachers in Osaka, Japan. J. Infect. Dis. 170, 439–443.

Palombo, E.A., Bishop, R.F., 1996. Annual incidence, serotype distribution, and genetic diversity of human astrovirus isolates from hospitalized children in Melbourne, Australia. J. Clin. Microbiol. 34, 1750–1753.

Pankovics, P., Boros, Á., Kiss, T., et al., 2015. Detection of a mammalian-like astrovirus in bird, European roller (*Coracias garrulus*). Infect. Genet. Evol. 34, 114–121.

Pantin-Jackwood, M.J., Strother, K.O., Mundt, E., et al., 2011. Molecular characterization of avian astroviruses. Arch. Virol. 156, 235–244.

Papaventsis, D.C., Dove, W., Cunliffe, N.A., et al., 2008. Human astrovirus gastroenteritis in children, Madagascar, 2004–2005. Emerg. Infect. Dis. 14, 844–846.

Pativada, M., Nataraju, S.M., Ganesh, B., et al., 2012. Emerging trends in the epidemiology of human astrovirus infection among infants, children and adults hospitalized with acute watery diarrhea in Kolkata, India. Infect. Genet. Evol. 12, 1685–1693.

Phan, T.G., Okame, M., Nguyen, T.A., et al., 2004. Human astrovirus, norovirus (GI, GII), and sapovirus infections in Pakistani children with diarrhea. J. Med. Virol. 73, 256–261.

Quan, P.L., Wagner, T.A., Briese, T., et al., 2010. Astrovirus encephalitis in boy with X-linked agammaglobulinemia. Emerg. Infect. Dis. 16, 918–925.

Reuter, G., Pankovics, P., Delwart, E., et al., 2012. Identification of a novel astrovirus in domestic sheep in Hungary. Arch. Virol. 157, 323–327.

Rice, M., Wilks, C.R., Jones, B.R., et al., 1993. Detection of astrovirus in the faeces of cats with diarrhoea. N. Z. Vet. J. 41, 96–97.

Rivera, R., Nollens, H.H., Venn-Watson, S., et al., 2010. Characterization of phylogenetically diverse astroviruses of marine mammals. J. Gen. Virol. 91, 166–173.

Schnagl, R.D., Belfrage, K., Farrington, R., et al., 2002. Incidence of human astrovirus in central Australia (1995 to 1998) and comparison of deduced serotypes detected from 1981 to 1998. J. Clin. Microbiol. 40, 4114–4120.

Sdiri-Loulizi, K., Gharbi-Khelifi, H., de Rougemont, A., et al., 2009. Molecular epidemiology of human astrovirus and adenovirus serotypes 40/41 strains related to acute diarrhea in Tunisian children. J. Med. Virol. 81, 1895–1902.

Sezen, F., Aval, E., Agkurt, T., et al., 2015. A large multi-pathogen gastroenteritis outbreak caused by drinking contaminated water from antique neighbourhood fountains, Erzurum city, Turkey, December 2012. Epidemiol. Infect. 143, 704–710.

Shan, T.L., Dai, X.Q., Guo, W., et al., 2009. Human astrovirus infection in children with gastroenteritis in Shanghai. J. Clin. Virol. 44, 248–249.

Silva, P.A., Stark, K., Mockenhaupt, F.P., et al., 2008. Molecular characterization of enteric viral agents from children in northern region of Ghana. J. Med. Virol. 80, 1790–1798.

Silva, P.A., Santos, R.A., Costa, P.S., et al., 2009. The circulation of human astrovirus genotypes in the Central West Region of Brazil. Mem. Inst. Oswaldo Cruz. 104, 655–658.

Smits, S.L., van Leeuwen, M., Kuiken, T., et al., 2010. Identification and characterization of deer astroviruses. J. Gen. Virol. 91, 2719–2722.

Snodgrass, D.R., Gray, E.W., 1977. Detection and transmission of 30 nm virus particles (astroviruses) in faeces of lambs with diarrhoea. Arch. Virol. 55, 287–291.

So, C.W., Kim, D.S., Yu, S.T., et al., 2013. Acute viral gastroenteritis in children hospitalized in Iksan, Korea during December 2010-June 2011. Korean J. Pediatr. 56, 383–388.

Stenglein, M.D., Velazquez, E., Greenacre, C., et al., 2012. Complete genome sequence of an astrovirus identified in a domestic rabbit (Oryctolagus cuniculus) with gastroenteritis. Virol. J. 9, 216.

Takano, T., Takashina, M., Doki, T., et al., 2015. Detection of canine astrovirus in dogs with diarrhea in Japan. Arch. Virol. 160, 1549–1553.

Tayeb, H.T., Al-Ahdal, M.N., Cartear, M.J., et al., 2010. Molecular epidemiology of human astrovirus infections in Saudi Arabia pediatric patients. J. Med. Virol. 82, 2038–2042.

Thongprachum, A., Takanashi, S., Kalesaran, A.F., et al., 2015. A four-year study of viruses that cause diarrhea in Japanese pediatric outpatients. J. Med. Virol. 87, 1141–1148.

Tran, A., Talmud, D., Lejeune, B., et al., 2010. Prevalence of rotavirus, adenovirus, norovirus, and astrovirus infections and coinfections among hospitalized children in northern France. J. Clin. Microbiol. 48, 1943–1946.

Tseng, W.C., Wu, F.T., Hsiung, C.A., et al., 2012. Astrovirus gastroenteritis in hospitalized children of less than 5 years of age in Taiwan, 2009. J. Microbiol. Immunol. Infect. 45, 311–317.

Tzipori, S., Menzies, J.D., Gray, E.W., 1981. Detection of astrovirus in the faeces of red deer. Vet. Rec. 108, 286.

Ulloa, J.C., Matiz, A., Lareo, L., et al., 2005. Molecular analysis of a 348 base-pair segment of open reading frame 2 of human astrovirus. A characterization of Colombian isolates. In Silico Biol. 5, 537–546.

van Hemert, F.J., Berkhout, B., Lukashov, V.V., 2007. Host-related nucleotide composition and codon usage as driving forces in the recent evolution of the Astroviridae. Virology 361, 447–454.

Walter, J.E., Mitchell, D.K., Guerrero, M.L., et al., 2001. Molecular epidemiology of human astrovirus diarrhea among children from a periurban community of Mexico City. J. Infect. Dis. 183, 681–686.

Wang, F., Wang, Y.H., Peng, J.S., et al., 2011. Genetic characterization of Human astrovirus infection in Wuhan, People's Republic of China, 2007-2008. Can. J. Microbiol. 57, 964–968.

Wei, H., Zeng, J., Deng, C., et al., 2013. A novel method of real-time reverse-transcription loop-mediated isothermal amplification developed for rapid and quantitative detection of human astrovirus. J. Virol. Methods 188, 126–131.

Willcocks, M.M., Carter, M.J., Laidler, F.R., et al., 1990. Growth and characterisation of human faecal astrovirus in a continuous cell line. Arch. Virol. 113, 73–81.

Williams, Jr., F.P., 1980. Astrovirus-like, coronavirus-like, and parvovirus-like particles detected in the diarrheal stools of beagle pups. Arch. Virol. 66, 215–226.

Woode, G.N., Bridger, J.C., 1978. Isolation of small viruses resembling astroviruses and caliciviruses from acute enteritis of calves. J. Med. Microbiol. 11, 441–452.

Yamaguchi, S., Imada, T., Kawamura, H., 1979. Characterization of a picornavirus isolated from broiler chicks. Avian. Dis. 23, 571–581.

Zhao, W., Zhu, A.L., Yu, Y., et al., 2011. Complete sequence and genetic characterization of pigeon avian nephritis virus, a member of the family Astroviridae. Arch. Virol. 156, 1559–1565.

Zhu, H.C., Chu, D.K., Liu, W., et al., 2009. Detection of diverse astroviruses from bats in China. J. Gen. Virol. 90, 883–887.

Chapter 5.1

Enteric Viral Metagenomics

N.J. Ajami, J.F. Petrosino

The Alkek Center for Metagenomics and Microbiome Research, Baylor College of Medicine, Houston, TX; Department of Molecular Virology and Microbiology, Baylor College of Medicine, Houston, TX, United States

1 BACKGROUND

The human commensal microbiota, also known as the human microbiome, is a complex ecosystem host to a milieu of organisms composed of bacteria, archaea, yeasts, fungi, and viruses. The functions encoded by these commensal organisms complement host-encoded capacities to provide a direct benefit to the host's overall health (Human Microbiome Project Consortium, 2012; Norman et al., 2014; Virgin, 2014). The microbiome in the gastrointestinal tract is thought to begin forming before birth and matures during the early years of life toward a more stable, "adult-like" community structure. Many factors contribute to the development of the microbiome including host genetics, environment, nutrition, and the use of antimicrobials (Koenig et al., 2011; Walter and Ley, 2011; Backhed et al., 2015; Lim et al., 2015; Nobel et al., 2015).

For many years, the detection and characterization of the members of the human microbiota, mainly as single entities, has been primarily achieved through microscopy, biochemical assays, serology, molecular amplification, cloning, and culture or animal infection studies. However, with the advent of high-throughput next-generation sequencing (NGS) technologies, it is now possible to probe and study the microbiome as a community rather than as single organisms without a need to culture them.

The depth with which one is able to sequence a specimen using NGS technologies, as compared to Sanger sequencing, has enabled the field of metagenomics to flourish. Through the ability of adding sequencing adapters to free DNA fragments of interest, NGS strategies eliminates the need for cloning prior to sequencing. This has resulted in massively parallel sequencing of unique and tagged DNA fragments, boosting the numbers of fragments sequenced by several orders of magnitude and thus facilitating the analysis of hundreds of samples in a single sequencing run. Due to NGS, the true diversity of the human microbiome and its components can now be evaluated.

The gastrointestinal tract contains one of the most diverse collections of microbiota in the human body (Human Microbiome Project Consortium, 2012). There are an estimated 10^{14} bacteria in the human intestine and although these organisms are the most studied, they are far from being the most abundant (Garrett et al., 2010). Human feces contain an estimated 10^9 viruses per gram. This viral collection, or virome, is dominated by viruses that infect bacteria (bacteriophages), plant viruses, and eukaryotic viruses, but the great majority still remains unidentified (Minot et al., 2012). Until recently, the gut virome has been generally considered to be comprised of viral pathogens some of which are widely known in the clinic including rotavirus, norovirus, adenovirus, and enterovirus (Dolin et al., 1971; Bishop et al., 1973; Appleton and Higgins, 1975). However, viral metagenomic data have led to the discovery of novel ones, including bacteriophages, pathogenic and nonpathogenic eukaryotic viruses, and dietary-associated viruses (Wylie et al., 2012).

2 METAGENOMICS FOR ENTERIC VIRUSES

Viral metagenomics emerged in 2002 with the publication of two uncultured marine viral communities (Breitbart et al., 2002). Since then, it has provided a different perspective to approach the study of viral communities. Prior to 2002, viral genomes were detected and characterized via a combination of restriction profiles, PCR, cloning, and Sanger sequencing (Sanger et al., 1977)

The metagenomic approach enables the detection and reconstitution of partial and complete viral genomes from the nucleic acids of samples collected from any environment. This provides a great advantage for the detection of low abundant or uncultivable viruses, some of which are found in the human gut, and also enables the processing of many samples at one time with high-throughput implementations of these strategies (Breitbart et al., 2008). The versatility of having an agnostic and untargeted method allows for the detection and characterization for both known and unknown viruses in common environments. As a result, an increasing number of diseases for which etiologies were not identified are now being recognized as of infectious origin (Wilson et al., 2014).

Gastroenteritis afflictions of viral origin are most common in industrialized nations where they are still a major cause of illness among people of all age groups (Glass et al., 2001). Among all diarrheal episodes, up to 40% are of unknown etiology suggesting that unrecognized infectious agents are yet to be identified (Kapikian, 1993; Simpson et al., 2003; Denno et al., 2007). Recent studies have turned to viral metagenomics to unravel the cause of idiopathic diarrheal diseases. Nonpathogenic enteric viruses, extra-intestinal viruses, plant viruses, and other novel viruses all were identified as associated with illness. These results compare favorably with early studies where viruses commonly associated with infectious diarrhea were detected after employing micromass sequencing, a low-throughput method that utilized a minimal starting sample quantity (<100 mg stool), with minimal sample purification, and limited sequencing (Finkbeiner et al., 2008). This approach identified common enteric viruses including rotavirus and norovirus, in the majority of cases. Furthermore, it generated data for possibly divergent strains of known viruses such as human picobirnavirus, human coxsackievirus A19, and human enterovirus 91 (Finkbeiner et al., 2008). Using the same metagenomic strategy, the same research group was able to identify a novel picornavirus in a child with acute diarrhea (Holtz et al., 2008). With the advent of high throughput technologies, studies such as this have served as a foundation for those in progress today. For instance, a study in the Netherlands using NGS, identified viruses of the families *Anelloviridae*, *Picobirnaviridae*, *Herpesviridae*, and *Picornaviridae,* in diarrheal stool samples obtained from patients with gastroenteritis whose etiology was not identified despite extensive testing. Some of these viruses might be associated with the development of gastroenteritis (Smits et al., 2014).

Another study found novel enteric eukaryotic viruses associated with acute diarrhea in children suffering from small bowel enteropathy in two different developing areas of Australia. In this study, many virus families were equally present in both cohorts including common diarrhea pathogens such as norovirus and astrovirus. However, other viruses with unknown pathogenic properties such as anelloviruses were frequently detected (Holtz et al., 2014). It remains unknown whether these viruses play a role in pathogenesis or are simply dietary passengers.

This research has added value to the field of molecular epidemiology by identifying viral causes of outbreaks where routine diagnostic methods fail (Moore et al., 2015), and also by providing a means to track viruses around the globe (Holtz et al., 2014).

3 METHODS FOR VIRAL METAGENOMICS

Viral metagenomics has been particularly suitable for providing an in-depth overview of the viral community structure (viral richness), and more recently, characterization of genes. In theory, the study of viruses through metagenomics allows for the identification of any virus regardless of its particular status, including low abundant, difficult to isolate, and uncultivable viruses. However, the state-of-the-art technology still has limitations in sample and library preparation, sequence strategy, and most importantly, taxonomic classification. Current strategies to improve the sensitivity of viral metagenomics include enrichment of viral template through subtraction of host nucleic acid via nuclease digestion, depletion of rRNA, semirandom amplification, probe-based capture, and alternative sequencing strategies. Although substantial progress has been made, more work needs to be done to before viral metagenomics become a point-of-care clinical tool.

Challenges begin with sample selection, collection, and preservation. For upper respiratory infections, the sensitivity for detecting respiratory viruses differs between nasal swabs and nasopharyngeal aspirates. Nasal swabs are generally optimal for the detection of most respiratory viruses except respiratory syncytial virus (RSV). RSV is known as a relatively labile virus and a larger-volume specimen, like the nasopharyngeal aspirate, is required for the detection of intact viruses. (Heikkinen et al., 2002). Similarly, to measure the gut microbiome, choices have to be made for sampling procedures because it is known that sampling methods, storage, and processing of samples can have an impact on the analysis (Budding et al., 2014).

For enteric virus detection, the choice of specimen collection has to be carefully selected (Table 5.1.1). The collection of raw primary samples followed by snap freezing is optimal. However, in some clinical settings and in remote locations, this is not practical. The use of stabilizing buffers has been proposed for remote places of sample collection but is yet to be validated thoroughly for known enteric viruses. Sample processing methods for metagenomic analysis rely on high-throughput assays and automated liquid handlers for the extraction and purification of viral nucleic acids. Prior to extraction, a homogenization step and viral enrichment processes are required to increase virus detection and avoid contaminating signals from free nucleic acids or from other organisms, respectively. This is critical for stool or intestinal samples where the presence of inhibitors is ubiquitous (Oikarinen et al., 2009). Clarification through filtration or sedimentation is also implemented for similar purposes. The average viral genome (~50 kb) is about 50 times smaller than the average microbial genome (~2.5 Mb), and about 60,000 times smaller than the human genome (~3 Gb), so any cellular contamination will overwhelm the viral signal (Edwards and Rohwer, 2005). Nonetheless, common processes for virus purification such as direct ultracentrifugation or ultracentrifugation in cesium chloride density gradients are not practical for large-scale metagenomic studies where hundreds or thousands of samples are to be examined, often with very limiting amounts of starting material (Bachrach and Friedmann, 1971; Huhti et al., 2010).

Following extraction, viral nucleic acids are converted into DNA/cDNA libraries. To detect RNA viruses, a step for reverse transcription is included to synthesize cDNA. The resulting viral DNA and cDNA molecules are often amplified to increase the sensitivity of detection. The choice of read length and sequencing depth (targeted reads per sample) will affect the mapping identity thresholds used for taxonomic classification which are mostly based on sequence-similarity algorithms similar to BLAST (Altschul et al., 1990) but adapted for massive datasets. Mapping strategies also include translated protein searches to increase sensitivity for detection. Quality control checkpoints are carefully evaluated after each step of the process to ensure the generation of high quality data.

Various initiatives are being developed to evaluate aspects of microbiome research including sample handling, sequencing, and analysis, aiming to evaluate methods for measuring the human microbiome. Such methods include tools for sampling human-associated microbes at different body sites, techniques and protocols for handling human samples, and computational pipelines for microbiome data processing. The overarching goal is to improve the state-of-the-art in each of these areas and to promote open sharing of standard operating procedures and best practices throughout the field.

TABLE 5.1.1 Considerations for a Viral Metagenomic Pipeline

Process	Description	Considerations
Collection	• Primary sample • Primary sample + stabilizer	• Primary sample collection in clinical setting is best, but not always practical
Storage	• −80°C • Room temperature + stabilizer	• Snap freezing (dry ice) followed by storage at −80°C is optimal
Processing	• Clarification • Extraction - DNA, RNA, DNA/RNA	• Clarification required for optimal extraction • Choice of extraction method depends on the type of viruses sought
Amplification	• Reverse transcription (RNA viruses) • Random amplification • Clean up • Library preparation	• Reverse transcription and second strand (cDNA) synthesis required for RNA viruses • Amplification results in better coverage of low-abundant viruses • Clean up step, removing undesired fragment sizes, required for optimal libraries
Sequencing	• Platform • Read length • Sequencing depth	• Platform choice requires consideration of balance between desired read length, sequencing depth, and acceptable error rate
Analysis	• Assembly • Nucleotide alignment • Translated nucleotide	• Reads may be assembled prior to mapping • Choice of assembly tools should take sequencing platform, sample complexity into account • Nucleotide + translated nucleotide alignments will result in a higher proportion of mapped reads • Independent validation of results of interest recommended (eg, PCR, RT-PCR)

4 VIRUS INTERACTION WITH GUT MICROBIOTA

Commensal intestinal bacteria aid host health by contributing with the metabolism of nutrients (He et al., 2015; Ramakrishna, 2013), the development of the immune system (Kau et al., 2011), and as a barrier for pathogen colonization

(Abt et al., 2012; Kamada et al., 2013; Smith and Nemerow, 2008). Disturbances that lead to the imbalance of the resident microbiota have been associated with enteric diseases such as *Clostridium difficile* infection (Chang et al., 2008) and systemic diseases like obesity, inflammatory bowel disease (Kostic et al., 2014), and diabetes (Kostic et al., 2015; Mejia-Leon et al., 2014) among others. Although commensal bacteria-host interactions have been established, it is less clear whether commensal microbiota have an effect on enteric viruses.

Recent research in murine models indicate that intestinal microbiota enhance infection of poliovirus, reovirus, Theiler's murine encephalomyelits virus (TMEV), and mouse mammary tumor virus (MMTV) (Kane et al., 2011; Kuss et al., 2011; Pullen et al., 1995). Although these are unrelated viruses, all seem to utilize bacterial components to increase their infectivity. MMTV binds to bacterial lipopolysaccharide (LPS) and induces immune tolerance though a Toll-like receptor 4-dependent mechanism (Kane et al., 2011). Similarly, poliovirus binds bacterial surface polysaccharides, which enhances virion stability and cell attachment by increasing the binding to the viral receptor resulting in increased infectivity (Kuss et al., 2011; Robinson et al., 2014). Reovirus also seem to exploit the intestinal microbiota as its infection is enhanced by the presence of resident microbes (Kuss et al., 2011). TMEV is another enteric virus and a common contaminant of mouse colonies, which its replication is enhanced with LPS treatment. In this case, data suggest that is likely due to increased inflammation in the central nervous system that results in an increased replication (Pullen et al., 1995). Components of enteric bacteria are instrumental in norovirus infection of Peyer's patch B cells in vitro and in vivo (Jones et al., 2014). (See also Chapter 5.2.)

In contrast, other studies suggest that commensal microbiota can also block viral infectivity. Research suggests that soluble factors obtained from cultures of *Bacteroides thetaiotaomicron* and *Lactobacillus casei* have the ability to modify cell-surface glycans and block rotavirus infection (Varyukhina et al., 2012). Another study sought to determine the mechanisms that contribute to the resolution of rotavirus gastroenteritis in response to probiotics in a mouse model. The data suggested *Lactobacillus reuteri* enhanced enterocyte proliferation and overall health of the intestine contributing to the resolution of diarrhea (Preidis et al., 2012). For many years studies have demonstrated the protective effects of bacteria, particularly lactic acid bacteria on rotavirus infection, and these results suggest a plausible explanation for the observed protective effects. (See also Chapter 2.7.)

Blockage of viral infections can also happen in an indirect way. For instance, influenza virus infection is enhanced in germfree and antibiotic treated mice and it is thought that certain gut bacteria may prime the immune system for influenza virus protection as neomycin-sensitive bacteria were associated with protective responses in the lung while stimulation with Toll-like receptor agonists was sufficient to restore immune responses (Ichinohe et al., 2011). In a similar study, antibiotic treated mice showed decreased influenza virus titers and pathogenesis. These mice showed reduced virus-specific CD8+ T-cell responses and IgG and IgM antibody levels, compared to infected conventional mice suggesting impaired adaptive immune responses in mice with depleted microbiota (Abt et al., 2012). In addition, the replication of adenoviruses was observed to be inhibited by antimicrobial peptides (defensins), which are produced by host cells in response to the microbiota. The inhibition is mediated by binding virions and limiting their replication (Smith and Nemerow, 2008). Altogether, these research indicate that commensal intestinal and upper airway microbiota may play important roles in limiting viral infections by constant stimulation of the immune system (Abt et al., 2012; Robinson and Pfeiffer, 2014)

For other viruses, including coxsackievirus B3 (CVB3), norovirus, and murine norovirus (MNV), the interactions with the microbiota remain unclear. In the case of CVB3, limited data generated using a germfree mouse revealed that mice were more susceptible to colonization after intraperitoneal injection (Schaffer et al., 1963) Follow-up studies are needed to elucidate the mechanisms behind this effect. Interestingly, CVB3 is closely related to poliovirus whose infectivity has been shown to increase in the presence of commensal microbiota (Kane et al., 2011; Kuss et al., 2011). For human noroviruses, it is still not clear whether infection disrupts the intestinal microbiota. A recent study showed outgrowth of *Escherichia coli* and reduced Bacteroidetes in a subset of individuals infected with norovirus. Although still inconclusive, the outgrowth of enteric pathogens following viral infection may elevate the risk for long-term health complications (Nelson et al., 2012). For MNV, it is established that infection does not alter the gut microbiota (Nelson et al., 2013). However, MNV-induced pathologies have been suggested to be associated with microbiota-induced immune mediators (eg, TNF-alpha and IFN-gamma) in a mouse model for Crohn's disease, providing an example of the dynamics in the host-bacteria-virus triad (Cadwell et al., 2010).

5 FUTURE DIRECTIONS

Metagenomics allows for the study of the different constituents of the human microbiome (including the virome) enabling evaluation of the impact of multiple associations between host and microbe, and between microbe and microbe. In addition, it provides a way to study observed phenomena without introducing the bias and limitations of targeted assays and gives way to the generation of multiple hypotheses. Nonetheless, a current limitation of the state-of-the-art is the lack of application for point-of-care testing. While in the past the limitations dealt with high technology costs, in this case sample preparation (presequencing) and bioinformatics analyses provide barriers to produce a result in adequate sample-to-answer turnaround times. A recent study showed progress toward narrowing this gap. By implementing nanopore sequencing, a third generation technology, coupled with a web-based pipeline for real-time bioinformatics analysis, the detection of chikungunya virus, Ebola virus, and hepatitis C virus directly form human blood samples was performed in under 6hrs (Greninger et al., 2015). Although challenges remain to decrease the time of library preparation, it is critical to have real-time sequence analysis for applications such as outbreak investigation and metagenomic diagnoses of life-threatening infections.

The diversity of viruses in the gut virome, some of which are not associated with pathologic states, is analogous to that vast bacterial diversity observed in the gut microbiome. This suggests that viral communities may behave in a fashion similar to the well-established bacterial associations where events of commensalism and mutualism are vital for the stability and resilience of the overall ecosystem. Nevertheless, further studies are necessary to elucidate the interactions between hosts and their microbiomes and translate these finding into clinical applications.

REFERENCES

Abt, M.C., Osborne, L.C., Monticelli, L.A., Doering, T.A., Alenghat, T., Sonnenberg, G.F., Paley, M.A., Antenus, M., Williams, K.L., Erikson, J., Wherry, E.J., Artis, D., 2012. Commensal bacteria calibrate the activation threshold of innate antiviral immunity. Immunity 37, 158–170.

Altschul, S.F., Gish, W., Miller, W., Myers, E.W., Lipman, D.J., 1990. Basic local alignment search tool. J. Mol. Biol. 215, 403–410.

Appleton, H., Higgins, P.G., 1975. Letter: Viruses and gastroenteritis in infants. Lancet 1, 1297.

Bachrach, U., Friedmann, A., 1971. Practical procedures for the purification of bacterial viruses. Appl. Microbiol. 22, 706–715.

Backhed, F., Roswall, J., Peng, Y., Feng, Q., Jia, H., Kovatcheva-Datchary, P., Li, Y., Xia, Y., Xie, H., Zhong, H., Khan, M.T., Zhang, J., Li, J., Xiao, L., Al-Aama, J., Zhang, D., Lee, Y.S., Kotowska, D., Colding, C., Tremaroli, V., Yin, Y., Bergman, S., Xu, X., Madsen, L., Kristiansen, K., Dahlgren, J., Wang, J., 2015. Dynamics and stabilization of the human gut microbiome during the first year of life. Cell Host Microbe 17, 852.

Bishop, R.F., Davidson, G.P., Holmes, I.H., Ruck, B.J., 1973. Virus particles in epithelial cells of duodenal mucosa from children with acute non-bacterial gastroenteritis. Lancet 2, 1281–1283.

Breitbart, M., Haynes, M., Kelley, S., Angly, F., Edwards, R.A., Felts, B., Mahaffy, J.M., Mueller, J., Nulton, J., Rayhawk, S., Rodriguez-Brito, B., Salamon, P., Rohwer, F., 2008. Viral diversity and dynamics in an infant gut. Res. Microbiol. 159, 367–373.

Breitbart, M., Salamon, P., Andresen, B., Mahaffy, J.M., Segall, A.M., Mead, D., Azam, F., Rohwer, F., 2002. Genomic analysis of uncultured marine viral communities. Proc. Natl. Acad. Sci. USA 99, 14250–14255.

Budding, A.E., Grasman, M.E., Eck, A., Bogaards, J.A., Vandenbroucke-Grauls, C.M., van Bodegraven, A.A., Savelkoul, P.H., 2014. Rectal swabs for analysis of the intestinal microbiota. PLoS One 9, e101344.

Cadwell, K., Patel, K.K., Maloney, N.S., Liu, T.C., Ng, A.C., Storer, C.E., Head, R.D., Xavier, R., Stappenbeck, T.S., Virgin, H.W., 2010. Virus-plus-susceptibility gene interaction determines Crohn's disease gene Atg16L1 phenotypes in intestine. Cell 141, 1135–1145.

Chang, J.Y., Antonopoulos, D.A., Kalra, A., Tonelli, A., Khalife, W.T., Schmidt, T.M., Young, V.B., 2008. Decreased diversity of the fecal Microbiome in recurrent Clostridium difficile-associated diarrhea. J. Infect. Dis. 197, 435–438.

Denno, D.M., Klein, E.J., Young, V.B., Fox, J.G., Wang, D., Tarr, P.I., 2007. Explaining unexplained diarrhea and associating risks and infections. Anim. Health Res. Rev. 8, 69–80.

Dolin, R., Blacklow, N.R., DuPont, H., Formal, S., Buscho, R.F., Kasel, J.A., Chames, R.P., Hornick, R., Chanock, R.M., 1971. Transmission of acute infectious nonbacterial gastroenteritis to volunteers by oral administration of stool filtrates. J. Infect. Dis. 123, 307–312.

Edwards, R.A., Rohwer, F., 2005. Viral metagenomics. Nat. Rev. Microbiol. 3, 504–510.

Finkbeiner, S.R., Allred, A.F., Tarr, P.I., Klein, E.J., Kirkwood, C.D., Wang, D, 2008. Metagenomic analysis of human diarrhea: viral detection and discovery. PLoS Pathog. 4, e1000011.

Garrett, W.S., Gordon, J.I., Glimcher, L.H., 2010. Homeostasis and inflammation in the intestine. Cell 140, 859–870.

Glass, R.I., Bresee, J., Jiang, B., Gentsch, J., Ando, T., Fankhauser, R., Noel, J., Parashar, U., Rosen, B., Monroe, S.S., 2001. Gastroenteritis viruses: an overview. Novartis Found. Symp. 238, 5–19.

Greninger, A.L., Naccache, S.N., Federman, S., Yu, G., Mbala, P., Bres, V., Stryke, D., Bouquet, J., Somasekar, S., Linnen, J.M., Dodd, R., Mulembakani, P., Schneider, B.S., Muyembe-Tamfum, J.J., Stramer, S.L., Chiu, C.Y., 2015. Rapid metagenomic identification of viral pathogens in clinical samples by real-time nanopore sequencing analysis. Genome Med. 7, 99.

He, B., Nohara, K., Ajami, N.J., Michalek, R.D., Tian, X., Wong, M., Losee-Olson, S.H., Petrosino, J.F., Yoo, S.H., Shimomura, K., Chen, Z., 2015. Transmissible microbial and metabolomic remodeling by soluble dietary fiber improves metabolic homeostasis. Sci. Rep. 5, 10604.

Heikkinen, T., Marttila, J., Salmi, A.A., Ruuskanen, O., 2002. Nasal swab versus nasopharyngeal aspirate for isolation of respiratory viruses. J. Clin. Microbiol. 40, 4337–4339.

Holtz, L.R., Cao, S., Zhao, G., Bauer, I.K., Denno, D.M., Klein, E.J., Antonio, M., Stine, O.C., Snelling, T.L., Kirkwood, C.D., Wang, D., 2014. Geographic variation in the eukaryotic virome of human diarrhea. Virology 468–470, 556–564.

Holtz, L.R., Finkbeiner, S.R., Kirkwood, C.D., Wang, D., 2008. Identification of a novel picornavirus related to cosaviruses in a child with acute diarrhea. Virol. J. 5, 159.

Huhti, L., Blazevic, V., Nurminen, K., Koho, T., Hytonen, V.P., Vesikari, T., 2010. A comparison of methods for purification and concentration of norovirus GII-4 capsid virus-like particles. Arch. Virol. 155, 1855–1858.

Human Microbiome Project, Consortium, 2012. Structure, function and diversity of the healthy human microbiome. Nature 486, 207–214.

Ichinohe, T., Pang, I.K., Kumamoto, Y., Peaper, D.R., Ho, J.H., Murray, T.S., Iwasaki, A., 2011. Microbiota regulates immune defense against respiratory tract influenza A virus infection. Proc. Natl. Acad. Sci. USA 108, 5354–5359.

Jones, M.K., Watanabe, M., Zhu, S., Graves, C.L., Keyes, L.R., Grau, K.R., Gonzalez-Hernandez, M.B., Iovine, N.M., Wobus, C.E., Vinje, J., Tibbetts, S.A., Wallet, S.M., Karst, S.M., 2014. Enteric bacteria promote human and mouse norovirus infection of B cells. Science 346, 755–759.

Kamada, N., Chen, G.Y., Inohara, N., Nunez, G., 2013. Control of pathogens and pathobionts by the gut microbiota. Nat. Immunol. 14, 685–690.

Kane, M., Case, L.K., Kopaskie, K., Kozlova, A., MacDearmid, C., Chervonsky, A.V., Golovkina, T.V., 2011. Successful transmission of a retrovirus depends on the commensal microbiota. Science 334, 245–249.

Kapikian, A.Z., 1993. Viral gastroenteritis. JAMA 269, 627–630.

Kau, A.L., Ahern, P.P., Griffin, N.W., Goodman, A.L., Gordon, J.I., 2011. Human nutrition, the gut microbiome and the immune system. Nature 474, 327–336.

Koenig, J.E., Spor, A., Scalfone, N., Fricker, A.D., Stombaugh, J., Knight, R., Angenent, L.T., Ley, R.E., 2011. Succession of microbial consortia in the developing infant gut microbiome. Proc. Natl. Acad. Sci. USA 108 (Suppl. 1), 4578–4585.

Kostic, A.D., Gevers, D., Siljander, H., Vatanen, T., Hyotylainen, T., Hamalainen, A.M., Peet, A., Tillmann, V., Poho, P., Mattila, I., Lahdesmaki, H., Franzosa, E.A., Vaarala, O., de Goffau, M., Harmsen, H., Ilonen, J., Virtanen, S.M., Clish, C.B., Oresic, M., Huttenhower, C., Knip, M., Xavier, R.J., Diabimmune Study Group, 2015. The dynamics of the human infant gut microbiome in development and in progression toward type 1 diabetes. Cell Host Microbe 17, 260–273.

Kostic, A.D., Xavier, R.J., Gevers, D., 2014. The microbiome in inflammatory bowel disease: current status and the future ahead. Gastroenterology 146, 1489–1499.

Kuss, S.K., Best, G.T., Etheredge, C.A., Pruijssers, A.J., Frierson, J.M., Hooper, L.V., Dermody, T.S., Pfeiffer, J.K., 2011. Intestinal microbiota promote enteric virus replication and systemic pathogenesis. Science 334, 249–252.

Lim, E.S., Zhou, Y., Zhao, G., Bauer, I.K., Droit, L., Ndao, I.M., Warner, B.B., Tarr, P.I., Wang, D., Holtz, L.R., 2015. Early life dynamics of the human gut virome and bacterial microbiome in infants. Nat. Med. 21 (10), 1228–1234.

Mejia-Leon, M.E., Petrosino, J.F., Ajami, N.J., Dominguez-Bello, M.G., de la Barca, A.M., 2014. Fecal microbiota imbalance in Mexican children with type 1 diabetes. Sci. Rep. 4, 3814.

Minot, S., Wu, G.D., Lewis, J.D., Bushman, F.D., 2012. Conservation of gene cassettes among diverse viruses of the human gut. PLoS One 7, e42342.

Moore, N.E., Wang, J., Hewitt, J., Croucher, D., Williamson, D.A., Paine, S., Yen, S., Greening, G.E., Hall, R.J., 2015. Metagenomic analysis of viruses in feces from unsolved outbreaks of gastroenteritis in humans. J. Clin. Microbiol. 53, 15–21.

Nelson, A.M., Elftman, M.D., Pinto, A.K., Baldridge, M., Hooper, P., Kuczynski, J., Petrosino, J.F., Young, V.B., Wobus, C.E., 2013. Murine norovirus infection does not cause major disruptions in the murine intestinal microbiota. Microbiome 1, 7.

Nelson, A.M., Walk, S.T., Taube, S., Taniuchi, M., Houpt, E.R., Wobus, C.E., Young, V.B., 2012. Disruption of the human gut microbiota following Norovirus infection. PLoS One 7, e48224.

Nobel, Y.R., Cox, L.M., Kirigin, F.F., Bokulich, N.A., Yamanishi, S., Teitler, I., Chung, J., Sohn, J., Barber, C.M., Goldfarb, D.S., Raju, K., Abubucker, S., Zhou, Y., Ruiz, V.E., Li, H., Mitreva, M., Alekseyenko, A.V., Weinstock, G.M., Sodergren, E., Blaser, M.J., 2015. Metabolic and metagenomic outcomes from early-life pulsed antibiotic treatment. Nat. Commun. 6, 7486.

Norman, J.M., Handley, S.A., Virgin, H.W., 2014. Kingdom-agnostic metagenomics and the importance of complete characterization of enteric microbial communities. Gastroenterology 146, 1459–1469.

Oikarinen, S., Tauriainen, S., Viskari, H., Simell, O., Knip, M., Virtanen, S., Hyoty, H., 2009. PCR inhibition in stool samples in relation to age of infants. J. Clin. Virol. 44, 211–214.

Preidis, G.A., Saulnier, D.M., Blutt, S.E., Mistretta, T.A., Riehle, K.P., Major, A.M., Venable, S.F., Barrish, J.P., Finegold, M.J., Petrosino, J.F., Guerrant, R.L., Conner, M.E., Versalovic, J., 2012. Host response to probiotics determined by nutritional status of rotavirus-infected neonatal mice. J. Pediatr. Gastroenterol. Nutr. 55, 299–307.

Pullen, L.C., Park, S.H., Miller, S.D., Dal Canto, M.C., Kim, B.S., 1995. Treatment with bacterial LPS renders genetically resistant C57BL/6 mice susceptible to Theiler's virus-induced demyelinating disease. J. Immunol. 155, 4497–4503.

Ramakrishna, B.S., 2013. Role of the gut microbiota in human nutrition and metabolism. J. Gastroenterol. Hepatol. 28 (Suppl. 4), 9–17.

Robinson, C.M., Jesudhasan, P.R., Pfeiffer, J.K., 2014. Bacterial lipopolysaccharide binding enhances virion stability and promotes environmental fitness of an enteric virus. Cell Host Microbe 15, 36–46.

Robinson, C.M., Pfeiffer, J.K., 2014. Viruses and the Microbiota. Annu. Rev. Virol. 1, 55–69.

Sanger, F., Air, G.M., Barrell, B.G., Brown, N.L., Coulson, A.R., Fiddes, C.A., Hutchison, C.A., Slocombe, P.M., Smith, M., 1977. Nucleotide sequence of bacteriophage phi X174 DNA. Nature 265, 687–695.

Schaffer, J., Beamer, P.R., Trexler, P.C., Breidenbach, G., Walcher, D.N., 1963. Response of germ-free animals to experimental virus monocontamination. I. Observation on Coxsackie B virus. Proc. Soc. Exp. Biol. Med. 112, 561–564.

Simpson, R., Aliyu, S., Iturriza-Gómara, M., Desselberger, U., Gray FJ., 2003. Infantile viral gastroenteritis: on the way to closing the diagnostic gap. J. Med. Virol. 70, 258–262.

Smith, J.G., Nemerow, G.R., 2008. Mechanism of adenovirus neutralization by human alpha-defensins. Cell Host Microbe 3, 11–19.

Smits, S.L., Schapendonk, C.M., van Beek, J., Vennema, H., Schurch, A.C., Schipper, D., Bodewes, R., Haagmans, B.L., Osterhaus, A.D., Koopmans, M.P., 2014. New viruses in idiopathic human diarrhea cases, the Netherlands. Emerg. Infect. Dis. 20, 1218–1222.

Varyukhina, S., Freitas, M., Bardin, S., Robillard, E., Tavan, E., Sapin, C., Grill, J.P., Trugnan, G., 2012. Glycan-modifying bacteria-derived soluble factors from Bacteroides thetaiotaomicron and Lactobacillus casei inhibit rotavirus infection in human intestinal cells. Microbes Infect. 14, 273–278.

Virgin, H.W., 2014. The virome in mammalian physiology and disease. Cell 157, 142–150.

Walter, J., Ley, R., 2011. The human gut microbiome: ecology and recent evolutionary changes. Annu. Rev. Microbiol. 65, 411–429.

Wilson, M.R., Naccache, S.N., Samayoa, E., Biagtan, M., Bashir, H., Yu, G., Salamat, S.M., Somasekar, S., Federman, S., Miller, S., Sokolic, R., Garabedian, E., Candotti, F., Buckley, R.H., Reed, K.D., Meyer, T.L., Seroogy, C.M., Galloway, R., Henderson, S.L., Gern, J.E., DeRisi, J.L., Chiu, C.Y., 2014. Actionable diagnosis of neuroleptospirosis by next-generation sequencing. N. Engl. J. Med. 370, 2408–2417.

Wylie, K.M., Weinstock, G.M., Storch, G.A., 2012. Emerging view of the human virome. Transl. Res. 160, 283–290.

Chapter 5.2

Interactions Between Enteric Viruses and the Gut Microbiota

R.R. Garg, S.M. Karst
College of Medicine, Department of Molecular Genetics and Microbiology, Emerging Pathogens Institute, University of Florida, Gainesville, FL, United States

1 INTRODUCTION

The human body is colonized by trillions of microorganisms collectively referred as the microbiota or the commensal flora. In fact, this microbiota is so immense that it out-numbers host cells by 10 times (Xu and Gordon, 2003). The intestinal lumen is home to a particularly rich microbial community. The density of non-pathogenic enteric microorganisms increases from the proximal to the distal region of intestine with the stomach containing 10^1 microbial cells/g of content, the duodenum 10^3 cells/g, the jejunum 10^4 cells/g, the ileum 10^7 cells/g, and the colon up to 10^{13} cells/g (Mowat and Agace, 2014). Resident gut microbes compete with invading pathogens for resources and space which can dramatically inhibit the growth of microbial pathogens. All viruses infecting along the gastrointestinal tract encounter this dense population of bacterial cells. Thus, one would expect there to be significant interaction between commensal bacteria and enteric viruses with potentially beneficial or inhibitory outcomes for the viral infection. Indeed, recent work by numerous groups has begun to reveal the complex influence of the microbiota on viral infection in the gut. For all enteric viruses studied to date including poliovirus (Kuss et al., 2011; Robinson et al., 2014), reovirus (Kuss et al., 2011), rotavirus (Uchiyama et al., 2014), mouse mammary tumor virus (Kane et al., 2011), and noroviruses (Jones et al., 2014; Baldridge et al., 2015; Kernbauer et al., 2014), commensal bacteria enhance viral infections although mechanistic insight into this stimulatory activity is in its infancy. The intent of this chapter is to discuss the state of understanding pertaining to mechanisms of bacterial stimulation of enteric viral infections and to highlight the potential relevance of understanding enteric virus–bacteria interactions in illuminating new treatment and prevention strategies. We provide a brief description of the nature and relevance of the gut microbiota within the mammalian host (Section 2) and then summarize findings regarding specific enteric viruses and the gut microbiota (Section 3).

Viral Gastroenteritis

2 THE GUT MICROBIOTA AND THE HOST—A SYMBIOTIC RELATIONSHIP

The mutually beneficial, or symbiotic, relationship existing between the host and its microbiota is nicely highlighted by the ability of metabolites produced by the microbial population to impact host physiology; in turn, microorganism-assisted digestion of food products within the gut lumen of the host also provides a rich source of nutrients for the survival of the microbial community (Nicholson et al., 2012). The composition of the gut microbiota is incredibly complex, consisting of 500–1000 bacterial species, and is influenced by environmental, dietary, and host genetic factors (Kau et al., 2011; Sommer and Bäckhed, 2013). Millions of genes encompassed within the overall microbiome encode for an immense collection of biosynthetic enzymes (eg, proteases and glycosidases) whose activities greatly expand the biochemical and metabolic potential of the host (Sommer and Bäckhed, 2013). This is best exemplified by the ability of commensal bacteria to degrade otherwise indigestible plant polysaccharides, complex carbohydrates, and other dietary substances, resulting in the production of essential energy sources and vitamins (Tremaroli and Bäckhed, 2012). A critical end product of this digestion is short-chain fatty acids (SCFAs) which can act as an energy source for cellular metabolism as well as modulators of immune responses and gut motility (Tremaroli and Bäckhed, 2012). Microbial enzymes can also metabolize bile acids and choline, providing beneficial effects to overall host metabolism (Nicholson et al., 2012; Tremaroli and Bäckhed, 2012).

The microbiota is also essential for shaping the development of the host immune system (Hooper et al., 2012; Round and Mazmanian, 2010). It is well-established that commensal bacteria play a critical role in immune development and functionality. For example, germ-free mice lacking commensal bacteria fail to develop mature lymphoid structures within the gastrointestinal tract and have severely reduced levels of secretory immunoglobulin A (IgA) and numbers of intestinal T cells. Molecular mechanisms responsible for this microbially driven immune homeostasis are beginning to be elucidated and have been reviewed recently elsewhere (Hooper et al., 2012; Round and Mazmanian, 2010; Kamada and Núñez, 2014). Again underscoring the cross-talk between the host and its microbiota, the immune system clearly regulates the composition and functioning of the gut flora as well (Hooper et al., 2012; Kamada and Núñez, 2014). For example, the host produces antimicrobial peptides and bacteria-specific IgA that are secreted into the gut lumen to restrict bacterial penetration of the small intestine (Vaishnava et al., 2011; Macpherson et al., 2000). There is also evidence for immune-driven regulation of the composition of the bacteria comprising the intestinal microbiota (Hooper et al., 2012).

A final benefit provided by the intestinal microbiota to the mammalian host is protection from infection by pathogenic microorganisms. For example, commensal bacteria-induced expression of antimicrobial proteins can target pathogens as well (Vaishnava et al., 2011; Hooper et al., 2003). Overall, the intestinal microbiota and the mammalian host have a multifaceted and complex symbiotic relationship that provides beneficial functions to host physiology, immune development and functioning, and protection from pathogenic bacterial infections. While these areas have been intensely studied, relatively little attention has been paid to the interactions between commensal bacteria and enteric viruses until very recently. The remainder of this chapter will focus on this emerging area of study.

3 THE INFLUENCE OF THE MICROBIOTA ON SPECIFIC ENTERIC VIRAL INFECTIONS

3.1 Poliovirus

Poliovirus, a nonenveloped, positive-sense, single-stranded RNA virus, is a member of the *Enterovirus* genus within the family *Picornaviridae*. Poliovirus is transmitted via the fecal–oral route and replicates within the gastrointestinal tract, rarely disseminating to the central nervous system and causing paralytic poliomyelitis. Poliovirus exhibits strict species specificity because of the highly divergent sequence of the poliovirus receptor (PVR) (Ida-Hosonuma et al., 2003). This limitation to studying poliovirus pathogenesis was overcome by the development of transgenic mice expressing the human PVR (Koike et al., 1991; Ren et al., 1990). PVR transgenic mice are not susceptible to oral poliovirus infection unless they are deficient in the type I interferon (IFN) response (Ohka et al., 2007; Ida-Hosonuma et al., 2005); so Kuss et al. (2011) utilized PVR-$IFNAR^{-/-}$ mice in order to address the impact of the intestinal microbiota on poliovirus infection. Specifically, PVR-$IFNAR^{-/-}$ mice were depleted of their commensal bacteria using a standard antibiotic (Abx) approach whereby mice are orally gavaged with a cocktail of four broad-spectrum Abx (ampicillin, vancomycin, neomycin, and metronidazole). Following oral poliovirus infection, a significant reduction in mortality was observed in Abx-treated mice compared to untreated control mice; this attenuation of poliovirus-induced mortality was reversed upon recolonization of Abx-treated mice with fecal bacteria (Kuss et al., 2011). These collective data demonstrated that commensal bacteria enhance the virulence of poliovirus in susceptible mice. This enhanced virulence correlated with increased viral replication in microbially colonized mice since peak infectious viral titers shed in the feces were higher in microbially colonized mice than in Abx-treated mice. Although Abx-treated mice continued to shed virus for prolonged periods compared to their colonized counterparts, Kuss et al. (2011) were able to demonstrate that this represented delayed shedding of inoculum virus and that in fact very little newly replicated virus was shed in bacterially-depleted mice. Confirming that bacterial depletion (in contrast to other effects of broad-spectrum Abx treatment) was responsible for impairment of poliovirus intestinal infection, colonization of Abx-treated mice with a single Abx-resistant bacterium restored poliovirus pathogenesis. Moreover, infection of cultured cells was unaffected by the presence of Abx and coadministration of poliovirus and Abx to mice did not impact viral intestinal infection (Kuss et al., 2011).

Two specific mechanisms of bacterial enhancement of poliovirus infection have been elucidated. First, commensal bacteria stabilize poliovirus particles (Fig. 5.2.1A). This was revealed by experiments demonstrating that virus incubated with feces from microbially colonized, but not Abx-treated or germ-free, mice displayed enhanced infection of cultured cells compared to virus incubated with PBS (Kuss et al., 2011). Moreover, viral incubation with a variety of Gram-negative or Gram-positive bacteria was sufficient to enhance viral infectivity; bacteria did not need to be viable for stimulatory activity (Kuss et al., 2011). Providing molecular insight into bacterial enhancement of poliovirus infectivity, *N*-acetylglucosamine (GlcNAc)-containing polysaccharides expressed on bacterial surfaces including lipopolysaccharide (LPS) and peptidoglycan (PG), also stabilized virus (Kuss et al., 2011; Robinson et al., 2014). Furthermore, polysaccharides composed of only GlcNAc were stimulatory although activity was regulated by length (requiring over six sugar residues) and dependent on acetylation (Robinson et al., 2014). Bacterial polysaccharides provided poliovirus with enhanced thermostability as well as resistance to bleach specifically by preventing a conformational change in the virion that results in viral genome release (Robinson et al., 2014). Second, bacterial polysaccharides promote poliovirus attachment to PVR expressed on permissive cells (Fig. 5.2.1B) (Robinson et al., 2014). Virus incubated with LPS or PG was more efficient at binding permissive cells than virus alone; and attachment was inhibited by pretreatment of cells with anti-PVR antibody irrespective of the presence of bacterial polysaccharide. In fact, poliovirus incubated with LPS was more efficient at binding purified PVR confirming that stimulatory activity occurs through a direct mechanism (Robinson et al., 2014).

Poliovirus directly binds LPS (Kuss et al., 2011). A specific surface-exposed residue in the viral VP1 capsid protein (a threonine at position 99) influences this interaction, underscored by the fact that the attenuated Sabin vaccine strain contains a lysine at this position and was not stabilized by LPS at physiological temperatures, although this residue is not involved in direct LPS binding (Robinson et al., 2014). While the T99K mutation impaired the ability of LPS to enhance viral stability, it did not affect LPS enhancement of viral attachment to permissive cells. Further supporting the idea that LPS functions to enhance poliovirus infection in two distinct manners, much greater LPS concentrations were required to stabilize poliovirus particles than to enhance viral attachment to permissive cells (Robinson et al., 2014). Although poliovirus containing the T99K mutation replicated comparably to wild-type virus in susceptible mice and displayed no loss in virulence, it did display reduced stability in the environment (Robinson et al., 2014). Collective data of Kuss et al. (2011) and

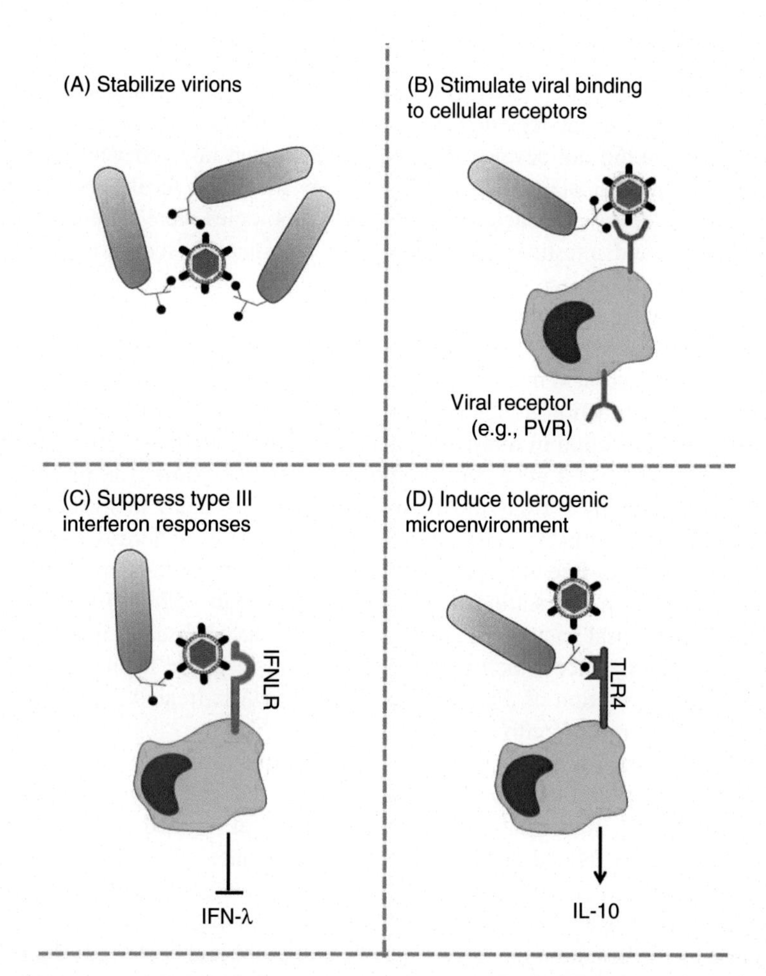

FIGURE 5.2.1 Mechanisms of commensal bacterial-mediated stimulation of enteric viral infections. (A) For poliovirus, commensal bacteria enhance thermostability of virions and even provide resistance to bleach. (B) For poliovirus and a HuNoV, bacterial surface-expressed polysaccharides (LPS and HBGA, respectively) stimulate viral attachment to the surface of permissive cells. For poliovirus, the cellular receptor is PVR. (C) For MNVs, commensal bacteria appear to suppress the antiviral activity of IFN-λ thereby promoting persistent viral infection. (D) For MMTV, the virus appears to bind bacterial LPS which is then recognized by TLR4 on host cells, resulting in secretion of IL-10. This leads to a tolerogenic microenvironment that suppresses immune clearance of MMTV.

Robinson et al. (2014) thus support a model whereby poliovirus binds bacterial surface components such as LPS, resulting in enhanced viral attachment to PVR on permissive cells and enhanced virion stability. This interaction confers a fitness advantage to the virus by enhancing its environmental stability, possibly facilitating transmission to new hosts.

3.2 Reovirus and Rotavirus

Similar to findings with poliovirus, reovirus infection is significantly attenuated in response to bacterial depletion (Kuss et al., 2011). Although immunocompetent adult mice did not develop overt symptoms when infected with reovirus, *IFNAR1*$^{-/-}$ mice displayed intestinal pathology including fecal abnormalities and enlarged Peyer's patches. Both of these pathologies were absent in Abx-treated mice, and intestinal virus titers were significantly reduced compared to microbially colonized mice (Kuss et al., 2011). Uchiyama et al., (2014) performed similar experiments testing the role of the intestinal microbiota in rotavirus infection. As with reovirus, particular rotavirus strains establish only asymptomatic infection in immunocompetent wild-type mice. Investigators in this study assessed infection in this context by viral antigen detection in feces and viral genome detection in intestinal tissue. Using both readouts, infection was reduced in Abx-treated mice compared to control mice. Moreover, fecal shedding was delayed by ~1 day in Abx-treated mice. This delay in shedding was also observed in germ-free mice, although peak levels of shed virus were actually higher in germ-free mice possibly because of the underdeveloped mucosal immune system in this setting (Round and Mazmanian, 2009). To determine whether commensal bacteria influence rotavirus-induced disease, investigators infected suckling mice which develop diarrhea in response to infection. The incidence and duration of diarrhea was reduced in Abx-treated suckling mice compared to controls (Uchiyama et al., 2014).

Because rotavirus titers were lower in Abx-treated mice, it seemed probable that the antiviral immune response would be reduced because of less antigen production. Remarkably however, the antiviral antibody response was enhanced in Abx-treated and germ-free mice compared to their microbially colonized counterparts (Uchiyama et al., 2014). The authors of this study offered two possible explanations for this surprising finding: First, rotavirus recognition by the host immune system may be more robust in bacterially depleted hosts due to reduced basal immune signaling in response to commensal bacterial recognition (ie, the immune system can focus more attention on viral antigens). Second, minimized activation of regulatory immune responses directed to commensal bacteria may promote a more inflammatory environment that is conducive to robust antiviral immune responses. When considering disparate rotavirus vaccine efficacy in industrialized versus impoverished parts of the world (see Chapter 2.11), it was further speculated that microbiota differences in these environments could be a contributing factor to vaccine response and that Abx treatment prior to administration of the rotavirus vaccine could boost immunogenicity. While this suggestion is extreme when considered in light of concerns over the emergence of Abx-resistant bacterial strains, it is possible that tailored probiotic therapy could be used to enhance vaccine responses in target populations (see Chapter 2.7). Consistent with the idea that the recognition of bacterial antigens during an enteric virus infection can shape the antiviral immune response, Zhang et al. (2014) recently reported that exposure to bacterial flagellin—an immunostimulatory bacterial ligand—protected mice from rotavirus infection. This protection required the flagellin sensors Toll-like receptor 5 (TLR5) and NOD-like receptor C4 (NLRC4) and the cytokines IL-22 and IL-18 (Zhang et al., 2014). Collectively, these data suggest that recognition of tolerogenic bacterial ligands can suppress antiviral immune responses whereas recognition of immunostimulatory bacterial ligands can promote antiviral immunity.

3.3 Noroviruses

The first clue that commensal bacteria may impact norovirus infections was provided by Basic et al. when they demonstrated that a murine norovirus (MNV) induced inflammation in mice deficient in the regulatory cytokine IL-10; this phenotype was dependent on commensal bacteria since germ-free IL-10$^{-/-}$ mice failed to develop inflammation when MNV-infected (Basic et al., 2014). In 2014–15, three studies were published in rapid succession confirming that noroviruses exploit commensal bacteria similarly to other enteric viruses described herein. First, Jones et al. (2014) reported that acute MNV titers in the distal ileum, mesenteric lymph nodes, and colon were significantly reduced at 1 day postinfection (1 dpi) in Abx-treated mice compared to their microbially colonized counterparts; experiments were performed to demonstrate that reduced titers reflected reduced in vivo viral replication. Second, Baldridge et al. (2015) revealed that persistent MNV infection in the intestine was dependent on the presence of commensal microbes: Abx-treated mice were resistant to persistent MNV infection, a phenotype that could be restored by fecal transplantation from microbially colonized (but not Abx-treated) mice. Importantly, systemic MNV infection was not affected by Abx treatment, demonstrating that commensal bacteria are specifically required to stimulate persistent MNV infection of the intestine. A variety of immune factors were examined for their role in Abx-mediated inhibition of MNV persistence establishment (Baldridge et al., 2015): TLR4, the LPS sensor, and its downstream signaling molecules MyD88 and Trif were not required nor were adaptive immune factors, type I IFN, or type II IFN. In contrast, Abx-mediated prevention of persistent MNV infection was regulated by type III IFN, or IFN-λ. These data suggest that commensal bacteria may normally suppress the antiviral effects of IFN-λ in the intestine during MNV infection (Fig. 5.2.1C). Third, Kernbauer et al. (2014) demonstrated that MNV infection of germ-free mice resulted in significantly less shedding of infectious virus compared to infection of microbially colonized mice. Remarkably, infection of germ-free mice with a MNV restored many of the defects observed in this setting pertaining to villus width, lymphocyte number and function, antibody levels, and basal suppression of innate lymphoid cell expansion; these phenotypes were dependent on type I IFN signaling, suggesting that an enteric virus can replace the intestinal microbiota in providing beneficial immune signals for a healthy gut (Kernbauer et al., 2014).

Recent data also revealed a stimulatory role for commensal bacteria in human norovirus (HuNoV) infections (Jones et al., 2014). Specifically, HuNoV-positive stool inoculum was able to initiate productive infection of B cells but filtration through a 0.2 μm membrane significantly reduced viral infectivity. This phenotype could be fully rescued by incubation of filtered stool with a single commensal bacteria, *Enterobacter cloacae*. Evidence suggests that *E. cloacae* stimulated HuNoV attachment to the surface of B cells via bacterial surface-expressed histo-blood group antigen (HBGA) molecules (Fig. 5.2.1B): (Miura et al. (2013) determined that *E. cloacae* expresses HBGA; non-HBGA-expressing bacteria and LPS both failed to restore infectivity (Jones et al., 2014); and synthetic HBGA fully restored infectivity and viral attachment to B cells (Jones et al., 2014). Collective observations thus revealed that both MNV and HuNoVs are stimulated by commensal bacteria but they do not bind LPS. Instead HuNoVs appear to bind to neutrally charged glycans, the HBGAs, while the bacterial ligands for MNV strains have yet to be identified.

3.4 Mouse Mammary Tumor Virus

Mouse mammary tumor virus (MMTV) is a retrovirus that is transmitted vertically from infected mothers to their pups via milk. Specifically, lymphocytes containing MMTV provirus in the milk are ingested by the pups where the virus can access intestinal Peyer's patches and initiate infection. A previous study revealed that MMTV persistence subsequent to ingestion of MMTV-laden milk requires the LPS receptor TLR4 and the antiinflammatory cytokine IL-10 (Jude et al., 2003). Based on these observations, Kane et al., (2011) tested the hypothesis that MMTV avoids immune-driven elimination by associating with bacterial LPS and inducing a tolerogenic immune response. Supporting this idea, Abx treatment of infected mothers prevented MMTV transmission to their offspring (Kane et al., 2011). Furthermore, germ-free infected mothers failed to transmit virus while their reconstitution with a defined gut flora fully restored transmission (Kane et al., 2011). Similar to poliovirus, MMTV appears to directly bind LPS: MMTV isolated from the stomachs of pups ingesting infected milk was associated with LPS; and LPS cofractionated with MMTV in ultracentrifugation density gradients. MMTV associated with LPS induced IL-10 secretion when inoculated onto splenocytes in culture whereas LPS-free MMTV isolated from germ-free animals failed to induce IL-10 secretion. Overall, these data support a model whereby an orally transmitted retrovirus avoids immune-driven elimination by binding commensal bacterial LPS that is recognized by TLR4 and ultimately induces IL-10 secretion and a tolerogenic immune environment (Fig. 5.2.1D). These results are in accordance with a well-established immune phenomenon referred to as bystander suppression (Miller et al., 1991).

4 CONCLUDING REMARKS

The exploitation of commensal bacteria by enteric viruses is a rapidly emerging area of study that has received significant attention only in the past 5 years. As summarized in this chapter, poliovirus, reovirus, rotavirus, murine and human noroviruses, and an enteric retrovirus have all now been revealed to rely on commensal bacteria for optimal infection of the intestine. While it is possible that examples of microbial inhibition of enteric viral infections will be discovered in future studies, all enteric viruses must circumvent the extremely dense population of microbes residing in the gut lumen so it is probable that they have uniformly evolved mechanisms to breach this barrier to infect the mammalian host. Although mechanisms of bacterial stimulation are unique for each virus studied to date, there are certain generalities. For example, all bacterial ligands that have been identified thus far are surface polysaccharides, including LPS and HBGAs. Defined mechanisms of bacterial stimulation segregate into two general categories: (1) bacteria can directly stimulate viral infection of cells by stabilizing the virion or facilitating attachment to the surface of permissive cells (Fig. 5.2.1A,B); and (2) bacteria can impact the antiviral immune response to promote establishment of persistent infections (Fig. 5.2.1C,D). In each case, one can envision designing novel therapeutic or preventative strategies that target virus-bacteria interactions appropriately. This will undoubtedly be an intensely studied area in the near future that will lead to a better understanding of the complex interplay between the host immune system, the intestinal microbiota, and pathogenic viral infections along the intestinal tract.

REFERENCES

Baldridge, M.T., Nice, T.J., McCune, B.T., et al., 2015. Commensal microbes and interferon-λ determine persistence of enteric murine norovirus infection. Science 347, 266–269.

Basic, M., Keubler, L.M., Buettner, MD rer. nat., et al., 2014. Norovirus triggered microbiota-driven mucosal inflammation in interleukin 10-deficient mice. Inflamm. Bowel Dis. 20, 431–443.

Hooper, L.V., Stappenbeck, T.S., Hong, C.V., Gordon, J.I., 2003. Angiogenins: a new class of microbicidal proteins involved in innate immunity. Nat. Immunol. 4, 269–273.

Hooper, L.V., Littman, D.R., Macpherson, A.J., 2012. Interactions between the microbiota and the immune system. Science 336, 1268–1273.

Ida-Hosonuma, M., Sasaki, Y., Toyoda, H., et al., 2003. Host range of poliovirus is restricted to simians because of a rapid sequence change of the poliovirus receptor gene during evolution. Arch. Virol. 148, 29–44.

Ida-Hosonuma, M., Iwasaki, T., Yoshikawa, T., et al., 2005. The alpha/beta interferon response controls tissue tropism and pathogenicity of poliovirus. J. Virol. 79, 4460–4469.

Jones, M.K., Watanabe, M., Zhu, S., et al., 2014. Enteric bacteria promote human and murine norovirus infection of B cells. Science 346, 755–759.

Jude, B.A., Pobezinskaya, Y., Bishop, J., et al., 2003. Subversion of the innate immune system by a retrovirus. Nat. Immunol. 4, 573–578.

Kamada, N., Núñez, G., 2014. Regulation of the immune system by the resident intestinal bacteria. Gastroenterology 146, 1477–1488.

Kane, M., Case, L.K., Kopaskie, K., et al., 2011. Successful transmission of a retrovirus depends on the commensal microbiota. Science 334, 245–249.

Kau, A.L., Ahern, P.P., Griffin, N.W., Goodman, A.L., Gordon, J.I., 2011. Human nutrition, the gut microbiome and the immune system. Nature 474, 327–336.

Kernbauer, E., Ding, Y., Cadwell, K., 2014. An enteric virus can replace the beneficial function of commensal bacteria. Nature 516, 94–98.

Koike, S., Taya, C., Kurata, T., et al., 1991. Transgenic mice susceptible to poliovirus. Proc. Natl. Acad. Sci. USA 88, 951–955.

Kuss, S.K., Best, G.T., Etheredge, C.A., et al., 2011. Intestinal microbiota promote enteric virus replication and systemic pathogenesis. Science 334, 249–252.

Macpherson, A.J., Gatto, D., Sainsbury, E., Harriman, G.R., Hengartner, H., Zinkernagel, R.M., Primitive T, A., 2000. Cell-independent mechanism of intestinal mucosal IgA responses to commensal bacteria. Science 288, 2222–2226.

Miller, A., Lider, O., Weiner, H.L., 1991. Antigen-driven bystander suppression after oral administration of antigens. J. Exp. Med. 174, 791–798.

Miura, T., Sano, D., Suenaga, A., et al., 2013. Histo-blood group antigen-like substances of human enteric bacteria as specific adsorbents for human noroviruses. J. Virol. 87, 9441–9451.

Mowat, A.M., Agace, W.W., 2014. Regional specialization within the intestinal immune system. Nat. Rev. Immunol.Nat. Rev. Immunol. 14, 667–685.

Nicholson, J.K., Holmes, E., Kinross, J., et al., 2012. Host–gut microbiota metabolic interactions. Science 336, 1262–1267.

Ohka, S., Igarashi, H., Nagata, N., et al., 2007. Establishment of a poliovirus oral infection system in human poliovirus receptor-expressing transgenic mice that are deficient in alpha/beta interferon receptor. J. Virol. 81, 7902–7912.

Ren, R., Costantini, F., Gorgacz, E.J., Lee, J.J., Racaniello, V.R., 1990. Transgenic mice expressing a human poliovirus receptor: A new model for poliomyelitis. Cell 63, 353–362.

Robinson, C.M., Jesudhasan, P.R., Pfeiffer, J.K., 2014. Bacterial lipopolysaccharide binding enhances virion stability and promotes environmental fitness of an enteric virus. Cell Host Microbe 15, 36–46.

Round, J.L., Mazmanian, S.K., 2009. The gut microbiota shapes intestinal immune responses during health and disease. Nat. Rev. Immunol.Nat. Rev. Immunol. 9, 313–323.

Round, J.L., Mazmanian, S.K., 2010. Inducible Foxp3+ regulatory T-cell development by a commensal bacterium of the intestinal microbiota. Proc. Natl. Acad. Sci. USA 107, 12204–12209.

Sommer, F., Bäckhed, F., 2013. The gut microbiota—masters of host development and physiology. Nat. Rev. Microbiol. 11, 227–238.

Tremaroli, V., Bäckhed, F., 2012. Functional interactions between the gut microbiota and host metabolism. Nature 489, 242–249.

Uchiyama, R., Chassaing, B., Zhang, B., Gewirtz, A.T., 2014. Antibiotic treatment suppresses rotavirus infection and enhances specific humoral immunity. J. Infect. Dis. 210, 171–182.

Vaishnava, S., Yamamoto, M., Severson, K.M., et al., 2011. The antibacterial lectin RegIIIγ promotes the spatial segregation of microbiota and host in the intestine. Science 334, 255–258.

Xu, J., Gordon, J.I., 2003. Honor thy symbionts. Proc. Natl. Acad. Sci. USA 100, 10452–10459.

Zhang, B., Chassaing, B., Shi, Z., et al., 2014. Prevention and cure of rotavirus infection via TLR5/NLRC4–mediated production of IL-22 and IL-18. Science 346, 861–865.

Index

A

B

H

I

O

P

R

W

X

Y

Z